Mathematics for Industry

Volume 40

Editor-in-Chief

Masato Wakayama, Kyushu University, NTT Institute for Fundamental Mathematics, Tokyo Japan, Fukuoka, Japan

Series Editors

Robert S. Anderssen, Commonwealth Scientific and Industrial Research Organisation, Canberra, ACT, Australia

Yuliy Baryshnikov, Department of Mathematics, University of Illinois at Urbana-Champaign, Urbana, IL, USA

Heinz H. Bauschke, University of British Columbia, Vancouver, BC, Canada

Philip Broadbridge, School of Engineering and Mathematical Sciences, La Trobe University, Melbourne, VIC, Australia

Jin Cheng, Department of Mathematics, Fudan University, Shanghai, China

Monique Chyba, Department of Mathematics, University of Hawaii at Mānoa, Honolulu, HI, USA

Georges-Henri Cottet, Joseph Fourier University, Grenoble, Isère, France

José Alberto Cuminato, University of São Paulo, São Paulo, Brazil

Shin-ichiro Ei, Department of Mathematics, Hokkaido University, Sapporo, Japan

Yasuhide Fukumoto, Kyushu University, Nishi-ku, Fukuoka, Japan

Jonathan R. M. Hosking, IBM T. J. Watson Research Center, Scarsdale, NY, USA

Alejandro Jofré, University of Chile, Santiago, Chile

Masato Kimura, Faculty of Mathematics & Physics, Kanazawa University, Kanazawa, Japan

Kerry Landman, The University of Melbourne, Victoria, Australia

Robert McKibbin, Institute of Natural and Mathematical Sciences, Massey University, Palmerston North, Auckland, New Zealand

Andrea Parmeggiani, Dir Partenariat IRIS, University of Montpellier 2, Montpellier, Hérault, France

Jill Pipher, Department of Mathematics, Brown University, Providence, RI, USA

Konrad Polthier, Free University of Berlin, Berlin, Germany

Osamu Saeki, Institute of Mathematics for Industry, Kyushu University, Fukuoka, Japan

Wil Schilders, Department of Mathematics and Computer Science, Eindhoven University of Technology, Eindhoven, The Netherlands

Zuowei Shen, Department of Mathematics, National University of Singapore, Singapore, Singapore

Kim Chuan Toh, Department of Analytics and Operations, National University of Singapore, Singapore, Singapur, Singapore

Evgeny Verbitskiy, Mathematical Institute, Leiden University, Leiden, The Netherlands

Nakahiro Yoshida, The University of Tokyo, Meguro-ku, Tokyo, Japan

Aims & Scope

The meaning of "Mathematics for Industry" (sometimes abbreviated as MI or MfI) is different from that of "Mathematics in Industry" (or of "Industrial Mathematics"). The latter is restrictive: it tends to be identified with the actual mathematics that specifically arises in the daily management and operation of manufacturing. The former, however, denotes a new research field in mathematics that may serve as a foundation for creating future technologies. This concept was born from the integration and reorganization of pure and applied mathematics in the present day into a fluid and versatile form capable of stimulating awareness of the importance of mathematics in industry, as well as responding to the needs of industrial technologies. The history of this integration and reorganization indicates that this basic idea will someday find increasing utility. Mathematics can be a key technology in modern society.

The series aims to promote this trend by 1) providing comprehensive content on applications of mathematics, especially to industry technologies via various types of scientific research, 2) introducing basic, useful, necessary and crucial knowledge for several applications through concrete subjects, and 3) introducing new research results and developments for applications of mathematics in the real world. These points may provide the basis for opening a new mathematics-oriented technological world and even new research fields of mathematics.

To submit a proposal or request further information, please use the PDF Proposal Form or contact directly: Smith Ahram Chae, Publishing Editor (smith.chae@springer.com).

Scientific Board Members

Robert S. Anderssen (Commonwealth Scientific and Industrial Research Organisation, Canberra, ACT, Australia)
Yuliy Baryshnikov (Department of Mathematics, University of Illinois at Urbana-Champaign, Urbana, IL, USA)
Heinz H. Bauschke (University of British Columbia, Vancouver, BC, Canada)
Philip Broadbridge (School of Engineering and Mathematical Sciences, La Trobe University, Melbourne, VIC, Australia)
Jin Cheng (Department of Mathematics, Fudan University, Shanghai, China)
Monique Chyba (Department of Mathematics, University of Hawaii at Mānoa, Honolulu, HI, USA)
Georges-Henri Cottet (Joseph Fourier University, Grenoble, Isère, France)
José Alberto Cuminato (University of São Paulo, São Paulo, Brazil)
Shin-ichiro Ei (Department of Mathematics, Hokkaido University, Sapporo, Japan)
Yasuhide Fukumoto (Kyushu University, Nishi-ku, Fukuoka, Japan)
Jonathan R. M. Hosking (IBM T. J. Watson Research Center, Scarsdale, NY, USA)
Alejandro Jofré (University of Chile, Santiago, Chile)
Masato Kimura (Faculty of Mathematics & Physics, Kanazawa University, Kanazawa, Japan)
Kerry Landman (The University of Melbourne, Victoria, Australia)
Robert McKibbin (Institute of Natural and Mathematical Sciences, Massey University, Palmerston North, Auckland, New Zealand)
Andrea Parmeggiani (Dir Partenariat IRIS, University of Montpellier 2, Montpellier, Hérault, France)
Jill Pipher (Department of Mathematics, Brown University, Providence, RI, USA)
Konrad Polthier (Free University of Berlin, Berlin, Germany)
Osamu Saeki (Institute of Mathematics for Industry, Kyushu University, Fukuoka, Japan)
Wil Schilders (Department of Mathematics and Computer Science, Eindhoven University of Technology, Eindhoven, The Netherlands)
Zuowei Shen (Department of Mathematics, National University of Singapore, Singapore, Singapore)
Kim Chuan Toh (Department of Analytics and Operations, National University of Singapore, Singapore, Singapur, Singapore)
Evgeny Verbitskiy (Mathematical Institute, Leiden University, Leiden, The Netherlands)
Nakahiro Yoshida (The University of Tokyo, Meguro-ku, Tokyo, Japan)

Tsuyoshi Takagi · Masato Wakayama ·
Noboru Kunihiro · Keisuke Tanaka ·
Kazufumi Kimoto · Momonari Kudo
Editors

Mathematical Foundations for Post-Quantum Cryptography

Crypto-Math CREST

 Springer

Editors
Tsuyoshi Takagi
Department of Mathematical Informatics
Graduate School of Information Science
and Technology
University of Tokyo
Bunkyo City, Tokyo, Japan

Noboru Kunihiro
Faculty of Engineering, Information
and Systems
University of Tsukuba
Tsukuba, Ibaraki, Tokyo, Japan

Kazufumi Kimoto
Department of Mathematical Sciences
University of the Ryukyus Senbaru
Nishihara, Nakagami, Okinawa, Japan

Masato Wakayama
NTT Institute for Fundamental
Mathematics
NTT Inc.
Musashino, Tokyo, Japan

Keisuke Tanaka
Department of Mathematical
and Computing Science
Institute of Science Tokyo
Meguro City, Tokyo, Japan

Momonari Kudo
Department of Information
and Communication Engineering
Fukuoka Institute of Technology
Higashi-ku, Fukuoka, Japan

ISSN 2198-350X   ISSN 2198-3518 (electronic)
Mathematics for Industry
ISBN 978-981-96-1217-8   ISBN 978-981-96-1218-5 (eBook)
https://doi.org/10.1007/978-981-96-1218-5

© The Editor(s) (if applicable) and The Author(s) 2026. This book is an open access publication.

Open Access This book is licensed under the terms of the Creative Commons Attribution 4.0 International License (http://creativecommons.org/licenses/by/4.0/), which permits use, sharing, adaptation, distribution and reproduction in any medium or format, as long as you give appropriate credit to the original author(s) and the source, provide a link to the Creative Commons license and indicate if changes were made.
The images or other third party material in this book are included in the book's Creative Commons license, unless indicated otherwise in a credit line to the material. If material is not included in the book's Creative Commons license and your intended use is not permitted by statutory regulation or exceeds the permitted use, you will need to obtain permission directly from the copyright holder.
The use of general descriptive names, registered names, trademarks, service marks, etc. in this publication does not imply, even in the absence of a specific statement, that such names are exempt from the relevant protective laws and regulations and therefore free for general use.
The publisher, the authors and the editors are safe to assume that the advice and information in this book are believed to be true and accurate at the date of publication. Neither the publisher nor the authors or the editors give a warranty, expressed or implied, with respect to the material contained herein or for any errors or omissions that may have been made. The publisher remains neutral with regard to jurisdictional claims in published maps and institutional affiliations.

This Springer imprint is published by the registered company Springer Nature Singapore Pte Ltd.
The registered company address is: 152 Beach Road, #21-01/04 Gateway East, Singapore 189721, Singapore

If disposing of this product, please recycle the paper.

Preface

The CREST Crypto-Math Project: "Creation and Development of Mathematical Foundations for Cryptography Required by the Post-Quantum Society" supported by the Japan Science and Technology Agency (JST) aims at realizing of cryptographic technology that is resistant to the compromise of various possible attackers such as attacks using quantum computers and side-channel attacks by power analysis. In this study, we attempt to develop cryptographic applications with a decentralized security function using blockchains for large-scale distributed systems. This book presents the mathematical foundations for cryptography securely used in the era of quantum computers. In particular, we aim to deepen the basic mathematics of post-quantum cryptography, model the strongest possible attacks such as side-channel attacks, and constructing cryptographic protocols that guarantee security against such attacks. Furthermore, we explore fundamental mathematical theories, such as the interaction between number theory, arithmetic geometry, and quantum computation, from a long-term perspective to design secure cryptography in the future.

This book is a collection of lectures from eight workshops held for the CREST Crypto-Math project starting from October 2021, which consists of 4 groups: cryptanalysis, quantum attacks, physical attacks, and cryptographic protocols, with 36 researchers and over 30 students participating. The homepage of the CREST Crypto-Math project is http://crypto.mist.i.u-tokyo.ac.jp/crest-cryptomath/.

This book comprises 24 papers categorized into four parts. Part I is devoted to the mathematical foundation and quantum theory, which address the ABC inequalities conjecture, Riemann zeta functions, quantum walks, quantum proof, non-commutative harmonic oscillators, and quantum Rabi models. Part II covers topics such as quantum attacks on ECDLP, lattice attacks on RSA cryptosystems, and the cryptanalysis of some quantum-resistant cryptosystems. Part III discusses the development of post-quantum cryptography such as lattice-based cryptography, multivariate public-key cryptography, code-based cryptography, and isogeny-based cryptography. Finally, Part IV discusses hash functions and digital signatures based on expander graphs, including Ramanujan graphs and group—subgroup pair graphs, and explains some applications of risk assessments, cryptocurrencies based on blockchains, and private information retrieval.

This book is suitable for graduate students and researchers interested in mathematical cryptography. We hope this book will be useful for promoting research on quantum theory and post-quantum cryptography.

Finally, we would like to thank Dr. Naonari Ueda, the Research Supervisor at the CREST Mathematical Information Platform supported by JPMJCR2113, for publishing this book.

Tokyo, Japan Tsuyoshi Takagi
July 2023 Masato Wakayama
Noboru Kunihiro
Keisuke Tanaka
Kazufumi Kimoto
Momonari Kudo

Contents

Mathematical Foundation and Quantum Theory

On New Interactions Between Quantum Theories and Arithmetic Geometry 3
Ivan Fesenko

Quantum Walks for the Working Mathematician 15
Hayato Saigo and Shingo Sugiyama

On the Hardness of Conversion from Entangled Proof into Separable One 35
Seiseki Akibue, Go Kato, and Seiichiro Tani

Representation Theory of $\mathfrak{sl}(2, \mathbb{R}) \simeq \mathfrak{su}(1, 1)$ and a Generalization of Non-commutative Harmonic Oscillators 51
Ryosuke Nakahama

The k-Photon Quantum Rabi Model 75
Daniel Braak

Meditations on the Farey Fractal 89
Shai Haran

Quantum and Physical Cryptanalysis

Shor's Quantum Algorithm for Solving the Binary ECDLP: A Survey 125
Ren Taguchi and Atsushi Takayasu

Proposal for Quantum GRS Algorithm and Cryptanalysis for ROLLO and RQC 143
Asuka Wakasugi and Mitsuru Tada

Full Key Recovery on RSA from Noisy Binary GCD Operation Sequences 163
Kenta Tani and Noboru Kunihiro

Performance Analysis of Fault Attack on UOV Multivariate Signature Scheme 183
Hiroki Furue, Tatsuya Nagasawa, and Tsuyoshi Takagi

On Hilbert–Poincaré Series of Affine Semi-regular Polynomial Sequences and Related Gröbner Bases 205
Momonari Kudo and Kazuhiro Yokoyama

Parallel DeepBKZ 2.0: Development of Parallel DeepBKZ Reduction with Large Blocksizes 233
Satoshi Nakamura, Nariaki Tateiwa, Masaya Yasuda, and Katsuki Fujisawa

Post-quantum Cryptography

A Survey on Middle-Product Learning with Errors Cryptography 251
Masayuki Tezuka and Keisuke Tanaka

Expanded Lattices for Solving Ring-Based LWE and NTRU Problems 273
Satoshi Nakamura and Masaya Yasuda

Analysis of (U,U+V)-Code Problem with Gramian Over Binary and Ternary Fields 295
Ichiro Iwata, Yusuke Yoshida, and Keisuke Tanaka

A Survey on Small Public Key Signature Schemes Derived from UOV Signature Scheme 313
Yasuhiko Ikematsu

A Survey of Attacks on Supersingular Isogeny with Torsion Problem and Its Variants 333
Hiroshi Onuki

An Optimization for Efficient Computation of Multiradical (3, 3)-isogenies on Jacobians 351
Masahiro Ishii and Daiki Hayashida

Hash Function, Graph Theory, and Applications

Hashing by Walking Over Expanders: A Recipe for Constructing Provably Secure Hash Functions 371
Yusuke Aikawa

Toward Hash Functions Based on Group-Subgroup Pair Graphs 389
Cid Reyes-Bustos

Improving Hash-Based Signature Schemes: From Theory to Practice 407
Quan Yuan

A Formal Approach for Secured Risk Assessment 425
Kengo Zenitani

SoK: A Taxonomy for Layer-2 Scalability Related Protocols for Cryptocurrencies 453
Maxim Jourenko, Mario Larangeira, Kanta Kurazumi, and Keisuke Tanaka

A Survey on Private Information Retrieval: Information-Theoretic Constructions and Error-Correction Techniques 475
Reo Eriguchi and Koji Nuida

Index 495

Mathematical Foundation and Quantum Theory

On New Interactions Between Quantum Theories and Arithmetic Geometry

Ivan Fesenko

Abstract One of the aims of this paper is to attract the attention of quantum theorists to certain areas of arithmetic geometry whose ideas and concepts and analogues of objects may find applications in quantum physics. Proposing or drawing these interdisciplinary links is novel and may open up new research directions. We discuss several analogies between some developments in arithmetic geometry, once stemming from Grothendieck and now including the IUT theory of Mochizuki, and some aspects of quantum theory including quantum computing. These analogies were spotted recently and it is hoped that related developments may be fruitful. In the appendix we also propose a new abc-ABC question about an asymptotic symmetry of the moduli space of Frey–Hellegouarch elliptic curves over rational numbers. This question goes beyond the standard abc inequalities conjectures/questions. We prove that the positive question to the question and the effective abc inequalities established in [27] using enhanced IUT theory imply the stronger version of the effective $(1 + \varepsilon)$-abc inequality.

Keywords Topos theory · Non-Boolean logic · Étale fundamental group · Group theoretic algorithms · Algorithmic Anabelian geometry · IUT theory · Abc inequalities · Effective abc inequalities · Elliptic curves over rational numbers · Improved foundations of quantum mechanics · Quantum computing · Stabiliser formalism in quantum computing

1 Introduction

Quantum theory in its standard formulation uses mathematics mostly known at the time of its creation, i.e. 100–60 years ago. Serious foundational problems and paradoxes of quantum theory affect its further developments and applications, including quantum computing and communication. It is reasonable to expect that the use

I. Fesenko (✉)
Institute of Theoretical Sciences, Westlake University, Hangzhou, China
e-mail: ivan.b.fesenko@westlake.edu.cn

of modern mathematics to produce mathematically improved versions of quantum theory may lead to the resolution of those foundational problems.

There is huge potential for utilising concepts and visions of modern arithmetic geometry in developments of quantum theory including quantum computing. The following examples will serve as a partial evidence for this opinion. In order to develop implications of these analogies, we need researchers with expertise in quantum theories and in Anabelian geometry and IUT.

In Sects. 2–4 we discuss, using a simple language in order not to frighten potential readers, analogies between some ideas and concepts in arithmetic geometry whose applications in quantum theory may be fruitful. Most of these analogies are not mentioned in the published literature. The author expresses his gratitude to E. Demler, N. Gisin, A. Lvovsky, I. Martin, A. Ustyuzhanin for interesting quantum discussions, to S. Mochizuki for numerous discussions of the IUT theory, and to the referee for valuable suggestions.

In the appendix we propose, using very simple mathematics, new questions about an asymptotic symmetry of the moduli space of Frey–Hellegouarch elliptic curves over rational numbers. These questions are motivated by the study of options to derive stronger effective abc-type inequalities from the already established ones. No knowledge of Anabelian geometry or the IUT theory is assumed.

2 Topos-Theoretical Approach to Quantum Theory

Non-locality and contextuality (Bell, Kochen–Specker) in quantum mechanics may help obtain quantum advantage over classical computational models in quantum computing. This is closely related to the use of non-Boolean logic in quantum theory.

The notion of topos was introduced in the early sixties by Grothendieck with the original first aim of bringing a topological or geometric intuition in parts of number theory where actual topological spaces do not occur [15, 20]. Grothendieck realised that many important properties of topological spaces X can be naturally formulated as properties of the categories $Sh(X)$ of sheaves of sets on the spaces. It is important that the map $X \longrightarrow Sh(X)$ is an embedding of continuous structures into categories which are discrete structures. The crucial unifying notion of topos is to provide the common geometric intuition for many areas of mathematics and to connect continuous with discrete. See the talk of Lafforgue [19] for more mathematical and historical details.

A topos has various features similar to the category of sets. Somehow similar to quantisation of a classical physical theory, constructions in topos theory can often be understood by looking at them first in the category of sets or geometrical categories and then lifting to the general case.

However, unlike sets, the law of excluded middle does not need to hold in a topos. Toposes incorporate non-Boolean logic in an organic way. A topos has an internal logical structure that is similar to the way in which Boolean algebra arises in set theory, but instead of two truth values 1 and 0, goes outside Boolean logic with truth values are in a larger set. Topos theory is a math theory that can 'speak' of indeterminism. It may also address the issue of 'real numbers are not real for physicists', mentioned in the papers and talks of Gisin [3, 14].

Isham, Döring and others proposed a partial reformulation of quantum theory in terms of topos theory [4–9, 12, 13, 17, 18]. The reason for their choice of topos theory is that the latter in the first approximation looks like sets theory and is equipped with an internal logic. Their topos theoretical approach to quantum theory builds locally on the topos of presheaves of commutative (hence classical) sub-algebras of the algebra of all bounded operators on the quantum theory's Hilbert space. This reformulation in several aspects looks like classical physics, propositions can be given truth values without using concepts of measurement or external observer, using the internal non-Boolean logic of appropriate toposes. The non-existence of classical explanations for quantum phenomena can be viewed as corresponding to the non-existence of global points or sections of certain toposes.

3 Etale Fundamental Group, Section Conjecture and Stabiliser Formalism in Quantum Computing

No quantities show up in category theory and topos theory. What matters is the form of a category and its structure. The notion of a geometric morphism in topos theory has allowed to build general cohomology theories which cannot be otherwise produced. The first example was étale cohomology theory. The Grothendieck definitions of étale sites, étale fundamental group and étale cohomology use toposes.

For any geometrically integral (quasi-compact) scheme X over a perfect field k one has its étale fundamental group $\pi_1(X)$. For example, if C is a complex irreducible smooth projective curve minus a finite set of its points, over an algebraically closed field of characteristic 0, then $\pi_1(C)$ is isomorphic to the profinite completion of the topological fundamental group of the Riemann surface associated to C.

A hyperbolic curve C over a field k of characteristic zero is a smooth projective geometrically connected curve of genus g minus r points such that the Euler characteristic $2 - 2g - r$ is negative. The étale fundamental group of a hyperbolic curve is highly (in appropriate strict mathematical sense) nonabelian, its centre is trivial. Grothendieck asked famous questions whether hyperbolic curves over number fields are Anabelian, i.e. whether one can restore the curve from its étale fundamental group. A partial positive answer was obtained by Tamagawa [29] and the full positive answer was obtained by Mochizuki [21–23]. This was the foundation of Anabelian geometry of hyperbolic curves over number fields and their non-archimedean completions. Most of the further development of this Anabelian geometry in the last 30 years has not been digested by mathematicians outside Japan.

A point x in $X(k)$ for a field k, i.e. a morphism $\text{Spec}(k) \longrightarrow X$, determines, in a functorial way, a continuous section from the absolute Galois group G_k of k to $\pi_1(X)$, well-defined up to composition with an inner automorphism, of the surjective map

$\pi_1(X) \longrightarrow G_k$. Grothendieck's section conjecture asks whether, for a geometrically connected smooth projective curve X over k, of genus >1, the map from rational points $X(k)$ to the set of conjugacy classes of sections, $x \mapsto D_x = \mathrm{Stab}(x)$, is surjective. Injectivity was already known. Various other similar conjectures, such as a combinatorial section conjecture, were established by Mochizuki and his collaborators. There are forthcoming discoveries of relations of the section conjecture to some fundamental open problems in number theory.

One of the key issues for quantum algorithms is whether they can run in polynomial time, instead of exponential time. Controlling loss of information/error correction is crucial. In quantum error correction one uses stabiliser groups in finite-dimensional complex spaces.

A map
$$s \mapsto D_s$$
from quantum states s in a $2n$-dimensional vector space over $\mathbb{C}$ to their stabiliser group D_s (unitary matrices acting trivially on s) is injective. However, D_s has too many generators, about 4^n.

Aaronson–Gottesman [1] considered the intersection $D_s \cap P_n$ where P_n is the group of n-qubit Pauli operators: all tensor products of n matrices
$$\begin{pmatrix} 1 & 0 \\ 0 & 1 \end{pmatrix} \begin{pmatrix} 0 & 1 \\ 1 & 0 \end{pmatrix} \begin{pmatrix} 0 & -i \\ i & 0 \end{pmatrix} \begin{pmatrix} 1 & 0 \\ 0 & -1 \end{pmatrix}$$
and their scalar products with roots of order 4. The number of elements of P_n is 4^{n+1}. This intersection $D_s \cap P_n$ has a much smaller number of generators than D_s.

It is easy to show that on the subset of quantum states that are stabilised by exactly 2^n elements of P_n the map
$$s \mapsto D_s \cap P_n$$
is still injective, [1]. This subset is further characterised as obtained from $|0\rangle^{\otimes n}$ by CNOT, Hadamard, and phase gates only.

Recently, Hoshi, Mochizuki and Tsujimura obtained results about the injectivity of the section map
$$x \mapsto D_x \cap G_L$$
from closed points x of hyperbolic curves over number fields to (conjugacy classes of) the intersection of their stabiliser groups (decomposition groups) with the absolute Galois group of various infinite extensions L of the number field k, [16]. Notice the similarity with the injectivity of the map $s \mapsto D_s \cap P_n$ in quantum stabiliser theory.

Note that the variety of decomposition groups and absolute Galois groups in number theory is much larger than the Clifford group (the group of unitary matrices that normalise P_n) and P_n in quantum computing. One perspective is to investigate whether analogues of the decomposition groups and absolute Galois groups in arithmetic geometry may provide new classes of groups useful for quantum computing and lead to new computational models. This may help to go beyond the Clifford ground in quantum computing, an important task indicated by the experts.

At the same time, there is a substantial difference between profinite decomposition groups and discrete stabiliser groups in quantum computing: the centre of the former is trivial whilst the centre of Clifford group is infinite but the quotient group by its centre is finite. However, when one works with those arithmetic stabiliser groups, one often considers them as the projective limit of their quotients which are extensions of a finite group by an infinite abelian group and such quotients modulo their centre are finite groups.

4 On Some Similarities Between IUT and Quantum Theory

Algebraic geometry involves locally the correspondence between affine varieties and commutative rings with two algebraic operations. Anabelian geometry for hyperbolic curves over number fields and other fields is a correspondence between these geometric objects and their arithmetic fundamental groups (or slightly more complicated objects). Fundamental groups are non-commutative, but they have one algebraic operation, not two. This opens the perspective to try to perform deformations of these geometric objects not seen by algebraic geometry, using the fact that there are more maps, group homomorphisms and variations of those between topological groups in comparison to ring homomorphisms between commutative rings.

The IUT theory, discovered by Mochizuki, provides, for certain hyperbolic curves (e.g. an elliptic curve minus a point), a new fundamental understanding of how to bound from above the deviation from commutativity of certain crucial diagrams associated to arithmetic deformation aspects of IUT. IUT is a new arithmetic deformation theory that is entirely unavailable in the standard arithmetic geometry dealing with ring structures. IUT is a non-linear algorithmic theory which addresses such fundamental aspects as to which extent the multiplication and addition on integers (or rings of algebraic integers) cannot be separated from one another. Arithmetic deformations in IUT are not compatible with ring structure. Deformations are coded in certain links, ring structures do not pass through the links, groups of symmetries such as Galois and étale fundamental groups do pass. To algorithmically restore certain rings from some groups that pass through a link in IUT, one uses deep results of Anabelian geometry about number fields and hyperbolic curves over number fields and their non-archimedean completions. The Anabelian algorithms produce from a given input data an object within a set of possible output objects. Several indeterminacies are used to weaken the input data in the algorithm, whilst mildly increasing the set of possible output objects (container) until that increased set includes the object in the original input data.

There are many analogies and relations between IUT and other areas of number theory, see 2.14 of [10]. We now list several intriguing similarities between IUT and quantum theory, by involving some basic concepts and ideas of IUT, those which can be stated in a relatively simple form and do not require an expertise in IUT. These similarities are even more amazing given the fact that representation theory plays no role in IUT and plays a fundamental role in quantum physics.

- In IUT, multiplication and addition are related to two symmetries, geometric and arithmetic symmetries, see, e.g. Sects. 3.6 of [24], 2.7 of [10]. These two dimensions are reminiscent of the two parameters, one of which is related to electricity, the other to magnetism. In particular, those two were employed in the experimental study of layers of hexagonal lattices in the growth of graphene and boron nitride layers.
- One of the key issues for quantum algorithms is whether they can run in polynomial time, instead of exponential time. The aspect of reducing exponential to polynomial is crucial for IUT.
- In mono-Anabelian geometry and IUT one algorithmically reconstructs objects from étale fundamental groups. IUT produces upper bounds on the change of the relevant data passing through the theta-link, using the action of étale fundamental groups which pass through the link unaffected. As discussed with practitioners, using appropriate group (that goes intact through the algorithmic process) action on a flow of information to control its loss may be useful in quantum computing and quantum computers.
- IUT works with two types of topological monoid structures: étale-like (coming from groups of symmetries) and frobenius-like (coming from 'ordered' objects) and their interactions, see, e.g. Sects. 2.7 of [24], 3.2 of [25]. A monoid is a set with binary associative operation and an identity element. For example, the multiplicative monoid $\mathbb{Z}_p\setminus\{0\}$ of non-zero elements of p-adic integers $\mathbb{Z}_p$ of the field $\mathbb{Q}_p$ of p-adic numbers is a submonoid of the group of invertible elements of $\mathbb{Q}_p$. It splits into the product of the group of units $\mathbb{Z}_p^\times$ and the monoid of non-negative integer powers of p. The latter monoid is totally ordered. The monoid $\mathbb{Z}_p\setminus\{0\}$ is an example of a frobenius-like object in IUT, p^n can be informally viewed as having 'mass' n. A more interesting example of a frobenius-like object comes at the level of algebraic closures. Denote by O the ring of integers of an algebraic closure K of $\mathbb{Q}_p$. The absolute Galois group G of $\mathbb{Q}_p$ is an example of étale-like object in IUT. Consider the monoid $(O\setminus\{0\}, G)$ of non-zero elements of O with respect to multiplication, under the action by the group G. By class field theory, the maximal abelian quotient of G is isomorphic to $\mathbb{Z}_p^\times$ times the procyclic group $\hat{\mathbb{Z}}$ generated by p. The totally ordered cyclic group $\mathbb{Z}$ is a subgroup of its profinite completion $\hat{\mathbb{Z}}$, the latter is not a totally ordered group, and the image of p^n in it has no longer non-zero 'mass'. The embedding of positive integers in $\hat{\mathbb{Z}}$ plays an important role in explicit class field theory, see, e.g. Sect. 2 Chap. III of [11].

One can prove that the map that sends $(O\setminus\{0\}, G)$ to G induces an isomorphism from the group of automorphisms of $(O\setminus\{0\}, G)$ to the group of automorphisms of G. If one replaces $O\setminus\{0\}$ with the group of units $O^\times$, the group of automorphisms of $(O^\times, G)$ is isomorphic to the product of $\hat{\mathbb{Z}}^\times$ and automorphisms of G, for more details see Examples 2.12.1 and 2.12.2 of [24].

Étale-like structures are functorial, rigid and invariant with respect to the links in IUT, whilst frobenius-like structures are used to construct the links. Relations between these two types of structures are of crucial importance in IUT. One of such relations is given by a generalised Kummer map. For example, there is the classical Kummer isomorphism $\mathbb{Q}_p^\times / \mathbb{Q}_p^{\times n} \simeq H^1(G, \mu_n)$ where μ_n are roots of unity of order n in K. The object on the right-hand side of the Kummer isomorphism is solely dependent on the étale-like object G and n, whilst the object on the left-hand side is a frobenius-like object which has the quotient generated by p modulo p^n. Generalised Kummer maps restore multiplication of sufficiently small fields such as number fields and their non-archimedean completions. The substance of Anabelian geometry is to further construct a non-trivial algorithm to restore the addition and the ring structure. The Kummer map embed the frobenius-like object $\mathbb{Z}_p \setminus \{0\}$ into the étale-like object $H^1(G, \mu)$ where μ are all roots of unity in K. This étale-like object can be viewed as a container for the frobenius-like object $\mathbb{Z}_p \setminus \{0\}$. An automorphism of $(O^\times, G)$ produces from a Kummer map $\mathbb{Z}_p^\times \to H^1(G, \mu)$ another one, the corresponding diagram of maps is commutative up to indeterminacy of the action by a suitable element of $\hat{\mathbb{Z}}^\times$, for more details see Example 2.12.2 of [24]. Involving appropriate indeterminacies to make diagrams commutative is important in IUT.

Étale-like objects, unlike frobenius-like objects, pass unmodified through certain non-ring-theoretical links in the IUT theory. For example, the theta-link rescales p to its fixed positive integer power and this is not compatible with the ring structure. In Sect. 2.2 of [24] frobenius-like and étale-like structures are compared with non-zero mass objects and zero mass objects. Depending on the context, the same structure can be a frobenius-like object or an étale-like object.

From a certain perspective, the analogues of these two non-archimedean mathematical structures are particles and waves in archimedean mechanics. Interaction of frobenius-like and étale-like structures via Kummer maps in IUT may be sometimes viewed a little analogous to the relation between particles and waves in quantum mechanics.

- In IUT passing to a set-theoretic subquotient by taking the log-volume at the very last stage of the algorithm, see (LVsQ) in Sect. 3.9, (Stp7) and (Stp8) in Sect. 3.10 of [25], sounds a little similar to a measurement of a quantum system with the wave function collapsing.

Appendix

The IUT theory [26] has applications to the proofs of non-effective abc inequalities for number fields. A recent paper [27] extends the IUT theory and deduces, for the first time, several effective abc inequalities. In the remaining part of the paper we set new questions about an asymptotic symmetry of the moduli space of Frey–Hellegouarch elliptic curves over rational numbers, in relation to how one can deduce from the established effective abc inequalities some stronger abc-type inequalities. These questions go beyond the previously asked abc inequalities conjectures. No knowledge of [27] is required.

For a non-zero integer its radical rad is the product of its prime divisors taken each with multiplicity one and its odd radical rad' is the product of its odd prime divisors taken each with multiplicity one.

One of the established effective abc inequalities in [27] is

for every $\varepsilon > 0$ there is an effectively described constant C'_ε such that for all relatively prime positive integer numbers a, b, the inequality

$$\log(a+b) < 1.5(1+\varepsilon) \cdot \log \text{rad}(ab(a+b)) + C'_\varepsilon$$

holds.

The constant C'_1 is slightly larger than $8.5 \cdot 10^{29}$. Note the constant 1.5 instead of the more common constant 1 in most abc inequality conjectures. The appearance of 1.5 is due to lack of some information at Archimedean places. A version of the previous inequality is also established in [27] over quadratic imaginary fields.

Another established effective abc inequality in [27] is

for every $\varepsilon > 0$ there is an effectively described constant C_ε such that for all relatively prime positive integer numbers a, b, the inequality

$$\log(ab(a+b)) < 3(1+\varepsilon) \cdot \log \text{rad}(ab(a+b)) + C_\varepsilon$$

holds.

The constant C_1 is slightly larger than $1.7 \cdot 10^{30}$. The number 10^{30} is approximately the ratio of the average diameter of the galaxy to the average diameter of atom.

The second abc inequality implies the first one. The second inequality was stated as a conjecture by Szpiro in [28] in 1990.

One can ask how to deduce an effective $(1 + \varepsilon)$-abc inequality from these effective abc inequalities without a further strengthening of IUT and its applications. Towards this aim, let's consider the following situation.

An elliptic curve over $\mathbb{Q}$ with all its 2-torsion points $\mathbb{Q}$-rational is isomorphic over an algebraic closure of $\mathbb{Q}$ to a (Frey–Hellegouarch) curve $E_{a,b}$ with affine equation

$$y^2 = x(x+a)(x-b)$$

for some coprime non-zero integers a, b. It can be written in the Weierstrass form as

$$Y^2 = X^3 - 27c_4 X - 54c_6, \quad c_4 = 16(a^2 + ab + b^2),$$
$$c_6 = 32(b-a)(2a+b)(a+2b).$$

Its discriminant $\Delta = (c_4^3 - c_6^2)/1728 = 16(ab(a+b))^2$. The minimal discriminant of $E_{a,b}$ is the same if 16 does not divide abc or if $a \equiv -1 \bmod 4$ and $b \equiv 0 \bmod 16$, and $16^{-2}(ab(a+b))^2$ if $a \equiv 1 \bmod 4$ and $b \equiv 0 \bmod 16$.

In particular,

$$(a^2 + ab + b^2)^3 = ((b-a)(2a+b)(a+2b)/2)^2 + 3^3(ab(a+b)/2)^2. \qquad (\dagger)$$

The j-invariant of the Weierstrass equation is

$$j_{a,b} = 2^8 \cdot \frac{(a^2+ab+b^2)^3}{(ab(a+b))^2} = 2^6 \cdot \frac{((b-a)(2a+b)(a+2b))^2}{(ab(a+b))^2} + 2^6 \cdot 3^3.$$

If 16 does not divide $ab(a+b)$ then $\mathrm{cond}(E_{a,b}) < 2^{12}\mathrm{rad}'(ab(a+b))$. If $16|ab(a+b)$ and say $4|(a-1)$, $16|b$ then $\mathrm{cond}(E_{a,b}) = \mathrm{rad}(2^{-4}ab(a+b)) \leqslant \mathrm{rad}(ab(a+b))$. If $16|ab(a+b)$ and say $4|(a+1)$, $16|b$ then $\mathrm{cond}(E_{a,b}) \leqslant 2^{4+2l}\mathrm{rad}'(ab(a+b))$ where l is the maximal power of 2 dividing b. All this is very well known. See, e.g. Sect. 12.5 of [2]. Note that the statement "Since E has multiplicative reduction at all primes $p|\Delta$" in the top line of its p. 434 is incorrect as the example of $E_{1,16}$ shows, but the inequality for the LHS and RHS of the next displayed inequality on that page is correct.

Now let in addition $0 < a < b$, a, b are still coprime. Put $c = a+b$. Define

$$A = (b-a)/d, \quad B = (2a+b)/d, \quad C = A+B = (a+2b)/d,$$

where $d = \gcd(b-a, 2a+b) (= 1 \text{ or } 3)$. Then $0 < A < B$, and A, B are coprime.

We have
$$a^2 + ab + b^2 = d^2(A^2 + AB + B^2)/3,$$
$$ab(a+b) = d^3(B-A)(A+2B)(2A+B)/3^3,$$
$$(b-a)(2a+b)(a+2b) = d^3 AB(A+B).$$

The map $\phi: (a,b) \mapsto (A,B)$ is an involution: $\phi^2 = \mathrm{id}$. The involution ϕ corresponds to $x \mapsto (2+x)/(x-1)$ on $\mathbb{P}^1$ sending b/a to B/A.

Thus we have an involution map on the moduli space of Frey–Hellegouarch elliptic curves: $E_{a,b} \mapsto E_{A,B}$. The map ϕ relates the two terms on the RHS of $(\dagger)$.

From $(\dagger)$ one gets

$$(A^2 + AB + B^2)^3 = 3^3(AB(A+B)/2)^2 + ((B-A)(2A+B)(A+2B)/2)^2.$$

We also have $j_{A,B} = 12^3 j_{a,b}/(j_{a,b} - 12^3) = (12^{-3} - j_{a,b}^{-1})^{-1}$.

Question (abc-ABC question). *Are the following equivalent statements true?*
1. $\mathrm{rad}(abc)$ and $\mathrm{rad}(ABC)$ are effectively asymptotically equal, i.e. for every $\varepsilon > 0$ there are constants $\mathfrak{c}_\epsilon, \mathfrak{c}'_\epsilon$, effectively depending on ϵ, such that for all relatively prime positive $a < b$

$$\mathrm{rad}(abc) < \mathfrak{c}_\epsilon \cdot \mathrm{rad}(ABC)^{1+\epsilon}, \quad \mathrm{rad}(ABC) < \mathfrak{c}'_\epsilon \cdot \mathrm{rad}(abc)^{1+\epsilon}.$$

2. For every $\epsilon > 0$ there is a positive constant κ_ϵ such that for all positive coprime integers $a < b$

$$\mathrm{rad}((b-a)(2a+b)(a+2b)) < \kappa_\epsilon \cdot \mathrm{rad}(ab(a+b))^{1+\epsilon}$$

with κ_ϵ effectively dependent on ϵ.
3. $\mathrm{rad}(\Delta(E_{a,b}))$ and $\mathrm{rad}(\Delta(E_{A,B}))$ are effectively asymptotically equivalent.
4. $\mathrm{rad}(c_6(E_{a,b}))$ and $\mathrm{rad}(\Delta(E_{a,b}))$ are effectively asymptotically equivalent.

The proof of the equivalences is immediate. The author does not know the proof of either a positive answer or a negative answer to this abc-ABC question, and several experts in arithmetic of elliptic curves were in a similar situation.

The positive answer to the question signifies a new asymptotic symmetry of the moduli space of elliptic curves over $\mathbb{Q}$ all of whose 2-torsion points are $\mathbb{Q}$-rational.

Fix a positive integer m. The second abc inequality above implies that *for every positive ε for all non-zero integers a, b, c such that $a + b + c = 0$ and $\gcd(a, b, c)$ divides m we have*

$$\log |abc| < 3(1+\epsilon) \cdot \log \mathrm{rad}(abc) + C_\epsilon + 3 \log m. \tag{$\sharp$}$$

In view of (†), consider the equation

$$x^3 = y^2 + 3^3 z^2$$

where x, y, z are positive integers such that $\gcd(x, y, z) | 3$. We use the standard notation $f \ll_\epsilon g$ for real functions on positive real numbers depending on parameter $\epsilon > 0$ which means $|f(\epsilon, x)| \leqslant C(\epsilon) g(\epsilon, x)$ for a real $C(\epsilon)$ depending on ϵ.

Applying ($\sharp$), we obtain $x^3 y^2 z^2 \ll_\epsilon \mathrm{rad}(xyz)^{3(1+\epsilon)}$. Since $y^2 \cdot 3^3 z^2 \leqslant x^6/4$, we deduce $yz \ll_\epsilon \mathrm{rad}(xyz)^{1+\epsilon}$. Assume that $y^2 \leqslant 3^3 z^2$, then we deduce $y \ll_\epsilon \mathrm{rad}(xyz)^{(1+\epsilon)/2}$. Since $x^3 \leqslant 2 \cdot 3^3 z^2$, we get $x^6 y^2 \leqslant x^3 \cdot 2 \cdot 3^3 z^2 \cdot y^2 \ll_\epsilon \mathrm{rad}(xyz)^{3(1+\epsilon)}$ and $x^6 y^6 \ll_\epsilon \mathrm{rad}(xyz)^{5(1+\epsilon)}$, so $xy \ll_\epsilon \mathrm{rad}(z)^{5(1+\epsilon)}$. Substituting the latter in the RHS of $y \ll_\epsilon \mathrm{rad}(xyz)^{(1+\epsilon)/2}$, we obtain $y \ll_\epsilon \mathrm{rad}(z)^{3(1+\epsilon)}$. From $x^6 y^2 \ll_\epsilon \mathrm{rad}(xyz)^{3(1+\epsilon)}$ we deduce $x^3 \ll_\epsilon y^{1+\epsilon} \cdot \mathrm{rad}(z)^{3(1+\epsilon)}$ so $x^3 \ll_\epsilon \mathrm{rad}(z)^{6(1+\epsilon)}$, hence $x \ll_\epsilon \mathrm{rad}(z)^{2(1+\epsilon)}$. Thus, ($\sharp$) implies: if $y^2 \leqslant 3^3 z^2$ then $x \ll_\epsilon \mathrm{rad}(z)^{2(1+\epsilon)}$. Similarly, if $y^2 \geqslant 3^3 z^2$ then $x \ll_\epsilon \mathrm{rad}(y)^{2(1+\epsilon)}$. All the implied constants are explicit functions of C_ϵ.

Now, for positive coprime $a < b$ denote $x = a^2 + ab + b^2$, $y = (b-a)(2a+b)(a+2b)/2$, $z = ab(a+b)/2$. Then $x^3 = y^2 + 3^3 z^2$. Note that since a and b are coprime, $\gcd(x, y, z)$ divides 3, so we can apply the previous paragraph to x, y, z. We deduce from the previous paragraph: if $((b-a)(2a+b)(a+2b))^2 \leqslant 3^3 (ab(a+b))^2$ then $3c^2/4 \leqslant a^2 + ab + b^2 \ll_\epsilon \mathrm{rad}(abc)^{2+\epsilon}$ and hence $c \ll_\epsilon \mathrm{rad}(abc)^{1+\epsilon}$; if $((b-a)(2a+b)(a+2b))^2 \geqslant 3^3 (ab(a+b))^2$, i.e. $((B-A)(2A+B)(A+2B))^2 \leqslant 3^3 (AB(A+B))^2$, then $A^2 + AB + B^2 \ll_\epsilon \mathrm{rad}(ABC)^{2+\epsilon}$ and hence $c \ll_\epsilon \mathrm{rad}(ABC)^{1+\epsilon}$. All the implied constants are explicit functions of C_ε.

Therefore, the inequality ($\sharp$) implies:

Theorem 1 *For every positive ε there is an effectively described constant K_ε such that for all coprime positive integers a, b and their sum $c = a + b$ and A, B, C defined for a, b as above*

$$\log c < (1+\epsilon) \cdot \log \max\{\mathrm{rad}(abc), \mathrm{rad}(ABC)\} + K_\varepsilon.$$

Using Theorem 1 we obtain.

Theorem 2 *Assume that the* abc-ABC Question *has positive answer. Then for every positive ε there is an effectively described constant L_ε such that for all coprime positive integers a, b and their sum $c = a + b$ the inequality*

$$\log c < (1+\varepsilon) \cdot \log \mathrm{rad}(abc) + L_\varepsilon$$

holds.

References

1. S. Aaronson, D. Gottesman, Improved simulation of stabilizer circuits, arXiv:quant-ph/0406196
2. E. Bombieri, W. Gubler, *Heights in Diophantine Geometry* (CUP, 2007)
3. F. Del Santo, N. Gisin, Physics without determinism: alternative interpretations of classical physics, arXiv:1909.03697 [quant-ph]
4. A. Döring, C.J. Isham, A topos foundation for theories of physics. I. Formal languages for physics. J. Math. Phys. **49**, 053515 (2008), arXiv:quant-ph/0703060 [quant-ph]
5. A. Döring, C.J. Isham, A topos foundation for theories of physics. II. Daseinisation and the liberation of quantum theory. J. Math. Phys.**49**, 053516 (2008), arXiv:quant-ph/0703062 [quant-ph]
6. A. Döring, C.J. Isham, A topos foundation for theories of physics. III. The representation of physical quantities with arrows. J. Math. Phys. **49**, 053517 (2008), arXiv:quant-ph/0703064 [quant-ph]
7. A. Döring, C.J. Isham, A topos foundation for theories of physics. IV. Categories of systems. J. Math. Phys. **49**, 053518 (2008), arXiv:quant-ph/0703066 [quant-ph]
8. A. Döring, C.J. Isham, 'What is a thing?': topos theory in the foundations of physics, arXiv: 0803.0417 [quant-ph]
9. A. Döring, C.J. Isham, Classical and quantum probabilities as truth values, arXiv: 1102.2213v1
10. I. Fesenko, Arithmetic deformation theory via fundamental groups and nonarchimedean theta-functions, notes on the work of Shinichi Mochizuki. Eur. J. Math. **1**, 405–440 (2015)
11. I. Fesenko, Core topics in number theory I, lecture notes (2023), https://ivanfesenko.org/wp-content/uploads/Q/C1/partI.pdf
12. C. Flori, *A First Course in Topos Quantum Theory*. Lecture Notes Physics, vol. 868 (Springer 2013)
13. C. Flori, *A Second Course in Topos Quantum Theory*. Lecture Notes Physics, vol. 944 (Springer 2018)
14. N. Gisin, Indeterminism in physics and intuitionistic mathematics, arXiv:2011.02348 [quant-ph]
15. R. Goldblatt, *Topoi: The Categorial Analysis of Logic* (North-Holland, London, 1984)
16. J. Hoshi, S. Mochizuki, S. Tsujimura, Combinatorial construction of the absolute Galois group of the field of rational numbers, preprint (2022), https://www.kurims.kyoto-u.ac.jp/~motizuki/Combinatorial%20absolute%20Galois%20groups.pdf
17. C.J. Isham, Topos theory and consistent histories: the internal logic of the set of all consistent sets. Int. J. Theor. Phys. **36**, 785 (1997), arXiv:gr-qc/9607069
18. C.J. Isham, Topos methods in the foundations of physics, arXiv:1004.3564 [quant-ph]

19. L. Lafforgue, Geometry according to Grothendieck: glimpses from "Récoltes et Semailles", in *Unifying Themes in Geometry*. Lake Como School of Advanced Studies, 27–30 September 2021, slides are available from https://www.dropbox.com/s/pnq81jcu8yhtlrj/LafforgueSlides.pdf?dl=0, video is available from https://www.youtube.com/watch?v=-CtzBL83bf0
20. S. MacLane, I. Moerdijk, *Sheaves in Geometry and Logic: A First Introduction to Topos Theory* (Springer, London, 1968)
21. S. Mochizuki, The profinite Grothendieck conjecture for closed hyperbolic curves over number fields. J. Math. Sci. Univ Tokyo **3**, 571–627 (1996)
22. S. Mochizuki, The local pro-p Grothendieck conjecture for hyperbolic curves. Invent. Math. **138**, 319–423 (1999)
23. S. Mochizuki, Topics surrounding the Anabelian geometry of hyperbolic curves, in *Galois Groups and Fundamental Groups* (MSRI Publications 41, CUP, 2003), pp. 119–165
24. S. Mochizuki, The mathematics of mutually alien copies: from Gaussian integrals to inter-universal Teichmüller theory, in *Inter-universal Teichmüller Theory Summit 2016*, RIMS Kōkyūroku Bessatsu B84 (2021), pp. 23–192, available from https://www.kurims.kyoto-u.ac.jp/~motizuki/Alien%20Copies,%20Gaussians,%20and%20Inter-universal%20Teichmuller%20Theory.pdf
25. S. Mochizuki, On the essential logical structure of inter-universal Teichmüller theory in terms of logical and/or relations: report on the occasion of the publication of the four main papers on inter-universal Teichmüller theory, preprint (2023), available from https://www.kurims.kyoto-u.ac.jp/~motizuki/Essential%20Logical%20Structure%20of%20Inter-universal%20Teichmuller%20Theory.pdf
26. S. Mochizuki, Inter-universal Teichmüller theory I: Constructions of Hodge theaters. Publ. Res. Inst. Math. Sci. **57**, 3–207 (2021); II: Hodge-Arakelov-theoretic evaluation. Publ. Res. Inst. Math. Sci. **57**, 209–401 (2021); III: Canonical splittings of the log-theta-lattice. Publ. Res. Inst. Math. Sci. **57**, 403–626 (2021); IV: Log-volume computations and set-theoretic foundations, Publ. Res. Inst. Math. Sci. **57**, 627–723 (2021)
27. S. Mochizuki, I. Fesenko, Y. Hoshi, A. Minamide, W. Porowski, Explicit estimates in inter-universal Teichmüller theory. Kodai Math. J. **45**, 175–236 (2022)
28. L. Szpiro, Discriminant et conducteur des courbes elliptiques. Astérisque **183**, 7–18 (1990)
29. A. Tamagawa, The Grothendieck conjecture for affine curves. Compositio Math. **109**, 135–194 (1997)

Open Access This chapter is licensed under the terms of the Creative Commons Attribution 4.0 International License (http://creativecommons.org/licenses/by/4.0/), which permits use, sharing, adaptation, distribution and reproduction in any medium or format, as long as you give appropriate credit to the original author(s) and the source, provide a link to the Creative Commons license and indicate if changes were made.

The images or other third party material in this chapter are included in the chapter's Creative Commons license, unless indicated otherwise in a credit line to the material. If material is not included in the chapter's Creative Commons license and your intended use is not permitted by statutory regulation or exceeds the permitted use, you will need to obtain permission directly from the copyright holder.

Quantum Walks for the Working Mathematician

Hayato Saigo and Shingo Sugiyama

Abstract This article is a survey on quantum walks for the working mathematician. We show some examples of quantum walks and introduce a candidate for a general definition of quantum walks to give a unified perspective for various concrete models. A part of this article is based on the second author's talk on "Quantum walks and zeta functions" at Crypto-Math CREST mini-workshop "Quantum computation and cryptography" on September 12th in 2022, under the Sponsorship of DWANGO Co., Ltd.

Keywords Quantum walks · Noncommutative probability theory · †–categories (dagger categories) · Rigs · Category algebras · States on categories

1 Introduction

The aim of this article is to introduce *quantum walks* to mathematicians. Quantum walks are, in short, quantum-like dynamics on discrete spacetimes. They have been studied in many aspects in terms of mathematics, physics, and quantum computers, etc. Two phenomena called localization and linear diffusion are remarkable features of quantum walks contrary to random walks, and therefore quantum walks enjoy potential for applications. As of now there are a lot of applications such as quantum search algorithms [5], quantum teleportations [27], isotope separations [4, 17, 18], topological insulators [8, 20], financial time series analysis [15], dressed photons [2, 22], and so on. For details of history and fundamental properties on quantum walks, see [6, 13], etc.

H. Saigo (✉)
Faculty of Social Informatics, ZEN University, Zushi, Kanagawa, Japan
e-mail: hayato_saigo@zen.ac.jp

S. Sugiyama (✉)
Faculty of Mathematics and Physics, Institute of Science and Engineering, Kanazawa University, Kanazawa, Ishikawa, Japan
e-mail: s-sugiyama@se.kanazawa-u.ac.jp

In spite of a lot of examples of quantum walks (cf. [6]), there are no unified definitions of quantum walks as of now. As a candidate for the unified definition of quantum walks, we introduce quantum walks on †-categories (dagger categories) in this article. The notion of †-categories gives us a general notion of spacetime and therefore it is suitable for unifying dynamics on spacetimes. Our definition of quantum walks on †-categories actually includes known examples of quantum walks. The authors hope that this article will be an opportunity to encourage mathematicians to apply quantum walks to other fields such as cryptography, etc., and contribute to develop a unified theory of quantum walks and related research areas such as quantum field theory.

This article is organized as follows. In Sect. 2, we review random walks since quantum walks are regarded as a quantization of random walks. For simplicity we treat quantum walks on the 1-dimensional lattice $\mathbb{Z}$. In Sect. 3, we consider 2-state quantum walks on $\mathbb{Z}$. We see that the normal distribution in random walks is replaced with the Konno distribution in quantum walks. In Sect. 4, we show three examples of quantum walks. First we consider the Hadamard walk and compare it with the random walk on $\mathbb{Z}$ via the return probability and the distribution of the random variables associated with the two walks. Furthermore, two phenomena called localization and linear diffusion are observed in this section. These two phenomena are not always observed in any quantum walks. We mainly show stationary quantum walks and free quantum walks as examples of localization and linear diffusion. In Sect. 5, as another example of quantum walks we introduce Grover walks on certain infinite graphs called jellyfish graphs, which have some application to a mathematical modeling of physical system called dressed photons. In Sect. 6, we exhibit a picture of general theory of quantum walks by using noncommutative probability theory and the concept of †-categories.

In this article, $\mathbb{Z}_{\geq 0}$ denotes the set of all non-negative integers.

2 Review of Random Walks

Before introducing quantum walks, let us first consider random walks on $\mathbb{Z}$ in a situation that a random walker moves at discrete times $n \in \mathbb{Z}_{\geq 0}$ due to the coin toss by the following rule:

- The random walker stands on $x = 0$ at time $n = 0$.
- At each time, the random walker goes to the left with probability p and to the right with probability q, respectively.

Here p and q are positive real numbers such that $p + q = 1$.

Let $X_n \in \mathbb{Z}$ be a random variable as the locus of the random walker at time n.

Question 2.1 *For $n \in \mathbb{Z}_{\geq 0}$ and $x \in \mathbb{Z}$, what is the probability that the random walker is on the locus x at time n? Namely, what is the value of $P(X_n = x)$?*

For example, we have
- $P(X_1 = 1) = q$, $P(X_1 = -1) = p$,
- $P(X_2 = 2) = q^2$, $P(X_2 = 0) = pq + qp = 2pq$, $P(X_2 = -2) = p^2$.

For the question above, we consider the coin toss. Suppose that the random walker moves to the left if the coin lands on heads and to the right otherwise. If a coin turns heads ℓ times and tails m times, then we have $\ell + m = n$ and $m - \ell = x$. By a direct computation, ℓ and m are determined as $\ell = \frac{n-x}{2}$ and $m = \frac{n+x}{2}$, and the answer to the question above is given as follows.

Proposition 2.2 *We have*

$$P(X_n = x) = \begin{cases} \binom{n}{\ell} p^\ell q^m = \binom{n}{\frac{n-x}{2}} p^{\frac{n-x}{2}} q^{\frac{n+x}{2}} & (n + x \in 2\mathbb{Z} \ \& \ x \in [-n, n]), \\ 0 & (otherwise). \end{cases}$$

Corollary 2.3 *The return probability at time $2n$ is given by*

$$P(X_{2n} = 0) = \binom{2n}{n} p^n q^n.$$

By combining the corollary for $p = q = 1/2$ with the Stirling formula

$$n! \sim \sqrt{2\pi n} \left(\frac{n}{e}\right)^n, \quad n \to \infty,$$

we have the asymptotics

$$P(X_{2n} = 0) = \binom{2n}{n} \left(\frac{1}{2}\right)^n \left(\frac{1}{2}\right)^n \sim \frac{1}{\sqrt{\pi n}}, \quad n \to \infty.$$

Let us recall the central limit theorem, which tells us that any probability distribution converges to the normal distribution (the Gaussian distribution). In the setting of our random walk on $\mathbb{Z}$ (the random variables $\{X_n\}_n$), the central limit theorem is stated as follows.

Theorem 2.4 (The central limit theorem) *For any $a, b \in \mathbb{R}$ with $a < b$, we have*

$$\lim_{n \to \infty} P\left(a \leq \frac{X_n - (q-p)n}{\sqrt{4pqn}} \leq b\right) = \int_a^b \frac{1}{\sqrt{2\pi}} e^{-x^2/2} dx.$$

Corollary 2.5 *Assume $p = q = 1/2$. For $a, b \in \mathbb{R}$ with $a < b$, we have*

$$\lim_{n \to \infty} P\left(a \leq \frac{X_n}{\sqrt{n}} \leq b\right) = \int_a^b \frac{1}{\sqrt{2\pi}} e^{-x^2/2} dx.$$

3 Introducing Quantum Walks

Next let us consider quantum walks. For simplicity, we discuss 2-state quantum walks on the 1-dimensional lattice $\mathbb{Z}$ at discrete time.[1] We assume a quantum walker, which is the probability of existence at every point as time passes. Hence the quantum walker stands on a lot of loci simultaneously!

The quantum walker has overlapping of two directions "left" and "right". These two directions are called chiralities (or spins). We define the chiralities $|L\rangle$ and $|R\rangle$ by $|L\rangle := \begin{bmatrix}1\\0\end{bmatrix}$ and $|R\rangle := \begin{bmatrix}0\\1\end{bmatrix}$, respectively. We have $\mathbb{C}^2 = \mathbb{C}|L\rangle + \mathbb{C}|R\rangle$ and this space is called a coin space.

Instead of $p + q = 1$ in the random walk, We prepare a unitary operator U on $\mathbb{C}^2$ and operators P, Q on $\mathbb{C}^2$ such that $P + Q = U$. If we take a unitary matrix

$$U = \begin{bmatrix} a & b \\ c & d \end{bmatrix},$$

then P and Q are defined by

$$P := \begin{bmatrix}1&0\\0&0\end{bmatrix}U = \begin{bmatrix}a&b\\0&0\end{bmatrix}, \qquad Q := \begin{bmatrix}0&0\\0&1\end{bmatrix}U = \begin{bmatrix}0&0\\c&d\end{bmatrix}.$$

The operator P means "Go to the left", and the operator Q means "Go to the right", respectively. Note the relations $|a|^2 + |b|^2 = |a|^2 + |c|^2 = |c|^2 + |d|^2 = |b|^2 + |d|^2 = 1$ and $a\bar{b} + c\bar{d} = a\bar{c} + b\bar{d} = 0$. In other words, we have

$$PP^\dagger + QQ^\dagger = P^\dagger P + Q^\dagger Q = \begin{bmatrix}1&0\\0&1\end{bmatrix},$$
$$PQ^\dagger = Q^\dagger P = QP^\dagger = P^\dagger Q = \begin{bmatrix}0&0\\0&0\end{bmatrix},$$

where we put $X^\dagger := {}^t\overline{X}$ for any square matrix X.

Let $\ell^2(\mathbb{Z})$ be the space of all $\mathbb{C}$-valued square-integrable functions on $\mathbb{Z}$ with respect to the counting measure. We regard $\ell^2(\mathbb{Z}) \otimes \mathbb{C}^2$ as the space of all functions $\Psi : \mathbb{Z} \to \mathbb{C}^2$ such that

$$\sum_{x \in \mathbb{Z}} \|\Psi(x)\|^2 < \infty$$

with $\|\ \|$ being the Euclidean norm on $\mathbb{C}^2$. We remark that $\ell^2(\mathbb{Z}) \otimes \mathbb{C}^2$ is a Hermitian space with a standard inner product $\langle \Phi, \Psi \rangle := \sum_{x \in \mathbb{Z}} {}^t\Phi(x)\overline{\Psi(x)}$. Any element of $\ell^2(\mathbb{Z}) \otimes \mathbb{C}^2$ is called *a state* (or *a probability amplitude*). For $\Psi \in \ell^2(\mathbb{Z}) \otimes \mathbb{C}^2$, we define $\Psi^L : \mathbb{Z} \to \mathbb{C}$ and $\Psi^R : \mathbb{Z} \to \mathbb{C}$ by

$$\Psi(x) = \begin{bmatrix} \Psi^L(x) \\ \Psi^R(x) \end{bmatrix} = \Psi^L(x)|L\rangle + \Psi^R(x)|R\rangle, \qquad x \in \mathbb{Z}.$$

---

[1] The term "state" should not be confused with that in Sect. 6. Although it is possible to consider the former as the special case of the latter, we will not discuss this point in the present article.

Let us take $\Psi_0 \in \ell^2(\mathbb{Z}) \otimes \mathbb{C}^2$ such that $\sum_{x \in \mathbb{Z}} \|\Psi_0(x)\|^2 = 1$. Define an operator $U^{(s)}$ on $\ell^2(\mathbb{Z}) \otimes \mathbb{C}^2$ by

$$U^{(s)}\Psi(x) := P\Psi(x+1) + Q\Psi(x-1), \quad x \in \mathbb{Z}, \quad \Psi \in \ell^2(\mathbb{Z}) \otimes \mathbb{C}^2.$$

The suffix "(s)" of $U^{(s)}$ is the initial of "system". The operator $U^{(s)}$ is unitary. Indeed, we observe

$$\begin{aligned}
&\langle U^{(s)}\Phi, U^{(s)}\Psi \rangle \\
&= \sum_{x \in \mathbb{Z}} {}^t\overline{\{P\Phi(x+1) + Q\Phi(x-1)\}}\{P\Psi(x+1) + Q\Psi(x-1)\} \\
&= \sum_{x \in \mathbb{Z}} {}^t\overline{\Phi(x+1)} {}^t\overline{P}P\,\Psi(x+1) + \sum_{x \in \mathbb{Z}} {}^t\overline{\Phi(x-1)} {}^t\overline{Q}Q\,\Psi(x-1) \\
&\quad + \sum_{x \in \mathbb{Z}} \{{}^t\overline{\Phi(x+1)} {}^t\overline{P}Q\,\Psi(x-1) + {}^t\overline{\Phi(x-1)} {}^t\overline{Q}P\,\Psi(x+1)\} \\
&= \sum_{x \in \mathbb{Z}} {}^t\overline{\Phi(x)}({}^t\overline{P}P + {}^t\overline{Q}Q)\Psi(x) = \sum_{x \in \mathbb{Z}} {}^t\overline{\Phi(x)}\Psi(x) = \langle \Phi, \Psi \rangle
\end{aligned}$$

for all $\Phi, \Psi \in \ell^2(\mathbb{Z}) \otimes \mathbb{C}^2$.

Definition 3.1 (*1-dimensional 2-state quantum walks*) Let us take $\Psi_0 \in \ell^2(\mathbb{Z}) \otimes \mathbb{C}^2$, a state at time 0, such that $\sum_{x \in \mathbb{Z}} \|\Psi_0(x)\|^2 = 1$. We call (U, P, Q, Ψ_0) a 1-dimensional 2-state quantum walk.

The operator $U^{(s)}$ is called a coin operator or a quantum coin, and the operators P and Q are called shift operators. The state Ψ_0 is called an initial state, and $\Psi_n := (U^{(s)})^n \Psi_0$ is called the state at time n, respectively.

Remark 3.2 For general positive integers d and n, d-dimensional n-state quantum walks are defined as dynamics of states on $\ell^2(\mathbb{Z}^d) \otimes \mathbb{C}^n$ in a general setting (cf. [25, 26]).

By definition, we have the following recurrence equation of states $\{\Psi_n\}_n$ for a 1-dimensional 2-state quantum walk (U, P, Q, Ψ_0):

$$\Psi_{n+1}(x) = P\Psi_n(x+1) + Q\Psi_n(x-1), \quad x \in \mathbb{Z}, \quad n \in \mathbb{Z}_{\geq 0}.$$

It is regarded as an analogue of the recurrence equation

$$P(X_{n+1} = x) = pP(X_n = x+1) + qP(X_n = x-1) \quad x \in \mathbb{Z}, \quad n \in \mathbb{Z}_{\geq 0}$$

in the case of the random walk in Sect. 2.

For a 1-dimensional 2-state quantum walk (U, P, Q, Ψ_0) and any $x \in \mathbb{Z}$, we define the probability $P(X_n = x)$ that the quantum walker stands on the locus x at time n by

$$P(X_n = x) := \|\Psi_n(x)\|^2 = |\Psi_n^L(x)|^2 + |\Psi_n^R(x)|^2.$$

Then we have $\sum_{x \in \mathbb{Z}} P(X_n = x) = 1$. The symbol X_n is regarded as a random variable expressing the locus of the quantum walker at time n.

4 Examples of Quantum Walks on $\mathbb{Z}$

In this section, we show three examples of quantum walks on $\mathbb{Z}$.

4.1 The Hadamard Walk

We consider *the Hadamard walk*. This is a quantum walk determined by

$$U := \begin{bmatrix} \frac{1}{\sqrt{2}} & \frac{1}{\sqrt{2}} \\ \frac{1}{\sqrt{2}} & -\frac{1}{\sqrt{2}} \end{bmatrix}.$$

This unitary matrix U is called *the Hadamard gate*. Then the shift operators are defined as

$$P := \begin{bmatrix} \frac{1}{\sqrt{2}} & \frac{1}{\sqrt{2}} \\ 0 & 0 \end{bmatrix}, \quad Q := \begin{bmatrix} 0 & 0 \\ \frac{1}{\sqrt{2}} & -\frac{1}{\sqrt{2}} \end{bmatrix}.$$

We have the relation $P + Q = U$. Define an initial state $\Psi_0 \in \ell^2(\mathbb{Z}) \otimes \mathbb{C}^2$ by

$$\Psi_0 := \delta_0 \otimes \begin{bmatrix} \frac{1}{\sqrt{2}} \\ \frac{i}{\sqrt{2}} \end{bmatrix}.$$

Here $\delta_n : \mathbb{Z} \to \mathbb{C}$ for $n \in \mathbb{Z}$ is the function satisfying $\delta_n(n) = 1$ and $\delta_n(x) = 0$ for all $x \in \mathbb{Z} - \{n\}$. The quantum walk (U, P, Q, Ψ_0) is called *the Hadamard walk*.

For $x \in \mathbb{Z}$ and $n \in \mathbb{Z}_{\geq 0}$, the probability $P(X_n = x)$ is computed as follows. We take unique $\ell, m \in \mathbb{Z}_{\geq 0}$ such that $\ell + m = n$ and $m - \ell = x$. We put

$$\Xi_n(\ell, m) = \sum_{(\ell_j)_j, (m_j)_j} P^{\ell_n} Q^{m_n} P^{\ell_{n-1}} Q^{m_{n-1}} \cdots P^{\ell_2} Q^{m_2} P^{\ell_1} Q^{m_1},$$

where $(\ell_j)_j$ and $(m_j)_j$ run over the set of element of $\mathbb{Z}_{\geq 0}^n$ such that $\ell_1 + \ell_2 + \cdots + \ell_{n-1} + \ell_n = \ell$, $m_1 + m_2 + \cdots + m_{n-1} + m_n = m$ and $\ell_j + m_j = 1$ for all $j = 1, \ldots, n$. Then $\Psi_n(x)$ and $P(X_n = x)$ are given by

$$\Psi_n(x) = \Xi_n(\ell, m) \begin{bmatrix} \frac{1}{\sqrt{2}} \\ \frac{i}{\sqrt{2}} \end{bmatrix}, \quad P(X_n = x) = \left\| \Xi_n(\ell, m) \begin{bmatrix} \frac{1}{\sqrt{2}} \\ \frac{i}{\sqrt{2}} \end{bmatrix} \right\|^2.$$

Table 1 Probability at x at time $n = 4$

x	-4	-2	0	2	4
(Hadamard walk) $P(X_4 = x)$	$\frac{1}{16}$	$\frac{6}{16}$	$\frac{2}{16}$	$\frac{6}{16}$	$\frac{1}{16}$
(Random walk) $P(X_4 = x)$	$\frac{1}{16}$	$\frac{4}{16}$	$\frac{6}{16}$	$\frac{4}{16}$	$\frac{1}{16}$

Table 1 shows the probabilities at any locus x at time $n = 4$ both in the case of the Hadamard walk and the random walk explained in Sect. 2 (cf. [10]).

The return probability for the Hadamard walk is given by Konno [12].

Proposition 4.1 *The return probability of the Hadamard walk is given as*

$$P(X_{2n+1} = 0) = 0, \quad (n \in \mathbb{Z}_{\geq 0}),$$

$$P(X_{4m} = 0) = P(X_{4m+2} = 0) = \frac{1}{2^{4m+1}}\binom{2m}{m}^2, \quad (m \in \mathbb{Z}, m \geq 1),$$

$P(X_0 = 0) = 1$ *and* $P(x_2 = 0) = 1/2$. *In particular, we have*

$$P(X_{2n} = 0) \sim \frac{1}{\pi n}, \quad n \to \infty$$

by the Stirling formula.

This proposition can be proved by the formula

$$P(X_{2n} = 0) = \frac{1}{2}\{p_{n-1}(0)^2 + p_n(0)^2\},$$

where

$$p_n(x) = \frac{1}{2^n n!}\frac{d^n}{dx^n}(x^2 - 1)^n$$

is the Legendre polynomial of degree n. We sketch a proof of this formula following [12]. By the explicit formula of $\Xi_n(\ell, m)$ in [12, Lemma 4.1] (see also [10, Lemma 2]), we have

$$P(X_{2n} = 0) = \left\|\Xi_{2n}(n,n)\begin{bmatrix}\frac{1}{\sqrt{2}}\\ \frac{i}{\sqrt{2}}\end{bmatrix}\right\|^2 = \left(\frac{1}{2}\right)^{2n}\Bigg[2n^2\Bigg\{\sum_{j=1}^{n}\frac{(-1)^j}{j}\binom{n-1}{j-1}^2\Bigg\}^2$$

$$- 4n\sum_{j=1}^{n}\sum_{k=1}^{n}\frac{(-1)^{j+k}}{j}\binom{n-1}{j-1}^2\binom{n-1}{k-1}^2$$

$$+ 4\sum_{j=1}^{n}\sum_{k=1}^{n}(-1)^{j+k}\binom{n-1}{j-1}^2\binom{n-1}{k-1}^2\Bigg].$$

By the formula $p_n(x) = {}_2F_1(-n, n+1; 1; \frac{1-x}{2})$ and the connection formula ${}_2F_1(a, b; c; z) = (1-z)^{-a} {}_2F_1(a, c-b; c; \frac{z}{z-1})$, of the Gaussian hypergeometric function, we obtain

$$P(X_{2n} = 0)$$
$$= \left(\frac{1}{2}\right)^{2n} \left[2n^2 \times \frac{2^{2(n-1)}}{n^2} \{p_{n-1}(0) - p_n(0)\}^2 \right.$$
$$\left. - 4n \times \frac{2^{2(n-1)}}{n} \{p_{n-1}(0) - p_n(0)\} p_{n-1}(0) + 4 \times 2^{2(n-1)} p_{n-1}(0)^2 \right.$$
$$= \frac{1}{2} \{p_{n-1}(0)^2 + p_n(0)^2\}.$$

From the formula of $P(X_{2n} = 0)$, we have

$$P(X_{4m} = 0) = P(X_{4m+2} = 0) = \frac{1}{2} p_{2m}(0) = \frac{1}{2^{4m+1}} \binom{2m}{m}^2$$

for integers $m \geq 2$.

Remark 4.2 The generating function of $\{P(X_n = 0)\}_n$ is expressed as

$$\sum_{n=0}^{\infty} P(X_n = 0) z^n = \frac{1+z^2}{\pi} K(z^2) + \frac{1}{2},$$

where $K(k)$ is the complete elliptic integral of the first kind (cf. [12]).

We note that the asymptotics $P(X_{2n} = 0) \sim \frac{1}{\pi n}$ of the Hadamard walk is quite differrent from $P(X_{2n} = 0) = \binom{2n}{n}(\frac{1}{2})^n(\frac{1}{2})^n \sim \frac{1}{\sqrt{\pi n}}$ in the random walk in Sect. 2.

The limit distribution of the Hadamard walk is given as the special case of the weak limit theorem of 1-dimensional 2-state quantum walks [9, 10].

Proposition 4.3 (The weak limit theorem) *For $a, b \in \mathbb{R}$ such that $a < b$, we have*

$$\lim_{n \to \infty} P\left(a \leq \frac{X_n}{n} \leq b\right) = \int_a^b \frac{1}{\pi(1-x^2)\sqrt{1-2x^2}} \operatorname{ch}_{[-\frac{1}{\sqrt{2}}, \frac{1}{\sqrt{2}}]}(x) dx,$$

where $\operatorname{ch}_{[-\frac{1}{\sqrt{2}}, \frac{1}{\sqrt{2}}]}$ *is the characteristic function of the interval $[-\frac{1}{\sqrt{2}}, \frac{1}{\sqrt{2}}]$ on $\mathbb{R}$.*

The distribution

$$\frac{1}{\pi(1-x^2)\sqrt{1-2x^2}} dx$$

on $[-\frac{1}{\sqrt{2}}, \frac{1}{\sqrt{2}}]$ is called the Konno distribution. Konno [9, 10] proved this weak limit theorem in a more general setting by combinatorial arguments. As for alternative proofs, Grimmett et al. [1] used Fourier analysis in spectral and scattering theory and their method is called the GJS method.

Recall the central limit theorem for the random walk in Sect. 2 when $p = q = 1/2$:

$$\lim_{n\to\infty} P\left(a \leq \frac{X_n}{\sqrt{n}} \leq b\right) = \int_a^b \frac{1}{\sqrt{2\pi}} e^{-x^2/2} dx.$$

Comparing two limit theorems, we conclude that the random walker moves asymptotically on the order of $\sqrt{n}$ but the Hadamard walker moves asymptotically on the order of n. Hence the Hadamard walker moves with *linear diffusion*.

In addition to linear diffusion, we show *localization* as a feature of quantum walks.

Definition 4.4 (*Localization*) Localization occurs (at $x = 0$) if we have $\limsup_{n\to\infty} P(X_n = 0) > 0$.

The localization does not occur in the random walk in Sect. 2 nor in the Hadamard walk.

In the next section, we will see some examples of localization of quantum walks.

4.2 Stationary Quantum Walks

Let us consider the *stationary quantum walks*. Set

$$U := \begin{bmatrix} 0 & 1 \\ 1 & 0 \end{bmatrix}, \quad P := \begin{bmatrix} 0 & 1 \\ 0 & 0 \end{bmatrix}, \quad Q := \begin{bmatrix} 0 & 0 \\ 1 & 0 \end{bmatrix}$$

and $\Psi_0 := \delta_0 \otimes \begin{bmatrix} \alpha \\ \beta \end{bmatrix} \in \ell^2(\mathbb{Z}) \otimes \mathbb{C}^2$, where $\alpha, \beta \in \mathbb{C}$ satisfy $|\alpha|^2 + |\beta|^2 = 1$. Then the quantum walk (U, P, Q, Ψ_0) is called a stationary quantum walk. We can compute the state Ψ_n at time n as follows.

Proposition 4.5 *For any* $n \in \mathbb{Z}_{\geq 0}$, *we have*

$$\Psi_{2n+1} = \delta_{-1} \otimes \begin{bmatrix} \beta \\ 0 \end{bmatrix} + \delta_1 \otimes \begin{bmatrix} 0 \\ \alpha \end{bmatrix}, \quad \Psi_{2n} = \delta_0 \otimes \begin{bmatrix} \alpha \\ \beta \end{bmatrix}.$$

From this, we obtain $P(X_{2n+1} = 0) = 0$ and $P(X_{2n} = 0) = |\alpha|^2 + |\beta|^2 = 1$. It deduces

$$\limsup_{n\to\infty} P(X_n = 0) = 1 > 0.$$

As a result, the localization occurs in the stationary quantum walk. In the stationary quantum walk, we can obtain $P(X_n \in \{-1, 0, 1\}) = 1$. Thus we observe no linear diffusion in that case.

4.3 Free Quantum Walks

Next let us consider *free quantum walks*. Set

$$U = \begin{bmatrix} 1 & 0 \\ 0 & 1 \end{bmatrix}, \quad P = \begin{bmatrix} 1 & 0 \\ 0 & 0 \end{bmatrix}, \quad Q = \begin{bmatrix} 0 & 0 \\ 0 & 1 \end{bmatrix},$$

$$\Psi_0 = \delta_0 \otimes \begin{bmatrix} \alpha \\ \beta \end{bmatrix} \in \ell^2(\mathbb{Z}) \otimes \mathbb{C}^2.$$

Then $P + Q = U$ holds. The quantum walk (U, P, Q, Ψ_0) is called a free quantum walk. We can check

$$\Psi_1 = \delta_{-1} \otimes \begin{bmatrix} \alpha \\ 0 \end{bmatrix} + \delta_1 \otimes \begin{bmatrix} 0 \\ \beta \end{bmatrix}, \quad \Psi_2 = \delta_{-2} \otimes \begin{bmatrix} \alpha \\ 0 \end{bmatrix} + \delta_2 \otimes \begin{bmatrix} 0 \\ \beta \end{bmatrix}.$$

In general, we can prove the following formula of the state Ψ_n at time n.

Proposition 4.6 *For any $n \in \mathbb{Z}_{\geq 0}$, we have*

$$\Psi_n = \delta_{-n} \otimes \begin{bmatrix} \alpha \\ 0 \end{bmatrix} + \delta_n \otimes \begin{bmatrix} 0 \\ \beta \end{bmatrix}.$$

This formula leads us to

$$\lim_{n \to \infty} P(X_n = 0) = 0$$

and

$$\lim_{n \to \infty} P\left(\frac{X_n}{n} \in \{-1, 1\}\right) = 1.$$

These limits mean that we can observe no localization but linear diffusion in the free quantum walk.

Before closing this section, we have several remarks on localization and linear diffusion on quantum walks. Inui et al. [7] proved that both linear diffusion and localization can occur in some *1-dimensional 3-state Grover walks*, which are given by certain unitary operators on $\ell^2(\mathbb{Z}) \otimes \mathbb{C}^3$. Konno [11] found the localization in a *one defect model*, which is given by a family $(U_x)_{x \in \mathbb{Z}}$ of unitary operators U_x on $\mathbb{C}^2$. If there exists a unitary operator U on $\mathbb{C}^2$ such that $U_x = U$ for all $x \in \mathbb{Z}$, then $(U_x)_{x \in \mathbb{Z}}$ gives a 1-dimensional 2-state quantum walk introduced in Sect. 3. Such a quantum walk is called homogeneous or uniform. On the other hand, the one defect model is inhomogeneous and defined by two distinct unitary operators U and U' such that $U_x = U$ for all $x \in \mathbb{Z} - \{0\}$ and $U_0 = U'$. Category algebras on (simple examples of) $\dagger$-categories introduced in Sect. 6 will play an important role to investigate inhomogeneous cases in general as group algebras do for homogeneous cases.

5 Grover Walks

In this section, we focus on Grover walks on graphs, which are dynamical systems considered as typical quantum walks. (The contents and description of this section are based on [2].) Let us first summarize basic concepts of graphs.

Definition 5.1 (*Directed graph*) A directed graph (or digraph) G is a quadruple (V_G, A_G, s_G, t_G) composed of a set V_G, a set A_G, a mapping $s_G : A_G \longrightarrow V_G$ and a mapping $t_G : A_G \longrightarrow V_G$. The elements of V_G and of A_G are called vertices and arcs, respectively. For an arc a, $s_G(a)$ is called the source of a and $t_G(a)$ is called the target of a, respectively.

Definition 5.2 (*Path*) A path in a directed graph G is a finite sequence of the form

$$(v_n, a_n, v_{n-1}, \ldots, a_1, v_0)$$

of vertices $v_0, \ldots, v_n$ and arcs $a_1, a_2, \ldots, a_n$ in G such that $s_G(a_i) = v_{i-1}$ and $t_G(a_i) = v_i$ holds for any $i = 1, \ldots, n$.[2] The vertex v_0 is called the source of the path and v_n is called the target of the path.

Definition 5.3 (*Connected*) A directed graph is called connected if for any two vertices v, v' there is some path whose source is v and whose target is v'.[3]

Definition 5.4 (*Symmetric, degree, locally finite*) A directed graph G is called symmetric if there exists a mapping $\overline{(\ )} : A_G \longrightarrow A_G$ such that $s_G(\overline{a}) = t_G(a)$, $t_G(\overline{a}) = s_G(a)$ and $\overline{\overline{a}} = a$ hold for all $a \in A_G$. For a symmetric graph, the cardinals of the set of arrows whose source is v and one whose target is v are coincide. This cardinal is called as the degree of the vertex v. If the degrees of the vertices are all finite, G is said to be locally finite.

Definition 5.5 (*Simple*) A directed graph G is called simple if for any $v, v' \in V_G$ there exists at most one $a \in A_G$ such that $v = s_G(a)$, $v' = t_G(a)$ hold and for any $a \in A_G$, $s_G(a) \neq t_G(a)$ holds.

Hereafter, in this section, "a graph" simply means "a symmetric simple directed graph". We also omit the indices indicating the name of a graph.

The main focus of this section is Grover walks on "jellyfish graphs" defined below, which are useful for constructing toy models of the dressed photon phenomena [21]. In short, a jellyfish graph is a graph composed of a finite connected graph and a finite number of "half-lines" attached to it. More precisely, it is defined as follows.

---

[2] In graph theory, many authors call this notion as "(finite) walk" and they keep the term "(finite) path" for a more restricted notion. However, we use the term "path" to avoid the confusion with "quantum walk" and keep the coherence of category theoretic terms as "Moore path category".

[3] Although the notion of connectivity defined here is usually called "strongly connected", no confusion will occur by this abuse of the term for the graphs treated in the present article.

Definition 5.6 A graph G is called a jellyfish graph if it is the union of a finite number of graphs
$$G^{(0)}, l^{(1)}, l^{(2)}, \ldots$$
which satisfy the following conditions:

- $G^{(0)}$ is a finite connected graph, i.e., a connected graph composed of a finite number of vertices and arcs. For simplicity, we identify the set of vertices with the set $\{1, 2, 3, \ldots, n\}$, where n denotes the number of vertices.
- Each $l^{(i)}$ is a half-line graph, i.e., a connected graph such that degree of any vertex v is 2 except for one vertex called the endvertex of $l^{(i)}$ which is a unique common vertex of $l^{(i)}$ and $G^{(0)}$.

A Grover walk on a locally finite graph is a discrete time dynamical system on a certain class of complex valued functions on the set of arcs such that the "amplitudes" $\alpha, \alpha', \alpha'', \ldots$, i.e., the values of each function, on any arcs $a, a', a'', \ldots$ with the common target in time t determines the ones $\beta, \beta', \beta'', \ldots$ in time $t+1$ of $\overline{a}, \overline{a'}, \overline{a''}, \ldots$ by

$$\begin{bmatrix} \beta \\ \beta' \\ \beta'' \\ \vdots \end{bmatrix} = \begin{bmatrix} 2/r - 1 & 2/r & \cdots & 2/r \\ 2/r & 2/r - 1 & \cdots & 2/r \\ \vdots & \vdots & \ddots & \vdots \\ 2/r & 2/r & \cdots & 2/r - 1 \end{bmatrix} \begin{bmatrix} \alpha \\ \alpha' \\ \alpha'' \\ \vdots \end{bmatrix},$$

where r denotes the degree of the common vertex.

Most studies have considered square-integrable functions as the class of functions for which a Grover walk is defined, and have achieved many results. On the other hand, it is important to consider different classes, such as bounded functions, when considering certain physical problems.

Higuchi and Segawa [3] showed the following limit theorem about the amplitudes of the Grover walks on jellyfish graphs, for the class of bounded functions.

Theorem 5.7 *Let $\alpha^{(i)}$ denote the constant amplitude of a in $l^{(i)}$ and let $\beta^{(i)} = \psi_\infty(\overline{a})$ denote the limit amplitude of a, which is shown to be constant for each $l^{(i)}$. The relation between these amplitudes is given as follows:*

$$\begin{bmatrix} \beta^{(1)} \\ \beta^{(2)} \\ \beta^{(3)} \\ \vdots \end{bmatrix} = \begin{bmatrix} 2/r - 1 & 2/r & \cdots & 2/r \\ 2/r & 2/r - 1 & \cdots & 2/r \\ \vdots & \vdots & \ddots & \vdots \\ 2/r & 2/r & \cdots & 2/r - 1 \end{bmatrix} \begin{bmatrix} \alpha^{(1)} \\ \alpha^{(2)} \\ \alpha^{(3)} \\ \vdots \end{bmatrix}.$$

In [3], the following theorem is also proved (the terms are modified here).

Theorem 5.8 *Let $J(a)$ be the quantity defined as*

$$J(a) := \psi_\infty(a) - \operatorname{ave}(\alpha^{(1)}, \alpha^{(2)}, \ldots, \alpha^{(m)}),$$

where $\mathrm{ave}(\alpha^{(1)}, \alpha^{(2)}, \ldots, \alpha^{(m)})$ denotes the average of $\alpha^{(1)}, \alpha^{(2)}, \ldots, \alpha^{(m)}$. The following equations hold:

- $J(a) + J(\bar{a}) = 0$.
- For any $v \in V_G$, $\sum_{a \in A_G, t(a)=v} J(a) = 0$.

This is nothing but "Kirchhoff's current law" for the quantity $J(a)$. Similarly, a law corresponding to "Kirchhoff's voltage law" is also shown, but it is omitted here. It is important to note that these laws on $J(a)$ allow us to calculate the limit amplitudes.

Hamano and the first author [2] analyzed the behavior of a physical system arising from the interaction between light and nano-scale matter, called "dressed photons" [21], based on the above theorem. Recently, there has been an increasing interplay between the research on quantum walks and the research on dressed photons (see [22], for example).

However, a more general concept of quantum walks is needed to deepen our understanding of interacting quantum fields such as dressed photons. In the next section, we will sketch an attempt to construct a general theory of quantum walks.

6 Toward a General Theory

To construct a general theory of quantum walks, it is necessary to consider at a very general level what are the essence of quantum nature and the essence of spacetime. In this section, we will clarify the essence of the mathematical structure of quantum theory in terms of "noncommutative probability spaces". Then we show that a noncommutative probability space can be constructed from †-categories, a generalized concept of spacetime. Based on these preparations, we will introduce a concept of quantum walks on †-categories and give a sketch of a direction toward a general theory of quantum walks.

6.1 Noncommutative Probability Theory

The fundamental concepts of quantum theory are the algebras of observables (physical quantities) and the states on them. These are defined as certain algebras called *-algebras over $\mathbb{C}$ and certain linear functionals on the *-algebras, respectively. A pair of a *-algebra and a state on it is said to be a "noncommutative probability space" over $\mathbb{C}$ and the study of these noncommutative probability spaces is called noncommutative probability theory.

From a noncommutative probability space, a Hilbert space and a representation of the algebra on it can be reconstructed through a procedure called the Gelfand-Naimark-Segal construction. A noncommutative probabilistic framework can also investigate quantum systems with infinite degrees of freedom that cannot be mathematically modeled in the usual Hilbert space-based framework. In this sense, noncommutative probability spaces over $\mathbb{C}$ can be said to be a sufficiently general mathematical framework for the conventional quantum theory.

On the other hand, using $\mathbb{C}$ as a scalar is not necessarily an absolute requirement from a mathematical perspective. In fact, the fundamental part of noncommutative probability theory remains unchanged even if we generalize the concepts for $\mathbb{C}$ to ones for a general "rig (ring without negatives)" defined below. In this subsection, we describe the basic concepts of this generalized version of noncommutative probability theory.

Definition 6.1 (*Rig*) A rig R is a set with two binary operations called addition and multiplication such that

1. Addition is associative, commutative, and R has the unit 0 for addition,
2. Multiplication is associative and R has the unit 1 for multiplication,
3. $r''(r' + r) = r''r' + r''r$ and $(r'' + r')r = r''r + r'r$ hold for any $r, r', r'' \in R$,
4. $0r = 0$ and $r0 = 0$ hold for any $r \in R$.

The concept of rigs is "a concept of rings minus the commutativity of multiplication and the reversibility of addition". Most of quantum walk studies focus on the cases that the scalar rig is $\mathbb{C}$. However, there are some interesting studies on "quantum walks" whose coefficients are quaternion [14] or max-plus algebra (the rig consisting of real numbers and $-\infty$ whose "addition" is "maximum" and "multiplication" is "sum") [28]. In the former, multiplication is not commutative, and in the latter, addition is not reversible. In order to include these examples and other possible models such as "operator valued" walks which will be important in the study of quantum physics, it is necessary to consider general rigs. Interestingly, considerable structures survive in this generalization.

One can naturally generalize basic notions such as the center (the set of the elements which commute with all elements) and bimodules (a generalizaton of vector spaces) for general rigs.[4] Moreover, we can define the notions of algebras over rigs, *-rigs, and *-algebras over *-rigs as follows [23]. (The reader can read below considering only the case where $R = \mathbb{C}$.)

Definition 6.2 (*Algebra over rig*) A bimodule A over a rig R is called an algebra over R if it is also a rig with respect to its own multiplication which is compatible with scalar multiplication, i.e.,

$$(r'a')(ar) = r'(a'a)r, \ (a'r)a = a'(ra)$$

for any $a, a' \in A$ and $r, r' \in R$.

---

[4] For details, see [23].

Definition 6.3 (**-Rig*) Let R be a rig. An operation $(\ )^*$ on R is said to be a (contravariant) involution on R if it satisfies

$$(a+b)^* = a^* + b^*, \ (ab)^* = b^*a^*, \ (a^*)^* = a$$

for any $a, b \in R$. A pair $(R, (\ )^*)$ of a rig R and a (contravariant) involution $(\ )^*$ on R is called a **-rig*, and just denoted as R when the (contravariant) involution is specified.

Definition 6.4 (**-Algebra over *-rig*) Let R be a *-rig with the involution $\overline{(\ )}$. A pair $(A, (\ )^*)$ of an algebra A over R (as a rig) and a (contravariant) involution $(\ )^*$ on A (as a rig) is said to be a **-algebra over R* if it is compatible with the scalar multiplication, i.e.,

$$(r'ar)^* = \overline{r}a^*\overline{r'}$$

holds for any $r, r' \in R$ and $a \in A$, and just denoted as A when the (contravariant) involution is specified.

A *-algebra can be considered as an algebra of observables. The next problem is how to define a concept of states. Intuitively, a state is a statistical law that associates each observable with its expectation. This notion of states can be defined as follows [23]:

Definition 6.5 (*Linear functional*) Let A be an algebra over a rig R. An R-valued linear function on A, i.e., a function preserving addition and scalar multiplication, is called a linear functional on A. A linear functional φ on A is said to be unital if $\varphi(\epsilon) = 1$, where ϵ and 1 denote the multiplicative units in A and in R, respectively.

Definition 6.6 (*Positivity*) A pair of *-rigs (R, R_+) is called a positivity structure on R if R_+ is a subrig of R such that $r, s \in R_+$ and $r + s = 0$ imply $r = s = 0$, and such that $a^*a \in R_+$ for any $a \in R$.

Definition 6.7 (*State*) Let R be a *-rig and (R, R_+) a positivity structure on R. A state φ on a *-algebra A over R with respect to (R, R_+) is a unital linear functional $\varphi : A \longrightarrow R$ which satisfies $\varphi(a^*a) \in R_+$ and $\varphi(a^*) = \overline{\varphi(a)}$ for any $a \in R$, where $(\ )^*$ and $\overline{(\ )}$ denote the involutions on A and on R, respectively.[5]

Definition 6.8 (*Noncommutative probability space*) A noncommutative probability space is a pair (A, φ) of a *-algebra A and a state φ on A.

Then the very abstract notion of quantum walks is defined [24]:

Definition 6.9 (*Quantum walk on *-algebra*) Let A be a *-algebra. A sequence of states given by

$$\varphi^t(\alpha) = \varphi((\omega^*)^t \alpha \omega^t) \quad t = 0, 1, 2, 3, \ldots$$

generated by a unitary element $\omega \in A$, i.e., an element satisfying $\omega^*\omega = \omega\omega^* = \epsilon$ is called a quantum walk on A.

---

[5] The last condition $\varphi(a^*) = \overline{\varphi(a)}$ follows from other conditions, if $R = \mathbb{C}$.

The above definition is useful to clarify the "quantum nature" of quantum walks. However, the aspects of dynamics "on spacetimes" are not clearly incorporated. In the next subsection, we consider the concept of spacetime as †-categories, and define "quantum walks on †-categories" as dynamics that reflect the "temporospatial" structures.

6.2 Quantum Walks on †-Categories

A category is a system composed of two kinds of entities called objects and arrows, equipped with domain/codomain, composition, and identity arrows, satisfying associative law and unit law (for details one can see any textbook on category theory such as [16]). A category $\mathcal{C}$ is called small if the collections of objects and arrows are sets. In this article, we assume that categories are small unless otherwise specified. The set of all arrows in a category $\mathcal{C}$ is identified with the category itself and is also denoted as $\mathcal{C}$. Note that the objects can be identified with the identity arrows. The set of all objects (identity arrows) in $\mathcal{C}$ is denoted as $|\mathcal{C}|$.

The first author proposed to consider the notion of †-categories (equipped with certain subcategory of "causal" arrows) as a mathematical counterpart of "spacetime" on which quantum fields are defined in the most general meaning [24]. A †-category is, in short, a category with involution structure ("reversing the direction of all the arrows"). More precisely, it is defined as follows.[6]

Definition 6.10 (†-category) Let $\mathcal{C}$ be a category. A contravariant endofunctor $(\ )^\dagger$ from $\mathcal{C}$ to $\mathcal{C}$ is said to be a (contravariant) involution on $\mathcal{C}$ when $(\ )^\dagger \circ (\ )^\dagger$ is equal to the identity functor on $\mathcal{C}$. A category with contravariant involution which is the identity on the objects is called a †-category.[7]

To understand why this concept is appropriate to be considered as general spacetime, let us consider the following simple example: We can construct a †-category called "the free category of the symmetric directed graph" from any symmetric directed graph by considering vertices as objects, paths as arrows, source/target as domain/codomain, concatenation as composition, trivial paths which can be identified with vertices as identity arrows, and † is given by

$$(v_n, a_n, v_{n-1}, \ldots, a_1, v_0)^\dagger = (v_0, \overline{a_1}, v_1, \ldots, \overline{a_n}, v_n).$$

A generalization of this category to continuous contexts is called Moore path categories [19], which are also important †-categories.

---

[6] See any category theory textbook for the meanings of terminologies used here.

[7] There are many large †-categories (such as the category of Hilbert spaces and the category of cobordism), and it would be interesting to apply the following discussion by appropriately restricting them, but we omit such discussions here.

As another kind of $\dagger$-categories, let us focus on "indiscrete categories", which have played major roles (although implicitly) in quantum walks studies so far.

Definition 6.11 (*Indiscrete category*) A category is said to be indiscrete if there is always a unique arrow whose domain is A and codomain is B for any objects A, B. (It becomes a $\dagger$-category equipped with $(\ )^\dagger$ as the correspondence between the unique arrow from A to B and one from B to A.)

Let us define the notion of a category algebra as an "observable algebra" that reflects the "spacetime" structure as $\dagger$-category as follows [23, 24] :

Definition 6.12 (*Category algebra*) Let $\mathcal{C}$ be a category and R be a rig. An R-valued function α defined on $\mathcal{C}$ is said to be of finite propagation if for any object C there are at most finite number of arrows whose codomain or domain is C in its support. The module over R consisting of all R-valued functions of finite propagation together with the multiplication defined by

$$(\alpha'\alpha)(c'') = \sum_{\{(c',c)|\ c''=c'\circ c\}} \alpha'(c')\alpha(c), \quad c, c', c'' \in \mathcal{C}$$

becomes an algebra over R with unit ϵ defined by

$$\epsilon(c) = \begin{cases} 1 & (c \in |\mathcal{C}|), \\ 0 & (otherwise), \end{cases}$$

and is called the category algebra of finite propagation of $\mathcal{C}$ over R, which is denoted as $R[\mathcal{C}]$. In the present article, we simply call $R[\mathcal{C}]$ the category algebra of $\mathcal{C}$.

Category algebras are noncommutative if there is some arrow connecting distinct objects ("quantization as categorification"). When a category $\mathcal{C}$ is an indiscrete category with n objects, the category algebra $R[\mathcal{C}]$ is isomorphic to the matrix algebra $M_n(R)$. Hence, category algebras can be considered as a generalization of matrix algebras, replacing "indices" with (maybe infinite) objects and "numbers" with linear combinations of arrows as indeterminates. In the authors' opinion, these features are essential to unify various quantum walk studies. Furthermore, the notion of the Hermite conjugate can be naturally generalized as follows [23].

Theorem 6.13 (*Category algebra as*-algebra*) *Let $\mathcal{C}$ be a category with a (contravariant) involution $(\ )^\dagger$, especially a $\dagger$-category, and let R be a *-rig. Then the category algebra $R[\mathcal{C}]$ becomes a *-algebra over R.*

Proof. The operation $(\ )^*$ is defined as $\alpha^*(c) = \overline{\alpha(c^\dagger)}$ for all $\alpha \in R[\mathcal{C}]$, where $\overline{(\ )}$ denotes the involution on R. Then $(\ )^*$ satisfies $(\alpha\beta)^* = \beta^*\alpha^*$ for any $\alpha, \beta \in R[\mathcal{C}]$. Indeed, for any $c \in \mathcal{C}$ we have

$$(\alpha\beta)^*(c) = \overline{\alpha\beta(c^\dagger)} = \overline{\sum_{c^\dagger = c'oc''} \alpha(c')\beta(c'')} = \sum_{c^\dagger = c'oc''} \overline{\alpha(c')\beta(c'')}$$

$$= \sum_{c^\dagger = c'oc''} \overline{\beta(c'')}\,\overline{\alpha(c')} = \sum_{c = c''^\dagger oc'^\dagger} \overline{\beta(c'')}\,\overline{\alpha(c')}.$$

By changing the labels of arrows, $\sum_{c = c''^\dagger oc'^\dagger} \overline{\beta(c'')}\,\overline{\alpha(c')}$ can be rewritten as

$$\sum_{c = c''^\dagger oc'^\dagger} \overline{\beta(c'')}\,\overline{\alpha(c')} = \sum_{c = c'oc''} \overline{\beta(c'^\dagger)}\,\overline{\alpha(c''^\dagger)} = \sum_{c = c'oc''} \beta^*(c')\alpha^*(c'') = \beta^*\alpha^*(c).$$

It is easy to check the other properties which ( )* should satisfy. □

Based on the above, we obtain the concepts of states and of quantum walks that reflect the structure of †-categories [23, 24].

Definition 6.14 (*State on category*) Let R be a *-rig, (R, R_+) a positivity structure on R, and C a category with involution, especially a †-category, respectively. A state on the category algebra $R[C]$ over R with respect to (R, R_+) is said to be a state on a category C with respect to (R, R_+).

Definition 6.15 (*Quantum walk on †-category*) Let C be a †-category and R a *-rig. A quantum walk on the *-algebra $R[C]$ is called as a quantum walk on the †-category C.

The notion of quantum walks on a †-category includes various dynamical models studied under the name of quantum walks. In fact, quantum walks presented in this article can be considered as quantum walks on indiscrete categories. Quantum walks on $\mathbb{Z}$ treated in Sects. 3 and 4 can be considered as quantum walks on the indiscrete category whose objects are elements of the product of $\mathbb{Z}$ and the set of "inner degrees of freedom". Grover walks can also be considered as quantum walks on the indiscrete category whose objects are arcs.

One may think that it would be sufficient to consider only indiscrete categories. However, if we want to model more general dynamics (for example, when the transition amplitude depends on the paths of transitions), we need to proceed to general †-categories. This is because information about the diversity of paths cannot be naturally written in terms of indiscrete categories. Path dependencies will become increasingly complex when considering quantum walks on graphs with multiple arcs between vertices or self-loops. Even in such cases, it is possible to understand them naturally as quantum walks on †-categories.

Another important point is that by considering quantum walks on a †-category, the mathematical approach to quantum fields and the theory of quantum walks are deeply connected. As a new approach to quantum fields, the first author [23] proposed that the category algebra on a †-category is an algebra of local observables, and the states on the category are identified with (local) states of the quantum field. Quantum walks on the †-category are expected to play an important role as simple models of dynamics of the quantum fields. The authors hope that the interplay between the mathematics of quantum information science (including quantum walks) and many other fields (including mathematical research of quantum fields) will be enriched by the ideas above.

Acknowledgements The authors would like to thank the anonymous referee for pointing out fruitful comments. The second author was supported by JSPS KAKENHI Grant Number JP20K14298 (Grant-in-Aid for Early-Career Scientists). This work was supported by JST CREST Grant Number JPMJCR2113, Japan.

References

1. G. Grimmett, S. Janson, P.F. Scudo, Weak limits for quantum random walks. Phys. Rev. E Stat. Nonlinear Soft Matter Phys. **69**(2), 026119 (2004)
2. M. Hamano, H. Saigo, Quantum walk and dressed photon, in *Proceedings 9th International Conference on Quantum Simulation and Quantum Walks*, ed. by G. Di Molfetta, V. Kendon, Y. Shikano. Electronic Proceedings in Theoretical Computer Science, vol. 315 (Open Publishing Association, 2020), pp. 93–99
3. Y. Higuchi, E. Segawa, A dynamical system induced by quantum walk. J. Phys. A: Math. Theor. **52**(39), 395202 (2019)
4. A. Ichihara, L. Matsuoka, Y. Kurosaki, K. Yokoyama, An analytic formula for describing the transient rotational dynamics of diatomic molecules in an optical frequency comb. Chin. J. Phys (2013)
5. Ide Y (2019) Partition of graphs and quantum walk based search algorithms. Nonlinear Theory Its Appl., IEICE 10(1):16–27
6. Y. Ide, N. Konno (eds.), *New Development of Quantum Walks* (Baifukan, Tokyo, 2019) (in Japanese)
7. N. Inui, N. Konno, E. Segawa, One-dimensional three-state quantum walk. Phys. Rev. E Stat. Nonlinear Soft Matter. Phys. **72**(5), 056112 (2005)
8. Kitagawa T (2012) Topological phenomena in quantum walks: elementary introduction to the physics of topological phases. Quantum Inf. Process. 11(5):1107–1148
9. Konno N (2002) Quantum random walks in one dimension. Quantum Inf. Process. 1(5):345–354
10. Konno N (2005) A new type of limit theorems for the one-dimensional quantum random walk. J. Math. Soc. Japan 57(4):1179–1195
11. Konno N (2010) Localization of an inhomogeneous discrete-time quantum walk on the line. Quantum Inf. Process. 9(3):405–418
12. Konno N (2010) Quantum walks and elliptic integrals. Math. Struct. Comput. Sci. 20(6):1091–1098
13. N. Konno, *Quantum Walk* (Morikita Shuppan, Tokyo, 2014) (in Japanese)
14. Konno N (2015) Quaternionic quantum walks. Quantum Stud. Math. Found. 2(1):63–76
15. Konno N (2019) A new time-series model based on quantum walk. Quantum Stud. Math. Found. 6:61–72
16. S. Mac Lane, *Categories for the Working Mathematician*, 2nd edn. Graduate Texts in Mathematics (Springer, New York, 2010)
17. L. Matsuoka, T. Kasajima, M. Hashimoto, K. Yokoyama, Numerical study on quantum walks implemented on cascade rotational transitions in a diatomic molecule. J. Korean Phys. Soc. **59**(4(1)), 2897–2900 (2011)
18. Matsuoka L, Yokoyama K (2013) Physical implementation of quantum cellular automaton in a diatomic molecule. J. Comput. Theor. Nanosci. 10(7):1617–1620

19. *n*Lab, *Moore path category*, https://ncatlab.org/nlab/show/Moore+path+category. Accessed 28 Oct 2023
20. H. Obuse, N. Kawakami, Topological phases and delocalization of quantum walks in random environments. Phys. Rev. B Condens. Matter **84**(19), 195139 (2011)
21. Ohtsu M (2013) *Dressed Photons: Concepts of Light-Matter Fusion Technology*. Nano-Optics and Nanophotonics (Springer, Berlin)
22. M. Ohtsu, E. Segawa, K. Yuki, *A Quantum Walk Model for the Energy Transfer of a Dressed Photon*. IEICE Proceedings Series, vol. 71, no. A3L-C-01 (2022)
23. H. Saigo, Category algebras and states on categories. Symmetry **13**(7), 1172 (2021)
24. H. Saigo, Quantum fields as category algebras. Symmetry **13**(9), 1727 (2021)
25. Saigo H, Sako H (2020) Space-homogeneous quantum walks on $\mathbb{Z}$ from the viewpoint of complex analysis. J. Math. Soc. Jpn. 72(4):1201–1237
26. H. Sako, Convergence theorems on multi-dimensional homogeneous quantum walks. Quantum Inf. Process. **20**(3), 24 (2021)
27. Y. Wang, Y. Shang, P. Xue, Generalized teleportation by quantum walks. Quantum Inf. Process. **16**(9), 13 (2017)
28. Watanabe S, Fukuda A, Segawa E, Sato I (2020) A walk on max-plus algebra. Linear Algebra Appl. 598:29–48

Open Access This chapter is licensed under the terms of the Creative Commons Attribution 4.0 International License (http://creativecommons.org/licenses/by/4.0/), which permits use, sharing, adaptation, distribution and reproduction in any medium or format, as long as you give appropriate credit to the original author(s) and the source, provide a link to the Creative Commons license and indicate if changes were made.

The images or other third party material in this chapter are included in the chapter's Creative Commons license, unless indicated otherwise in a credit line to the material. If material is not included in the chapter's Creative Commons license and your intended use is not permitted by statutory regulation or exceeds the permitted use, you will need to obtain permission directly from the copyright holder.

On the Hardness of Conversion from Entangled Proof into Separable One

Seiseki Akibue, Go Kato, and Seiichiro Tani

Abstract A quantum channel whose image approximates the set of separable states is called a disentangler, which plays a prominent role in the investigation of variants of the computational model called Quantum Merlin Arthur games and has potential applications in classical and quantum algorithms for the separability testing and NP-complete problems. So far, two types of a disentangler, constructed based on ϵ-nets and the quantum de Finetti theorem, have been known; however, both of them require an exponentially large input system. Moreover, in 2008, John Watrous conjectured that any disentangler requires an exponentially large input system, called the disentangler conjecture. In this paper, we show that both of the two known disentanglers can be regarded as examples of a *strong disentangler*, which is a disentangler approximately breaking entanglement between one output system and the composite system of another output system and the arbitrarily large environment. Note that the strong disentangler is essentially an approximately entanglement-*breaking* channel while the original disentangler is an approximately entanglement-*annihilating* channel, and the set of strong disentanglers is a subset of disentanglers. As a main result, we show that the disentangler conjecture is true for this subset, the set of strong disentanglers, for a wide range of approximation parameters without any computational hardness assumptions.

Keywords Quantum entanglement · Quantum interactive proof · Entanglement-breaking channel · Entanglement-annihilating channel · Quantum de Finetti theorem

S. Akibue (✉) · S. Tani
NTT Communication Science Laboratories, NTT Corporation, Atsugi, Kanagawa, Japan
e-mail: seiseki.akibue@ntt.com

S. Tani
e-mail: seiichiro.tani@ntt.com

G. Kato
Advanced ICT Research Institute, NICT, Koganei, Tokyo, Japan
e-mail: go.kato@nict.go.jp

1 Introduction

Entanglement is an essential resource that provides non-classical phenomena in quantum mechanics and advantages in quantum information processing over classical one. Thus, testing whether a given quantum state is entangled or separable is a fundamental task for investigating and utilizing quantum nature. One of the primitive ways to detect entanglement is using an entanglement witness [1]. However, it is known that the number of entanglement witnesses represented by positive maps necessary for detecting any (even robustly) entangled state in $\mathbb{C}^d \otimes \mathbb{C}^d$ is $\exp(\Omega(d^3/\log d))$ [2]. Moreover, if we formalize the quantum separability testing as a promise problem via the weak membership problem within an inverse polynomial precision, it has been shown to be NP-hard [3, 4]. On the other hand, such complex structures of separable states provide benefits to the computation when we use a separable state as *quantum proof* in the computational model called Quantum Merlin Arthur games (QMA) [5, 6]. Indeed, proof encoded in a log-size separable state is sufficient for solving NP-complete problems, 3-COL [7] and 3-SAT [8, 9], whereas it seems impossible to solve such NP-complete problems with proof encoded in a log-size entangled state [10].

The remarkable computational power provided by proof encoded in a separable state has induced the *disentangler conjecture*, which states the difficulty of converting the set of entangled states into that of separable states [6]. More precisely, the conjecture states that exponentially large input dimension D, i.e., $\log D = \Omega(d)$ with respect to dimension d of one output system, is necessary for realizing the quantum channel called an (ϵ, δ)-disentangler, whose output state is an almost separable state within precision ϵ and approximates an arbitrary separable state within precision δ as shown in Fig. 1. Thus, the disentangler is an approximated-entanglement-*annihilating* channel, whose exact version is defined in [11], in contrast to the well-known entanglement-*breaking* channel [12, 13]. Despite its simplicity and importance, the conjecture is far from a complete proof or a falsification. Indeed, there exist only a few known ways to construct a disentangler including a construction based on ϵ-nets and that based on the quantum de Finetti theorem [6]. Moreover, the only nonexistence proofs without assuming any computational hardness assumption are given by [6] for the nonexistence of $(0, 0)$-disentanglers on a finite-dimensional Hilbert space and by [14] for the nonexistence of (ϵ, δ)-disentanglers with $\epsilon + \delta = O\left(\frac{1}{d^2}\right)$ having a polynomial input dimension, i.e., it requires $\log D = \Omega\left(\frac{(\log d)^2}{poly \log \log d}\right)$.

In this paper, we investigate a new subclass of disentanglers that includes the two known disentanglers as specific examples. We define this subclass as the set of quantum channels, called (ϵ, δ)-*strong disentanglers*, that are (ϵ, δ)-disentanglers and break entanglement between one output system and the composite system of another output system and the arbitrarily large environment within precision ϵ as shown in Fig. 2. Then, we verify that the two known disentanglers are examples of this strong disentangler. Note that the strong disentangler is essentially an approximately entanglement-*breaking* channel while the original disentangler is an approximately

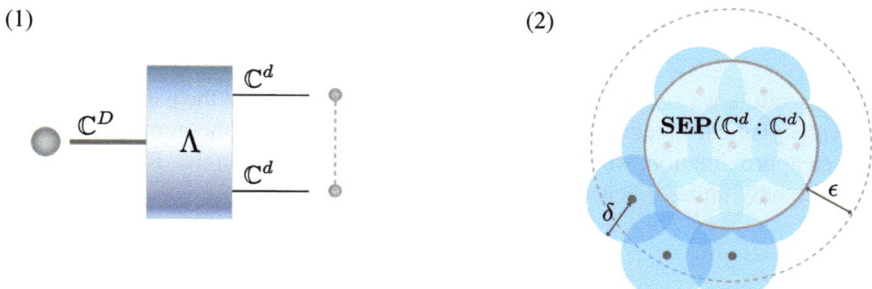

Fig. 1 Graphical representations of an (ϵ, δ)-disentangler Λ. **a** Λ produces only approximately separable states. **b** The points and small disks surrounding them represent producible states by Λ and their δ-neighborhoods, respectively. The large disk surrounded by the solid circle and that surrounded by the dashed circle represent the set of separable states $\mathbf{SEP}\left(\mathbb{C}^d : \mathbb{C}^d\right)$ and its ϵ-neighborhood, respectively. The first condition of the disentangler requires that all the producible states by Λ reside in the ϵ-neighborhood of $\mathbf{SEP}\left(\mathbb{C}^d : \mathbb{C}^d\right)$. The second condition requires that δ-neighborhoods of the producible states cover $\mathbf{SEP}\left(\mathbb{C}^d : \mathbb{C}^d\right)$. More precise definitions are given in Sect. 3

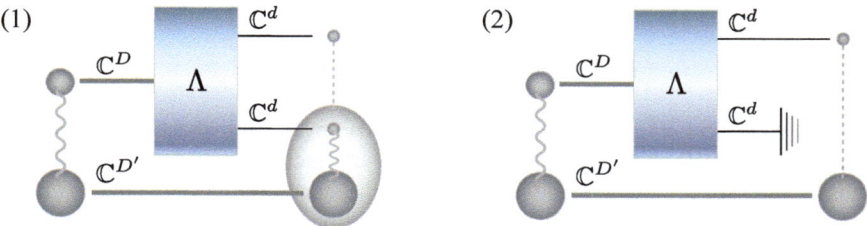

Fig. 2 Graphical representations of strong disentangler Λ. **a** Λ satisfies not only the conditions of the disentangler but also the condition such that Λ approximately breaks entanglement between one output system and the composite system of another output system and the arbitrarily large environment. **b** If we discard one output system of Λ, it is an approximately entanglement-breaking channel. More precise definitions are given in Sect. 3

entanglement-*annihilating* channel, as shown in Fig. 2b. As a main result, we show, without any computational hardness assumption, that the disentangler conjecture is true for the strong disentangler for a wide range of ϵ and δ. More precisely, we obtain the following theorem.

Theorem 1 (informal version) *For any constants $\epsilon, \delta \geq 0$ such that $\epsilon + \sqrt{\delta} < 1$, every (ϵ, δ)-strong disentangler satisfies $\log D = \Omega(d)$ as $d \to \infty$.*

2 Notations and Reviews

In this section, we summarize basic notations used throughout the paper and review the entanglement-breaking channel, the quantum de Finetti theorem, ϵ-net, and a convex approximation, which are deeply related to the disentangler. Remark that we consider only finite-dimensional Hilbert spaces.

2.1 Basic Notations

L $(\mathcal{H})$, **Herm** $(\mathcal{H})$ and **Pos** $(\mathcal{H})$ represent the set of linear operators, hermitian operators and positive semidefinite operators on Hilbert space $\mathcal{H}$, respectively. The set of quantum states is represented by that of density operators, defined as

$$\mathbf{S}(\mathcal{H}) := \{\rho \in \mathbf{Pos}(\mathcal{H}) : tr[\rho] = 1\}. \tag{1}$$

The set of pure quantum states is represented by

$$\mathbf{P}(\mathcal{H}) := \{\rho \in \mathbf{S}(\mathcal{H}) : tr[\rho^2] = 1\}. \tag{2}$$

It is known that $\mathbf{S}(\mathcal{H})$ and $\mathbf{P}(\mathcal{H})$ are compact and convex. Note that the compactness of subsets in $\mathbf{L}(\mathcal{H})$ and the (uniform) continuity of functions on subsets in $\mathbf{L}(\mathcal{H})$ (or their product space) are defined with respect to the standard topology on $\mathbf{L}(\mathcal{H})$, i.e., the topology induced by a norm on $\mathbf{L}(\mathcal{H})$. Sometimes, a pure state is alternatively represented by complex unit vector $|\phi\rangle \in \mathcal{H}$ such that $\langle\phi|\phi\rangle = 1$. Unnormalized complex vector is denoted with a tilde, e.g., $|\tilde{\eta}\rangle$. For pure state $|\phi\rangle$ (or complex vector $|\tilde{\eta}\rangle$), we denote its density operator (or rank-1 linear operator) as $\phi := |\phi\rangle\langle\phi| \in \mathbf{P}(\mathcal{H})$ (or $\tilde{\eta} := |\tilde{\eta}\rangle\langle\tilde{\eta}|$.) We sometimes denote a subscript to emphasize the system where a state resides, e.g., ρ_A implies $\rho_A \in \mathbf{S}(\mathcal{H}_A)$. A reduced density operator of pure state $|\Phi\rangle_{AB} \in \mathcal{H}_A \otimes \mathcal{H}_B$ is denoted as

$$\Phi_A := tr_B[\Phi], \tag{3}$$

where the partial trace $tr_B : \mathbf{L}(\mathcal{H}_A \otimes \mathcal{H}_B) \to \mathbf{L}(\mathcal{H}_A)$ is defined by $tr_B = id_A \otimes tr$ by using the identity map $id : \mathbf{L}(\mathcal{H}_A) \to \mathbf{L}(\mathcal{H}_A)$ and the trace map $tr : \mathbf{L}(\mathcal{H}_B) \to \mathbb{C}$. We use a subscript to identify the tracing system and to emphasize the system where the linear map acts.

The set of separable states is denoted by

$$\mathbf{SEP}(\mathcal{H}_A : \mathcal{H}_B) := \mathrm{conv}(\{\phi_A \otimes \psi_B : \phi_A \in \mathbf{P}(\mathcal{H}_A), \psi_B \in \mathbf{P}(\mathcal{H}_B)\}). \tag{4}$$

It is known that $\mathbf{SEP}(\mathcal{H}_A : \mathcal{H}_B)$ is compact and convex.

Any physical process can be represented by a quantum channel, defined as a linear completely positive and trace-preserving (CPTP) map $\Gamma : \mathbf{L}(\mathcal{H}_A) \to \mathbf{L}(\mathcal{H}_B)$, where the initial state of the process is regarded as input state $\rho \in \mathbf{S}(\mathcal{H}_A)$ of quantum channel Γ and the final state of the process is given by output state $\Gamma(\rho) \in \mathbf{S}(\mathcal{H}_B)$. For any linear map $\Gamma : \mathbf{L}(\mathcal{H}_A) \to \mathbf{L}(\mathcal{H}_B)$, we can define corresponding Choi operator $J(\Gamma) \in \mathbf{L}(\mathcal{H}_A \otimes \mathcal{H}_B)$ such as

$$J(\Gamma) := \sum_{i,j} |i\rangle\langle j|_A \otimes \Gamma(|i\rangle\langle j|_A), \tag{5}$$

where $\{|i\rangle\}_i$ is an orthonormal basis of $\mathcal{H}_A$.

We often measure the distinguishability of quantum states by using a norm on $\mathbf{L}(\mathcal{H})$ called the trace distance, defined by

$$\|M\|_{tr} := \frac{1}{2}\|M\|_1, \tag{6}$$

where $\|M\|_1 := tr\left[\sqrt{MM^\dagger}\right]$. Note that for any states $\rho, \sigma \in \mathbf{S}(\mathcal{H})$, it holds that $\|\rho - \sigma\|_{tr} \leq 1$, and the equality holds if and only if ρ and σ are perfectly distinguishable.

We also use the fidelity function to measure the distinguishability, defined by

$$F(\rho, \sigma) := \max |\langle \Phi^\rho | \Phi^\sigma \rangle|^2 \tag{7}$$

for any quantum states $\rho, \sigma \in \mathbf{S}(\mathcal{H}_A)$, where pure states $|\Phi^\rho\rangle_{AB}$ and $|\Phi^\sigma\rangle_{AB}$ on $\mathcal{H}_A \otimes \mathcal{H}_B$ represent purifications of ρ and σ respectively, i.e., $\Phi^\rho_A = \rho$ and $\Phi^\sigma_A = \sigma$, and the maximization is taken over all the purifications. The trace distance and the fidelity of two states $\rho, \sigma \in \mathbf{S}(\mathcal{H})$ are related as follows:

$$1 - \sqrt{F(\rho, \sigma)} \leq \|\rho - \sigma\|_{tr} \leq \sqrt{1 - F(\rho, \sigma)}, \tag{8}$$

where the right equality holds when ρ and σ are pure states.

Both the trace distance and the fidelity function satisfy the *monotone* property that any physical process cannot increase the distinguishability of quantum states as follows: for any two states $\rho, \sigma \in \mathbf{S}(\mathcal{H}_A)$ and any quantum channel $\Gamma : \mathbf{L}(\mathcal{H}_A) \to \mathbf{L}(\mathcal{H}_B)$,

$$\|\rho - \sigma\|_{tr} \geq \|\Gamma(\rho) - \Gamma(\sigma)\|_{tr} \,\wedge\, F(\rho, \sigma) \leq F(\Gamma(\rho), \Gamma(\sigma)). \tag{9}$$

2.2 Entanglement-Breaking Channel

$\Lambda : \mathbf{S}(\mathcal{H}_E) \to \mathbf{S}(\mathcal{H}_B)$ is called an entanglement-breaking channel if Λ is a linear CPTP map and $(\Lambda \otimes id_{E'})(\rho) \in \mathbf{SEP}\left(B : E'\right)$ for all input states $\rho \in \mathbf{S}(\mathcal{H}_E \otimes \mathcal{H}_{E'})$. The equivalent condition can be written as $\frac{1}{\dim \mathcal{H}_E} J(\Lambda) \in \mathbf{SEP}\,(E : B)$ by using the Choi operator $J(\Lambda)$ of Λ. Define the set of Choi operators of entanglement-breaking channels by $\mathcal{E}_{E \to B}$, which is equivalent to

$$\mathcal{E}_{E \to B} = \left\{ (\dim \mathcal{H}_E)\sigma : \sigma \in \mathbf{SEP}\,(E : B) \wedge tr_B[\sigma] = \frac{1}{\dim \mathcal{H}_E} \mathbb{I}_E \right\}. \tag{10}$$

By definition, it is clear that $\mathcal{E}_{E \to B}$ is convex. It is also obvious that $\mathcal{E}_{E \to B}$ is compact since it is the intersection of a compact set and a closed set. Owing to the Caratheodory's theorem, we can represent the set as

$$\mathcal{E}_{E \to B} = \left\{ \sum_{i=1}^{s} \psi_B^{(i)} \otimes \tilde{\eta}_i : \{\tilde{\eta}_i\}_i \text{ is rank}-1 \text{ POVM} \right\}, \tag{11}$$

with $s \leq (\dim \mathcal{H}_B \dim \mathcal{H}_E)^2$.

2.3 Quantum de Finetti Theorem

Quantum de Finetti theorem is obtained by extending the (classical) de Finetti theorem and asserts that a reduced state of any quantum state ρ on symmetric subspace $\bigvee_n \mathbb{C}^d \subseteq (\mathbb{C}^d)^{\otimes n}$ is approximately a probability mixture of independent and identically distributed (i.i.d.) states $\int d\mu(\phi)\phi^{\otimes k}$ for some probability measure μ [15, 16], i.e.,

$$\left\| tr_{[n-k]}[\rho] - \int d\mu(\phi)\phi^{\otimes k} \right\|_{tr} < \frac{kd}{n}, \tag{12}$$

where $[n] = \{1, 2, \cdots, n\}$ and $tr_{[n-k]}$ represents the partial trace of the first $(n - k)$ system. The theorem not only provides a basis for "information-based interpretations" of quantum mechanics [17] but also has several applications to information processing tasks where an i.i.d. state is favorable, including quantum key distribution [18], quantum tomography [18] and quantum hypothesis testing [19].

Note that in Eq. (12), we refer an improved bound given in [16] comparing to the original bound given in [15]. Furthermore, in [16, Theorem 3], the quantum de Finetti theorem has been generalized to the following form: for any finite-dimensional Hilbert space $\mathcal{H}$ and any quantum state ρ on $\bigvee_n \mathbb{C}^d \otimes \mathcal{H}$, there exists some probability measure μ and quantum state $\sigma : \mathbf{P}\left(\mathbb{C}^d\right) \to \mathbf{S}(\mathcal{H})$ such that

$$\left\| tr_{[n-k]}[\rho] - \int d\mu(\phi)\phi^{\otimes k} \otimes \sigma(\phi) \right\|_{tr} < \frac{kd}{n}. \tag{13}$$

Remark that Eq. (13) implies Eq. (12) when $\mathcal{H} = \mathbb{C}$. This generalized theorem tells us how well $tr_{[n-k]}[\rho]$ can be approximated by a (not necessarily i.i.d.) separable state, which ensures the completeness of the separability testing based on the k-extendibility [20]. In the next section, we construct a disentangler based on these quantum de Finetti theorems.

2.4 ϵ-Net

The ϵ-net is a subset of a set which can approximate any element of the set within precision ϵ with respect to some distance. In this paper, we use ϵ-net $I \subseteq \mathbf{P}(\mathbb{C}^d)$ of $\mathbf{P}(\mathbb{C}^d)$ with respect to the trace distance. That is, I satisfies

$$\forall \phi \in \mathbf{P}(\mathbb{C}^d), \exists \hat{\phi} \in I, \left\| \hat{\phi} - \phi \right\|_{tr} \leq \epsilon. \tag{14}$$

An estimation of the minimum size $|I|$ of the ϵ-net was given by us as follows.

Lemma 1 ([21, Lemma 5]) For any $\epsilon \in (0, 1]$ and an integer $d \geq 2$, the minimum size $|I|$ of the ϵ-net of $\mathbf{P}(\mathbb{C}^d)$ is bounded by

$$2(d-1)\log_2\left(\frac{1}{\epsilon}\right) \leq \log_2 |I| \leq 2(d-1)\log_2\left(\frac{1}{\epsilon}\right) + \log_2(5d \ln d). \tag{15}$$

2.5 Convex Approximation

We use the following lemmas proven by us [21] to prove the disentangler conjecture for strong disentanglers.

Lemma 2 ([21, Theorem 1]) For any finite subset $\{\hat{\phi}_x\}_{x \in X}$ of pure states, it holds that

$$\max_\phi \min_p \left\| \phi - \sum_{x \in X} p(x)\hat{\phi}_x \right\|_{tr} = \max_\phi \min_{x \in X} \left\| \phi - \hat{\phi}_x \right\|_{tr}^2, \tag{16}$$

where the maximization of ϕ is taken over the set of pure states.

Lemma 3 ([21, Lemma 3 (extended version)]) For any compact and convex subset $\{\hat{\rho}_x \in \mathbf{S}(\mathcal{H})\}_{x \in X}$ of mixed states, it holds that

$$\min_{x \in X} \left\| \phi - \hat{\rho}_x \right\|_{tr} = \max_{\psi \in \mathbf{P}(\mathcal{H})} \left(tr\left[\psi\phi\right] - \max_{x \in X} tr\left[\psi \hat{\rho}_x\right] \right). \tag{17}$$

3 Disentangler and Strong Disentangler

In this section, we review the definition of the disentangler and give two explicit constructions of it. We also define a strong disentangler and verify that the two disentanglers are examples of the strong disentangler.

3.1 Definitions

Linear CPTP map $\Lambda : \mathbf{S}\left(\mathbb{C}^D\right) \to \mathbf{S}\left(\mathbb{C}^d \otimes \mathbb{C}^d\right)$ is called an (ϵ, δ)-disentangler if it satisfies the following two conditions:

1. $\forall \rho \in \mathbf{S}\left(\mathbb{C}^D\right)$, $\min_{\sigma \in \mathbf{SEP}(\mathbb{C}^d:\mathbb{C}^d)} \|\Lambda(\rho) - \sigma\|_{tr} \leq \epsilon$
2. $\forall \sigma \in \mathbf{SEP}\left(\mathbb{C}^d : \mathbb{C}^d\right)$, $\min_{\rho \in \mathbf{S}(\mathbb{C}^D)} \|\Lambda(\rho) - \sigma\|_{tr} \leq \delta$.

Linear CPTP map $\Lambda : \mathbf{S}\left(\mathbb{C}^D\right) \to \mathbf{S}\left(\mathbb{C}^d \otimes \mathbb{C}^d\right)$ is called an (ϵ, δ)-strong disentangler if it satisfies the following two conditions:

1. $\forall \mathcal{H}_R, \forall \rho \in \mathbf{S}\left(\mathbb{C}^D \otimes \mathcal{H}_R\right)$, $\min_{\sigma \in \mathbf{SEP}(\mathbb{C}^d:\mathbb{C}^d \otimes \mathcal{H}_R)} \|(\Lambda \otimes id_R)(\rho) - \sigma\|_{tr} \leq \epsilon$
2. $\forall \sigma \in \mathbf{SEP}\left(\mathbb{C}^d : \mathbb{C}^d\right)$, $\min_{\rho \in \mathbf{S}(\mathbb{C}^D)} \|\Lambda(\rho) - \sigma\|_{tr} \leq \delta$.

By definition, an (ϵ, δ)-strong disentangler is an (ϵ, δ)-disentangler.

The disentangler conjecture is the following conjecture:

Disentangler conjecture [6] For any constants $\epsilon, \delta \geq 0$ such that $\epsilon + \delta < 1$, every (ϵ, δ)-disentangler satisfies $\log D = \Omega(d)$ as $d \to \infty$.

3.2 Constructions

We review two types of a disentangler, constructed based on ϵ-nets and the quantum de Finetti theorem. which were suggested in [6]. We provide their explicit construction with bounds for the input dimension. By slightly modifying the proof, we can verify that these two disentanglers are also strong disentanglers with the same approximation parameters ϵ and δ.

3.2.1 Disentangler Based on ϵ-Net

Let I and $\{|e_{\hat{\phi}}\rangle \in \mathbb{C}^{|I|}\}_{\hat{\phi} \in I}$ be a $\sqrt{\delta}$-net of $\mathbf{P}(\mathbb{C}^d)$ and an orthonormal basis, respectively. Let $D = |I|d$. Define

$$\Lambda(\rho) := \sum_{\hat{\phi} \in I} \hat{\phi} \otimes tr_1 \left[|e_{\hat{\phi}}\rangle\langle e_{\hat{\phi}}| \otimes \mathbb{I}_d \rho \right]. \tag{18}$$

Since the image of Λ contains only separable states, Λ satisfies the first condition of the disentangler with $\epsilon = 0$. We can verify the second condition as follows: Let $\sigma = \sum_j p(j) \phi^{(j)} \otimes \psi^{(j)}$. Since Lemma 2 implies that the convex hull of a $\sqrt{\delta}$-net forms a δ-net of $\mathbf{P}(\mathbb{C}^d)$, we can find probability distribution $q(\hat{\phi}|j)$ such that $\left\| \phi^{(j)} - \sum_{\hat{\phi} \in I} q(\hat{\phi}|j) \hat{\phi} \right\|_{tr} \leq \delta$. By letting the input state be $\rho = \sum_j p(j) \sum_{\hat{\phi} \in I} q(\hat{\phi}|j) |e_{\hat{\phi}}\rangle\langle e_{\hat{\phi}}| \otimes \psi^{(j)}$, we can verify that

$$\|\Lambda(\rho) - \sigma\|_{tr} \leq \sum_j p(j) \left\| \sum_{\hat{\phi} \in I} q(\hat{\phi}|j) \hat{\phi} \otimes \psi^{(j)} - \phi^{(j)} \otimes \psi^{(j)} \right\|_{tr} \tag{19}$$

$$= \sum_j p(j) \left\| \sum_{\hat{\phi} \in I} q(\hat{\phi}|j) \hat{\phi} - \phi^{(j)} \right\|_{tr} \leq \delta, \tag{20}$$

where we use the triangle inequality in the first inequality.

By using Lemma 1, this construction provides a $(0, \delta)$-disentangler with $\log_2 D \leq (d-1)\log_2\left(\frac{1}{\delta}\right) + \log_2(5d \ln d) + \log_2 d$.

3.2.2 Disentangler Based on Quantum de Finetti Theorem

Let $U : \mathbb{C}^{\dim(\bigvee_n \mathbb{C}^d)} \to \bigvee_n \mathbb{C}^d$ be an isometry operator and $D = \dim(\bigvee_n \mathbb{C}^d)d$. Define

$$\Lambda(\rho) := tr_{[n-1]} \left[(U \otimes \mathbb{I}_d) \rho (U^\dagger \otimes \mathbb{I}_d) \right]. \tag{21}$$

Since the image of Λ contains any separable states, Λ satisfies the second condition with $\delta = 0$. We can verify the first condition as follows: for any ρ, by letting $\rho' = (U \otimes \mathbb{I}_d)\rho(U^\dagger \otimes \mathbb{I}_d) \in \mathbf{S}\left(\bigvee_n \mathbb{C}^d \otimes \mathbb{C}^d\right)$ and applying Eq. (13), we obtain

$$\min_{\sigma \in \mathbf{SEP}(\mathbb{C}^d : \mathbb{C}^d)} \|\Lambda(\rho) - \sigma\|_{tr} = \min_{\sigma \in \mathbf{SEP}(\mathbb{C}^d : \mathbb{C}^d)} \left\|tr_{[n-1]}[\rho'] - \sigma\right\|_{tr} < \frac{d}{n}. \tag{22}$$

Thus, the second condition is satisfied with $\epsilon = \frac{d}{n}$. This construction provides a $(\epsilon, 0)$-disentangler with $\log_2 D < (d-1) \log_2 \left(e(1 + \frac{d}{d-1}\left(\frac{1}{\epsilon} + \frac{1}{d}\right))\right) + \log_2 d$, where we use $\dim(\bigvee_n \mathbb{C}^d) = \binom{n+d-1}{d-1}$ and an inequality $\binom{n}{k} < \left(\frac{ne}{k}\right)^k$.

4 Proof of the Conjecture for Strong Disentanglers

In this section, we prove the following main theorem.

Theorem 1 *For any constants $\epsilon, \delta \geq 0$ such that $\epsilon + \sqrt{\delta} < 1$, every (ϵ, δ)-strong disentangler satisfies $\log_2 D \geq \frac{d-1}{2} \log_2\left(\frac{1}{\Delta}\right) - 2\log_2 d$ as $d \to \infty$, where $\Delta = 1 - \left(1 - \epsilon - \sqrt{\delta}\right)^2$.*

Proof Define induced CPTP map $\Gamma : \mathbf{S}\left(\mathbb{C}^D\right) \to \mathbf{S}\left(\mathbb{C}^d\right)$ by $\Gamma(\rho) = tr_2[\Lambda(\rho)]$ with an (ϵ, δ)-strong disentangler $\Lambda : \mathbf{S}\left(\mathbb{C}^D\right) \to \mathbf{S}\left(\mathbb{C}^d \otimes \mathbb{C}^d\right)$. Then the conditions of the strong disentangler imply that

1. $\forall \mathcal{H}_R, \forall \rho \in \mathbf{S}\left(\mathbb{C}^D \otimes \mathcal{H}_R\right), \min_{\sigma \in \mathbf{SEP}(\mathbb{C}^d : \mathcal{H}_R)} \|(\Gamma \otimes id_R)(\rho) - \sigma\|_{tr} \leq \epsilon$,
2. $\max_{\phi \in \mathbf{P}(\mathbb{C}^d)} \min_{\rho \in \mathbf{S}(\mathbb{C}^D)} \|\Gamma(\rho) - \phi\|_{tr} \leq \delta$.

Since $\{\Gamma(\rho) : \rho \in \mathbf{S}\left(\mathbb{C}^D\right)\}$ is a compact and convex subset of $\mathbf{S}\left(\mathbb{C}^d\right)$, by using Lemma 3, the second condition is equivalent to

$$\max_{\phi, \psi \in \mathbf{P}(\mathbb{C}^d)} \left(tr[\phi\psi] - \max_{\rho \in \mathbf{S}(\mathbb{C}^D)} tr[\psi\Gamma(\rho)]\right) \leq \delta \tag{23}$$

$$\Leftrightarrow \min_{\psi \in \mathbf{P}(\mathbb{C}^d)} \max_{\rho \in \mathbf{S}(\mathbb{C}^D)} F(\psi, \Gamma(\rho)) \geq 1 - \delta. \tag{24}$$

Let $V : \mathcal{H}_A \to \mathcal{H}_B \otimes \mathcal{H}_E$ with $\dim \mathcal{H}_A = D$, $\dim \mathcal{H}_B = d$, and $\dim \mathcal{H}_E = Dd$ be an Stinespring dilation of Γ, i.e., $\Gamma(\rho) = tr_E\left[V\rho V^\dagger\right]$. Let $\mathcal{V}$ be the range of V. Equation (24) implies that

$$\forall |\phi\rangle \in \mathcal{H}_B, \exists |\psi\rangle \in \mathcal{H}_E, \exists |\Phi\rangle \in \mathcal{V}, \langle \phi\psi|\Phi\rangle \geq \sqrt{1-\delta}. \tag{25}$$

Let $\{\phi^{(i)} \in \mathbf{P}(\mathcal{H}_B)\}_i$, $\{\psi^{(i)} \in \mathbf{P}(\mathcal{H}_E)\}_i$ and $\{\Phi^{(i)} \in \mathbf{P}(\mathcal{V})\}_i$ be an ϵ'-net of $\mathbf{P}(\mathcal{H}_B)$ and the corresponding states satisfying in Eq. (25). Let $\rho = \Psi \in \mathbf{P}(\mathcal{H}_A \otimes \mathcal{H}_R)$

with $(V \otimes \mathbb{I}_R)|\Psi\rangle = \sum_i \sqrt{p(i)} |\Phi^{(i)}\rangle_{BE} |i\rangle_R$ be an input state of Γ. Then, the first condition of the strong disentangler implies that

$$\max_p \min_{\sigma \in \mathbf{SEP}(\mathcal{H}_B : \mathcal{H}_R)} \left\| tr_E \left[(V \otimes \mathbb{I}_R) \Psi (V^\dagger \otimes \mathbb{I}_R) \right] - \sigma \right\|_{tr} \leq \epsilon. \quad (26)$$

Since $\langle \hat{\Psi} | (V \otimes \mathbb{I}_R) | \Psi \rangle \geq \sqrt{1-\delta}$ with $|\hat{\Psi}\rangle = \sum_i \sqrt{p(i)} |\phi^{(i)}\rangle_B |\psi^{(i)}\rangle_E |i\rangle_R$, Eqs. (8) implies that

$$\left\| (V \otimes \mathbb{I}_R) \Psi (V^\dagger \otimes \mathbb{I}_R) - \hat{\Psi} \right\|_{tr} \leq \sqrt{\delta}. \quad (27)$$

By using the triangle inequality and the monotonicity of the trace distance, Eqs. (26) and (27) imply that

$$\max_p \min_{\sigma \in \mathbf{SEP}(\mathcal{H}_B : \mathcal{H}_R)} \left\| tr_E \left[\hat{\Psi} \right] - \sigma \right\|_{tr} \leq \epsilon + \sqrt{\delta}. \quad (28)$$

From now on, we assume that $\epsilon + \sqrt{\delta} < 1$. By using Eq. (8), we obtain

$$\min_p \max_{\sigma \in \mathbf{SEP}(\mathcal{H}_B : \mathcal{H}_R)} F\left(tr_E \left[\hat{\Psi} \right], \sigma \right) \geq \left(1 - \epsilon - \sqrt{\delta} \right)^2. \quad (29)$$

By using Lemma 4 shown in Appendix, we obtain

$$\min_p \max_{J(\mathcal{E}) \in \mathcal{E}_{E \to B}} tr \left[J(\mathcal{E}) tr_R \left[\hat{\Psi} \right] \right] \geq \left(1 - \epsilon - \sqrt{\delta} \right)^2 \quad (30)$$

$$\Leftrightarrow \min_p \max_{J(\mathcal{E}) \in \mathcal{E}_{E \to B}} \sum_i p(i) tr \left[J(\mathcal{E}) \phi_B^{(i)} \otimes \psi_E^{(i)} \right] \geq \left(1 - \epsilon - \sqrt{\delta} \right)^2. \quad (31)$$

Since the domain of p and $J(\mathcal{E})$ is convex and compact sets and the target function is bilinear, we can apply the minimax theorem. After applying the theorem, we obtain

$$\max_{J(\mathcal{E}) \in \mathcal{E}_{E \to B}} \min_i tr \left[J(\mathcal{E}) \phi_B^{(i)} \otimes \psi_E^{(i)} \right] \geq \left(1 - \epsilon - \sqrt{\delta} \right)^2. \quad (32)$$

By taking the limit of $\epsilon' \to 0$, we obtain

$$\max_{J(\mathcal{E}) \in \mathcal{E}_{E \to B}} \min_{\phi \in P(\mathcal{H}_B)} \max_{\psi \in P(\mathcal{H}_E)} tr \left[J(\mathcal{E}) \phi_B \otimes \psi_E \right] \geq \left(1 - \epsilon - \sqrt{\delta} \right)^2 \quad (33)$$

$$\Leftrightarrow \max_{J(\mathcal{E}) \in \mathcal{E}_{E \to B}} \min_{\phi \in P(\mathcal{H}_B)} \max_{\psi \in P(\mathcal{H}_E)} tr \left[\mathcal{E}(\psi) \phi \right] \geq \left(1 - \epsilon - \sqrt{\delta} \right)^2. \quad (34)$$

$$\Leftrightarrow \max_{J(\mathcal{E}) \in \mathcal{E}_{E \to B}} \min_{\phi \in P(\mathcal{H}_B)} \max_{\psi \in P(\mathcal{H}_E)} F(\phi, \mathcal{E}(\psi)) \geq 1 - \Delta, \quad (35)$$

where $\Delta = 1 - \left(1 - \epsilon - \sqrt{\delta}\right)^2$. By using Lemma 3 in the similar way to derive Eq. (24), we can verify that Eq. (35) is equivalent to

$$\min_{J(\mathcal{E}) \in \mathcal{E}_{E \to B}} \max_{\phi \in \mathbf{P}(\mathcal{H}_B)} \min_{\rho \in \mathbf{S}(\mathcal{H}_E)} \|\mathcal{E}(\rho) - \phi\|_{tr} \leq \Delta. \tag{36}$$

Let $\mathcal{E}(\rho) = \sum_{\hat{\phi} \in E} tr\left[M_{\hat{\phi}}\rho\right]\hat{\phi}$ maximize Eq. (36). Equation (36) implies that

$$\max_{\phi \in \mathbf{P}(\mathcal{H}_B)} \min_{p} \left\|\sum_{\hat{\phi} \in E} p(\hat{\phi})\hat{\phi} - \phi\right\|_{tr} \leq \Delta. \tag{37}$$

Since this implies that the convex hull of E forms a Δ-net of $\mathbf{P}(\mathcal{H}_B)$, Lemma 2 implies that E is a $\sqrt{\Delta}$-net of $\mathbf{P}(\mathcal{H}_B)$, i.e.,

$$\max_{\phi \in \mathbf{P}(\mathcal{H}_B)} \min_{\hat{\phi} \in E} \|\hat{\phi} - \phi\|_{tr} \leq \sqrt{\Delta}. \tag{38}$$

By using Lemma 1, we obtain

$$\log_2(Dd^2)^2 = \log_2(\dim \mathcal{H}_B \dim \mathcal{H}_E)^2 \geq \log_2 |E| \geq (d-1)\log_2\left(\frac{1}{\Delta}\right). \tag{39}$$

□

5 Discussion

We investigated how the input dimension of a strong disentangler relates to the output dimension and approximation parameters. Firstly, we presented explicit constructions of the strong disentangler based on ϵ-net and quantum de Finetti theorem. This shows the existence of a strong disentangler with an exponential input dimension compared to the output dimension. Secondly, we proved that any strong disentangler with certain approximation parameters must have such an exponential input dimension. We achieved this by reducing the strong disentangler into an approximately entanglement-breaking channel. This provides an important partial proof to the original disentangler conjecture since the class of strong disentanglers is a subset of that of disentanglers that is wide enough to contain all the known disentanglers. As mentioned in the introduction, the disentangler is an approximately entanglement-annihilating channel, which differs from an approximately entanglement-breaking channel in general. Therefore, it is important to conduct further research on finding more connections between these two types of channels in order to resolve the disentangler conjecture.

Fidelity Distance from the Set of Separable States

We derive an alternative formulation of the maximal fidelity to separable states, which is essentially the minimal Bures distance to separable states, as following.

Lemma 4 *For given $\rho \in \mathbf{S}(\mathcal{H}_A \otimes \mathcal{H}_B)$, by letting $|\Phi\rangle_{ABE}$ be a purification of ρ, it holds that*

$$\max_{\sigma \in \mathrm{SEP}(A:B)} F(\rho, \sigma) = \max_{J(\mathcal{E}) \in \mathcal{E}_{E \to B}} tr\left[J(\mathcal{E})\Phi_{BE}\right]. \tag{40}$$

Proof We can find the closest separable state to ρ in the form of $\sigma = \sum_{i=1}^{s} p_i \phi_A^{(i)} \otimes \psi_B^{(i)}$ with $s \geq (\dim \mathcal{H}_B \dim \mathcal{H}_E)^2$. Thus, the left-hand side is equivalent to

$$\max_{\{p_i, \phi_A^{(i)}, \psi_B^{(i)}\}_{i=1}^{s}} F\left(\rho, \sum_{i=1}^{s} p_i \phi_A^{(i)} \otimes \psi_B^{(i)}\right) \tag{41}$$

$$= \max_{\{p_i, \phi_A^{(i)}, \psi_B^{(i)}\}_{i=1}^{s}} \max_{V: \mathcal{H}_E \to \mathcal{H}_{E'}} \left|\sum_{i=1}^{s} \sqrt{p_i} \langle\Phi|\phi_A^{(i)}\rangle|\psi_B^{(i)}\rangle V^\dagger |i\rangle_{E'}\right|^2 \tag{42}$$

$$= \max_{\{p_i, \phi_A^{(i)}, \psi_B^{(i)}, \tilde{\eta}_i\}_{i=1}^{s}} \left|\sum_{i=1}^{s} \sqrt{p_i} \langle\Phi|\phi_A^{(i)}\rangle|\psi_B^{(i)}\rangle|\tilde{\eta}_i\rangle_E\right|^2 \tag{43}$$

$$= \max_{\{p_i, \phi_A^{(i)}, \psi_B^{(i)}, \tilde{\eta}_i\}_{i=1}^{s}} \left(\sum_{i=1}^{s} \sqrt{p_i} \left|\langle\Phi|\phi_A^{(i)}\rangle|\psi_B^{(i)}\rangle|\tilde{\eta}_i\rangle_E\right|\right)^2, \tag{44}$$

where $\{|i\rangle_{E'}\}_{i=1}^{s}$ is an orthonormal basis of $\mathcal{H}_{E'}$, V is chosen from isometries, $\{\tilde{\eta}_i\}_{i=1}^{s}$ is chosen from a set of rank-1 POVMs. By using Cauchy-Schwarz inequality,

$$(44) = \max_{\{\phi_A^{(i)}, \psi_B^{(i)}, \tilde{\eta}_i\}_{i=1}^{s}} \sum_{i=1}^{s} \left|\langle\Phi|\phi_A^{(i)}\rangle|\psi_B^{(i)}\rangle|\tilde{\eta}_i\rangle_E\right|^2 \tag{45}$$

$$= \max_{\{\psi_B^{(i)}, \tilde{\eta}_i\}_{i=1}^{s}} \sum_{i=1}^{s} tr\left[(\psi_B^{(i)} \otimes \tilde{\eta}_i)\Phi_{BE}\right]. \tag{46}$$

Equation (46) is upper bounded by the right hand side of the lemma. This implies (LHS) $\leq$ (RHS). On the other hand, since the Choi operator of any entanglement-breaking channel can be described by $\sum_{i=1}^{s} \psi_B^{(i)} \otimes \tilde{\eta}_i$, we obtain (RHS) $\leq$ (LHS). This completes the proof. □

Acknowledgements We thank Yoshifumi Nakata, Takaya Matsuura, Mio Murao, Koji Azuma, Hayata Yamasaki, Tomoyuki Morimae, Ryuhei Mori, Takuya Ikuta, Yuki Takeuchi, and Yasuhiro Takahashi for their helpful discussions.

Competing Interests This work was partially supported by JST Moonshot R&D MILLENNIA Program (Grant no. JPMJMS2061). SA was partially supported by JST, PRESTO Grant no. JPMJPR2111 and MEXT Q-LEAP Grant no. JPMXS0120319794. GK was supported in part by the Grant-in-Aid for Scientific Research (C) no. 20K03779, (C) no. 21K03388, and (S) no.18H05237 of JSPS, and CREST (Japan Science and Technology Agency) Grant no.JPMJCR1671. ST was partially supported by JSPS KAKENHI Grant nos. JP20H05966 and JP22H00522. The authors have no conflicts of interest to declare that are relevant to the content of this chapter.

References

1. M. Horodecki, P. Horodecki, R. Horodecki, Separability of mixed states: necessary and sufficient conditions. Phys. Lett. A **223**(1), 1–8 (1996)
2. G. Aubrun, S. Szarek, Dvoretzky's theorem and the complexity of entanglement detection. Discrete Anal. **1**, 20 (2017)
3. L. Gurvits, Classical complexity and quantum entanglement. J. Comput. Syst. Sci. **69**(3), 448–484 (2004). Special Issue on STOC 2003
4. G. Sevag, Strong NP-hardness of the quantum separability problem. Quantum Info. Comput. **10**(3), 343–360 (2010)
5. H. Kobayashi, K. Matsumoto, T. Yamakami, Quantum certificate verification: single versus multiple quantum certificates (2001), ArXiv arXiv:quant-ph/0110006
6. S. Aaronson, S. Beigi, A. Drucker, B. Fefferman, P. Shor, The power of unentanglement. Theory Comput. **5**, 1–42 (2009)
7. H. Blier, A. Tapp, All languages in NP have very short quantum proofs, in *2009 Third International Conference on Quantum, Nano and Micro Technologies* (2009), pp. 34–37
8. B. Salman, NP vs $QMA_{log}(2)$. Quantum Info. Comput. **10**(1), 141–151 (2010)
9. L.G. François, N. Shota, N. Harumichi, On QMA protocols with two short quantum proofs. Quantum Info. Comput. **12**(7–8), 589–600 (2012)
10. M. Chris, W. John, Quantum Arthur-Merlin games. Comput. Complex. **14**(2), 122–152 (2005)
11. L. Moravčíková, M. Ziman, Entanglement-annihilating and entanglement-breaking channels. J. Phys. A Math. Theor. **43**(27), 275306 (2010)
12. A.S. Holevo, Coding theorems for quantum communication channels, in *Proceedings. 1998 IEEE International Symposium on Information Theory (Cat. No.98CH36252)* (1998), p. 84
13. M. Horodecki, P.W. Shor, M.B. Ruskai, Entanglement breaking channels. Rev. Math. Phys. **15**(06), 629–641 (2003)
14. A.W. Harrow, A. Natarajan, X. Wu, Limitations of semidefinite programs for separable states and entangled games. Commun. Math. Phys. **366**(2), 423–468 (2019)
15. M. Christandl, R. König, G. Mitchison, R. Renner, One-and-a-half quantum de Finetti theorems. Commun. Math. Phys. **273**(2), 473–498 (2007)
16. G. Chiribella, On quantum estimation, quantum cloning and finite quantum de Finetti theorems, in *Theory of Quantum Computation, Communication, and Cryptography*, ed. by W. van Dam, V.M. Kendon, S. Severini (Springer, Berlin, Heidelberg), pp. 9–25
17. C.M. Caves, C.A. Fuchs, R. Schack, Unknown quantum states: the quantum de Finetti representation. J. Math. Phys. **43**(9), 4537–4559 (2002)
18. R. Renner, Symmetry of large physical systems implies independence of subsystems. Nat. Phys. **3**(9), 645–649 (2007)
19. F.G.S.L. Brandão, M.B. Plenio, A Generalization of quantum Stein's lemma. Commun. Math. Phys. **295**(3), 791–828 (2010)

20. A.C. Doherty, P.A. Parrilo, F.M. Spedalieri, Complete family of separability criteria. Phys. Rev. A **69**, 022308 (2004)
21. S. Akibue, G. Kato, S. Tani, Probabilistic state synthesis based on optimal convex approximation. npj Quantum Inform. **10**(1), 3 (2024)

Open Access This chapter is licensed under the terms of the Creative Commons Attribution 4.0 International License (http://creativecommons.org/licenses/by/4.0/), which permits use, sharing, adaptation, distribution and reproduction in any medium or format, as long as you give appropriate credit to the original author(s) and the source, provide a link to the Creative Commons license and indicate if changes were made.

The images or other third party material in this chapter are included in the chapter's Creative Commons license, unless indicated otherwise in a credit line to the material. If material is not included in the chapter's Creative Commons license and your intended use is not permitted by statutory regulation or exceeds the permitted use, you will need to obtain permission directly from the copyright holder.

Representation Theory of $\mathfrak{sl}(2,\mathbb{R}) \simeq \mathfrak{su}(1,1)$ and a Generalization of Non-commutative Harmonic Oscillators

Ryosuke Nakahama

Abstract The non-commutative harmonic oscillator (NCHO) was introduced as a specific Hamiltonian operator on $L^2(\mathbb{R}) \otimes \mathbb{C}^2$ by Parmeggiani and Wakayama. Then it was proved by Ochiai and Wakayama that the eigenvalue problem for NCHO is reduced to a Heun differential equation. In this article, we consider some generalization of NCHO for $L^2(\mathbb{R}^n) \otimes \mathbb{C}^p$ as a rotation-invariant differential equation. Then by applying a representation theory of $\mathfrak{sl}(2,\mathbb{R}) \simeq \mathfrak{su}(1,1)$, we check that its restriction to the space of products of radial functions and homogeneous harmonic polynomials is reduced to a holomorphic differential equation on the unit disk, which is generically Fuchsian.

Keywords Non-commutative harmonic oscillator · Representation of $\mathfrak{sl}(2,\mathbb{R})$ · Unitary highest weight representation · Fuchsian differential equation

1 Introduction

In [14, 15], Parmeggiani and Wakayama introduced the non-commutative harmonic oscillator (NCHO) on $L^2(\mathbb{R}) \otimes \mathbb{C}^2$. For the η-shifted version of the NCHO [16], its Hamiltonian is given by the self-adjoint operator

$$H_{\text{NCHO}} := \mathbf{A}\left(-\frac{1}{2}\frac{d^2}{dx^2} + \frac{1}{2}x^2\right) + \mathbf{J}\left(x\frac{d}{dx} + \frac{1}{2}\right) + 2\eta i \det(\mathbf{A} \pm i\mathbf{J})^{1/2}\mathbf{J},$$

where $\mathbf{A}, \mathbf{J} \in M(2,\mathbb{R})$, $\mathbf{A} = {}^t\mathbf{A}$, $\mathbf{J} = -{}^t\mathbf{J}$, and $\eta \in \mathbb{R}$. The original NCHO corresponds to $\eta = 0$ case. Then in [11, 12, 16, 18], Ochiai, Reyes-Bustos and Wakayama proved that its eigenvalue problem $H_{\text{NCHO}}\psi = \lambda\psi$ is reduced to a Heun differential equation on a certain simply-connected domain in $\mathbb{C}$. In the proof they used (quasi) intertwining operators for (reducible) principal series representations of $\mathfrak{sl}(2, \mathbb{R})$.

In this article, we consider a generalization of this problem, that is, we want to find a function $\psi \in L^2(\mathbb{R}^n) \otimes \mathbb{C}^p$ satisfying

$$\left[-\mathbf{A}_1 \sum_{j=1}^n \frac{\partial^2}{\partial x_j^2} + i\mathbf{A}_2 \sum_{j=1}^n \left(x_j \frac{\partial}{\partial x_j} + \frac{1}{2} \right) + \mathbf{A}_3 \sum_{j=1}^n x_j^2 \right] \psi = 2\mathbf{C}\psi, \qquad (1)$$

where $\mathbf{A}_k, \mathbf{C} \in M(p, \mathbb{C}), \mathbf{A}_k = \mathbf{A}_k^\dagger, \mathbf{C} = \mathbf{C}^\dagger$. Here $\mathbf{C}^\dagger = {}^t\overline{\mathbf{C}}$. This equation commutes with the rotation of $\mathbb{R}^n$ by orthogonal matrices $g \in O(n)$. Then according to the decomposition of $L^2(\mathbb{R}^n)$ under $\mathfrak{sl}(2, \mathbb{R}) \times O(n)$ in terms of harmonic polynomials,

$$L^2(\mathbb{R}^n) \simeq \sum_{k \in \mathbb{Z}_{\geq 0}}^{\oplus} L^2_{k+\frac{n}{2}}(\mathbb{R}^+) \otimes \mathcal{HP}_k(\mathbb{R}^n)$$

(where notations are explained later), we check that the restriction of (1) on $L^2_{k+\frac{n}{2}}(\mathbb{R}^+) \otimes \mathcal{HP}_k(\mathbb{R}^n) \otimes \mathbb{C}^p$ is reduced to a holomorphic differential equation on the unit disk $\mathbf{D}$, which has some $SU(1, 1)$-covariance and is generically Fuchsian. As a tool we use intertwining operators for unitary highest weight representations of $\mathfrak{sl}(2, \mathbb{R})$, between $L^2_{k+\frac{n}{2}}(\mathbb{R}^+)$ and the space of holomorphic functions on $\mathbf{D}$.

This paper is organized as follows. In Sect. 2 we begin with a review on usual harmonic oscillators. In Sect. 3 we review representation theory for the Lie algebra $\mathfrak{sl}(2, \mathbb{R}) \simeq \mathfrak{su}(1, 1)$. In Sect. 4 we apply representation theory of $\mathfrak{sl}(2, \mathbb{R}) \simeq \mathfrak{su}(1, 1)$ for the analysis of the generalized NCHO.

2 Usual Harmonic Oscillator and Harmonic Polynomials

First we recall the usual harmonic oscillator of n-variables. For details about this and the next section, see, e.g., [4, 6, 10]. Let H_{HO} be the self-adjoint operator on $L^2(\mathbb{R}^n)$ given by

$$H_{\text{HO}} := \frac{1}{2} \sum_{j=1}^n \left(-\frac{\partial^2}{\partial x_j^2} + x_j^2 \right).$$

To find eigenfunctions of H_{HO}, we consider the creation and annihilation operators

$$a_j^\dagger := \frac{1}{\sqrt{2}} \left(x_j - \frac{\partial}{\partial x_j} \right), \qquad a_j := \frac{1}{\sqrt{2}} \left(x_j + \frac{\partial}{\partial x_j} \right),$$

so that

$$H_{\text{HO}} = \sum_{j=1}^n a_j^\dagger a_j + \frac{n}{2}, \qquad [a_j, a_k^\dagger] = a_j a_k^\dagger - a_k^\dagger a_j = \delta_{jk}.$$

hold. Also let
$$h_0(x) := \pi^{-n/4} e^{-\sum_{j=1}^n x_j^2/2} \in L^2(\mathbb{R}^n),$$

and for $\mathbf{m} = (m_1, \ldots, m_n) \in (\mathbb{Z}_{\geq 0})^n$ let

$$h_\mathbf{m}(x) := \frac{1}{\sqrt{\mathbf{m}!}} (a_1^\dagger)^{m_1} \cdots (a_n^\dagger)^{m_n} h_0(x)$$
$$= \frac{\pi^{-n/4}}{\sqrt{\mathbf{m}!}} \left(-\frac{1}{\sqrt{2}}\right)^{|\mathbf{m}|} \prod_{j=1}^n e^{x_j^2/2} \left(\frac{\partial}{\partial x_j}\right)^{m_j} e^{-x_j^2},$$

where $\mathbf{m}! := m_1! \cdots m_n!$, $|\mathbf{m}| := m_1 + \cdots + m_n$. These $h_\mathbf{m}(x)$ are called the Hermite functions. Then we can show

$$a_j^\dagger h_\mathbf{m}(x) = \sqrt{m_j + 1} h_{\mathbf{m}+\mathbf{e}_j}(x), \quad a_j h_\mathbf{m}(x) = \begin{cases} \sqrt{m_j} h_{\mathbf{m}-\mathbf{e}_j}(x) & (m_j \geq 1), \\ 0 & (m_j = 0), \end{cases}$$

$$\langle h_\mathbf{m}(x), h_\mathbf{n}(x) \rangle_{L^2(\mathbb{R}^n)} = \int_{\mathbb{R}^n} h_\mathbf{m}(x) \overline{h_\mathbf{n}(x)} \, dx = \delta_{\mathbf{mn}},$$

where $\mathbf{e}_j = (0, \ldots, 0, \overset{j\text{th}}{1}, 0, \ldots, 0)$. Especially $h_\mathbf{m}(x)$ is an eigenfunction of H_{HO} with the eigenvalue $|\mathbf{m}| + \frac{n}{2}$. Now since

$$L^2(\mathbb{R}^n)_{\text{fin}} := \bigoplus_{\mathbf{m} \in (\mathbb{Z}_{\geq 0})^n} \mathbb{C} h_\mathbf{m}(x) = \mathcal{P}(\mathbb{R}^n) h_0(x)$$

is dense in $L^2(\mathbb{R}^n)$, where $\mathcal{P}(\mathbb{R}^n)$ is the space of polynomials on $\mathbb{R}^n$, the set of eigenvalues of H_{HO} is $\mathbb{Z}_{\geq 0} + \frac{n}{2}$, and for each $k \in \mathbb{Z}_{\geq 0}$, the corresponding eigenspaces are given by

$$\left\{ \psi \in L^2(\mathbb{R}^n) \mid H_{\text{HO}} \psi = \left(k + \frac{n}{2}\right) \psi \right\} = \bigoplus_{\substack{\mathbf{m} \in (\mathbb{Z}_{\geq 0})^n \\ |\mathbf{m}| = k}} \mathbb{C} h_\mathbf{m}(x).$$

The set $\{h_\mathbf{m}(x)\}_{\mathbf{m} \in (\mathbb{Z}_{\geq 0})^n}$ forms an orthonormal basis of $L^2(\mathbb{R}^n)$.

The space $L^2(\mathbb{R}^n)$ is unitarily equivalent to the following *Fock space*

$$\mathcal{F}(\mathbb{C}^n) := \left\{ f \in O(\mathbb{C}^n) \,\middle|\, \|f\|_{\mathcal{F}(\mathbb{C}^n)}^2 := \frac{1}{\pi^n} \int_{\mathbb{C}^n} |f(w)|^2 e^{-\sum_{j=1}^n |w_j|^2} \, dw < \infty \right\},$$

where $O(\mathbb{C}^n)$ denotes the space of holomorphic functions on $\mathbb{C}^n$, via the *Bargmann transform*

$$\mathcal{B} \colon L^2(\mathbb{R}^n) \longrightarrow \mathcal{F}(\mathbb{C}^n),$$

$$(\mathcal{B}\psi)(w) := \pi^{-n/4} \int_{\mathbb{R}^n} \psi(x) e^{\sum_{j=1}^n \left(\sqrt{2}x_j w_j - \frac{x_j^2}{2} - \frac{w_j^2}{2}\right)} dx.$$

Via $\mathcal{B}$, the creation and annihilation operators are transformed as

$$(\mathcal{B}(a_j^\dagger \psi))(w) = w_j (\mathcal{B}\psi)(w), \qquad (\mathcal{B}(a_j \psi))(w) = \frac{\partial}{\partial w_j}(\mathcal{B}\psi)(w).$$

Especially we have

$$\mathcal{B} \circ H_{\mathrm{HO}} \circ \mathcal{B}^{-1} = \mathcal{B} \circ \left(\sum_{j=1}^n a_j^\dagger a_j + \frac{n}{2}\right) \circ \mathcal{B}^{-1} = \sum_{j=1}^n w_j \frac{\partial}{\partial w_j} + \frac{n}{2},$$

$$(\mathcal{B} h_{\mathbf{m}})(w) = \frac{1}{\sqrt{\mathbf{m}!}} w_1^{m_1} \cdots w_n^{m_n}.$$

The structure of eigenspaces of $\sum_{j=1}^n w_j \frac{\partial}{\partial w_j} + \frac{n}{2}$ coincides with that of H_{HO}.

Next we recall harmonic polynomials. For $k \in \mathbb{Z}_{\geq 0}$, let $\mathcal{P}_k(\mathbb{R}^n)$ be the space of homogeneous polynomials of degree k, and let $\mathcal{HP}_k(\mathbb{R}^n)$ be the space of homogeneous harmonic polynomials of degree k.

$$\mathcal{P}_k(\mathbb{R}^n) := \{h(x) \in \mathcal{P}(\mathbb{R}^n) \mid h(tx) = t^k h(x) \ (t \in \mathbb{R})\},$$

$$\mathcal{HP}_k(\mathbb{R}^n) := \left\{h(x) \in \mathcal{P}_k(\mathbb{R}^n) \ \bigg| \ \sum_{j=1}^n \frac{\partial^2 h}{\partial x_j^2}(x) = 0\right\}.$$

We note that if $n = 1$, then $\mathcal{HP}_0(\mathbb{R}) = \mathbb{C}$, $\mathcal{HP}_1(\mathbb{R}) = \mathbb{C}x$ and $\mathcal{HP}_k(\mathbb{R}) = \{0\}$ for $k \geq 2$. Then it is well-known that we have

$$\mathcal{P}_k(\mathbb{R}^n) = \mathcal{HP}_k(\mathbb{R}^n) \oplus \mathcal{P}_{k-2}(\mathbb{R}^n) r^2,$$

where $r^2(x) := x_1^2 + \cdots + x_n^2$, and hence we have

$$\mathcal{P}(\mathbb{R}^n) = \bigoplus_{k=0}^\infty \mathbb{C}[r^2] \mathcal{HP}_k(\mathbb{R}^n),$$

$$L^2(\mathbb{R}^n)_{\mathrm{fin}} = \bigoplus_{k=0}^\infty \mathbb{C}[r^2] \mathcal{HP}_k(\mathbb{R}^n) e^{-r^2/2},$$

where $\mathbb{C}[r^2] := \bigoplus_{j=0}^\infty \mathbb{C}r^{2j}$. Also, for $\phi(t) \in \mathbb{C}[t]e^{-t}$, $h(x) \in \mathcal{HP}_k(\mathbb{R}^n)$, we have

$$\int_{\mathbb{R}^n} \left|\phi\left(\frac{r^2(x)}{2}\right) h(x)\right|^2 dx = \iint_{\mathbb{R}^+ \times S^{n-1}} \left|\phi\left(\frac{r^2}{2}\right) h(r\omega)\right|^2 r^{n-1} dr d\omega$$

$$= \int_0^\infty \left|\phi\left(\frac{r^2}{2}\right)\right|^2 r^{2k+n-1}\, dr \int_{S^{n-1}} |h(\omega)|^2\, d\omega$$

$$= 2^{k+\frac{n}{2}-1} \int_0^\infty |\phi(t)|^2 t^{k+\frac{n}{2}-1}\, dt \int_{S^{n-1}} |h(\omega)|^2\, d\omega \cdot \frac{2}{\Gamma(k+\frac{n}{2})} \int_0^\infty e^{-s^2} s^{2k+n-1}\, ds$$

$$= \frac{2^{k+\frac{n}{2}}}{\Gamma(k+\frac{n}{2})} \int_0^\infty |\phi(t)|^2 t^{k+\frac{n}{2}-1}\, dt \iint_{\mathbb{R}^+ \times S^{n-1}} |h(s\omega)|^2 e^{-s^2} s^{n-1}\, ds d\omega$$

$$= \frac{2^{k+\frac{n}{2}}}{\Gamma(k+\frac{n}{2})} \int_0^\infty |\phi(t)|^2 t^{k+\frac{n}{2}-1}\, dt \int_{\mathbb{R}^n} |h(x)|^2 e^{-r^2(x)}\, dx,$$

where $\mathbb{R}^+ := \{t \in \mathbb{R} \mid t > 0\}$, $S^{n-1} := \{x \in \mathbb{R}^n \mid r^2(x) = 1\}$. Now, for $\mu > 0$ we define a Hilbert space $L^2_\mu(\mathbb{R}^+)$ by

$$L^2_\mu(\mathbb{R}^+) := \left\{ \phi \colon \mathbb{R}^+ \longrightarrow \mathbb{C} \;\Big|\; \|\phi\|^2_{\mu,\mathbb{R}^+} := \frac{2^\mu}{\Gamma(\mu)} \int_0^\infty |\phi(t)|^2 t^{\mu-1}\, dt < \infty \right\},$$

and define an inner product on $\mathcal{HP}_k(\mathbb{R}^n)$ by

$$\langle h_1, h_2 \rangle_{\mathcal{HP}} := \int_{\mathbb{R}^n} h_1(x)\overline{h_2(x)} e^{-r^2(x)}\, dx.$$

Then we have a Hilbert direct sum decomposition

$$L^2(\mathbb{R}^n) \simeq \sum_{k \in \mathbb{Z}_{\geq 0}}^{\oplus} L^2_{k+\frac{n}{2}}(\mathbb{R}^+) \otimes \mathcal{HP}_k(\mathbb{R}^n). \tag{2}$$

Especially, if $n = 1$ then $L^2(\mathbb{R}) = L^2_{\mathrm{even}}(\mathbb{R}) \oplus L^2_{\mathrm{odd}}(\mathbb{R}) \simeq L^2_{1/2}(\mathbb{R}^+) \oplus L^2_{3/2}(\mathbb{R}^+)x$ holds. In addition, for $\phi(t) \in L^2_{k+\frac{n}{2}}(\mathbb{R}^+)_{\mathrm{fin}} := \mathbb{C}[t]e^{-t} \subset L^2_{k+\frac{n}{2}}(\mathbb{R}^+)$, $h(x) \in \mathcal{HP}_k(\mathbb{R}^n)$, we have

$$H_{\mathrm{HO}}\left(\phi\left(\frac{r^2(x)}{2}\right)h(x)\right) = h(x)\left(-t\frac{d^2}{dt^2} - \left(k+\frac{n}{2}\right)\frac{d}{dt} + t\right)\phi(t)\bigg|_{t=r^2(x)/2}.$$

That is, the eigenvalue problem for H_{HO} on $L^2(\mathbb{R}^n)$ restricted to $L^2_{k+\frac{n}{2}}(\mathbb{R}^+) \otimes \mathcal{HP}_k(\mathbb{R}^n)$ is equivalent to that for

$$H_{\mathrm{HO}}^{(\mu)} := -t\frac{d^2}{dt^2} - \mu\frac{d}{dt} + t \tag{3}$$

on $L^2_\mu(\mathbb{R}^+)$ for $\mu = k + \frac{n}{2}$. We note that the orthogonal group $O(n) := \{g \in M(n, \mathbb{R}) \mid {}^t g g = I_n\}$ acts irreducibly on $\mathcal{HP}_k(\mathbb{R}^n)$ by $h(x) \mapsto h(g^{-1}x)$. In the next section we review the $\mathfrak{sl}(2, \mathbb{R})$-module structure on $L^2_\mu(\mathbb{R}^+)$. Especially, Eq. (2) gives a decomposition as representations of $\mathfrak{sl}(2, \mathbb{R}) \times O(n)$, which is a special case of *Howe's dual pair correspondence* (see [5, 8]).

3 Representation Theory of $\mathfrak{sl}(2, \mathbb{R}) \simeq \mathfrak{su}(1, 1)$

In this section we consider the Lie groups

$$SL(2, \mathbb{R}) := \left\{ g = \begin{bmatrix} a & b \\ c & d \end{bmatrix} \,\middle|\, \begin{array}{c} a, b, c, d \in \mathbb{R}, \\ \det(g) = ad - bc = 1 \end{array} \right\},$$

$$SU(1, 1) := \left\{ g = \begin{bmatrix} a & b \\ \overline{b} & \overline{a} \end{bmatrix} \,\middle|\, \begin{array}{c} a, b \in \mathbb{C}, \\ \det(g) = |a|^2 - |b|^2 = 1 \end{array} \right\}.$$

Then their corresponding Lie algebras are given by

$$\mathfrak{sl}(2, \mathbb{R}) := \left\{ \begin{bmatrix} a & b \\ c & -a \end{bmatrix} \,\middle|\, a, b, c \in \mathbb{R} \right\}, \quad \mathfrak{su}(1, 1) := \left\{ \begin{bmatrix} ia & b \\ \overline{b} & -ia \end{bmatrix} \,\middle|\, a \in \mathbb{R}, b \in \mathbb{C} \right\}.$$

That is, for $X, Y \in \mathfrak{sl}(2, \mathbb{R})$ (resp. $\mathfrak{su}(1, 1)$), $[X, Y] := XY - YX \in \mathfrak{sl}(2, \mathbb{R})$ (resp. $\mathfrak{su}(1, 1)$) holds, and $e^{tX} = \sum_{n=0}^\infty (tX)^n/n! \in SL(2, \mathbb{R})$ (resp. $SU(1, 1)$) holds for all $t \in \mathbb{R}$. Here $i := \sqrt{-1}$. Their complexification is given by

$$\mathfrak{sl}(2, \mathbb{R}) \otimes_\mathbb{R} \mathbb{C} = \mathfrak{su}(1, 1) \otimes_\mathbb{R} \mathbb{C} = \mathfrak{sl}(2, \mathbb{C}) := \left\{ \begin{bmatrix} a & b \\ c & -a \end{bmatrix} \,\middle|\, a, b, c \in \mathbb{C} \right\}.$$

Let $C := \begin{bmatrix} 1 & i \\ i & 1 \end{bmatrix}$. Then we have isomorphisms

$$\text{Int}(C) \colon SU(1, 1) \longrightarrow SL(2, \mathbb{R}), \qquad g \mapsto CgC^{-1},$$
$$\text{Ad}(C) \colon \mathfrak{su}(1, 1) \longrightarrow \mathfrak{sl}(2, \mathbb{R}), \qquad X \mapsto CXC^{-1}.$$

We take two bases $\{H, E, F\}, \{{}^c H, {}^c E, {}^c F\} \subset \mathfrak{sl}(2, \mathbb{C})$ by

$$H := \begin{bmatrix} 1 & 0 \\ 0 & -1 \end{bmatrix}, \qquad E := \begin{bmatrix} 0 & 1 \\ 0 & 0 \end{bmatrix}, \qquad F := \begin{bmatrix} 0 & 0 \\ 1 & 0 \end{bmatrix},$$
$${}^c H := \text{Ad}(C)H, \qquad {}^c E := \text{Ad}(C)E, \qquad {}^c F := \text{Ad}(C)F,$$

so that

$$[H, E] = 2E, \qquad [H, F] = -2F, \qquad [E, F] = H,$$

$$[{}^c H, {}^c E] = 2{}^c E, \qquad [{}^c H, {}^c F] = -2{}^c F, \qquad [{}^c E, {}^c F] = {}^c H.$$

A representation (τ, V) of $\mathfrak{sl}(2, \mathbb{C})$ is a pair of a $\mathbb{C}$-vector space V and a $\mathbb{C}$-linear map $\tau \colon \mathfrak{sl}(2, \mathbb{C}) \to \mathrm{End}_{\mathbb{C}}(V) := \{\mathbb{C}\text{-linear maps } V \to V\}$ satisfying $\tau([X, Y]) = [\tau(X), \tau(Y)] := \tau(X)\tau(Y) - \tau(Y)\tau(X)$ for all $X, Y \in \mathfrak{sl}(2, \mathbb{C})$. We say (τ, V) is $\mathfrak{sl}(2, \mathbb{R})$-(infinitesimally) unitary (resp. $\mathfrak{su}(1, 1)$-(infinitesimally) unitary) if there exists an inner product $\langle \cdot, \cdot \rangle$ on V satisfying $\langle \tau(X)v, w \rangle = -\langle v, \tau(X)w \rangle$ for all $v, w \in V$ and $X \in \mathfrak{sl}(2, \mathbb{R})$ (resp. $\mathfrak{su}(1, 1)$), or equivalently,

$$\langle \tau(H)v, w \rangle = -\langle v, \tau(H)w \rangle, \quad \langle \tau(E)v, w \rangle = -\langle v, \tau(E)w \rangle, \quad \langle \tau(F)v, w \rangle = -\langle v, \tau(F)w \rangle,$$

resp. $\quad \langle \tau(H)v, w \rangle = \langle v, \tau(H)w \rangle, \quad \langle \tau(E)v, w \rangle = -\langle v, \tau(F)w \rangle.$

In the following we give some examples of representations of $\mathfrak{sl}(2, \mathbb{C})$.

First we consider the spaces of holomorphic functions on the unit disk and the upper half plane. Let

$$\mathbf{D} := \{z \in \mathbb{C} \mid |z| < 1\}, \qquad \mathbf{H} := \{y \in \mathbb{C} \mid \mathrm{Im}\, y > 0\},$$

and for $\mu > 1$, let $\mathcal{H}_\mu(\mathbf{D})$, $\mathcal{H}_\mu(\mathbf{H})$ be the Hilbert spaces given by

$$\mathcal{H}_\mu(\mathbf{D}) := \left\{ f(z) \in \mathcal{O}(\mathbf{D}) \;\middle|\; \|f\|_{\mu,\mathbf{D}}^2 := \frac{\mu-1}{\pi} \int_{\mathbf{D}} |f(z)|^2 (1-|z|^2)^{\mu-2}\, dz < \infty \right\},$$

$$\mathcal{H}_\mu(\mathbf{H}) := \left\{ f(y) \in \mathcal{O}(\mathbf{H}) \;\middle|\; \|f\|_{\mu,\mathbf{H}}^2 := \frac{\mu-1}{4\pi} \int_{\mathbf{H}} |f(y)|^2 (\mathrm{Im}\, y)^{\mu-2}\, dy < \infty \right\}.$$

These spaces are called *weighted Bergman spaces*. These are unitarily isomorphic by

$$C_\mu \colon \mathcal{H}_\mu(\mathbf{D}) \longrightarrow \mathcal{H}_\mu(\mathbf{H}), \qquad (C_\mu f)(y) := \left(\frac{-iy+1}{2} \right)^{-\mu} f\left(\frac{y-i}{-iy+1} \right),$$

$$C_\mu^{-1} \colon \mathcal{H}_\mu(\mathbf{H}) \longrightarrow \mathcal{H}_\mu(\mathbf{D}), \qquad (C_\mu^{-1} f)(z) := (iz+1)^{-\mu} f\left(\frac{z+i}{iz+1} \right),$$

where we choose the branches as $\left(\frac{-iy+1}{2} \right)^{-\mu}\big|_{y=i} = (iz+1)^{-\mu}\big|_{z=0} = 1$. The norm $\|f\|_{\mu,\mathbf{D}}$ is explicitly computed by using the Taylor expansion of f as

$$\left\| \sum_{m=0}^{\infty} a_m z^m \right\|_{\mu,\mathbf{D}}^2 = \sum_{m=0}^{\infty} \frac{m!}{(\mu)_m} |a_m|^2,$$

where $(\mu)_m := \mu(\mu+1)(\mu+2) \cdots (\mu+m-1)$ for $m \in \mathbb{Z}_{>0}$ and $(\mu)_0 := 1$. Then this is positive definite for $\mu > 0$, and we can redefine $\mathcal{H}_\mu(\mathbf{D})$, $\mathcal{H}_\mu(\mathbf{H})$ for $\mu > 0$ by

$$\mathcal{H}_\mu(\mathbf{D}) := \left\{ f(z) = \sum_{m=0}^{\infty} a_m z^m \in \mathcal{O}(\mathbf{D}) \,\middle|\, \|f\|_{\mu,\mathbf{D}}^2 := \sum_{m=0}^{\infty} \frac{m!}{(\mu)_m} |a_m|^2 < \infty \right\},$$

$$\mathcal{H}_\mu(\mathbf{H}) := C_\mu(\mathcal{H}_\mu(\mathbf{D})).$$

Also, for $\mu > 0$ let

$$\mathcal{H}_\mu(\mathbf{D})_{\text{fin}} := \mathbb{C}[z] \subset \mathcal{H}_\mu(\mathbf{D}), \qquad \mathcal{H}_\mu(\mathbf{H})_{\text{fin}} := C_\mu(\mathcal{H}_\mu(\mathbf{D})_{\text{fin}}),$$

where $\mathbb{C}[z]$ denotes the space of holomorphic polynomials on $\mathbb{C}$.

Next we consider the representations on $\mathcal{H}_\mu(\mathbf{D}), \mathcal{H}_\mu(\mathbf{H})$. When $\mu \in \mathbb{Z}_{>0}$, the Lie group $SU(1,1)$ acts on $\mathcal{H}_\mu(\mathbf{D})$ and $SL(2,\mathbb{R})$ acts on $\mathcal{H}_\mu(\mathbf{H})$ unitarily by

$$\tau_\mu\left(\begin{bmatrix} a & b \\ c & d \end{bmatrix}^{-1} \right) f(z) := (cz+d)^{-\mu} f\left(\frac{az+b}{cz+d} \right).$$

We note that this is not well-defined if $\mu \notin \mathbb{Z}_{>0}$, since $(cz+d)^{-\mu}$ is not single-valued. However, we can make this well-defined by considering the universal covering groups of $SU(1,1)$ and $SL(2,\mathbb{R})$ for general $\mu > 0$. The representations $(\tau_\mu, \mathcal{H}_\mu(\mathbf{D}))$, $(\tau_\mu, \mathcal{H}_\mu(\mathbf{H}))$ for $\mu > 1$ are called *holomorphic discrete series representations*, since they correspond to discrete spectra in $L^2(SU(1,1))$ and $L^2(SL(2,\mathbb{R}))$. By differentiating τ_μ, we get the Lie algebra representation of $\mathfrak{su}(1,1)$ on $\mathcal{H}_\mu(\mathbf{D})_{\text{fin}}$ and that of $\mathfrak{sl}(2,\mathbb{R})$ on $\mathcal{H}_\mu(\mathbf{H})_{\text{fin}}$ as

$$\tau_\mu(X) f(z) := \frac{d}{dt} \tau_\mu(e^{tX}) f(z) \bigg|_{t=0}$$

(we use the same symbol τ_μ), and by extending this complex-linearly, we get the representations of $\mathfrak{sl}(2,\mathbb{C})$ on $\mathcal{H}_\mu(\mathbf{D})_{\text{fin}}, \mathcal{H}_\mu(\mathbf{H})_{\text{fin}}$. The resulting representation is given by

$$\tau_\mu(H) f(z) = -2z \frac{df}{dz}(z) - \mu f(z), \qquad \tau_\mu(E) f(z) = -\frac{df}{dz}(z),$$

$$\tau_\mu(F) f(z) = z^2 \frac{df}{dz}(z) + \mu z f(z).$$

This is well-defined for all $\mu > 0$. $\mathcal{H}_\mu(\mathbf{D})_{\text{fin}}$ is $\mathfrak{su}(1,1)$-infinitesimally unitary and $\mathcal{H}_\mu(\mathbf{H})_{\text{fin}}$ is $\mathfrak{sl}(2,\mathbb{R})$-infinitesimally unitary. When we distinguish $\mathbf{D}$ and $\mathbf{H}$, we write $\tau_\mu^{\mathbf{D}}, \tau_\mu^{\mathbf{H}}$ instead of τ_μ. These are related as

$$C_\mu \circ \tau_\mu^{\mathbf{D}}(g) = \tau_\mu^{\mathbf{H}}(\text{Int}(C)g) \circ C_\mu \qquad (g \in SU(1,1)),$$

$$C_\mu \circ \tau_\mu^{\mathbf{D}}(X) = \tau_\mu^{\mathbf{H}}(\text{Ad}(C)X) \circ C_\mu \qquad (X \in \mathfrak{sl}(2,\mathbb{C})).$$

$\mathcal{H}_\mu(\mathbf{D})$ has an orthogonal basis $\{z^m\}_{m=0}^\infty$, and $\mathcal{H}_\mu(\mathbf{H})$ has an orthogonal basis $\{C_\mu(z^m) = \left(\frac{-iy+1}{2}\right)^{-\mu}\left(\frac{y-i}{-iy+1}\right)^m\}_{m=0}^\infty$. These satisfy

$$\langle z^m, z^n \rangle_{\mu,\mathbf{D}} = \langle C_\mu(z^m), C_\mu(z^n) \rangle_{\mu,\mathbf{H}} = \frac{m!}{(\mu)_m}\delta_{mn}, \tag{4a}$$

$$\tau_\mu^{\mathbf{D}}(H)z^m = -(2m+\mu)z^m, \quad \tau_\mu^{\mathbf{H}}({}^cH)C_\mu(z^m) = -(2m+\mu)C_\mu(z^m), \tag{4b}$$

$$\tau_\mu^{\mathbf{D}}(E)z^m = -mz^{m-1}, \quad \tau_\mu^{\mathbf{H}}({}^cE)C_\mu(z^m) = -mC_\mu(z^{m-1}), \tag{4c}$$

$$\tau_\mu^{\mathbf{D}}(F)z^m = (m+\mu)z^{m+1}, \quad \tau_\mu^{\mathbf{H}}({}^cF)C_\mu(z^m) = (m+\mu)C_\mu(z^{m+1}). \tag{4d}$$

The representations $(\tau_\mu^{\mathbf{D}}, \mathcal{H}_\mu(\mathbf{D})_{\text{fin}})$, $(\tau_\mu^{\mathbf{H}}, \mathcal{H}_\mu(\mathbf{H})_{\text{fin}})$ for $\mu > 0$ are called *unitary highest weight representations*.

Second, for $\mu > 0$ we consider the Hilbert space and its dense subspace

$$L_\mu^2(\mathbb{R}^+) := \left\{\phi: \mathbb{R}^+ \longrightarrow \mathbb{C} \;\middle|\; \|\phi\|_{\mu,\mathbb{R}^+}^2 := \frac{2^\mu}{\Gamma(\mu)}\int_0^\infty |\phi(t)|^2 t^{\mu-1}\,dt < \infty\right\},$$

$$L_\mu^2(\mathbb{R}^+)_{\text{fin}} := \mathbb{C}[t]e^{-t} \subset L_\mu^2(\mathbb{R}^+).$$

Let $\mathcal{L}_\mu$ be the *weighted Laplace transform* given by

$$\mathcal{L}_\mu: L_\mu^2(\mathbb{R}^+) \longrightarrow \mathcal{H}_\mu(\mathbf{H}), \quad (\mathcal{L}_\mu \phi)(y) := \frac{2^\mu}{\Gamma(\mu)}\int_0^\infty \phi(t)e^{iyt}t^{\mu-1}\,dt.$$

For $m \in \mathbb{Z}_{\geq 0}$ let

$$l_m^{(\mu)}(t) := i^m \sum_{j=0}^m \frac{(-m)_j}{(\mu)_j j!}(2t)^j e^{-t} = \frac{i^m}{(\mu)_m}t^{-\mu+1}e^t \frac{d^m}{dt^m}t^{\mu+m-1}e^{-2t} \in L_\mu^2(\mathbb{R}^+)_{\text{fin}}.$$

This is also written as $l_m^{(\mu)}(t) = i^m \frac{m!}{(\mu)_m} L_m^{(\mu-1)}(2t)e^{-t}$ by using the generalized Laguerre polynomial $L_m^{(\alpha)}(t)$. Then we can directly show that $\{l_m^{(\mu)}(t)\}_{m=0}^\infty$ forms an orthogonal basis of $L_\mu^2(\mathbb{R}^+)$ satisfying

$$\langle l_m^{(\mu)}(t), l_n^{(\mu)}(t) \rangle_{\mu,\mathbb{R}^+} = \frac{m!}{(\mu)_m}\delta_{mn}, \quad (\mathcal{L}_\mu l_m^{(\mu)})(y) = i^m \sum_{j=0}^m \frac{(-m)_j}{j!}\left(\frac{1-iy}{2}\right)^{-\mu-j}$$

$$= i^m \left(\frac{1-iy}{2}\right)^{-\mu}\left(1 - \frac{2}{1-iy}\right)^m = C_\mu(z^m)(y).$$

Especially $\mathcal{L}_\mu$ is a unitary isomorphism. Next we consider the following representation of $\mathfrak{sl}(2,\mathbb{C})$ on $L_\mu^2(\mathbb{R}^+)_{\text{fin}}$.

$$\tau_\mu^{\mathbb{R}^+}(H)\phi(t) := 2t\frac{d\phi}{dt}(t) + \mu\phi(t), \qquad \tau_\mu^{\mathbb{R}^+}(E)\phi(t) := -it\phi(t),$$

$$\tau_\mu^{\mathbb{R}^+}(F) := -i\left(t\frac{d^2\phi}{dt^2}(t) + \mu\frac{d\phi}{dt}(t)\right).$$

Then this representation is $\mathfrak{sl}(2,\mathbb{R})$-infinitesimally unitary, and $\mathcal{L}_\mu$ intertwines the $\mathfrak{sl}(2,\mathbb{C})$-action, that is, we have

$$\mathcal{L}_\mu \circ \tau_\mu^{\mathbb{R}^+}(X) = \tau_\mu^{\mathbf{H}}(X) \circ \mathcal{L}_\mu \qquad (X \in \mathfrak{sl}(2,\mathbb{C})).$$

Especially we have

$$\tau_\mu^{\mathbb{R}^+}({}^cH)l_m^{(\mu)} = -(2m+\mu)l_m^{(\mu)},$$
$$\tau_\mu^{\mathbb{R}^+}({}^cE)l_m^{(\mu)} = -m l_{m-1}^{(\mu)}, \quad \tau_\mu^{\mathbb{R}^+}({}^cF)l_m^{(\mu)} = (m+\mu)l_{m+1}^{(\mu)}.$$

Since $\tau_\mu^{\mathbb{R}^+}({}^cH) = -i\tau_\mu^{\mathbb{R}^+}(E-F) = -H_{\mathrm{HO}}^{(\mu)}$ in (3), this solves the eigenvalue problem of $H_{\mathrm{HO}}^{(\mu)}$ on $L_\mu^2(\mathbb{R}^+)$. Now we have gotten three different pictures of the unitary highest weight representation of $\mathfrak{sl}(2,\mathbb{R}) \simeq \mathfrak{su}(1,1)$,

$$L_\mu^2(\mathbb{R}^+) \xrightarrow[\sim]{\mathcal{L}_\mu} \mathcal{H}_\mu(\mathbf{H}) \xleftarrow[\sim]{C_\mu} \mathcal{H}_\mu(\mathbf{D})$$
$$l_m^{(\mu)}(t) \longmapsto (\mathcal{L}_\mu l_m^{(\mu)})(y) = (C_\mu z^m)(y) \longleftarrow z^m.$$

Remark 3.1 By replacing $\mathbb{R}^+$, $\mathbf{H}$ and $\mathbf{D}$ with a symmetric cone $\Omega \subset V$, a tube domain $\Omega + iV \subset V \otimes_{\mathbb{R}} \mathbb{C}$ and a bounded symmetric domain $D \subset V \otimes_{\mathbb{R}} \mathbb{C}$ associated with a Euclidean Jordan algebra V, these equivalences for $SL(2,\mathbb{R}) \simeq SU(1,1)$ are generalized to those for Hermitian Lie groups of tube type, $Sp(n,\mathbb{R})$, $SU(n,n)$, $SO^*(4n)$, $SO_0(2,n)$ and $E_{7(-25)}$. See [3].

Third, we consider a representation of $\mathfrak{sl}(2,\mathbb{C})$ on

$$L^2(\mathbb{R}^n) \supset L^2(\mathbb{R}^n)_{\mathrm{fin}} := \mathcal{P}(\mathbb{R}^n)e^{-r^2/2},$$

given by

$$\tau^{\mathbb{R}^n}(H)\psi(x) = \sum_{j=1}^n x_j \frac{\partial \psi}{\partial x_j}(x) + \frac{n}{2}\psi(x), \qquad \tau^{\mathbb{R}^n}(E)\psi(x) = -\frac{i}{2}\sum_{j=1}^n x_j^2 \psi(x),$$

$$\tau^{\mathbb{R}^n}(F)\psi(x) = -\frac{i}{2}\sum_{j=1}^n \frac{\partial^2 \psi}{\partial x_j^2}(x).$$

This representation is $\mathfrak{sl}(2,\mathbb{R})$-infinitesimally unitary. In terms of ${}^cH, {}^cE, {}^cF$, this representation is rewritten as

$$\tau^{\mathbb{R}^n}(^cH) = -i\tau^{\mathbb{R}^n}(E-F) = -\frac{1}{2}\sum_{j=1}^{n}\left(x_j^2 - \frac{\partial^2}{\partial x_j^2}\right) = -\sum_{j=1}^{n}a_j^\dagger a_j - \frac{n}{2},$$

$$\tau^{\mathbb{R}^n}(^cE) = \frac{1}{2}\tau^{\mathbb{R}^n}(-iH+E+F) = -\frac{i}{4}\sum_{j=1}^{n}\left(2x_j\frac{\partial}{\partial x_j} + 1 + x_j^2 + \frac{\partial^2}{\partial x_j^2}\right)$$

$$= -\frac{i}{2}\sum_{j=1}^{n}(a_j)^2,$$

$$\tau^{\mathbb{R}^n}(^cF) = \frac{1}{2}\tau^{\mathbb{R}^n}(iH+E+F) = -\frac{i}{4}\sum_{j=1}^{n}\left(-2x_j\frac{\partial}{\partial x_j} - 1 + x_j^2 + \frac{\partial^2}{\partial x_j^2}\right)$$

$$= -\frac{i}{2}\sum_{j=1}^{n}(a_j^\dagger)^2.$$

According to the decomposition (2), for $\phi(t) \in L^2_{k+\frac{n}{2}}(\mathbb{R}^+)_{\text{fin}}$, $h(x) \in \mathcal{HP}_k(\mathbb{R}^n)$, we have

$$\tau^{\mathbb{R}^n}(X)\left(\phi\left(\frac{r^2(x)}{2}\right)h(x)\right) = (\tau^{\mathbb{R}^+}_{k+\frac{n}{2}}(X)\phi)\left(\frac{r^2(x)}{2}\right)h(x) \qquad (X \in \mathfrak{sl}(2,\mathbb{C})).$$

4 A Generalization of NCHO

In this section we consider a generalization of the non-commutative harmonic oscillator, with the Hamiltonian given by the self-adjoint operator on $L^2(\mathbb{R}) \otimes \mathbb{C}^2$,

$$H_{\text{NCHO}} := \mathbf{A}\left(-\frac{1}{2}\frac{d^2}{dx^2} + \frac{1}{2}x^2\right) + \mathbf{J}\left(x\frac{d}{dx} + \frac{1}{2}\right) + 2\eta i \det(\mathbf{A} \pm i\mathbf{J})^{1/2}\mathbf{J}$$

$$= \mathbf{A}\left(a^\dagger a + \frac{1}{2}\right) + \frac{1}{2}\mathbf{J}(a^2 - (a^\dagger)^2) + 2\eta i \det(\mathbf{A} \pm i\mathbf{J})^{1/2}\mathbf{J},$$

where $\mathbf{A}, \mathbf{J} \in M(2, \mathbb{R})$, $\mathbf{A} = {}^t\mathbf{A}$, $\mathbf{J} = -{}^t\mathbf{J}$, and $\eta \in \mathbb{R}$. That is, we want to find a function $\psi \in L^2(\mathbb{R}^n) \otimes \mathbb{C}^p$ satisfying

$$\left[-\mathbf{A}_1\sum_{j=1}^{n}\frac{\partial^2}{\partial x_j^2} + i\mathbf{A}_2\sum_{j=1}^{n}\left(x_j\frac{\partial}{\partial x_j} + \frac{1}{2}\right) + \mathbf{A}_3\sum_{j=1}^{n}x_j^2\right]\psi = 2\mathbf{C}\psi,$$

where $\mathbf{A}_k, \mathbf{C} \in M(p, \mathbb{C})$, $\mathbf{A}_k = \mathbf{A}_k^\dagger$, $\mathbf{C} = \mathbf{C}^\dagger$. Here $\mathbf{C}^\dagger = {}^t\overline{\mathbf{C}}$. Then by putting

$$\mathbf{A}_1 + \mathbf{A}_3 =: \mathbf{A}, \qquad \frac{1}{2}(-i\mathbf{A}_1 + \mathbf{A}_2 + i\mathbf{A}_3) =: \mathbf{B} \qquad (5)$$

so that $\mathbf{A}, \mathbf{B} \in M(p, \mathbb{C})$, $\mathbf{A} = \mathbf{A}^\dagger$, this is rewritten as

$$\left[\mathbf{A}\sum_{j=1}^{n}\left(a_j^\dagger a_j + \frac{1}{2}\right) - i\mathbf{B}\sum_{j=1}^{n}(a_j^\dagger)^2 + i\mathbf{B}^\dagger\sum_{j=1}^{n}(a_j)^2\right]\psi = 2\mathbf{C}\psi. \qquad (6)$$

In terms of $\tau^{\mathbb{R}^n}$, this becomes

$$\left[-\mathbf{A}\tau^{\mathbb{R}^n}({}^cH) + 2\mathbf{B}\tau^{\mathbb{R}^n}({}^cF) - 2\mathbf{B}^\dagger\tau^{\mathbb{R}^n}({}^cE)\right]\psi = 2\mathbf{C}\psi.$$

Then according to the decomposition (2), on $L^2_{k+\frac{n}{2}}(\mathbb{R}^+) \otimes \mathcal{HP}_k(\mathbb{R}^n) \otimes \mathbb{C}^p$, the above equation is reduced to an ordinary differential equation on $\mathcal{H}_{k+\frac{n}{2}}(\mathbf{D}) \otimes \mathbb{C}^p$.

Theorem 4.1 *Let* $h(x) \in \mathcal{HP}_k(\mathbb{R}^n)$, $\{u_m\}_{m=0}^\infty \subset \mathbb{C}^p$, *and let*

$$\psi(x) := \sum_{m=0}^{\infty} u_m l_m^{(k+\frac{n}{2})}\left(\frac{r^2(x)}{2}\right) h(x).$$

Then $\psi(x) \in L^2(\mathbb{R}^n) \otimes \mathbb{C}^p$ *and* (6) *hold if and only if*

$$f(z) := \sum_{m=0}^{\infty} u_m z^m \in \mathcal{H}_{k+\frac{n}{2}}(\mathbf{D}) \otimes \mathbb{C}^p,$$

$$\left[-\mathbf{A}\tau^{\mathbf{D}}_{k+\frac{n}{2}}(H) + 2\mathbf{B}\tau^{\mathbf{D}}_{k+\frac{n}{2}}(F) - 2\mathbf{B}^\dagger\tau^{\mathbf{D}}_{k+\frac{n}{2}}(E)\right]f = 2\mathbf{C}f$$

hold.

In the following, for general $\mu > 0$ we consider

$$\left[-\mathbf{A}\tau^{\mathbf{D}}_\mu(H) + 2\mathbf{B}\tau^{\mathbf{D}}_\mu(F) - 2\mathbf{B}^\dagger\tau^{\mathbf{D}}_\mu(E)\right]f = 2\mathbf{C}f \qquad (7)$$

on $\mathcal{H}_\mu(\mathbf{D}) \otimes \mathbb{C}^p$, that is,

$$\left[\mathbf{A}\left(2z\frac{d}{dz} + \mu\right) + 2\mathbf{B}\left(z^2\frac{d}{dz} + \mu z\right) + 2\mathbf{B}^\dagger\frac{d}{dz}\right]f = 2\mathbf{C}f,$$

or equivalently,

$$(\mathbf{B}z^2 + \mathbf{A}z + \mathbf{B}^\dagger)\frac{df}{dz} = \left(-\mu\mathbf{B}z - \frac{\mu}{2}\mathbf{A} + \mathbf{C}\right)f. \qquad (8)$$

Remark 4.2 We consider the space of functions on $\sqrt{\mu}\mathbf{D} := \{z \in \mathbb{C} \mid |z| < \sqrt{\mu}\}$,

$$\mathcal{H}_\mu(\sqrt{\mu}\mathbf{D}) := \left\{f(w) \in \mathcal{O}(\sqrt{\mu}\mathbf{D}) \ \bigg|\ \frac{\mu-1}{\mu\pi}\int_{\sqrt{\mu}\mathbf{D}} |f(w)|^2\left(1 - \frac{|w|^2}{\mu}\right)^{\mu-2} dw < \infty\right\},$$

and the map $\mathcal{H}_\mu(\mathbf{D}) \to \mathcal{H}_\mu(\sqrt{\mu}\mathbf{D})$, $f(z) \mapsto f(w/\sqrt{\mu})$. Then the Eq. (8) on $\mathcal{H}_\mu(\mathbf{D}) \otimes \mathbb{C}^p$ is equivalent to the equation

$$\left(\frac{1}{\sqrt{\mu}}\mathbf{B}w^2 + \mathbf{A}w + \sqrt{\mu}\mathbf{B}^\dagger\right)\frac{df}{dw} = \left(-\sqrt{\mu}\mathbf{B}w - \frac{\mu}{2}\mathbf{A} + \mathbf{C}\right)f$$

on $\mathcal{H}_\mu(\sqrt{\mu}\mathbf{D}) \otimes \mathbb{C}^p$. Then by putting $\sqrt{\mu}\mathbf{B} =: \widetilde{\mathbf{B}}$, $\mathbf{C} - \frac{\mu}{2}\mathbf{A} =: \widetilde{\mathbf{C}}$ and taking the limit $\mu \to \infty$, this "converges" to the equation

$$(\mathbf{A}w + \widetilde{\mathbf{B}}^\dagger)\frac{df}{dw} = (-\widetilde{\mathbf{B}}w + \widetilde{\mathbf{C}})f$$

on

$$\mathcal{F}(\mathbb{C}) \otimes \mathbb{C}^p := \left\{f(w) \in O(\mathbb{C}) \;\Big|\; \frac{1}{\pi}\int_{\mathbb{C}} |f(w)|^2 e^{-|w|^2}\, dw < \infty\right\} \otimes \mathbb{C}^p,$$

and via the Bargmann transform, this is equivalent to the equation

$$(\mathbf{A}a^\dagger a + \widetilde{\mathbf{B}}^\dagger a + \widetilde{\mathbf{B}}a^\dagger)\psi = \widetilde{\mathbf{C}}\psi$$

on $L^2(\mathbb{R}) \otimes \mathbb{C}^p$. Especially, let $p = 2$, let $\sigma_1, \sigma_2, \sigma_3 \in M(2, \mathbb{C})$ be the Pauli matrices, let $\mathbf{I} \in M(2, \mathbb{C})$ be the identity, and let $\omega, \Delta, g, \varepsilon \in \mathbb{R}$. If $\mathbf{A} = \omega\mathbf{I}$, $\widetilde{\mathbf{B}} = g\sigma_1$ and $\widetilde{\mathbf{C}} = -\Delta\sigma_3 - \varepsilon\sigma_1 + \lambda\mathbf{I}$, then this is the asymmetric quantum Rabi model [1, 2, 9, 19],

$$(\omega\mathbf{I}a^\dagger a + g\sigma_1(a + a^\dagger) + \Delta\sigma_3 + \varepsilon\sigma_1)\psi = \lambda\psi. \tag{9}$$

That is, the non-commutative harmonic oscillator is regarded as a "covering model" of the quantum Rabi model, as is pointed out in [16, 18]. Similarly, if $\mathbf{A} = \omega\mathbf{I}$, $\widetilde{\mathbf{B}} = g(\sigma_1 - i\sigma_2)/2 =: g\sigma^-$, $\widetilde{\mathbf{B}}^\dagger = g(\sigma_1 + i\sigma_2)/2 =: g\sigma^+$ and $\widetilde{\mathbf{C}} = -\Delta\sigma_3 + \lambda\mathbf{I}$, then this is the Jaynes–Cummings model [7],

$$(\omega\mathbf{I}a^\dagger a + g(\sigma^+ a + \sigma^- a^\dagger) + \Delta\sigma_3)\psi = \lambda\psi.$$

We return to the analysis of (8). The inverse of $\mathbf{B}z^2 + \mathbf{A}z + \mathbf{B}^\dagger$ is written as

$$(\mathbf{B}z^2 + \mathbf{A}z + \mathbf{B}^\dagger)^{-1} = \frac{\mathrm{adj}(\mathbf{B}z^2 + \mathbf{A}z + \mathbf{B}^\dagger)}{\det(\mathbf{B}z^2 + \mathbf{A}z + \mathbf{B}^\dagger)},$$

where adj is the adjugate matrix. Its denominator and numerator have degrees at most $2p$ and $2p - 2$ respectively. Now we assume that all poles of $(\mathbf{B}z^2 + \mathbf{A}z + \mathbf{B}^\dagger)^{-1}$ are of order 1, and let

$$(\mathbf{B}z^2 + \mathbf{A}z + \mathbf{B}^\dagger)^{-1} = \sum_{j=1}^{N} \frac{1}{z - \alpha_j} \mathbf{P}_{\alpha_j} + \sum_{k=0}^{M} \mathbf{Q}_k z^k \tag{10}$$

be its partial fraction decomposition, where $\alpha_j \in \mathbb{C}$ are distinct, and $\mathbf{P}_{\alpha_j}, \mathbf{Q}_k \in M(p, \mathbb{C})$, $\mathbf{P}_{\alpha_j} \neq 0$. Then the following holds.

Lemma 4.3 1. *If $\alpha \neq 0$ is a pole of $(\mathbf{B}z^2 + \mathbf{A}z + \mathbf{B}^\dagger)^{-1}$, then $\overline{\alpha}^{-1}$ is also a pole of it.*
2. *For a pole $\alpha \neq 0$, $\mathbf{P}_{\overline{\alpha}^{-1}} = -\mathbf{P}_\alpha^\dagger$, and $\mathbf{Q}_k = 0$.*
3. *If $\det(\mathbf{B}) \neq 0$, then $\sum_{j=1}^{N} \mathbf{P}_{\alpha_j} = 0$, $\sum_{j=1}^{N} \alpha_j \mathbf{P}_{\alpha_j} \mathbf{B} = \mathbf{I}$.*
4. *If $\det(\mathbf{B}) = 0$, then $\sum_{j=1}^{N} \mathbf{P}_{\alpha_j} = \mathbf{P}_0^\dagger$, $\sum_{j=1}^{N} \mathbf{P}_{\alpha_j} \mathbf{B} = 0$, $\sum_{j=1}^{N} \alpha_j \mathbf{P}_{\alpha_j} \mathbf{B} = \mathbf{I} - \mathbf{P}_0^\dagger \mathbf{A}$.*
5. $\operatorname{rank} \mathbf{P}_{\alpha_j} \leq \dim \operatorname{Ker}(\alpha_j^2 \mathbf{B} + \alpha_j \mathbf{A} + \mathbf{B}^\dagger)$.
6. $\mathbf{P}_{\alpha_j}(2\alpha_j \mathbf{B} + \mathbf{A}) \mathbf{P}_{\alpha_j} = \mathbf{P}_{\alpha_j}$.

Proof 1. If $\alpha \neq 0$ is a pole of $(\mathbf{B}z^2 + \mathbf{A}z + \mathbf{B}^\dagger)^{-1}$, then $\alpha^2 \mathbf{B} + \alpha \mathbf{A} + \mathbf{B}^\dagger$ is non-invertible, and hence

$$\overline{\alpha}^{-2} \mathbf{B} + \overline{\alpha}^{-1} \mathbf{A} + \mathbf{B}^\dagger = \overline{\alpha}^{-2} (\alpha^2 \mathbf{B} + \alpha \mathbf{A} + \mathbf{B}^\dagger)^\dagger$$

is also non-invertible. Therefore $\overline{\alpha}^{-1}$ is also a pole of $(\mathbf{B}z^2 + \mathbf{A}z + \mathbf{B}^\dagger)^{-1}$.

2. By (10) we have

$$(\mathbf{B}z^2 + \mathbf{A}z + \mathbf{B}^\dagger)^{-1} = z^{-2}((\mathbf{B}\overline{z}^{-2} + \mathbf{A}\overline{z}^{-1} + \mathbf{B}^\dagger)^{-1})^\dagger$$
$$= z^{-2}\left(\sum_{j=1}^{N} \frac{1}{z^{-1} - \overline{\alpha_j}} \mathbf{P}_{\alpha_j}^\dagger + \sum_{k=0}^{M} \mathbf{Q}_k^\dagger z^{-k}\right) = \sum_{j=1}^{N} \left(\frac{1}{z} + \frac{-\overline{\alpha_j}}{\overline{\alpha_j} z - 1}\right) \mathbf{P}_{\alpha_j}^\dagger + \sum_{k=0}^{M} \mathbf{Q}_k^\dagger z^{-k-2}.$$

Then by the simple-pole assumption, we have $\mathbf{Q}_k = 0$, and comparing the residues at $\alpha_j = \overline{\alpha_{j'}}^{-1}$ with (10), we get

$$\mathbf{P}_\alpha = -\mathbf{P}_{\overline{\alpha}^{-1}}^\dagger, \qquad \sum_{j=1}^{N} \mathbf{P}_{\alpha_j} = \begin{cases} 0 & (0 \text{ is not a pole}), \\ \mathbf{P}_0^\dagger & (0 \text{ is a pole}). \end{cases}$$

3, 4. By (10) we have

$$\sum_{j=1}^{N} \frac{1}{z(z - \alpha_j)} \mathbf{P}_{\alpha_j} (\mathbf{B}z^2 + \mathbf{A}z + \mathbf{B}^\dagger) = \frac{1}{z} \mathbf{I},$$

and taking the limit $z \to \infty$, we get $\sum_{j=1}^{N} \mathbf{P}_{\alpha_j} \mathbf{B} = 0$. Similarly, we have

$$\mathbf{I} = \sum_{j=1}^{N} \frac{1}{z-\alpha_j} \mathbf{P}_{\alpha_j}(\mathbf{B}z^2 + \mathbf{A}z + \mathbf{B}^\dagger)$$

$$= \sum_{j=1}^{N} \left(\left(1 + \frac{\alpha_j}{z-\alpha_j}\right) \mathbf{P}_{\alpha_j}(\mathbf{B}z + \mathbf{A}) + \frac{1}{z-\alpha_j} \mathbf{P}_{\alpha_j} \mathbf{B}^\dagger \right)$$

$$= \sum_{j=1}^{N} \left(\mathbf{P}_{\alpha_j} \mathbf{B}z + \frac{\alpha_j z}{z-\alpha_j} \mathbf{P}_{\alpha_j} \mathbf{B} + \mathbf{P}_{\alpha_j} \mathbf{A} + \frac{1}{z-\alpha_j} \mathbf{P}_{\alpha_j}(\alpha_j \mathbf{A} + \mathbf{B}^\dagger) \right).$$

Then since $\sum_{j=1}^{N} \mathbf{P}_{\alpha_j} \mathbf{B} = 0$, by taking the limit $z \to \infty$, we have

$$\sum_{j=1}^{N} \alpha_j \mathbf{P}_{\alpha_j} \mathbf{B} = \mathbf{I} - \sum_{j=1}^{N} \mathbf{P}_{\alpha_j} \mathbf{A} = \begin{cases} \mathbf{I} & (0 \text{ is not a pole}), \\ \mathbf{I} - \mathbf{P}_0^\dagger \mathbf{A} & (0 \text{ is a pole}). \end{cases}$$

5. By comparing the residues at $z = \alpha_j$ of

$$\sum_{j=1}^{N} \frac{1}{z-\alpha_j} (\mathbf{B}z^2 + \mathbf{A}z + \mathbf{B}^\dagger) \mathbf{P}_{\alpha_j} = \mathbf{I},$$

we get $(\alpha_j^2 \mathbf{B} + \alpha_j \mathbf{A} + \mathbf{B}^\dagger) \mathbf{P}_{\alpha_j} = 0$, and the desired formula holds. $\square$

6. By differentiating (10) with respect to z, we get

$$-\sum_{j=1}^{N} \frac{1}{(z-\alpha_j)^2} \mathbf{P}_{\alpha_j} = -(\mathbf{B}z^2 + \mathbf{A}z + \mathbf{B}^\dagger)^{-1}(2\mathbf{B}z + \mathbf{A})(\mathbf{B}z^2 + \mathbf{A}z + \mathbf{B}^\dagger)^{-1}$$

$$= -\sum_{j=1}^{N}\sum_{k=1}^{N} \frac{1}{(z-\alpha_j)(z-\alpha_k)} \mathbf{P}_{\alpha_j}(2\mathbf{B}z + \mathbf{A}) \mathbf{P}_{\alpha_k}.$$

Then multiplying $(z-\alpha_j)^2$ and substituting $z = \alpha_j$, we get the desired formula.

From this lemma we easily get the following.

Theorem 4.4 *Suppose that all poles of $(\mathbf{B}z^2 + \mathbf{A}z + \mathbf{B}^\dagger)^{-1}$ are of order 1, and we consider the partial fraction decomposition (10).*

1. *The Eq. (8) is equal to the Fuchsian equation*

$$\frac{df}{dz} = \sum_{j=1}^{N} \frac{1}{z-\alpha_j} \mathbf{P}_{\alpha_j}\left(-\mu\left(\alpha_j \mathbf{B} + \frac{1}{2}\mathbf{A}\right) + \mathbf{C}\right) f. \quad (11)$$

2. *If (8) is singular at $z = \alpha$ ($\neq 0, \infty$), then it is also singular at $z = \overline{\alpha}^{-1}$.*

3. We have $-\sum_{j=1}^{N} \mathbf{P}_{\alpha_j}\left(-\mu\left(\alpha_j \mathbf{B} + \frac{1}{2}\mathbf{A}\right) + \mathbf{C}\right) = \begin{cases} \mu \mathbf{I} & (\det(\mathbf{B}) \neq 0), \\ \mu \mathbf{I} - \mathbf{P}_0^{\dagger}\left(\frac{\mu}{2}\mathbf{A} + \mathbf{C}\right) & (\det(\mathbf{B}) = 0). \end{cases}$

4. We have $\operatorname{rank}\left(\mathbf{P}_{\alpha_j}\left(-\mu\left(\alpha_j \mathbf{B} + \frac{1}{2}\mathbf{A}\right) + \mathbf{C}\right)\right) \le \dim \operatorname{Ker}(\alpha_j^2 \mathbf{B} + \alpha_j \mathbf{A} + \mathbf{B}^{\dagger})$. That is, $\mathbf{P}_{\alpha_j}\left(-\mu\left(\alpha_j \mathbf{B} + \frac{1}{2}\mathbf{A}\right) + \mathbf{C}\right)$ has an eigenvalue 0 with multiplicity at least $p - \dim \operatorname{Ker}(\alpha_j^2 \mathbf{B} + \alpha_j \mathbf{A} + \mathbf{B}^{\dagger})$.

5. For $u \in \operatorname{Im} \mathbf{P}_{\alpha_j} \subset \mathbb{C}^p$, we have $\mathbf{P}_{\alpha_j}\left(-\mu\left(\alpha_j \mathbf{B} + \frac{1}{2}\mathbf{A}\right) + \mathbf{C}\right)u = \mathbf{P}_{\alpha_j}\mathbf{C}u - \frac{\mu}{2}u$. That is, non-zero eigenvalues of $\mathbf{P}_{\alpha_j}\left(-\mu\left(\alpha_j \mathbf{B} + \frac{1}{2}\mathbf{A}\right) + \mathbf{C}\right)$ are equal to $-\frac{\mu}{2}$ plus those of $\mathbf{P}_{\alpha_j}\mathbf{C}$.

Suppose that all singularities α_j are not in $S^1 := \{z \in \mathbb{C} \mid |z| = 1\}$. Then if a solution f of (8) is holomorphic at all the singularities $\alpha_1, \ldots, \alpha_{N'}$ inside the unit disk $\mathbf{D}$, then f automatically has the radius of convergence strictly larger than 1 at the origin, and hence f belongs to $\mathcal{H}_{\mu}(\mathbf{D}) \otimes \mathbb{C}^p$. That is, we do not need to estimate the norm of f. Especially, if $\mathbf{A}, \mathbf{B}$ (or $\mathbf{A}_1, \mathbf{A}_2, \mathbf{A}_3$ in (5)) satisfy the condition

$$\mathbf{B}z + \mathbf{A} + \mathbf{B}^{\dagger}\overline{z} = 2(\mathbf{A}_1 \cos^2 \theta + \mathbf{A}_2 \cos \theta \sin \theta + \mathbf{A}_3 \sin^2 \theta)$$

is positive definite for all $z = ie^{-2i\theta} \in S^1$, (12)

then a solution f of (8) in $O(\mathbf{D}) \otimes \mathbb{C}^p$ automatically belongs $\mathcal{H}_{\mu}(\mathbf{D}) \otimes \mathbb{C}^p$.

Remark 4.5 If the left-hand side of (7) is bounded below on $\mathcal{H}_{\mu}(\mathbf{D}) \otimes \mathbb{C}^p$, then $\mathbf{B}z + \mathbf{A} + \mathbf{B}^{\dagger}\overline{z}$ is positive semidefinite for all $z \in S^1$. Indeed,

$$\left\langle \left[-\mathbf{A}\tau_{\mu}^{\mathbf{D}}(H) + 2\mathbf{B}\tau_{\mu}^{\mathbf{D}}(F) - 2\mathbf{B}^{\dagger}\tau_{\mu}^{\mathbf{D}}(E)\right] \sum_{m=0}^{\infty} u_m z^m, \sum_{m=0}^{\infty} v_m z^m \right\rangle_{\mu,\mathbf{D}}$$
$$= \sum_{m=0}^{\infty} \frac{(m+1)!}{(\mu)_m}\left(\frac{m+1}{\mu+m} v_{m+1}^{\dagger}\mathbf{A}u_{m+1} + 2v_{m+1}^{\dagger}\mathbf{B}u_m + 2v_m^{\dagger}\mathbf{B}^{\dagger}u_{m+1} + \frac{\mu+m}{m+1} v_m^{\dagger}\mathbf{A}u_m\right)$$

holds for all $\{u_m\}, \{v_m\} \subset \mathbb{C}^p$ by (4), and especially if the left-hand side of (7) has the lower bound $c \in \mathbb{R}$, then for $u \in \mathbb{C}^p, w \in \mathbf{D}$, we have

$$0 \le \left\langle \left[-\mathbf{A}\tau_{\mu}^{\mathbf{D}}(H) + 2\mathbf{B}\tau_{\mu}^{\mathbf{D}}(F) - 2\mathbf{B}^{\dagger}\tau_{\mu}^{\mathbf{D}}(E) - c\mathbf{I}\right](1 - z\overline{w})^{-\mu}u, (1 - z\overline{w})^{-\mu}u\right\rangle_{\mu,\mathbf{D}}$$
$$= \left\langle \left[-\mathbf{A}\tau_{\mu}^{\mathbf{D}}(H) + 2\mathbf{B}\tau_{\mu}^{\mathbf{D}}(F) - 2\mathbf{B}^{\dagger}\tau_{\mu}^{\mathbf{D}}(E) - c\mathbf{I}\right]\sum_{m=0}^{\infty}\frac{(\mu)_m}{m!}\overline{w}^m z^m u, \sum_{m=0}^{\infty}\frac{(\mu)_m}{m!}\overline{w}^m z^m u\right\rangle_{\mu,\mathbf{D}}$$
$$= \sum_{m=0}^{\infty}\frac{(\mu)_{m+1}}{m!}|w|^{2m} u^{\dagger}(\mathbf{A}|w|^2 + 2\mathbf{B}w + 2\mathbf{B}^{\dagger}\overline{w} + \mathbf{A})u - c\sum_{m=0}^{\infty}\frac{(\mu)_m}{m!}|w|^{2m} u^{\dagger}u$$
$$= \mu(1 - |w|^2)^{-\mu-1} u^{\dagger}(\mathbf{A}|w|^2 + 2\mathbf{B}w + 2\mathbf{B}^{\dagger}\overline{w} + \mathbf{A})u - c(1 - |w|^2)^{-\mu}u^{\dagger}u.$$

Then by taking the limit $|w| \nearrow 1$, we get $u^{\dagger}(\mathbf{B}w + \mathbf{A} + \mathbf{B}^{\dagger}\overline{w})u \ge 0$ for all $w \in S^1$. Also, by putting $w = 0$ we get $\mathbf{A} \ge \frac{c}{\mu}\mathbf{I}$.

Next we consider the action of the Lie group $SU(1, 1)$. In the following we assume $\mu \in \mathbb{Z}_{>0}$, but we may consider general $\mu > 0$ by taking the universal covering group of $SU(1, 1)$. We take an arbitrary $g \in SU(1, 1)$. Then the Eq. (7) is transformed as

$$\tau_\mu^D(g)[-\mathbf{A}\tau_\mu^D(H) + 2\mathbf{B}\tau_\mu^D(F) - 2\mathbf{B}^\dagger\tau_\mu^D(E)]\tau_\mu^D(g)^{-1}f$$
$$= [-\mathbf{A}\tau_\mu^D(\mathrm{Ad}(g)H) + 2\mathbf{B}\tau_\mu^D(\mathrm{Ad}(g)F) - 2\mathbf{B}^\dagger\tau_\mu^D(\mathrm{Ad}(g)E)]f = 2\mathbf{C}f$$

on $\mathcal{H}_\mu(\mathbf{D}) \otimes \mathbb{C}^p$, where $\mathrm{Ad}(g)X := gXg^{-1}$. Let ${}^g\mathbf{A}, {}^g\mathbf{B} \in M(p, \mathbb{C})$ be the matrices satisfying

$$-\mathbf{A} \otimes \mathrm{Ad}(g)H + 2\mathbf{B} \otimes \mathrm{Ad}(g)F - 2\mathbf{B}^\dagger \otimes \mathrm{Ad}(g)E$$
$$= -{}^g\mathbf{A} \otimes H + 2{}^g\mathbf{B} \otimes F - 2{}^g\mathbf{B}^\dagger \otimes E \in \mathfrak{sl}(2, \mathbb{C}) \otimes M(p, \mathbb{C}),$$

namely, for $g = \begin{bmatrix} a & b \\ \bar{b} & \bar{a} \end{bmatrix} \in SU(1, 1)$, let

$$({}^g\mathbf{A}, {}^g\mathbf{B}) := \left((|a|^2 + |b|^2)\mathbf{A} - 2(\bar{a}b\mathbf{B} + a\bar{b}\mathbf{B}^\dagger), -\bar{a}b\mathbf{A} + \bar{a}^2\mathbf{B} + \bar{b}^2\mathbf{B}^\dagger\right).$$

Then this equation is equivalent to

$$({}^g\mathbf{B}z^2 + {}^g\mathbf{A}z + {}^g\mathbf{B}^\dagger)\frac{df}{dz} = \left(-\mu{}^g\mathbf{B}z - \frac{\mu}{2}{}^g\mathbf{A} + \mathbf{C}\right)f.$$

On the other hand, the Eq. (11) is transformed as

$$\tau_\mu^D(g)\frac{d}{dz}\tau_\mu^D(g)^{-1}f = \tau_\mu^D(g)\sum_{j=1}^N \frac{1}{z - \alpha_j}\mathbf{P}_{\alpha_j}\left(-\mu\left(\alpha_j\mathbf{B} + \frac{1}{2}\mathbf{A}\right) + \mathbf{C}\right)\tau_\mu^D(g)^{-1}f,$$

and by direct computation this is equivalent to

$$\frac{df}{dz} = \bigg(\sum_{\substack{1 \le j \le N \\ g.\alpha_j \ne \infty}} \frac{1}{z - g.\alpha_j}\mathbf{P}_{\alpha_j}\left(-\mu\left(\alpha_j\mathbf{B} + \frac{1}{2}\mathbf{A}\right) + \mathbf{C}\right)$$
$$- \frac{1}{z - g.\infty}\bigg(\sum_{1 \le j \le N}\mathbf{P}_{\alpha_j}\left(-\mu\left(\alpha_j\mathbf{B} + \frac{1}{2}\mathbf{A}\right) + \mathbf{C}\right) + \mu\mathbf{I}\bigg)\bigg)f$$
$$= \bigg(\sum_{\substack{1 \le j \le N \\ g.\alpha_j \ne \infty}} \frac{1}{z - g.\alpha_j}\mathbf{P}_{\alpha_j}\left(-\mu\left(\alpha_j\mathbf{B} + \frac{1}{2}\mathbf{A}\right) + \mathbf{C}\right) - \frac{1}{z - g.\infty}\mathbf{P}_0^\dagger\left(\frac{\mu}{2}\mathbf{A} + \mathbf{C}\right)\bigg)f,$$

where we have used Theorem 4.4, and we set $\mathbf{P}_0 := 0$ if $\det(\mathbf{B}) \ne 0$. Here, for $g = \begin{bmatrix} a & b \\ \bar{b} & \bar{a} \end{bmatrix} \in SU(1, 1)$ we write

$$g.\alpha := \begin{cases} \dfrac{a\alpha + b}{\bar{b}\alpha + \bar{a}} & (\alpha \ne -\frac{\bar{a}}{\bar{b}}), \\ \infty & (\alpha = -\frac{\bar{a}}{\bar{b}}), \end{cases} \qquad g.\infty := \frac{a}{\bar{b}}.$$

Then these two equations must coincide. That is, we have

$$\frac{df}{dz} = ({}^g\mathbf{B}z^2 + {}^g\mathbf{A}z + {}^g\mathbf{B}^\dagger)^{-1}\left(-\mu {}^g\mathbf{B}z - \frac{\mu}{2}{}^g\mathbf{A} + \mathbf{C}\right)f$$

$$= \left(\sum_{\substack{1 \leq j \leq N \\ g.\alpha_j \neq \infty}} \frac{1}{z - g.\alpha_j} P_{\alpha_j}\left(-\mu\left(\alpha_j \mathbf{B} + \frac{1}{2}\mathbf{A}\right) + \mathbf{C}\right) - \frac{1}{z - g.\infty} P_0^\dagger\left(\frac{\mu}{2}\mathbf{A} + \mathbf{C}\right)\right) f.$$

Also, the positivity condition (12) is stable under $(\mathbf{A}, \mathbf{B}) \mapsto ({}^g\mathbf{A}, {}^g\mathbf{B})$, since for $z \in S^1$, $g = \begin{bmatrix} a & b \\ \bar{b} & \bar{a} \end{bmatrix} \in SU(1,1)$, we have

$${}^g\mathbf{B}z + {}^g\mathbf{A} + {}^g\mathbf{B}^\dagger \bar{z} = |-\bar{b}z + a|^2 \left(\mathbf{B}(g^{-1}.z) + \mathbf{A} + \mathbf{B}^\dagger\overline{(g^{-1}.z)}\right).$$

In the following we consider the case $p = 2$. We assume $(\mathbf{A}, \mathbf{B})$ satisfies the positivity condition (12), and that $\det(\mathbf{B}z^2 + \mathbf{A}z + \mathbf{B}^\dagger)$ has at least 3 distinct roots. If it has 4 distinct roots, then by (12) the roots are not on S^1, and by Lemma 4.3, they are of the form $\{\beta, \gamma, \overline{\beta}^{-1}, \overline{\gamma}^{-1}\}$ with $\beta, \gamma \in D$. Then by considering $g \in SU(1,1)$ satisfying $g.\beta = 0$, $g.\overline{\beta}^{-1} = \infty$, $g.\gamma =: \alpha \in D$ and replacing $(\mathbf{A}, \mathbf{B})$ with $({}^g\mathbf{A}, {}^g\mathbf{B})$, without loss of generality we may assume $\det(\mathbf{B}) = 0$ and $\det(\mathbf{B}z^2 + \mathbf{A}z + \mathbf{B}^\dagger)$ has the roots $\{0, \alpha, \overline{\alpha}^{-1}\}$ with $\alpha \in D$. Next, since $\mathbf{A}$ is positive definite by (12), by replacing $\mathbf{B}$ and $\mathbf{C}$ with $\mathbf{A}^{-\frac{1}{2}}\mathbf{B}\mathbf{A}^{-\frac{1}{2}}$ and $\mathbf{A}^{-\frac{1}{2}}\mathbf{C}\mathbf{A}^{-\frac{1}{2}}$, we may assume $\mathbf{A} = \mathbf{I}$. In addition, by taking a gauge transform by a unitary matrix, we may assume

$$\mathbf{A} = \begin{bmatrix} 1 & 0 \\ 0 & 1 \end{bmatrix}, \quad \mathbf{B} = \begin{bmatrix} b_1 & b_2 \\ 0 & 0 \end{bmatrix}, \quad \mathbf{C} = \begin{bmatrix} c_1 & c_2 \\ \overline{c_2} & c_3 \end{bmatrix} \quad \begin{pmatrix} c_1, c_3 \in \mathbb{R}, \ b_1, b_2, c_2 \in \mathbb{C}, \\ b_1 \neq 0, \ 2|b_1| + |b_2|^2 < 1 \end{pmatrix}. \quad (13)$$

Then (8) becomes

$$\begin{bmatrix} b_1 z^2 + z + \overline{b_1} & b_2 z^2 \\ \overline{b_2} & z \end{bmatrix} \frac{df}{dz} = \begin{bmatrix} -\mu b_1 z - \frac{\mu}{2} + c_1 & -\mu b_2 z + c_2 \\ \overline{c_2} & -\frac{\mu}{2} + c_3 \end{bmatrix}. \quad (14)$$

Let

$$\begin{bmatrix} b_1 z^2 + z + \overline{b_1} & b_2 z^2 \\ \overline{b_2} & z \end{bmatrix}^{-1} \begin{bmatrix} -\mu b_1 z - \frac{\mu}{2} + c_1 & -\mu b_2 z + c_2 \\ \overline{c_2} & -\frac{\mu}{2} + c_3 \end{bmatrix}$$

$$= \frac{1}{z(b_1 z^2 + (1 - |b_2|^2)z + \overline{b_1})} \begin{bmatrix} z & -b_2 z^2 \\ -\overline{b_2} & b_1 z^2 + z + \overline{b_1} \end{bmatrix} \begin{bmatrix} -\mu b_1 z - \frac{\mu}{2} + c_1 & -\mu b_2 z + c_2 \\ \overline{c_2} & -\frac{\mu}{2} + c_3 \end{bmatrix}$$

$$=: \frac{1}{z}\begin{bmatrix} 0 & 0 \\ X_0 & Y_0 \end{bmatrix} + \frac{1}{(z - \alpha)(z - \overline{\alpha}^{-1})}\begin{bmatrix} V_1 z + V_2 & W_1 z + W_2 \\ X_1 z + X_2 & Y_1 z + Y_2 \end{bmatrix}$$

$$= \frac{1}{z}\begin{bmatrix} 0 & 0 \\ X_0 & Y_0 \end{bmatrix} + \frac{1}{z - \alpha_+}\begin{bmatrix} V_+ & W_+ \\ X_+ & Y_+ \end{bmatrix} + \frac{1}{z - \alpha_-}\begin{bmatrix} V_- & W_- \\ X_- & Y_- \end{bmatrix},$$

where we write $\alpha =: \alpha_+$, $\overline{\alpha}^{-1} =: \alpha_-$, so that

Representation Theory of $\mathfrak{sl}(2, \mathbb{R}) \simeq \mathfrak{su}(1, 1)$ ...

$$\alpha_{\pm} = \frac{-1 + |b_2|^2 \pm \sqrt{(1 - |b_2|^2)^2 - 4|b_1|^2}}{2b_1},$$

$$\begin{bmatrix} 0 & 0 \\ X_0 & Y_0 \end{bmatrix} = \frac{1}{\overline{b_1}} \begin{bmatrix} 0 & 0 \\ -\overline{b_2} & \overline{b_1} \end{bmatrix} \begin{bmatrix} c_1 - \frac{\mu}{2} & c_2 \\ \overline{c_2} & c_3 - \frac{\mu}{2} \end{bmatrix},$$

$$\begin{bmatrix} V_{\pm} & W_{\pm} \\ X_{\pm} & Y_{\pm} \end{bmatrix} = \frac{\pm 1}{b_1(\alpha_+ - \alpha_-)} \begin{bmatrix} 1 & -b_2\alpha_{\pm} \\ -\overline{b_2}\alpha_{\pm}^{-1} & |b_2|^2 \end{bmatrix} \begin{bmatrix} -\mu b_1\alpha_{\pm} - \frac{\mu}{2} + c_1 & -\mu b_2\alpha_{\pm} + c_2 \\ \overline{c_2} & -\frac{\mu}{2} + c_3 \end{bmatrix}.$$

We note that $V_{\pm}Y_{\pm} - W_{\pm}X_{\pm} = 0$ holds by Theorem 4.4. Then (14) is rewritten as

$$\frac{df_1}{dz} = \frac{V_1 z + V_2}{(z - \alpha)(z - \overline{\alpha}^{-1})} f_1 + \frac{W_1 z + W_2}{(z - \alpha)(z - \overline{\alpha}^{-1})} f_2,$$

$$\frac{df_2}{dz} = \left(\frac{X_0}{z} + \frac{X_1 z + X_2}{(z - \alpha)(z - \overline{\alpha}^{-1})} \right) f_1 + \left(\frac{Y_0}{z} + \frac{Y_1 z + Y_2}{(z - \alpha)(z - \overline{\alpha}^{-1})} \right) f_2.$$

By transforming this into a single higher-order differential equation, we get

$$\left(\frac{d}{dz} - \frac{Y_0}{z} - \frac{Y_1 z + Y_2}{(z - \alpha)(z - \overline{\alpha}^{-1})} \right) \frac{(z - \alpha)(z - \overline{\alpha}^{-1})}{W_1 z + W_2} \left(\frac{d}{dz} - \frac{V_1 z + V_2}{(z - \alpha)(z - \overline{\alpha}^{-1})} \right) f_1$$
$$= \left(\frac{X_0}{z} + \frac{X_1 z + X_2}{(z - \alpha)(z - \overline{\alpha}^{-1})} \right) f_1,$$

that is,

$$\left[\frac{d^2}{dz^2} - \left(\frac{Y_0}{z} + \frac{V_+ + Y_+ - 1}{z - \alpha} + \frac{V_- + Y_- - 1}{z - \overline{\alpha}^{-1}} + \frac{W_1}{W_1 z + W_2} \right) \frac{d}{dz} \right.$$
$$\left. + \frac{V_1(Y_0 + Y_1) - W_1(X_0 + X_1)}{(z - \alpha)(z - \overline{\alpha}^{-1})} + \frac{V_2 Y_0 - W_2 X_0}{z(z - \alpha)(z - \overline{\alpha}^{-1})} + \frac{V_2 W_1 - V_1 W_2}{(z - \alpha)(z - \overline{\alpha}^{-1})(W_1 z + W_2)} \right] f_1$$
$$= 0.$$

Therefore we get the following.

Proposition 4.6 *The Eq. (14) is equivalent to the single differential equation*

$$\left[\frac{d^2}{dz^2} + \left(\frac{-\kappa_0 + \frac{\mu}{2}}{z} + \frac{1 - \kappa_1 + \frac{\mu}{2}}{z - \alpha} + \frac{1 + \overline{\kappa_1} + \frac{\mu}{2}}{z - \overline{\alpha}^{-1}} - \frac{1}{z - \epsilon} \right) \frac{d}{dz} \right.$$
$$\left. + \frac{\mu(-\overline{\kappa_0} + \frac{\mu}{2})}{(z - \alpha)(z - \overline{\alpha}^{-1})} + \frac{q_1}{z(z - \alpha)(z - \overline{\alpha}^{-1})} + \frac{q_2}{(z - \alpha)(z - \overline{\alpha}^{-1})(z - \epsilon)} \right] f_1 = 0,$$

where

$$\alpha = \frac{-1 + |b_2|^2 + \sqrt{(1 - |b_2|^2)^2 - 4|b_1|^2}}{2b_1}, \qquad \kappa_0 = \frac{\overline{b_1} c_3 - \overline{b_2} c_2}{\overline{b_1}} = \frac{\mathrm{tr}(\mathrm{adj}(\mathbf{B}^{\dagger})\mathbf{C})}{\mathrm{tr}(\mathrm{adj}(\mathbf{A})\mathbf{B}^{\dagger})},$$

$$\kappa_1 = \frac{1}{b_1(\alpha - \overline{\alpha}^{-1})} \mathrm{tr}\left(\begin{bmatrix} 1 & -b_2\alpha \\ -\overline{b_2}\alpha^{-1} & |b_2|^2 \end{bmatrix} \begin{bmatrix} c_1 & c_2 \\ \overline{c_2} & c_3 \end{bmatrix} \right) = \frac{\mathrm{tr}(\mathrm{adj}(\mathbf{B}\alpha + \mathbf{A} + \mathbf{B}^{\dagger}\alpha^{-1})\mathbf{C})}{(\alpha - \overline{\alpha}^{-1}) \mathrm{tr}(\mathrm{adj}(\mathbf{A})\mathbf{B})},$$

$$\epsilon = \frac{c_2}{b_2(c_3 + \frac{\mu}{2})}, \quad q_1 = \frac{(c_1 - \frac{\mu}{2})(c_3 - \frac{\mu}{2}) - |c_2|^2}{b_1} = \frac{\det(\mathbf{C} - \frac{\mu}{2}\mathbf{A})}{\text{tr}(\text{adj}(\mathbf{A})\mathbf{B})},$$

$$q_2 = \frac{\mu(b_2(c_1 - \frac{\mu}{2}) - b_1 c_2) + b_2((c_1 - \frac{\mu}{2})(c_3 - \frac{\mu}{2}) - |c_2|^2)}{b_1 b_2(c_3 + \frac{\mu}{2})}.$$

This equation has original regular singularities at $\{0, \alpha, \overline{\alpha}^{-1}, \infty\}$ and an additional apparent singularity at $z = \epsilon$. The characteristic exponents at each singularity are given as

$$\left\{ \begin{array}{ccccc} z=0 & z=\alpha & z=\overline{\alpha}^{-1} & z=\epsilon & z=\infty \\ 0 & 0 & 0 & 0 & \mu \\ 1+\kappa_0 - \frac{\mu}{2} & \kappa_1 - \frac{\mu}{2} & -\overline{\kappa_1} - \frac{\mu}{2} & 2 & -\overline{\kappa_0} + \frac{\mu}{2} \end{array} \right\}.$$

Such an equation appears in the context of isomonodromic deformations and Painlevé equations (see, e.g., [13]). If the space of solutions for the Eq. (14) in $\mathcal{H}_\mu(\mathbf{D}) \otimes \mathbb{C}^2$ is 2-dimensional, then all local solutions around the singularities $0, \alpha \in \mathbf{D}$ must be holomorphic, and this holds only if the corresponding characteristic exponents are positive integers. That is, the following holds.

Corollary 4.7 *If the space of solutions for the Eq. (14) in $\mathcal{H}_\mu(\mathbf{D}) \otimes \mathbb{C}^2$ is 2-dimensional, then $1 + \kappa_0 - \frac{\mu}{2}, \kappa_1 - \frac{\mu}{2} \in \mathbb{Z}_{>0}$.*

We note that this condition is not sufficient since there may exist solutions with logarithmic terms.

As a special case, if $\mathbf{B} = \mathbf{B}^\dagger$ holds, or equivalently if $\mathbf{A}_1 = \mathbf{A}_3$ holds in (5), then by a suitable $g \in SO(1,1) := SU(1,1) \cap M(2, \mathbb{R})$ and by a suitable gauge transform, we can change $(\mathbf{A}, \mathbf{B}, \mathbf{C})$ to a form in (13) with $b_1 \in \mathbb{R}, b_2 = 0$. Then by transforming (14) into a single higher-order differential equation, we get a Heun equation [17],

$$\left[\frac{d^2}{dz^2} + \left(\frac{-\kappa_0 + \frac{\mu}{2}}{z} + \frac{1 - \kappa_1 + \frac{\mu}{2}}{z - \alpha} + \frac{1 + \kappa_1 + \frac{\mu}{2}}{z - \alpha^{-1}}\right)\frac{d}{dz} + \frac{\mu(1 - \kappa_0 + \frac{\mu}{2})z + q_1}{z(z-\alpha)(z-\alpha^{-1})}\right] f_1 = 0, \quad (15)$$

with regular singularities at $\{0, \alpha, \alpha^{-1}, \infty\}$ and with no additional apparent singularity. The characteristic exponents at each singularity are given as

$$\left\{ \begin{array}{cccc} z=0 & z=\alpha & z=\alpha^{-1} & z=\infty \\ 0 & 0 & 0 & \mu \\ 1+\kappa_0 - \frac{\mu}{2} & \kappa_1 - \frac{\mu}{2} & -\kappa_1 - \frac{\mu}{2} & 1 - \kappa_0 + \frac{\mu}{2} \end{array} \right\}.$$

Example 4.8 The original η-shifted non-commutative harmonic oscillator $H_{\text{NCHO}} \psi = \lambda \psi$ treated in [16], in the standard form, corresponds to (6) with $n = 1, p = 2$ and

$$\mathbf{A} = \begin{bmatrix} \beta & 0 \\ 0 & \gamma \end{bmatrix}, \quad \mathbf{B} = \frac{1}{2}\begin{bmatrix} 0 & i \\ -i & 0 \end{bmatrix}, \quad 2\mathbf{C} = \begin{bmatrix} \lambda & 2\eta i \sqrt{\beta\gamma - 1} \\ -2\eta i \sqrt{\beta\gamma - 1} & \lambda \end{bmatrix}.$$

The positivity condition (12) is equivalent to $\beta, \gamma > 0$, $\beta\gamma > 1$, which makes the spectra of H_{NCHO} discrete and bounded below. Its restriction to $L^2_{\text{even}}(\mathbb{R}) \otimes \mathbb{C}^2$ and $L^2_{\text{odd}}(\mathbb{R}) \otimes \mathbb{C}^2$ are equivalent to the differential equation (8) with $\mu = \frac{1}{2}$ and $\mu = \frac{3}{2}$ respectively. Let $l := \begin{bmatrix} \sqrt{\beta i}^{-1} & 0 \\ 0 & \sqrt{-\gamma i}^{-1} \end{bmatrix} \in GL(2, \mathbb{C})$, and we put

$$\mathbf{A}' := l\mathbf{A}l^\dagger = \begin{bmatrix} 1 & 0 \\ 0 & 1 \end{bmatrix}, \qquad \mathbf{B}' := l\mathbf{B}l^\dagger = \frac{1}{2\sqrt{\beta\gamma}} \begin{bmatrix} 0 & 1 \\ 1 & 0 \end{bmatrix} = \begin{bmatrix} 0 & g \\ g & 0 \end{bmatrix},$$

$$\mathbf{C}' := l\mathbf{C}l^\dagger = \frac{\lambda}{2} \begin{bmatrix} \beta^{-1} & 0 \\ 0 & \gamma^{-1} \end{bmatrix} + \eta \frac{\sqrt{\beta\gamma - 1}}{\sqrt{\beta\gamma}} \begin{bmatrix} 0 & 1 \\ 1 & 0 \end{bmatrix} = \begin{bmatrix} \lambda' - \Delta & -\varepsilon \\ -\varepsilon & \lambda' + \Delta \end{bmatrix},$$

with $g, \lambda', \Delta, \varepsilon \in \mathbb{R}$, $|g| < \frac{1}{2}$, as an analogue of the Rabi model (9) (with $\omega = 1$). Let $\tanh(\theta) := \sqrt{\beta\gamma}^{-1} = 2g$ so that (8) has 5 regular singularities at $\pm \tanh(\theta/2)^{\pm 1}$ and ∞, and let $h_+ := \begin{bmatrix} \cosh(\theta/2) & \sinh(\theta/2) \\ \sinh(\theta/2) & \cosh(\theta/2) \end{bmatrix}$, $h_- := \begin{bmatrix} i\cosh(\theta/2) & -i\sinh(\theta/2) \\ i\sinh(\theta/2) & -i\cosh(\theta/2) \end{bmatrix} \in SU(1,1)$. Then by replacing $\mathbf{A}, \mathbf{B}, \mathbf{C}$ with

$$h_\pm \mathbf{A}' = (\cosh\theta)\mathbf{A}' \mp 2(\sinh\theta)\mathbf{B}' = \frac{1}{\sqrt{1-4g^2}} \begin{bmatrix} 1 & \mp 4g^2 \\ \mp 4g^2 & 1 \end{bmatrix},$$

$$h_\pm \mathbf{B}' = -\frac{1}{2}(\sinh\theta)\mathbf{A}' \pm (\cosh\theta)\mathbf{B}' = \frac{g}{\sqrt{1-4g^2}} \begin{bmatrix} -1 & \pm 1 \\ \pm 1 & -1 \end{bmatrix},$$

and $\mathbf{C}'$, we get equivalent differential equations. Transforming this into a single differential equation, we get the Heun differential equation (15) with

$$\alpha = 2g = \sqrt{\beta\gamma}^{-1}, \quad \kappa_0 = \kappa_\pm, \quad \kappa_1 = \kappa_\mp, \quad q_1 = q_\pm, \qquad \text{where}$$

$$\kappa_\pm = \frac{\lambda' \mp \varepsilon}{\sqrt{1-4g^2}} = \frac{\lambda}{4} \frac{\beta + \gamma}{\sqrt{\beta\gamma(\beta\gamma - 1)}} \pm \eta,$$

$$q_\pm = -\frac{1}{2g}\left(\lambda'^2 - \frac{\lambda'\mu}{\sqrt{1-4g^2}} + \frac{\mu^2}{4}(1+4g^2) \pm \frac{4g^2\mu\varepsilon}{\sqrt{1-4g^2}} - \varepsilon^2 - \Delta^2 \right)$$

$$= -\frac{1}{\sqrt{\beta\gamma}}\left(\frac{\lambda^2}{4} - \frac{\lambda\mu\sqrt{\beta\gamma}(\beta+\gamma)}{4\sqrt{\beta\gamma-1}} + \frac{\mu^2}{4}(\beta\gamma + 1) \mp \eta\mu - \eta^2(\beta\gamma - 1) \right).$$

For details on analysis of this equation, see [16].

Next we consider a confluence process as in Remark 4.2. Let $\tilde{g} := \sqrt{\mu}g$, $\tilde{\lambda} := \lambda' - \frac{\mu}{2}$, $z = w/\sqrt{\mu}$. Then by taking the limit $\mu \to \infty$, we get

$$\left[\frac{d^2}{dw^2} + \left(\frac{-\tilde{\kappa}_\pm}{w} + \frac{1-\tilde{\kappa}_\mp}{w - 2\tilde{g}} - 2\tilde{g} \right) \frac{d}{dw} - \frac{2\tilde{g}(1-\tilde{\kappa}_\pm)w - \tilde{q}_\pm}{w(w - 2\tilde{g})} \right] f_1 = 0, \quad \text{where}$$

$$\tilde{\kappa}_\pm := \tilde{\lambda} + \tilde{g}^2 \mp \varepsilon, \quad \tilde{q}_\pm := (\tilde{\lambda} + \tilde{g}^2)(\tilde{\lambda} - 3\tilde{g}^2) \pm 4\tilde{g}^2\varepsilon - \varepsilon^2 - \Delta^2,$$

since $\kappa_\pm = \tilde{\kappa}_\pm + \frac{\mu}{2} + O(\mu^{-1})$, $q_\pm = -\frac{\sqrt{\mu}}{2g}(\tilde{q}_\pm + O(\mu^{-1}))$ hold. This gives the confluent Heun picture of the asymmetric quantum Rabi model (9). For details see [9, 16].

Acknowledgements This work was supported by JST CREST Grant Number JPMJCR2113, Japan.

References

1. D. Braak, Integrability of the Rabi model. Phys. Rev. Lett. **107**, 100401 (2011)
2. D. Braak, A generalized G-function for the quantum Rabi model. Ann. Phys. **525**(3), L23–L28 (2013)
3. J. Faraut, A. Koranyi, *Analysis on Symmetric Cones*. Oxford Mathematical Monographs Oxford Science Publications (The Clarendon Press, Oxford University Press, New York, 1994), xii+382 pp
4. G.B. Folland, *Harmonic Analysis in Phase Space*. Annals of Mathematics Studies, vol. 122 (Princeton University Press, Princeton, NJ, 1989), x+277 pp
5. R. Howe, Remarks on classical invariant theory. Trans. Amer. Math. Soc. **313**(2), 539–570 (1989)
6. R. Howe, E.C. Tan, *Non-abelian Harmonic Analysis: Applications of SL(2, **R**)*. Universitext (Springer, New York, 1992), xvi+257 pp
7. E.T. Jaynes, F.W. Cummings, Comparison of quantum and semiclassical radiation theories with application to the beam maser. Proc. IEEE **51**, 89–109 (1963)
8. M. Kashiwara, M. Vergne, On the Segal-Shale-Weil representations and harmonic polynomials. Invent. Math. **44**(1), 1–47 (1978)
9. K. Kimoto, C. Reyes-Bustos, M. Wakayama, Determinant expressions of constraint polynomials and the spectrum of the asymmetric quantum Rabi model. Int. Math. Res. Not. IMRN (12), 9458–9544 (2021)
10. T. Nomura, *Spherical Harmonics and Group Representations* (processinlinefigureincludegraphics *in Japanese*) (Nippon-Hyoron-Sha, 2018), x+358 pp
11. H. Ochiai, Non-commutative harmonic oscillators and Fuchsian ordinary differential operators. Comm. Math. Phys. **217**(2), 357–373 (2001)
12. H. Ochiai, Non-commutative harmonic oscillators and the connection problem for the Heun differential equation. Lett. Math. Phys. **70**(2), 133–139 (2004)
13. K. Okamoto, Polynomial Hamiltonians Associated with Painlevé Equations. I. Proc. Jpn. Acad. Ser. A Math. Sci. **56**(6), 264–268 (1980)
14. A. Parmeggiani, M. Wakayama, Non-commutative harmonic oscillators. I. Forum Math. **14**(4), 539–604 (2002)
15. A. Parmeggiani, M. Wakayama, Non-commutative harmonic oscillators. II. Forum Math. **14**(5), 669–690 (2002)
16. C. Reyes-Bustos, M. Wakayama, Covering families of the asymmetric quantum Rabi model: η-shifted non-commutative harmonic oscillators. Comm. Math. Phys. **403**(3), 1429–1476 (2023)
17. A. Ronveaux (ed.), *Heun's Differential Equations*. With contributions by F. M. Arscott, S. Yu. Slavyanov, D. Schmidt, G. Wolf, P. Maroni, A. Duval. Oxford Science Publication (The Clarendon Press, Oxford University Press, New York, 1995), xxiv+354 pp
18. M. Wakayama, Equivalence between the eigenvalue problem of non-commutative harmonic oscillators and existence of holomorphic solutions of Heun differential equations, eigenstates degeneration, and the Rabi model. Int. Math. Res. Not. IMRN (3), 759–794 (2016)
19. Q. Xie, H. Zhong, M.T. Batchelor, C. Lee, The quantum Rabi model: solution and dynamics. J. Phys. A **50**(11), 113001, 40 pp. (2017)

Open Access This chapter is licensed under the terms of the Creative Commons Attribution 4.0 International License (http://creativecommons.org/licenses/by/4.0/), which permits use, sharing, adaptation, distribution and reproduction in any medium or format, as long as you give appropriate credit to the original author(s) and the source, provide a link to the Creative Commons license and indicate if changes were made.

The images or other third party material in this chapter are included in the chapter's Creative Commons license, unless indicated otherwise in a credit line to the material. If material is not included in the chapter's Creative Commons license and your intended use is not permitted by statutory regulation or exceeds the permitted use, you will need to obtain permission directly from the copyright holder.

The k-Photon Quantum Rabi Model

Daniel Braak

Abstract A generalization of the quantum Rabi model is obtained by replacing the linear (dipole) coupling between the two-level system and the radiation mode by a non-linear expression in the creation and annihilation operators, corresponding to multi-photon excitations. If each spin flip involves k photons, it is called the "k-photon" quantum Rabi model. While the formally symmetric Hamilton operator is self-adjoint in the case $k = 2$, it is demonstrated here that the Hamiltonian is not self-adjoint for $k \geq 3$. Therefore it does not generate a unitary time evolution and is unphysical. This result cannot be obtained by numerical calculations in finite-dimensional spaces which attempt to approximate an unbounded operator by a finite-rank operator.

Keywords Spectral theory · Self-adjointness · Quantum optics · Light-matter interaction · Integrable systems · Rabi model

1 Introduction

The quantum Rabi model with linear coupling is a well-understood elementary model of quantum optics (see, e.g., [11]) with Hamiltonian

$$H_R = \omega \hat{a}^\dagger \hat{a} + g\left(\hat{a} + \hat{a}^\dagger\right)\sigma_x + \Delta\sigma_z, \tag{1}$$

coupling a bosonic mode (described by bosonic creation/annihilation operators $\hat{a}, \hat{a}^\dagger$) to a two-level system or "pseudospin", described by the Pauli matrices σ_x, σ_z. One sees from (1) that a spin flip, generated by σ_x, leads to a concomitant emission (effected by $\hat{a}^\dagger$) or absorption ($\hat{a}$) of a single light quantum. One would therefore assume that the following generalization of H_R,

---

D. Braak (✉)
TP III and Center for Electronic Correlations and Magnetism, Institute of Physics, University of Augsburg, Augsburg, Germany
e-mail: daniel.braak@uni-a.de

© The Author(s) 2026
T. Takagi et al. (eds.), *Mathematical Foundations for Post-Quantum Cryptography*, Mathematics for Industry 40, https://doi.org/10.1007/978-981-96-1218-5_5

$$H_{kp} = \omega \hat{a}^\dagger \hat{a} + g\left(\hat{a}^k + \hat{a}^{\dagger k}\right)\sigma_x + \Delta\sigma_z, \qquad (2)$$

corresponds to a physical process in which each spin flip is accompanied by the emission or absorption of k photons. Indeed, processes involving k-photon resonance have been observed and are of great interest for various applications in quantum optics [1, 14], but those processes are not described by (2), but by the k-photon Jaynes-Cummings model,

$$H_{JCk} = \omega \hat{a}^\dagger \hat{a} + g\left(\hat{a}^k \sigma^+ + \hat{a}^{\dagger k} \sigma^-\right) + \Delta\sigma_z, \qquad (3)$$

with $\sigma^\pm = (\sigma_x \pm i\sigma_y)/2$, the spin raising/lowering operators. The Hamiltonian H_{JCk} is easily diagonalizable and has analytical properties completely different from H_{kp}. It can be directly derived for each experimental implementation if the coupling g is small enough [2, 10]. In this sense H_{JCk} is *not* the "rotating-wave" approximation of H_{kp}, in contrast to the case $k = 1$, where the Jaynes-Cummings model appears as the weak coupling (and close to resonance) approximation of H_R which in turn is obtained directly from the dipole limit of the atomic light-matter interaction [11].

Nevertheless, the model H_{kp} has been studied from a theoretical and mathematical viewpoint for several years, especially the case $k = 2$ [6], which features the "spectral collapse" phenomenon for sufficiently strong coupling, $g > g_c = \omega/2$. For values $g < g_c$ the model can be physically implemented via trapped ions [7]. For the case $g > g_c$, it was conjectured in [15] that H_{2p} is not self-adjoint: It is easily seen that for $g > \omega/2$ and $\Delta = 0$, the Hamiltonian (2) can be mapped to a harmonic oscillator with inverted potential. Therefore, it has no ground state and indeed no eigenvectors, the pure point spectrum is empty. However, the harmonic oscillator with inverted potential is known to be self-adjoint [8, 9, 24] and the continuous spectrum spans the whole real axis, similar to the position operator $\hat{x}$. The same applies to H_{2p} [4]. The case $k \geq 3$ has been studied in [13, 25]. Lo et al., as well as Zhang, come to the conclusion that the operator H_{kp} is "ill-defined". While Zhang only states that the pure point spectrum is empty, which does not exclude self-adjointness, it is correctly claimed in [13] that H_{kp} is not self-adjoint for $k \geq 3$. To show this, Lo et al. use Nelson's theorem in the reverse, i.e., it is assumed that an operator $\hat{A}$ is not self-adjoint if some dense set of vectors $\{\phi_n\}$ in the Hilbert space $\mathcal{H}$ yields no convergence of the expression

$$\sum_{j=0} \frac{t^j}{j!} \| \hat{A}^j \phi_n \| \qquad (4)$$

for any $t > 0$ and all ϕ_n. Nelson's theorem states that a symmetric operator $\hat{A}$ is essentially self-adjoint if $\mathcal{D}(\hat{A})$ contains at least one such set of vectors, which are then called analytic vectors for $\hat{A}$ [19]. Therefore, in order to prove the negative ($\hat{A}$ is not essentially self-adjoint), one has to show that for *all* dense sets $\{\phi_n\}$ in $\mathcal{D}(\hat{A})$ the expression (4) does not converge. It is not sufficient to prove it for just one set (in [13] this was shown for the eigenvectors of the harmonic oscillator). Indeed, the usual way to show lack of self-adjointness for $\hat{A}$ is constructing the domain of $\hat{A}^\dagger$, which is larger than $\mathcal{D}(\hat{A})$ if $\hat{A}$ is symmetric, a highly non-trivial task [20, 24].

In the present paper, it is proven with Bargmann space methods and asymptotic analysis that H_{kp} has a continuous spectrum (in fact, all of $\mathbb{C}$), while the eigenvectors are normalizable.[1] However, these elements of $L^2(\mathbb{R})$ are not located in $\mathcal{D}(\hat{A}) = \mathcal{D}(\hat{a}^k) \cap \mathcal{D}(\hat{a}^{\dagger k})$.

H_{kp} is formally symmetric because it contains the sum of operators $\hat{a}^k$ and $\hat{a}^{\dagger k}$ which are mutually adjoint in $L^2(\mathbb{R})$. The total Hilbert space of H_{kp} is $\mathbb{C}^2 \otimes L^2(\mathbb{R})$. In the following, we shall study the problem in the Bargmann space of analytic functions $\phi(z)$, $\mathcal{B}$, which is isomorphic to $L^2(\mathbb{R})$ [3, 23]. In this space, $\hat{a}$ is given by the derivative ∂_z and $\hat{a}^\dagger$ by multiplication with z. The operators z and ∂_z are mutually adjoint in $\mathcal{B}$ with the scalar product

$$\langle \phi | \psi \rangle = \frac{1}{\pi} \int dz \bar{z}\, e^{-z\bar{z}} \bar{\phi}(\bar{z}) \psi(z) \tag{5}$$

A function $\phi(z)$ is an element of $\mathcal{B}$ if $\phi(z)$ is analytic in $\mathbb{C}$ and $\langle \phi | \phi \rangle < \infty$. After a trivial transformation of (2) which interchanges σ_x and σ_z, we obtain the eigenvalue equation (setting $g = 1$),

$$\begin{aligned}
\left(\omega z \partial_z + z^k + \partial_z^k\right) \phi_1(z) + \Delta \phi_2(z) &= E \phi_1(z), \\
\left(\omega z \partial_z - z^k - \partial_z^k\right) \phi_2(z) + \Delta \phi_1(z) &= E \phi_2(z),
\end{aligned} \tag{6}$$

for an element $\left(\phi_1(z), \phi_2(z)\right)^T$ in $\mathbb{C}^2 \otimes \mathcal{B}$.

One could attempt now a shortcut by studying the decoupled system by setting $\Delta = 0$, arguing that Δ, as a bounded operator, cannot influence the self-adjointness properties of $\mathcal{H}_{kp}$. But we shall see in the following that in this way only part of the actually needed information (namely only one of the exponents of first kind, see below) is gained. Therefore, we shall first eliminate $\phi_2(z)$ to obtain an equation for $\phi_1(z)$,

$$\left[(\omega z \partial_z - E)^2 - \left(z^k + \partial_z^k\right)^2 + \omega k \left(z^{k-1} - \partial_z^{k-1}\right) - \Delta^2 \right] \phi_1 = 0. \tag{7}$$

Normal ordering of the term $\left(z^k + \partial_z^k\right)^2$ yields

$$\left(z^k + \partial_z^k\right)^2 = \partial_z^{2k} + z^{2k} + 2 z^k \partial_z^k + \sum_{j=1}^{k} a_j^{(k)} z^{k-j} \partial_z^{k-j}, \tag{8}$$

with known functions $a_j^{(k)}$. Equation (7) is an ordinary differential equation in the complex domain with no singular points in $\mathbb{C}$ except at $z = \infty$, where it has an

---

[1] This seems to contradict Zhang's statement in [25]. But Zhang's scalar product differs from (5), so it cannot be compared to the present calculation.

unramified irregular singular point of s-rank three and class $2k$ [4, 22]. Therefore all formal solution of (7) are analytic in $\mathbb{C}$ with an essential and isolated singularity at $z = \infty$. Thus, $\phi_1(z)$ will be an element of $\mathcal{B}$ if it is normalizable with respect to the measure (5). The formal solutions of (7) have the asymptotic expansion for $|z| \to \infty$,

$$\psi(z) = e^{\frac{\gamma}{2}z^2 + \beta z} z^\rho \sum_{n=0}^{\infty} c_n z^{-n} \qquad (9)$$

for all $k \geq 2$ [4]. In the following sections we shall use the method applied in [4] to the case $k = 2$ to study H_{kp} for all integer $k \geq 3$.

2 The Cases $k = 3, 4$

In this section, we treat the cases with $k = 3, 4$ separately, because the exponents behave differently for these low values of k from the case $k \geq 5$, studied in Sect. 3. Especially, they depend on ω (albeit not on E as for $k = 2$ [4]).

2.1 $k = 3$

A direct calculation yields for the coefficients $a_j^{(3)}$, $a_1^{(3)} = 9$, $a_2^{(3)} = 18$, $a_3^{(3)} = 6$. Then (7) reads for $k = 3$,

$$\begin{aligned}\left[-\partial_z^6 - 2z^3 \partial_z^3 + (\omega^2 - 9)z^2 \partial_z^2 + (\omega^2 - 2E\omega - 18)z\partial_z \right. \\ \left. -z^6 + 3\omega z^2 + E^2 - \Delta^2 - 6\right] \phi_1 = 0. \end{aligned} \qquad (10)$$

The equation for the exponent of second kind with order two, γ, is obtained by plugging the expansion (9) into (10),

$$\gamma^6 + 2\gamma^3 - 1 = (\gamma^3 + 1)^2 = 0. \qquad (11)$$

There are three different, each doubly degenerate value for γ, located on the unit circle (sixth roots of unity). This entails that the solutions of (10) are conditionally normalizable, determined by the subdominant exponents β and ρ which lift the remaining two-fold degeneracy of the asymptotic expansion. A direct calculation yields for $\beta(\gamma)$,

$$\beta = \pm \frac{\omega}{3\gamma}. \qquad (12)$$

The two possible values of β for each γ give altogether six linearly independent asymptotic forms for the formal solution of (10). Each such solution is only valid in a certain sector of the complex plane due to the Stokes phenomenon [22]. Which expansion is asymptotically valid in a given Stokes sector cannot be computed from the values of $\phi_1(z)$ in any bounded domain of the complex plane. However, we shall see that the condition of normalizability does not depend on the knowledge of the Stokes phenomenon. The critical line for the exponent γ with $|\gamma| = 1$, where the integral in $\langle \phi_1 | \phi_1 \rangle$ is not controlled by the factor $\exp(-z\bar{z})$ in the Bargmann measure (5) is given by the ray $z(r) = r \exp(-i\theta/2)$ with $\theta = \arg(\gamma)$ [4]. The solution is normalizable if for all γ the γ-critical lines are not in the Stokes sector where γ is valid. On the other hand, if the critical line for some γ lies in the Stokes sector of γ, we find

$$\phi_1(t) \sim \exp\left(\frac{1}{2}r^2 \pm \frac{\omega}{3}e^{-i3\theta/2}r\right) z^\rho (c_0 + O(r^{-1})).$$

Because $\arg(\gamma) \in \{\pi, \pm\pi/3\}$, the factor $\exp(\beta z)$ has unit modulus along the critical line and does not determine the finiteness of $|\phi_1|$. The convergence of the integral in $\langle \phi_1 | \phi_1 \rangle$ is then given by the value of $\rho(\gamma, \beta)$. The explicit calculation yields $\rho(\gamma, \beta) = -2$ for all possible γ, β. Because $\Re(\rho) < -1/2$, it follows that $|\phi_1| < \infty$. This is true whether a Stokes sector contains a critical line or not, thus knowledge of the Stokes multipliers is not needed, as proven in [4]. This means that the eigenvalue equation has normalizable solutions for all $E \in \mathbb{C}$ and all six combinations of γ and β. This entails that H_{3p} is not self-adjoint, although it may have self-adjoint extensions because the defect indices are $\{6, 6\}$ [19]. However, these extensions are not unique and it is not possible to define the extended domain of H_{3p} by restricting the behavior of $\phi_1(z)$ for $z \to \infty$ in a physically meaningful way. For example, just demanding a finite photon content, $\langle \phi_1 | \hat{a}^\dagger \hat{a} | \phi_1 \rangle \overset{!}{<} \infty$, does not remove enough of the uncountably many, not mutually orthogonal eigenstates.

2.2 $k = 4$

In this case, we have $a_1^{(4)} = 16$, $a_2^{(4)} = 72$, $a_3^{(4)} = 96$ and $a_4^{(4)} = 24$. The equation for $\phi_1(z)$ is

$$\hat{A}_4[\phi_1](z) = \Big[-\partial_z^8 - 2z^4\partial_z^4 - 16z^3\partial_z^3 + (\omega^2 - 72)z^2\partial_z^2 \\ + (\omega^2 - 2\omega E - 96)z\partial_z + (-z^8 + 4\omega z^3 + E^2 - \Delta^2 - 24) \Big]\phi_1 = 0 \tag{13}$$

The equation for γ reads

$$(\gamma^4 + 1)^2 = 0, \tag{14}$$

the exponents of second kind of order two are located on the unit circle. In this case, the two-fold degeneracy of the four different values of γ is not lifted by the exponent of second kind of order one, β, because $\beta(\gamma) = 0$ for all γ. The equation determining the exponent of first kind, ρ, is obtained from the term b_4 in the expansion

$$\hat{A}_4[\phi_1](z) = e^{\frac{\gamma}{2}z^2 + \beta z} z^\rho \sum_{l=0}^{\infty} b_{8-l} z^{8-l}, \qquad (15)$$

and reads

$$\rho^2 + 5\rho + \frac{\omega^2}{16} + \frac{21}{4} = 0, \qquad (16)$$

for all γ with $\gamma^4 = -1$. The two solutions of (16) are

$$\rho_\pm = -\frac{5}{2} \pm \frac{\sqrt{16 - \omega^2}}{4}, \qquad (17)$$

whose real part is always less than $-3/2$. We conclude again that all formal solutions of (13) are normalizable and H_{4p} is therefore not self-adjoint. Moreover, the argument of Sect. 2.1 about the feasability of possible self-adjoint extensions applies here as well: The Hamiltonian H_{4p} does not describe a physically realizable system.

3 The Case $k \geq 5$

We have seen in Sect. 2.2 that ρ lifts the degeneracy of γ and at the same time determines the normalizability properties of the formal solutions if $\beta = 0$. We shall show below that $\beta = 0$ for all $k > 4$. Obviously, ρ is given for $k \geq 4$ by the condition $b_{2k-4} = 0$ in the expansion

$$\hat{A}_k[\phi_1](z) = e^{\frac{\gamma}{2}z^2 + \beta z} z^\rho \sum_{l=0}^{\infty} b_{2k-l} z^{2k-l}. \qquad (18)$$

Because the terms depending on ω in (7) contribute at most to the term b_4, it is sufficient to consider the reduced operator

$$\hat{A}_k^{red} = \partial_z^{2k} + 2z^k \partial_z^k + a_1^{(k)} z^{k-1} \partial_z^{k-1} + a_2^{(k)} z^{k-2} \partial_z^{k-2}, \qquad (19)$$

which does not depend on ω, Δ, E. We begin with $b_{2k} = 0$ which reads

$$(\gamma^k + 1)^2 = 0, \qquad (20)$$

as in the previous cases (in fact, γ may differ form a root of unity only for $k = 2$).
The term b_{2k-1} contains $2k-1$ factors γz and one factor β coming from the operator ∂_z^{2k}, with multiplicity $2k$, likewise the operator ∂_z^k (from $2z^k \partial_z^k$) yields $k -$

1 factors γz and one factor β with multiplicity k. The maximal power of z produced by $z^{k-1}\partial_z^{k-1}$ is $2k-2$, so this part of $\hat{A}_k^{red}$ does not contribute to b_{2k-1}. Apart from these terms in b_{2k-1} multiplying c_0 (see (9)), there appear terms proportional to c_1 which, however, vanish for any γ in the solution set because they are produced by the operator $\partial_z^{2k} + 2z^k\partial_z^k + z^{2k}$. In a similar fashion, all terms proportional to c_j, $j \geq 1$ in b_{2k-l} for $l \leq 4$ are redundant because they do not fix the exponents. After the determination of γ, β, ρ, the c_j are computed recursively depending on $c_0 \neq 0$ which can be arbitrarily chosen.

The terms $\propto c_0$ in b_{2k-1} read

$$2k\beta(\gamma^{2k-1} + \gamma^{k-1}) = 2k\beta\gamma^{k-1}(\gamma^k + 1), \tag{21}$$

which vanishes due to $\gamma^k = -1$, so b_{2k-1} does not determine β.

The three operators in $\hat{A}_k^{red}$ contributing to b_{2k-2} are $\partial_z^{2k}, 2z^k\partial_z^k$ and $a_1^{(k)}z^{k-1}\partial_z^{k-1}$. To find the $a_j^{(k)}$, we note the recurrence relation

$$a_j^{(n)} = a_j^{(n-1)} + (2n - 2j + 1)a_{j-1}^{(n-1)} + (n - j + 1)^2 a_{j-2}^{(n-1)}, \tag{22}$$

with $a_0^{(n)} = 1$ and $a_l^{(n)} = 0$ for $l < 0$. We obtain then

$$a_1^{(n)} = \sum_{l=1}^{n}(2l - 1) = n^2, \tag{23}$$

and

$$a_2^{(n)} = 2\sum_{l=1}^{n-1}l^3 = \frac{(n-1)^2 n^2}{2}. \tag{24}$$

The three relevant operators produce the following terms in b_{2k-2}:
∂_z^{2k}: $z^{2k-2}\gamma^{2k-2}\beta^2$ with multiplicity $\binom{2k}{2}$ and $z^{2k-1}\gamma^{2k-1}\rho z^{-1}$ with multiplicity $2k$.
$2z^k\partial_z^k$: $2z^k z^{k-2}\gamma^{k-2}\beta^2$ with multiplicity $\binom{k}{2}$ and $2z^k z^{k-1}\gamma^{k-1}\rho z^{-1}$ with multiplicity k.
$a_1^{(k)}z^{k-1}\partial_z^{k-1}$: $a_1^{(k)}z^{k-1}z^{k-1}\gamma^{k-1}$ with multiplicity 1.

Apart from the terms proportional to β^2 and ρ, there are terms coming from the split $\partial_z^n = \partial_z \partial_z^{n-1}$ where $n-1$ operators ∂_z act on $\exp(\gamma z^2/2)$ and one ∂_z on the ensuing factor $\gamma^{n-1}z^{n-1}$ yielding z^{n-2}. Counting all possibilities gives a combinatorial factor of $1 + 2 + 3 \ldots + (n-1) = n(n-1)/2$.

Collecting all terms gives for the coefficient of z^{2k-2} in $b_{2k-2}|_{c_0}$,

$$k(2k-1)\gamma^{2k-1} + k(k-1)\gamma^{k-1} + a_1^{(k)}\gamma^{k-1}$$
$$+ \beta^2\left(\binom{2k}{2}\gamma^{2k-2} + 2\binom{k}{2}\gamma^{k-2}\right) \tag{25}$$
$$+ (2k\gamma^{2k-1} + 2k\gamma^{k-1})\rho.$$

The first line in (25) is $(2k^2 - k)\gamma^{k-1}(\gamma^k + 1)$ because $a_1^{(k)} = k^2$ and thus vanishes like the last line. Therefore ρ is not determined by b_{2k-2}. The middle line gives

$$k\gamma^{k-2}((2k-1)\gamma^k + k - 1)\beta^2 = -k^2\gamma^{k-2}\beta^2. \tag{26}$$

It follows $\beta = 0$ for all $k \geq 5$.

Setting $\beta = 0$ from now on, we consider next $b_{2k-3}|_{c_0}$. The factor z^{2k-3} has odd exponent and cannot be produced by the even operators in $\hat{A}_k^{red}$ if $\beta = 0$. It therefore vanishes for all $k \geq 5$.

The term $b_{2k-4}|_{c_0}$ determines ρ, but the combinatorial factors are more complicated because the differential operator ∂_z^n splits now as $\partial_z \partial_z \partial_z^{n-2}$, meaning that two derivatives acting on factors $(\gamma z)^j$ are interspersed between derivatives acting on $\exp(\gamma z^2/2)$.

In a first step, we shall derive a generating function for the terms proportional to $\gamma^{2k-2} z^{2k-4}$ in $\partial_z^{2k} \exp(\gamma z^2/2)$. To this end, we use the representation of $SL_2(\mathbb{C})$ in $\mathcal{B}$ [4]. The generators of $\mathfrak{sl}_2(\mathbb{C})$ are

$$K_+ = \frac{z^2}{2}, \quad K_- = \frac{1}{2}\partial_z^2, \quad K_0 = z\partial_z + \frac{1}{2}. \tag{27}$$

Then

$$\exp\left(\frac{\lambda}{2}\partial_z^2\right) \exp\left(\frac{\gamma}{2}z^2\right) = \exp(\lambda K_-) \exp(\gamma K_+)$$
$$= \exp(\gamma' K_+) \exp(\alpha K_0) \exp(\lambda' K_-), \tag{28}$$

with

$$\alpha = -\ln(1 - \lambda\gamma), \quad \lambda' = \frac{-\lambda}{1 - \lambda\gamma}, \quad \gamma' = \frac{\gamma}{1 - \lambda\gamma}, \tag{29}$$

which can be obtained via the two-dimensional representation of the group $SL_2(\mathbb{C})$. We apply now the normal ordered form of the operator in (28) to the constant function and obtain

$$e^{\frac{\lambda}{2}\partial_z^2} e^{\frac{\gamma}{2}z^2} = \frac{1}{\sqrt{1-\lambda\gamma}} \exp\left(\frac{\gamma}{1-\lambda\gamma} \frac{z^2}{2}\right) = G(\lambda, \gamma, z). \tag{30}$$

For a function $f(\gamma, z)$ one has the expansion

$$\exp\left(\frac{\lambda}{2}\partial_z^2\right) f(\gamma, z) = \sum_{n=0}^{\infty} \lambda^n \Gamma_n(\gamma, z), \quad \Gamma_n(\gamma, z) = \frac{1}{n! 2^n} \partial_z^{2n} f(\gamma, z). \tag{31}$$

We expand now $G(\lambda, \gamma, z)$ in powers of λ, γ and z (for $|\lambda| < 1$),

$$G(\lambda, \gamma, z) = e^{\frac{\gamma}{2}z^2} \sum_{\{n_j\}} \eta_{n_0} \lambda^{n_0} \gamma^{n_0} \frac{1}{\prod_{j=1} n_j! 2^{\sum_{j=1} n_j}} \lambda^{\sum_{j=1} j n_j} \gamma^{\sum_{j=1}(j+1)n_j} z^{2\sum_{j=1} n_j}, \quad (32)$$

where the η_l are given as

$$\frac{1}{\sqrt{1-\lambda\gamma}} = \sum_{l=0} \eta_l (\lambda\gamma)^l. \quad (33)$$

We wish to extract the coefficient of the term $\lambda^k \gamma^{2k-2} z^{2k-4}$ in (32) which yields the term $C_0^{(2k)}$ proportional to $\gamma^{2k-2} z^{2k-4} e^{\gamma z^2/2}$ in $\Gamma_k(\gamma, z)$ for $f(\gamma, z) = \exp(\gamma z^2/2)$. This leads to the conditions

$$n_0 + \sum_{j=1} n_j = k,$$

$$n_0 + \sum_{j=1}(j+1)n_j = 2k - 2, \quad (34)$$

$$2\sum_{j=1} n_j = 2k - 4.$$

They entail $n_j = 0$ for $j \geq 4$. The remaining four possibilities for the set $\{n_0, n_1, n_2, n_3\}$ leads to

$$e^{-\frac{\gamma}{2}z^2} G(\lambda, \gamma, z)|_{\lambda^k \gamma^{2k-2} z^{2k-4}} = \eta_0 \left[\frac{1}{(k-4)!2!0!2^{k-4+2}} + \frac{1}{(k-3)!0!1!2^{k-3+1}} \right]$$
$$+ \frac{\eta_1}{(k-3)!1!0!2^{k-3+1}} + \frac{\eta_2}{(k-2)!0!0!2^{k-2}}. \quad (35)$$

The right-hand side of (35) gives $C_0^{(2k)}/(2^k k!)$ and we find

$$C_0^{(2k)} = \frac{3}{2}k(k-1) + 6k(k-1)(k-2) + 2k(k-1)(k-2)(k-3). \quad (36)$$

To compute the combinatorial factor for the operator ∂_z^k appearing in $2z^k \partial_z^k$, we can apply the same formula if k is even (setting $k = 2l$). For odd $k = 2l + 1$, we obtain instead

$$C_0^{(2l+1)} = (2l-2)(2l-1)l + C_0^{(2l)}. \quad (37)$$

It turns out that both expressions are the same if written in terms of k,

$$C_0^{(k)} = \frac{k}{8}(k^3 - 6k^2 + 11k - 6). \quad (38)$$

The operator $a_1^{(k)}z^{k-1}\partial_z^{k-1}$ contributes to z^{2k-4} if one ∂_z acts on z^{k-2}, the combinatorial factor is thus $(k-2)(k-1)/2$. The operator $a_2^{(k)}z^{k-2}\partial_z^{k-2}$ appears for the first time in b_{2k-4}, so it has multiplicity 1. The terms not containing ρ in b_{2k-4} are then

$$\gamma^{2k-2}C_0^{(2k)} + \gamma^{k-2}\left(2C_0^{(k)} + a_1^{(k)}\frac{(k-1)(k-2)}{2} + a_2^{(k)}\right). \tag{39}$$

The terms proportional to ρ and ρ^2 in b_{2k-4} created by ∂_z^{2k} are related to two insertions of ∂_z in the factor $\gamma^{2k-2}z^{2k-2}z^\rho$. If both of them act on z^ρ, they produce the coefficient $\rho(\rho-1)$ with multiplicity $\binom{2k-2+2}{2}$. If only one of the ∂_z acts on z^ρ, the other acts on z^{2k-2}, yielding the coefficient ρ. One has to discern two cases, namely whether the first inserted derivative acts on z^ρ or the second. In the first case, the combinatorial factor reads

$$(2k-1)(2k-2) + (2k-3)(2k-1)(k-1)$$
$$+ \frac{1}{2}((k-1)(2k-1) - 1) - \frac{1}{2}\sum_{j=2}^{2k-2} j^2, \tag{40}$$

and in the second case

$$(2k-1)^2(k-1) - \sum_{j=1}^{2k-2} j^2. \tag{41}$$

Finally, we obtain for the coefficient $C_{\rho^2}^{(2k)}$ of $\gamma^{2k-2}\rho^2$ and for the coefficient $C_\rho^{(2k)}$ of $\gamma^{2k-2}\rho$,

$$C_{\rho^2}^{(2k)} = k(2k-1), \qquad C_\rho^{(2k)} = 4k^3 - 8k^2 + 3k. \tag{42}$$

The corresponding factors coming from ∂_z^k in $2z^k\partial_z^k$ are computed in the same way to yield the coefficients of $\gamma^{k-2}\rho^2$ and $\gamma^{k-2}\rho$, respectively,

$$C_{\rho^2}^{(k)} = \frac{k(k-1)}{2}, \qquad C_\rho^{(k)} = \frac{1}{2}k(k^2+3) - 2k^2. \tag{43}$$

The operator $a_1^{(k)}z^{k-1}\partial_z^{k-1}$ has a single insertion of ∂_z in the factor $\gamma^{k-2}z^{k-2}z^\rho$ and produces $\gamma^{k-2}\rho$ with multiplicity $k-1$. The operator $a_2^{(k)}z^{k-2}\partial_z^{k-2}$ does not yield ρ-dependent terms in b_{2k-4}.

Collecting all terms, we find finally for the coefficient of z^{2k-4} in $b_{2k-4}|_{c_0}$,

$$\gamma^{2k-2}\left[C_{\rho^2}^{(2k)}\rho^2 + C_\rho^{(2k)}\rho + C_0^{(2k)}\right]$$
$$+ \gamma^{k-2}\left[2C_{\rho^2}^{(k)}\rho^2 + \left(2C_\rho^{(k)} + (k-1)a_1^{(k)}\right)\rho + 2C_0^{(k)} + \frac{(k-2)(k-1)}{2}a_1^{(k)} + a_2^{(k)}\right].$$
$$\tag{44}$$

Further simplifications arise when we use now $\gamma^k = -1$ and (23) and (24). The final equation for ρ has the remarkably simple form

$$\rho^2 + (2k-3)\rho + \frac{3}{4}k^2 - 2k + \frac{5}{4} = 0. \tag{45}$$

It has the solutions

$$\rho_+ = -\frac{k}{2} + \frac{1}{2}, \quad \rho_- = -\frac{3k}{2} + \frac{5}{2}. \tag{46}$$

Both exponents are less than $-1/2$ for $k > 2$. Together with the results from Sects. 2.1 and 2.2, it follows that H_{kp} is not self-adjoint in $\mathcal{B}$ for $k \geq 3$. The constructed solutions of the eigenvalue equation are elements of $\mathcal{H} = \mathcal{B} \otimes \mathbb{C}^2$, but not of $\mathcal{D}(H_{kp})$, as seen as follows. The expectation value of $z^k = \hat{a}^{\dagger k}$,

$$|\langle \phi_1 | z^k | \phi_1 \rangle| > C_1 \int_R^\infty dr \, r^k |\phi_1(re^{-i\theta/2})|^2 = C_2 \int_R^\infty dr \, r^{k+2\rho}, \tag{47}$$

with certain positive constants C_1, C_2 and R large enough, is not finite for $\rho = \rho_+$ and as $\mathcal{D}(H_{kp}) \subset \mathcal{D}(\hat{a}^{\dagger k})$, the eigenfunctions must be located in some (not unique) self-adjoint extension of H_{kp}.

4 Conclusions

We have shown that the k-photon quantum Rabi model is unphysical in the strict sense for $k \geq 3$, because its Hamilton operator H_{kp} is, though apparently hermitian, not self-adjoint. Therefore, according to Stone's theorem [20], it cannot generate a unitary time development of quantum states. This feature of H_{kp} is not detectable by any numerical evaluation which necessarily operates in a truncated, finite-dimensional Hilbert space. Any such calculation is misleading, even if a calculation of the spectrum seems to converge [15]. The employed method makes use of the analytically accessible asymptotics of formal solutions to the eigenvalue equation, although it contains arbitrarily high derivatives and is quite different from the commonly studied case where the (d-dimensional) Laplacian is augmented by some potential. Remarkably, the Stokes phenomenon, though certainly present, has no bearing on the conclusions. This is due to the vanishing of the exponent of the second kind with order one for $k \geq 4$. If β does not vanish, it may vary from Stokes sector to Stokes sector. Thus the usual lateral connection problem reappears which is hard to solve by analytic means. Finally, we note that the k-photon coupling in (2) may acquire physical meaning in the Dicke models with more than one qubit. Under certain conditions on the parameters of the Dicke models, there exist finite-dimensional invariant sub-spaces with bounded photon content, so-called "dark-like states". The first such state was discovered numerically in [5] and mathematically identified as exact eigenstate in [16]. Later on, these states have been found in several generalizations of the Dicke model [17] and are highly relevant for applications in quantum information technology because they can be used for the fast generation of W-states [18]. For these states to exist, several qubits are required. If the qubit frequencies are different and fine-tuned to the mode frequency (but *not* to the coupling), a condition easily realizable in current experimental platforms, there are exact eigenstates of the Dicke Hamiltonian with finite photon number and energy which is independent from the coupling between qubits and radiation mode [17]. These states bear some resemblance to the quasi-exact states in the asymmetric quantum Rabi model [21] and hint at another type of hidden symmetry. Very recently, there has been an attempt to identify this symmetry in the asymmetric generalization of the two-qubit Dicke model [12]. The k-photon Dicke model possesses also such exact eigenstates with finite photon number. If restricted to just these states, the Hamiltonian becomes clearly a finite-dimensional diagonal matrix with real entries and is trivially self-adjoint. But the analysis above demonstrates that the generic eigenstates of a system with k-photon coupling cannot be meaningfully associated with any self-adjoint extension of its Hamilton operator.

Acknowledgements The author wishes to thank for enlightening discussions with Fumio Hiroshima, Daniel Burgarth and Davide Lonigro. This work was funded by the Deutsche Forschungsgemeinschaft through grant no. 439943572.

References

1. E.O. Akeweje, G. Bader, A.M. Dikandé, P. Kameni Nteutse, Femtosecond laser inscriptions in Kerr nonlinear transparent media: dynamics in the presence of K-photon absorptions, radiative recombinations and electron diffusions. J. Mod. Opt. **68**(21), 1211–1220 (2021)
2. H.R. Baghshahi, M.K. Tavassoly, Dynamics of different entanglement measures of two three-level atoms interacting nonlinearly with a single-mode field. Eur. Phys. J. Plus **130**(3), 37 (2015)
3. V. Bargmann, On a Hilbert space of analytic functions and an associated integral transform part I. Commun. Pure Appl. Math. **14**(3), 187–214 (1961)
4. D. Braak, Spectral determinant of the two-photon quantum Rabi model. Annalen der Physik **535**(3), 2200519 (2023)
5. S.A. Chilingaryan, B.M. Rodríguez-Lara, The quantum Rabi model for two qubits. J. Phys. A: Math. Theor. **46**(33), 335301 (2013)
6. L. Duan, Y.-F. Xie, D. Braak, Q.-H. Chen, Two-photon Rabi model: analytic solutions and spectral collapse. J. Phys. A: Math. Theor. **49**(46), 464002 (2016)
7. S. Felicetti, J.S. Pedernales, I.L. Egusquiza, G. Romero, L. Lamata, D. Braak, E. Solano, Spectral collapse via two-phonon interactions in trapped ions. Phys. Rev. A **92**(3), 033817 (2015)
8. B. Hellwig, Ein Kriterium für die Selbstadjungiertheit elliptischer Differentialoperatoren im $\mathbb{R}^n$. Math Z **86**(3), 255–262 (1964)
9. H. Kalf, Self-adjointness for strongly singular potentials with a $-|x|^2$ fall-off at infinity. Math. Z **133**(3), 249–255 (1973)

10. A.B. Klimov, A. Navarro, L.L. Sanchez-Soto, Lie-type transformations and effective Hamiltonians in nonlinear quantum optics: applications to multilevel systems, February 2001
11. J. Larson, T. Mavrogordatos, *The Jaynes–Cummings Model and Its Descendants* (IOP Publishing Ltd., 2021)
12. Z.-F. Lei, J. Tian, J. Peng, Dark-state solution and hidden symmetries of the two-qubit multimode asymmetric quantum Rabi model, October 2023
13. C.F. Lo, K.L. Liu, K.M. Ng, The multiquantum Jaynes-Cummings model with the counter-rotating terms. EPL **42**(1), 1 (1998)
14. G.C. Ménard, A. Peugeot, C. Padurariu, C. Rolland, B. Kubala, Y. Mukharsky, Z. Iftikhar, C. Altimiras, P. Roche, H. Le Sueur, P. Joyez, D. Vion, D. Esteve, J. Ankerhold, F. Portier, Emission of photon multiplets by a DC-biased superconducting circuit. Phys. Rev. X **12**(2), 021006 (2022)
15. K.M. Ng, C.F. Lo, K.L. Liu, Exact eigenstates of the two-photon Jaynes-Cummings model with the counter-rotating term. Eur. Phys. J. D **6**(1), 119–126 (1999)
16. J. Peng, Z. Ren, D. Braak, G. Guo, G. Ju, X. Zhang, X. Guo, Solution of the two-qubit quantum Rabi model and its exceptional eigenstates. J. Phys. A: Math. Theor. **47**(26), 265303 (2014)
17. J. Peng, Z. Ren, H. Yang, G. Guo, X. Zhang, G. Ju, X. Guo, C. Deng, G. Hao, Algebraic structure of the two-qubit quantum Rabi model and its solvability using Bogoliubov operators. J. Phys. A: Math. Theor. **48**(28), 285301 (2015)
18. J. Peng, J. Zheng, J. Yu, P. Tang, G. Alvarado Barrios, J. Zhong, E. Solano, F. Albarrán-Arriagada, L. Lamata, One-photon solutions to the multiqubit multimode quantum Rabi model for fast W-state generation. Phys. Rev. Lett. **127**(4), 043604 (2021)
19. M. Reed, B. Simon, *Methods of Modern Mathematical Physics II: Fourier Analysis, Self-adjointness* (Elsevier, 1975)
20. M. Reed, B. Simon, *Methods of Modern Mathematical Physics I: Functional Analysis* (Academic Press, 1981)
21. C. Reyes-Bustos, M. Wakayama, Degeneracy and hidden symmetry for the asymmetric quantum Rabi model with integral bias. Comm. Num. Theory Phys. **16**(3), 615–672 (2022)
22. S.Y. Slavyanov, W. Lay, *Special Functions: A Unified Theory Based on Singularities* (Oxford University Press, 2000)
23. A. Vourdas, Analytic representations in quantum mechanics. J. Phys. A: Math. Gen. **39**(7), R65–R141 (2006)
24. E. Wienholtz, Bemerkungen über elliptische Differentialoperatoren. Arch. Math **10**(1), 126–133 (1959)
25. Y.-Z. Zhang, On the 2-mode and k-photon quantum Rabi models. Rev. Math. Phys. **29**(04), 1750013 (2017)

Open Access This chapter is licensed under the terms of the Creative Commons Attribution 4.0 International License (http://creativecommons.org/licenses/by/4.0/), which permits use, sharing, adaptation, distribution and reproduction in any medium or format, as long as you give appropriate credit to the original author(s) and the source, provide a link to the Creative Commons license and indicate if changes were made.

The images or other third party material in this chapter are included in the chapter's Creative Commons license, unless indicated otherwise in a credit line to the material. If material is not included in the chapter's Creative Commons license and your intended use is not permitted by statutory regulation or exceeds the permitted use, you will need to obtain permission directly from the copyright holder.

Meditations on the Farey Fractal

Shai Haran

Abstract We define the "coronas", which are especially spiky paths in the Farey graph going from $\underline{0} = (1, 0)$ to $\underline{\infty} = (0, 1)$. We show that for $R \geq 2$, $\{(x, y) \in \mathbb{N}^+ \times \mathbb{N}^+, \gcd(x, y) = 1, x + y \leq R\}$ is a corona.

Keywords Riemann hypothesis · Farey tree · Operad structure

Preface

For me, the greatest mystery of mathematics was André Weil's "Roseta Stone" [16, 18]: the analogies between number fields and function fields.

Regarding the Riemann hypothesis, initially I followed Weil's approach [22] to Tate's thesis [15], viewing it as the harmonic analysis of the action of $\mathbb{A}_K^\star / K^\star$ on $\mathbb{A}_K / K^\star$ (or on distribution on $\mathbb{A}_K$ that are $K^\star$-invariant), and on $\mathbb{P}^1 (\mathbb{A}_k) / K^\star$. This suggested that an (real valued) index-theorem, analogue of the Riemann-Roch for the associated surface, will give a proof of the Riemann-Hypothesis along the lines of Weil's proof, see [4]. The formula [3] for Weil's explicit-sums-distribution [21] was also the starting point for the program of Alain Connes and collaborates (cf. [1] appendix, where the formula [3] is stated in an asymptotic form). But note that there is still not even a proof of Weil's Riemann-Hypothesis for a function field K [19] using the "non-commutative" space $\mathbb{A}_K / K^\star$!

The mysterious analogy between number fields and function fields is clarified by the concept of generalized-ring (see [6] for a quick introduction). The language of generalized-rings can be used as the foundation of algebraic geometry in the style of Grothendieck (see [8, 10]):

- The final object of geometry is the absolute-point $\operatorname{spec}(\mathbb{F})$, where $\mathbb{F}$ is the initial object of generalized-rings, the "Field with one element"

S. Haran (✉)
Department of Mathematics Technion, Israel Institute of Technology, Haifa, Israel
e-mail: haran@technion.ac.il

- The real $\mathbb{R}$ and complex $\mathbb{C}$ numbers, when viewed as (topological) generalized-rings, have (maximal compact topological)-sub-generalized-rings $\mathbb{Z}_\mathbb{R} \subseteq \mathbb{R}$, and $\mathbb{Z}_\mathbb{C} \subseteq \mathbb{C}$, (analogous to $\mathbb{Z}_p \subseteq \mathbb{Q}_p$); and spec($\mathbb{Z}$), and spec($O_K$), K a number field, have natural compactifications.
- There are non-trivial Arithmetical surfaces, and higher arithmetical dimensions, as the tensor-product (=the categorical sum) does not reduce to its diagonal:

$$\mathbb{Z} \otimes_\mathbb{F} \mathbb{Z} \neq \mathbb{Z}.$$

- There is a natural generalization of homological algebra, and of the derived category of quasi-coherent sheaves of O_X-modules, and the derived functors of direct and inverse images, [9].

However, we are still missing an arithmetical analogue of the Frobenius correspondence. For the tropical examples (see [6, 8], p. 29):

$$\mathcal{B} = \{0, 1\}_t \subseteq \mathcal{J} = [0, 1]_t \subseteq \mathcal{R} = [0, \infty)_t$$

where the subscript "t" indicates that addition is $x + y := \max\{x, y\}$, we have that $\mathcal{R}$ is a generalized-field, and the (multiplicative) group $\mathbb{R}^+$ acts on $\mathcal{R}$ by automorphism $x \mapsto x^p$, with fixed field the Boolean-field $\mathcal{B}$. This resembles the Frobenius-automorphism $x \mapsto x^p$ of the field $\overline{\mathbb{F}_p}$ with fixed field $\mathbb{F}_p$. Unfortunately, the tensor-product (=categorical sum, in the categories of generalized-rings or of semi-rings) vanishes: $\mathbb{Z} \otimes \mathcal{B} = \{0\}$ the zero=final object.

As Weil suggested [20], the arithmetical analogue of extending scalars to $\overline{\mathbb{F}_p}$, is the cyclotomic extension obtained by adding (all!) roots of unity μ_∞ (this idea, in the p-power cyclotomic extension, $\mathbb{Z}[\mu_{p^\infty}]$, was developed by K. Iwasawa, who related it to the Kobuta-Leopoldt p-adic L-function [11]).

All this makes the prospect of seeing, in our life-time, a proof of the Riemann-Hypothesis, along the lines of Weil's proof for a function field, unrealistic! But perhaps there is an alternative route: after all, the field of rational numbers $\mathbb{Q}$ is the analogue of the field of rational functions $\mathbb{F}_p(T)$, and the Riemann-Hypothesis for $\mathbb{F}_p(T)$, that is for $\mathbb{P}^1/\mathbb{F}_p$, is a triviality: there are no zeros of the zeta function, only the poles, and we can constructively generate all the primes, $\overline{\mathbb{F}_p}/(x \sim x^p)$, and we can count exactly the number of points,

$$\#\mathbb{P}^1\left(\mathbb{F}_{p^d}\right) = \frac{p^{2d} - 1}{p^d - 1} = p^d + 1$$

(and the Riemann-Hypothesis for a general function field follows from this by the Bombieri-Stepanov argument). So perhaps, for the basic number field of rational numbers $K = \mathbb{Q}$ there is a constructive proof.

1 Introduction

The non-zero natural numbers, $\mathbb{N}^+ = \mathbb{N}\setminus\{0\}$, are the free commutative, unital, monoid on the set of primes,

$$\text{Prime} = \mathbb{N}^{++}\setminus\left(\mathbb{N}^{++} \bullet \mathbb{N}^{++}\right), \quad \mathbb{N}^{++} = \mathbb{N}^+\setminus\{1\}. \tag{1}$$

Using the notation

$$\{\alpha\} = \begin{cases} 1 & \text{unconditionally} \\ \alpha & \text{if the Riemann Hypothesis holds} \end{cases}$$

the convergence of

$$\frac{1}{\zeta(s)} = \prod_{p \in \text{Prime}} (1 - p^{-s}) = \sum_{n \geq 1} \frac{\mu(n)}{n^s} \tag{2}$$

for $\Re(s) > \{\tfrac{1}{2}\}$ is equivalent to

$$\left|\sum_{n=1}^{R} \mu(n)\right| = \left|\sum_{\substack{1 \leq x \leq y \leq R \\ \gcd(x,y)=1}} e^{2\pi i x/y}\right| = O\left(R^{\{1/2\}} \log R\right). \tag{3}$$

Similarly, the positive rational numbers $\mathbb{Q}^+$ are the free abelian group on the primes

$$\mathbb{Q}^+ = \mathbb{Z}\,\text{Prime} = \bigoplus_{p \in \text{Prime}} p^{\mathbb{Z}} \tag{4}$$

While this does not determine the set of primes as in (1), we do have that every $v \in \mathbb{Q}^+$ can be written uniquely as $v = y/x$ with $x, y \in \mathbb{N}^+$, and $\gcd(x, y) = 1$; we write $\underline{v} = (x, y)$, so that

$$\underline{\mathbb{Q}^+} \equiv \{(x, y) \in \mathbb{N}^+ \mid \gcd(x, y) = 1\} \equiv (\mathbb{N}^+ \times \mathbb{N}^+)\setminus\mathbb{N}^{++} \bullet (\mathbb{N}^+ \times \mathbb{N}^+) \tag{5}$$

We have now quite similarly to (3), with $\Phi(n) = \#(\mathbb{Z}/n\mathbb{Z})^*$,

$$\left|\sum_{n=1}^{R} \Phi(n)\right| = 1 + \#\left\{(x, y) \in \underline{\mathbb{Q}}^+ \mid x + y \leq R\right\}$$

$$= 1 + \sum_{n \geq 1} \mu(n) \cdot \#\left\{(x, y) \in \mathbb{N}^+ \times \mathbb{N}^+ \mid x + y \leq \tfrac{R}{n}\right\} \qquad (6)$$

$$\approx \sum_{n \geq 1} \mu(n) \cdot \tfrac{1}{2}\left(\tfrac{R}{n}\right)^2$$

$$= \tfrac{R^2}{2\zeta(2)} + O\left(R^{\{\frac{1}{2}\}} \log R\right)$$

Indeed, on the analytic side we have, since Φ is multiplicative

$$\sum_{n \geq 1} \frac{\Phi(n)}{n^s} = \prod_{p \in \text{Prime}} \left(1 + (1 - p^{-1}) \sum_{n \geq 1} p^{n(1-s)}\right) \qquad (7)$$

$$= \prod_{p \in \text{Prime}} \frac{1 - p^{-s}}{1 - p^{1-s}} = \frac{\zeta(s-1)}{\zeta(s)}.$$

This has a simple pole at $s = 2$, with residue $\frac{1}{\zeta(2)}$, and otherwise is analytic for $\Re(s) > \{\frac{1}{2}\}$.

The functions $\mu(n)$ and $\Phi(n)$ are as mysterious as the primes, for instance,

$$\text{Prime} \equiv \{n \in \mathbb{N}^+, \Phi(n) = n - 1\}.$$

But the sum in (6) is better than the sum in (3), because it can be made <u>constructive</u>. Our purpose here is to <u>linearize</u> this constructive approach so as to have an explicit recursive formula for the sum in (6) and to explore some of the structures behind it. Curiously, there is some kind of interaction between the binary and the Fibonacci expansions of integers.

2 The Farey Graph

The group $\underline{\mathbb{Q}}^+$ embeds as a dense subgroup of the multiplicative group of positive real numbers

$$\underline{\mathbb{Q}}^+ \hookrightarrow \mathbb{R}^+ = (0, \infty) \subseteq [0, \infty] \qquad (8)$$

and we have an induced total order $\leq$ on $\underline{\mathbb{Q}}^+$. We add to $\underline{\mathbb{Q}}^+$ the points in the plane

$$\underline{0} := (1, 0), \qquad \underline{\infty} := (0, 1), \qquad (9)$$

and we have the Farey Graph $\mathcal{G}$ with vertices $\mathcal{G}_0 = \{\infty\} \sqcup \mathbb{Q}^+ \sqcup \{\underline{0}\}$ and edges

$$\mathcal{G}_1 = \left\{ \begin{array}{l} (v_- = (x_-, y_-), v_+ = (x_+, y_+)) \in \mathcal{G}_0 \times \mathcal{G}_0, \\ \det \begin{pmatrix} v_- \\ v_+ \end{pmatrix} = x_- y_+ - y_- x_+ = 1 \end{array} \right\} \quad (10)$$

Every edge $(v_-, v_+) \in \mathcal{G}_1$, gives a parallelogram

$$P_v = \{t_+ v_+ + t_- v_- \mid t_\pm \in [0, 1]\}$$
$$\text{Area } (P_v) = \det \begin{pmatrix} v_- \\ v_+ \end{pmatrix} = 1 \, , \quad P_v \cap (\mathbb{N} \times \mathbb{N}) = \{(0,0), v_-, v, v_+\} \quad (11)$$

and a triangle

$$\Delta_v = \{t_+ v_+ + t_- v_- \mid t_\pm \in [0, 1], \ t_+ + t_- \geq 1\}$$
$$\text{Area } (\Delta_v) = \tfrac{1}{2} \, , \quad \Delta_v \cap (\mathbb{N} \times \mathbb{N}) = \{v_+ > v > v_-\} \quad (12)$$

where we denote them using the mediant

$$v = v_+ + v_- = (x_+ + x_-, y_+ + y_-) \quad (13)$$

(Obtained as the vector addition in the plane; not addition in $\mathbb{Q}^+$!).
If we add to the triangles $\{\Delta_v\}_{v \in \mathbb{Q}^+}$ the triangle

$$\Delta_0 = \{(t_+, t_-) \mid t_\pm \in [0, 1], \ t_+ + t_- \leq 1\} \quad (14)$$

we get a triangulation of the first quadrant of the plane minus the multiples $t \cdot v$, $v \in \mathbb{Q}^+ \sqcup \{\underline{0}, \infty\}, t > 1$:

$$[0, \infty) \times [0, \infty) \setminus (1, \infty) \bullet (\mathbb{Q}^+ \sqcup \{\underline{0}, \infty\}) \equiv \Delta_0 \sqcup \coprod_{v \in \mathbb{Q}^+} \Delta_v. \quad (15)$$

3 The Binary Tree

The monoid

$$\text{SL}_2(\mathbb{N}) := \left\{ \begin{pmatrix} a & b \\ c & d \end{pmatrix} \in \text{SL}_2(\mathbb{Z}) \, \middle| \, a, b, c, d \geq 0 \right\}$$
$$\equiv \left\{ \begin{pmatrix} v_- \\ v_+ \end{pmatrix} \, \middle| \, (v_-, v_+) \in \mathcal{G}_1 \right\} \equiv \mathcal{G}_1 \quad (16)$$

acts on $\underline{\mathbb{Q}}^+$, (and on $\underline{\mathbb{Q}}^+ \amalg \{0, \infty\}$), on the right:

$$(x, y) \begin{pmatrix} v_- \\ v_+ \end{pmatrix} = x \cdot v_- + y \cdot v_+ \tag{17}$$

$$\underline{0} \begin{pmatrix} v_- \\ v_+ \end{pmatrix} = v_-, \qquad \underline{\infty} \begin{pmatrix} v_- \\ v_+ \end{pmatrix} = v_+$$

This action preserves the graph structure $\mathcal{G}$, and the triangles Δ_v:

$$\Delta_v \cdot g = \Delta_{vg}, \quad v \in \underline{\mathbb{Q}}^+, \quad g \in \mathrm{SL}_2(\mathbb{N}). \tag{18}$$

For $g = \begin{pmatrix} a & b \\ c & d \end{pmatrix} \in \mathrm{SL}_2(\mathbb{N}) \setminus \left\{ \begin{pmatrix} 1 & 0 \\ 0 & 1 \end{pmatrix} \right\}$ with $ad = bc + 1$, we have

$$g \cdot \begin{pmatrix} 1 & 1 \\ 0 & 1 \end{pmatrix}^{-1} = \begin{pmatrix} a & b-a \\ c & d-c \end{pmatrix} \in \mathrm{SL}_2(\mathbb{N})$$

$$\text{or } g \cdot \begin{pmatrix} 1 & 0 \\ 1 & 1 \end{pmatrix}^{-1} = \begin{pmatrix} a-b & b \\ c-d & d \end{pmatrix} \in \mathrm{SL}_2(\mathbb{N}), \tag{19}$$

that is, either ($b \geq a$ and $d \geq c$) or ($b \leq a$ and $d \leq c$).
It follows that $\mathrm{SL}_2(\mathbb{N})$ is the <u>free</u> monoid on these two generators

$$\mathrm{SL}_2(\mathbb{N}) = \langle g_+, g_- \rangle, \quad g_+ = \begin{pmatrix} 1 & 1 \\ 0 & 1 \end{pmatrix}, \; g_- = \begin{pmatrix} 1 & 0 \\ 1 & 1 \end{pmatrix} \tag{20}$$

and every $g \in \mathrm{SL}_2(\mathbb{N})$ has a unique representation as a "<u>word</u>"

$$g = g_\delta^{a_\ell} \cdots g_+^{a_2} g_-^{a_1} g_+^{a_0}, \tag{21}$$

with $a_0 \geq 0$, $a_{1+i} \geq 1$ and $\delta = \delta(g) = (-1)^\ell$.

The action of $\mathrm{SL}_2(\mathbb{N})$ on $\underline{\mathbb{Q}}^+$ is free, and we get an <u>identification</u>

$$\mathrm{SL}_2(\mathbb{N}) \xleftrightarrow{\sim} \underline{\mathbb{Q}}^+$$

$$\begin{pmatrix} v_- \\ v_+ \end{pmatrix} \longleftrightarrow v$$

$$g = \begin{pmatrix} v_- \\ v_+ \end{pmatrix} \longmapsto v = (1,1)g = v_+ + v_- \tag{22}$$

$$g_v = \begin{pmatrix} v_- \\ v_+ \end{pmatrix} \longleftarrow v$$

Meditations on the Farey Fractal

Under this identification we have

$$g_v = g_\delta^{a_\ell} \cdots g_-^{a_1} g_+^{a_0} \longleftrightarrow v = (x, y) \qquad (23)$$

where ℓ continued fraction expansion

$$\frac{y}{x} = [\![a_0, \cdots, a_\ell]\!] := a_0 + \cfrac{1}{a_1 + \cfrac{1}{\ddots + \cfrac{1}{a_{\ell-1} + \cfrac{1}{a_\ell + 1}}}}, \qquad (24)$$

and $\delta = \delta(v) = (-1)^\ell \in \{\pm 1\}$.

We get the maps of <u>upper</u> and <u>lower</u> bounds:

$$t_+ : \underline{\mathbb{Q}^+} \longrightarrow \underline{\mathbb{Q}^+} \amalg \{\underline{\infty}\}, \quad t_+(v) = v_+ = \underline{\infty} \, g_v$$

$$t_- : \underline{\mathbb{Q}^+} \longrightarrow \underline{\mathbb{Q}^+} \amalg \{\underline{0}\}, \quad t_-(v) = v_- = \underline{0} \, g_v \qquad (25)$$

$$t_+^{-1}(\underline{\infty}) = \{(1, n) \mid n \geq 1\} = f_\infty^-, \quad t_-^{-1}(\underline{0}) = \{(n, 1) \mid n \geq 1\} = f_{\underline{0}}^+$$

We get a structure of a binary tree on $\underline{\mathbb{Q}^+}$, the Stern-Brocot tree, with root $\underline{1} = (1, 1)$, and each $v \in \underline{\mathbb{Q}^+}$ has the two <u>offsprings</u>

$$v \begin{array}{c} \nearrow v + v_+ \\ \\ \searrow v + v_- \end{array} \qquad (26)$$

In terms of the identification (22), this can be written as

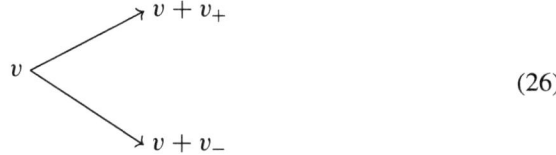

$$g = \begin{pmatrix} v_- \\ v_+ \end{pmatrix} \begin{array}{c} \nearrow g_+ \cdot g = \begin{pmatrix} v \\ v_+ \end{pmatrix} \\ \\ \searrow g_- \cdot g = \begin{pmatrix} v_- \\ v \end{pmatrix} \end{array}, \quad v = v_+ + v_-. \qquad (27)$$

cf. Fig. 1.

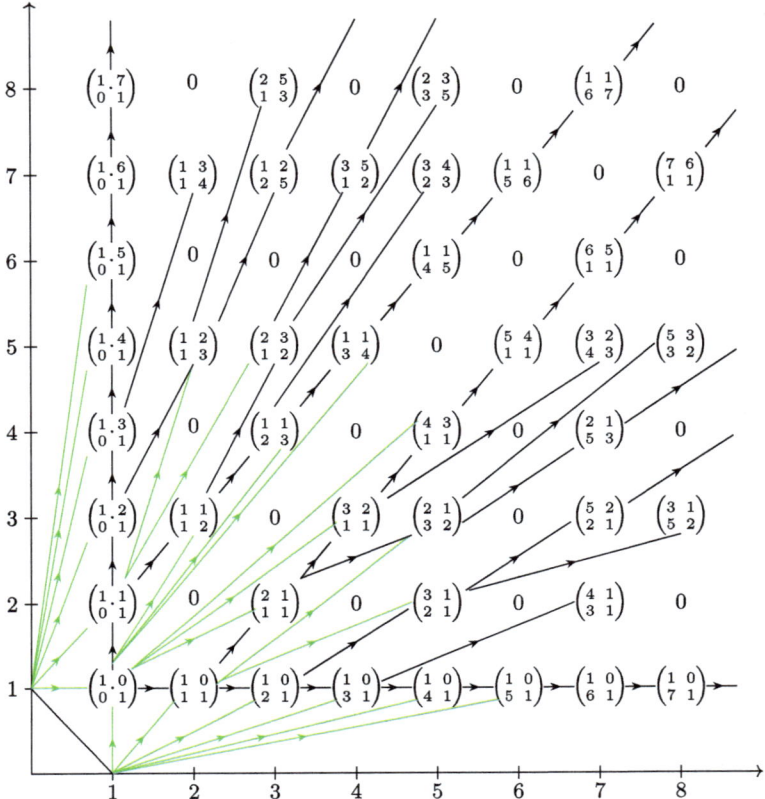

Fig. 1 The Stern-Brocot tree

Thus each $v \in \underline{\mathbb{Q}}^+ \setminus \{(1,1)\}$ has a unique <u>Mother</u> in $\underline{\mathbb{Q}}^+$ given by

$$M(v) = t_{-\delta(v)}(v) = \begin{cases} v_- & \text{if } \delta(v) = +1 \\ v_+ & \text{if } \delta(v) = -1 \end{cases} \tag{28}$$

$$M^{-1}(v) = \{v + v_+, v + v_-\}$$

We also define the <u>Father</u> of $v \in \underline{\mathbb{Q}}^+ \setminus \{(1,1)\}$ to be the element in $\underline{\mathbb{Q}}^+ \amalg \{\underline{0}, \infty\}$ defined by

$$F(v) = t_{\delta(v)}(v) = \begin{cases} v_+ & \text{if } \delta(v) = +1 \\ v_- & \text{if } \delta(v) = -1 \end{cases} \tag{29}$$

For all $v \in \underline{\mathbb{Q}}^+ \setminus \{(1,1)\}$, we have

$$\{F(v), M(v)\} \equiv \{v_+, v_-\} \tag{30}$$

and if $\delta(v) = +1$ we say that v is "convex":

$$F(v) = v_+, \quad M(v) = v_- \qquad (31)$$

and if $\delta(v) = -1$ we say v is "concave":

$$F(v) = v_-, \quad M(v) = v_+, \qquad (32)$$

(cf. Fig. 2).

For the root $v = \underline{1} = (1, 1)$, we have $g_v = \begin{pmatrix} 1 & 0 \\ 0 & 1 \end{pmatrix}$, $v_+ = \underline{\infty}$, $v_- = \underline{0}$, and we consider these two points, $\underline{\infty}$ and $\underline{0}$, to be both the mothers and fathers of $v = \underline{1}$. If $v = (x, y)$ has continued fraction expansion (24), then for $a_\ell > 1$ we have

$$M(v) = [\![a_0, \cdots, a_\ell - 1]\!] \qquad (33)$$
$$F(v) = [\![a_0, \cdots, a_{\ell-1}]\!],$$

for $a_\ell = 1$ and $\ell \geq 2$,

$$M(v) = [\![a_0, \cdots, a_{\ell-1}]\!] \qquad (34)$$
$$F(v) = [\![a_0, \cdots, a_{\ell-2}]\!]$$

if $\ell = 1$, $F(v) = a_0$ and if $\ell = 0$, $F(a_0) = \infty$.

Note that if $F(v) = v_\delta$, there is a unique $m = m(v) \geq 1$, such that

$$v_\delta = F(v) = M^{1+m}(v) \qquad (35)$$

i.e., every father is an "m-th great grandmother", and more precisely we have

$$v = \begin{cases} (1 + m)v_- + t_+(v_-) & \text{if } \delta(v) = -1 \\ (1 + m)v_+ + t_-(v_+) & \text{if } \delta(v) = +1 \end{cases} \qquad (36)$$

cf. Fig. 2.

Definition 1 For $v \in \mathbb{Q}^+$, the <u>fin f_v</u> around v consists of the offsprings $M^{-1}(v) = \{v + v_+, v + v_-\}$ as well as the the elements of $F^{-1}(v)$. Concretely,

$$f_v = M^{-1}(v) \amalg F^{-1}(v) \qquad (37)$$
$$= f_v^+ \amalg f_v^-,$$

with the positive and negative parts $f_v^\pm$ are defined by

$$f_v^+ := \{v_+ + nv \mid n \geq 1\}, \quad f_v^- := \{v_- + nv \mid n \geq 1\}. \qquad (38)$$

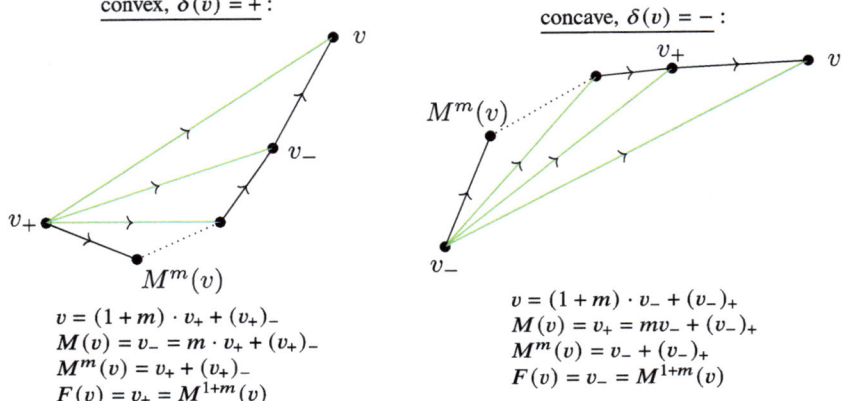

Fig. 2 Concavity and convexity of $\delta(v)$

We put (Fig. 3).

$$f_{\underline{\infty}} = f_{\underline{\infty}}^- = F^{-1}(\infty) = \{(1, n)\}_{n \geq 1}, \tag{39}$$
$$f_{\underline{0}} = f_{\underline{0}}^+ = F^{-1}(0) = \{(n, 1)\}_{n \geq 1}.$$

4 The Collection of ◁-Sets

The binary tree structure on $\mathbb{Q}^+$ gives a partial order ◁ on $\mathbb{Q}^+$,

$$v' \triangleleft v \quad \text{if and only if} \quad v' = M^n(v), \tag{40}$$

for some $n \geq 0$.

Definition 2 A ◁-set c is a finite subset of $\mathbb{Q}^+$, such that

$$v \in c, \quad v' \triangleleft v \Longrightarrow v' \in c$$

or equivalently, $M(c) \subseteq c$ and c is a <u>finite subtree</u> of $\mathbb{Q}^+$.

We let C denote the collection of all ◁-sets, it is a lattice:

$$c, c' \in C \Longrightarrow c \cap c', \quad c \cup c' \in C \tag{41}$$

$$C = \coprod_{m \geq 1} C_m, \quad C_m = \{c \in C \mid \#c = m - 1\} \tag{42}$$

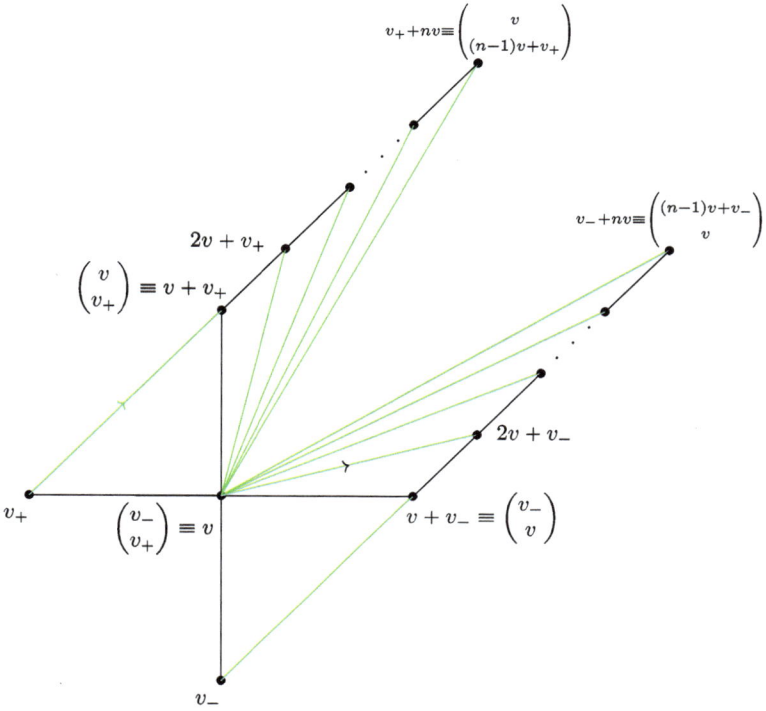

Fig. 3 Fin around v

For $v \in \mathbb{Q}^+$ we have the $\triangleleft$-set c_v consisting of the path from the root $\underline{1} = (1, 1)$ to v,

$$c_v = \{v' \in \mathbb{Q}^+ \mid v' \trianglelefteq v\} = \{M^n(v)\}_{n \geq 0}. \tag{43}$$

We define the operation $\wedge : \mathbb{Q}^+ \times \mathbb{Q}^+ \to \mathbb{Q}^+$ by

$$c_{v_1} \cap c_{v_2} = c_{v_1 \wedge v_2}. \tag{44}$$

It is immediate to verify that $\wedge$ is associative, commutative and

$$\begin{aligned} v_1 \triangleleft v_2 &\iff v_1 \wedge v_2 = v_1, \\ v_1 < v_2 &\iff v = v_1 \wedge v_2 \quad \text{satisfies} \quad v + v_+ \triangleleft v_2 \text{ or } v + v_- \triangleleft v_1. \end{aligned} \tag{45}$$

Every $\triangleleft$-set c is determined by its set of $\triangleleft$-maximal elements $c^{\max}$:

$$c = \bigcup_{v \in c^{\max}} c_v \tag{46}$$

For a ◁-set $c \in C_m$, we write its elements in increasing $\leq$ order

$$c = \{c_{m-1} > \cdots > c_2 > c_1\}, \quad m = 1 + \#c, \tag{47}$$

and we put

$$c_m = \underline{\infty}, \quad c_0 = \underline{0}$$
$$\partial c = \{[c_{i-1}, c_i]\}_{i=1}^m \tag{48}$$

The collection of edges $\partial c = \{[c_{i-1}, c_i]\}$ forms a <u>polygonal path</u> in the Farey graph going from $\underline{0}$ to $\underline{\infty}$. Conversely, every path in the Farey graph $\{[c_{i-1}, c_i]\}$, with $c_0 = \underline{0}, c_m = \underline{\infty}, \det\begin{pmatrix} c_{i-1} \\ c_i \end{pmatrix} = 1$, forms a ◁-set $c = \{c_i\}_{i=1}^{m-1} \in C_m$.

Thus: the finite subtrees of $\underline{\mathbb{Q}^+}$ are precisely the paths in $\mathcal{G}$ from $\underline{0}$ to $\underline{\infty}$.

5 Structure of ◁-Sets

For $c = \{c_i\} \in C$, we have

$$c^{\max} = \{c_i \mid c_{i+1} \triangleleft c_i, c_{i-1} \triangleleft c_i\}$$
$$= \{c_i \mid c_i = c_{i+1} + c_{i-1}\} \tag{49}$$
$$= \{v \in c \mid v + v_+ \notin c \text{ and } v + v_- \notin c\},$$

that is, $c^{\max}$ is the sets of leaves of the tree c.

We also define the <u>local ◁-minima</u>

$$\Phi c = \{c_j \mid c_j \triangleleft c_{j+1} \text{ and } c_j \triangleleft c_{j-1}\}$$
$$= \{v \in c \mid v + v_+ \in c \text{ and } v + v_- \in c\} \tag{50}$$

and we put

$$c^{\min} = \Phi c \amalg \{\underline{0}, \underline{\infty}\}. \tag{51}$$

These sets are intertwined:

$$c^{\max} = \{c_{i_\ell} > \cdots > c_{i_1} > c_{i_0}\}, \tag{52}$$
$$c^{\min} = \{\underline{\infty} > c_{j_\ell} > \cdots > c_{j_1} > \underline{0}\}$$

with

$$m > i_\ell > j_\ell > \cdots > i_1 > j_1 > i_0 > 0.$$

There are still more perspectives on ◁-sets.

Definition 3 For $c \in C_m$ we have the triangulated polygon

$$\Delta(c) := \bigcup_{v \in c} \Delta_v$$

and the associated Friez pattern

$$f(c) := \{f(c)_0, f(c)_1, \cdots, f(c)_m\},$$

with

$$f(c)_j = \#\{v \in c \mid c_j \in \Delta_v\}.$$

Clearly, the Friez pattern $f(c)$ determines c and we easily verify that

$$c^{\max} = \{c_j \mid f(c)_j = 1\}.$$

We also have the associated bipartite graph $\mathbb{B}(c)$ given by

$$\mathbb{B}(c) := \{(v', v) \mid v' \in \Delta_v, v \in c\},$$

with natural projections

$$c \amalg \{0, \infty\} \xleftarrow{\pi_0} \mathbb{B}(c) \xrightarrow{\pi_1} c$$

satisfying

$$\#\pi_0^{-1}(c_j) = f(c)_j, \qquad j = 0, \cdots, m,$$
$$\#\pi_1^{-1}(c_j) \equiv 3, \qquad j = 1, \cdots, m-1.$$

It is also useful to consider Friez indices associated with a Friez pattern $f(c)$.

Definition 4 For $c \in C$, with $c_i = v \in c$, we define the Friez indices $n(c)_i^\pm \geq 0$ as the integers such that

$$c_{i-1} \in f_v^- \amalg \{v_-\} = \{v_- + n \cdot v \mid n \geq 0\}, \quad c_{i-1} = v_- + n(c)_i^- \cdot v,$$
$$c_{i+1} \in f_v^+ \amalg \{v_+\} = \{v_+ + n \cdot v \mid n \geq 0\}, \quad c_{i+1} = v_+ + n(c)_i^+ \cdot v,$$

then

$$f(c)_i = 1 + n(c)_i^+ + n(c)_i^-, \quad i = 1, \cdots, m-1.$$

With the foregoing notation, we see that

$$c_i \in c^{\max} \iff n(c)_i^- = n(c)_i^+ = 0$$

and

$$c_i \in \Phi c \iff n(c)_i^- > 0 \text{ and } n(c)_i^+ > 0.$$

We also note that if $n = n(c)_i^- > 0$, then there exists

$$i_0 < i_1 < \cdots < i_n \equiv i - 1, \tag{53}$$

with $c_{i_k} = v_- + k \cdot v$ for $k = 0, \cdots n$.

Similarly, if $m = n(c)_i^+ > 0$, then there exists

$$j_0 > j_1 > \cdots > j_m \equiv i + 1, \tag{54}$$

with $c_{j_k} = v_+ + k \cdot v$ for $k = 0, \cdots, m$.

Remark 1 The monoid $SL_2(\mathbb{N})$ has two commuting <u>involutions</u>. One is the automorphism (outer in $SL_2(\mathbb{Z})$, inner in $GL_2(\mathbb{Z})$),

$$g = \begin{pmatrix} x_- & y_- \\ x_+ & y_+ \end{pmatrix} \mapsto g^\star := \begin{pmatrix} 0 & 1 \\ 1 & 0 \end{pmatrix} g \begin{pmatrix} 0 & 1 \\ 1 & 0 \end{pmatrix} = \begin{pmatrix} y_+ & x_+ \\ y_- & x_- \end{pmatrix}$$

$$(g_\pm)^\star = g_\mp \quad , \quad (g_1 \cdot g_2)^\star = g_1^\star \cdot g_2^\star \quad , \quad g^{\star\star} = g.$$

The other is the <u>anti</u>-automorphism

$$g = \begin{pmatrix} x_- & y_- \\ x_+ & y_+ \end{pmatrix} \mapsto g^t := \begin{pmatrix} x_- & x_+ \\ y_- & y_+ \end{pmatrix}$$

$$(g_\pm)^t = g_\mp \quad , \quad (g_1 \cdot g_2)^t = g_2^t \cdot g_1^t \quad , \quad g^{tt} = g.$$

In terms of the identification $SL_2(\mathbb{N}) \equiv \mathbb{Q}^+$ these read:

$$v^\star = (y/x)^\star = \left(\frac{y_+ + y_-}{x_+ + x_-}\right)^\star = \frac{x_- + x_+}{y_- + y_+} = x/y = v^{-1}$$

$$v^t = \left(\frac{y_+ + y_-}{x_+ + x_-}\right)^t = \frac{y_+ + x_+}{y_- + x_-} = \frac{|v_+|_1}{|v_-|_1}.$$

5.1 Creation and Annihilation Operators

For $c = \{c_i\}_{i=1}^{m-1} \in C_m$, we have $c \cup \{c_i + c_{i-1}\} \in C_{m+1}, i = 1, \ldots, m$, and we define the creation operator

$$d^\star : \mathbb{Z}C_m \longrightarrow \mathbb{Z}C_{m+1}$$

$$d^\star[c] := \sum_{i=1}^{m} \left[c \cup \{c_i + c_{i-1}\} \right]. \tag{55}$$

Similarly, for $c_j \in c^{\max}$, we have $c \backslash \{c_j\} \in C_{m-1}$, and we obtain the annihilation operator

$$d : \mathbb{Z}C_m \longrightarrow \mathbb{Z}C_{m-1}$$

$$d[c] := \sum_{c_j \in c^{\max}} \left[c \backslash \{c_j\} \right] \tag{56}$$

Since the operation of adding a mediant, and of removing a (different) maximal point commute, we see that the Number operator

$$N = d \circ d^\star - d^\star \circ d : \mathbb{Z}C_m \longrightarrow \mathbb{Z}C_m \tag{57}$$

is diagonalizable in the basis of $\triangleleft$-sets, and we have

$$N[c] = (m - \#c^{\max}) \cdot [c], \quad c \in C_m. \tag{58}$$

5.2 The Operad Structure

Let $c = \{c_i\}_{i=1}^{m-1} \in C_m$ and $b^{(i)} = \left\{ b_j^{(i)} \right\}_{j=1}^{n_i - 1} \in C_{n_i}, i = 1, \cdots, m$. Here, as usual, we set $c_m = b_{n_i}^{(i)} = \infty, c_0 = b_0^{(i)} = 0$.

Recall that

$$\begin{pmatrix} c_{i-1} \\ c_i \end{pmatrix} \in \text{SL}_2(\mathbb{N}),$$

thus, we may define paths

$$I_i = \left\{ b_j^{(i)} \begin{pmatrix} c_{i-1} \\ c_i \end{pmatrix} \right\}_{j=0}^{n_i}$$

from $\underline{0} \begin{pmatrix} c_{i-1} \\ c_i \end{pmatrix} = c_{i-1}$ to $\underline{\infty} \begin{pmatrix} c_{i-1} \\ c_i \end{pmatrix} = c_i$, for $i = 1, \cdots, m$. The union of paths $\bigcup_i I_i$ gives a new path from $\underline{0}$ to $\underline{\infty}$ in $\mathcal{G}$, denoted by $c \circ b$.

Proposition 1 *The set* $C = \coprod_{m \geq 1} C_m$ *is an* <u>operad</u> *via*

$$C_m \times C_{n_1} \times \cdots \times C_{n_m} \longrightarrow C_{n_1 + \cdots + n_m}$$

$$c \, , \, b^{(1)} \, , \, \ldots \, , \, b^{(m)} \longmapsto c \circ b$$

$$c \circ b := \left\{ b_j^{(i)} \begin{pmatrix} c_{i-1} \\ c_i \end{pmatrix} \right\} \quad 1 \leq i \leq m, \ 1 \leq j \leq n_i$$

$$(c \circ b) \circ a = c \circ (b \circ a) \, , \quad \{\phi\} \circ c = c = c \circ \{\phi\}^m.$$

The unit is the empty $\triangleleft$-set $\emptyset$, $C_1 = \{\emptyset\}$, $\partial \emptyset = \{[0, \infty]\}$.
The root $\underline{1} = (1, 1)$ satisfies $\partial \{\underline{1}\} = \{[0, 1], [\underline{1}, \infty]\}$, $\Phi \{\underline{1}\} = \emptyset$, and we put $\Phi(\emptyset) = \emptyset$.

6 Coronas

We shall identify the integers $\mathbb{Z}$ with $\triangleleft$-set via

$$\mathbb{Z} \ni m \longleftrightarrow \nu_m = \begin{cases} \{(1,1), (1,2), \cdots, (1, m+1)\}, \, m \geq 0 \\ \{(1,1)\} \quad m = 0 \\ \{(1,1), (2,1), \cdots, (|m|+1, 1)\}, \, m \leq 0 \end{cases} \in C_{|m|+2} \quad (59)$$

Note that the $\triangleleft$-set ν_m has a unique $\triangleleft$-maximal point, and the path $\partial \nu_m$ consists of a straight line from $\underline{0}$ to the maximal point, followed by a straight line from the maximal point to $\underline{\infty}$; this properly characterizes the $\triangleleft$-sets $\{\nu_m\}$, $m \in \mathbb{Z}$. Thus for a $\triangleleft$-set $c \in C_m$, and vector $\nu = \sum_{i=1}^m n_i \cdot [c_{i-1}, c_i] \in \mathbb{Z}\partial c$, we obtain the $\triangleleft$-set

$$c \circ \nu := c \circ \{\nu_{n_i}\} \in C_{2m+|\nu|} \, , \quad |\nu| = \sum_{i=1}^m |n_i|. \quad (60)$$

Explicitly, the $\triangleleft$-set $c \circ \nu$ is obtained by replacing $\{c_i > c_{i-1}\}$ in c by

$$\{c_i > (n+1) \cdot c_i + c_{i-1} > \cdots > c_i + c_{i-1} > c_{i-1}\}, \quad \text{if } n_i = n \geq 0,$$

$$\{c_i > c_i + c_{i-1} > c_{i-1}\}, \quad \text{if } n_i = 0, \quad (61)$$

$$\{c_i > c_i + c_{i-1} > \cdots > c_i + (|n|+1) \cdot c_{i-1} > c_{i-1}\}, \quad \text{if } n_i = n \leq 0.$$

We have
$$\Phi(c \circ \nu) = c. \tag{62}$$

Definition 5 A $\star$-set $c \in C_m^\star$ is a $\triangleleft$-set $c \in C_m$ such that for any consecutive edges of ∂c, $c_{i+1} > c_i > c_{i-1}$, one of the conditions

- $c_i \in c^{\max}$, that is, $c_i = c_{i+1} + c_{i-1}$,
- $c_i \in c^{\min}$, that is, $c_i \triangleleft c_{i+1}$ and $c_i \triangleleft c_{i-1}$,
- they form a straight line, $c_i - c_{i-1} = c_{i+1} - c_i$, or equivalently, $c_i = \frac{1}{2}(c_{i+1} + c_{i-1})$,

holds.

Clearly, the $\triangleleft$-set $c \circ \nu$ is a $\star$-set according to the definition. We write $C^\star = \bigsqcup_{m \geq 1} C_m^\star$ for the collection of $\star$-sets.

Remark 2 For $c \in C^\star$, and for $c_{i'} > c_i$ two consecutive points of $c^{\min}$, let c_j be the unique point of $c^{\max}$ between them, $i' > j > i$, we have either a positive or a negative fin between them: there exists $m > n \geq 0$ such that either

$$(f_{c_i}^+) : c \cap [c_i, c_{i'}] \equiv \begin{cases} c_{i'} \equiv (c_i)_+ + n \cdot c_i > (c_i)_+ + (n+1)c_i > \cdots \\ \cdots > c_j = c_{i+1} = (c_i)_+ + mc_i > c_i \end{cases} \tag{63}$$

(note that $c_{i'} \triangleleft c_i$; unless $n = 0$ and $c_i \triangleleft c_{i'}$); put $\lambda_c([c_i, c_{i'}]) = n - m + 1$; or

$$(f_{c_{i'}}^-) : c \cap [c_i, c_{i'}] \equiv \begin{cases} c_{i'} > c_j = c_{i'-1} = (c_{i'})_- + mc_{i'} > \cdots \\ \cdots > (c_{i'})_- + nc_{i'} = c_i \end{cases}. \tag{64}$$

(note that $c_i \triangleleft c_{i'}$, unless $n = 0$ and $c_{i'} \triangleleft c_i$); put $\lambda_c([c_i, c_{i'}]) = m - n - 1$. Thus in any case $[c_i, c_{i'}]$ is an edge of the Farey graph and $\partial c^{\min}$ is again a path from $\underline{0}$ to ∞ in $\mathcal{G}$, and $\Phi c \in C$ is again a $\triangleleft$-set.

Thus any $\star$-set c can be written uniquely as
$$c = \Phi c \circ \lambda_c. \tag{65}$$

We obtain the fibration
$$\Phi : C^\star \twoheadrightarrow C$$
$$c \mapsto \Phi c = c^{\min} \setminus \{\underline{0}, \infty\} \tag{66}$$
$$\mathbb{Z}\partial \overline{c} \equiv \Phi^{-1}(\overline{c})$$

We get the pull-back diagram

$$\begin{array}{ccc}
C^\star & \xrightarrow{\Phi} & C \\
\cup| & & \cup| \\
\Phi^{-1}(C^\star) & \xrightarrow{\Phi} C^\star \xrightarrow{\Phi} & C \\
\cup| & \cup| & \cup| \\
\Phi^{-2}(C^\star) \xrightarrow{\Phi} & \Phi^{-1}(C^\star) \xrightarrow{\Phi} C^\star \xrightarrow{\Phi} & C \\
\vdots & &
\end{array} \qquad (67)$$

Definition 6 The set of <u>Coronas</u> is

$$\mathrm{Cor} := \bigcap_{n \geq 0} \Phi^{-n}(C^\star) = \coprod_{m \geq 1} \mathrm{Cor}_m ,$$

$$\mathrm{Cor}_m = \{c \in \mathrm{Cor}, \#c = m - 1\}.$$

We get the fibration

$$\Phi : \mathrm{Cor} \longrightarrow \mathrm{Cor}$$

$$c \longmapsto \Phi c \qquad (68)$$

$$\mathbb{Z}\partial \overline{c} \equiv \Phi^{-1}(\overline{c})$$

Given $c \in \mathrm{Cor}_m$ and any vector $v = \sum_{i=1}^{m} n_i [c_{i-1}, c_i] \in \mathbb{Z}\partial c$, we get $c \circ v \in \mathrm{Cor}_{2m+|v|}$, $|v| = \sum_{i=1}^{m} |n_i|$, and $\Phi(c \circ v) = c$.

Conversely, any $c \in \mathrm{Cor}_m$ can be written uniquely as

$$c = \Phi(c) \circ \lambda_c , \qquad (69)$$

with $\Phi(c) \in \mathrm{Cor}_{\frac{1}{2}(m-|\lambda_c|)}$ and $\lambda_c \in \mathbb{Z}\partial \Phi c$.

7 Structure of Coronas

We give next a constructive approach to coronas based on the <u>Inductive Principle</u>.

Proposition 2 (Inductive principle) *For $m > 1$ and $c \in \mathrm{Cor}_m$, then there exists $c_j \in c^{\max}$ such that*

$$c \setminus \{c_j\} \in \mathrm{Cor}_{m-1} .$$

Proof Writing $c = \Phi c \circ \lambda_c$ with $\lambda_c = \sum n_i \left[\bar{c}_{i-1}, \bar{c}_i\right]$ and $\Phi c = \bar{c} = \{\bar{c}_i\}$, if $n_{i_0} \neq 0$ we can take c_j to be the $\lhd$-maximal element in $c \cap \left[\bar{c}_{i_0-1}, \bar{c}_{i_0}\right]$, then $\Phi(c \setminus \{c_j\}) \equiv \Phi c$.

Otherwise, $\lambda_c \equiv 0$, $\bar{c} = \Phi c \in \text{Cor}_{\frac{m}{2}}$, by induction there is $\bar{c}_i \in \bar{c}^{\max}$ such that $\bar{c} \setminus \{\bar{c}_i\} \in \text{Cor}_{\frac{m}{2}-1}$, and we take c_j to be either $\bar{c}_{i+1} + \bar{c}_i$ or $\bar{c}_i + \bar{c}_{i-1}$. □

Thus, every $c \in \text{Cor}_m$ is obtained from the empty corona $\phi \in \text{Cor}_1$ by adding one point at a time, and the set $\text{Cor} = \bigsqcup_{m \geq 1} \text{Cor}_m$ forms the vertices of a connected rooted graph with edges

$$\text{Cor}^1 \equiv \left\{(c, c') \in \text{Cor} \times \text{Cor}, c \subseteq c', \#c' = \#c + 1\right\} \tag{70}$$

as shown in Figs. 4 and 5.

7.1 Creation and Annihilation Operators

For $c \in \text{Cor}_m$, define the set of <u>closed points</u> of c, $\text{Cl}(c) \subseteq c^{\max}$, by declaring all $c_j \in c^{\max}$ to be closed except when $c_{j+1}, c_{j-1} \in c^{\min}$, and either

- $c_{j+1} \lhd c_{j-1} \lhd c_{j-2}$ and $c_{j-1} \neq \frac{1}{2}(c_{j+1} + c_{j-2})$, or
- $c_{j-1} \lhd c_{j+1} \lhd c_{j+2}$ and $c_{j+1} \neq \frac{1}{2}(c_{j+2} + c_{j-1})$.

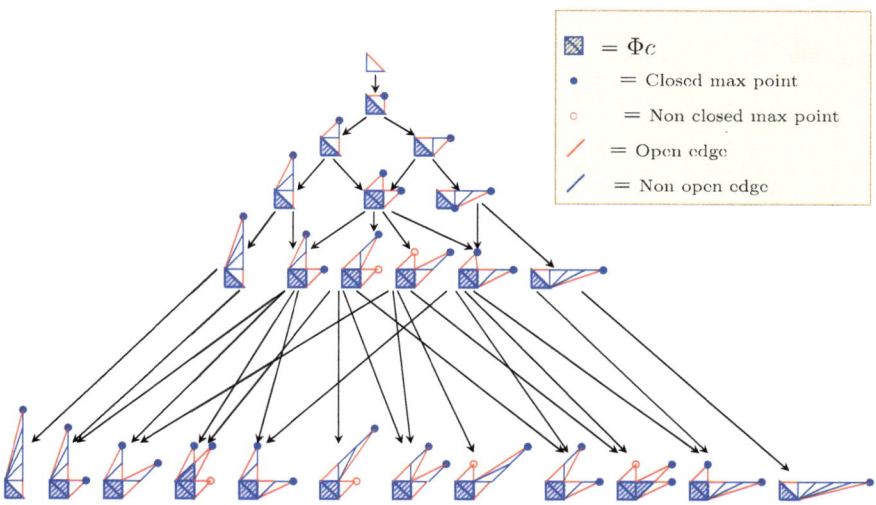

Fig. 4 First six levels of Coronas

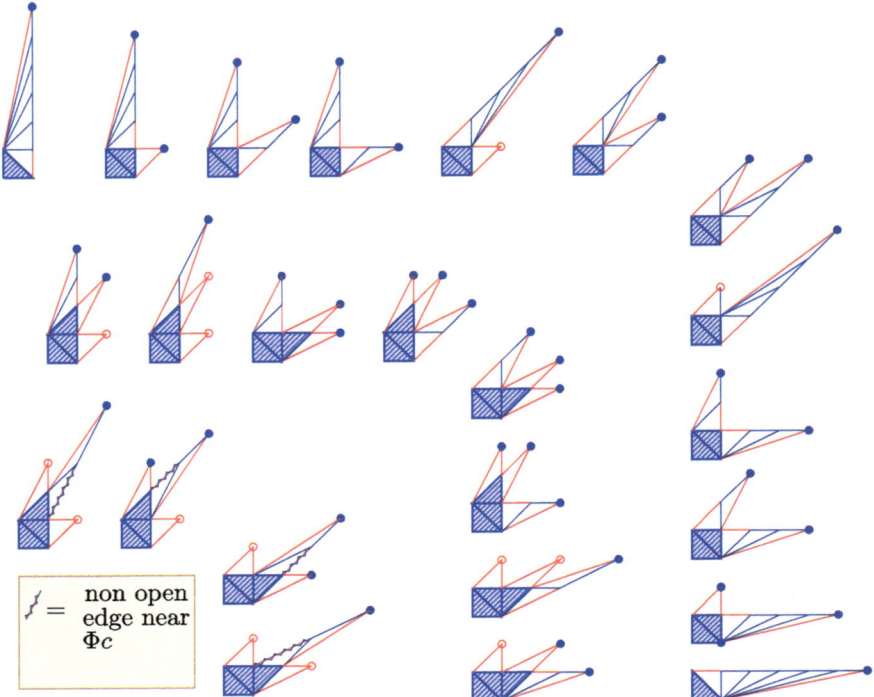

Fig. 5 Level seven of the corona tree

and, thus
$$\mathrm{Cl}(c) = \{c_j \in c^{\max} \mid c \setminus \{c_j\} \in \mathrm{Cor}_{m-1}\}$$
$$\#\mathrm{Cl}(c) = 1 + \#\Phi c - h^\circ(c), \quad 0 \le h^\circ(c) \le \#\Phi c. \tag{71}$$

We define the <u>annihilation operator</u>
$$d : \mathbb{Z}\,\mathrm{Cor}_m \longrightarrow \mathbb{Z}\,\mathrm{Cor}_{m-1}$$
$$d\,[c] = \sum_{c_j \in \mathrm{Cl}(c)} [c \setminus \{c_j\}] \tag{72}$$

Similarly, for $c \in \mathrm{Cor}_m$ define the set of <u>open edges</u> of c
$$\mathrm{Op}(c) = \{[c_{i-1}, c_i] \in \partial c \mid c \cup \{c_i + c_{i-1}\} \in \mathrm{Cor}_{m+1}\}$$
$$\#\mathrm{Op}(c) = 2 + 2 \cdot \#\Phi c - h^1(c), \quad 0 \le h^1(c) \le \#\Phi c. \tag{73}$$

Indeed, for consecutive points $c_{i'} > c_i$ of $c^{\min}$ with $c_j \in c^{\max}$ the $\triangleleft$-maximal point between them (cf. Eq. (64)), the edge $[c_i, c_j] \equiv [c_i, c_{i+1}]$ (resp. $[c_j, c_{i'}] \equiv [c_{i'-1}, c_{i'}]$) is always open in case of $f^+_{c_i}$ (resp. $f^-_{c_{i'}}$). The only other possibly open edge in $c \cap [c_i, c_{i'}]$ is the edge $[c_{i'-1}, c_{i'}]$ (resp. $[c_i, c_{i+1}]$), and this edge is open if and only if $[c_i, c_{i'}]$ is open in Φc.

We define the <u>creation operator</u>

$$d^\star : \mathbb{Z}\,\mathrm{Cor}_m \longrightarrow \mathbb{Z}\,\mathrm{Cor}_{m+1}$$

$$d^\star[c] = \sum_{[c_{i-1}, c_i] \in \mathrm{Op}(c)} \left[c \cup \{c_i + c_{i-1}\}\right] \tag{74}$$

Since the operations of adding a mediant and that of removing a (different) maximal point commute, we see that the <u>Number operator</u>

$$N = d \circ d^\star - d^\star \circ d : \mathbb{Z}\,\mathrm{Cor}_m \longrightarrow \mathbb{Z}\,\mathrm{Cor}_m \tag{75}$$

is diagonalizable in the basis of coronas with eigenvalues

$$N[c] = e_c \cdot [c], \tag{76}$$

given by

$$e_c = \#\mathrm{Op}(c) - \#\mathrm{Cl}(c) = 1 + \#\Phi c + h^0(c) - h^1(c), \tag{77}$$

with $1 \le e_c \le 1 + 2 \cdot \#\Phi c$. We remark that when $h^0(c) = h^1(c)$, we have

$$e_c = 1 + \#\Phi c = \#c^{\max}.$$

7.2 The D.N.A. of a Corona

For $c \in \mathrm{Cor}_m$ define its <u>height</u> to be

$$\mathrm{ht}(c) = \min\{n \mid \Phi^n c = \phi\}. \tag{78}$$

For $n = 1, 2, \cdots, \mathrm{ht}(c)$ we have $\Phi^n c \in \mathrm{Cor}_{\ell_n}$, and

$$\Phi^{n-1} c = \Phi^n c \circ \lambda^n_c, \quad \lambda^n_c \in \mathbb{Z}\partial \Phi^n c \cong \mathbb{Z}^{\ell_n}. \tag{79}$$

We therefore have

$$c = \phi \circ \lambda^{\mathrm{ht}(c)}_c \circ \cdots \lambda^n_c \circ \cdots \circ \lambda^1_c, \tag{80}$$

and thus c is determined by the $\mathbb{Z}$-valued vectors

$$\lambda_c^n = \left(\lambda_1^n, \lambda_2^n, \ldots, \lambda_{\ell_n}^n\right) \in \mathbb{Z}^{\ell_n}, \quad \lambda_j^n \in \mathbb{Z}, \tag{81}$$

of length $\ell_n = 2 \cdot \ell_{n+1} + |\lambda_c^{n+1}|$, $|\lambda_c^{n+1}| = \sum_j |\lambda_j^{n+1}|$ and $\ell_{\text{ht}(c)} = 1$.

We refer to the set of vectors $\lambda_c^{\text{ht}(c)} \in \mathbb{Z}, \ldots, \lambda_c^n \in \mathbb{Z}^{\ell_n}, \ldots, \lambda_c^1 \in \mathbb{Z}^{\ell_1}$ as the <u>d.n.a. of c</u>.
We have

$$\begin{aligned} m = \#c + 1 &= 2^{\text{ht}(c)} + 2^{\text{ht}(c)-1} \cdot |\lambda_c^{\text{ht}(c)}| + \cdots 2^n \cdot |\lambda_c^{n+1}| + \cdots |\lambda_c^1| \\ &= 2 \cdot \ell_1 + |\lambda_c^1| =: \ell_0. \end{aligned} \tag{82}$$

8 The Main Examples

Besides the real total order $\leq$ and the tree partial order $\lhd$ we shall use the following two partial orders on $\mathbb{Q}^+$. We have the <u>pointwise order</u> $\prec$:

$$\begin{aligned} v \prec v' &\iff v' - v \in \mathbb{N} \times \mathbb{N} \\ &\text{or} \\ (x, y) \prec (x', y') &\iff x \leq x' \text{ and } y \leq y', \end{aligned} \tag{83}$$

and the <u>fundamental order</u> $\ll$:

$$v \ll v' \iff v_+ \prec v'_+ \quad \text{and} \quad v_- \prec v'_- \tag{84}$$

Note that we have the strict implications

$$v \lhd v' \implies v \ll v' \implies v \prec v'. \tag{85}$$

Let $|\ | : \mathbb{Q}^+ \to \mathbb{R}$ be any map that is <u>fundamentally monotone</u>, that is,

$$v \ll v' \implies |v| \leq |v'|, \tag{86}$$

then, for any $R \in \mathbb{R}$ put

$$c\left(|\_| \leq R\right) := \left\{v \in \mathbb{Q}^+ \mid |v| \leq R\right\} \tag{87}$$

Theorem 1 *Assuming $c\left(|\_| \leq R\right)$ is finite, then it is a corona.*

Note that $v \lhd v'$ implies $|v| \leq |v'|$, (85) and (86), so that $c(|\_| \leq R)$ is a $\lhd$-set.

To prove the theorem we make use of a lemma.

Meditations on the Farey Fractal

Lemma 1 $c\left(\left|\_\right| \leq R\right) \in C$ is a $\star$-set.

Proof Write $c = c(|\_| \leq R) = \{c_i\}_{i=1}^{m-1}$ and let $c_i = v \in c \setminus \left(c^{\max} \sqcup c^{\min}\right)$.
We have, by Definition 4, that either

- negative part f_v^- of fin: $n(c)_i^+ = 0$ and $n(c)_i^- = m > 0$, therefore
  $c_{i+1} = (c_i)_+ = v_+$ and $c_{i-1} = v_- + m \cdot v$, or
- positive part f_v^+ of fin: $n(c)_i^+ = m > 0$ and $n(c)_i^- = 0$, therefore
  $c_{i+1} = v_+ + m \cdot v$ and $c_{i-1} = (c_i)_- = v_-$.

If $m = 1$ then $c_{i+1} - c_i = c_i - c_{i-1}$ and $\{c_{i-1} < c_i < c_{i+1}\}$ are on a straight line, so assume $m > 1$. Then, for the case f_v^- we have

$$|c_{i+1} + c_i| = |v_+ + v| = \left|\begin{pmatrix} v \\ v_+ \end{pmatrix}\right| \leq \left|\begin{pmatrix} (m-1)v + v_- \\ v \end{pmatrix}\right|$$
$$= |m \cdot v + v_-| = |c_{i-1}| \leq R$$

but this is a contradiction since $c_{i+1} + c_i \notin c$. Similarly, for the case of f_v^+, we have

$$|c_i + c_{i-1}| = |v + v_-| = \left|\begin{pmatrix} v_- \\ v \end{pmatrix}\right| \leq \left|\begin{pmatrix} v \\ (m-1)v + v_+ \end{pmatrix}\right|$$
$$= |mv + v_+| = |c_{i+1}| \leq R,$$

a contradiction since $c_i + c_{i-1} \notin c$. □

Note that

$$v \in \Phi c(|\_| \leq R) \iff v + v_+, \ v + v_- \in c(|\_| \leq R) \tag{88}$$
$$\iff |v|_1 := \sup\{|v + v_+|, |v + v_-|\} \leq R$$

The map $|\_|_1$ is again fundamentally monotone so that by the lemma, $\Phi c(|\_| \leq R) = c(|\_|_1 \leq R) \in C^\star$ is again a $\star$-set.
Denoting the <u>Fibonacci numbers</u> by

$$a_1 = a_2 = 1, \quad a_n = a_{n-1} + a_{n-2} = \frac{1}{\sqrt{5}}\left[\left(\frac{1+\sqrt{5}}{2}\right)^n - \left(\frac{1-\sqrt{5}}{2}\right)^n\right] \tag{89}$$

for $n \geq 0$, we define

$$|v|_n := \sup\{|a_{n+2}v_+ + a_{n+1}v_-|, |a_{n+1}v_+ + a_{n+2}v_-|\} \equiv |a_{n+2}M(v) + a_{n+1}F(v)|. \tag{90}$$

Then $|\_|_n$ is again fundamentally monotone, and from the lemma we may deduce inductively that

$$\Phi^n c\left(|\_| \leq R\right) \equiv c\left(|\_|_n \leq R\right) \in C^\star \tag{91}$$

and thus $c\left(|\_| \leq R\right)$ is indeed a corona. We get (91) by induction that $v \in \Phi^n c(|\_| \leq R)$ holds if and only if $v + v_+, v + v_- \in \Phi^{n-1}c\left(|\_| \leq R\right)$.

Equivalently, if

$$R \geq \sup\{|v + v_+|_{n-1}, |v + v_-|_{n-1}\}$$
$$= \sup\{|a_n v_+ + a_{n+1} v|, |a_{n+1} v + a_n v_-|\}$$
$$= \sup\begin{cases} |(a_{n+1} + a_n)v_+ + a_{n+1} v_-|, \\ |a_{n+1} v_+ + (a_{n+1} + a_n)v_-| \end{cases}$$
$$= \sup\{|a_{n+2} v_+ + a_{n+1} v_-|, |a_{n+1} v_+ + a_{n+2} v_-|\}$$
$$= |v|_n$$

This complete the proof of Theorem 1.

Note that if $|\_|$ is pointwise-monotone (i.e., $v \prec v'$ implies $|v| \leq |v'|$), hence a-posteriori fundamentally monotone, the norms $|\_|_n$, $n \geq 1$, need not be pointwise-monotone.

Let $|\_|$ be pointwise-monotone map $|\_| : \mathbb{N} \times \mathbb{N} \to \mathbb{R}$ that is also homogeneous, that is, $|a \cdot v| = a \cdot |v|$ for $a \in \mathbb{N}^+$, then we have

$$|v|_n = \sup\{|a_{n+2} v_+ + a_{n+1} v_-|, |a_{n+1} v_+ + a_{n+2} v_-|\} \leq |a_{n+2} \cdot v| = a_{n+2} \bullet |v| \quad (92)$$

Moreover, if $|\_|$ comes from a norm, i.e., satisfies the triangle inequality, we have for $v \in \Phi^n c(|\_| \leq R)$

$$a_{n+3} \bullet |v| = |a_{n+3} \cdot v| \leq |a_{n+2} v_+ + a_{n+1} v_-| + |a_{n+1} v_+ + a_{n+2} v_-| \leq 2 \cdot R \quad (93)$$

Together we have the exponential decay of $\Phi^n c(|\_| \leq R)$. For a norm $|\_|$:

$$c\left(|\_| \leq \frac{1}{a_{n+2}} \cdot R\right) \subseteq \Phi^n c\left(|\_| \leq R\right) \subseteq c\left(|\_| \leq \frac{2}{a_{n+3}} \cdot R\right) \quad (94)$$

Examples of such coronas are given by, $p \geq 1$, $R \geq 2$,

$$c_R^{(p)} := c\left(x^p + y^p \leq R^p\right), \quad c_R^{(\infty)} := c\left(\max\{x, y\} \leq R\right) \quad (95)$$

We have

$$c_R^{(1)} \subseteq c_R^{(2)} \subseteq c_R^{(\infty)} \subseteq c_{2R}^{(1)} \subseteq c_{2R}^{(2)} \subseteq c_{2R}^{(\infty)} \subseteq c_{4R}^{(1)} \subseteq \cdots \quad (96)$$

For a positive real matrix $A = \begin{pmatrix} a_{11} & a_{12} \\ a_{21} & a_{22} \end{pmatrix}$ with $a_{ij} > 0$, we define the norm $|\_|_A$ by

Meditations on the Farey Fractal

$$|g|_A = \left|\begin{pmatrix} x_- & y_- \\ x_+ & y_+ \end{pmatrix}\right|_A := \operatorname{tr}(g \cdot A^t) = x_- \cdot a_{11} + y_- \cdot a_{12} + x_+ \cdot a_{21} + y_+ \cdot a_{22}. \tag{97}$$

In particular taking $A = \begin{pmatrix} \alpha & \beta \\ \alpha & \beta \end{pmatrix}$ with $\alpha, \beta > 0$, we have the <u>linear-norms</u>

$$|(x, y)|_{(\alpha,\beta)} := \alpha \cdot x + \beta \cdot y$$

$$|v|_{(\alpha,\beta)} = |v_+ + v_-|_{(\alpha,\beta)} = |v_+|_{(\alpha,\beta)} + |v_-|_{(\alpha,\beta)} \tag{98}$$

$$|vg|_{(\alpha,\beta)} = \operatorname{tr}\left(g_v \cdot g \cdot \begin{pmatrix} \alpha & \beta \\ \alpha & \beta \end{pmatrix}^t\right) = \operatorname{tr}\left(g_v \cdot \left(\begin{pmatrix} \alpha & \beta \\ \alpha & \beta \end{pmatrix} \cdot g^t\right)^t\right) = |v|_{(\alpha,\beta)g^t}$$

Thus we have the coronas

$$c_R^{(\alpha,\beta)} := \{(x, y) \in \mathbb{Q}^+, \alpha \cdot x + \beta \cdot y \leq R\}. \tag{99}$$

We have for $g \in \operatorname{SL}_2(\mathbb{N})$, with $\underline{0}g, \underline{\infty}g \in c_R^{(\alpha,\beta)}$,

$$c_R^{(\alpha,\beta)} \cap (\underline{0}g, \underline{\infty}g) = \left(c_R^{(\alpha,\beta)g^t}\right)g \tag{100}$$

9 The D.N.A of $c_R^{(\alpha,\beta)}$

Fix a linear-norm $|\_| = |\_|_{(\alpha,\beta)}$ with $\alpha, \beta > 0$, $R \in \mathbb{R}$ and $n \geq 1$, and define $c^{n-1}, c^n \in C$ by

$$c^{n-1} := \Phi^{n-1} c_R^{(\alpha,\beta)} = \left\{v \in \mathbb{Q}^+ \;\middle|\; a_{n+1}|v_+| + a_n|v_-|, a_n|v_+| + a_{n+1}|v_-| \leq R\right\},$$

$$c^n := \Phi^n c_R^{(\alpha,\beta)} = \left\{v \in \mathbb{Q}^+ \;\middle|\; a_{n+2}|v_+| + a_{n+1}|v_-|, a_{n+1}|v_+| + a_{n+2}|v_-| \leq R\right\},$$

and $\lambda^n \in \mathbb{Z}^m$ such that $c^{n-1} = c^n \circ \lambda^n$, that is,

$$\lambda^n = \sum_{i=1}^{m} \lambda_i^n [c_{i-1}, c_i] \in \mathbb{Z}\partial c^n, \quad c^n = \{c_i\}_{i=1}^{m-1}.$$

Let $v \in (c^{n-1})^{\max}$, and let $c_i, c_{i-1} \in (c^{n-1})^{\min}$ be the points immediately above and below it, that is, $c_i > v > c_{i-1}$. Then, via (64) (cf. Remark 2), we see that there exists $k_1 > k_0 \geq 0$ such that either, we have the $\left(f_{c_{i-1}}^+\right)$ case, that is,

$$c^{n-1} \cap [c_{i-1}, c_i] = \left\{ \begin{array}{l} c_i = (c_{i-1})_+ + k_0 c_{i-1} > \cdots \\ \\ \cdots > v = (c_{i-1})_+ + k_1 c_{i-1} > c_{i-1} \end{array} \right\}, \quad (101)$$

and $\lambda_i^n = -(k_1 - k_0 - 1)$, or we have the $\left(f_{c_i}^-\right)$ case, that is,

$$c^{n-1} \cap [c_{i-1}, c_i] = \left\{ \begin{array}{l} c_i > v = (c_i)_- + k_1 c_i > \cdots \\ \\ \cdots > (c_i)_- + k_0 c_i = c_{i-1} \end{array} \right\}, \quad (102)$$

with $\lambda_i^n = (k_1 - k_0 - 1)$.

Therefore, for the case of $\left(f_{c_{i-1}}^+\right)$ we have

$$c_i = (c_{i-1})_+ + k_0 \cdot c_{i-1} \in c^n, \quad c_i + c_{i-1} = (c_{i-1})_+ + (k_0 + 1) c_{i-1} \notin c^n$$
$$v = (c_{i-1})_+ + k_1 \cdot c_{i-1} \in c^{n-1}, \quad v + c_{i-1} = (c_{i-1})_+ + (k_1 + 1) \cdot c_{i-1} \notin c^{n-1}, \quad (103)$$

it follows that

$$\left|(c_{i-1})_+ + k_0 c_{i-1}\right|_n \leq R < \left|(c_{i-1})_+ + (k_0 + 1) c_{i-1}\right|_n \quad (104)$$

$$\left|(c_{i-1})_+ + k_1 c_{i-1}\right|_{n-1} \leq R < \left|(c_{i-1})_+ + (k_1 + 1) c_{i-1}\right|_{n-1}$$

whence

$$a_{n+2}\left|(c_{i-1})_+ + (k_0 - 1) c_{i-1}\right| + a_{n+1}|c_{i-1}| \leq R < a_{n+2}\left|(c_{i-1})_+ + k_0 c_{i-1}\right|$$
$$+ a_{n+1}|c_{i-1}|,$$

$$a_{n+1}|(c_{i-1})_+ + (k_1 - 1)c_{i-1}| + a_n|c_{i-1}| \leq R < a_{n+1}|(c_{i-1})_+ + k_1 c_{i-1}|$$
$$+ a_n|c_{i-1}|. \quad (105)$$

Thus $k = |\lambda_i^n| = k_1 - k_0 - 1$ is the maximal integer such that

$$(a_{n+1} \cdot k + a_n) \cdot |c_{i-1}| + a_{n+1}|c_i| = a_{n+1}|(c_{i-1})_+ + (k_1 - 1)c_{i-1}| + a_n|c_{i-1}| \leq R,$$

and we obtain $c_i = k_0 c_{i-1} + (c_{i-1})_+$. Therefore,

$$|\lambda_i^n| = \left\lfloor \frac{R - a_{n+1}|c_i| - a_n|c_{i-1}|}{a_{n+1}|c_{i-1}|} \right\rfloor \quad (106)$$
$$= \left\lfloor \frac{R - a_{n+1}|(c_{i-1})_+| - a_n|c_{i-1}|}{a_{n+1}|c_{i-1}|} \right\rfloor - k_0,$$

and, when $k_0 > 0$, we obtain

$$|\lambda_i^n| = \left\lfloor \frac{R}{a_{n+1}|c_{i-1}|} - \frac{|(c_{i-1})_+|}{|c_{i-1}|} - \frac{a_n}{a_{n+1}} \right\rfloor - \left\lfloor \frac{R}{a_{n+2}|c_{i-1}|} - \frac{|(c_{i-1})_+|}{|c_{i-1}|} - \frac{a_{n+1}}{a_{n+2}} \right\rfloor - 1. \tag{107}$$

Similarly in the case of $\left(f_{c_i}^-\right)$ we have $c_{i-1} = k_0 c_i + (c_i)_-$, and

$$\lambda_i^n = \left\lfloor \frac{R - a_{n+1}|c_{i-1}| - a_n|c_i|}{a_{n+1}|c_i|} \right\rfloor \tag{108}$$

$$= \left\lfloor \frac{R - a_{n+1}|(c_i)_-| - a_n|c_i|}{a_{n+1}|c_i|} \right\rfloor - k_0,$$

and, when $k_0 > 0$, we see that

$$\lambda_i^n = \left\lfloor \frac{R}{a_{n+1}|c_i|} - \frac{|(c_i)_-|}{|c_i|} - \frac{a_n}{a_{n+1}} \right\rfloor - \left\lfloor \frac{R}{a_{n+2}|c_i|} - \frac{|(c_i)_-|}{|c_i|} - \frac{a_{n+1}}{a_{n+2}} \right\rfloor - 1. \tag{109}$$

Note that, in the case of $\left(f_{c_{i-1}}^+\right)$, we have $\lambda_i^n \neq 0$ if and only if

$$\frac{R - a_{n+1}|c_i| - a_n|c_{i-1}|}{a_{n+1}|c_{i-1}|} \geq 1, \tag{110}$$

or, equivalently,

$$a_{n+2}|c_{i-1}| + a_{n+1}|c_i| \leq R. \tag{111}$$

We cannot have $k_0 = 0$, because then $c_i = (c_{i-1})_+$, $k_1 \geq 2$, so that

$$a_{n+1}|(c_{i-1})_+ + c_{i-1}| + a_n|c_{i-1}| = |(c_{i-1})_+ + 2 \cdot c_{i-1}|_{n-1} \leq R$$
$$< |(c_{i-1})_+ + c_{i-1}|_n$$
$$= a_{n+2}|c_{i-1}| + a_{n+1}|(c_{i-1})_+|$$

a contradiction. Therefore $k_0 \geq 1$, and so $k_1 = |\lambda_i^n| + k_0 + 1 \geq 3$. We see that in this case $c_{i-1} \in (c^n)^{\min}$, that is $c_{i-1} + (c_{i-1})_- \in c^n$, for otherwise we get

$$a_{n+2}|c_{i-1}| + a_{n+1}|(c_{i-1})_-| = |c_{i-1} + (c_{i-1})_-|_n$$
$$> R$$
$$\geq |3 \cdot c_{i-1} + (c_{i-1})_+|_{n-1} \tag{112}$$
$$= a_{n+1}|2 \cdot c_{i-1} + (c_{i-1})_+| + a_n|c_{i-1}|$$
$$= a_{n+2}|c_{i-1}| + a_{n+1}|c_{i-1} + (c_{i-1})_+|$$

a contradiction.

Similarly in the case of $(f_{c_i}^-)$ we have $\lambda_i^n \neq 0$ if and only if

$$a_{n+2}|c_i| + a_{n+1}|c_{i-1}| \leq R \tag{113}$$

and this implies $|c_{i-1}| > |c_i|$, $k_0 \geq 1$, $k_1 \geq 3$, and $c_i \in (c^n)^{\min}$.

We summarize the discussion in the following description of the "d.n.a. of $c_R^{(\alpha,\beta)}$".

Theorem 2 *For a linear-norm* $|x, y| := \alpha x + \beta y$, $\alpha, \beta > 0$, *we have for* $n \geq 1$,

$$\Phi^{n-1} c_R^{(\alpha,\beta)} = \Phi^n c_R^{(\alpha,\beta)} \circ \lambda_R^n$$

$$\lambda_R^n = \sum_{i=1}^m \lambda_i^n \cdot [c_{i-1}, c_i] \in \mathbb{Z} \partial \Phi^n c_R^{(\alpha,\beta)} \quad, \quad \Phi^n c_R^{(\alpha,\beta)} = \{c_i\}_{i=1}^{m-1},$$

and

$$\lambda_R^n = \lambda_\infty^- \cdot [c_{m-1}, \infty] + \sum_{c_i \in \Phi^{n+1} c_R^{(\alpha,\beta)}} \left(\lambda_{c_i}^+ \cdot [c_i, c_{i+1}] + \lambda_{c_i}^-[c_{i-1}, c_i] \right) + \lambda_0^+ \cdot [0, c_1]$$

$$|\lambda_{c_i}^+| = \begin{cases} \left\lfloor \dfrac{R}{a_{n+1}|c_i|} - \dfrac{|(c_i)_+|}{|c_i|} - \dfrac{a_n}{a_{n+1}} \right\rfloor - \left\lfloor \dfrac{R}{a_{n+2}|c_i|} - \dfrac{|(c_i)_+|}{|c_i|} - \dfrac{a_{n+1}}{a_{n+2}} \right\rfloor - 1, \\ 0 \quad \text{if} \quad a_{n+2}|c_i| + a_{n+1}|c_{i+1}| > R; \end{cases}$$

$$\lambda_{c_i}^- = \begin{cases} \left\lfloor \dfrac{R}{a_{n+1}|c_i|} - \dfrac{|(c_i)_-|}{|c_i|} - \dfrac{a_n}{a_{n+1}} \right\rfloor - \left\lfloor \dfrac{R}{a_{n+2}|c_i|} - \dfrac{|(c_i)_-|}{|c_i|} - \dfrac{a_{n+1}}{a_{n+2}} \right\rfloor - 1, \\ 0 \quad \text{if} \quad a_{n+2}|c_i| + a_{n+1}|c_{i-1}| > R; \end{cases}$$

$$|\lambda_0^+| = \begin{cases} \left\lfloor \dfrac{R}{a_{n+1}\alpha} - \dfrac{\beta}{\alpha} - \dfrac{a_n}{a_{n+1}} \right\rfloor - \left\lfloor \dfrac{R}{a_{n+2}\alpha} - \dfrac{\beta}{\alpha} - \dfrac{a_{n+1}}{a_{n+2}} \right\rfloor - 1, \\ 0 \quad \text{if} \quad a_{n+2}\alpha + a_{n+1}|c_1| > R; \end{cases}$$

$$\lambda_\infty^- = \begin{cases} \left\lfloor \dfrac{R}{a_{n+1}\beta} - \dfrac{\alpha}{\beta} - \dfrac{a_n}{a_{n+1}} \right\rfloor - \left\lfloor \dfrac{R}{a_{n+2}\beta} - \dfrac{\alpha}{\beta} - \dfrac{a_{n+1}}{a_{n+2}} \right\rfloor - 1, \\ 0 \quad \text{if} \quad a_{n+2}\beta + a_{n+1}|c_{m-1}| > R. \end{cases}$$

Thus if $\Phi^n c_R^{(\alpha,\beta)} = \{c_i\}_{i=1}^{m-1}$, we obtain $\Phi^{n-1} c_R^{(\alpha,\beta)}$ from it by adding all the mediants $c_i + c_{i-1}$, and for those $c_i \in \Phi^{n+1} c_R^{(\alpha,\beta)}$, as well as $c_m = \underline{\infty}$, $c_0 = \underline{0}$, we add the fin around c_i whose length is given by the $\lambda_{c_i}^\pm$.

Corollary 1 *We have, with* $\lfloor R \rfloor_+ := \max\{0, \lfloor R \rfloor\}$,

$$\#\Phi^{n-1}c_R^{(\alpha,\beta)} = 2 \cdot \#\Phi^n c_R^{(\alpha,\beta)}$$

$$+ \sum_{c_i = v \in \Phi^{n+1}c_R^{(\alpha,\beta)}} \left\lfloor \frac{R}{a_{n+1}|v|} - \frac{|c_{i+1}|}{|v|} - \frac{a_n}{a_{n+1}} \right\rfloor_+ + \left\lfloor \frac{R}{a_{n+1}|v|} - \frac{|c_{i-1}|}{|v|} - \frac{a_n}{a_{n+1}} \right\rfloor_+$$

$$+ \left\lfloor \frac{R}{a_{n+1}\alpha} - \frac{|c_1|}{\alpha} - \frac{a_n}{a_{n+1}} \right\rfloor_+ + \left\lfloor \frac{R}{a_{n+1}\beta} - \frac{|c_{m-1}|}{\beta} - \frac{a_n}{a_{n+1}} \right\rfloor_+$$

The formula of Theorem 2 show an interaction between the binary and Fibonacci bases.

Recall that every integer $R \in \mathbb{N}$ has a binary expansion

$$R = 2^{n_1} + \cdots + 2^{n_j} + \cdots + 2^{n_\ell} \ , \quad n_j \geq 0. \tag{114}$$

This expansion is unique if we require $n_j > n_{j+1}$. We can add such numbers and bring them to the canonical form using the "carry-reminders" rule $2^n + 2^n = 2^{n+1}$. We can multiply numbers using the simple rule $2^n \cdot 2^m = 2^{n+m}$.
Rewriting the Fibonacci numbers as

$$\varphi^n := a_{1+n} = \frac{1}{\sqrt{5}} \left[\left(\frac{1+\sqrt{5}}{2}\right)^{n+1} - \left(\frac{1-\sqrt{5}}{2}\right)^{n+1} \right]$$
$$= \frac{1}{2^n} \sum_{k=0}^{n} (1+\sqrt{5})^k (1-\sqrt{5})^{n-k} \tag{115}$$

Similarly, every $R \in \mathbb{N}$ has a Fibonacci or Zeckendorf expansion as a sum

$$R = \varphi^{n_1} + \cdots + \varphi^{n_j} + \cdots + \varphi^{n_\ell} \tag{116}$$

This expansion is unique if we require that $n_j > n_{j+1} + 1$, i.e., we can represent R by a sequence of zeros and ones, where no two ones are neighbors. We can add numbers in this representation, and bring them to the canonical form using the "carry-reminder" rules: $\varphi^n + \varphi^{n+1} = \varphi^{n+2}$, and $\varphi^n + \varphi^n = \varphi^{n+1} + \varphi^{n-2}$. We can also multiply numbers using the rule, for $m \geq n$:

$$(\star)_{n,m} \ \varphi^n \cdot \varphi^m = \varphi^{n+m} + \varphi^{n+m-4} + \cdots + \varphi^{n+m-4j} + \cdots$$

$$+ \begin{cases} \varphi^{m-n+4} + \varphi^{m-n} & \text{if } n \equiv 0 \pmod{2} \\ \varphi^{m-n+2} + \varphi^{m-n-1} & \text{if } n \equiv 1 \pmod{2} \text{ and } n < m \\ \varphi^2 + \varphi^0 & \text{if } m = n \equiv 1 \pmod{2} \end{cases}$$

One prove $(\star)_{n,m}$ by induction, via

$$(\star)_{n,n} + (\star)_{n-1,n} \Longrightarrow (\star)_{n,n+1}$$

$$(\star)_{n-1,n} + (\star)_{n-2,n} \Longrightarrow (\star)_{n,n}$$

$$(\star)_{n,n} + (\star)_{n,n+1} \Longrightarrow (\star)_{n,m}, \quad m \geq n.$$

Note the curious 4-periodicity of $(\star)_{n,m}$.

Perhaps this interaction of the binary and Fibonacci expansions should come as no surprise since our very approach to $\mathbb{Q}^+$ is as a binary tree of Fibonacci growth.

10 Equidistribution

We end with some remarks on equidistribution. First, we define the potential function

$$h : \mathcal{G}_1 = \text{SL}_2(\mathbb{N}) \to [0, 1]$$
$$h(v_-, v_+) := \frac{1}{|v_-| \cdot |v_+|}, \quad |(x, y)| = x + y. \tag{117}$$

For each triangle Δ_v we have the exactness

$$h(v_-, v_+) = h(v_-, v) + h(v, v_+). \tag{118}$$

We get the function

$$H : \{\underline{\infty}\} \amalg \underline{\mathbb{Q}^+} \amalg \{\underline{0}\} = \mathcal{G}_0 \longrightarrow [0, 1]$$
$$H(v) = \int_{\underline{0}}^{v} h\,(dg) = \text{sum of } h \text{ along (any) path from } \underline{0} \text{ to } v. \tag{119}$$

We have
$$H(x, y) = \frac{y}{x + y}. \tag{120}$$

Indeed, it follows from

Meditations on the Farey Fractal

$$\partial H \begin{pmatrix} x_- & y_- \\ x_+ & y_+ \end{pmatrix} = \frac{y_+}{x_+ + y_+} - \frac{y_-}{x_- + y_-}$$

$$= \frac{1}{(x_+ + y_+)(x_- + y_-)} \tag{121}$$

$$= h \begin{pmatrix} x_- & y_- \\ x_+ & y_+ \end{pmatrix}.$$

For a ◁-set $c = \{c_i\}_{i=1}^{m-1} \in C_m$, we get the function

$$R_c : \{\infty\} \sqcup c \sqcup \{\underline{0}\} \longrightarrow [0, 1]$$

$$R_c(v) = \begin{cases} 1 & v = c_m = \infty \\ j/m & v = c_j \\ 0 & v = c_0 = \underline{0} \end{cases} = \frac{1}{m} \int_{\underline{0}}^{v} 1, \tag{122}$$

the length of the path c from $\underline{0}$ to v divided by the total length of c.
Put for $p \geq 1$,

$$\delta_p(c) = \|R_c - H\|_{\ell_{p(c)}}^p = \sum_{j=1}^{m-1} \left| \frac{j}{m} - \frac{y_j}{x_j + y_j} \right|^p, \quad c = \{c_j = (x_j, y_j)\}. \tag{123}$$

For $c = c_R = c_R^{(1,1)} = \{(x, y) \in \mathbb{Q}^+, x + y \leq R\}$, we have that the following estimates imply Riemann-Hypothesis,

$$\text{Franel [2]:} \quad \delta_2(c_R) = O\left(\frac{\log R}{R}\right) \tag{124}$$

$$\text{Landau [12]:} \quad \delta_1(c_R) = O\left(R^{1/2} \log R\right) \tag{125}$$

To obtain this using an inductive procedure, one will need a good estimation of $\delta_p(\Phi^{n-1} c_R)$ in terms of $\delta_p(\Phi^n c_R)$. One can try to do this "locally", by dissecting $\Phi^n c_R$ to intervals.

A partial ◁-set, or a ◁-interval, is a path $c = \{c_i\}_{i=0}^m$ in the Farey Graph, $\det \begin{pmatrix} c_{i-1} \\ c_i \end{pmatrix} \equiv 1$, from the initial-point c_0 to the end-point c_m (and similarly one can define a partial ⋆-set and partial corona).
For such ◁-interval $c = \{c_i = (x_i, y_i)\}_{i=0}^m$, we can define

$$\delta_p(c) := \sum_i \left| \frac{y_0}{x_0 + y_0} + \frac{i}{m}\left(\frac{y_m}{x_m + y_m} - \frac{y_0}{x_0 + y_0}\right) - \frac{y_i}{x_i + y_i} \right|^p \tag{126}$$

This agrees with (123) when $(x_0, y_0) = \underline{0} = (1, 0)$. $(x_m, y_m) = \underline{\infty} = (0, 1)$.
One can also demand that $\det \begin{pmatrix} c_0 \\ c_m \end{pmatrix} = 1$, so that

$$\left(\frac{y_m}{x_m + y_m} - \frac{y_0}{x_0 + y_0} \right) = \frac{1}{(x_m + y_m) \cdot (x_0 + y_0)},$$

and $c = \left\{ c_i = \check{c}_i \begin{pmatrix} c_0 \\ c_m \end{pmatrix} \right\}$ where $\check{c} = \{\check{c}_i\}$ is a usual ◁-set (or corona).

Example 1 For a partial ◁-set (or corona) $c = \{c_i\}_{i=0}^m$ one has the associated partial ◁-set (corona) $\tilde{c} := c \cup \{c_i + c_{i-1}\}_{i=1}^m$ with the same initial and end points, obtained by adding all mediants. There is an elementary estimation

$$\delta_1(\tilde{c}) \leq 2 \cdot \delta_1(c) + \tfrac{1}{2} h(c) \tag{127}$$

$h(c) = \frac{y_m}{x_m + y_m} - \frac{y_0}{x_0 + y_0} \in [0, 1]$, the real length of c.

Along the fins one can estimate δ_1 using the Euler-MacLaurin formula. Also, if $c = \coprod_j c_j$, where the end point of c_{j-1} is the initial port of c_j, we have the elementary estimate

$$\left| \delta_1(c) - \sum_j \delta_1(c_j) \right| \leq \tfrac{1}{2} \sum_{j_1 < j_2} \left| h(c_{j_1}) m(c_{j_2}) - h(c_{j_2}) m(c_{j_1}) \right| \tag{128}$$

$m(c) = \#c + 1 = m \in \mathbb{N}$, the degree or length of the path ∂c.

But it is important to note that $H(c_R) \subseteq [0, 1]$ is <u>not</u> equidistributed, it is only on average so (124) and (125): the real distance between $H(v) = H(c_i)$ and $H(c_{i \pm 1})$, for an "old" v, so $c_{i \pm 1} \in f_v^{\pm}$, is of the order $O\left(\frac{1}{|v| \cdot (R + |v_{\pm}|)} \right)$, while the real distance between the elements of f_v^+, or of f_v^-, is smaller

$$|H(c_{i \pm 1}) - H(c_{i \pm 2})| = O\left(\frac{1}{R \cdot (R + |v|)} \right).$$

Example 2 For $v = \underline{0} = (1, 0) = c_0$, we have $c_1 = (R, 1)$, $c_2 = (R - 1, 1)$, and $|H(c_1) - H(c_0)| = \frac{1}{(R+1)}$, while

$$|H(c_2) - H(c_1)| = \frac{1}{R} - \frac{1}{R+1} = \frac{1}{R \cdot (R+1)}.$$

Thus it is important that the d.n.a. of $\Phi^n c_R$ adds extra points just around such old v's (cf. Theorem 2).

Remark 3 Let $c(n)$ denote the corona of height n, with d.n.a. identically 0, so that $c(n+1)$ is obtained from $c(n)$ just by adding all mediants, and $\#c(n) = 2^n - 1$, so $c(n) = \{(x_i, y_i)\}_{i=1}^{2^n-1}$ in increasing real order, and let

$$S_n = \sum_{i=1}^{2^n-1} \left| \frac{i}{2^n} - \frac{y_i}{x_i + y_i} \right|^2. \tag{129}$$

The sum S_n appears in all the even places of the sum S_{n+1}, so that

$$S_1 = 0 < S_2 = \frac{2}{144} < S_3 = \frac{668}{14400} < \cdots < \cdots$$

is monotone increasing, and does not converge to 0! Comparing this to (124), we see that it is the d.n.a. that is responsible for the uniform distribution of the rationals within the continuum.

References

1. A. Connes, Trace formula in noncommutative geometry and the zeros of the Riemann zeta function. Selecta Math. (N.S.) **5**(1), 29–106 (1999)
2. J. Franel, Les suites de Farey et le probleme des nombres premiers. Nachrichten von der Gesellschaft der Wissenschaften zu Göttingen, Mathematisch-Physikalische Klasse **1924**, 198–201 (1924)
3. S. Haran, Riesz potentials and explicit sums in arithmetic. Invent. Math. **101**(3), 697–703S (1990)
4. S. Haran, Index theory, potential theory, and the Riemann hypothesis, in *L-functions and arithmetic (Durham, 1989)* (Cambridge University Press, Cambridge, 1989), pp. 257–270
5. S. Haran, *The Mysteries of the Real Prime*. London Mathematical Society Monographs. New Series, vol. 25 (The Clarendon Press, Oxford University Press, New York, 2001)
6. S. Haran, *Invitation to Nonadditive Arithmetical Geometry*. Casimir Force. Casimir Operators and the Riemann Hypothesis (Walter de Gruyter, Berlin, 2010), pp 249–265
7. S. Haran, Geometry over F1 (2017), ArXiv Preprint arXiv:1709.05831
8. S. Haran, *New Foundations for Geometry—Two Non-additive Languages for Arithmetical Geometry*. Memoirs of the American Mathematical Society, vol. 246 (American Mathematical Society, 2017)
9. S. Haran, Algebra over generalized rings (2020), ArXiv Preprint arXiv:2006.15613
10. S. Haran, Non Additive Geometry. World Scientific Publications (2024)
11. K. Iwasawa, *Lectures on p-adic L-functions*. Annals of Mathematics Studies, No. 74 (Princeton University Press, Princeton, NJ; University of Tokyo Press, Tokyo, 1972)
12. E. Landau, Bemerkungen zu der obenstehenden Abhandlung von J. Franel. Nachrichten von der Gesellschaft der Wissenschaften zu Göttingen, Mathematisch-Physikalische Klasse **1924**, 202–206 (1924)
13. L. Lindroos, A. Sills, H. Wang, Odd Fibbinary numbers and the golden ratio. Fibonacci Quart. **52**(1), 61–65 (2014)
14. T.J. Stieltjes, Sur la réduction en fraction continue d'une série procédant suivant les puissances descendantes d'une variable. Ann. Fac. Sci. Toulouse Sci. Math. Sci. Phys. **3**, H1–H17 (1889)
15. J. Tate, Fourier analysis in number fields and Hecke's zeta-functions. Thesis (Ph.D.), Princeton University, ProQuest LLC, Ann Arbor, MI, 1950

16. A. Weil, Sur les analogie entre les corps de nombres algébriques et les corps de fonctions algébriques. Oeuvres Scient. **I**, 236–240 (1980)
17. A. Weil, Sur les fonctions algébriques à corps de constantes fini. C. R. Acad. Sci. Paris **210**, 592–594 (1940)
18. A. Weil, Une lettre et un extrait de lettre à Simone Weil. March 26 1940. Collected Pap. **1979**, 244–255 (1940)
19. A. Weil, On the Riemann hypothesis in function-fields. Proc. Nat. Acad. Sci. U.S.A. **27**, 345–347 (1941)
20. A. Weil, Lettre a E. Artin, July 10th, in *Oeuvres Scientifiques* vol. 1 (1942)
21. A. Weil, Sur les "formules explicites" de la théorie des nombres premiers. Comm. Sém. Math. Univ. Lund [Medd. Lunds Univ. Mat. Sem.] **1952** (Tome Supplémentaire), 252–265 (1952)
22. A. Weil, Fonction zêta et distributions. Séminaire Bourbaki **9**(312), 523–531 (1966)

Open Access This chapter is licensed under the terms of the Creative Commons Attribution 4.0 International License (http://creativecommons.org/licenses/by/4.0/), which permits use, sharing, adaptation, distribution and reproduction in any medium or format, as long as you give appropriate credit to the original author(s) and the source, provide a link to the Creative Commons license and indicate if changes were made.

The images or other third party material in this chapter are included in the chapter's Creative Commons license, unless indicated otherwise in a credit line to the material. If material is not included in the chapter's Creative Commons license and your intended use is not permitted by statutory regulation or exceeds the permitted use, you will need to obtain permission directly from the copyright holder.

Quantum and Physical Cryptanalysis

Shor's Quantum Algorithm for Solving the Binary ECDLP: A Survey

Ren Taguchi and Atsushi Takayasu

Abstract Shor's quantum algorithm can solve the factorization problem and the elliptic curve discrete logarithm problem (ECDLP) in polynomial time. Due to the physical barrier for realizing large-scale reliable quantum computer, designs of concrete quantum circuits and their resource estimates have been actively discussed. In this paper, we focus on the case of the binary ECDLP. In particular, since the quantum resource for solving the binary ECDLP heavily depends on the quantum inversion algorithm over $\mathbb{F}_{2^n}$, we summarize the known quantum inversion algorithms and compare the quantum resource.

Keywords Shor's algorithm · ECDLP · GCD-based inversion · FLT-based inversion · Quantum resource estimate

1 Introduction

1.1 Shor's Algorithm and Factorization

Due to the popularity of RSA [31] and elliptic curve cryptography (ECC) [19, 24], the computational hardness of the factorization problem and the elliptic curve discrete logarithm problem (ECDLP) have been extensively studied. As a result, there seem to be no *classical* polynomial time algorithms for solving the problems. In fact, the largest composite number that has been experimentally factorized so far is 829-bit [4], while 2048-bit composite numbers are currently recommended. Therefore, RSA and ECC are believed to be secure in practice by setting appropriate cryptographic parameters, e.g., bit length of composite numbers.

R. Taguchi (✉) · A. Takayasu
Graduate School of Information Science and Technology, The University of Tokyo, Tokyo, Japan
e-mail: rtaguchi-495@g.ecc.u-tokyo.ac.jp

A. Takayasu
e-mail: takayasu-a@g.ecc.u-tokyo.ac.jp

In 1994, this situation changed drastically since Shor [32] proposed a *quantum polynomial time algorithm* for solving the factorization problem and the ECDLP. Thus, RSA and ECC are no longer secure when large-scale reliable quantum computers are physically realized. Due to the situation, cryptographic research in two directions is now paid much attention. The first direction is designing post-quantum public key cryptosystems (PQC), while the second one is a quantum resource estimate of Shor's algorithm. The first one should be arguably important to maintain a secure information society. Then, various PQC schemes have been already proposed so far. The second direction enables us to determine when we should replace RSA and ECC with PQC schemes in the actual information systems. In this paper, we focus on the topic hereafter.

Shor's quantum algorithm can efficiently solve the factorization problem and the ECC in theory, while it is still hard to complete the task in practice. Indeed, although various physical implementations of Shor's algorithm have been reported [1, 7, 21–23, 25–27, 33, 37], most papers focus on the factorization of small composite numbers such as $15 = 3 \times 5, 21 = 3 \times 7, 35 = 5 \times 7$. Thus, many researchers believe that the factorization of practically used 2048-bit composite numbers is still hard. The situation is caused by the hardness of the physical realization of large-scale reliable quantum computers. Briefly speaking, it is hard to realize quantum computers such that they have many qubits, their gate operations are reliable, and their coherence time is long. To factorize large composite numbers, there should be sufficiently many qubits to encode the numbers, gate operations should be sufficiently reliable to compute correct solutions, and coherence time should be sufficiently long to complete the computations. Although quantum error corrections may be able to solve the second problem, they require more qubits and longer computation. Since it may be physically hard to realize quantum computers that satisfy all the requirements perfectly, cryptographic researchers have designed as efficient quantum circuits as possible to run Shor's algorithm so that the circuits require smaller number of qubits, smaller number of quantum gates, and smaller depth. There have been several papers [3, 8, 9, 12, 14, 15, 20, 34, 36, 38] that designed and improved quantum circuits to run Shor's algorithm for solving the factorization problem. For example, Gidney and Ekerå [9] claimed that Shor's algorithm can factorize the 2048-bit numbers with 20 million qubits, 2.7 billion quantum gates, and eight hours under the assumption that the error rate per gate operation is about 0.1% by taking into account quantum error corrections.

1.2 Shor's Algorithm and ECDLP

As we mentioned, there have been several physical implementations and circuit designs of Shor's algorithm; however, their target is the factorization problem. Compared with the factorization problem, the same research for the ECDLP has not yet progressed. To the best of our knowledge, there are no physical implementations of Shor's algorithm for solving the ECDLP. Although there have been a few papers that study designs of quantum circuits for solving the ECDLP, they assumed that gate operations do not fail and quantum error corrections are not considered. Therefore, it is indispensable to develop the topic and the aim of this paper is the summary of known results for the ECDLP.

It is known that the most dominant part of running Shor's algorithm for solving the ECDLP is an elliptic curve point addition. Proos and Zalka [29] first studied the topic; however, they did not provide actual quantum circuits to perform elliptic curve point additions that are the most dominant part of running Shor's algorithm. Roetteler et al. [30] proposed the first concrete quantum circuits for solving the ECDLP over a prime field $\mathbb{F}_q$. When q is an n-bit prime number, Roetteler et al.'s quantum circuit requires $9n + 2\lceil \log_2 n \rceil + 10$ qubits and $448n^3 \log_2 n + 4090n^3$ Toffoli gates. Although they did not provide the depth of quantum circuits asymptotically, they showed that the quantum circuit requires 4,719 qubits, $1.14 \cdot 10^{12}$ Toffoli gates, and $1.05 \cdot 10^{12}$ Toffoli depth for $n = 521$. Then, Häner et al. [13] improved the quantum circuits for achieving various trade-offs. For example, Häner et al.'s quantum circuits require $8n + 10.2\lceil \lg n \rceil - 1$ qubits, $436n^3 - 1.05 \cdot 2^{26}$ T-gates, and $120n^3 - 1.67 \cdot 2^{22}$ T-depth by optimizing the number of qubits and $10n + 7.4\lceil \lg n \rceil + 1.3$ qubits, $1115n^3/\lg n - 1.08 \cdot 2^{24}$ T-gates, and $389n^3/\lg n - 1.70 \cdot 2^{22}$ T-depth by optimizing the number of T-gates.

1.3 Shor's Algorithm and Binary ECDLP

Recently, concrete designs of quantum circuits for solving the ECDLP over a binary field $\mathbb{F}_{2^n}$ (binary ECDLP) have been actively studied, where degrees $n = 163, 233, 283, 409$, and 571 are recommended [6]. The first result was given by Banegas et al. [2]. The above Roetteler et al.'s result [30] indicates that quantum resources of whole circuits mainly depend on those of quantum inversion over a prime field $\mathbb{F}_q$. Then, Banegas et al. proposed two quantum inversion algorithms over $\mathbb{F}_{2^n}$. Their first quantum inversion algorithm is a GCD-based method which is the same as Roetteler et al., while their second quantum inversion algorithm is an FLT-based method[1] which is effective in a binary case. Their GCD-based quantum circuit requires $7n + \lfloor \log_2 n \rfloor + 9$ qubits and $48n^3 + 8n^{\log_2 3+1} + 352n^2 \log_2 n + 512n^2 + o(n^{\log_2 3})$ Toffoli gates, where $\log_2 3 = 1.58 \cdots$. Concretely, for $n = 571$, their GCD-based quantum circuit requires $4,015$ qubits, $1.0 \cdot 10^{10}$ Toffoli gates, and depth $1.2 \cdot 10^{10}$, while their FLT-based quantum circuit requires $9,137$ qubits, $2.4 \cdot 10^8$ Toffoli gates, and $1.0 \cdot 10^{10}$ depth. Thus, their GCD-based quantum circuit is effective in optimizing the number of qubits, while their FLT-based quantum circuit is effective in optimizing the number of Toffoli gates and the depth.

Kim and Hong [18] proposed a quantum GCD-based inversion algorithm with fewer qubits and more Toffoli gates compared with Banegas et al.'s quantum GCD-based inversion algorithm. In particular, Kim-Hong's quantum circuit to run Shor's

---

[1] FLT is an abbreviation of Fermat's little theorem.

algorithm requires $6n + \lfloor n/2 \rfloor + \lfloor \log_2 n \rfloor + 10$. For $n = 571$, Kim-Hong's quantum circuit requires $3,727$ qubits.

Putranto et al. [28] proposed a quantum FLT-based inversion algorithm with more qubits and less depth compared with Banegas et al.'s quantum FLT-based inversion algorithm. In particular, for $n = 571$, Putranto et al. to run Shor's algorithm requires $14,276$ qubits, $2.4 \cdot 10^8$ Toffoli gates, and $8.2 \cdot 10^9$ depth. Taguchi and Takayasu [35] proposed two improved quantum FLT-based inversion algorithms. The core of their improvement is an addition chain which indicates the sequence of computation. If we use specific addition chains, Taguchi-Takayasu's quantum inversion algorithms become the same as Banegas et al.'s algorithm and Putranto et al.'s algorithm. Taguchi and Takayasu showed that their FLT-based inversion algorithm are improvement of previous ones since there exist more suitable addition chains. When $n = 571$, their improvement of Banegas et al.'s algorithm requires $8,566$ qubits, $2.2 \cdot 10^8$ Toffoli gates, and $8.0 \cdot 10^9$ depth, while their improvement of Putranto et al.'s algorithm requires $10,850$ qubits, $2.2 \cdot 10^8$ Toffoli gates, and $8.0 \cdot 10^9$ depth.

1.4 Organization

Since concrete designs of quantum circuits for solving the binary ECDLP have been actively studied, we summarize the known results. In Sect. 2, we review the basic of quantum computation and Shor's algorithm. In Sect. 3, we summarize quantum computations over a binary field $\mathbb{F}_{2^n}$. In Sects. 4 and 5, we review quantum FLT-based and GCD-based inversion algorithms, respectively. In Sect. 6, we compare the quantum resources of Shor's algorithm for solving the binary ECDLP with various quantum inversion algorithms.

2 Shor's Algorithm

In Sect. 2.1, we review basic of quantum computation. In Sect. 2.2, we define the binary elliptic curve discrete logarithm problem (binary ECDLP). In Sect. 2.3, we briefly explain Shor's algorithm for solving the binary ECDLP.

2.1 Quantum Computation

Let $\{|0\rangle, |1\rangle\}$ be an orthonormal basis of $\mathbb{C}^2$ called a computational basis. A *qubit* is represented as a two-dimensional unit complex vector $\alpha_0|0\rangle + \alpha_1|1\rangle \in \mathbb{C}^2$ such that $\alpha_0, \alpha_1 \in \mathbb{C}$ and $|\alpha_0|^2 + |\alpha_1|^2 = 1$. When we measure the qubit, we observe 0 and 1 with probability $|\alpha_0|^2$ and $|\alpha_1|^2$, respectively. For $i_1, i_2, \ldots, i_n \in \{0, 1\}$, we use $|i_1 i_2 \cdots i_n\rangle$ to denote a 2^n-dimensional unit complex vector $|i_1\rangle \otimes |i_2\rangle \otimes \cdots \otimes |i_n\rangle \in$

$\mathbb{C}^{2^n}$, where "$\otimes$" denotes a tensor product. As the case of one qubit, an orthonormal basis $\{|i_1 i_2 \cdots i_n\rangle\}_{i_1, i_2, \ldots, i_n \in \{0,1\}}$ of $\mathbb{C}^{2^n}$ is a computational basis and n qubits are represented as a 2^n-dimensional unit complex vector $\sum_{i_1, i_2, \ldots, i_n \in \{0,1\}} \alpha_{i_1, i_2, \ldots, i_n} |i_1 i_2 \cdots i_n\rangle$, where $\alpha_{i_1, i_2, \ldots, i_n} \in \mathbb{C}$ and $\sum_{i_1, i_2, \ldots, i_n \in \{0,1\}} |\alpha_{i_1, i_2, \ldots, i_n}|^2 = 1$ hold. When we measure the qubits, we observe $i_1, i_2, \ldots, i_n$ with probability $|\alpha_{i_1, i_2, \ldots, i_n}|^2$.

We use a quantum gate model to denote quantum computations and every quantum gate is a unitary matrix. In short, quantum gates take n qubits as input and outputs n qubits multiplied by a $2^n \times 2^n$ unitary matrix. Most quantum algorithms use a Hadamard gate denoted by H. When we apply the Hadamard gate H to a qubit $|0\rangle$, we have $\frac{1}{\sqrt{2}}(|0\rangle + |1\rangle)$. When we apply the Hadamard gate H to all n qubits $|0 \cdots 0\rangle$, we have $\frac{1}{\sqrt{2}}(|0\rangle + |1\rangle) \otimes \cdots \otimes \frac{1}{\sqrt{2}}(|0\rangle + |1\rangle) = \frac{1}{\sqrt{2^n}} \sum_{i_1, i_2, \ldots, i_n \in \{0,1\}} |i_1 i_2 \cdots i_n\rangle$. Thus, the Hadamard gate H maps $|0 \cdots 0\rangle$ to a sum of all vectors of a computational basis with the same coefficients. Next, we review elementary quantum gates called CNOT gates, Toffoli (TOF) gates, and SWAP gates, where the CNOT and SWAP gates are two-qubit gates and the TOF gate is a three-qubit gate. For $i, j, k \in \{0, 1\}$, the gates work as follows:

$$\text{CNOT}(i, j) = (i, i \oplus j), \quad \text{TOF}(i, j, k) = (i, j, k \oplus (i \cdot j)),$$
$$\text{SWAP}(i, j) = (i, j).$$

Among them, the TOF gate is the most expensive. Figure 1 describes the quantum circuit of the CNOT, TOF, and SWAP gates. Observe that the CNOT gate does not change the input if the first bit satisfies $i = 0$ and changes only the second bit as $i \oplus j$ otherwise. Thus, the first and second bits are called a controlled bit and a target bit, respectively. As illustrated in Fig. 1, the controlled bit is represented as a black dot. Similarly, since the TOF gate does not change the input if the first and second bits satisfy $i = j = 0$, the first two bits are controlled bits and represented as black dots in Fig. 1.

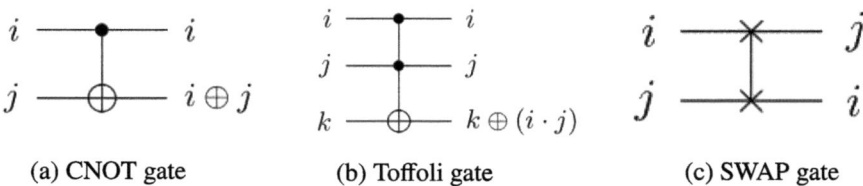

(a) CNOT gate (b) Toffoli gate (c) SWAP gate

Fig. 1 Quantum gates

2.2 Elliptic Curve Discrete Logarithm Problem

For a positive integer n, a binary elliptic curve of degree n is given by

$$y^2 + xy = x^3 + ax^2 + b,$$

where $a \in \mathbb{F}_{2^n}$ and $b \in \mathbb{F}_{2^n}^*$. A set of rational points on an elliptic curve and a special point O forms a group under elliptic curve point addition, where O is called a point at infinity and an identity element in the group. For two points $P = (x_1, y_1)$ and $Q = (x_2, y_2)$ on a binary elliptic curve, we define an elliptic curve point addition in two ways. If $P \neq Q$, we have $P + Q = (x_3, y_3)$ such that

$$x_3 = \lambda^2 + \lambda + x_1 + x_2 + a, \quad y_3 = (x_2 + x_3)\lambda + x_3 + y_2,$$

where

$$\lambda = \frac{y_1 + y_2}{x_1 + x_2}.$$

Otherwise, if $P = Q$, $P + P = (x_3, y_3)$ such that

$$x_3 = \lambda^2 + \lambda + a, \quad y_3 = x_1^2 + (\lambda + 1)x_3,$$

where

$$\lambda = x_1 + \frac{y_1}{x_1}.$$

Thus, we have to compute an inversion over $\mathbb{F}_{2^n}$ when we compute an elliptic curve point addition.

We use $[k]P$ to denote an elliptic curve point addition of k P's, i.e., $[k]P = \underbrace{P + \cdots + P}_{k\ P\text{'s}}$. The task of the binary ECDLP is given $(P, [k]P)$ and computes k for randomly selected k.

2.3 Shor's Algorithm for Solving the Binary ECDLP

Figure 2 briefly describes the quantum circuit of Shor's algorithm. The input of the circuit consists of two n-qubit quantum states $|0 \cdots 0\rangle$ and a quantum state of the point at infinity on the elliptic curve. We first apply the Hadamard gates H to all top two n-qubit $|0 \cdots 0\rangle$. Next, the quantum gate indicated by "$+[2^i]P$" is a controlled point addition gate that adds $+[2^i]P$ to the bottom states if the controlled bit is 1 and adds nothing otherwise. Thus, we apply $2n + 2$ controlled point addition gates

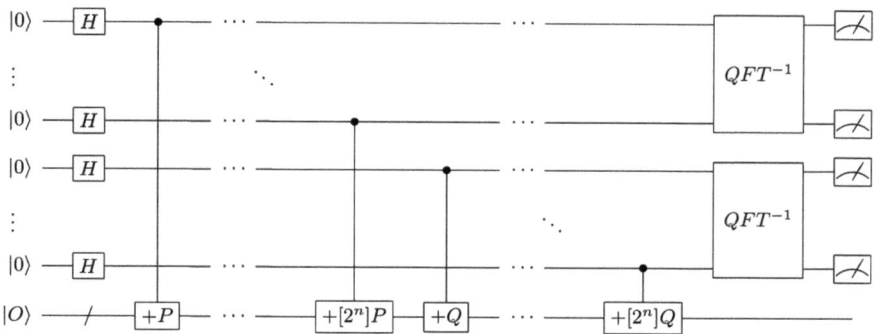

Fig. 2 A quantum circuit of Shor's algorithm for solving a binary ECDLP

$+P, \ldots, +[2^n]P, +Q, \ldots, +[2^n]Q$. Finally, we apply the inverse quantum Fourier transform to the top two n-qubit quantum states and measure them. Although we do not explain how to compute the discrete logarithm in detail, the quantum circuit enables us to solve the binary ECDLP.

As Fig. 2 illustrates, the first Hadamard step requires only $O(n)$ quantum gates. It is known that the point addition and the inverse QFT steps require $O(n^3)$ and $O(n^2)$ quantum gates, respectively. Therefore, the point addition step is the most expensive step of Shor's algorithm. Moreover, as we explained in Sect. 2.2, an inversion over $\mathbb{F}_{2^n}$ has to be computed to perform an elliptic curve point addition. Thus, the quantum resource of Shor's algorithm heavily depends on that of a quantum inversion over $\mathbb{F}_{2^n}$.

3 Quantum Basic Arithmetics

Hereafter, we represent an element of a binary field $f \in \mathbb{F}_{2^n}$ as a polynomial $f = a_0 + a_1 x + \cdots + a_{n-1} x^{n-1}$ of degree at most $n - 1$ with coefficients $a_0, a_1, \ldots, a_{n-1} \in \mathbb{F}_2$, where we use the fact that $\mathbb{F}_{2^n} \simeq \mathbb{F}_2[x]/(m(x))$ holds for an irreducible polynomial $m(x) \in \mathbb{F}_2[x]$ of degree n. Then, we can represent f as an n-qubit quantum state $|a_0, a_1, \ldots, a_{n-1}\rangle$, while we may use f itself to denote the quantum state for simplicity. In this section, we review quantum basic arithmetics, i.e., quantum addition, squaring, and multiplication over $\mathbb{F}_{2^n}$. Hereafter, let $f, g, h \in \mathbb{F}_{2^n}$.

Quantum Addition. We use Banegas et al.'s algorithm [2] denoted by ADD to perform quantum addition, where it works as follows:

$$\mathrm{ADD}(f, g) = (f, f + g).$$

Since ADD performs quantum addition over $\mathbb{F}_{2^n}$, it holds that

$$\mathrm{ADD}(f, f + g) = (f, g).$$

A quantum circuit of ADD consists only of CNOT gates. Specifically, the number of CNOT gates is n and the depth is 1.

Quantum Squaring. We use Banegas et al.'s algorithm [2] denoted by SQUARE and spSQUARE to perform quantum squaring, where they work as follows:

$$\text{SQUARE}(f) = f^2, \quad \text{spSQUARE}(f, g) = (f, f^2 + g).$$

In this paper, we also use SQUARE^{-1} that works as follows:

$$\text{SQUARE}^{-1}(f^2) = f.$$

Quantum circuits of SQUARE and spSQUARE consist only of CNOT gates. Specifically, the number of CNOT gates and the depth of SQUARE (resp. spSQUARE) is at most $n(n-1)/2$ (resp. $n(n+1)/2$) although $O(n)$ may be sufficient for specific irreducible polynomial $m(x)$. SQUARE^{-1} is a reverse circuit of SQUARE.

Quantum Multiplication. We use Hoof's algorithm [16] denoted by MODMULT to perform quantum multiplication, where it works as follows:

$$\text{MODMULT}(f, g, h) = (f, g, h + fg).$$

A quantum circuit of MODMULT consists of TOF gates and CNOT gates. Specifically, the numbers of TOF and CNOT gates are $O(n^{\log_2 3})$ and $O(n^2)$, respectively, where $\log_2 3 = 1.58 \cdots$.

4 Quantum FLT-Based Inversion Algorithm

Let f be an element of $\mathbb{F}_{2^n}^*$. For simplicity, we use a notation

$$\langle \alpha \rangle := f^\alpha$$

in this section. Based on the extended Fermat's little theorem, the FLT-based inversion method takes $\langle 1 \rangle$ as input and outputs

$$\langle 2^n - 2 \rangle = \langle -1 \rangle.$$

We explain how the FLT-based inversion method computes $\langle 2^n - 2 \rangle$ by using an addition chain for $n - 1$.

Definition 1 (*Addition chain*) Let ℓ and N be non-negative integers. An addition chain for N of length ℓ is a sequence $p_0 = 1, p_1, p_2, \ldots, p_\ell = N$ which satisfies

- for all $s = 1, 2, \ldots, \ell$, there exist i and j which satisfy $p_s = p_i + p_j$, where $0 \leq i, j < s$.

For all s, we call p_s a doubled term if there are no i and j which satisfy three conditions $0 \leq i, j < s$, $p_s = p_i + p_j$, and $p_i \neq p_j$, and an added term otherwise. At first, the FLT-based inversion method takes $\langle 1 \rangle = \langle 2^1 - 1 \rangle = \langle 2^{p_0} - 1 \rangle$ as input and computes $\langle 2^{p_1} - 1 \rangle, \langle 2^{p_2} - 1 \rangle, \ldots, \langle 2^{p_\ell} - 1 \rangle = \langle 2^{n-1} - 1 \rangle$ sequentially. Then, the method outputs

$$\langle 2^{n-1} - 1 \rangle^2 = \langle 2^n - 2 \rangle = \langle -1 \rangle.$$

We explain how the FLT-based method computes the above sequence. For a doubled term $p_s = 2p_i$, the method uses $\langle 2^{p_i} - 1 \rangle$ and computes

$$\langle 2^{p_i} - 1 \rangle \times \langle 2^{p_i} - 1 \rangle^{2^{p_i}} = \langle 2^{p_i} - 1 \rangle \times \langle 2^{2p_i} - 2^{p_i} \rangle = \langle 2^{2p_i} - 1 \rangle = \langle 2^{p_s} - 1 \rangle.$$

Similarly, for an added term $p_s = p_i + p_j$, the method uses $\langle 2^{p_i} - 1 \rangle$ and $\langle 2^{p_j} - 1 \rangle$ and computes

$$\langle 2^{p_i} - 1 \rangle \times \langle 2^{p_j} - 1 \rangle^{2^{p_i}} = \langle 2^{p_i + p_j} - 1 \rangle = \langle 2^{p_s} - 1 \rangle.$$

Taguchi-Takayasu's Algorithm. Taguchi-Takayasu's quantum FLT-based inversion algorithm [35] is a quantum realization of the above procedure. To be precise, Taguchi and Takayasu proposed two algorithms which they call the Basic algorithm and the Extended algorithm. Hereafter, we only describe the Extended algorithm since it requires fewer qubits than the Basic algorithm and call it simply Taguchi-Takayasu's algorithm. Taguchi-Takayasu's algorithm takes $\langle 1 \rangle = \langle 2^1 - 1 \rangle = \langle 2^{p_0} - 1 \rangle$ and an addition chain $\{p_s\}_{s=0}^{\ell}$ [2] for $n - 1$ of length ℓ as input and computes the sequence $\langle 2^{p_1} - 1 \rangle, \langle 2^{p_2} - 1 \rangle, \ldots, \langle 2^{p_\ell} - 1 \rangle$ in two distinct ways for all $1 \leq s \leq \ell$ depending on whether p_s is a doubled term or an added term.

Case of a doubled term: The algorithm takes $\langle 2^{p_i} - 1 \rangle$ as input and computes $\langle 2^{p_s} - 1 \rangle$. The algorithm first applies $\mathrm{ADD}(\langle 2^{p_i} - 1 \rangle, 0) = (\langle 2^{p_i} - 1 \rangle, \langle 2^{p_i} - 1 \rangle)$ and computes a copy of $\langle 2^{p_i} - 1 \rangle$. Next, the algorithm applies $\mathrm{SQUARE}(\langle 2^{p_i} - 1 \rangle^{2^k}) = (\langle 2^{p_i} - 1 \rangle^{2^{k+1}})$ p_i times to the copied $\langle 2^{p_i} - 1 \rangle$ and computes $\langle 2^{p_i} - 1 \rangle^{2^{p_i}}$. Finally, the algorithm applies $\mathrm{MODMULT}(\langle 2^{p_i} - 1 \rangle, \langle 2^{p_i} - 1 \rangle^{2^{p_i}}, 0) = (\langle 2^{p_i} - 1 \rangle, \langle 2^{p_i} - 1 \rangle^{2^{p_i}}, \langle 2^{p_s} - 1 \rangle)$ and obtain $\langle 2^{p_s} - 1 \rangle$. Afterward, the algorithm applies $\mathrm{SQUARE}^{-1}(\langle 2^{p_i} - 1 \rangle^{2^k}) = (\langle 2^{p_i} - 1 \rangle^{2^{k-1}})$ p_i times to $\langle 2^{p_i} - 1 \rangle^{2^{p_i}}$ and computes $\langle 2^{p_i} - 1 \rangle$. Then, the algorithm applies $\mathrm{ADD}(\langle 2^{p_i} - 1 \rangle, \langle 2^{p_i} - 1 \rangle) = (\langle 2^{p_i} - 1 \rangle, 0)$ and uncomputes a copied $\langle 2^{p_i} - 1 \rangle$.

Case of an added term: The algorithm takes $\langle 2^{p_i} - 1 \rangle$ and $\langle 2^{p_j} - 1 \rangle$ as input and computes $\langle 2^{p_s} - 1 \rangle$. The algorithm applies $\mathrm{SQUARE}(\langle 2^{p_j} - 1 \rangle^{2^k}) = (\langle 2^{p_j} - 1 \rangle^{2^{k+1}})$ p_i times to $\langle 2^{p_j} - 1 \rangle$ and computes $\langle 2^{p_j} - 1 \rangle^{2^{p_i}}$. Then, the algorithm applies $\mathrm{MODMULT}(\langle 2^{p_i} - 1 \rangle, \langle 2^{p_j} - 1 \rangle^{2^{p_i}}, 0) = (\langle 2^{p_i} - 1 \rangle, \langle 2^{p_j} - 1 \rangle^{2^{p_i}}, \langle 2^{p_s} - 1 \rangle)$ and obtain $\langle 2^{p_s} - 1 \rangle$.

---

[2] p_ℓ has to be an added term although we omit the details.

Finally, the algorithm applies $\text{SQUARE}(\langle 2^{p_\ell} - 1\rangle) = \text{SQUARE}(\langle 2^{n-1} - 1\rangle) = (\langle 2^n - 2\rangle)$ and obtains $\langle 2^n - 2\rangle = \langle -1\rangle$. Theorem 1 describes the quantum resources for Taguchi-Takayasu's algorithm.

Theorem 1 ([35], Theorem 2) *Let f be an element in $\mathbb{F}_{2^n}^*$ and $\{p_s\}_{s=0}^\ell$ be an addition chain for $n - 1$ of length ℓ with p_ℓ is an added term. Taguchi-Takayasu's algorithm takes $f = \langle 1\rangle$ and $\{p_s\}_{s=0}^\ell$ as input and outputs $\langle -1\rangle = \langle 2^n - 2\rangle$ with ℓ ancillary registers and ℓ multiplications.*

Algorithm 1 Taguchi-Takayasu's Algorithm

Input: An irreducible polynomial $m(x) \in \mathbb{F}_{2^n}^*$ of degree n, an addition chain $\{p_s\}_{s=0}^\ell$ for $n - 1$ of length ℓ (composed of d doubled terms and m added terms) and related $\{a_s\}_{s=1}^\ell, \{b_s\}_{s=1}^\ell, \{Q_s\}_{s=1}^\ell, \{c\ell_t\}_{t=0}^d$, a polynomial $g_0 = f \in \mathbb{F}_{2^n}^*$ of degree up to $n - 1$, polynomials $g_1, \ldots, g_{d+m-1}$, h initialized to an all-$|0\rangle$ state.
Output: $h = f^{2^n-2}$
1: $dcount \leftarrow 0$
2: **for** $s = 1, \ldots, d + m - 1$ **do**
3:   **if** $s \in D$ **then**
4:     $\text{ADD}(g_{a_s}, h)$
5:     **for** $i = 1, \ldots, Q_s$ **do**
6:       $\text{SQUARE}(h)$
7:     $\text{MODMULT}(g_{a_s}, h, g_s)$
8:     $\text{GARBAGECLEAR}(c\ell_d, a_s)$
9:     $dcount \leftarrow dcount + 1$
10:   **else** $\{s \in M\}$
11:     **for** $i = 1, \ldots, Q_s$ **do**
12:       $\text{SQUARE}(g_{a_s})$
13:     $\text{MODMULT}(g_{a_s}, g_{b_s}, g_s)$
14: **for** $i = 1, \ldots, Q_{d+m}$ **do**
15:   $\text{SQUARE}(g_{a_{d+m}})$
16: $\text{MODMULT}(g_{a_{d+m}}, g_{b_{d+m}}, h)$
17: $\text{SQUARE}(h)$

Remark 1 In the above explanation, Taguchi-Takayasu's algorithm uses SQUARE to compute $\text{SQUARE}(\langle 2^{p_i} - 1\rangle^{2^k}) = (\langle 2^{p_i} - 1\rangle^{2^{k+1}})$. We can replace SQUARE with spSQUARE and complete the computation.

Remark 2 In the above explanation, we explain the procedure of Taguchi-Takayasu's *Extended* algorithm. Their Basic algorithm is the same except that the copied $\langle 2^{p_i} - 1\rangle$ is not uncomputed while computing $\langle 2^{p_s} - 1\rangle$ for a doubled term p_s.

Remark 3 Banegas et al.'s and Putranto et al.'s quantum FLT-based inversion algorithms are the same as Taguchi-Takayasu's Extended and Basic algorithms, respectively, except that the fixed addition chains are used.

Algorithm 1 illustrates Taguchi-Takayasu's quantum FLT-based inversion algorithm. Algorithm 1 takes $\{c\ell_t\}_{t=0}^d$ as an additional input. The sequence $\{c\ell_t\}_{s=0}^t$

describe the number of times to apply SQUARE or SQUARE^{-1} for clearing garbages, i.e., uncomputation. More precisely, we apply SQUARE $c\ell_t$ times if $c\ell_t > 0$ and SQUARE^{-1} $-c\ell_t$ times if $c\ell_t < 0$. We set $c\ell_0 = 0$. Garbages are stored in h and clearing is performed by initializing it to 0. We describe the algorithm for clearing garbages in Algorithm 2.

Algorithm 2 GARBAGECLEAR (c, k)

Input: Integers c, k.
1: **if** $c > 0$ **then**
2:    **for** $i = 1, \ldots, c$ **do**
3:       SQUARE (h)
4: **if** $c < 0$ **then**
5:    **for** $i = 1, \ldots, -c$ **do**
6:       SQUARE$^{-1}(h)$
7: ADD (g_k, h)

---

5 Quantum GCD-Based Inversion Algorithm

Let deg f be a degree of a polynomial $f(x) \in \mathbb{F}_2[x]$. A GCD-based inversion method takes $f(x)$ as input and outputs $V(x) = f^{-1}$ such that

$$f(x)V(x) \equiv 1 \mod m(x)$$

by using an extended-Euclid's greatest-common-divisor-based algorithm. Banegas et al.'s quantum GCD-based inversion algorithm [2] is based on Bernstein and Yang's classical GCD-based inversion algorithm [5].

Bernstein-Yang's Classical GCD-based Inversion Algorithm. Let K denote a field. The following divstep takes $F(x), G(x) \in K[x]$, and $\delta \in \mathbb{Z}$ as input and reduces deg $F(x)$ or deg $G(x)$:

divstep $(\delta, F(x), G(x))$
$$= \begin{cases} (1 - \delta, G(x), (G(0)F(x) - F(0)G(x))/x) & \text{if } \delta > 0 \wedge G(0) \neq 0, \\ (1 + \delta, F(x), (F(0)G(x) - G(0)F(x))/x) & \text{otherwise.} \end{cases}$$

Given $\delta_0 \in \mathbb{Z}$, $F_0(x), G_0(x) \in K[x]$, we define $\delta_s, F_s(x), G_s(x)$ for $s = 0, 1, 2, \ldots$ which satisfy

$$(\delta_s, F_s(x), G_s(x)) = \text{divstep}^s(\delta_0, F_0(x), G_0(x)).$$

We note that $\delta_s = \deg F_s(x) - \deg G_s(x)$ holds when $\delta_0 = \deg F_0(x) - \deg G_0(x)$. Then, there exist matrices $\mathcal{T}_0, \ldots, \mathcal{T}_{s-1}$ such that

$$\begin{pmatrix} F_s(x) \\ G_s(x) \end{pmatrix} = \mathcal{T}_0 \cdots \mathcal{T}_{s-1} \begin{pmatrix} F_0(x) \\ G_0(x) \end{pmatrix},$$

where we use the notation $\mathcal{T}_{s-1} \cdots \mathcal{T}_0 = \begin{pmatrix} u_s & v_s \\ q_s & r_s \end{pmatrix}$ for $u_s, v_s, q_s, r_s \in K[1/x]$. Bernstein and Yang proved the following fact.

Theorem 2 ([5], Theorem 6.2.) *Let $f(x) \in \mathbb{F}_2[x]$ of degree at most $n-1$ and $m(x)$ denote an irreducible polynomial of degree n described in Sect. 2.1. Let $F(x) = x^n m(1/x)$, $G(x) = x^{n-1} f(1/x)$. $V(x) = x^{-n+1} v_{2n-1}(1/x)$ satisfies*

$$f(x)V(x) \equiv 1 \mod m(x).$$

By Theorem 2, an inverse of f over $\mathbb{F}_{2^n}$ determined by $m(x)$ is V.

Banegas et al.'s Quantum GCD-based Inversion Algorithm. Algorithm 3 illustrates Banegas et al.'s quantum GCD-based inversion algorithm based on Theorem 2, where $h[i]$ is a coefficient of x^i of a polynomial h and $H[i]$ is i-th qubit of a register H.

Algorithm 3 Banegas et al.'s Qunatum GCD-based inversion Algorithm
---
Input: an irreducible polynomial $m(x)$, an n-qubit register G which stores $x^{n-1} f(1/x)$, $n+1$-qubit registers F, r, v, g initialized to an all-$|0\rangle$ state, a $\lceil \log n \rceil + 2$-qubit register δ initialized to an all-$|0\rangle$ state, 2 qubits a, $G[n]$ initialized to $|0\rangle$
Output: $v = f^{-1}$
1: **for** $i = 0, \ldots, n$ **do**
2:    **if** $m[s] = 1$ **then**
3:       $X(F[n-s])$
4: $X(sign)$
5: $X(r[0])$
6: **for** $s = 0, \ldots, 2n-2$ **do**
7:    RIGHTSHIFT(v)
8:    TOF$(sign, G[0], a)$
9:    **for** $i = 0, \ldots, \lceil \log n \rceil + 1$ **do**
10:       CNOT$(a, \delta[i])$
11:    **for** $i = 0, \ldots, \Phi$ **do**
12:       ctrl_SWAP$_a(F[i], G[i])$
13:    **for** $i = 0, \ldots, \phi$ **do**
14:       ctrl_SWAP$_a(r[i], v[i])$
15:    INC$_{1+a}(\delta)$
16:    CNOT$(v[0], a)$
17:    CNOT$(G[0], g[s])$
18:    **for** $i = 0, \ldots, \Phi$ **do**
19:       TOF$(g[s], F[i], G[i])$
20:    **for** $i = 0, \ldots, \phi$ **do**
21:       TOF$(g[s], r[i], v[i])$
22:    LEFTSHIFT$(G[0, \ldots, \Phi])$
23: **for** $i = 0, \ldots, \lfloor n/2 \rfloor - 1$ **do**
24:    SWAP$(v[i], v[n-1-i])$

Let H_1, H_2 be N-qubit registers and a be a single qubit. `ctrl_SWAP`$_a$($H_1[i]$, $H_2[i]$) applies SWAP($H_1[i]$, $H_2[i]$) if $a = 1$ and does nothing if $a = 0$. X(a) is applying X gate to a. RIGHTSHIFT(H_1) and LEFTSHIFT(H_1) are constructed by only swap gates. δ describes an integer and $sign = \delta[\lceil \log n \rceil + 1]$. We note that δ describes $\delta[0]2^0 + \cdots + \delta[\lceil \log n \rceil]2^{\lceil \log n \rceil} + 1$ if $sign = 1$ and $-(\delta[0]2^0 + \cdots + \delta[\lceil \log n \rceil]2^{\lceil \log n \rceil})$ if $sign = 0$. INC$_{1+a}(\delta)$ is a quantum computation proposed by Gidney [10] which adds 1 to δ if $a = 0$, and requires $22\lfloor \log n \rfloor + 26$ TOF gates and $2\lfloor \log n \rfloor + 3$ CNOT gates. Moreover, we define $\Phi = \min(2n - 2 - s, n)$, $\phi = \min(s + 1, n)$ for $s = 0, 1, \ldots, 2n - 2$. By [5], it holds that $\Phi = \max(\deg F_s, \deg G_s)$ and $\phi = \max(\deg v_s, \deg r_s)$. Since $G[2n - 2 - s + 1], G[2n - 2 - s + 2], \ldots, G[n - 1] = 0$ for all $s = n, n + 1, \ldots, 2n - 2$, we use $G[2n + 1 - s]$ as $g[s]$ for all $s = n, n + 1, \ldots, 2n - 2$.

Briefly speaking, Algorithm 3 compute `divstep`(δ_s, F_s, G_s) and v_s, r_s for each $s = 0, 1, \ldots, 2n - 2$ in a loop from line 6 to 17. In Algorithm 3, we do not hold u_s, q_s. More precisely, it follows that $v_{s+1} = r_s, r_{s+1} = (G_s(0)v_s - F_s(0)r_s)/x$ or $v_{s+1} = v_s, r_{s+1} = (-G_s(0)v_s + F_s(0)r_s)/x$. Thus, it is sufficient for computing f^{-1} to hold v_s and r_s since we just need v_{2n-1} by Theorem 2. However, we treat v, r as in $\mathbb{F}_2[x]$ while they are in $\mathbb{F}_2[1/x]$.

Kim et al.'s quantum GCD-based inversion algorithm does not prepare g, while it requires ($\lfloor \log n \rfloor + 1$)-qubit register *mask* and other ancillary $2\lfloor \log n \rfloor + 1$ qubits for unary iterations. Thus, they reduce $n - 3\lfloor \log n \rfloor - 2$ qubits compared to Algorithm 3. A naive unary iteration requires $2(n + \min(s, n - 1) - 2\max(\lfloor (s - 1)/2 \rfloor, 0))$ TOF gates for each $s = 0, 1, \ldots, 2n - 2$, however, Kim et al. replaced a TOF gate to a Hadamard gate, a measurement, and a ctrl-X gate and halved the number of TOF gates for a unary iteration. Then, the total number of TOF gates for Kim et al.'s GCD-based algorithm is fewer than Banegas et al.'s GCD-based algorithm. On the other hand, Kim et al.'s GCD-based algorithm requires $O(n^3)$ measurements for Shor's algorithm, while other all GCD-based and FLT-based algorithms require $O(n)$ measurements.

6 Comparison

In this section, we compare the quantum resources of all known methods for Shor's algorithm, i.e., four FLT-based methods as follows:

- BBHL21-FLT: Banegas et al.'s FLT-based algorithm
- PWLK22-FLT: Putranto et al.'s FLT-based algorithm
- TT23-FLT basic: Taguchi-Takayasu's FLT-based basic algorithm
- TT23-FLT extended: Taguchi-Takayasu's FLT-based extended algorithm described in Algorithm 1

and two GCD-based methods as follows:

Table 1 The number of Hadamard gates and rotation gates for Shor's algorithm with semiclassical Fourier transform

n	Hadamard gates	Rotation gates
163	656	328
233	936	468
283	1, 136	568
571	2, 288	1, 144

- BBHL21-GCD: Banegas et al.'s GCD-based algorithm described in Algorithm 3
- KH23-GCD: Kim-Hong's GCD-based algorithm

in terms of the number of qubits, TOF, depth, and CNOT.

First, we explain common quantum resources in all methods, i.e., quantum resources for a quantum Fourier transform. In this paper, we apply semiclassical Fourier transform [11] in Shor's algorithm since it requires only 1 qubit. Shor's algorithm with semiclassical Fourier transform requires $4n + 4$ Hadamard gates and $2n + 2$ rotation gates. Table 1 describes the concrete number of Hadamard gates and rotation gates for all n. Then, we estimate the quantum resources for point additions. As we describe in Sect. 2.3, Shor's algorithm requires $2n + 2$ point additions. As Roetteler et al. [30] mentioned, we can ignore the special cases of point addition since it does not affect the quantum Fourier transform. For estimating the resources of TT23-FLT basic and extended, we use addition chains that Taguchi and Takayasu [35] used. Values of the depth are upper bounds because we do not completely consider parallel quantum computation. Moreover, we compute the concrete number of CNOT gates of SQUARE, SQUARE^{-1}, and spSQUARE by calculating a matrix T_n for all n. In a quantum point addition algorithm by Banegas et al. [2], we compute const_ADD, ctrl_ADD, and ctrl_const_ADD while we omit the details. We assume that const_ADD requires $n/2$ X gates on average, ctrl_ADD requires $n/2$ TOF gates on average, and ctrl_const_ADD requires $n/2$ CNOT gates on average. When we estimate the upper bound of the depth for Shor's algorithm, we simply add the upper bound of the depth for each distinct quantum computation.

Table 2 compares the number of qubits, TOF gates, depth, and CNOT gates in all cases for all n.

Comparison between FLT-based methods and GCD-based methods. The number of qubits for GCD-based methods is half or less of the number for FLT-based methods. More precisely, BBHL21-GCD and KH23-GCD requires $7n + \lfloor \log n \rfloor + 9$ and $6n + 4\lfloor \log n \rfloor + 11$ qubits, respectively. On the other hand, BBHL21-FLT and PWLK22-FLT requires $(\lfloor \log (n-1) \rfloor + w + 1)n + 1$ and $(2\lfloor \log (n-1) \rfloor + w)n + 1$ qubits, where w is an Hamming weight of $n - 1$, respectively, and TT23-FLT basic and TT23-FLT extended requires $(\ell + d + 3)n + 1$ and $(\ell + 3)n + 1$ qubits, where ℓ is a length of a shortest addition chain for $n - 1$ and d is a number of doubled terms, respectively. Thus, GCD-based methods require $O(n)$ qubits while FLT-based

Table 2 Comparison of the number of TOF gates, qubits, depth, and CNOT gates in Shor's algorithm

n	BBHL21-FLT			
	Qubits	TOF	Depth	CNOT
163	1, 957	57, 717, 832	360, 418, 208	616, 115, 528
233	3, 030	130, 530, 348	740, 504, 700	1, 448, 957, 952
283	3, 963	280, 565, 304	1, 847, 583, 584	3, 168, 486, 328
571	9, 137	1, 998, 898, 616	16, 005, 278, 432	25, 605, 409, 544
n	PWLK22-FLT			
	Qubits	TOF	Depth	CNOT
163	3, 098	57, 717, 832	305, 423, 104	559, 632, 616
233	4, 661	130, 530, 348	693, 893, 772	1, 399, 306, 896
283	6, 227	280, 565, 304	1, 508, 639, 808	2, 824, 416, 920
571	14, 276	1, 998, 898, 616	12, 130, 614, 496	21, 707, 270, 728
n	TT23-FLT basic			
	Qubits	TOF	Depth	CNOT
163	2, 772	57, 717, 832	305, 420, 480	559, 204, 904
233	3, 962	130, 530, 348	693, 891, 900	1, 398, 870, 720
283	4, 812	280, 565, 304	1, 446, 164, 352	2, 760, 019, 352
571	10, 850	1, 856, 260, 120	11, 360, 835, 200	20, 244, 689, 608
n	TT23-FLT extended			
	Qubits	TOF	Depth	CNOT
163	1, 957	57, 717, 832	346, 126, 592	601, 186, 280
233	3, 030	130, 530, 348	776, 460, 204	1, 484, 479, 152
283	3, 963	280, 565, 304	1, 721, 305, 824	3, 039, 005, 048
571	8, 566	1, 856, 260, 120	14, 120, 867, 904	23, 020, 372, 232
n	BBHL21-GCD			
	Qubits	TOF	Depth	CNOT
163	1, 157	293, 095, 880	352, 060, 768	271, 056, 904
233	1, 647	781, 231, 932	969, 740, 460	755, 393, 184
283	1, 998	1, 378, 745, 592	1, 716, 407, 392	1, 338, 808, 376
571	4, 015	10, 281, 586, 744	13, 237, 503, 136	10, 383, 826, 024
n	KH23-GCD			
	Qubits	TOF	Depth	CNOT
163	1, 017	247, 502, 568	329, 381, 536	340, 341, 000
233	1, 437	703, 401, 660	923, 032, 188	957, 771, 360
283	1, 741	1, 256, 914, 136	1, 638, 850, 400	1, 700, 794, 776
571	3, 473	10, 068, 075, 160	12, 882, 835, 680	13, 357, 283, 368

methods require $O(n \log n)$ qubits since $w, \ell, d = O(\log n)$. The number of TOF gates for FLT-based methods is much fewer than the number for GCD-based methods. While we omit the details, FLT-based methods require $O(n^{\log_2 3+1} \log n)$ TOF gates and GCD-based methods require $O(n^3)$ TOF gates.

Comparison among FLT-based Methods. BBHL21-GCD requires much fewer qubits than PWLK22-FLT, while the former requires more depth and CNOT gates than the latter. TT23-FLT basic reduces qubits, depth, and CNOT gates compared to PWLK22-FLT without sacrificing TOF gates for all n. Similarly, TT23-FLT extended reduces depth and CNOT gates compared to BBHL21-FLT without sacrificing qubits and TOF gates for all n. Moreover, when $n = 571$, TT23-FLT basic and TT23-FLT extended reduce all quantum resources compared to PWLK22-FLT and BBHL21-FLT, respectively. TT23-FLT algorithms use an arbitrary addition chain for $n - 1$ while BBHL21-FLT and PWLK22-FLT use a specific addition chain by Itoh-Tsujii [17] for $n - 1$. Since we find better addition chains for all NIST-recommended n than Itoh-Tsujii's addition chain, TT23-FLT algorithms achieve the above improvements.

Comparison among GCD-based Methods. KH23-GCD achieves fewer resources than BBHL21-GCD except for the number of CNOT gates. KH23-GCD reduces qubits by applying unary iteration processes which require an amount of TOF gates. However, Kim-Hong halved TOF gates by replacing a specific TOF gate with a Hadamard gate, a measurement, and a ctrl-X gate. Then, KH23-GCD achieves fewer TOF gates compared to BBHL21-GCD.

References

1. M. Amico, Z.H. Saleem, M. Kumph, Experimental study of Shor's factoring algorithm using the IBM Q experience. Phys. Rev. A **100**, 012305 (2019)
2. G. Banegas, D.J. Bernstein, I. van Hoof, T. Lange, Concrete quantum cryptanalysis of binary elliptic curves. IACR Trans. CHES **2021**(1), 451–472 (2020)
3. S. Beauregard, Circuit for Shor's algorithm using $2n + 3$ qubits. Quantum Inf. Comput. **3**, 175–185 (2003)
4. F. Boudot, P. Gaudry, A. Guillevic, N. Heninger, E. Thomé, P. Zimmermann, Factorization of RSA-250 (2020)
5. D.J. Bernstein, B.-Y. Yang, Fast constant-time GCD computation and modular inversion. IACR Trans. CHES **2019**(3), 340–398 (2019)
6. C.F. Kerry, P.D. Gallagher, FIPS PUB 186-4 Digital Signature Standard (DSS), in *NIST* (2013), pp. 92–101
7. Z.-C. Duan, J.-P. Li, J. Qin, Y. Yu, Y.-H. Huo, S. Hofling, C.-Y. Lu, N.-L. Liu, K. Chen, J.-W. Pan, Proof-of-principle demonstration of compiled Shor's algorithm using a quantum dot single-photon source. Optics Express **28**, 18917–18930 (2020)
8. A.G. Fowler, M. Mariantoni, J.M. Martinis, A.N. Cleland, Surface codes: towards practical large-scale quantum computation. Phys. Rev. A **86**, 032324 (2012)
9. C. Gidney, M. Ekerå, How to factor 2048 bit RSA integers in 8 hours using 20 million noisy qubits. Quantum **5**, 433 (2021)
10. C. Gidney, Constructing large increment gates (2015)
11. R.B. Griffiths, C.-S. Niu, Semiclassical Fourier transform for quantum computation. Phys. Rev. Lett. **76**(17), 3228–3231 (1996)

12. É. Gouzien, N. Sangouard, Factoring 2048-bit RSA integers in 177 days with 13,436 qubits and a multimode memory. Phys. Rev. Lett. **127**, 140503 (2021)
13. T. Häner, S. Jaques, M. Naehrig, M. Roetteler, M. Soeken, Improved quantum circuits for elliptic curve discrete logarithms, in *Post-Quantum Cryptography*, ed. by J. Ding, J.-P. Tillich (Springer, Cham, 2020), pp. 425–444
14. J. Ha, J. Lee, J. Heo, Resource analysis of quantum computing with noisy qubits for Shor's factoring algorithms. Quantum Inf. Process. **21**(2), 60 (2022)
15. T. Haener, M. Roetteler, K.M. Svore, Factoring using $2n+2$ qubits with Toffoli based modular multiplication. Quantum Inf. Comput. **18**(7–8), 673–684 (2017)
16. I. van Hoof, Space-efficient quantum multiplication of polynomials for binary finite fields with sub-quadratic Toffoli gate count. CoRR, abs/1910.02849 (2019)
17. T. Itoh, S. Tsujii, A fast algorithm for computing multiplicative inverses in GF(2^m) using normal bases. Inf. Comput. **78**(3), 171–177 (1988)
18. H. Kim, S. Hong, New space-efficient quantum algorithm for binary elliptic curves using the optimized division algorithm. Quantum Inf. Process. **22**(6) (2023)
19. N. Koblitz, Elliptic curve cryptosystems. Math. Comput. **48**(177), 203–209 (1987)
20. N. Kunihiro, Exact analyses of computational time for factoring in quantum computers. IEICE Trans. Fundam. Electron. Commun. Comput. Sci. **88**-A(1), 105–111 (2005)
21. E. Lucero, R. Barends, Y. Chen, J. Kelly, M. Mariantoni, A. Megrant, P. O'Malley, D. Sank, A. Vainsencher, J. Wenner, T. White, Y. Yin, A.N. Cleland, J.M. Martinis, Computing prime factors with a Josephson phase qubit quantum processor. Nat. Phys. **8**, 719–723s (2012)
22. C.-Y. Lu, D.E. Browne, T. Yang, J.-W. Pan, Demonstration of a compiled version of Shor's quantum factoring algorithm using photonic qubits. Phys. Rev. Lett. **99**, 250504 (2007)
23. B.P. Lanyon, T.J. Weinhold, N.K. Langford, M. Barbieri, D.F.V. James, A. Cilchrist, A.G. White, Experimental demonstration of a compiled version of Shor's algorithm with quantum entanglement. Phys. Rev. Lett. **99**, 250505 (2007)
24. V.S. Miller, Use of elliptic curves in cryptography, in *CRYPTO '85*, Lecture Notes in Computer Science, vol. 218, ed. by H.C. Williams (Springer, Cham, 1985), pp. 417–426
25. E. Martin-Lopez, A. Laing, T. Lawson, R. Alvarez, X.-Q. Zhou, J.L. O'Brien, Experimental realisation of Shor's quantum factoring algorithm using qubit recycling. Nat. Photon **6**, 773–776 (2012)
26. T. Monz, D. Nigg, E.A. Martinez, M.F. Brandl, P. Schindler, R. Rines, S.X. Wang, I.L. Chuang, R. Blatt, Realization of a scalable Shor algorithm. Science **351**, 1068–1070 (2016)
27. A. Politi, J.C.F. Matthews, J.L. O'Brien, Shor's quantum factoring algorithm on a photonic chip. Science **325**, 1221 (2009)
28. D.S.C. Putranto, R.W. Wardhani, H.T. Larasati, H. Kim, Another concrete quantum cryptanalysis of binary elliptic curves. Cryptology ePrint Archive, Paper 2022/501 (2022)
29. J. Proos, C. Zalka, Shor's discrete logarithm quantum algorithm for elliptic curves. Quantum Inf. Comput. **3**(4) (2003)
30. M. Roetteler, M. Naehrig, K.M. Svore, K. Lauter, Quantum resource estimates for computing elliptic curve discrete logarithms, in *ASIACRYPT*, vol. 2017 (2017), pp. 241–270
31. R.L. Rivest, A. Shamir, L.M. Adleman, A method for obtaining digital signatures and public-key cryptosystems. Commun. ACM **21**(2), 120–126 (1978)
32. P.W. Shor, Algorithms for quantum computation: discrete logarithms and factoring, in *FOCS 1994* (1994), pp. 124–134
33. J.A. Smolin, G. Smith, A. Vargo, Oversimplifying quantum factoring. Nature **499**, 163–165 (2013)
34. Y. Takahashi, N. Kunihiro, A quantum circuit for Shor's factoring algorithm using $2n+2$ qubits. Quantum Inf. Comput. **6**(2), 184–192 (2006)
35. R. Taguchi, A. Takayasu, Concrete quantum cryptanalysis of binary elliptic curves via addition chain, in *Topics in Cryptology - CT-RSA, 2023* (2023), pp. 57–83
36. V. Vedral, A. Barenco, A. Ekert, Quantum networks for elementary arithmetic operations. Phys. Rev. A **54**, 147–153 (1996)

37. L. Vandersypen, M. Steffen, G. Breyta, C.S. Yannoni, M.H. Sherwood, I.L. Chuang, Experimental realization of Shor's quantum factoring algorithm using nuclear magnetic resonance. Nature **414**, 883–887 (2001)
38. C. Zalka, Fast versions of Shor's quantum factoring algorithm (1998)

Open Access This chapter is licensed under the terms of the Creative Commons Attribution 4.0 International License (http://creativecommons.org/licenses/by/4.0/), which permits use, sharing, adaptation, distribution and reproduction in any medium or format, as long as you give appropriate credit to the original author(s) and the source, provide a link to the Creative Commons license and indicate if changes were made.

The images or other third party material in this chapter are included in the chapter's Creative Commons license, unless indicated otherwise in a credit line to the material. If material is not included in the chapter's Creative Commons license and your intended use is not permitted by statutory regulation or exceeds the permitted use, you will need to obtain permission directly from the copyright holder.

Proposal for Quantum GRS Algorithm and Cryptanalysis for ROLLO and RQC

Asuka Wakasugi and Mitsuru Tada

Abstract Code-based cryptosystem (CBC) is one of the candidates for post-quantum cryptosystems (PQCs). Its security is primarily based on the syndrome decoding problem (SDP). In this paper, we focus on the rank CBC, whose security relies on the rank SDP. The Gaborit–Ruatta–Schrek (GRS) algorithm is well known as the current best decoding algorithm for the rank SDP. Therefore, we propose the quantum version of the GRS algorithm. Then, we introduce an attack strategy using this quantum algorithm for the previous rank CBCs that remained at the 2nd Round of the NIST's PQC standardization project and consider the quantum security for those cryptosystems. We present results that according to the currently recommended key-size parameter, our approach does not threaten the security of ROLLO, but that it is effective and can be a threat to the RQC encryption scheme. Then, we give new RQC instances that are secure against such an attack.

Keywords Rank code-based cryptography · GRS algorithm · Grover's algorithm · ROLLO · RQC

1 Introduction

Public-key cryptosystems are currently being used for our secure communication. Their security is largely based on number theory problems, such as the integer factorization problem. The Rivest–Shamir–Adleman (RSA) cryptosystem relies on this problem. In 1994, Shor [26] proposed the polynomial-time quantum algorithm to solve this problem. Once a quantum computer of an appropriate size is built, the

---

A. Wakasugi (✉)
Graduate School of Science and Engineering, Chiba University, Chiba, Japan
e-mail: a_wakasugi@eaglys.co.jp

EAGLYS Inc, Research and Development, Tokyo, Japan

M. Tada
Graduate School of Science, Chiba University, Chiba, Japan
e-mail: m.tada@faculty.chiba-u.jp

© The Author(s) 2026
T. Takagi et al. (eds.), *Mathematical Foundations for Post-Quantum Cryptography*, Mathematics for Industry 40, https://doi.org/10.1007/978-981-96-1218-5_8

RSA cryptosystems shall be vulnerable to Shor's algorithm. Therefore, we need to consider post-quantum cryptosystems (PQCs) and code-based cryptosystem (CBC) as among the promising candidates for PQCs.

1.1 Rank-Metric Code

Let q be a prime power, and m and n be positive integers. For $x \in \mathbb{F}_{q^m}^n$, let φ_x be the linear transformation such that $\varphi_x : \mathbb{F}_q^n \ni v \mapsto vx \in \mathbb{F}_{q^m}$, and $\text{rank}_{\mathbb{F}_q}(x)$ denotes the rank of x, that is, the dimension of the image of φ_x. Also, an $\mathbb{F}_{q^m}$-linear code C of n-length and k-dimension is a linear subspace with k-dimension in $\mathbb{F}_{q^m}^n$. Such a C is called an $[n, k]_{q^m}$-linear code. A rank-metric code $C \subseteq \mathbb{F}_{q^m}^n$ is an $[n, k]_{q^m}$-linear code with rank-metric. For example, the Gabidulin code [6] and the low-rank-parity-check code [7] are the rank-metric codes. These codes correspond to the rank version of the Reed-Solomon code [23] and the low-density-parity-check code [10] with the Hamming metric, respectively.

1.2 Rank Syndrome Decoding Problem (SDP)

Definition 1 (*Rank SDP*) Let q, m, n, k, and w be positive integers, H be a matrix in $\mathbb{F}_{q^m}^{(n-k) \times n}$, and s be a vector in $\mathbb{F}_{q^m}^{n-k}$. Then, the rank SDP is the problem to find an $e \in \mathbb{F}_{q^m}^n$ such that $He^T = s$ and $\text{rank}_{\mathbb{F}_{q^m}}(e) = w$ on input q, m, n, k, w, H and s.

The rank SDP is not known to be NP-complete, but it is known that there is a UR reduction [13] between the rank SDP and the SDP with Hamming metric, which is NP-complete [9]. In this paper, we consider rank CBC, which uses the rank SDP as the basis for its security. NIST has been standardizing PQCs since 2016, and the project is currently in its 4th Round. Two rank CBCs, ROLLO [19] and RQC [18], remained in the PQC project until its 2nd Round, but not in the 4th Round.

1.3 Attack Method for Rank CBC

Rank CBC is the McEliece cryptosystem [17] using the rank-metric code. The Gaborit–Ruatta–Schrek (GRS) algorithm [8] is the best-known decoding algorithm to solve the rank SDP. In addition, an algebraic attack is one of the attack methods for solving equations with multivariate polynomials derived from the public-key in multivariate polynomial-based public-key cryptosystems, and can be applied to rank CBC. Bardet et al. [2] proposed the algebraic attack for ROLLO and RQC. In this paper, we focus on the GRS algorithm as a previous attack method for rank CBC. We review the GRS algorithm in Sect. 2.

1.4 Previous Study

CBC is derived from the McEliece cryptosystem. Since the McEliece cryptosystem appeared in 1978, there are some studies of security from a classical view. For example, the Information Set Decoding (ISD) algorithm is the algorithm to solve the SDP with Hamming metric. Prange [22] proposed the ISD algorithm with its subsequent derivation, such as May–Meurer-Thomae (MMT) [16] and Becker–Joux–May–Meurer (BJMM) [3]. In addition, Bernstein [4] proposed a quantum version of the Prange algorithm, and Kachigar et al. [14] proposed a quantum version of the MMT/BJMM algorithm. The computational complexity of these quantum ISD algorithms can be much smaller than those of corresponding classical ISD algorithms, by combining previous classical ISD algorithms and Grover's algorithm. For example, the computational complexity of Bernstein's algorithm is about half of that of Prange's algorithm, by using Grover's algorithm. Perriello et al. [21] introduced an analysis strategy for the previous Hamming CBCs using the Bernstein algorithm. As seen in Sect. 1.3, we can see some studies on rank CBC from a classical view, but there are few ones from a quantum view. In PQC other than CBC, several studies are known to report that combining Grover's algorithm with existing classical algorithms helps to reduce computational complexity. For example, Schwabe and Westerbaan [24] applied Grover's algorithm to the multivariate quadratic problem that is the basis for the security of a multivariate polynomial cryptosystem. Also, Biasse and Pring [5], by Grover's algorithm, provided the cryptanalysis for the supersingular isogeny cryptosystem SIKE.

1.5 Contribution and Organization

In this paper, we propose the quantum version of the GRS algorithm, and find the computational complexity of the quantum GRS algorithm is reduced compared to that of the classical GRS algorithm. In the classical GRS algorithm, an exhaustive search is done with respect to the whole basis in some field ($B_{m,r}$ to appear in Sect. 2). Grover's algorithm can reduce the computational cost for such a search, and hence in this paper, we make use of such an advantage. Then, we compute the computational costs of the quantum GRS algorithm over the quantum circuit. Finally, we can find this quantum attack method is effective for RQC by applying it to rank CBCs that remained at the NIST PQC project 2nd Round. One reason why it is effective is that the value of a certain instance is small. Therefore, we propose a value at which RQC is secure against this attack method.

This paper is organized as follows. First, we have already seen the definition of the rank-metric code and the rank SDP. In Sect. 2, we review the abstract of the classical GRS algorithm. In Sect. 3, we introduce Grover's algorithm [11]. In Sect. 4, we propose a quantum GRS algorithm based on the above preparations. In Sect. 5, we consider the attack strategy using the quantum GRS algorithm and the results for ROLLO and RQC. We conclude this paper in Sect. 6.

2 Classical GRS Algorithm

Suppose that the rank SDP's inputs q, m, n, k, w, H, and s are as in Definition 1. Let r be a positive integer, and let $V_{m,r}$ be an r-dimension entire subspace over $\mathbb{F}_{q^m}$. For $v_m \in V_{m,r}$, let b_m be one of the bases that generate v_m, and $B_{m,r}$ be such the whole b_m, that is, $B_{m,r} = \{b_m \mid v_m \in V_{m,r}\}$. Furthermore, let $\beta = (\beta_1, \ldots, \beta_m)$ be one of the bases over $\mathbb{F}_{q^m}$. For $x \in \mathbb{F}_{q^m}$ and every $i \in [1, m]$, there exists $x_i \in \mathbb{F}_q$ such that $x = \sum_{i=1}^{m} x_i \beta_i$. Then, let $p_i : \mathbb{F}_{q^m} \ni x \mapsto x_i \in \mathbb{F}_q$. In the following, we consider only the case $n \geq m$.[1]

We randomly choose r from the closed interval $[w, m - \lceil km/n \rceil]$, and randomly take $F = (F_1, \ldots, F_r)$ from $(B_{m,r})^r$ without duplication. Then, we assume that, for $\ell' \in [1, n]$, there exists $\lambda_{\ell'} = (\lambda_{\ell',1}, \ldots, \lambda_{\ell',r}) \in \mathbb{F}_q^r$ such that $e_{\ell'} = \sum_{j=1}^{r} \lambda_{\ell',j} F_j$, where $e_{\ell'}$ means the ℓ'-th element of $e = (e_1, \ldots, e_n)$. We define $\hat{H} \in \mathbb{F}_q^{m(n-k) \times nr}$ as follows ($\ell \in [1, N]$ where $N = n - k$):

$$\hat{H} = \begin{pmatrix} \hat{H}[1][1] & \cdots & \hat{H}[1][n] \\ \vdots & \ddots & \vdots \\ \hat{H}[N][1] & \cdots & \hat{H}[N][n] \end{pmatrix} \text{ where } \hat{H}[\ell][\ell'] = \begin{pmatrix} p_1(H_{\ell,\ell'} F_1) & \cdots & p_1(H_{\ell,\ell'} F_r) \\ \vdots & \ddots & \vdots \\ p_m(H_{\ell,\ell'} F_1) & \cdots & p_m(H_{\ell,\ell'} F_r) \end{pmatrix}.$$

Also, let $\hat{s}[\ell] = (p_1(s_\ell), \cdots, p_m(s_\ell)) \in \mathbb{F}_q^m$, where s_ℓ means the ℓ-th element in s, and we define $\hat{s} = (\hat{s}[1], \ldots, \hat{s}[n-k])$. Then, there exists $\lambda \in \mathbb{F}_q^{nr}$ such that $\hat{H}\lambda^T = \hat{s}$, where $\lambda[\ell'] = (\lambda_{\ell',1}, \ldots, \lambda_{\ell',r})$ and $\lambda = (\lambda[1], \ldots, \lambda[n])$. $\hat{H}\lambda^T = \hat{s}$ is equivalent to $He^T = s$ in the rank SDP. It is necessary that $nr \geq m(n-k)$ holds for the size of $\hat{H}$ to have a solution at least for this simultaneous equation with an unknown λ. We estimate the range of r from this condition. We evaluate e from λ using Gaussian elimination over $\mathbb{F}_q$ for $\hat{H}$ and $\hat{s}$. Hence, the classical GRS algorithm if $n \geq m$ is given in Algorithm 1. $F = (F_1, \ldots, F_r) \xleftarrow{\$} (B_{m,r})^r$ described in Line 3 refers to randomly choosing r elements from $B_{m,r}$ without duplication. GE described in Line 11 refers to the subroutine that receives $(\hat{H}, \hat{s})$, executes Gaussian elimination for those, and computes the solution for the simultaneous equation.

Let q and x be positive integers. We define $[x]_q := 1 + q + \cdots + q^{x-1}$ and $[x]_q! = \prod_{k=1}^{x} [k]_q$. Then, let y be a positive integer, and $\binom{x}{y}_q = \frac{[x]_q!}{[x-y]_q! [y]_q!}$.

---

[1] This is because of lack of space. The discussion is available also for the case $n < m$.

Algorithm 1 Classical GRS algorithm ($n \geq m$)

Input: $q, m, n, k, w, H, s, \beta = (\beta_1, \ldots, \beta_m), r, B_{m,r}$
Output: e
1: $e \leftarrow 0^n$
2: **while** $\text{rank}_{\mathbb{F}_q}(e)! = w$ **do**
3:      $F = (F_1, \ldots, F_r) \xleftarrow{\$} (B_{m,r})^r$
4:      **for** $\ell := 1$ to $n - k$ **do**
5:          **for** $\ell' := 1$ to n **do**
6:              $\hat{H}[\ell][\ell'] \leftarrow ()$ // $\hat{H}[\ell][\ell'] : (m \times r)$ matrix
7:              **for** $j := 1$ to r **do**
8:                  $H_{\ell,\ell',j} \leftarrow H_{\ell',\ell} \cdot F_r$
9:                  $\hat{H}[\ell][\ell'] \leftarrow \hat{H}[\ell][\ell'] \cup (p_1(H_{\ell,\ell',j}), \ldots, p_m(H_{\ell,\ell',j}))^T$
10:      $\hat{s}[\ell] \leftarrow (p_1(s_\ell), \ldots, p_m(s_\ell))$
11:      $(\lambda_{\ell',j})_{1 \leq \ell' \leq n, 1 \leq j \leq r} \leftarrow \text{GE}(\hat{H}, \hat{s})$
12:      **for** $\ell' := 1$ to n **do**
13:          $e_{\ell'} \leftarrow \sum_{j=1}^{r} \lambda_{\ell',j} F_j$
14: **return** e

Then, let $\ell_{\text{CGRS},n \geq m}$ be the expected number of loop times in the while sentence in Algorithm 1, and $\ell_{\text{CGRS},n \geq m}$ is given in $\ell_{\text{CGRS},n \geq m} = \binom{m}{w}_q / \binom{r}{w}_q$ [1]. Moreover, the computational complexity of Gaussian elimination for the simultaneous equation with $m(n-k)$ unknowns is $O(m^3(n-k)^3)$. Thus, the computational complexity in the classical GRS algorithm is as follows:

$$\begin{cases} O\left((n-k)^3 m^3 \binom{m}{w}_q / \binom{r}{w}_q\right), & n \geq m \\ O\left((n-k)^3 m^3 \binom{n}{w}_q / \binom{r}{w}_q\right), & n < m \end{cases}$$

3 Quantum Gates and Grover's Algorithm

In this section, we introduce Grover's algorithm used for efficient searching. Before we see the algorithm, we first review some quantum gate representation.

3.1 Quantum Gate Representation

We give the matrix representation for some quantum gates.

$$H = \frac{1}{\sqrt{2}}\begin{pmatrix} 1 & 1 \\ 1 & -1 \end{pmatrix}, \; S = \begin{pmatrix} 1 & 0 \\ 0 & i \end{pmatrix}, \; \text{CNOT} = \begin{pmatrix} 1 & 0 & 0 & 0 \\ 0 & 1 & 0 & 0 \\ 0 & 0 & 0 & 1 \\ 0 & 0 & 1 & 0 \end{pmatrix}, \; T = \begin{pmatrix} 1 & 0 \\ 0 & e^{\frac{i\pi}{4}} \end{pmatrix},$$

$$X = \begin{pmatrix} 0 & 1 \\ 1 & 0 \end{pmatrix}, \; Z = \begin{pmatrix} 1 & 0 \\ 0 & -1 \end{pmatrix} \text{ and Toffoli} = \begin{pmatrix} 1 & 0 & 0 & 0 & 0 & 0 & 0 & 0 \\ 0 & 1 & 0 & 0 & 0 & 0 & 0 & 0 \\ 0 & 0 & 1 & 0 & 0 & 0 & 0 & 0 \\ 0 & 0 & 0 & 1 & 0 & 0 & 0 & 0 \\ 0 & 0 & 0 & 0 & 1 & 0 & 0 & 0 \\ 0 & 0 & 0 & 0 & 0 & 1 & 0 & 0 \\ 0 & 0 & 0 & 0 & 0 & 0 & 0 & 1 \\ 0 & 0 & 0 & 0 & 0 & 0 & 1 & 0 \end{pmatrix}.$$

3.2 Grover's Algorithm

Let n be a positive integer, V be $\{0, 1\}^n$, and M be a non-empty subset of V. Let $f : V \to \{0, 1\}$ be the function such that $f(v)$ is 1 if $v \in M$ and 0 otherwise. Grover's algorithm is the quantum algorithm used to search $x_0 \in M$ taking (V, f) as inputs. This algorithm has the computational complexity of $O\left(\sqrt{\frac{|V|}{|M|}}\right)$. Let H^V be the Hilbert space associated with V. U_o and U_d are the unitary operators over H^V and are defined as follows:

$$U_o(|x\rangle) := \begin{cases} -|x\rangle & x \in M \Leftrightarrow f(x) = 1; \\ |x\rangle & \text{o.w.}; \end{cases}$$

$$U_d(|x\rangle) := (2H^{\otimes n}|0\rangle\langle 0|H^{\otimes n} - I_n)|x\rangle.$$

$H^{\otimes n}$ denotes $\underbrace{H \otimes \cdots \otimes H}_{n}$, that is, the Tensor products of n H gates. U_o is called the oracle operator, and U_d is the unitary operator called diffuser. Then, Grover's algorithm is written in Algorithm 2.

4 Quantum GRS Algorithm

In this section, we propose a quantum GRS algorithm that combines the classical GRS algorithm and Grover's algorithm. In the quantum GRS algorithm, by using Grover's algorithm, we can efficiently probe a set F of bases.

Algorithm 2 Grover's algorithm

Input: $V \subset \{0,1\}^n$, $f : V \to \{0,1\}$
Output: $x_0 \in \{0,1\}^n$ s.t. $f(x_0) = 1$
1: $|\psi\rangle \leftarrow |0^n\rangle$
2: $|\psi\rangle \leftarrow H^{\otimes n}|\psi\rangle$
3: **for** $i := 1$ **to** $\left\lfloor \dfrac{\pi}{4\arcsin(\sqrt{\frac{|M|}{|V|}})} \right\rfloor$ **do**
4: $\quad |\psi\rangle \leftarrow U_o|\psi\rangle$
5: $\quad |\psi\rangle \leftarrow U_d|\psi\rangle$
6: **return** $|\psi\rangle$

Suppose $n \geq m$. Denote, by $\text{Grover}_{n \geq m}$, the subroutine using Grover's algorithm. Let $B_{m,r}$ be V in Grover's algorithm. Then, $|B_{m,r}| = \binom{m}{r}_q$ holds. Also, let F_{m-w} be a subspace over $\mathbb{F}_{q^m}$ whose dimension is $m - w$, and V_{m-w} be such a whole F_{m-w}. Moreover, let $V_{m-w,r-w}$ be an $(r - w)$-dim entire subspace included in V_{m-w}. Then, let $F_{m-w,r-w}$ be one of the subspaces in $V_{m-w,r-w}$, and $b_{m-w,r-w}$ be one of the basis of $F_{m-w,r-w}$. That is, $B_{m-w,r-w} = \{b_{m-w} \mid F_{m-w,r-w} \in V_{m-w,r-w}\}$, and it implies $|B_{m-w,r-w}| = \binom{m-w}{r-w}$. Let M in Grover's algorithm be $B_{m-w,r-w}$, and we take f as follows. For $v \in V$, we construct $\hat{H}$ and $\hat{s}$ from (H, s, β, v) and evaluate $(\lambda_{\ell',j})_{\ell',j}$ as in Line 11 in Algorithm 1, and evaluate e as in Line 13. Then, if $\text{rank}_{\mathbb{F}_q}(e) = w$, f returns 1 to indicate $v \in M$, and returns 0 otherwise. Then, $|M|$ is equal to the number of r-dim subspaces including the subspace F_e, generated by e, whose dimension is w. Such a number is equal to the number of $(n - w)$-dimension subspaces in the subspace from $\mathbb{F}_{q^m}^m$ minus F_e, so $|M| = \binom{m-w}{r-w}$. Here,

$$\frac{\binom{m}{r}_q}{\binom{m-w}{r-w}_q} = \frac{\frac{[m]_q!}{[m-r]_q![r]_q!}}{\frac{[m-w]_q!}{[m-r]_q![r-w]_q!}} = \frac{\frac{[m]_q!}{[r]_q!}}{\frac{[m-w]_q!}{[r-w]_q!}} = \frac{\frac{[m]_q!}{[m-w]_q!}}{\frac{[r]_q!}{[r-w]_q!}} = \frac{\frac{[m]_q!}{[m-w]_q![w]_q!}}{\frac{[r]_q!}{[r-w]_q![w]_q!}} = \frac{\binom{m}{w}_q}{\binom{r}{w}_q}.$$

Hence, by $|V|/|M| = \binom{m}{w}_q / \binom{r}{w}_q$, $|V|/|M|$ coincides with the number of iteration in the while sentence in Algorithm 1. That is, in $\text{Grover}_{n \geq m}$, the operations corresponding to the while sentence in Algorithm 1 are executed in $\ell_{\text{QGRS},n \geq m} (= \sqrt{\ell_{\text{CGRS},n \geq m}})$ times. Thus, $\text{Grover}_{n \geq m}$ can be constructed with V, M and f shown above, and given by Algorithm 3.

QRA appearing in Line 6 refers to the quantum random access [12]. Let $A = (a_1, \ldots, a_n)$ be an n-element array each of whose element is of m bits. For A, $1 \leq i \leq m$ and $b \in \mathbb{F}_2^m$, we think about the quantum circuit to compute $b \oplus' a_i$, where $\oplus'$ means the addition in $\mathbb{F}_2^m$. That is, the quantum circuit which returns $(|i\rangle, |b \oplus' a_i\rangle, |A\rangle)$ for an input $(|i\rangle, |b\rangle, |A\rangle)$. Such a quantum circuit is called quantum random access. In particular, when $b = 0^m$, we define $\text{QRA}(|A\rangle, |i\rangle) = |a_i\rangle$ using $(|i\rangle, |0^m\rangle, |A\rangle) \mapsto (|i\rangle, |a_i\rangle, |A\rangle)$. In Line 10, $|H_{\ell',\ell}\rangle \cdot |F_{\psi,j}\rangle$ is the quantum state corresponding to the multiplication of $H_{\ell',\ell}$ and $F_{\psi,j}$ over $\mathbb{F}_{2^m}$. Also in Line 10, for each $i \in [1, m]$, $|p_i(|H_{\ell',\ell}\rangle \cdot |F_{\psi,j}\rangle)\rangle$ represents the quantum state corresponding

Algorithm 3 $\text{Grover}_{n \geq m}$

Input: $w, H, s, \beta, B_{m,r}$
Output: F
1: $|\psi\rangle \leftarrow |0^{|B_{m,r}|}\rangle$
2: $|\psi\rangle \leftarrow H^{\otimes |B_{m,r}|}|\psi\rangle$
3: **for** $x := 1$ **to** $\sqrt{\dfrac{\binom{m}{w}_q}{\binom{r}{w}_q}}$ **do**
4:     $|e\rangle = |e_1\rangle \cdots |e_r\rangle \leftarrow |0^{\log q}\rangle \cdots |0^{\log q}\rangle$
5:     **for** $j := 1$ **to** r **do**
6:        $|F_{\psi,j}\rangle \leftarrow \text{QRA}((\text{QRA}(|B_{m,r}\rangle, |\psi\rangle)), |j\rangle)$
7:     **for** $\ell := 1$ **to** $n - k$ **do**
8:        **for** $\ell' := 1$ **to** n **do**
9:           **for** $j := 1$ **to** r **do**
10:              $|\hat{H}_{\ell,\ell',j}\rangle \leftarrow |p_1(|H_{\ell',\ell}\rangle \cdot |F_{\psi,j}\rangle)\rangle \cdots |p_m(|H_{\ell',\ell}\rangle \cdot |F_{\psi,j}\rangle)\rangle$
11:     **for** $\ell := 1$ **to** $n - k$ **do**
12:        $|\hat{s}_\ell\rangle \leftarrow |p_1(|s_\ell\rangle)\rangle \cdots |p_m(|s_\ell\rangle)\rangle$
13:     $(|\lambda_{\ell',j}\rangle)_{1 \leq \ell' \leq n, 1 \leq j \leq r} \leftarrow \text{GE}(|\hat{H}\rangle, |\hat{s}\rangle)$
14:     **for** $\ell' := 1$ **to** n **do**
15:        $|e_{\ell'}\rangle \leftarrow \displaystyle\sum_{j=1}^{r} |\lambda_{\ell',j}\rangle \cdot |F_{\psi,j}\rangle$
16:     **if** $\text{rank}_{\mathbb{F}_q}(|e\rangle)! = w$ **then**
17:        $|\psi\rangle \leftarrow -|\psi\rangle$
18:     $|\psi\rangle \leftarrow U_d |\psi\rangle$
19: **return** $\text{QRA}(|B_{m,r}\rangle, |\psi\rangle)$

Algorithm 4 Quantum GRS algorithm ($n \geq m$)

Input: $q, m, n, k, w, H, s, \beta = (\beta_1, \ldots, \beta_m), r, B_{m,r}$
Output: $|e\rangle$
1: $|e\rangle = |e_1\rangle \cdots |e_r\rangle \leftarrow |0^{\log q}\rangle \cdots |0^{\log q}\rangle$
2: $|F\rangle = |F_1\rangle \cdots |F_r\rangle \leftarrow \text{Grover}_{n \geq m}(w, H, s, \beta, B_{m,r})$
3: **for** $\ell := 1$ **to** $n - k$ **do**
4:     **for** $\ell' := 1$ **to** n **do**
5:        **for** $j := 1$ **to** r **do**
6:           $|\hat{H}_{\ell,\ell',j}\rangle \leftarrow |p_1(|H_{\ell',\ell}\rangle \cdot |F_j\rangle)\rangle \cdots |p_m(|H_{\ell',\ell}\rangle \cdot |F_j\rangle)\rangle$
7: **for** $\ell := 1$ **to** $n - k$ **do**
8:     $|\hat{s}_\ell\rangle \leftarrow |p_1(|s_\ell\rangle)\rangle \cdots |p_m(|s_\ell\rangle)\rangle$
9: $(|\lambda_{\ell',j}\rangle)_{1 \leq \ell' \leq n, 1 \leq j \leq r} \leftarrow \text{GE}(|\hat{H}\rangle, |\hat{s}\rangle)$
10: **for** $\ell' := 1$ **to** n **do**
11:     **for** $j := 1$ **to** r **do**
12:        $|e_{\ell'}\rangle \leftarrow |e_{\ell'}\rangle + |\lambda_{\ell',j}\rangle \cdot |F_j\rangle$
13: **return** $|e\rangle$

---

to $p_i(H_{\ell',\ell} \odot_m F_{\psi,j})$, where $\odot_m$ represents the multiplication over $\mathbb{F}_{2^m}$. Moreover, GE in Line 13 refers to the subroutine that obtains the solution $\lambda_{\ell',j}$ for the simultaneous equation derived from $|H\rangle$ and $|s\rangle$ applying Gaussian elimination. The quantum GRS algorithm in the case $n \geq m$ is given in Algorithm 4 using this subroutine.

The arguments in the quantum GRS algorithm are the same as those in the classical one. Hence, the computational complexity in the quantum GRS algorithm is as follows:

$$\begin{cases} O\left((n-k)^3 m^3 \sqrt{\binom{m}{w}_q / \binom{r}{w}_q}\right), & n \geq m; \\ O\left((n-k)^3 m^3 \sqrt{\binom{n}{w}_q / \binom{r}{w}_q}\right), & n < m. \end{cases}$$

5 Analysis

We discuss the attack method using the quantum GRS algorithm and its result for the rank SDP with the parameters given in Table 1. Suppose $q = 2$ and β is the standard basis over $\mathbb{F}_{q^m}$, that is, $\beta = (\beta_0, \beta_1, \ldots, \beta_{m-1}) = (1, 2, \ldots, 2^{m-1})$. Also, Table 1 shows the rank SDP instances corresponding to ROLLO and RQC, the cryptosystems remained at the 2nd Round of the NIST PQC standardization project, with security levels of 128, 192 and 256.

In this section, we introduce the G-cost, D-cost, and W-cost to estimate the computational costs of the quantum GRS algorithm and the improved quantum GRS algorithm. Then, we consider quantum circuits that execute various operations using these quantum algorithms. We can combine these circuits to form quantum circuits that execute quantum GRS algorithms. Therefore, we can evaluate the computational costs of the quantum GRS algorithm and its improved version by finding the various operations used in the quantum GRS algorithms and the number of times each is executed. In addition, we can compute the computational costs of the attack algorithms, that is the quantum GRS algorithm and the improved quantum GRS algorithm, for each cryptosystem using the instances (m, n, k, w) in Table 1. Hence, we can determine whether the cryptosystem in Table 1 is secure or not, by comparing the computational costs of the attack algorithms for each cryptosystem with each security level at the computational costs for the circuit corresponding to that bit security.

Table 1 Targeted cryptosystem and bit security

Cryptosystem	Bit security	m	n	k	w
ROLLO	128	67	166	83	7
	192	79	194	97	8
	256	97	226	113	9
RQC	128	127	113	3	7
	192	151	149	5	8
	256	181	179	3	9

5.1 Introducing New Computational Costs

We introduce the computational costs explained in [12]. Let C be a quantum circuit that consists of Clifford+T gates. The G-cost means the number of all the quantum gates appearing in C. The D-cost represents the depth of C, and the W-cost refers to the number of qubits in C. These computational costs are evaluated by $\log_2$. In this paper, we mainly compare the G-cost for a circuit corresponding to the bit security with the G-cost obtained by our strategy from the parameters (m, n, k, w) in Table 1. We can estimate the computational costs (G, D, and W) of quantum GRS algorithms simulated over classical circuits or quantum circuits. The latter ones are the computational costs of quantum circuits consisting of Clifford+T gate and executing quantum GRS algorithms. The former are the computational costs restricted to classical PRAM operations for the latter quantum circuit. As seen in [12], and in this paper, in the case of estimating the computational costs of the quantum GRS algorithms over classical circuits, we do not consider a superposition of quantum states. This is because a superposition that can be executed in quantum RAM operations cannot be simulated in classical ones. Therefore, we do not take into account of the computational costs of Grover's algorithm itself. Then, in the improved quantum GRS algorithm, since we alter only the part concerned with quantum RAM operations, the computational costs of the classical circuit for the quantum GRS algorithm remain the same with those of the classical circuit for the improved version of this algorithm. We consider the G-cost, D-cost, and W-cost of the operations in the quantum GRS algorithms. In addition, we consider the numbers of input qubits and ancilla bits to estimate the W-cost. In the following, let m and n be positive integers, and let $a, b \in \mathbb{F}_{2^m}$ and $e \in \mathbb{F}_{2^m}^n$. Also, let β be the standard basis over $\mathbb{F}_{2^m}$.

Here, we introduce the computational costs for the quantum circuits that execute Gaussian elimination. In the following, let x and y be positive integers, for a matrix $A \in \mathbb{F}_2^{x \times y}$, let $|A\rangle$ be a quantum state corresponding to A. For $\hat{H} \in \mathbb{F}_2^{x \times y}$, let $U \in \mathbb{F}_2^{x \times x}$ and $Q = U\hat{H} \in \mathbb{F}_2^{x \times y}$, where U is a matrix to execute Gaussian elimination for $\hat{H}$. Then, we consider the quantum circuit to take $|\hat{H}\rangle$ and return $|U\rangle$ and $|Q\rangle$. Perriello et al. [21] introduced the quantum circuit which executes Gaussian elimination, and such a quantum circuit has xy input qubits, $\frac{3}{2}(x-1)x$ ancilla qubits, the G-cost of $(x-1)\{x(36y - 20x + \frac{43}{2}) + 8\}$ and the D-cost of $\frac{8}{3}(x-1)x(9y - 4x + 5)$.

5.2 Addition of Quantum Bits Over $\mathbb{F}_{2^m}$

In the following, for $a, b \in \mathbb{F}_{2^m}$, we expand a and b with β, and consider $a = a_1 \ldots a_m, b = b_1 \ldots b_m$, where β is the m-dimensional standard basis as mentioned in Sect. 5.1. We define $|a\rangle \in \mathcal{H}^{\otimes m}$ as the quantum state corresponding to $a \in \mathbb{F}_{2^m}$, therefore, $|a\rangle = |a_1 \ldots a_m\rangle = |a_1\rangle \cdots |a_m\rangle$ holds. Then, we define $|a\rangle + |b\rangle := |a \oplus_m b\rangle$, where $\oplus_m$ means the addition over $\mathbb{F}_{2^m}$. That is, for $a, b \in \mathbb{F}_2^m$, the

Algorithm 5 Addition for classical bits

Input: $a = a_1 \cdots a_m, b = b_1 \cdots b_m$
Output: $c = c_1 \cdots c_m$
1: **for** $i := 1$ to m **do**
2:    $c_i \leftarrow a_i \oplus_1 b_i$
3: **return** c

sum of m-quantum bits $|a\rangle$ and $|b\rangle$ corresponds to $|a \oplus_m b\rangle$, the quantum state of $a \oplus_m b$. In a classical circuit, let $c = a \oplus_m b$, and the algorithm that takes a and b and outputs c is described in Algorithm 5. Similarly, for $1 \leq i \leq m$, we consider the quantum circuit to compute $|a_i \oplus_1 b_i\rangle$ from $|a_i\rangle$ and $|b_i\rangle$. Such a quantum circuit can be constructed with two CNOT gates as shown in Fig. 1. That is, the quantum circuit to realize the sum $|a\rangle + |b\rangle$ of m-quantum bits $|a\rangle$ and $|b\rangle$, has $2m$ input qubits, m ancilla bits, the G-cost of $2m$ and the D-cost of 2.

5.3 Product of Quantum Bits Over $\mathbb{F}_{2^m}$

We define $|a\rangle \cdot |b\rangle := |a \odot_m b\rangle$, where $\odot_m$ means the multiplication over $\mathbb{F}_{2^m}$. In a classical circuit, $\odot_m$ can be realized with m^2 times $\odot_1$ and $(m^2 - 1)$ times $\oplus_1$. Then, for $1 \leq i \leq m$ and $|a_i\rangle, |b_i\rangle \in \mathcal{H}$, $|a_i \odot_1 b_i\rangle$ can be constructed with one Toffoli gate. The construction for the Toffoli gate by Clifford+T gates is given in [25] by Shende. That quantum circuit that executes the Toffoli gate has 3 input qubits, no ancilla bit, the G-cost of 24, and the D-cost of 16. Therefore, the quantum circuit that executes $|a \odot_m b\rangle$ can be constructed with m^2 Toffoli gates and $(m^2 - 1)$ CNOT gates. Such a quantum circuit has $2m$ input qubits, m ancilla bits, the G-cost of $24m^2 + 16(m - 1) = 40m^2 - 16$, and the D-cost of $16m + 2(m - 1) = 18m - 2$ at most.

5.4 Estimating $\mathrm{rank}_{\mathbb{F}_2}(e)$

For $e = (e_1, \ldots, e_n) \in \mathbb{F}_{2^m}^n$ and $1 \leq \ell' \leq n$, by expanding $e_{\ell'}$ with β, we see $e_{\ell'}$ as $e_{\ell'} = (e_{\ell',1}, \ldots, e_{\ell',m}) \in \mathbb{F}_2^m$ and e as $e \in \mathbb{F}_2^{n \times m}$. By the definition of $\mathrm{rank}_{\mathbb{F}_2}(e)$, $\mathrm{rank}_{\mathbb{F}_2}(e)$ is $\dim(\mathrm{Im}(\varphi_e))$, and $\dim(\mathrm{Im}(\varphi_e))$ for $e \in \mathbb{F}_{2^m}^n$ coincides with $\mathrm{rank}(e)$, the

Fig. 1 Addition for one qubit

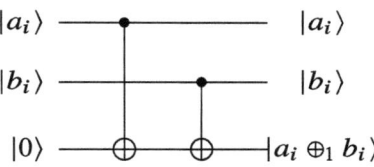

rank of the matrix $e \in \mathbb{F}_2^{n \times m}$. From the above, for $e \in \mathbb{F}_{2^m}^n$, to compute $\mathrm{rank}_{\mathbb{F}_2}(e)$, we consider e as the $n \times m$ matrix over $\mathbb{F}_2$, and execute Gaussian elimination for $e \in \mathbb{F}_2^{n \times m}$. We need nm ancilla bits and nm CNOT gates as the copy of the input qubits because the values of the input qubits are altered during the execution of Gaussian elimination. Therefore, that quantum circuit has nm input qubits, nm ancilla bits, the G-cost of $\mathrm{GE}(n, m) + nm$, and the D-cost of $16(m + 1)$.

5.5 Quantum Random Access (QRA)

For $i \in [1, n]$ and $A = (a_1, \ldots, a_n) \in (\mathbb{F}_2^m)^n$, we think the quantum circuit that executes the quantum random access. By [12], such a quantum circuit has $\log n + m + mn$ input qubits, $nm + n \log n$ ancilla bits, the G-cost of $nm + n \log n$ and the D-cost of $\log m + \log n$.

5.6 Superposition

A superposition can be realized with $|B_{m,r}| = \binom{m}{r}_q$-H gates. That is, the quantum circuit that executes the superposition has $|B_{m,r}|$ input qubits, no ancilla bits, the G-cost of $|B_{m,r}|$, and the D-cost of 1.

5.7 Oracle Phase Flipping (OPF)

In Grover's algorithm, V and M denote the set of all elements and targeted elements, respectively. Then, U_o can be constructed with some X gates. Therefore, the quantum circuit that executes U_o has $\log |V|$ input qubits, no ancilla bit, the G-cost of $4 \log |M|$, and the D-cost of 4.

5.8 Diffuser (Dif)

For the quantum state $|\psi\rangle \in \mathcal{H}^{\otimes |V|}$, let $x = \log |V|$, and the diffuser can be constructed with one $C^{x-1}(Z)$ gate.[2] Therefore, the quantum circuit that performs U_d can be realized with $2(x - 1)$ Toffoli gates and one Z gate. Hence, such a

---

[2] For a positive integer x, $C^x(Z)$ gate means the quantum gate that makes Z gate on the target bit if x-controlled bits are $|1^x\rangle$. Also, $C^x(Z)$ gate can be realized with $2(x - 1)$ Toffoli gates and one Z gate.

quantum circuit has $\log|V|$ input qubits, $(\log|V| - 1)$ ancilla bit, the G-cost of $48(\log|V| - 1) - 46 = 48\log|V| - 94$ and the D-cost of 32.

5.9 Computational Costs of Quantum GRS Algorithm

We discuss the computational costs of the quantum GRS algorithm and the improved version over the classical circuit and the quantum one. Table 2 shows the result up to the previous subsection, where GE represents Gaussian elimination. Here, $G_{GE}(x, y) = (x - 1)\{x(36y - 20x + \frac{43}{2}) + 8\}$ and $D_{GE}(x, y) = \frac{8}{3}(x - 1)x(9y - 4x + 5)$. Also, '—' in Table 2 denotes that there is no such qubits.

For the classical circuit, in estimating the computational costs of the classical circuit to simulate the quantum GRS algorithm, we do not consider the costs for Grover's algorithm. Suppose $n \geq m$. We think about realizing the while sentence in Algorithm 1 over the quantum circuit. In one iteration, $\mathrm{rank}_{\mathbb{F}_q}(e)$ is evaluated once, the multiplication of $H_{\ell,\ell'}F_j$ over $\mathbb{F}_{2^m}$ is $(n-k)nr$ times, the Gaussian elimination for the $nr \times m(n-k)$ matrix over $\mathbb{F}_2$ is once, the multiplication of $\lambda_{\ell',j}F_j$ over $\mathbb{F}_{2^m}$ is $(n-k)nr$ times, and the addition over $\mathbb{F}_{2^m}$ is $(n-1)r$ times. These operations are repeated in ℓ_{QGRS} times.

Moreover, to estimate the computational costs over the quantum circuit, we have to take into account of four extra operations: quantum random access, superposition, the oracle process, and the diffuser per one iteration in Algorithm 4, besides the computational costs in what we have just mentioned above.

In the following, we present the improved quantum GRS algorithm given as Algorithm 7 using the subroutine described below. Thereby the improved one can be more efficient by the improvement for the original quantum GRS algorithm. When $n \geq m$, let $B'_{m,r}$ be a set of $|B_{m,r}|/|B_{m-w,r-w}|$-element basis randomly selected from $B_{m,r}$ without duplication. Let V in Grover's algorithm be $B'_{m,r}$. Define M and f to be the same in Algorithm 4, and we can get the expected value of $|M|$

Table 2 Computational cost for each operation

Operation	G-cost	D-cost	Input	Ancilla
GE	$G_{GE}(x, y)$	$D_{GE}(x, y)$	xy	$\frac{3}{2}(x - 1)x$
Addition	$2m$	2	$2m$	m
Product	$40m^2 - 16$	$18m - 2$	$2m$	m
Rank	$G_{GE}(x, y) + xy$	$16(y + 1)$	xy	xy
QRA	$nm + n\log n$	$\log m + \log n$	$\log n + m + mn$	$nm + n\log n$
Superposition	$|B_{m,r}|$	1	$|B_{m,r}|$	—
OPF	$4\log|M|$	4	$\log|V|$	—
Dif	$48\log|V| - 94$	32	$\log|V|$	$\log|V| - 1$

Algorithm 6 Improved_Grover$_{n \geq m}$

Input: $w, H, s, \beta, B'_{m,r}$
Output: F
1: $|\psi\rangle \leftarrow |0^{|B'_{m,r}|}\rangle$
2: $|\psi\rangle \leftarrow H^{\otimes |B'_{m,r}|}|\psi\rangle$
3: **for** $x := 1$ **to** $\sqrt{\dfrac{\binom{m}{w}_q}{\binom{r}{w}_q}}$ **do**
4:     $|e\rangle = |e_1\rangle \cdots |e_r\rangle \leftarrow |0^{\log q}\rangle \cdots |0^{\log q}\rangle$
5:     **for** $j := 1$ **to** r **do**
6:       $|F_{\psi,j}\rangle \leftarrow \text{QRA}((\text{QRA}(|B'_{m,r}\rangle, |\psi\rangle)), |j\rangle)$
7:     **for** $\ell := 1$ **to** $n - k$ **do**
8:       **for** $\ell' := 1$ **to** n **do**
9:          **for** $j := 1$ **to** r **do**
10:            $|\hat{H}_{\ell,\ell',j}\rangle \leftarrow |p_1(|H_{\ell',\ell}\rangle \cdot |F_{\psi,j}\rangle)\rangle \cdots |p_m(|H_{\ell',\ell}\rangle \cdot |F_{\psi,j}\rangle)\rangle$
11:     **for** $\ell := 1$ **to** $n - k$ **do**
12:       $|\hat{s}_\ell\rangle \leftarrow |p_1(|s_\ell\rangle))\rangle \cdots |p_m(|s_\ell\rangle)\rangle$
13:     $(|\lambda_{\ell',j}\rangle)_{1 \leq \ell' \leq n, 1 \leq j \leq r} \leftarrow \text{GE}(|\hat{H}\rangle, |\hat{s}\rangle)$
14:     **for** $\ell' := 1$ **to** n **do**
15:       $|e_{\ell'}\rangle \leftarrow \sum_{j=1}^{r} |\lambda_{\ell',j}\rangle \cdot |F_{\psi,j}\rangle$
16:     **if** $\text{rank}_{\mathbb{F}_q}(|e\rangle)! = w$ **then**
17:       $|\psi\rangle \leftarrow -|\psi\rangle$
18:     $|\psi\rangle \leftarrow U_d |\psi\rangle$
19: **return** $\text{QRA}(|B_{m,r}\rangle, |\psi\rangle)$

Algorithm 7 Improved quantum GRS algorithm $(n \geq m)$

Input: $q, m, n, k, w, H, s, \beta = (\beta_1, \ldots, \beta_m), r, B'_{m,r}$
Output: e
1: $|e\rangle = |e_1\rangle \cdots |e_r\rangle \leftarrow |0^{\log q}\rangle \cdots |0^{\log q}\rangle$
2: $|F\rangle = |F_1\rangle \cdots |F_r\rangle \leftarrow \text{Improved\_Grover}_{n \geq m}(w, H, s, \beta, B'_{m,r})$
3: **for** $\ell := 1$ **to** $n - k$ **do**
4:     **for** $\ell' := 1$ **to** n **do**
5:       **for** $j := 1$ **to** r **do**
6:          $|\hat{H}_{\ell,\ell',j}\rangle \leftarrow |p_1(|H_{\ell',\ell}\rangle \cdot |F_j\rangle)\rangle \cdots |p_m(|H_{\ell',\ell}\rangle \cdot |F_j\rangle)\rangle$
7: **for** $\ell := 1$ **to** $n - k$ **do**
8:     $|\hat{s}_\ell\rangle \leftarrow |p_1(|s_\ell\rangle))\rangle \cdots |p_m(|s_\ell\rangle)\rangle$
9: $(|\lambda_{\ell',j}\rangle)_{1 \leq \ell' \leq n, 1 \leq j \leq r} \leftarrow \text{GE}(|\hat{H}\rangle, |\hat{s}\rangle)$
10: **for** $\ell' := 1$ **to** n **do**
11:     **for** $j := 1$ **to** r **do**
12:       $|e_{\ell'}\rangle \leftarrow |e_{\ell'}\rangle + |\lambda_{\ell',j}\rangle \cdot |F_j\rangle$
13: **return** $|e\rangle$

is 1. Define Improved_Grover$_{n \geq m}$ as the above subroutine, and we can describe Improved_Grover$_{n \geq m}$ as in Algorithm 6.

We estimate the computational costs for the improved quantum GRS algorithm. We use the denotation of the computational costs given in Table 2. The operations from Line 4 to Line 18 in Algorithm 6 are shown in Table 3.

Table 3 Each operation in Improved Grover$_{n \geq m}$

Line	Operation
6	QRA$(m, \log|V'_{m,r}|)$ and QRA(m, r)
10	pro(m)
13	GE$(m(n-k), nr)$
15	pro(m) and add(m)
16	rank(m, n)
17	OPF(V, M)
18	dif(V)

Here, in Line 17 and 18, $|V| = |V'_{m,r}|$ and $|M| = 1$. Also, $G_{\text{IGroloop},n \geq m}$, $D_{\text{IGroloop},n \geq m}$ and $A_{\text{IGroloop},n \geq m}$ denote the G-cost, D-cost and the number of the ancilla bits from Line 4 to Line 18 in Algorithm 6. Then,

$$G_{\text{IGroloop},n \geq m} = r(G_{\text{QRA}}(m, \log|V'_{m,r}|) + G_{\text{QRA}}(m, r)) + (n-k)nr G_{\text{pro}}(m)$$
$$+ G_{\text{GE}}(m(n-k), nr) + nr(G_{\text{add}}(m) + G_{\text{pro}}(m))$$
$$+ G_{\text{rank}}(m, n) + G_{\text{OPF}}(V, M) + G_{\text{dif}}(V),$$
$$D_{\text{IGroloop},n \geq m} = \max\{r \cdot \max\{D_{\text{QRA}}(m, \log|V'_{m,r}|), D_{\text{QRA}}(m, r)\}$$
$$, (n-k)nr D_{\text{pro}}(m), D_{\text{GE}}(m(n-k), nr), nr \cdot D_{\text{add}}(m)$$
$$, nr \cdot D_{\text{pro}}(m), D_{\text{rank}}(m, n), D_{\text{OPF}}(V, M), D_{\text{dif}}(V)\},$$
$$A_{\text{IGroloop},n \geq m} = r(A_{\text{QRA}}(m, \log|V'_{m,r}|) + A_{\text{QRA}}(m, r)) + (n-k)nr A_{\text{pro}}(m)$$
$$+ A_{\text{GE}}(m(n-k), nr) + nr(A_{\text{add}}(m) + A_{\text{pro}}(m))$$
$$+ A_{\text{rank}}(m, n) + A_{\text{OPF}}(V, M) + A_{\text{dif}}(V).$$

The for sentence is executed in $\ell_{\text{QGRS},n \geq m}$ times. $G_{\text{IGro},n \geq m}$, $D_{\text{IGro},n \geq m}$ and $A_{\text{IGro},n \geq m}$ denote the G-cost, D-cost and the number of the ancilla bits in Algorithm 6. Hence, p_{Igro} processors are in parallel, the following holds:

$$G_{\text{IGro},n \geq m} = G_{\text{IGroloop},n \geq m} \cdot \ell_{\text{QGRS},n \geq m} \cdot \sqrt{p_{\text{Igro}}} + G_{\text{QRA}}(m, \log|V'_{m,r}|),$$
$$D_{\text{IGro},n \geq m} = \max\{D_{\text{IGroloop},n \geq m} \cdot \ell_{\text{QGRS},n \geq m} \cdot \frac{1}{\sqrt{p_{\text{Igro}}}}, D_{\text{QRA}}(m, \log|V'_{m,r}|)\},$$
$$A_{\text{IGro},n \geq m} = A_{\text{IGroloop},n \geq m} \cdot \ell_{\text{QGRS},n \geq m} \cdot p_{\text{Igro}} + A_{\text{QRA}}(m, \log|V'_{m,r}|).$$

Therefore, $G_{\text{IQGRS},n \geq m}$, $D_{\text{IQGRS},n \geq m}$ and $A_{\text{IQGRS},n \geq m}$ denote the G-cost, D-cost, and the number of the ancilla bits in Algorithm 7. Then,

$$G_{\text{IQGRS},n \geq m} = G_{\text{IGro},n \geq m} + (n-k)nr G_{\text{pro}}(m)$$
$$+ G_{\text{GE}}(m(n-k), nr) + nr(G_{\text{add}}(m) + G_{\text{pro}}(m)),$$
$$D_{\text{IQGRS},n \geq m} = \max\{D_{\text{IGro},n \geq m}, (n-k)nr D_{\text{pro}}(m)$$
$$, D_{\text{GE}}(m(n-k), nr), nr \max\{D_{\text{add}}(m), D_{\text{pro}}(m)\},$$
$$A_{\text{IQGRS},n \geq m} = A_{\text{IGro},n \geq m} + (n-k)nr A_{\text{pro}}(m)$$
$$+ A_{\text{GE}}(m(n-k), nr) + nr(A_{\text{add}}(m) + A_{\text{pro}}(m)).$$

In addition, $I_{\text{IGro},n\geq m}$ and $W_{\text{IGro},n\geq m}$ denote the number of input qubits and the W-cost in Algorithm 7, respectively. Here, $I_{\text{IGro},n\geq m} = m(n-k) \cdot mn + m(n-k) + m^2 + |B'_{m,r}| \cdot m^2$ because Algorithm 7 is the algorithm for H, s, β and $B'_{m,r}$. We estimate the W-cost by $W_{\text{IGro},n\geq m} = I_{\text{IGro},n\geq m} + A_{\text{IGro},n\geq m}$.

5.10 Evaluation Criteria and Parallelizing the Quantum Algorithm

NIST [20] states that the classical circuit corresponding to the 128, 192, and 256-bit security is equivalent to the classical circuit having the 2^{143}, 2^{207}, and 2^{272} classical gates, respectively. Also, the quantum circuit having the 128, 192, and 256 security levels is equivalent to the quantum circuit whose sums of the G-cost and the D-cost are 170, 233, and 298, respectively. Therefore, by considering a classical gate as a RAM operation by a classical computer, we can directly compare the G-cost with the above numbers. Table 4 shows the security evaluation criteria simulated over classical and quantum circuits. For example, for a cryptosystem with a 128 security level, if its G-cost is greater than 143, it is secure against attacks.

Also, the D-cost is limited to 96 or less under the condition from NIST [20]. If the D-cost exceeds 96, we use the parallel Grover in [12]. This is the technique parallelizing Algorithm 2. Let G, D, and W be the G-cost, D-cost, and W-cost, respectively, for the entire statement with one processor. Then G-cost, D-cost and W-cost for the entire statement with p processors are $\sqrt{p}G$, $\frac{1}{\sqrt{p}}D$, and pW, respectively. Therefore, if the D-cost exceeds 96 with one processor, it can be reduced to less than 96 using an appropriate number of processors for parallelizing.

5.11 Result

We made a C implementation with the GNU Multiple Precision arithmetic library of the costs of the previous section for the instances in Table 1. The implementation code is available on https://github.com/AsukaWakasugi/quantum_cryptanalysis/tree/main/rank_CBC/quantum_GRS.

Table 4 Security criteria

Bit security	Classical circuit	Quantum circuit
128	G-cost $\geq$ 143	G-cost + D-cost $\geq$ 170
192	G-cost $\geq$ 207	G-cost + D-cost $\geq$ 233
256	G-cost $\geq$ 272	G-cost + D-cost $\geq$ 298

Table 5 Computational costs for each cryptosystem and bit security

Cryptosystem	Bit security	G-cost	D-cost	W-cost
ROLLO	128	478	95	654
		1344	95	1413
		448	95	258
	192	647	95	908
		1868	95	1980
		616	95	341
	256	894	95	1277
		2784	95	2958
		862	95	464
RQC	128	82	74	71
		492	60	492
		83	**60**	**83**
	192	105	95	95
		926	72	926
		116	**72**	**116**
	256	94	86	82
		764	67	764
		99	**68**	**99**

Table 5 lists the computational costs for each encryption scheme and security level. The upper row for each security level in the table shows the computational costs of the quantum GRS algorithm over the classical circuit. The middle row shows the computational costs of the quantum GRS algorithm over the quantum circuit. The lower row shows the computational costs of the improved quantum GRS algorithm over the quantum circuit.

From the upper and middle rows, in the quantum GRS algorithm, we find that the computational costs over the classical circuit are less than those over the quantum one. There is a difference between the middle row and the lower row because the size of $|V|$ is small in the improved quantum GRS algorithm. In addition, the bolded rows in Table 5 shows that this attack method using the improved quantum GRS algorithm is effective for RQC. In addition, in RQC, the computational costs in the 256 security level are less than those in the 192 security level. One of the factors contributing to this is that the value of the parameter k in the 256 security level is less than that in the 192 security level. In general, for $r \in [w, n-k]$, $\binom{n}{r}_q = \binom{n}{n-r}_q$ holds. Thus, in Table 1, $\binom{n}{n-k}_q = \binom{n}{k}_q < \binom{n}{w}_q$ holds. Therefore, $\binom{n}{r}_q = \binom{n}{k}_q = \binom{179}{3}_2 \approx 2^{530}$ in the 256 security level is smaller than $\binom{n}{r}_q = \binom{149}{5}_2 \approx 2^{722}$. Based on the above discussion, one reason why our attack method is effective for RQC is that the value of the parameter k is small. Then, Table 6 shows the parameter k and

Table 6 Proposal for RQC's parameter k

Bit security	k	G-cost	D-cost	W-cost
128	5	104	67	104
192	8	152	84	152
256	11	215	95	223

Table 7 Computational costs for 128 security level RQC

k	G-cost	D-cost	G-cost + D-cost
3	83	60	143
4	94	63	157
5	104	67	171

Table 8 Computational costs for 192 security level RQC

k	G-cost	D-cost	G-cost + D-cost
5	116	72	188
6	128	76	204
7	140	80	220
8	152	84	236

Table 9 Computational costs for 256 security level RQC

k	G-cost	D-cost	G-cost + D-cost
3	99	67	166
4	113	72	185
5	126	76	202
6	140	81	221
7	153	85	238
8	167	90	257
9	180	94	274
10	198	95	293
11	215	95	310

the computational costs for RQC to be secure against this attack by the improved quantum GRS algorithm when the instances except for k are the same in Table 1.

In moving only the parameter k, we show Tables 7, 8 and 9 for the computational costs of 128, 192, and 256 security level RQC, respectively. In each table, if the value of k is in the lower row, the sum of the corresponding G-cost and D-cost exceeds the criteria given in Table 4.

6 Conclusion

In this paper, for rank CBC, we have proposed the quantum GRS algorithm, the best-known algorithm for the rank SDP. In addition, we have proposed an attack method using the quantum GRS algorithm for rank CBCs that remained at the 2nd Round of the NIST PQC standardization project. As a result, this attack method is effective for RQC, so we have recommended the value of k for RQC that is secure against this attack.

References

1. N. Aragon, P. Gaborit, A. Hauteville, J.P. Tillich, A new algorithm for solving the rank syndrome decoding problem, in *The 2018 IEEE International Symposium on Information Theory (ISIT)* (IEEE, 2018), pp. 2421–2425
2. M. Bardet, P. Briaud et al., An algebraic attack on rank metric code-based cryptosystems, in *Annual International Conference on the Theory and Applications of Cryptographic Techniques* (Springer, 2020), pp. 64–93
3. A. Becker, A. Joux, A. May, A. Meurer, Decoding random binary linear codes in $2^{n/20}$: how 1 + 1 = 0 improves information set decoding, in *Annual International Conference on the Theory and Applications of Cryptographic Techniques* (Springer, 2012), pp. 520–536
4. D.J. Bernstein, Grover vs. McEliece, in *International Workshop on Post-Quantum Cryptography* (Springer, 2010), pp. 73–80
5. J.-F. Biasse, B. Pring, A framework for reducing the overhead of the quantum oracle for use with Grover's algorithm with applications to cryptanalysis of SIKE. J. Math. Cryptol. **15**(1), 143–156 (2020)
6. E.M. Gabidulin, Theory of codes with maximum rank distance. Problemy peredachi informatsii **21**(1), 3–16 (1985)
7. P. Gaborit, G. Murat et al., Low rank parity check codes and their application to cryptography, in *Proceedings of the Workshop on Coding and Cryptography WCC*, vol. 2013 (2013)
8. P. Gaborit, O. Ruatta, J. Schrek, On the complexity of the rank syndrome decoding problem. IEEE Trans. Inf. Theory **62**(2), 1006–1019 (2015)
9. P. Gaborit, G. Zémor, On the hardness of the decoding and the minimum distance problems for rank codes. IEEE Trans. Inf. Theory **62**(12), 7245–7252 (2016)
10. R. Gallager, Low-density parity-check codes. IRE Trans. Inf. Theory **8**(1), 21–28 (1962)
11. L.K. Grover, A fast quantum mechanical algorithm for database search, in *Proceedings of the Twenty-Eighth Annual ACM Symposium on Theory of Computing* (1996), pp. 212–219
12. S. Jaques, J.M. Schanck, Quantum cryptanalysis in the RAM model: claw-finding attacks on SIKE, in *Annual International Cryptology Conference* (Springer, 2019), pp. 32–61
13. D.S. Johnson, A catalog of complexity classes, in *Algorithms and Complexity* (Elsevier, 1990), pp. 67–161
14. G. Kachigar, J.P. Tillich, Quantum information set decoding algorithms, in *International Workshop on Post-Quantum Cryptography*, Lecture Notes in Computer Science, vol. 10346 (Springer, 2017), pp. 69–89
15. F. Levy-dit-Vehel, L. Perret, Algebraic decoding of rank metric codes, in *Proceedings of YACC* (2006), pp. 142–152
16. A. May, A. Meurer, E. Thomae, Decoding random linear codes in $\tilde{O}(2^{0.054n})$, in *International Conference on the Theory and Application of Cryptology and Information Security* (Springer, 2011), pp. 107–124

17. R.J. McEliece, A public-key cryptosystem based on algebraic coding theory. Coding Thv **4244**, 114–116 (1978)
18. C.A. Melchor, N. Aragon et al., Hauteville: Rank quasi cyclic (RQC), in *Second Round Submission to the NIST Post-quantum Cryptography Call* (2019)
19. C.A. Melchor, N. Aragon et al., ROLLO - Rank-Ouroboros, LAKE & LOCKER, in *Second Round Submission to the NIST Post-quantum Cryptography Call* (2019)
20. NIST: Post-Quantum Cryptography, Security (Evaluation Criteria), https://csrc.nist.gov/projects
21. S. Perriello, A. Barenghi, G. Pelosi, A complete quantum circuit to solve the information set decoding problem, in *2021 IEEE International Conference on Quantum Computing and Engineering (QCE)* (IEEE, 2021), pp. 366–377
22. E. Prange, The use of information sets in decoding cyclic codes. IRE Trans. Inf. Theory **8**(5), 5–9 (1962)
23. I.S. Reed, G. Solomon, Polynomial codes over certain finite fields. J. Soc. Ind. Appl. Math. **8**(2), 300–304 (1960)
24. P. Schwabe, B. Westerbaan, Solving binary with Grover's algorithm, in *International Conference on Security, Privacy, and Applied Cryptography Engineering* (Springer, Cham, 2016), pp. 303–322
25. V.V. Shende, I.L. Markov, On the cnot-cost of toffoli gates, arXiv preprint arXiv:0803.2316 (2008)
26. P.W. Shor, Algorithms for quantum computation: discrete logarithms and factoring, in *Proceedings 35th Annual Symposium on Foundations of Computer Science* (IEEE, 1994), pp. 124–134

Open Access This chapter is licensed under the terms of the Creative Commons Attribution 4.0 International License (http://creativecommons.org/licenses/by/4.0/), which permits use, sharing, adaptation, distribution and reproduction in any medium or format, as long as you give appropriate credit to the original author(s) and the source, provide a link to the Creative Commons license and indicate if changes were made.

The images or other third party material in this chapter are included in the chapter's Creative Commons license, unless indicated otherwise in a credit line to the material. If material is not included in the chapter's Creative Commons license and your intended use is not permitted by statutory regulation or exceeds the permitted use, you will need to obtain permission directly from the copyright holder.

Full Key Recovery on RSA from Noisy Binary GCD Operation Sequences

Kenta Tani and Noboru Kunihiro

Abstract In CHES2019, Aldaya et al. reported a vulnerability of the binary GCD algorithm used in RSA key generation in OpenSSL. Furthermore, they proposed an attack that exploits this vulnerability. The attack consists roughly of (1) collecting the sequences of operations performed in the binary GCD algorithm using a side-channel attack, (2) error correction to generate candidate solutions for the LSBs of the secret key, and (3) full key recovery using the Coppersmith algorithm. Tani and Kunihiro proposed novel error correction algorithms in ACNS2023. In ACNS2023, they presented no experimental results about the overall attack, concretely the full key recovery phase. In this paper, we first optimize the parameters of the Coppersmith algorithm. Using the optimized parameters, we apply the Coppersmith algorithm to the candidate solutions and evaluate the success rate and execution time. This allows for a more rigorous evaluation of the success rate and execution time of the overall attack.

Keywords Binary GCD algorithm · Coppersmith algorithm · Full RSA key recovery

1 Introduction

Side-channel attacks are one of the threats to public-key cryptography. Side channel attacks include attacks that observe the power consumption of a computer and attacks that analyze the sound emitted by a computer. When implementing cryptographic schemes, it is necessary to prevent the leakage of secret information due to side-channel attacks. Aldaya et al. pointed out a vulnerability in the algorithm used in OpenSSL to calculate the greatest common divisor [1]. Some versions of OpenSSL use the binaryGCD algorithm to compute the greatest common divisor.

K. Tani · N. Kunihiro (✉)
Department of Computer Science, University of Tsukuba, Tsukuba, Japan
e-mail: kunihiro@cs.tsukuba.ac.jp

This algorithm computes the greatest common divisor by repeatedly dividing by 2 and subtracting. While the binary GCD algorithm is fast, its behavior depends on the input, so the input may be recovered by observing the execution.

Aldaya et al. proposed an attack to recover the secret key of an RSA scheme by exploiting a vulnerability in the binary GCD algorithm. Their attack consists of the following three steps. First, a side channel attack is performed against the binary GCD algorithm used for prime number generation in the RSA scheme to obtain information on the operation sequence. The obtained information contains errors. Next, using the error correction algorithm (AGTB algorithm), multiple candidate solutions for the lower bits of the secret key are output from the operation sequence with errors. Finally, the whole bits of the secret key are recovered. For each candidate solution, we apply the Coppersmith algorithm [3]. The attack is successful if the whole bits of the secret key are obtained by applying the Coppersmith algorithm. To improve the success rate of the Aldaya et al. attack, it is essential to propose an error correction algorithm that outputs correct candidate solutions even under noisy situations. If a more efficient error correction algorithm can be proposed, the overall execution time of the attack will be reduced.

Tani and Kunihiro proposed a novel error correction algorithm at ACNS2023 [6]. Their algorithm is classified as a branch and bound method like Aldaya et al., which repeats the process of generating candidate solutions (Expand) and removing them (Prune). Their proposed algorithm can recover the RSA secret key from the more noisy sequences than the AGTB algorithm. Their algorithm successfully corrects errors with higher probability by using information on the probability distribution of errors. In their evaluation of the error correction algorithm, they assumed that the attack was successful if the candidate solution set output by the algorithm contained a correct answer. Therefore, the evaluation was not rigorous enough to consider the full key recovery of the RSA secret key. In this study, we evaluate the entire attack, including the whole bit recovery, using the Coppersmith algorithm.

1.1 Our Contributions

In this paper, we optimize the parameters in the Coppersmith method. This allows Coppersmith's method to run in less time than the theoretical parameters. In addition, we evaluate the overall attack, including error correction and the whole bit recovery, using our proposed error correction algorithm. Our algorithm recovers the full RSA key more efficiently than the attack of Aldaya et al.

2 Preliminaries

2.1 RSA

Let (N, e) be a public key and (p, q, d) be a secret key. The integers p and q are distinct odd primes, and N is a product of p and q. Furthermore, it holds that $ed \equiv 1 \pmod{(p-1)(q-1)}$.

We first explain the RSA key generation in OpenSSL. The public exponent e is set to be $e = 2^{16} + 1$. We generate p and q such that $\gcd(p-1, e) = 1$ and $\gcd(q-1, e)$, which ensures that there exists an integer d which holds $ed \equiv 1 \pmod{(p-1)(q-1)}$. Versions 1.1.0-1.1.0h and 1.0.2b-1.0.2o in OpenSSL employ the binary GCD algorithm, which includes the vulnerability in computing the greatest common divisors.

2.2 Binary GCD Algorithm

We show the binary GCD algorithm in Algorithm 1.

We follow the same notations as [1, 6]. The division by 2 ($u \leftarrow u/2$ or $v \leftarrow v/2$) is denoted by 'L' and the subtraction ($u \leftarrow u - v$ or $v \leftarrow v - u$) by 'S' in the binary GCD algorithm. We refer to the operation sequence performed by the binary GCD algorithm as LS-sequence. We also denote by Z_i the number of Ls between the i-th S and the $(i + 1)$-th S in the LS-sequence. Z_1 is the number of Ls from the beginning of the LS-sequence to the first S. For an LS-sequence, a Z-sequence $(Z_i)_i$ is obtained. Owing to the nature of the binary GCD algorithm, S is never executed consecutively. Therefore, Z_i always takes a value greater than or equal to 1. We give a small example. Suppose that the LS-sequence is LLSLLLSLS. The Z-sequence is then given by $Z_1 = 2$, $Z_2 = 3$, $Z_3 = 1$.

2.3 Lattice

For m-dimension vector space $\mathbf{R}^m$, define the norm $\|\mathbf{v}\|$ of vector $\mathbf{v} = (v_1, \ldots, v_m) \in \mathbf{R}^m$ by $\sqrt{v_1^2 + \cdots + v_m^2}$. We define a lattice L spanned by linearly independent vectors $\mathbf{b}_1, \ldots, \mathbf{b}_n \in \mathbf{R}^m$ by $L := \{\sum_{i=1}^n x_i \mathbf{b}_i \mid x_1, \ldots, x_n \in \mathbf{Z}\}$. We call $\mathbf{b}_1, \ldots, \mathbf{b}_n$ the basis of lattice L. We say that n is the lattice dimension and write it as $\dim(L)$. We say that L is full-rank if $n = m$.

We construct the basis matrix B of the lattice L whose row vectors are basis vectors $\mathbf{b}_1, \ldots, \mathbf{b}_n \in \mathbf{R}^m$ of the lattice L. In other words, $B := (\mathbf{b}_1 \cdots \mathbf{b}_n)^\top$. The lattice L can be defined by using basis matrix B as $L := \{\mathbf{x}B \mid \mathbf{x} \in \mathbf{Z}^n\}$. We define $\mathrm{vol}(L)$ of lattice L by the volume of parallelepiped $\{\sum_{i=1}^n x_i \mathbf{b}_i \mid x_1, \ldots, x_n \in [0, 1]\}$. It holds

Algorithm 1: binary gcd algorithm

Input : Integers a, b such that $0 < b < a$
Output: gcd (a, b)

1 **begin**
2     $u \leftarrow a$, $v \leftarrow b$, $i \leftarrow 0$
3   **while** even(u) **and** even(v) **do**
4         $u \leftarrow u/2$, $v \leftarrow v/2$, $i \leftarrow i+1$
5   **while** $u \neq 0$ **do**
6   **while** even(u) **do**
7             $u \leftarrow u/2$ /* u-loop */
8   **while** even(v) **do**
9             $v \leftarrow v/2$ /* v-loop */
10   **if** $u \geq v$ **then**
11           $u \leftarrow u - v$ /* sub-step */
12   **else**
13           $v \leftarrow v - u$ /* sub-step */
14   **return** $v \cdot 2^i$

that $\mathrm{vol}(L) = \sqrt{\det(BB^\top)}$ for the basis matrix B. If the lattice L is full-rank, it holds that $\mathrm{vol}(L) = |\det(B)|$.

Regarding the lattice-related problems, the following shortest vector problem is well known.

Problem 1 (*Shortest Vector Problem*) Given the basis of lattice L, find a non-trivial shortest vector in the lattice L.

Under the Gaussian heuristic, the length of the shortest vector is estimated by $\sqrt{n/(2\pi e)}\mathrm{vol}(L)^{\frac{1}{n}}$ for lattice L with dimension n. Note that e is a base of natural logarithm. The shortest vector problem is known as a hard problem, and it is difficult to find the shortest vector for large lattice dimensions in practical time. However, we can effectively find a relatively short vector by using a lattice reduction algorithm. The LLL algorithm is one of the well-known lattice reduction algorithms. Regarding the vector that the LLL algorithm outputs, the following is known.

Theorem 1 *For the basis of n-dimension lattice L, let $\mathbf{v}$ be the shortest vector that the LLL algorithm output. Then, it holds that $\|\mathbf{v}\| \leq 2^{\frac{n-1}{4}} \mathrm{vol}(L)^{\frac{1}{n}}$.*

2.4 Coppersmith Algorithm

Coppersmith algorithm [3] efficiently finds the small solution for univariate modular equations. Formally, the Coppersmith algorithm solves the following problem.

Problem 2 Let N be the composite integer with unknown factors, b be a divisor of N and X be a positive integer. Suppose that it holds that $b \geq N^\beta$, $0 < \beta \leq 1$. Let $f(x)$ be a monic univariate polynomial with degree δ. Find all the solutions x_0 satisfying $|x_0| \leq X$ of the equation

$$f(x_0) \equiv 0 \pmod{b}.$$

Before obtaining the small root x_0 of polynomial $f(x)$ shown in Problem 2, we first obtain a polynomial with root x_0. We can efficiently find an integer solution of a polynomial equation using standard methods, like Newton's method. How to set $g(x)$ is crucial for Coppersmith's algorithm.

Coppersmith algorithm employs Howgrave-Graham's Theorem to obtain $g(x)$. Define the polynomial norm by $\|\sum_i a_i x^i\| := \sqrt{\sum_i a_i^2}$ for a polynomial $\sum_i a_i x^i$.

Theorem 2 (Howgrave-Graham's Theorem [5]) *Let $g(x)$ be a univariate polynomial with n monomials. Let X be a positive integer. If*

1. $|x_0| < X$ and $g(x_0) = 0 \bmod b^m$
2. $\|g(xX)\| \leq \frac{b^m}{\sqrt{n}}$

for a positive integer m, it holds that $g(x_0) = 0$ over integers.

Next, we show how to obtain $g(x)$. Letting t be a positive integer, we define shift-polynomials as follows:

$$\begin{aligned} g_{i,j}(x) &= x^j N^{m-i} f(x)^i \quad (0 \leq i \leq m-1, \, 0 \leq j \leq \delta - 1) \\ h_i(x) &= x^i f(x)^m \quad (0 \leq i \leq t-1) \end{aligned} \quad (1)$$

Let $\mathbf{b}_1, \ldots, \mathbf{b}_{m\delta+t}$ be the vectors with coefficients of $g_{i,j}(xX)$ and $h_i(xX)$. We can easily verify that the vectors $\mathbf{b}_1, \ldots, \mathbf{b}_{m\delta+t}$ are linearly independent. Let $\mathbf{v}$ be the shortest vector which the LLL algorithm outputs for these vectors. We assume that $\|\mathbf{v}\| \leq b^m/\sqrt{m\delta+t}$.

Let $g(x)$ be a polynomial whose coefficient is the element of the vector $\mathbf{v}$. Since $\mathbf{v}$ is an integer linear combination of $\mathbf{b}_1, \ldots, \mathbf{b}_{m\delta+t}$ from the property of LLL algorithm, the polynomial $g(x)$ is also an integer linear combination of polynomials defined in Eq. (1). Then, the polynomial $g(x)$ satisfies Condition (1) in Theorem 2. Furthermore, $g(x)$ also satisfies Condition (2) in Theorem 2 from the assumption. Then, if it satisfies that $\|\mathbf{v}\| \leq b^m/\sqrt{m\delta+t}$, we can obtain the polynomial $g(x)$.

2.4.1 Full Key Recovery from Known Lower Bits

By applying the Coppersmith algorithm, we can recover the secret prime p if half the lower bits of p are known. Assuming $N^{1/2} < p < 2N^{1/2}$, we can reduce the recover problem into Problem 2 with $b = p$, $\beta = 1/2$, $\delta = 1$. Now, we assume that the l

least significant bits of p are known. Let p_{high} be unknown part of p and p_{low} be known part of p. Supposing p_{low} is of l-bit, we have the following equation:

$$p = p_{high} \cdot 2^l + p_{low}.$$

Now, if we can find the small solution $x_0 = p_{high}$ of the equation:

$$2^l x + p_{low} \mod p,$$

we can recover p. By multiplying $\alpha = (2^l)^{-1} \mod N$, we can obtain the monic polynomial $f(x) = x + \alpha p_{low}$.

We derive the appropriate parameters X, m, t that make the Coppersmith algorithm work well. From the assumption, it holds that $\sqrt{N} < p < 2\sqrt{N}$. Then, we have

$$p_{high} = \frac{p - p_{low}}{2^l} < \frac{2\sqrt{N} - p_{low}}{2^l} < \frac{2\sqrt{N}}{2^l}.$$

It is sufficient to set $X = 2\sqrt{N}/2^l$.

We next discuss the setting of m, t. Let L be a lattice spanned by $\{\mathbf{b}_1, \ldots, \mathbf{b}_{m+t}\}$. Noting that the degree δ of f is 1, we can easily obtain that

$$\text{vol}(L) = N^{\frac{1}{2}m(m+1)} X^{\frac{1}{2}(m+t)(m+t-1)} \qquad (2)$$

and the dimension of L is $m + t$. The vector $\mathbf{v}$ satisfies the condition $\|\mathbf{v}\| \leq b^m/\sqrt{m\delta + t} = p^m/\sqrt{m+t}$ from Theorem 1 if it holds that

$$2^{\frac{m+t-1}{4}} \left(N^{\frac{1}{2}m(m+1)} X^{\frac{1}{2}(m+t)(m+t-1)} \right)^{\frac{1}{m+t}} \leq \frac{N^{\frac{1}{2}m}}{\sqrt{m+t}}.$$

By simplifying the above equation, we have

$$X \leq 2^{-\frac{1}{2}} \cdot (m+t)^{-\frac{1}{m+t-1}} \cdot N^{\frac{m(t-1)}{(m+t)(m+t-1)}}. \qquad (3)$$

Then, it is sufficient to choose m, t satisfying Eq. (3).

2.5 The Attack of Aldaya et al.

The attack of Aldaya et al. consists of three steps, as described above. Each of them is briefly described below. First, a side channel attack is performed against the binary GCD algorithm used for prime number generation in RSA cryptosystem to collect LS sequences. This side channel attack consists of a Flush+Reload attack [7] and a Performance Degradation attack [2]. Next, an error correction algorithm is used to output multiple candidate solutions for the lower bits of the prime. Finally, the Coppersmith's method is used to recover the whole bits of the prime.

Sequences obtained by side-channel attacks contain errors. AGTB algorithm addresses errors that change the number of Ls between S and S in the LS series with errors. Their algorithm consists of Expand that generates candidate solutions and Prune that eliminates candidate solutions. In Expand, the bits held by each candidate solution are expanded. In Prune, the number of candidate solutions generated by Expand is reduced. In this case, each candidate solution is removed depending on how the bits are extended in Expand.

For whole prime bit recovery, Coppersmith's method is applied to the set of candidate solutions output by the error correction algorithm. They then proposed a round-robin method for the order in which Coppersmith's method is applied to each candidate solution [1]. In their method, the order in which Coppersmith's method is applied depends on how the bits are expanded in the Expand.

2.6 Error Correction Algorithm in ACNS2023

Tani and Kunihiro proposed the error correction algorithm at [6]. Their algorithm employs Expand generating candidate solutions and Prune omitting candidate solutions. It is classified into branch and bound method. Algorithm 2 shows the algorithm. They employed the generation algorithm of the candidate solution proposed by Heninger–Shacham [4] as Expand. They proposed a novel Prune algorithm.

In this paper, we discuss attacks in which the only errors contained in an LS series with errors are errors that change the number of L's between S and S. We first explain the error correction algorithm described in [6].

Suppose that we have candidate solution $\mathbf{b} := (p', q') \in \mathcal{B}$ of primes p and q from 0-th bit to $(j-1)$-th bit. In Expand of the HS algorithm [4], the j-th bit is expanded using the following equation.

$$p[j] + q[j] \equiv (N - p'q')[j] \pmod{2} \qquad (4)$$

Next, we obtain a Z-sequence from the bit sequence of the candidate solution. For a candidate solution p' of p from 0-th bit to $(j-1)$-th bit, let x'_1 be the following.

$$x'_1 \equiv e^{-1}(p' - 1) \pmod{2^j}$$

We can show that the binary representation of x'_1 has the following structure.

$$x'_1 = 2^{Z_1} + 2^{Z_1 + Z_2} + \cdots + 2^{Z_1 + \cdots + Z_{l'_p}}$$

where $l_{p'}$ is the Hamming weight of $e^{-1}(p'-1) \bmod 2^n$. From the value of x'_1, we can obtain the Z-sequence $(Z_i^{(p')})_{i=1}^{l_{p'}}$ for the candidate solution p'.

Algorithm 2: Error correction algorithm shown in [6]

Input : (N, e), $(\tilde{Z}_i^p)_i$, $(\tilde{Z}_i^q)_i$, m and L.
Output: Set of candidate solutions $\mathcal{B}$.

1 **begin**
2     $p_0 \leftarrow 1$, $q_0 \leftarrow 1$
3     $\mathcal{B} \leftarrow \{(p_0, q_0)\}$
4   **for** *Iterate $m-1$ times* **do**
5        $\mathcal{B}' \leftarrow \emptyset$
6   **foreach** $b \in \mathcal{B}$ **do**
7           $\mathcal{B}' \leftarrow \mathcal{B}' \cup \text{Expand}(b, N)$
8        $\mathcal{B} \leftarrow \text{Prune}(\mathcal{B}', e, L, \tilde{LS}_p, \tilde{LS}_q)$
9   **return** $\mathcal{B}$

In Prune, we discard the candidate solutions by following procedures.

1. Compute the loss value for each candidate solution using the loss function.
2. Keep L candidate solutions with the lowest loss values and remove the remainder.

We then present a framework for the loss functions. Let **x** denote the sequence observed by the side-channel attack and $\mathbf{b} \in \mathcal{B}$ denote the candidate solution generated by Expand. Let $(\tilde{Z}_i^{(p)})_{i=1}^{\tilde{l}_p}$ and $(\tilde{Z}_i^{(q)})_{i=1}^{\tilde{l}_q}$ be Z-sequences obtained by side channel attacks. Let $(Z_i^{(p')})_{i=1}^{l_{p'}}$ and $(Z_i^{(q')})_{i=1}^{l_{q'}}$ be the Z-sequences held by the candidate solution $\mathbf{b} \in \mathcal{B}$. For a function $d : \mathbb{N} \times \mathbb{N} \to \mathbb{R}^+$, we define the loss function as follows.

$$\text{Loss}(\mathbf{x}, \mathbf{b}) := \frac{1}{l_{p'}} d((\tilde{Z}_i^p)_{i=1}^{\tilde{l}_p}, (Z_i^{p'})_{i=1}^{l_{p'}}) + \frac{1}{l_{q'}} d((\tilde{Z}_i^q)_{i=1}^{\tilde{l}_q}, (Z_i^{q'})_{i=1}^{l_{q'}})$$

By properly defining the function d, we can introduce "distance" between **x** and **b**. The function d for the likelihood-based loss function is defined as follows:

$$d((\tilde{Z}_i)_{i=1}^{\tilde{l}}, (Z_i')_{i=1}^{l'}) := -\sum_{i=1}^{l'} \Big(\log \Pr(\tilde{Z}_i = \tilde{z} \mid Z_i' = z) + \log \Pr(Z_i' = z)\Big),$$

where $\log x$ is the logarithm with base 2. Since it is assumed that $\Pr(Z_i = z) = (1/2)^z$ in [6], we know $\Pr(Z_i = z)$. Hence, it is necessary to estimate $\Pr(\tilde{Z}_i = \tilde{z} \mid Z_i = z)$ when we use the likelihood-based loss function. In the norm-based loss function, the function d is as follows:

$$d((\tilde{Z}_i)_{i=1}^{\tilde{l}}, (Z_i')_{i=1}^{l'}) := \left(\sum_{i=1}^{l'} |\tilde{Z}_i - Z_i'|^k\right)^{\frac{1}{k}}.$$

We can calculate the loss values without knowledge about distributions of errors.

3 Parameter Optimization for Coppersmith's Algorithm

In this section, we discuss optimizing parameters involved in the Coppersmith algorithm, which recovers the whole bit of prime. For applying the Coppersmith attack, we must adequately set the parameters (X, m, t). We will optimize the parameters by focusing on the situation described in Sect. 2.4.1. First, we obtain the optimal parameters by theoretical analysis. Then, by modifying their values, we obtain the optimal parameters by experiments.

Suppose that the lower l-bit of the prime p is known. It is sufficient to fix $X = 2\sqrt{N}/2^l$ from the discussion of Sect. 2.4.1. The conditions which parameters X, m, t should satisfy are given by Eq. (3). Here, the term

$$N^{\frac{m(t-1)}{(m+t)(m+t-1)}}$$

is dominant in Eq. (3). Letting the function $k(t)$ be $k(t) := m(t-1)/((m+t)(m+t-1))$, we can rewrite the term into $N^{k(t)}$.
The derivation of $k(t)$ with t is given as follows:

$$\frac{dk}{dt} = \frac{-mt^2 + 2mt + m^3 + m^2 - m}{(m+t)^2(m+t-1)^2}$$

If and only is $-\sqrt{m^2 + m} + 1 \leq t \leq \sqrt{m^2 + m} + 1$, it holds that $dk/dt \geq 0$ and otherwise $dk/dt < 0$. Then, the function is maximum when $t = \sqrt{m^2 + m} + 1$ for fixed integer m. Since t should be an integer, $t = \lceil \sqrt{m^2 + m} + 1 \rceil, \lfloor \sqrt{m^2 + m} + 1 \rfloor = m+1, m+2$. Then, $m(t-1)/((m+t)(m+t-1))$ is maximum for fixed m. For the both case of $t = m+1, m+2$, the values fall into $N^{\frac{m}{2(2m+1)}}$. For simplicity, hereafter, we assume that $t = m+1$. If we substitute $X = 2\sqrt{N}/2^l, t = m+1$ into Eq. (3), we have

$$2^{-l} \leq 2^{-\frac{3}{2}} \cdot (2m+1)^{-\frac{1}{2m}} \cdot N^{\frac{m}{2(2m+1)} - \frac{1}{2}}$$

$$2^{-l} \leq 2^{-\frac{3}{2}} \cdot 2^{-\frac{1}{2m}\log(2m+1)} \cdot 2^{\left(\frac{m}{2(2m+1)} - \frac{1}{2}\right)\log N}.$$

Summing up, the condition for the number of known bits l and m is given by

$$l \geq \frac{3}{2} + \frac{1}{2m}\log(2m+1) + \frac{m+1}{2(2m+1)}\log N.$$

Let $\lfloor \log N \rfloor = 2047$ for 2048-bit RSA. Then, the following inequation should hold

$$l \geq \frac{3}{2} + \frac{1}{2m}\log(2m+1) + \frac{2047(m+1)}{2(2m+1)}. \quad (5)$$

Then, it is enough to choose the smallest m satisfying Eq. (5) when the l least significant bits are known in the prime p. Table 1 shows the smallest m satisfying

Table 1 The smallest m satisfying Eq. (5) and the lattice dimension

l	512	517	522	527	532	537	542
m	–	69	30	19	14	11	9
Dimension ($m+t$)	–	139	61	39	29	23	19

Eq. (5) and the lattice dimension ($:= m + t = 2m + 1$) for concrete values l. Since each prime is 1024-bit long for 2048-bit RSA, the Coppersmith algorithm theoretically recovers the whole bit of each prime for $l \geq 512$.

3.1 Parameter Optimization for Experiments

It is well-known that there is a gap between theoretical analysis and experimental performance. Concretely, in most cases, LLL algorithm outputs a shorter vector than one described in Theorem 1. Hence, we can expect to recover the whole bit of prime even if we use small m, which does not satisfy Eq. (5). In this section, we obtain optimal parameters X, m, t by modifying those derived from theoretical analysis. We can expect to reduce the total computational time.

We give how to determine the optimal parameters experimentally. Regarding with X, set $X = c \cdot 2\sqrt{N}/2^l$ for parameters $c = 1.0, 0.95, \ldots, 0.1, 0.05$ and conduct experiments for each X. Theoretical derived X corresponds to $c = 1.0$. Aldaya et al. [1] conducted experiments under $c = 0.5$. In this section, we conduct experiments with more choices of c than Aldaya et al. and obtain the optimal value of c. We set $t = m + 1$, which is the same value as theoretically obtained. We conduct experiments for adequately setting m, c for $522 \leq l \leq 542$. The flow of experiments is given as follows.

1. Set l, m, c.
2. Generate 1024-bit prime.
3. Apply Coppersmith algorithm.

We consider that the attack succeeds if we recover the upper part of the prime, and consequently, we can recover the whole of the prime. We set m as follows: for each l, first obtain the smallest m satisfying Eq. (5) and then decrease m by 1.

We conduct experiments and find the parameters c, m with the success rate 100% (the number of trials is 100), and the computational time is minimum for each l. Table 2 shows such parameters. Figure 1 shows the average computational time for parameters derived from theoretical analysis and parameters shown in Table 2. We conducted experiments by using computers with Intel Xeon Silver 4110 CPU@2.10GHz. We implement the program by using SageMath. In particular, we use the function LLL() in SageMath for the LLL algorithm.

Table 2 Optimal parameters c, m derived from experiments

l	(c, m)	Dimension	Time [s]	l	(c, m)	Dimension	Time [s]
542	(0.6, 9)	19	1.5	532	(0.3, 13)	27	7.4
541	(0.3, 9)	19	1.4	531	(0.6, 14)	29	10.8
540	(0.6, 9)	19	1.4	530	(0.6, 14)	29	10.6
539	(0.6, 10)	21	2.2	529	(0.55, 15)	31	15.6
538	(0.55, 10)	21	2.2	528	(0.6, 16)	33	22.2
537	(0.45, 10)	21	2.2	527	(0.5, 17)	35	31.9
536	(0.6, 11)	23	3.3	526	(0.55, 18)	37	44.5
535	(0.6, 11)	23	3.3	525	(0.4, 20)	41	86.9
534	(0.25, 12)	25	5.0	524	(0.35, 22)	45	157.2
533	(0.6, 12)	25	4.9	523	(0.4, 24)	49	265.6
				522	(0.5, 26)	53	421.4

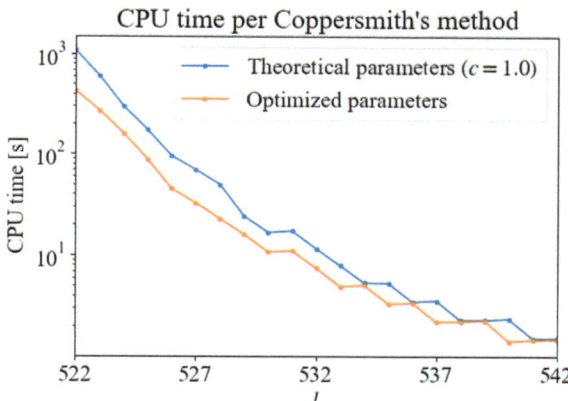

Fig. 1 Average computational time for each parameter

4 Evaluation for Full Key Recovery

We will present the whole computational time for full key recovery and success rate of the attack of Aldaya et al. and our algorithm. We first apply an error correction algorithm to artificially generated sequences and then obtain the set of candidate solutions. Then, we apply the Coppersmith algorithm to each candidate solution in an adequate order. We evaluate the computational time for the full key recovery phase, including the error correction phase and the success rate.

4.1 How to Artificially Generate Z-Sequences with Errors

We define the error probability distribution as

$$\Pr(\tilde{z} \mid z) = \begin{cases} 1 - \varepsilon & \text{if } \tilde{z} = z \\ \varepsilon/2 & \text{if } \tilde{z} = z - 1 \\ \varepsilon/2 & \text{if } \tilde{z} = z + 1, \\ 0 & \text{otherwise} \end{cases} \quad (6)$$

where $0 \leq \varepsilon \leq 1$. We sample Z-sequence following the above distribution.

We explain how to generate artificially error-generated Z-sequence in detail.

1. Generate 2048-bit RSA key (N, p, q, d) by using OpenSSL.
2. Obtain the Z-sequence for each p and q.
3. Fix ε and sample errors following $\Pr(\tilde{z} \mid z)$.

By following the above procedure, we sample noisy Z-sequence of one hundred pair of (p, q) with $\varepsilon = 0.05, 0.10, 0.20$.

Algorithm 3 presents the flow of experiments. It consists of an error correction phase and full key recovery phase, executed by applying the Coppersmith algorithm.

For the whole bit of candidate solution p_{cand}, if $\gcd(N, p_{\text{cand}}) \neq 1, N$, we find the prime factors of public key N. In that case, we consider that the attack succeeds.

4.2 Setting of Experiments

The experiments in this section follow those conducted by Aldaya et al. [1]. The attack flow is shown in Algorithm 3. It consists of an error correction phase and a phase in which Coppersmith's method is applied to the candidate solutions output by the error correction algorithm. First, the error correction algorithm takes the Z-sequence with errors as input and outputs a set of candidate solutions for the lower l bits of prime numbers. Next, Coppersmith's method is applied to each candidate solution in the candidate solution set to recover the whole bits. Since each candidate solution holds a bit sequence for two primes, the Coppersmith algorithm is applied twice to one candidate solution. Let p_{cand} be the whole bit of candidate output from the Coppersmith algorithm. We can consider the attack succeeds if it holds that $\gcd(N, p_{\text{cand}}) \neq 1, N$.

For the error correction phase, we use error correction algorithms proposed in [6] and AGTB algorithm [1]. First, we explain the setting related with [6]. We use the following four types of the loss function in the error correction algorithm: L1 norm-based loss function, L2 norm-based loss function, and two types of likelihood-based with $\varepsilon' = \varepsilon$ and $\varepsilon' = 0.10$. In all experiments related with [6], we set $L = 2^{10}$.

Algorithm 3: Attack Flow
Input : (N, e), $(\tilde{\mathbf{Z}}_p, \tilde{\mathbf{Z}}_q)$, c and m.
Output: Success or Failure.
1 begin
// Error correction phase
2 $\mathcal{B} \leftarrow \text{ErrorCorrection}(\tilde{\mathbf{Z}}_p, \tilde{\mathbf{Z}}_q)$
// Full-recovery phase
3   **foreach** $b \in \mathcal{B}$ **do**
4   **foreach** $p' \in b$ **do**
5 $p_{\text{candidate}} \leftarrow \text{Coppersmith}(p', c, m)$
6   **if** $\gcd(N, p_{\text{cand}}) \neq 1$, N **then**
7   **return** *Success*
8   **return** *Failure*

Next, we explain the setting related with [1]. The parameters shown in [1] are given by g, G, cons, th. The number of candidate solutions is at most $g \times G$. Here, we consider the following three setting ($g = 2^2$, $G = 2^{14}$, cons = 3, th = 200), ($g = 2^3$, $G = 2^{13}$, cons = 3, th = 200), ($g = 2^4$, $G = 2^{12}$, cons = 3, th = 200). Note that the number of candidate solutions is at most $g \times G = 2^{16}$. The details of the parameter setting are given in Appendix A. We explain the order of candidate solutions in applying the Coppersmith algorithm. For the full key recovery phase, we apply the Coppersmith algorithm to the candidate solution in decreasing order of loss function. For experiments for the attack of Aldaya et al., we apply the Coppersmith algorithm in the round-robin method shown in Sect. 2.5. In our experiments, we stop the execution if the computational time for full key recovery phase exceeds 10 hours.

4.3 Experimental Results

We show the success rate for each ε. Tables 3, 4, 5 show the success rates for $\varepsilon = 0.05, 0.10, 0.20$, respectively. We conducted experiments on the computer with an Intel Xeon Silver 4110 CPU @ 2.10GHz. Success rates for both our attack and the attack of Aldaya et al. change little if l varies. Further, the success rate for our attack is higher for every choice of l.

Table 3 Success rate for $\varepsilon = 0.05$

l	522	526	530	534	538	542
L1 norm	99%	99%	99%	99%	99%	99%
L2 norm	99%	99%	99%	99%	99%	99%
Likelihood ($\varepsilon' = \varepsilon$)	100%	100%	100%	100%	100%	100%
Likelihood ($\varepsilon' = 0.10$)	100%	100%	100%	100%	100%	100%
Aldaya ($g = 2^2$)	97%	97%	97%	97%	97%	97%
Aldaya ($g = 2^3$)	97%	97%	97%	97%	97%	97%
Aldaya ($g = 2^4$)	97%	97%	97%	97%	97%	97%

Table 4 Success rate for $\varepsilon = 0.10$

l	522	526	530	534	538	542
L1 norm	100%	100%	100%	100%	100%	100%
L2 norm	100%	100%	100%	100%	100%	100%
Likelihood ($\varepsilon' = \varepsilon$)	100%	100%	100%	100%	100%	100%
Likelihood ($\varepsilon' = 0.10$)	100%	100%	100%	100%	100%	100%
Aldaya ($g = 2^2$)	78%	78%	78%	78%	77%	77%
Aldaya ($g = 2^3$)	77%	77%	77%	77%	76%	76%
Aldaya ($g = 2^4$)	77%	77%	77%	77%	76%	76%

Table 5 Success rate for $\varepsilon = 0.20$

l	522	526	530	534	538	542
L1 norm	96%	96%	97%	97%	97%	96%
L2 norm	100%	99%	100%	100%	100%	100%
Likelihood ($\varepsilon' = \varepsilon$)	100%	99%	100%	100%	100%	100%
Likelihood ($\varepsilon' = 0.10$)	100%	99%	100%	100%	100%	100%
Aldaya ($g = 2^2$)	30%	30%	30%	30%	28%	28%
Aldaya ($g = 2^3$)	37%	37%	37%	37%	34%	34%
Aldaya ($g = 2^4$)	33%	34%	34%	35%	31%	31%

Next, we show the computational time for the error correction step and the full key recovery step. Figures 2, 3, 4 show the average computational time for $\varepsilon = 0.05, 0.10, 0.20$. From the figures, we can find that the computational time for our proposed attack is smaller than that of Aldaya et al. for the error correction phase. The reason is that our algorithm is enough to keep fewer solution candidates. We can see that the computational time for the attack of Aldaya et al. varies if we choose the different pairs of g, G under the restriction $g \times G = 2^{16}$. The computational time for

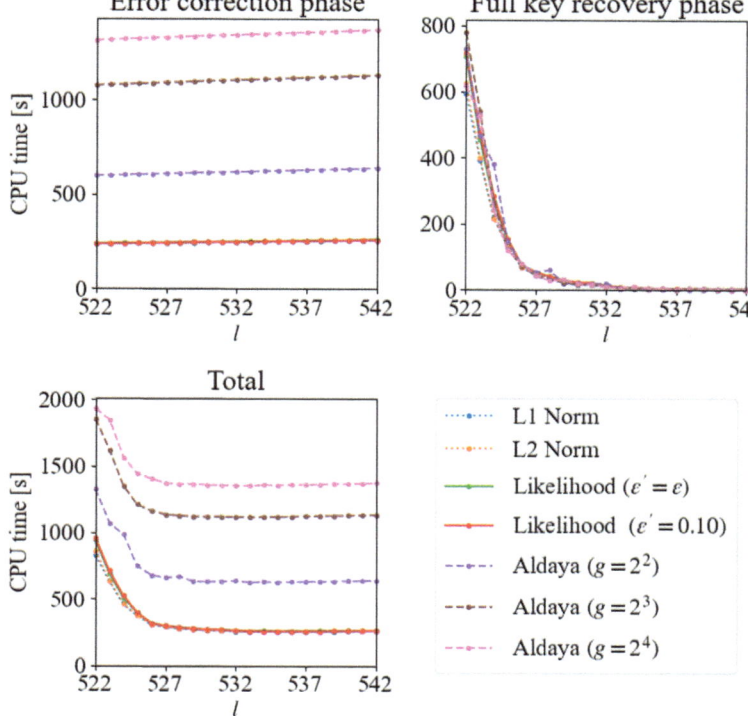

Fig. 2 Computational time for $\varepsilon = 0.05$

the full key recovery phase for our algorithm is smaller for any l. We can conclude that our algorithm succeeds in ordering the candidate solutions appropriately. On the other hand, the computational time for the attack of Aldaya et al. is unstable, especially for larger ε.

We conclude the above discussion. Under $l \geq 532$, our algorithm achieves a high success rate while keeping the overall attack execution time small. Furthermore, our attack achieves a higher success rate than that of Aldaya et al. with a smaller number of candidate solutions and a shorter execution time. This indicates that our algorithm can be used to attack more efficiently in more noisy situations.

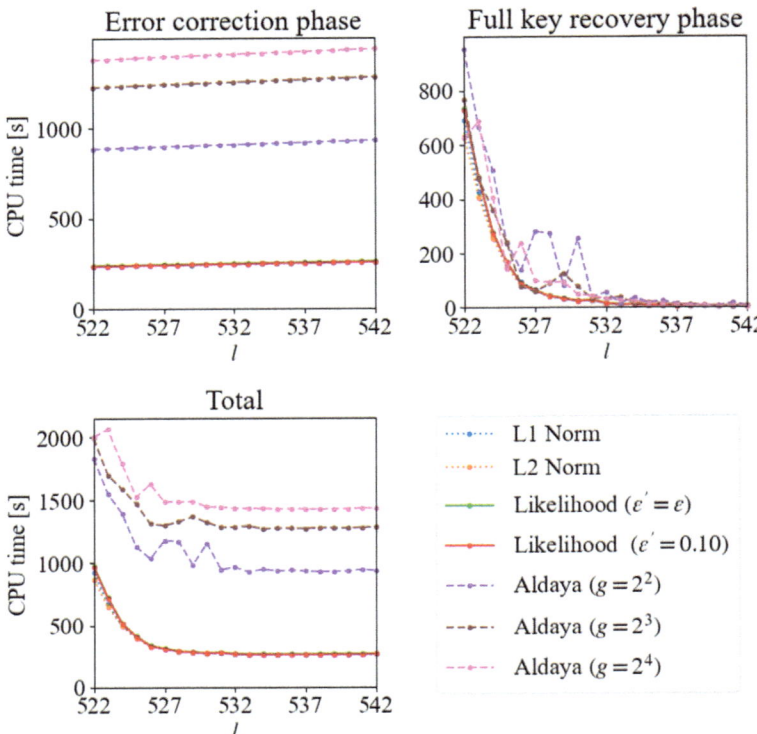

Fig. 3 Computational time for $\varepsilon = 0.10$

5 Conclusions

In this paper, we optimized the parameters of the Coppersmith algorithm in RSA key recovery from the noisy binary GCD operation sequences. We evaluated our previously proposed error correction algorithm. We showed that our algorithm can efficiently correct errors with a small number of candidate solutions and a small running time. This shows that our algorithm can successfully recover the RSA secret key under noisy situations.

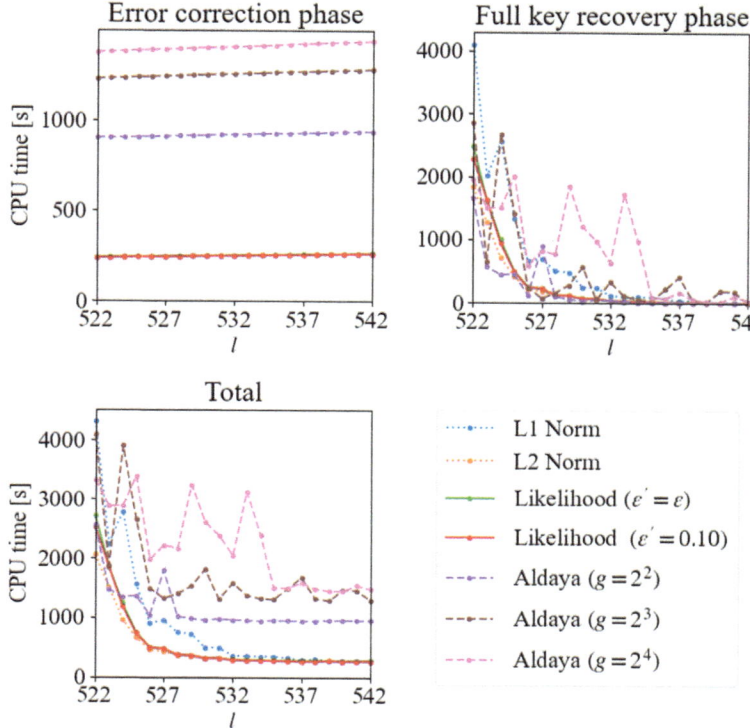

Fig. 4 Computational time for $\varepsilon = 0.20$

Acknowledgements This work was supported by JST CREST Grant Number JPMJCR2113 and JSPS KAKENHI Grant Number 21H03440.

Appendix A: Details of Experiments for AGTB Algorithm

We modify the AGTB algorithm to deal with errors that occur according to Eq. (6). The AGTB algorithm cannot perform error correction when the value $\tilde{Z}_i$ of Z_i, an element of the Z-sequence with error, is 0. On the other hand, in Eq. (6), the noisy value $\tilde{Z}_i$ may be zero. Due to the properties of the binary GCD algorithm, $Z_i \geq 1$ always holds since the subtraction is never performed in succession. Therefore, as a preprocess, the operation $Z_i \leftarrow 1$ is applied to all elements for which $\tilde{Z}_i = 0$ for the input Z sequence. This preprocess allows the AGTB algorithm to accommodate the error defined by Eq. (6). Moreover, this preprocess does not work against the AGTB algorithm.

We will explain the parameter setting for the error correction algorithm. The AGTB algorithm performs error correction based on the parameters g, G, cons, th.

In this appendix, we evaluate how the success rate of the results changes when these parameters are varied. The noisy sequences used is generated in the same way as the noisy sequences in the Sect. 4.3. The AGTB algorithm is considered successful if the candidate solution set output by the algorithm contains a candidate solution that matches the lower 512 bits of the correct prime and is unsuccessful if it does not. The experimental results of 100 experiments for each ε are shown in Figs. 5, 6, 7.

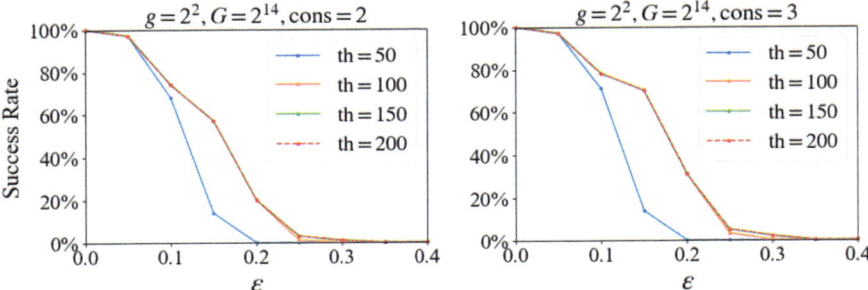

Fig. 5 Success rate for $g = 2^2$, $G = 2^{14}$

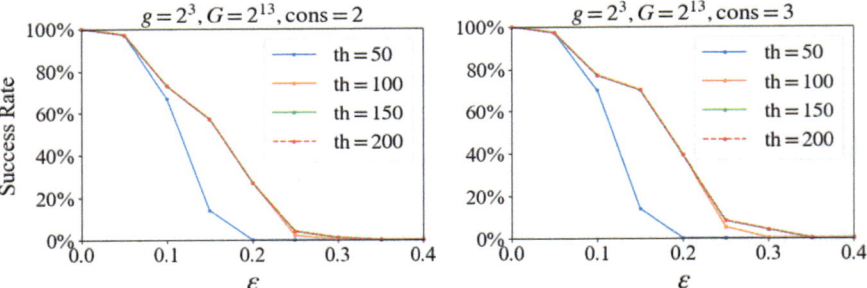

Fig. 6 Success rate for $g = 2^3$, $G = 2^{13}$

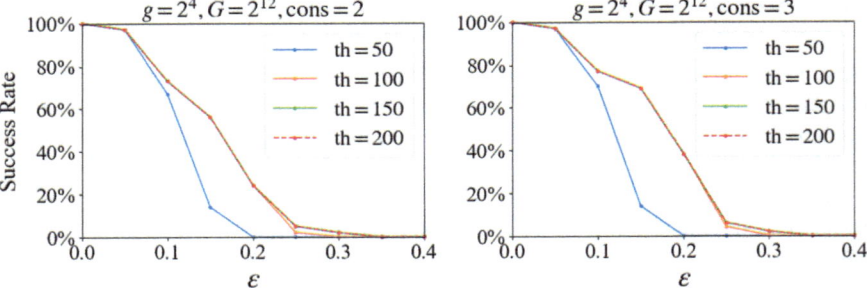

Fig. 7 Success rate for $g = 2^4$, $G = 2^{12}$

From these experimental results, in Sect. 4.3, we use $(g = 2^2, G = 2^{14}, \text{cons} = 3, \text{th} = 200)$, $(g = 2^3, G = 2^{13}, \text{cons} = 3, \text{th} = 200)$, $(g = 2^4, G = 2^{12}, \text{cons} = 3, \text{th} = 200)$ for the parameters in the AGTB algorithm.

References

1. A.C. Aldaya, C.P. García, L.M.A. Tapia, B.B. Brumley, Cache-timing attacks on RSA key generation. IACR Trans. Cryptogr. Hardw. Embed. Syst. **2019**(4), 213–242 (2019)
2. T. Allan, B.B. Brumley, K.E. Falkner, J. van de Pol, Y. Yarom, Amplifying side channels through performance degradation, in *Proceedings of the 32nd Annual Conference on Computer Security Applications, ACSAC 2016*, Los Angeles, CA, USA, Dec 5–9, 2016, ed. by S. Schwab, W.K. Robertson, D. Balzarotti (ACM, 2016), pp. 422–435
3. D. Coppersmith, Finding a small root of a univariate modular equation, *Advances in Cryptology - EUROCRYPT '96, International Conference on the Theory and Application of Cryptographic Techniques*, Saragossa, Spain, May 12–16, 1996, Proceeding, ed. by U.M. Maurer, Lecture Notes in Computer Science, vol. 1070 (Springer, 1996), pp. 155–165
4. N. Heninger, H. Shacham, Reconstructing RSA, private keys from random key bits, *Advances in Cryptology - CRYPTO, (2009) 29th Annual International Cryptology Conference*, Santa Barbara, CA, USA, August 16–20, 2009. Proceedings, ed. by S. Halevi, Lecture Notes in Computer Science, vol. 5677 (Springer 2009), pp. 1–17
5. N. Howgrave-Graham, Finding small roots of univariate modular equations revisited, *Cryptography and Coding, 6th IMA International Conference*, Cirencester, UK, Dec 17–19 (1997) Proceedings, ed. by M. Darnell, Lecture Notes in Computer Science, vol. 1355 (Springer, 1997), pp. 131–142
6. K. Tani, N. Kunihiro, Hs-based error correction algorithm for noisy binary gcd side-channel sequences, *Applied Cryptography and Network Security (Cham)*, ed. by M. Tibouchi, X. Wang (Springer Nature Switzerland, 2023), pp. 59–88
7. Y. Yarom, K. Falkner, FLUSH+RELOAD: a high resolution, low noise, L3 cache side-channel attack, *Proceedings of the 23rd USENIX Security Symposium*, San Diego, CA, USA, Aug 20–22, 2014, ed. by K. Fu, J. Jung (USENIX Association, 2014), pp. 719–732

Open Access This chapter is licensed under the terms of the Creative Commons Attribution 4.0 International License (http://creativecommons.org/licenses/by/4.0/), which permits use, sharing, adaptation, distribution and reproduction in any medium or format, as long as you give appropriate credit to the original author(s) and the source, provide a link to the Creative Commons license and indicate if changes were made.

The images or other third party material in this chapter are included in the chapter's Creative Commons license, unless indicated otherwise in a credit line to the material. If material is not included in the chapter's Creative Commons license and your intended use is not permitted by statutory regulation or exceeds the permitted use, you will need to obtain permission directly from the copyright holder.

Performance Analysis of Fault Attack on UOV Multivariate Signature Scheme

Hiroki Furue, Tatsuya Nagasawa, and Tsuyoshi Takagi

Abstract The unbalanced oil and vinegar signature scheme (UOV), which is one of the multivariate signature schemes, is expected to be secure against quantum attacks. To achieve cryptosystem security in a practical manner, we need to deal with security against physical attacks such as fault attacks, which generate computational errors to lead to security failures. We simulate the performance of a fault attack on UOV proposed by Furue et al. at PQCrypto 2022, which uses faults occurring on the secret key. This attack first recovers a part of the linear map of the secret key by utilizing faults occurring on the secret key, and then transforms the public key system. In this paper, we simulate the fault attack by Furue et al. in the following two conditions: the case where the attack is completed and the case where the number of faults is limited. For two practical parameter sets satisfying 100-bit security, in the first case, we confirmed that the attack breaks the parameters by smaller than 100-bit manipulations with approximately 90% probability and by smaller than 50-bit manipulations with approximately 30% probability. In the second case, our simulation shows that the attack breaks the claimed security level at least 60% probability with only two faults on the central map.

Keywords Post-quantum cryptography · Multivariate public key cryptography · UOV · Fault attack

H. Furue (✉) · T. Nagasawa · T. Takagi
Department of Mathematical Informatics, The University of Tokyo, Tokyo, Japan
e-mail: hiroki.furue@ntt.com

T. Takagi
e-mail: takagi@g.ecc.u-tokyo.ac.jp

1 Introduction

Currently used public key cryptosystems such as RSA and ECC can be broken in polynomial time using Shor's algorithm [23], which is a quantum computer algorithm. Thus, the amount of research conducted on post-quantum cryptography (PQC), which is secure against quantum computing attacks, has been accelerating. Indeed, the U.S. National Institute for Standards and Technology (NIST) has initiated a PQC standardization project [20]. Among various PQC candidates, multivariate public-Multivariate public key cryptography key cryptography (MPKC) is one of the main categories. MPKCs are cryptosystems constructed based on the difficulty of solving a system of multivariate quadratic polynomial equations over a finite field (the multivariate quadratic (MQ) problem). The MQ problem is NP-complete [14] and is thus likely to be secure against quantum computers.

The unbalanced oil and vinegar signature scheme (UOV) is a multivariate signature scheme proposed by Kipnis et al. [16]. UOV has withstood various types of attacks since 1999 and thus is considered to achieve sufficient security. In addition, UOV is a well-established signature scheme owing to its short signature and brief execution time. Indeed, several studies [5, 7, 8, 25] have presented efficient implementations of UOV and Rainbow [9], the latter of which is a multilayer UOV variant selected as a third-round finalist in the NIST PQC project [21]. Concretely, the public key of UOV is a quadratic map $\mathcal{P}$ constructed using $\mathcal{P} = \mathcal{F} \circ \mathcal{T}$ with an invertible quadratic map $\mathcal{F}$ and a linear map $\mathcal{T}$, which make up the secret key. Beullens recently proposed some attacks [2, 3] on UOV and Rainbow, the intersection, rectangular MinRank, and simple attacks. The rectangular MinRank and simple attacks do not work on the plain UOV, and by choosing the parameters properly, UOV can be secured against the intersection attack.

To achieve the security of a cryptosystem in a practical manner, we need to consider not only the weaknesses of the cryptographic protocols but also those of the devices implementing the cryptosystems. Physical attacks utilize information gained from the implementation of a computer system. As an example of such attacks on UOV-type signature schemes, a correlation power analysis attack was proposed at CHES 2018 [22]. Furthermore, among various physical attacks, fault attacks aim to stress the cryptographic device and generate computational errors, leading to a security failure of the cryptosystem. As fault attacks on UOV or its variants, some attacks combining information leakage through computational errors and algebraic measures have been proposed. Hashimoto et al. proposed two fault attacks at PQCrypto 2011 [15]. One attack is applicable to MPKCs with the public key $\mathcal{P} = S \circ F \circ T$, where an invertible quadratic map $\mathcal{F}$ and two linear maps S and $\mathcal{T}$ make up the secret key. This attack utilizes a fault that changes the coefficients of unknown terms in $\mathcal{F}$ and recovers the secret linear map S. This can be applied to Rainbow owing to its construction of the public and secret keys. The other fault attack proposed by Hashimoto et al. [15] is one in which the attacker fixes parameters chosen at random during the signature generation step. This attack weakens the security of UOV-like schemes such as UOV and Rainbow. These attacks proposed by Hashimoto et al. were complemented by some studies [18, 24], who analyzed how to apply them to UOV and Rainbow. Furthermore, at ACM CCS 2020 [19], Mus et al. proposed *QuantumHammer*, which is a fault attack on LUOV [4], a variant of UOV using a subfield. This attack utilizes faults occurring on the secret key, similar to the first attack proposed by Hashimoto et al. [15]. The authors demonstrated a full key recovery for LUOV-7-57-197 with a bit-tracing phase of less than 4 h and an algebraic phase of 49 h. Note that these proposed fault attacks on UOV do not utilize the recently proposed intersection algebraic attack.

Although several fault attacks on UOV or its variants have been proposed, a fault attack on plain UOV using faults occurring on the secret key $\mathcal{F}$ and $\mathcal{T}$ has yet to be proposed. The attack on the central map proposed by Hashimoto et al. [15] cannot be applied to UOV because the secret key of the scheme is composed of a quadratic map $\mathcal{F}$ and a linear map $\mathcal{T}$, unlike Rainbow. Indeed, at COSADE 2019 [18], the authors confirmed that the first attack proposed by Hashimoto et al. [15] does not work on UOV. Furthermore, the attack by Mus et al. [19] also does not work on UOV because it utilizes the fact that the secret key of LUOV is over the finite field of two elements.

At PQCrypto 2022 [13], Furue et al. proposed a new fault attack on the plain UOV scheme.[1] To the best of our knowledge, this is the first fault attack on UOV that causes faults on the secret key $\mathcal{F}$ and $\mathcal{T}$. The proposed attack transforms the public key system into a UOV system with fewer variables than the original. The secret key can then be fully recovered from this smaller system by using existing key recovery attacks on UOV. It should be noted that the attack model mainly follows the model used in the attack by Hashimoto et al. [15]. In the proposed attack, we assume that the fault caused on the secret key randomly changes the coefficient of the secret key, which is composed of the central map $\mathcal{F}$ and linear map $\mathcal{T}$. In UOV with n variables and m equations, numbers of coefficients of $\mathcal{F}$ and $\mathcal{T}$ are estimated as $O(m \cdot n^2)$ and $O(n^2)$, respectively, and thus we utilize the faults caused on the central map $\mathcal{F}$. The proposed attack is mainly composed of two steps:

1. Recover a part of $\mathcal{T}$ utilizing faults.
2. Transform the public key system $\mathcal{P}$.

The first step utilizes a fault occurring on the quadratic map $\mathcal{F}$. Some of the rows of the representation matrix of $\mathcal{T}$ are recovered from the computational error generated using the signing oracle several times, which generates a signature using the secret key with a fault. This step is iterated for each new fault unless a new fault is generated on $\mathcal{T}$. Subsequently, using a part of $\mathcal{T}$ recovered in the first step, the second step reduces the given public key system into one with fewer variables than the original. For the resulting smaller system obtained by executing the proposed attack, existing key recovery attacks can clearly be conducted with a smaller complexity than that of the key recovery attack on the original system.

---

[1] The paper published at PQCrypto 2022 by Furue et al. is a preliminary version of this paper.

The purpose of this work is to analyze the performance of the fault attack by Furue et al. on practical parameter sets. We will test the effectiveness of the fault attack by simulating it in the following two cases:

- The fault attack is completed. (The number of faults is unlimited.)
- The number of faults is limited.

More specifically, for two parameter sets of UOV $(q, v, m) = (16, 60, 39)$ and $(256, 50, 33)$ satisfying 100-bit security, we estimate the size of the resulting system of the proposed attack and the complexity of existing key recovery attacks on the resulting system.

2 Preliminaries

First, we recall the construction of UOV and explain some algebraic attacks on UOV. Second, we recall some fault attacks on UOV and its variants. We finally explain the fault attack on UOV proposed by Furue et al. [13], which we will mainly deal with in Sect. 3.

2.1 Unbalanced Oil and Vinegar Signature Schemes

In this section, we first describe the MQ problem and general signature schemes based on this problem. Subsequently, we review the construction of UOV [16] and some attacks applied to it.

2.1.1 Multivariate Signature Schemes

Let $\mathbb{F}_q$ be a finite field with q elements, and let n and m be two positive integers. For a system of quadratic polynomials $\mathcal{P} = (p_1(\mathbf{x}), \ldots, p_m(\mathbf{x}))$ in n variables over $\mathbb{F}_q$, the problem of obtaining a solution $\mathbf{x} \in \mathbb{F}_q^n$ to $\mathcal{P}(\mathbf{x}) = \mathbf{0}$ is called the MQ problem. Garey and Johnson [14] proved that this problem is NP-complete if $n \approx m$, and is thus considered to have the potential to resist quantum computer attacks.

Next, we briefly describe the construction of general multivariate signature schemes. First, an easily invertible quadratic map $\mathcal{F} = (f_1, \ldots, f_m) : \mathbb{F}_q^n \to \mathbb{F}_q^m$, called a *central map*, is generated. Next, two invertible linear maps $\mathcal{T} : \mathbb{F}_q^n \to \mathbb{F}_q^n$ and $\mathcal{S} : \mathbb{F}_q^m \to \mathbb{F}_q^m$ are randomly chosen to hide the structure of $\mathcal{F}$. These two linear maps $\mathcal{S}$ and $\mathcal{T}$ can be seen as two matrices in $\mathbb{F}_q^{m \times m}$ and $\mathbb{F}_q^{n \times n}$. The public key $\mathcal{P}$ is then provided as a polynomial map,

$$\mathcal{P} = \mathcal{S} \circ \mathcal{F} \circ \mathcal{T} : \mathbb{F}_q^n \to \mathbb{F}_q^m, \tag{1}$$

and the secret key comprises $\mathcal{S}$, $\mathcal{F}$, and $\mathcal{T}$. The signature is generated as follows: Given a message $\mathbf{m} \in \mathbb{F}_q^m$ to be signed, compute $\mathbf{m}_1 = \mathcal{S}^{-1}(\mathbf{m})$, and obtain a solution $\mathbf{m}_2$ to the equation $\mathcal{F}(\mathbf{x}) = \mathbf{m}_1$. This gives the signature $\mathbf{s} = \mathcal{T}^{-1}(\mathbf{m}_2) \in \mathbb{F}_q^n$ for the message $\mathbf{m}$. Verification is applied by confirming whether $\mathcal{P}(\mathbf{s}) = \mathbf{m}$.

2.1.2 Unbalanced Oil and Vinegar Signature Scheme

In this section, UOV is described. Let v be a positive integer and $n = v + m$. For variables $\mathbf{x} = (x_1, \ldots, x_n)$ over $\mathbb{F}_q$, we call $x_1, \ldots, x_v$ *vinegar variables* and $x_{v+1}, \ldots, x_n$ *oil variables*.

In UOV, a central map $\mathcal{F} = (f_1, \ldots, f_m) : \mathbb{F}_q^n \to \mathbb{F}_q^m$ is designed such that each f_k ($k = 1, \ldots, m$) is a quadratic polynomial of the form:

$$f_k(x_1, \ldots, x_n) = \sum_{i=1}^{v} \sum_{j=i}^{n} \alpha_{i,j}^{(k)} x_i x_j \qquad (2)$$

where $\alpha_{i,j}^{(k)} \in \mathbb{F}_q$. By using a randomly chosen linear map $\mathcal{T} : \mathbb{F}_q^n \to \mathbb{F}_q^n$, the public key map $\mathcal{P} : \mathbb{F}_q^n \to \mathbb{F}_q^m$ is computed using $\mathcal{P} = \mathcal{F} \circ \mathcal{T}$. The linear map $\mathcal{S}$ in Eq. (1) is not required because it does not help hide the structure of $\mathcal{F}$ in UOV. Indeed, the secret key is composed of only two maps $\mathcal{F}$ and $\mathcal{T}$. It should be noted that we here omit linear and constant terms of $\mathcal{F}$ and constant terms of $\mathcal{T}$ for simplicity.

Next, we describe the inversion of the central map $\mathcal{F}$. When we find $\mathbf{x} \in \mathbb{F}_q^n$ satisfying $\mathcal{F}(\mathbf{x}) = \mathbf{y}$ for a given $\mathbf{y} \in \mathbb{F}_q^m$, we first choose random values $a_1, \ldots, a_v$ in $\mathbb{F}_q$ as the values of the vinegar variables. We can then easily obtain a solution for the equation $\mathcal{F}(a_1, \ldots, a_v, x_{v+1}, \ldots, x_n) = \mathbf{y}$, because this is a linear system of m equations in m oil variables. If there is no solution to this equation, we choose new random values $a'_1, \ldots, a'_v$, and repeat the above procedure. By using this inversion approach, we can execute the signing process described in Sect. 2.1.1.

Finally, we introduce some matrices representing the public and secret keys of UOV. For each polynomial p_i of the public key $\mathcal{P}$, there exists an $n \times n$ matrix P_i such that $p_i(\mathbf{x}) = \mathbf{x}^\top \cdot P_i \cdot \mathbf{x}$. Similarly, an $n \times n$ matrix F_i can be taken for each f_i with $1 \leq i \leq m$, and an $n \times n$ matrix T is defined to satisfy $\mathcal{T}(\mathbf{x}) = T \cdot \mathbf{x}$. In general, these matrices P_i and F_i are taken as symmetric matrices if q is odd, and are taken as upper triangular matrices if q is even. For these representation matrices, based on Eq. (2), F_i can be considered as follows:

$$\begin{pmatrix} *_{v \times v} & *_{v \times m} \\ *_{m \times v} & 0_{m \times m} \end{pmatrix}.$$

Furthermore, from $\mathcal{P} = \mathcal{F} \circ \mathcal{T}$, we have

$$P_i = T^\top F_i T, \quad (i = 1, \ldots, m).$$

2.1.3 Attacks on UOV

A straightforward approach to attacking UOV is finding a signature **s** satisfying $\mathcal{P}(\mathbf{s}) = \mathbf{m}$ for a given public key $\mathcal{P}$ and a message **m**. A direct attack tries to solve this MQ system using an algorithm such as XL [6] or a Gröbner-based approach such as F_4 [11] and F_5 [12].

Several attacks that aim to recover the secret key have also been proposed. The attacker can sign any message after the secret key is recovered, and we call such attacks *key recovery attacks*. In the following, we review three key recovery attacks, the Kipnis-Shamir attack [17], a reconciliation attack [10], and an intersection attack [2]. All of these attacks are constructed for the purpose of obtaining the subspace $\mathcal{T}^{-1}(O)$ of $\mathbb{F}_q^n$, where O is the oil subspace defined as

$$O := \{(0, \ldots, 0, \alpha_1, \ldots, \alpha_m)^\top \mid \alpha_i \in \mathbb{F}_q\}.$$

This subspace $\mathcal{T}^{-1}(O)$ can induce the secret linear map $\mathcal{T}$ or an equivalent linear map $\mathcal{T}' : \mathbb{F}_q^n \to \mathbb{F}_q^n$ such that every component of $\mathcal{P} \circ \mathcal{T}'^{-1}$ has the form of Eq. (2).

Kipnis-Shamir Attack

To obtain $\mathcal{T}^{-1}(O)$, the Kipnis-Shamir attack chooses two invertible matrices W_i, W_j from the set of linear combinations of the representation matrices $P_1, \ldots, P_m$ for the public key. We then compute the invariant subspace of $W_i^{-1} W_j$, and a part of the subspace $\mathcal{T}^{-1}(O)$ is probabilistically recovered. The complexity of the Kipnis-Shamir attack is estimated as

$$O\left(q^{v-m-1} \cdot m^4\right). \tag{3}$$

Reconciliation Attack

The reconciliation attack treats a vector y of $\mathcal{T}^{-1}(O)$ as variables and solves the quadratic system $y^\top P_i y = 0$ $(i = 1, \ldots, m)$. Here, the number of dimensions of $\mathcal{T}^{-1}(O)$ is m, and thus if we impose affine constraints, we then solve a system of m equations in $n - m = v$ variables and still have a solution with high probability. However, the parameters of UOV are set to satisfy $v > m$ for the security against the Kipnis-Shamir attack, and in this case the system of $y^\top P_i y = 0$ has a large number of solutions. Therefore, to determine a solution uniquely, we need to solve the following system to find multiple vectors $y_1, \ldots, y_k$ of $\mathcal{T}^{-1}(O)$:

$$\begin{cases} y_j^\top P_i y_j = 0 & (1 \le i \le m, 1 \le j \le k) \\ y_j^\top P_i y_\ell = 0 & (1 \le i \le m, 1 \le j < \ell \le k) \end{cases}.$$

However, this attack will usually not outperform a direct attack because the number of variables to be solved increases.

Intersection Attack

Beullens [2] proposed a new attack against UOV, called an intersection attack. For an integer k satisfying $k < \frac{v}{v-m}$, let $L_1, \ldots, L_k$ be k invertible matrices randomly chosen from a set of linear combinations of the representation matrices $P_1, \ldots, P_m$ for the public key. This attack then solves the following equations for $\mathbf{y} \in \mathbb{F}_q^n$:

$$\begin{cases} (L_j^{-1}\mathbf{y})^\top P_i(L_j^{-1}\mathbf{y}) = 0 & (1 \le i \le m, 1 \le j \le k) \\ (L_j^{-1}\mathbf{y})^\top P_i(L_\ell^{-1}\mathbf{y}) = 0 & (1 \le i \le m, 1 \le j < \ell \le k) \end{cases}.$$

The solution space obtained from the above equation has $km - (k-1)v$ dimensions. Thus, its complexity is equivalent to that of solving the quadratic system with $n - (km - (k-1)v) = kv - (k-1)m$ variables and $\binom{k+1}{2}m - 2\binom{k}{2}$ equations owing to its linear dependency. The value of k is generally chosen such that the complexity of solving the above system takes the minimum value under the condition of $k < \frac{v}{v-m}$. Note that, when $v \ge 2m$, the intersection attack becomes a probabilistic algorithm.

2.2 Fault Attacks on UOV and Its Variant

This subsection mainly recalls two fault attacks on multivariate signatures proposed by Hashimoto et al. [15] and the attack on LUOV [4] proposed by Mus et al. [19].

Hashimoto et al. proposed a fault attack on the stepwise triangular system (STS) and big field (BF)-type signatures that have the public and secret keys of the form (1). This attack causes a fault to change the coefficients of unknown terms in the secret key $\mathcal{S}$, $\mathcal{F}$, and $\mathcal{T}$. In the case of the STS type, if the fault is caused on the central map $\mathcal{F}$, the attacker can recover a part of $\mathcal{S}$ directly from a pair of a message and a signature given by the faulty central map. The authors also show that the probability that the fault will successfully change $\mathcal{F}$ is sufficiently high. In [18], the authors complementarily investigated the behavior of this attack on UOV and Rainbow and confirmed it does not work on UOV owing to the construction of the secret key.

Furthermore, Hashimoto et al. proposed another attack that creates faults on the randomly chosen values in the signature generation. For UOV and its variant, this attack can recover a part of $\mathcal{T}$. Regarding this attack on UOV and Rainbow, Krämer and Loiero discussed a special case in [18], whereas Shim and Koo investigated the algebraic part in detail in [24].

In [19], Mus et al. proposed *QuantumHammer* for use on LUOV [4], whose secret keys, like those of UOV, are composed of a central map $\mathcal{F}$ and a linear map $\mathcal{T}$. This attack combines a bit-tracing phase, which causes faults on the linear map $\mathcal{T}$ and recovers a part of $\mathcal{T}$, with an algebraic phase, which recovers the remaining part of $\mathcal{T}$. The authors demonstrated a full key recovery for LUOV with $(q, v, m) =$ (7, 197, 57) using a bit-tracing phase of less than 4 h and an algebraic phase of 49 h.

In Sect. 2.3, we will explain a new fault attack on the plain UOV proposed by Furue et al. at PQCrypto 2022 [13]. This attack is the first attack that recovers the secret keys of the plain UOV utilizing faults caused on the secret key $\mathcal{F}$ and $\mathcal{T}$. From a technical point of view, they use a way of transforming matrices used in the result of Hashimoto et al., but other main algebraic techniques ware the original ones proposed in the fault attack by Furue et al.

2.3 Fault Attack on UOV by Furue et al.

This section explains a new fault attack on UOV proposed by Furue et al. The proposed attack utilizes faults generated on the secret key, particularly on the central map $\mathcal{F}$. After describing the attack model in Sect. 2.3.1, we describe the details of the proposed attack in Sect. 2.3.2. We finally explain how to apply existing key recovery attacks to the resulting smaller system of the proposed attack.

2.3.1 Attack Model

This section describes the attack model of the fault attack by Furue et al. We suppose that the coefficients of the secret key, $\mathcal{F}$ and $\mathcal{T}$, are stored in the device as fixed parameters used in UOV. In the proposed attack, we deal with the case in which the attacker causes a fault to change a coefficient, $\mathcal{F}$ or $\mathcal{T}$, of the secret key. The proposed attack is then constructed under the following assumptions:

- One fault changes one coefficient of the secret key, $\mathcal{F}$ or $\mathcal{T}$.
- A coefficient changed by a fault is randomly chosen.
- The attacker cannot know the location of the fault.
- Coefficients changed by the faults do not return to the original values (even if new faults are injected).

It should be noted that this attack model mainly follows that used in fault attacks on the secret key by Hashimoto et al. [15].

2.3.2 Description

This section describes the details of the proposed attack. The proposed attack mainly consists of the following two steps:

1. Some rows of the representation matrix T for the secret key are recovered utilizing faults.

2. The public key system $\mathcal{P}$ is reduced to a UOV public key system $\bar{\mathcal{P}}$ with fewer variables than the original system.

The resulting system with fewer variables can be broken with smaller complexity using existing key recovery attacks. In the following, for any vector $\mathbf{v} \in \mathbb{F}_q^a$ with $a \in \mathbb{N}$, $(\mathbf{v})_b$ with $b \in \{1, \ldots, a\}$ denotes the b-th element of $\mathbf{v}$.

Partial Recovery of $\mathcal{T}$

Herein, because, the data sizes of $\mathcal{F}$ and $\mathcal{T}$ are estimated as $O(\log(q) \cdot m \cdot n^2)$ and $O(\log(q) \cdot n^2)$, respectively, and thus a random fault is generated on the central map $\mathcal{F}$ with high probability, we consider a case in which the first fault is generated on $\mathcal{F}$. Concretely, we suppose that the coefficient $\alpha_{i,j}^{(k)}$ in Eq. (2) is changed into $\bar{\alpha}_{i,j}^{(k)}$ with $\alpha_{i,j}^{(k)} \neq \bar{\alpha}_{i,j}^{(k)}$. The central map with the fault is then denoted by $\mathcal{F}'$.

For each fault occurring on $\mathcal{F}$, the row vector recovery step is described as follows:

1. Iterate the following three steps N times:

    a. Randomly choose $\mathbf{m}_\ell \in \mathbb{F}_q^m$.
    b. Obtain $\mathbf{s}_\ell = \mathcal{T}^{-1} \circ \mathcal{F}'^{-1}(\mathbf{m}_\ell)$ using the signing oracle with the fault.
    c. $\delta_\ell = \mathcal{P}(\mathbf{s}_\ell) - \mathbf{m}_\ell$.

2. Solve the following linear system in $\{y_{p,r}\}_{1 \leq p \leq r \leq n}$:

$$\sum_{p \leq r} (\mathbf{s}_\ell)_p (\mathbf{s}_\ell)_r y_{p,r} = \bar{\delta}_\ell \quad (1 \leq \ell \leq N), \tag{4}$$

   where $\bar{\delta}_\ell$ is the only nonzero element of δ_ℓ. (If $\delta_\ell = \mathbf{0}$, then $\bar{\delta}_\ell = 0$.)
3. For the solution $\{z_{p,r}\}_{1 \leq p \leq r \leq n}$ of Eq. (4), find two vectors $(a_1, \ldots, a_n)$ and $(b_1, \ldots, b_n)$ satisfying

$$\begin{cases} a_i b_j + a_j b_i = z_{i,j} & (i < j) \\ a_i b_i = z_{i,i} \end{cases}. \tag{5}$$

As a result, two vectors $(a_1, \ldots, a_n)$ and $(b_1, \ldots, b_n)$ obtained in the third step correspond to the row vectors of the secret key $\mathcal{T}$.

The first step generates N pairs of two vectors $\mathbf{s}_\ell \in \mathbb{F}_q^n$ and $\delta_\ell \in \mathbb{F}_q^m$. These pairs are generated through the following three steps: First, a message $\mathbf{m}_\ell$ is randomly chosen in $\mathbb{F}_q^m$. Second, we input $\mathbf{m}_\ell$ to the signing oracle with the faulty central map $\mathcal{F}'$ and receive $\mathbf{s}_\ell$ as an output. Third, we compute the difference δ_ℓ between $\mathcal{P}(\mathbf{s}_\ell)$ and $\mathbf{m}_\ell$. These manipulations are iterated N times. For each $1 \leq \ell \leq N$, we then have

$$\delta_\ell = (\mathcal{F} \circ \mathcal{T})(\mathbf{s}_\ell) - (\mathcal{F}' \circ \mathcal{T})(\mathbf{s}_\ell)$$
$$= (\mathcal{F} - \mathcal{F}') \circ \mathcal{T}(\mathbf{s}_\ell)$$
$$= \left(0, \ldots, 0, \left(\alpha_{i,j}^{(k)} - \bar{\alpha}_{i,j}^{(k)}\right) \cdot (\mathcal{T}(\mathbf{s}_\ell))_i \cdot (\mathcal{T}(\mathbf{s}_\ell))_j, 0, \ldots, 0\right),$$

and thus, if $(\mathcal{T}(\mathbf{s}_\ell))_i \neq 0$ and $(\mathcal{T}(\mathbf{s}_\ell))_j \neq 0$, then δ_ℓ has the only nonzero k-th element. This is because $\mathcal{F}$ and $\mathcal{F}'$ only differ in the coefficient of $x_i x_j$ of f_k. Here, let $\bar{\delta}_\ell \in \mathbb{F}_q$ be the nonzero element of δ_ℓ (if $\delta_\ell = \mathbf{0}$, then $\bar{\delta}_\ell = 0$), and $t_{i,j}$ be the (i, j)-element of the representation matrix T for the secret key. We then hold

$$\bar{\delta}_\ell = \left(\alpha_{i,j}^{(k)} - \bar{\alpha}_{i,j}^{(k)}\right) \cdot \left(\sum_p t_{i,p}(\mathbf{s}_\ell)_p\right) \cdot \left(\sum_r t_{j,r}(\mathbf{s}_\ell)_r\right)$$
$$= \left(\alpha_{i,j}^{(k)} - \bar{\alpha}_{i,j}^{(k)}\right) \sum_{p \leq r}(\mathbf{s}_\ell)_p (\mathbf{s}_\ell)_r \begin{cases} (t_{i,p}t_{j,r} + t_{j,p}t_{i,r}) & (p \neq r) \\ t_{i,p}t_{j,p} & (p = r) \end{cases}. \quad (6)$$

In the second step, we introduce new $n(n+1)/2$ variables $\{y_{p,r}\}_{1 \leq p \leq r \leq n}$ such that every component $y_{p,r}$ corresponds to $(t_{i,p}t_{j,r} + t_{j,p}t_{i,r})$ in the case of $p \neq r$ and $t_{i,p}t_{j,p}$ in the case of $p = r$, as in Eq. (6). We then generate a linear system of N Eq. (4) in $n(n+1)/2$ variables $\{y_{p,r}\}_{1 \leq p \leq r \leq n}$. Here, although Eq. (4) is clearly deduced from Eq. (6), we omit the part of $\alpha_{i,j}^{(k)} - \bar{\alpha}_{i,j}^{(k)}$ from Eq. (6) because recovering a multiple of row vectors of T is sufficient. If we set $N = n(n+1)/2$, then the solution is determined uniquely with high probability. When a linear system has some solutions, we add a new linear equation obtained using a new pairing of $\mathbf{s}_\ell$ and δ_ℓ until the solution is uniquely determined.

Subsequently, from the solution $\{z_{p,r}\}_{1 \leq p \leq r \leq n}$ of Eq. (4), we generate two vectors $(a_1, \ldots, a_n)$ and $(b_1, \ldots, b_n)$ satisfying Eq. (5), as in the definition of $\{y_{p,r}\}_{1 \leq p \leq r \leq n}$. Specifically, these two vectors can be found as follows:

1. $a_1 = 1$.
2. $b_1 = z_{1,1}$.
3. Find a_2 and b_2 solving
$$\begin{cases} a_1 b_2 + a_2 b_1 = z_{1,2} \\ a_2 b_2 = z_{2,2} \end{cases}.$$

4. For $3 \leq i \leq n$, find a_i and b_i solving
$$\begin{cases} a_1 b_i + a_i b_1 = z_{1,i} \\ a_2 b_i + a_i b_2 = z_{2,i} \end{cases}.$$

Then, $(a_1, \ldots, a_n)$ and $(b_1, \ldots, b_n)$ clearly correspond to constant multiples of two row vectors of T from Eq. (6).

After executing the manipulations described above for the first fault, we cause another fault on the secret key and recover two row vectors of T again by applying a similar method if a new fault also occurs on $\mathcal{F}$. As the main difference from the first fault, δ_ℓ may have several nonzero elements owing to the previous faults. Herein, we suppose that, for the i-th fault with $1 \leq i < \bar{i}$ caused on $\mathcal{F}$, we obtain $\{z_{p,r}^{(i)}\}_{1 \leq p \leq r \leq n}$ by solving Eq. (4), and k_i then denotes the index in which δ_ℓ has a nonzero element. Then, after generating the $\bar{i}$-th fault, for a new pairing of $\mathbf{s}_\ell$ and δ_ℓ, by subtracting $\left(\sum_{p \leq r}(\mathbf{s}_\ell)_p(\mathbf{s}_\ell)_r z_{p,r}^{(i)}\right)\mathbf{e}_{k_i}$ from δ_ℓ for every $1 \leq i < \bar{i}$, δ_ℓ becomes a vector with one nonzero element, as in the first fault case. Based on this manipulation, the same recovery approach as the first fault case can be applied. Note that if a recovered row vector of T is dependent on one of the row vectors already recovered, it indicates that the same vector is recovered in a duplicate manner.

These manipulations are iterated until n independent row vectors are recovered or a new fault is caused on $\mathcal{T}$, which can be easily confirmed because δ_ℓ has many nonzero elements even after subtracting $\left(\sum_{p \leq r}(\mathbf{s}_\ell)_p(\mathbf{s}_\ell)_r z_{p,r}^{(i)}\right)\mathbf{e}_{k_i}$.

Reduction to Smaller UOV

Herein, we assume that α row vectors $(a_1^{(i)}, \ldots, a_n^{(i)})$ with $1 \leq i \leq \alpha$ are recovered in the first step, and each $(a_1^{(i)}, \ldots, a_n^{(i)})$ corresponds to the β_i-th row of T. This reduction step is then described as follows:

1. For $i = 1, \ldots, \alpha$ do

   a. $(b_1, \ldots, b_n) := (a_1^{(i)}, \ldots, a_n^{(i)}) \cdot T_1 \cdots T_{i-1}$,
   b. Choose k_i from $\{k' \notin \{k_1, \ldots, k_{i-1}\} \mid b_{k'} \neq 0\}$,
   c. $T_i \in \mathbb{F}_q^{n \times n}$ is taken as follows:

$$T_i := \begin{pmatrix} I & & & & & O \\ & & & k_i & & \\ -\frac{b_1}{b_{k_i}} & \cdots & -\frac{b_{k_i-1}}{b_{k_i}} & 1 & -\frac{b_{k_i+1}}{b_{k_i}} & \cdots & -\frac{b_n}{b_{k_i}} \\ & & & & & \\ O & & & & & I \end{pmatrix}. \quad (7)$$

2. Substitute 0 for $\{x_k \mid k \in \{k_1, \ldots, k_\alpha\}\}$ in $\mathcal{P}(T_1 \cdots T_\alpha \cdot \mathbf{x})$.

It should be noted that the first part of this reduction step mainly originates from the result of Hashimoto et al. [15].

First, for $(a_1^{(1)}, \ldots, a_n^{(1)})$ obtained in the partial recovery step, we take the matrix T_1 to transform the β_1-th row of T. We then choose one nonzero element $a_{k_1}^{(1)}$ from $(a_1^{(1)}, \ldots, a_n^{(1)})$ and take a matrix T_1 following Eq. (7). We then have

$$T \cdot T_1 = \beta_1 \begin{pmatrix} & & & k_1 & & & \\ & & & * & & & \\ 0 & \cdots & 0 & * & 0 & \cdots & 0 \\ & & & * & & & \end{pmatrix}.$$

We iterate such processes for all α row vectors obtained in the first step. It should be noted that, because the β_i-th row of T has already been transformed by $T_1 \cdots T_{i-1}$, we need to multiply $T_1 \cdots T_{i-1}$ to $(a_1^{(i)}, \ldots, a_n^{(i)})$ from the right side before choosing k_i and setting T_i. As a result, when we let $T' := T_1 \cdots T_\alpha$, the $\beta_1, \ldots, \beta_\alpha$ rows of T are eliminated by multiplying T' from the right side. Indeed, if we take two $n \times n$ permutation matrices A_1 and A_2 satisfying

$$A_1 \cdot (1 \cdots m)^\top = (\beta_1 \cdots \beta_\alpha * \cdots *)^\top$$
$$(1 \cdots m) \cdot A_2 = (k_1 \cdots k_\alpha * \cdots *),$$

then we have

$$A_1 \cdot (T \cdot T') \cdot A_2 = \begin{pmatrix} I_\alpha & 0_{\alpha \times (n-\alpha)} \\ B_1 & B_2 \end{pmatrix}, \tag{8}$$

where B_1 and B_2 are $(n-\alpha) \times \alpha$ and $(n-\alpha) \times (n-\alpha)$ matrices, respectively.

Subsequently, the second part reduces the transformed public key system $\mathcal{P}(T' \cdot \mathbf{x})$ into a UOV public key system with fewer variables than the original. In this part, we substitute 0 for $x_{k_1}, \ldots, x_{k_\alpha}$ in $\mathcal{P}(T' \cdot \mathbf{x})$. Here, we denote the remaining $n - \alpha$ variables $\bar{\mathbf{x}} = (\bar{x}_1, \ldots, \bar{x}_{n-\alpha})$ such that

$$A_2^{-1} \cdot \mathbf{x} = A_2^\top \cdot \mathbf{x}$$
$$= (x_{\beta_1} \cdots x_{\beta_\alpha} \bar{x}_1 \cdots \bar{x}_{n-\alpha})^\top,$$

and denote by $\bar{\mathcal{P}} = (\bar{p}_1(\bar{\mathbf{x}}), \ldots, \bar{p}_m(\bar{\mathbf{x}}))$ the resulting public key system obtained by executing the second part. With two permutation matrices in Eq. (8), we assume that $A_1^{-\top} \cdot F_i \cdot A_1^{-1}$ is written in the following forms:

$$A_1^{-\top} \cdot F_i \cdot A_1^{-1} = \begin{pmatrix} C_1 & C_2 \\ C_3 & C_4 \end{pmatrix},$$

where C_1, C_2, C_3, and C_4 are $\alpha \times \alpha$, $\alpha \times (n-\alpha)$, $(n-\alpha) \times \alpha$, and $(n-\alpha) \times (n-\alpha)$ matrices, respectively. We then have

$p_i(T' \cdot \mathbf{x})$
$= (T \cdot T' \cdot \mathbf{x})^\top \cdot F_i \cdot (T \cdot T' \cdot \mathbf{x}),$
$= (A_2^{-1} \cdot \mathbf{x})^\top \cdot (A_1 \cdot T \cdot T' \cdot A_2)^\top \cdot (A_1^{-\top} \cdot F_i \cdot A_1^{-1}) \cdot (A_1 \cdot T \cdot T' \cdot A_2) \cdot (A_2^{-1} \cdot \mathbf{x}),$

and thus it holds that

$$\begin{aligned}
\bar{p}_i(\bar{\mathbf{x}}) &= (0,\ldots,0,\bar{\mathbf{x}}^\top)\begin{pmatrix} I_\alpha & 0_{\alpha\times(n-\alpha)} \\ B_1 & B_2 \end{pmatrix}^\top \begin{pmatrix} C_1 & C_2 \\ C_3 & C_4 \end{pmatrix}\begin{pmatrix} I_\alpha & 0_{\alpha\times(n-\alpha)} \\ B_1 & B_2 \end{pmatrix}(0,\ldots,0,\bar{\mathbf{x}}^\top)^\top \\
&= \bar{\mathbf{x}}^\top \begin{pmatrix} 0_{\alpha\times(n-\alpha)} \\ B_2 \end{pmatrix}^\top \begin{pmatrix} C_1 & C_2 \\ C_3 & C_4 \end{pmatrix}\begin{pmatrix} 0_{\alpha\times(n-\alpha)} \\ B_2 \end{pmatrix}\bar{\mathbf{x}} \\
&= \bar{\mathbf{x}}^\top \cdot B_2 \cdot C_4 \cdot B_2 \cdot \bar{\mathbf{x}}.
\end{aligned}$$

In the above equation, because A_1^{-1} is also a permutation matrix, C_4 can be seen as a representation matrix of the central map of UOV with $n - \alpha$ variables. Therefore, we can regard $\bar{\mathcal{P}}$ as the public key system of UOV in $n - \alpha$ variables. Furthermore, if, among $\beta_1,\ldots,\beta_\ell$, the v' elements are in the set of $\{1,\ldots,v\}$ and the m' elements are in the set of $\{v+1,\ldots,n\}$, then $\bar{\mathcal{P}}$ is the UOV public key with $v - v'$ vinegar variables and $m - m'$ oil variables owing to the structure of A_1^{-1}.

2.3.3 Application of Key Recovery Attacks

In this section, we describe how to recover the secret key from the resulting smaller system $\bar{\mathcal{P}}$ with $v - v'$ vinegar variables and $m - m'$ oil variables obtained by the proposed attack described in Sect. 2.3.2. In the resulting system $\bar{\mathcal{P}}$, key recovery attacks on UOV can be executed with a smaller complexity in most cases. As the reason why the direct attack cannot be conducted with a smaller complexity, no pre-image of $\bar{\mathcal{P}}$ exists with high probability because the number $m - m'$ of the oil variables is smaller than the number m of equations if $m' > 0$.

As stated in Sect. 2.1.3, key recovery attacks on UOV aim to recover $\mathcal{T}^{-1}(O)$, where O is the oil subspace. If we obtain a vector $\mathbf{a} \in \mathbb{F}_q^{n-\alpha}$ by applying key recovery attacks on $\bar{\mathcal{P}}$, then we can obtain a vector in $\mathcal{T}^{-1}(O)$ by concatenating $\mathbf{a}$ with 0 for $x_{k_1},\ldots,x_{k_\alpha}$ and multiplying $T_1 \cdots T_\alpha$ from the construction of $\bar{\mathcal{P}}$. Note that, using key recovery attacks on $\bar{\mathcal{P}}$, we only obtain at most $m - m'$ independent vectors of $\mathcal{T}^{-1}(O)$. However, in most cases, after some basis of $\mathcal{T}^{-1}(O)$ is obtained, it becomes easy to recover the remaining basis of $\mathcal{T}^{-1}(O)$. The proposed attack described in Sect. 2.2.2 can be clearly executed in polynomial time, and thus the complexity of the attack is dominated by that of the key recovery attacks. In the following, we apply two key recovery attacks, the Kipnis-Shamir attack [17] and intersection attack [2], because they are particularly effective for our chosen parameter sets in Sect. 3. Note that the attacker here cannot know the number of vinegar and oil variables of $\bar{\mathcal{P}}$.

Kipnis-Shamir Attack

The Kipnis-Shamir attack can be executed without the number of vinegar and oil variables, and thus we can simply apply the Kipnis-Shamir attack on $\bar{\mathcal{P}}$. From Eq. (3), the complexity of the Kipnis-Shamir attack on $\bar{\mathcal{P}}$ is estimated as

$$O\left(q^{(v-v')-(m-m')-1} \cdot (m-m')^4\right). \tag{9}$$

Because v' is larger than m' with high probability, as stated in Sect. 3.1, this attack will be applied with a smaller complexity than that of the Kipnis-Shamir attack on the original system.

Intersection Attack

By contrast, the intersection attack generally requires the number of vinegar and oil variables. As described in Sect. 2.1.3, for UOV with v vinegar variables, o oil variables, and m equations, the intersection attack solves an MQ system of $\binom{k+1}{2}m - 2\binom{k}{2}$ equations in $kv - (k-1)o$ variables for an integer $k < \frac{v}{v-o}$. (We here distinguish the number o of oil variables and the number m of equations unlike Sect. 2.1.3.) The above value $kv - (k-1)o$ is determined by subtracting the dimension $ko - (k-1)v$ of the solution subspace from the number $n = v + o$ of variables to be solved. However, the attacker here only knows the value of $\alpha = v' + m'$ and does not know each value of v' and m', and thus the dimensions of the solution space and the value of k for the optimal complexity cannot be correctly conjectured before solving the system.

In the following, we introduce a way to execute the intersection attack on $\bar{\mathcal{P}}$. First, the attacker assumes that $v' = \alpha$ and $m' = 0$, and then conducts the intersection attack by choosing the optimal $k < \frac{v-\alpha}{(v-\alpha)-m}$ and supposing that the dimension of the solution space is $km - (k-1)(v-\alpha)$, namely, the number of variables to be solved is

$$(v+m-\alpha) - (km - (k-1)(v-\alpha)) = k(v-\alpha) - (k-1)m.$$

If there exists no solution for the above system, we then assume that $v' = \alpha - 1$ and $m' = 1$, and solve the system by supposing that the solution space is $k(m-1) - (k-1)(v-\alpha+1)$ for $k < \frac{v-\alpha+1}{(v-\alpha+1)-(m-1)}$. In this way, we iterate the decrease in v' and increase in m' until we obtain a solution. We can then find a solution in which we correctly assume the values of v' and m', which will dominate the complexity because the number of variables is larger than that of the previous case. In conclusion, the complexity of the intersection attack on $\bar{\mathcal{P}}$ is estimated as that of solving an MQ system of $\binom{k+1}{2}m - 2\binom{k}{2}$ equations in $k(v-v') - (k-1)(m-m')$ variables for $k < \frac{v-v'}{(v-v')-(m-m')}$. If we set $N = k(v-v') - (k-1)(m-m')$ and $M = \binom{k+1}{2}m - 2\binom{k}{2}$, by considering the hybrid approach [1, 26] with Wiedemann XL [27], the complexity of solving the above system is given as

$$\min_{k} \left(O\left(q^k \cdot 3 \cdot \binom{N-k}{2} \cdot \binom{D+N-k}{D}^2 \right) \right), \tag{10}$$

where D is the degree of the first non-positive term of $(1-z)^{M-N+k}(1+z)^M$.

3 Results

In this section, we simulate the fault attack by Furue et al. in the cases where the attack is fully completed and the case where the number of faults is limited to small values. As we described in Sect. 2.2, we obtain a smaller UOV public key system $\bar{\mathcal{P}}$ with $v - v'$ vinegar and $m - m'$ oil variables with $v' + m' = \alpha$ using the proposed attack. In our simulation, we first estimate the number of rows of $\mathcal{T}$ recovered in the first step of the proposed attack by assuming random faults. From this number of rows, we then estimate the complexity of the key recovery attacks. We leave the performance analysis by using the physical device and by implementing UOV as our future work.

This section applies the proposed attack to two practical parameter sets $(q, v, m) = (16, 60, 39)$ and $(256, 50, 33)$ satisfying 100-bit security. These parameter sets are chosen such that the public key and signature sizes reach the smallest values unless the complexity of the direct attack, Kipnis-Shamir attack [17], reconciliation attack [10], or intersection attack [2] is smaller than 100 bits.

3.1 Simulations of Fault Attack by Furue et al.

By simulating the proposed attack 1000 times for each parameter set, Table 1 shows the occurrence probabilities, the average rate of v' to $v' + m'$, and the average number of faults occurring in the attack for each difference in the numbers of variables of $\mathcal{P}$ and $\bar{\mathcal{P}}$, which is equal to $v' + m'$. For example, with the parameter set $(q, v, m) = (16, 60, 39)$, $v' + m'$ falls within the range of 5 to 9 with a 9.5% probability, and in this case, the average value of $v'/(v' + m')$ and the number of faults are 0.68 and 4.8, respectively. From this table, we can confirm that, by using the proposed fault attack, we can obtain $\bar{\mathcal{P}}$ with 5 fewer variables than the original system with a probability of approximately $80 \sim 90\%$, and $\bar{\mathcal{P}}$ with 10 fewer variables with a probability of approximately 70%. Furthermore, Table 1 shows that v' is larger than m' in most cases because, for each fault, either one of the first v rows and one of the last m rows, or two of the first v rows are recovered during the recovery step described in Sect. 2.2.2 owing to Eq. (2). In addition, the number of faults used in the proposed attack clearly increases with the increase in the difference in the numbers of variables of $\mathcal{P}$ and $\bar{\mathcal{P}}$.

Table 1 With two 100-bit secure parameters $(q, v, m) = (16, 60, 39)$ and $(256, 50, 33)$, for each difference between the numbers of variables of the public key system $\mathcal{P}$ and the resulting system $\bar{\mathcal{P}}$ of the proposed attack ($= v' + m'$), the probability of its occurrence, the average rate of v' to $v' + m'$, and the average number of faults used in the attack

$(q, v, m) = (16, 60, 39)$	The difference between the numbers of variables of $\mathcal{P}$ and $\bar{\mathcal{P}}$				
	$0 \sim 4$	$5 \sim 9$	$10 \sim 19$	$20 \sim 29$	$30 \sim$
Probability	15.4%	9.5%	22.6%	19.8%	32.7%
$v'/(v' + m')$	–	0.68	0.70	0.71	0.69
The number of faults	1.9	4.8	8.7	14.8	33.9
$(q, v, m) = (256, 50, 33)$	The difference between the numbers of variables of $\mathcal{P}$ and $\bar{\mathcal{P}}$				
	$0 \sim 4$	$5 \sim 9$	$10 \sim 19$	$20 \sim 29$	$30 \sim$
Probability	18.8%	13.2%	21.8%	20.0%	26.2%
$v'/(v' + m')$	–	0.73	0.71	0.70	0.68
The number of faults	1.9	4.8	8.7	15.7	32.1

Table 2 With two 100-bit secure parameters $(q, v, m) = (16, 60, 39)$ and $(256, 50, 33)$, the complexities of the Kipnis-Shamir (KS) and intersection (Int) attacks breaking the resulting system $\bar{\mathcal{P}}$ of the proposed attack when the parameters (v', m') of the proposed attack are set to $(4, 1)$, $(7, 3)$, $(14, 6)$, or $(21, 9)$

Parameter (100 bits)(q, v, m)	Attack	$(v' + m', v', m')$			
		$(5, 4, 1)$	$(10, 7, 3)$	$(20, 14, 6)$	$(30, 21, 9)$
(16, 60, 39)	KS Eq. (9)	89.0	84.7	68.2	51.6
	Int Eq. (10)	105.8	97.6	75.8	59.9
(256, 50, 33)	KS Eq. (9)	124.0	115.6	83.0	50.3
	Int Eq. (10)	89.5	81.3	63.2	40.6

Subsequently, for the two parameter sets satisfying 100-bit security, herein we estimate the complexity of the Kipnis-Shamir and intersection attacks on the resulting system $\bar{\mathcal{P}}$ of the proposed attack. As indicated in Table 2, we choose four values, 5, 10, 20, and 30, for $v' + m'$, which is equal to the difference between the numbers of variables of $\mathcal{P}$ and $\bar{\mathcal{P}}$. In addition, we set (v', m') as $(4, 1)$, $(7, 3)$, $(14, 6)$, and $(21, 9)$ in each case owing to the rate of v' to $v' + m'$ shown in Table 1. Table 2 shows the bit complexities of the Kipnis-Shamir and intersection attacks, where, for example, the complexity of the Kipnis-Shamir attack on $\bar{\mathcal{P}}$ with $(q, v, m) = (16, 60, 39)$ and $(v', m') = (4, 1)$ is 89.0 bits. This table indicates that, by choosing the optimal attack among the two key recovery attacks, the secret key can be recovered from $\bar{\mathcal{P}}$ with less complexity than the claimed security level in each case.

Table 3 estimates the complexity of the Kipnis-Shamir or intersection attack on the resulting system $\bar{\mathcal{P}}$ of the proposed attack through 1000 times simulations. We here choose the Kipnis-Shamir attack for $(q, v, m) = (16, 60, 39)$ and the intersection attack for $(q, v, m) = (256, 50, 33)$ from the result of Table 2.

Table 3 With two 100-bit secure parameters $(q, v, m) = (16, 60, 39)$ and $(256, 50, 33)$, the bit complexity of the Kipnis-Shamir or intersection attack on the resulting system $\bar{\mathcal{P}}$ of the proposed attack

$(q, v, m) = (16, 60, 39)$	The bit complexity of the Kipnis-Shamir attack on $\bar{\mathcal{P}}$						
	~ 50	$50 \sim 60$	$60 \sim 70$	$70 \sim 80$	$80 \sim 90$	$90 \sim 100$	$100 \sim$
Probability	33.2%	8.7%	11.1%	11.6%	14.9%	10.1%	10.4%
$(q, v, m) = (256, 50, 33)$	The bit complexity of the intersection attack on $\bar{\mathcal{P}}$						
	~ 50	$50 \sim 60$	$60 \sim 70$	$70 \sim 80$	$80 \sim 90$	$90 \sim 100$	$100 \sim$
Probability	34.6%	7.5%	10.9%	10.8%	12.0%	10.6%	13.6%

For $(q, v, m) = (16, 60, 39)$, the Kipnis-Shamir attack can be applied with smaller manipulations than 100 bits with 89.6% probability, and 50 bits with 33.2% probability. Similarly, for the case of $(q, v, m) = (256, 50, 33)$, the secret key can be recovered with smaller manipulations than 100 bits with 86.4% probability, and 50 bits with 34.6% probability. These results show a trade-off between the complexity of recovering the secret key and the probability that it will occur.

3.2 Limited Faults Cases

In this section, we consider applying the proposed attack in a case in which the number of faults is limited. Table 4 shows the results of the proposed attack on the parameter set $(q, v, m) = (16, 60, 39)$ under the conditions in which a fault occurs once, twice, and three times. For each case, we show the probability of occurrence and the complexity of the Kipnis-Shamir attack on the resulting system with $v - v'$

Table 4 For the 100-bit secure parameter $(q, v, m) = (16, 60, 39)$, the values of v' and m', the probability of such occurrence, and the complexity of the Kipnis-Shamir attack on the resulting system $\bar{\mathcal{P}}$ obtained from Eq. (9) when a fault occurs once, twice, and three times, respectively

Faults							
1	(v', m')	(1,1)	(2,0)	(0,0)			
	Probability	52.9%	41.4%	5.7%			
	$\log_2$(complexity)	101.0	**93.1**	101.1			
2	(v', m')	(3,1)	(2,2)	(4,0)	(0,0)	Other	
	Probability	43.8%	28.0%	17.1%	5.7%	5.4%	
	$\log_2$(complexity)	**93.0**	100.8	**85.1**	101.1		
3	(v', m')	(4,2)	(5,1)	(3,3)	(6,0)	Other	
	Probability	34.8%	27.2%	14.8%	7.1%	16.1%	
	$\log_2$(complexity)	**92.8**	**85.0**	100.7	**77.1**		

Table 5 With two 100-bit secure parameters $(q, v, m) = (16, 60, 39)$ (above) and $(256, 50, 33)$ (below), the complexities of the Kipnis-Shamir (KS) and intersection (Int) attacks breaking the resulting system $\bar{\mathcal{P}}$ of the proposed attack for the parameters (v', m') with $v' + m' = 4$

(v', m')	(4, 0)	(3, 1)	(2, 2)	(1, 3)	(0, 4)
Probability	15.5%	50.3%	34.2%	0.0%	0.0%
KS Eq. (9)	85.1	93.0	100.8	108.7	116.5
Int Eq. (10)	101.5	124.3	134.5	142.7	152.8
(v', m')	(4, 0)	(3, 1)	(2, 2)	(1, 3)	(0, 4)
Probability	16.6%	46.7%	36.7%	0.0%	0.0%
KS Eq. (9)	116.2	132.0	147.8	163.6	179.4
Int Eq. (10)	82.4	90.6	112.9	119.6	131.0

vinegar variables and $m - m'$ oil variables for each set of v' and m'. Herein, we choose the Kipnis-Shamir attack because it is more efficient than the intersection attack, as indicated from Table 2. Note that we dismiss the case in which rows of T are recovered in duplicate during the recovery step of the proposed attack, which occurs with negligible probability. As a result, in a case in which a fault occurs once, the secret key is recovered with a smaller complexity of 93.1 bits in comparison to the claimed security level when $(v', m') = (2, 0)$ with a probability of 41.4%. In the case of two faults, the system is less secure when $(v', m') = (3, 1)$ and $(4, 0)$ with a total probability of 60.9%, and in the case of three faults, the system is less secure when $(v', m') = (4, 2), (5, 1)$, and $(6, 0)$ with a total probability of 69.1%. These results indicate that the proposed approach weakens the security of UOV even when the number of faults is limited.

In Table 5, in the case where the sum of v' and m' is equal to 4, we estimate the probability that the set of (v', m') has each value and the complexity of the Kipnis-Shamir and intersection attacks on the resulting system $\bar{\mathcal{P}}$ for each case. Such a case with $v' + m' = 4$ will happen with high probability when two faults are caused on the central map $\mathcal{F}$. For the parameter $(q, v, m) = (16, 60, 39)$, the Kipnis-Shamir attack breaks its security criteria when (v', m') is $(4, 0)$ and $(3, 1)$, which happens with a probability of 65.8%. Similarly, for the parameter $(q, v, m) = (256, 50, 33)$, the intersection attack breaks its security criteria when (v', m') is $(4, 0)$ and $(3, 1)$, which happens with a probability of 63.3%. These facts show that the fault attack by Furue et al. reduces the security level with sufficiently high probability with only two faults.

4 Conclusion

In this work, we simulated the fault attack proposed by Furue et al. on some parameter sets of UOV, which is a multivariate signature scheme. This fault attack is the first attack on UOV that uses faults occurring on the secret key. Given a UOV public key system with v vinegar variables and m oil variables, the resulting system of the proposed attack is a UOV system with $v - v'$ vinegar variables and $m - m'$ oil variables, where v' and m' are determined based on the faults. For the resulting system, the existing key recovery attacks on UOV, i.e., Kipnis-Shamir and intersection attacks, can be applied with smaller complexity than that on the original system.

In this paper, we analyze the performance of the fault attack by Furue et al. in the following two conditions: the case where the attack is completed and the case where the number of faults is limited. More specifically, the simulation is performed on the two parameter sets $(q, v, m) = (16, 60, 39)$ and $(256, 50, 33)$ satisfying 100-bit security. In the first case, we confirmed that the proposed fault attack obtains the system with 5 fewer variables than the original system with a probability of approximately 80 ~ 90%, and the system with 10 fewer variables with a probability of approximately 70%. Furthermore, the attack breaks the parameters by smaller than 100-bit manipulations with approximately 90% probability and by smaller than 50-bit manipulations with approximately 30% probability. In the second case, our simulation shows that the attack breaks the claimed security level at least 60% probability with only two faults on the central map. These results show that, in the both cases, the fault attack reduces the security level with sufficiently high probability.

It should be noted that a naive countermeasure against the proposed attack would be to check whether the secret key is faulty, and if so, to avoid generating the signature [15]. This countermeasure will be practical in online scenario since the verification of UOV is known to be so efficient.

Acknowledgements This work was supported by JST CREST Grant Number JPMJCR2113, Japan, and JSPS KAKENHI Grant Number JP22KJ0554, Japan.

References

1. L. Bettale, J.-C. Faugère, L. Perret, Hybrid approach for solving multivariate systems over finite fields. J. Math. Cryptol. **3**, 177–197 (2009)
2. W. Beullens, Improved cryptanalysis of UOV and Rainbow, in *EUROCRYPT 2021*, LNCS, vol. 12696 (Springer, 2021), pp. 348–373
3. W. Beullens, Breaking Rainbow takes a weekend on a laptop, in *CRYPTO 2022*, LNCS, vol. 13508 (Springer, 2022), pp. 464–479
4. W. Beullens, B. Preneel, Field lifting for smaller UOV public keys, in *INDOCRYPT 2017*, LNCS, vol. 10698 (Springer, 2017), pp. 227–246
5. A. Bogdanov, T. Eisenbarth, A. Rupp, C. Wolf, Time-area optimized public-key engines: MQ-cryptosystems as replacement for elliptic curves? in *CHES 2008*, LNCS, vol. 5154 (Springer, 2008), pp. 45–61
6. N. Courtois, A. Klimov, J. Patarin, A. Shamir, Efficient algorithms for solving overdefined systems of multivariate polynomial equations, in *EUROCRYPT 2000*, LNCS, vol. 1807 (Springer, 2000), pp. 392–407
7. P. Czypek, S. Heyse, E. Thomae, Efficient implementations of MQPKS on constrained devices, in *CHES 2012*, LNCS, vol. 7428 (Springer, 2012), pp. 374–389

8. A.I.-T. Chen, M.-S. Chen, T.-R. Chen, C.-M. Cheng, J. Ding, E.L.-H. Kuo, F.Y.-S. Lee, B.-Y. Yang, SSE implementation of multivariate PKCs on modern x86 CPUs, in *CHES 2009*, LNCS, vol. 5747 (Springer, 2009), pp. 33–48
9. J. Ding, D. Schmidt, Rainbow, a new multivariable polynomial signature scheme, in *ACNS 2005*, LNCS, vol. 3531 (Springer, 2005), pp. 164–175
10. J. Ding, B. Yang, C.-O. Chen, M. Chen, C. Cheng, New differential-algebraic attacks and reparametrization of Rainbow, in *ACNS 2008*, LNCS, vol. 5037 (Springer, 2008), pp. 242–257
11. J.-C. Faugère, A new efficient algorithm for computing Gröbner bases (F4). J. Pure Appl. Algebra **139**(1–3), 61–88 (1999)
12. J.-C. Faugère, A new efficient algorithm for computing Gröbner bases without reduction to zero (F5), in *ISSAC 2002* (ACM, 2002), pp. 75–83
13. H. Furue, Y. Kiyomura, T. Nagasawa, T. Takagi, A new fault attack on UOV multivariate signature scheme, in *PQCrypto 2022*, LNCS, vol. 13512 (Springer, 2022), pp. 124–143
14. M.-R. Garey, D.-S. Johnson, Computers and intractability: a guide to the theory of NP-completeness (W. H. Freeman, 1979)
15. Y. Hashimoto, T. Takagi, K. Sakurai, General fault attacks on multivariate public key cryptosystems, in *PQCrypto 2011*, LNCS, vol. 7071 (Springer, 2011), pp. 1–18
16. A. Kipnis, J. Patarin, L. Goubin, Unbalanced oil and vinegar signature schemes, in *EUROCRYPT 1999*, LNCS, vol. 1592 (Springer, 1999), pp. 206–222
17. A. Kipnis, A. Shamir, Cryptanalysis of the oil and vinegar signature scheme, in *CRYPTO 1998*, LNCS, vol. 1462 (Springer, 1998), pp. 257–266
18. J. Krämer, M. Loiero, Fault attacks on UOV and Rainbow, in *COSADE 2019*, LNCS, vol. 11421 (Springer, 2019), pp. 193–214
19. K. Mus, S. Islam, B. Sunar, QuantumHammer: a practical hybrid attack on the LUOV signature scheme, in *CCS 2020* (ACM, 2020), pp. 1071–1084
20. NIST: Post-quantum cryptography CSRC. https://csrc.nist.gov/Projects/post-quantum-cryptography/post-quantum-cryptography-standardization
21. NIST: Status report on the second round of the NIST post-quantum cryptography standardization process. NIST Internal Report 8309, NIST (2020)
22. A. Park, K.-A. Shim, N. Koo, D.-G. Han, Side-channel attacks on post-quantum signature schemes based on multivariate quadratic equations - Rainbow and UOV -. IACR Trans. Cryptogr. Hardw. Embed. Syst. **2018**(3), 500–523 (2018)
23. P.W. Shor, Polynomial-time algorithms for prime factorization and discrete logarithms on a quantum computer. SIAM J. Comput. **26**(5), 1484–1509 (1997)
24. K.-A. Shim, N. Koo, Algebraic fault analysis of UOV and Rainbow with the leakage of random vinegar values. IEEE Trans. Inf. Forensics Secur. **15**, 2429–2439 (2020)
25. K.-A. Shim, S. Lee, N. Koo, Efficient implementations of Rainbow and UOV using AVX2. IACR Trans. Cryptogr. Hardw. Embed. Syst. **2022**(1), 245–269 (2022)
26. B.-Y. Yang, J.-M. Chen, N. Courtois, On asymptotic security estimates in XL and Gröbner bases-related algebraic cryptanalysis, in *ICICS 2004*, LNCS, vol. 3269 (Springer, 2004), pp. 401–413
27. B.-Y. Yang, O.C.-H. Chen, D.J. Bernstein, J.-M. Chen, Analysis of QUAD, in *FSE 2007*, LNCS, vol. 4593 (Springer, 2007), pp. 290–308

Open Access This chapter is licensed under the terms of the Creative Commons Attribution 4.0 International License (http://creativecommons.org/licenses/by/4.0/), which permits use, sharing, adaptation, distribution and reproduction in any medium or format, as long as you give appropriate credit to the original author(s) and the source, provide a link to the Creative Commons license and indicate if changes were made.

The images or other third party material in this chapter are included in the chapter's Creative Commons license, unless indicated otherwise in a credit line to the material. If material is not included in the chapter's Creative Commons license and your intended use is not permitted by statutory regulation or exceeds the permitted use, you will need to obtain permission directly from the copyright holder.

On Hilbert–Poincaré Series of Affine Semi-regular Polynomial Sequences and Related Gröbner Bases

Momonari Kudo and Kazuhiro Yokoyama

Abstract Gröbner bases are nowadays central tools for solving various problems in commutative algebra and algebraic geometry. A typical use of Gröbner bases is the multivariate polynomial system solving, which enables us to construct algebraic attacks against post-quantum cryptographic protocols. Therefore, the determination of the complexity of computing Gröbner bases is very important both in theory and in practice: one of the most important cases is the case where input polynomials compose an (overdetermined) affine semi-regular sequence. The first part of this paper aims to present a survey on Gröbner basis computation and its complexity. In the second part, we shall give an explicit formula on the (truncated) Hilbert–Poincaré series associated to the homogenization of an affine semi-regular sequence. Based on the formula, we also study (reduced) Gröbner bases of the ideals generated by an affine semi-regular sequence and its homogenization. Some of our results are considered to give mathematically rigorous proofs of the correctness of methods for computing Gröbner bases of the ideal generated by an affine semi-regular sequence.

Keywords Gröbner bases · Hilbert–Poincaré series · Semi-regular sequences · Solving degrees · Koszul complex · Homogenization

1 Introduction

Let K be a field, and $R = K[x_1, \ldots, x_n]$ the polynomial ring in n variables over K. For a polynomial f in $R \smallsetminus \{0\}$, let f^{top} denote its maximal total degree part which is called the *top part* of f here, and let f^h denote its homogenization in $R' = R[y]$ by an extra variable y, see Sect. 3.1.1 for details. We denote by $\langle F \rangle_R$ (or $\langle F \rangle$ simply)

M. Kudo (✉)
Fukuoka Institute of Technology, Fukuoka, Japan
e-mail: m-kudo@fit.ac.jp

K. Yokoyama
Rikkyo University, Tokyo, Japan
e-mail: kazuhiro@rikkyo.ac.jp

© The Author(s) 2026
T. Takagi et al. (eds.), *Mathematical Foundations for Post-Quantum Cryptography*, Mathematics for Industry 40, https://doi.org/10.1007/978-981-96-1218-5_11

the ideal generated by a non-empty subset F of R. For a finitely generated graded R-(or R'-)module M, we also denote by HF_M and HS_M its Hilbert function and its Hilbert–Poincaré series, respectively. A *Gröbner basis* of an ideal I in R is defined as a special kind of generating set for I, and it gives a computational tool to determine many properties of the ideal I. A typical application of computing Gröbner bases is solving the multivariate polynomial (MP) problem: given m polynomials $f_1, \ldots, f_m$ in R, find $(a_1, \ldots, a_n) \in K^n$ such that $f_i(a_1, \ldots, a_n) = 0$ for all i with $1 \le i \le m$. A particular case where polynomials are all quadratic is called the MQ problem, and its hardness is applied to constructing public-key cryptosystems and digital signature schemes that are expected to be quantum resistant. Therefore, analyzing the complexity of computing Gröbner bases is one of the most important problems both in theory and in practice.

An algorithm for computing Gröbner bases was found first by Buchberger [6], and so far a number of its improvements such as the F_4 [18] and F_5 [19] algorithms have been proposed, see Sect. 3.1 for a summary. In general, it is very difficult to determine the complexity of computing Gröbner bases, but in some cases, we can estimate it with several algebraic invariants such as the solving degree, the degree of regularity, the Castelnuovo–Mumford regularity, and the first and last fall degrees; we refer to [8] for the relations between these invariants.

The first part of this paper aims to survey Gröbner basis computation, and to review its complexity in the case where input polynomials generate a zero-dimensional ideal. For this, in Sect. 2, we first recall foundations in commutative algebra such as Koszul complex, Hilbert–Poincaré series, and semi-regular sequence, which are useful ingredients to estimate the complexity of computing Gröbner bases. Then, we overview existing Gröbner basis algorithms in Sect. 3.1. Subsequently, it will be described in Sect. 3.2 how to estimate the complexity of computing the reduced Gröbner basis of a zero-dimensional ideal, with the notion of homogenization.

In the second part, we focus on *affine semi-regular* polynomial sequences, where a sequence $\boldsymbol{F} = (f_1, \ldots, f_m) \in R^m$ of (not necessarily homogeneous) polynomials is said to be affine (cryptographic) semi-regular if $\boldsymbol{F}^{\mathrm{top}} = (f_1^{\mathrm{top}}, \ldots, f_m^{\mathrm{top}})$ is (cryptographic) semi-regular, see Definitions 4, 7, and 8 for details. Note that homogeneous semi-regular sequences are conjectured by Pardue [32, Conjecture B] to be generic sequences of polynomials, and affine (cryptographic) semi-regular sequences are often appearing in the construction of multivariate public-key cryptosystems and digital signature schemes. In Sect. 4, we relate the Hilbert–Poincaré series of $R'/\langle \boldsymbol{F}^h \rangle$ with that of $R/\langle \boldsymbol{F}^{\mathrm{top}} \rangle$. As a corollary, we obtain an explicit formula of the truncation at degree $D - 1$ of the Hilbert–Poincaré series of $R'/\langle \boldsymbol{F}^h \rangle$, where D is the degree of regularity for $\langle \boldsymbol{F}^{\mathrm{top}} \rangle$. The following theorem summarizes these results:

Theorem 1 (Theorem 7, Corollaries 1 and 2) *With notation as above, assume that $\boldsymbol{F}$ is affine cryptographic semi-regular. Then* $\mathrm{HF}_{R'/\langle \boldsymbol{F}^h \rangle}(d) = \sum_{i=0}^{d} \mathrm{HF}_{R/\langle \boldsymbol{F}^{\mathrm{top}} \rangle}(i)$ *and* $(\langle \mathrm{LM}(\langle \boldsymbol{F}^h \rangle) \rangle_{R'})_d = (\langle \mathrm{LM}(\langle \boldsymbol{F}^{\mathrm{top}} \rangle) \rangle_{R'})_d$ *for each d with $d < D$, where we use a degree reverse lexicographic (DRL) ordering on the set of monomials in R and its homogenization on that in R'. Hence, we also obtain* $\mathrm{HS}_{R'/\langle \boldsymbol{F}^h \rangle}(z) \equiv \prod_{i=1}^{m}(1 -$

$z^{d_i})/(1-z)^{n+1}$ (mod z^D), *so that $\boldsymbol{F}^h$ is D-regular (see Definition 4 for the definition of d-regularity).*

As an application of this theorem, we explore reduced Gröbner bases of $\langle \boldsymbol{F} \rangle$, $\langle \boldsymbol{F}^h \rangle$, and $\langle \boldsymbol{F}^{\mathrm{top}} \rangle$ in Sect. 5, dividing the cases into the degree less than D or not. In particular, we rigorously prove some existing results, which are often used for analyzing the complexity of computing Gröbner bases, and moreover extend them to our case.

2 Preliminaries

In this section, we recall definitions of Koszul complex, Hilbert–Poincaré series, and semi-regular polynomial sequences, and collect some known facts related to them. Throughout this section, let $R = K[X] = K[x_1, \ldots, x_n]$ be the polynomial ring of n variables $X = \{x_1, \ldots, x_n\}$ over a field K. As a notion, for a non-zero polynomial f in R, we denote its total degree by $\deg(f)$. As R is a graded ring with respect to total degree, for a non-zero polynomial f, its maximal total degree part, denoted by f^{top}, is defined as its graded component of $\deg(f)$, that is, the sum of all terms of f whose total degree equals to $\deg(f)$.

2.1 Koszul Complex and Its Homology

Let $f_1, \ldots, f_m \in R \smallsetminus \{0\}$ be homogeneous polynomials of degrees $d_1, \ldots, d_m$, and put $d_{j_1 \cdots j_i} := \sum_{k=1}^{i} d_{j_k}$. For each $0 \le i \le m$, we define a free R-module of rank $\binom{m}{i}$

$$K_i(f_1, \ldots, f_m) := \begin{cases} \bigoplus_{1 \le j_1 < \cdots < j_i \le m} R(-d_{j_1 \cdots j_i}) \mathbf{e}_{j_1 \cdots j_i} & (i \ge 1) \\ R & (i = 0), \end{cases}$$

where $\mathbf{e}_{j_1 \cdots j_i}$ is a standard basis. We also define a graded homomorphism

$$\varphi_i : K_i(f_1, \ldots, f_m) \longrightarrow K_{i-1}(f_1, \ldots, f_m)$$

of degree 0 by

$$\varphi_i(\mathbf{e}_{j_1 \cdots j_i}) := \sum_{k=1}^{i} (-1)^{k-1} f_{j_k} \mathbf{e}_{j_1 \cdots \hat{j}_k \cdots j_i}.$$

Here, $\hat{j}_k$ means to omit j_k. For example, we have $\mathbf{e}_{1\hat{2}3} = \mathbf{e}_{13}$. To simplify the notation, we set $K_i := K_i(f_1, \ldots, f_m)$. Then,

$$K_\bullet : 0 \xrightarrow{\varphi_{m+1}} K_m \xrightarrow{\varphi_m} \cdots \xrightarrow{\varphi_3} K_2 \xrightarrow{\varphi_2} K_1 \xrightarrow{\varphi_1} K_0 \xrightarrow{\varphi_0} K_{-1} := 0 \quad (1)$$

is a complex, and we call it the *Koszul complex* on $(f_1, \ldots, f_m)$. The i-th homology group of $K_\bullet$ is given by

$$H_i(K_\bullet) = \mathrm{Ker}(\varphi_i)/\mathrm{Im}(\varphi_{i+1}).$$

In particular, we have
$$H_0(K_\bullet) = R/\langle f_1, \ldots, f_m \rangle_R.$$

The kernel and the image of a graded homomorphism are both graded submodules in general, so that $\mathrm{Ker}(\varphi_i)$ and $\mathrm{Im}(\varphi_{i+1})$ are graded R-modules, and so is the quotient module $H_i(K_\bullet)$. In the following, we denote by $H_i(K_\bullet)_d$ the degree-d homogeneous part of $H_i(K_\bullet)$.

Note that $\mathrm{Ker}(\varphi_1) = \mathrm{syz}(f_1, \ldots, f_m)$ (the right-hand side is the module of syzygies), and that $\mathrm{Im}(\varphi_2) \subset K_1 = \bigoplus_{j=1}^m R(-d_j)\mathbf{e}_j$ is generated by

$$\{\mathbf{t}_{i,j} := f_i \mathbf{e}_j - f_j \mathbf{e}_i : 1 \le i < j \le m\}.$$

Hence, putting
$$\mathrm{tsyz}(f_1, \ldots, f_m) := \langle \mathbf{t}_{i,j} : 1 \le i < j \le m \rangle_R,$$

we have
$$H_1(K_\bullet) = \mathrm{syz}(f_1, \ldots, f_m)/\mathrm{tsyz}(f_1, \ldots, f_m). \quad (2)$$

Definition 1 (*Trivial syzygies*) With notation as above, we call each generator $\mathbf{t}_{i,j}$ (or each element of $\mathrm{tsyz}(f_1, \ldots, f_m)$) a *trivial syzygy* for $(f_1, \ldots, f_m)$. We also call $\mathrm{tsyz}(f_1, \ldots, f_m)$ the *module of trivial syzygies*.

We also note that $H_m(K_\bullet) = 0$, since φ_m is clearly injective by definition.

Remark 1 When $K = \mathbb{F}_q$, a vector of the form $f_i^{q-1}\mathbf{e}_i$ is also referred to as a trivial syzygy, in the context of Ding-Schmidt's definition for *first fall degree* [15] (see [7, Sect. 4.2] or [30, Sect. 3.2] for reviews). More concretely, putting $B := R/\langle x_1^q, \ldots, x_n^q \rangle_R$ and $\overline{f}_i := f_i \bmod \langle x_1^q, \ldots, x_n^q \rangle$, we define the Koszul complex on $(\overline{f}_1, \ldots, \overline{f}_m) \in B^m$ similar to that on $(f_1, \ldots, f_m) \in R^m$, and denote it by $\overline{K}_\bullet = \overline{K}_\bullet(\overline{f}_1, \ldots, \overline{f}_m)$. Then, the vectors $\overline{f}_i \mathbf{e}_j - \overline{f}_j \mathbf{e}_i$ and $\overline{f}_i^{q-1}\mathbf{e}_i$ in $\bigoplus_{j=1}^m B(-d_j)$ for $1 \le i < j \le m$ are syzygies for $(\overline{f}_1, \ldots, \overline{f}_m)$. Each $\overline{f}_i \mathbf{e}_j - \overline{f}_j \mathbf{e}_i$ is called a Koszul syzygy, and the Koszul syzygies together with $\overline{f}_i^{q-1}\mathbf{e}_i$'s are referred to as trivial syzygies for $(\overline{f}_1, \ldots, \overline{f}_m)$. The *first fall degree* $d_\mathrm{ff}(f_1, \ldots, f_m)$ is defined as the minimal integer d with $\mathrm{syz}(\overline{f}_1, \ldots, \overline{f}_m)_d \supsetneq \mathrm{tsyz}^+(\overline{f}_1, \ldots, \overline{f}_m)_d$ in $\bigoplus_{j=1}^m B(-d_j)_d$, where $\mathrm{tsyz}^+(\overline{f}_1, \ldots, \overline{f}_m)$ denotes the submodule in $\bigoplus_{j=1}^m B(-d_j)$ generated by the trivial syzygies for $(\overline{f}_1, \ldots, \overline{f}_m)$.

Note that, for each i, a homomorphism $H_i(K_\bullet) \to H_i(\overline{K}_\bullet)$ is canonically induced by taking modulo $\langle x_1^q, \ldots, x_n^q \rangle_R$. In particular, we have the following composite

K-linear map:

$$\eta_d : H_1(K_\bullet)_d \to H_1(\overline{K}_\bullet)_d \to \mathrm{syz}(\overline{f}_1, \ldots, \overline{f}_m)_d / \mathrm{tsyz}^+(\overline{f}_1, \ldots, \overline{f}_m)_d.$$

for each d. Putting $d = d_{\mathrm{ff}}(f_1, \ldots, f_m)$ and letting D to be the minimal integer with $H_1(K_\bullet)_D \neq 0$, it is straightforward to verify the following:

- If $q > D$, then η_D is injective (in fact, bijective), and $\mathrm{syz}(\overline{f}_1, \ldots, \overline{f}_m)_D \supsetneq \mathrm{tsyz}^+(\overline{f}_1, \ldots, \overline{f}_m)_D$, whence $D \geq d$.
- If $q > d$, then η_d is surjective (in fact, bijective). In this case, $H_1(K)_d \neq 0$, so that $D \leq d$.

See [30, Lemmas 4.2 and 4.3] for a proof. Therefore, we have $d = D$ if $q > \min\{d, D\}$.

2.2 Hilbert–Poincaré Series and Semi-regular Sequences

Definition 2 (*Hilbert–Poincaré series*) For a finitely generated graded R-module M, we define the *Hilbert function* HF_M of M, given by

$$\mathrm{HF}_M(d) = \dim_K M_d$$

for each $d \in \mathbb{Z}_{\geq 0}$. The *Hilbert–Poincaré series* HS_M of M is defined as the formal power series

$$\mathrm{HS}_M(z) = \sum_{d=0}^{\infty} \mathrm{HF}_M(d) z^d \in \mathbb{Z}[\![z]\!].$$

Theorem 2 (cf. [4, Chap. 10]) *Let I be a homogeneous ideal of R generated by a set $G \subset R \setminus \{0\}$ of homogeneous elements of degree not greater than a non-negative integer d. Let $\mathrm{LM}(f)$ denote the leading monomial of $f \in R \setminus \{0\}$ with respect to a graded ordering $\prec$ on the set of monomials in R. For a non-empty subset $F \subset R \setminus \{0\}$, put $\mathrm{LM}(F) := \{\mathrm{LM}(f) : f \in F\}$. Then, the following are equivalent:*

1. $\langle \mathrm{LM}(G) \rangle_{\leq d} = \langle \mathrm{LM}(I) \rangle_{\leq d}$.
2. *Every $f \in I_{\leq d}$ is reduced to zero modulo G.*
3. *For every pair of $f, g \in G$ with $\deg(\mathrm{LCM}(\mathrm{LM}(f), \mathrm{LM}(g))) \leq d$, the S-polynomial $S(f, g)$ is reduced to zero modulo G.*

In this case, G is called a d-Gröbner basis of I with respect to $\prec$.

We also review the notion of semi-regular sequence defined by Pardue [32].

Definition 3 (*Semi-regular sequences*, [32, Definition 1]) Let I be a homogeneous ideal of R. A degree-d homogeneous element $f \in R$ is said to be *semi-regular* on I if the multiplication map $(R/I)_{t-d} \longrightarrow (R/I)_d \; ; \; g \mapsto gf$ is injective or surjective, for every t with $t \geq d$. A sequence $(f_1, \ldots, f_m) \in R^m$ of homogeneous polynomials is said to be *semi-regular* on I if f_i is semi-regular on $I + \langle f_1, \ldots, f_{i-1} \rangle_R$, for every i with $1 \leq i \leq m$. We simply say that $(f_1, \ldots, f_m)$ is semi-regular if it is semi-regular on the zero ideal.

Throughout the rest of this subsection, let $f_1, \ldots, f_m \in R$ be homogeneous elements of positive degree $d_1, \ldots, d_m$, respectively, and put $I = \langle f_1, \ldots, f_m \rangle_R$. Furthermore, put $I^{(0)} := \{0\}$ and $A^{(0)} := R/I^{(0)} = R$. For each i with $1 \leq i \leq m$, we also set $I^{(i)} := \langle f_1, \ldots, f_i \rangle_R$ and $A^{(i)} := R/I^{(i)}$. The degree-d homogeneous part $A_d^{(i)}$ of each $A^{(i)}$ is given by $A_d^{(i)} = R_d / I_d^{(i)}$, where $I_d^{(i)} = I^{(i)} \cap R_d$. We denote by ψ_{f_i} the multiplication map

$$A^{(i-1)} \longrightarrow A^{(i-1)} \; ; \; g \mapsto g f_i,$$

which is a graded homomorphism of degree d_i. For every $t \geq d_i$, the restriction map

$$\psi_{f_i}|_{A_{t-d_i}^{(i-1)}} : A_{t-d_i}^{(i-1)} \longrightarrow A_t^{(i-1)}$$

is a K-linear map. On the other hand, as for the surjective homomorphism

$$\phi_{i-1} : A^{(i-1)} \longrightarrow A^{(i)} \; ; \; f + I^{(i-1)} \mapsto f + I^{(i)},$$

it is straightforward to see that for each t with $0 \leq t \leq d_i - 1$, the restriction map

$$\phi_{i-1}|_{A_t^{(i-1)}} : A_t^{(i-1)} \longrightarrow A_t^{(i)}$$

is an isomorphism of K-linear spaces, whence

$$\dim_K A_t^{(i-1)} = \dim_K A_t^{(i)} \quad (0 \leq t \leq d_i - 1).$$

Lemma 1 *With notation as above, for each $1 \leq i \leq m$ and for each $t \geq d_i$, we have the following equalities:*

$$\dim_K A_t^{(i)} = \dim_K A_t^{(i-1)} - \dim_K \mathrm{Im}\left(A_{t-d_i}^{(i-1)} \xrightarrow{\times f_i} A_t^{(i-1)} \right), \tag{3}$$

$$\dim_K \mathrm{Im}\left(A_{t-d_i}^{(i-1)} \xrightarrow{\times f_i} A_t^{(i-1)} \right) = \dim_K A_{t-d_i}^{(i-1)} - \dim_K (0 : f_i)_{t-d_i}, \tag{4}$$

where we set $(0 : f_i) := \mathrm{Ker}(\psi_{f_i}) = \{g \in A^{(i-1)} : g f_i = 0 \text{ in } A^{(i-1)}\}$. Hence,

- *The multiplication map $A_{t-d_i}^{(i-1)} \xrightarrow{\times f_i} A_t^{(i-1)}$ is injective if and only if*

$$\dim_K A_t^{(i)} = \dim_K A_t^{(i-1)} - \dim_K A_{t-d_i}^{(i-1)}. \tag{5}$$

In this case, one has $\dim_K A_{t-d_i}^{(i-1)} \leq \dim_K A_t^{(i-1)}$.

- The multiplication map $A_{t-d_i}^{(i-1)} \xrightarrow{\times f_i} A_t^{(i-1)}$ is surjective if and only if

$$\dim_K A_t^{(i)} = 0. \tag{6}$$

In this case, one has $\dim_K A_{t-d_i}^{(i-1)} \geq \dim_K A_t^{(i-1)}$.

Proof Let i and t be integers such that $1 \leq i \leq m$ and $t \geq d_i$. Since we have $(0 : f_i)_{t-d_i} = \{g \in A_{t-d_i}^{(i-1)} : gf_i = 0\}$, the sequence

$$0 \longrightarrow (0 : f_i)_{t-d_i} \longrightarrow A_{t-d_i}^{(i-1)} \xrightarrow{\times f_i} A_t^{(i-1)} \longrightarrow A_t^{(i)} \longrightarrow 0$$

of K-linear maps is exact, where $(0 : f_i)_{t-d_i} \to A_{t-d_i}^{(i-1)}$ is an inclusion map. The exactness of this sequence implies the desired equalities (3) and (4). □

The semi-regularity is characterized by equivalent conditions in Proposition 1. In particular, the fourth condition enables us to compute the Hilbert–Poincaré series of each $A^{(i)}$ easily.

Proposition 1 (cf. [32, Proposition 1]) *With notation as above, the following are equivalent:*

1. *The sequence* $(f_1, \ldots, f_m) \in R^m$ *is semi-regular.*
2. *For each* $1 \leq i \leq m$ *and for each* $t \geq d_i$, *we have (5) or (6), namely,*

$$\dim_K A_t^{(i)} = \max\{0, \dim_K(A_t^{(i-1)}) - \dim_K(A_{t-d_i}^{(i-1)})\}.$$

3. *For each* i *with* $1 \leq i \leq m$, *we have*

$$\mathrm{HS}_{A^{(i)}}(z) = [\mathrm{HS}_{A^{(i-1)}}(z)(1 - z^{d_i})],$$

where $[\cdot]$ *means truncating a formal power series over* $\mathbb{Z}$ *after the last consecutive positive coefficient.*
4. *For each* i *with* $1 \leq i \leq m$, *we have*

$$\mathrm{HS}_{A^{(i)}}(z) = \left[\frac{\prod_{j=1}^{i}(1 - z^{d_j})}{(1 - z)^n} \right].$$

When K is an infinite field, Pardue also conjectured in [32, Conjecture B] that generic polynomial sequences are semi-regular.

2.3 Cryptographic Semi-regular Sequences

We here review the notion of *cryptographic semi-regular* sequence, which is defined by a condition weaker than one for semi-regular sequences. The notion of cryptographic semi-regular sequence is introduced first by Bardet et al. (e.g., [2, 3]) motivated to analyze the complexity of computing Gröbner bases. Diem [13] also formulated cryptographic semi-regular sequences, in terms of commutative and homological algebra. The terminology "cryptographic" was named by Bigdeli et al. in their recent work [5], in order to distinguish such a sequence from a semi-regular one defined by Pardue (see Definition 3 in the previous subsection).

Definition 4 ([2, Definition 3]; *see also* [13, Definition 1]) Let $f_1, \ldots, f_m \in R$ be homogeneous polynomials of positive degrees $d_1, \ldots, d_m$, respectively, and put $I = \langle f_1, \ldots, f_m \rangle_R$. The notations $I^{(i)}$ and $A^{(i)}$ are also the same as in the previous subsection. For each integer d with $d \geq \max\{d_i : 1 \leq i \leq m\}$, we say that a sequence $(f_1, \ldots, f_m) \in R^m$ of homogeneous polynomials is d-*regular* if it satisfies the following condition:

- For each i with $1 \leq i \leq m$, if a homogeneous polynomial $g \in R$ satisfies $gf_i \in \langle f_1, \ldots, f_{i-1} \rangle_R$ and $\deg(gf_i) < d$, then we have $g \in \langle f_1, \ldots, f_{i-1} \rangle_R$. In other word, the multiplication map $A_{t-d_i}^{(i-1)} \longrightarrow A_t^{(i-1)}$; $g \mapsto gf_i$ is injective for every t with $d_i \leq t < d$.

Diem [13] determined the (truncated) Hilbert–Poincaré series of d-regular sequences as in the following proposition:

Theorem 3 (cf. [13, Theorem 1]) *With the same notation as in Definition 4, the following are equivalent for each d with $d \geq \max\{d_i : 1 \leq i \leq m\}$:*

1. *The sequence $(f_1, \ldots, f_m) \in R^m$ is d-regular. Namely, for each (i, t) with $1 \leq i \leq m$ and $d_i \leq t < d$, the equality (5) holds.*
2. *We have*

$$\mathrm{HS}_{A^{(m)}}(z) \equiv \frac{\prod_{j=1}^m (1 - z^{d_j})}{(1-z)^n} \pmod{z^d}.$$

3. $H_1(K_\bullet(f_1, \ldots, f_m))_{\leq d-1} = 0.$

Proposition 2 ([13, Proposition 2 (a)]) *With the same notation as in Definition 4, let D and i be natural numbers. Assume that $H_i(K(f_1, \ldots, f_m))_{\leq D} = 0$. Then, for each j with $1 \leq j < m$, we have $H_i(K(f_1, \ldots, f_j))_{\leq D} = 0$.*

Definition 5 A finitely generated graded R-module M is said to be *Artinian* if there exists a sufficiently large $D \in \mathbb{Z}$ such that $M_d = 0$ for all $d \geq D$.

Definition 6 ([2, Definition 4], [3, Definition 5]) For a homogeneous ideal I of R, we define its *degree of regularity* $d_{\mathrm{reg}}(I)$ as follows: If the finitely generated graded R-module R/I is Artinian, we set $d_{\mathrm{reg}}(I) := \min\{d : R_d = I_d\}$, and otherwise we set $d_{\mathrm{reg}}(I) := \infty$.

As for an upper bound on the degree of regularity, we refer to [22, Theorem 21].

Remark 2 In Definition 6, since R/I is Noetherian, it is Artinian if and only if it is of finite length. In this case, the degree of regularity $d_{\mathrm{reg}}(I)$ is equal to the *Castelnuovo–Mumford regularity* $\mathrm{reg}(I)$ of I (see, e.g., [16, Sect. 20.5] for the definition), whence $d_{\mathrm{reg}}(I) = \mathrm{reg}(I) = \mathrm{reg}(R/I) + 1$.

Definition 7 ([2, Definition 5], [3, Definition 5]; *see also* [14, Sect. 2]) A sequence $(f_1, \ldots, f_m) \in R^m$ of homogeneous polynomials is said to be *cryptographic semi-regular* if it is $d_{\mathrm{reg}}(I)$-regular, where we set $I = \langle f_1, \ldots, f_m \rangle_R$.

The cryptographic semi-regularity is characterized by equivalent conditions in Proposition 3. In particular, the second condition enables us to compute the Hilbert–Poincaré series of R/I easily.

Proposition 3 ([13, Proposition 1 (d)]; see also [3, Proposition 6]) *With the same notation as in Definition 4, we put $D = d_{\mathrm{reg}}(I)$. Then, the following are equivalent:*

1. $(f_1, \ldots, f_m) \in R^m$ *is cryptographic semi-regular.*
2. *We have*
$$\mathrm{HS}_{R/I}(z) = \left[\frac{\prod_{j=1}^{m}(1 - z^{d_j})}{(1 - z)^n} \right]. \tag{7}$$

3. $H_1(K_\bullet(f_1, \ldots, f_m))_{\leq D-1} = 0.$

Remark 3 By the definition of *degree of regularity*, if $(f_1, \ldots, f_m) \in R^m$ is cryptographic semi-regular, $d_{\mathrm{reg}}(I)$ coincides with $\deg(\mathrm{HS}_{R/I}(z)) + 1$, where we set $I = \langle f_1, \ldots, f_m \rangle$.

In 1985, Fröberg already conjectured in [21] that, when K is an infinite field, a generic sequence of homogeneous polynomials $f_1, \ldots, f_m \in R$ of positive degree $d_1, \ldots, d_m$ generates an ideal I with the Hilbert–Poincaré series of the form (7), namely, $(f_1, \ldots, f_m)$ is cryptographic semi-regular. It can be proved (cf. [32]) that Fröberg's conjecture is equivalent to Pardue's one [32, Conjecture B] introduced in Sect. 2.2. We also note that Moreno-Socías conjecture [29] is stronger than the above two conjectures, see [32, Theorem 2] for a proof.

It follows from the fourth condition of Proposition 1 together with the second condition of Proposition 3 that the semi-regularity implies the cryptographic semi-regularity. (The converse does not hold in general, see [24] or [25] for a counterexample.)

Definition 8 (*Affine semi-regular sequences*) A sequence $\boldsymbol{F} = (f_1, \ldots, f_m) \in R^m$ of not necessarily homogeneous polynomials $f_1, \ldots, f_m$ is said to be semi-regular (resp. cryptographic semi-regular) if $\boldsymbol{F}^{\mathrm{top}} = (f_1^{\mathrm{top}}, \ldots, f_m^{\mathrm{top}})$ is semi-regular (resp. cryptographic semi-regular). In this case, we call $\boldsymbol{F}$ an *affine semi-regular (resp. affine cryptographic semi-regular)* sequence.

Remark 4 For an affine cryptographic semi-regular sequence $\boldsymbol{F} = (f_1, \ldots, f_m) \in R^m$ with $K = \mathbb{F}_q$, it follows from Proposition 3 that $d_{\mathrm{reg}}(\langle \boldsymbol{F}^{\mathrm{top}}\rangle) \leq d_{\mathrm{ff}}(f_1^{\mathrm{top}}, \ldots, f_m^{\mathrm{top}})$ for $q \gg 0$, where $d_{\mathrm{ff}}(f_1^{\mathrm{top}}, \ldots, f_m^{\mathrm{top}})$ is the first fall degree defined in Remark 1.

3 Quick Review on the Computation of Gröbner Basis

In this section, we first review previous studies on the computation of Gröbner bases for polynomial ideals.

3.1 Overview of Existing Gröbner Basis Algorithms

Since Buchberger [6] discovered the notion of Gröbner basis and a fundamental algorithm for computing Gröbner bases, many efforts have been done for improving the efficiency of Gröbner basis computation based on Buchberger's algorithm. In his algorithm, S-polynomials play an important role for Gröbner basis computation and give a famous termination criterion called Buchberger's criterion, that is, for a given ideal I of a polynomial ring over a field, its finite generating subset G is a Gröbner basis of I with respect to a monomial ordering if and only if the S-polynomial $S(g, g')$ for any distinct pair $g, g' \in G$ is reduced to 0 modulo G. For details on Buchberger's algorithm and monomial orderings, see, e.g., [4].

In the below, we list effective improvements for algorithms which are, at the same time, very useful to analyze the complexity of Gröbner basis computation. Here we note that the choice of a monomial ordering is also very crucial for the efficiency of Gröber basis computation, but we here do not discuss about its choice. (In general, the degree reverse lexicographical (DRL) ordering[1] is considered as the most efficient ordering for the computation.)

(1) **Related to S-polynomial:**

(1-1) **Strategy for selecting S-polynomial:** It is considered to be very effective to apply the *normal selection strategy*, where we choose a pair (g, g') for which the least common multiple (LCM) of the leading monomials $LM(g)$ and $LM(g')$ with respect to the fixed ordering $\prec$ as smaller as possible. (See [4, Chap. 5.5].) The strategy is very suited for a homogeneous ideal with a *graded*[2] ordering such as DRL, as we can utilize the graded structure of a homogeneous ideal. Also, the *sugar strategy* is designed for a non-homogeneous ideal generated by F to make the computational behavior very close to that for the ideal generated by the *homogenization* F^h. See Sect. 3.1.1 for some details on homogenization. (See also [11, Chap. 2.10].)

(1-2) **Avoiding unnecessary S-polynomial computation:** In Buchberger's algorithm, we add a polynomial to a generating set G which is computed from an S-polynomial by possible reduction of elements in G. Since the construction of S-polynomials and their reduction dominate the whole computation, S-polynomials which are reduced to 0 are very harmful for the efficiency. Thus, it is highly desired to avoid such unnecessary S-polynomials as many as possible.

---

[1] This ordering is also called the graded reverse lexicographical (grevlex) ordering.

[2] We also call a graded ordering a *degree-compatible* ordering.

(A) **Based on simple rules:** At earlier stages, there are easily computable rules, called Buchberger's criterion and Gebauer-Möller's one. Those are using the relation of the LMs of a pair and those of a triple, see [4, Chap. 5.5]. Then, in 2002, Faugére [19] introduced the notion of *signature* and proposed his F_5 algorithm based on a general rule among signatures. We call algorithms using signatures including variants of F_5 *signature-based algorithms (SBA)*. See a survey [17] and Sect. 3.1.2 for details.

(B) **Using invariants of polynomial ideal:** For a homogeneous ideal I of a polynomial ring R, when its Hilbert function $\mathrm{HF}_{R/I}(z)$ is known before the computation, we can utilize the value $\mathrm{HF}_{R/I}(d)$ for each $d \in \mathbb{N}$ (cf. [38]). Because, by the value $\mathrm{HF}_{R/I}(d)$, we can check whether we can stop the computation of S-polynomials of degree d or not. We call an algorithm using Hilbert functions a *Hilbert-driven* (Buchberger's) algorithm. See [38], [11, Chap. 10.2], or [12, Sect. 3.5].

(2) **Efficient computation of S-polynomial reduction:** Since the computation of S-polynomial reduction is a dominant step in the whole Gröbner basis computation, its efficiency heavily affects the total efficiency. As the reduction for a polynomial by elements of G is sequentially applied, we can transform the whole reduction to a Gaussian elimination of a matrix. This approach was suggested in form of Macaulay matrices by Lazard [27] and the first efficient algorithm was given by Faugére [18], which is called the F_4 algorithm. Of course, we can combine the F_4 and F_5 algorithms effectively, which is called the *matrix-F_5 algorithm*.

(3) **Solving coefficient growth:** For a polynomial ideal I over the rational number field $\mathbb{Q}$, the computation may be suffered by certain growth of coefficients in polynomials appearing during Gröbner basis computation. To resolve this problem, several modular methods were proposed. As a typical one, we can use Chinese remainder algorithm (CRA), where we first compute the reduced Gröbner bases G_p over several finite fields $\mathbb{F}_p$ and then recover the reduced Gröbner basis of I from G_p's by CRA. See [31] for details about choosing primes p.

Remark 5 For several public key cryptosystems based on polynomial ideals over finite fields or the elliptic curve discrete logarithm problem, estimating the cost of finding zeros of polynomial ideals is important to analyze the security of those systems, where the computation of their Gröbner bases is a fundamental tool. In this situation, the F_5 algorithm and matrix-F_5 algorithm as its efficient variant with an efficient DRL ordering are considered, since not only those can attain efficient computation but also they are suited for estimating the computational complexity.

In the following, we introduce the notion of *homogenization* and an algorithm for Gröbner basis computation based on signatures (F_5 or its variants), which will be used for our study in Sect. 5.

3.1.1 Homogenization of Polynomials and Monomial Orderings

We begin with recalling the notion of homogenization. (See [23, Chap. 4] for details.) Let K be a field, $X = \{x_1, \ldots, x_n\}$ a set of variables, and $\mathcal{T}$ the set of all monomials in X.[3]

(1) For a non-zero polynomial $f = \sum_{t \in \mathcal{T}} c_t t$ in $K[X]$ with $c_t \in K$, its *homogenization* f^h is defined, by introducing a new variable y, since

$$f^h = \sum_{t \in \mathcal{T}} c_t t y^{\deg(f) - \deg(t)}.$$

Thus f^h is a homogeneous polynomial in $X \cup \{y\}$ over K with total degree $d = \deg(f)$. Also for a set F (or a sequence $\boldsymbol{F} = (f_1, \ldots, f_m) \in K[X]^m$) of polynomials, its *homogenization* F^h (or $\boldsymbol{F}^h$) is defined as $F^h = \{f^h \mid f \in F\}$ (or $\boldsymbol{F}^h = (f_1^h, \ldots, f_m^h) \in K[X \cup \{y\}]^m$). We also write X^h for $X \cup \{y\}$.

(2) Conversely, for a polynomial h in $K[X \cup \{y\}]$, its *dehomogenization* h^{deh} is defined by substituting y with 1, that is, $h^{\mathrm{deh}} = h(X, 1)$. (It is also denoted by $h|_{y=1}$.) For a set H of homogeneous polynomials in $K[X \cup \{y\}]$, its *dehomogenization* H^{deh} (or $H|_{y=1}$) is defined as $H^{\mathrm{deh}} = \{h^{\mathrm{deh}} \mid h \in H\}$. We also apply the dehomogenization to sequences of polynomials.

(3) For an ideal I of $K[X]$, its homogenization I^h, as an ideal, is defined as $\langle I^h \rangle_{K[X \cup \{y\}]}$. We remark that, for a set F of polynomials in $K[X]$, we have $\langle F^h \rangle_{K[X^h]} \subset I^h$ with $I = \langle F \rangle_{K[X]}$, and the equality does not hold in general.

(4) For a homogeneous ideal J in $K[X \cup \{y\}]$, its dehomogenization J^{deh}, as a set, is an ideal of $K[X]$. We note that if a homogeneous ideal J is generated by H, then $J^{\mathrm{deh}} = \langle H^{\mathrm{deh}} \rangle_{K[X]}$ and for an ideal I of $K[X]$, we have $(I^h)^{\mathrm{deh}} = I$.

(5) For a monomial (term) ordering $\prec$ on the set of *monomials* $\mathcal{T}$ in X, its *homogenization* $\prec^h$ on the set of *monomials* $\mathcal{T}^h$ in $X \cup \{y\}$ is defined as follows: For two monomials $X^\alpha y^a$ and $X^\beta y^b$ in $\mathcal{T}^h$, we say $X^\alpha y^a \prec^h X^\beta y^b$ if and only if one of the following holds: (i) $a + |\alpha| < b + |\beta|$, or (ii) $a + |\alpha| = b + |\beta|$ and $X^\alpha \prec X^\beta$, where $\alpha = (\alpha_1, \ldots, \alpha_n) \in \mathbb{Z}_{\geq 0}^n$ and, $|\alpha| = \alpha_1 + \cdots + \alpha_n$, and where X^α denotes $x_1^{\alpha_1} \cdots x_n^{\alpha_n}$. Here, for a monomial $X^\alpha y^a$, we call X^α and y^a its *X-part* and its $\{y\}$-part (or *y-part* simply), respectively. If a monomial ordering $\prec$ is *graded*, the restriction $\prec^h |_{\mathcal{T}}$ of $\prec^h$ on $\mathcal{T}$ coincides with $\prec$.

It is well known that for a Gröbner basis H of $\langle F^h \rangle$ with respect to $\prec^h$, its dehomogenization $\{h^{\mathrm{deh}} \mid h \in H\}$ is also a Gröbner basis of $\langle F \rangle$ with respect to $\prec$.

3.1.2 Signature and F_5 Algorithm

Here we briefly outline the F_5 algorithm, which is an improvement of Buchberger's algorithm. For details, see a survey [17]. Let $F = \{f_1, \ldots, f_m\} \subset R = K[X]$ be a

---

[3] As the symbol m is used for the size of a generating set, we use $\mathcal{T}$ instead of $\mathcal{M}$.

given generating set. For each polynomial h constructed during the Gröbner basis computation of $\langle F \rangle$, the F_5 algorithm attaches a *special label called a signature* as follows: Since h belongs to $\langle F \rangle$, it can be written as

$$h = a_1 f_1 + a_2 f_2 + \cdots + a_m f_m \tag{8}$$

for some $a_1, \ldots, a_m \in R$. Then, we assign h to $a_1 \mathbf{e}_1 + \cdots + a_m \mathbf{e}_m \in R^m$ and we call its leading monomial $t\mathbf{e}_i$ with respect to a monomial (module) ordering in R^m the *signature* of h. As the expression (8) is not unique, in order to determine the signature, we construct the expression procedurally or use the uniquely determined residue in $R^m/\mathrm{syz}(f_1, \ldots, f_m)$ by a module Gröbner basis of $\mathrm{syz}(f_1, \ldots, f_m)$. (For the latter case, we call it the *minimal* signature.) Here we denote the signature of h by $\mathrm{sig}(h)$. Anyway, in the F_5 algorithm, we can meet the both by carefully choosing S-polynomials and by applying restricted reduction steps (called Σ-reductions) for S-polynomials without any change of the signature. (So, we need not compute a module Gröbner basis of $\mathrm{syz}(f_1, \ldots, f_m)$.) We note that for the S-polynomial $S(h_1, h_2) = c_1 t_1 h_1 - c_2 t_2 h_2$ with $c_1, c_2 \in K$ and $t_1, t_2 \in \mathcal{T}$, the signature $\mathrm{sig}(S(h_1, h_2))$ is determined as the largest one between $\mathrm{sig}(c_1 t_1 h_1)$ and $\mathrm{sig}(c_2 t_2 h_2)$. Then, we have the following criteria, which are very useful to avoid the computation of unnecessary S-polynomials. (The latter one is called the *syzygy criterion*.)

Proposition 4 (cf. [11, 17]) *In the F_5 algorithm, we need not compute an S-polynomial if some S-polynomial of the same signature was already proceeded, since both are reduced to the same polynomial. Moreover, we need not compute an S-polynomial of signature s if there is a signature s' such that s' divides s and some S-polynomial with the signature s' is reduced to 0.*

3.2 Complexity of the Gröbner Basis Computation

In general, determining the complexity of computing a Gröbner basis is very hard; in the worst case, the complexity is doubly exponential in the number of variables, see, e.g., [9, 28, 33] for surveys. It is well known that a Gröbner basis with respect to a graded monomial ordering (in particular, DRL ordering) can be computed quite more efficiently than ones with respect to other orderings in general. Moreover, in the case where the input polynomials generate a zero-dimensional ideal, once a Gröbner basis with respect to an efficient monomial ordering is computed, one with respect to any other ordering can be computed easily by the FGLM basis conversion algorithm [20]. From this, we focus on the case where the monomial ordering is graded, and if necessary we also assume that the ideal generated by the input polynomials is zero-dimensional.

A typical way to estimate the complexity of computing a Gröbner basis for a sequence $F = (f_1, \ldots, f_m)$ of polynomials is to count the number of S-polynomials that are reduced during the Gröbner basis computation. In the case where the chosen monomial ordering is graded, the most efficient strategy to compute Gröbner bases is the normal selection strategy, on which we proceed *degree by degree*, namely, increase the degree of critical pairs defining S-polynomials, as in the F_4 and F_5 algorithms. For an algorithm adopting this strategy, several S-polynomials are dealt with consecutively at the same degree, which is called the *step degree*. The highest step degree through the execution of an algorithm adopting the normal selection strategy is called the *solving degree* of the algorithm, and it is denoted by $\text{sd}_\prec^{\text{hsd}}(F)$. Determining (or finding a tight bound for) the solving degree is difficult without computing any Gröbner basis. Once it is specified, we may estimate the complexity of the algorithm, as in [37].

On the other hand, for a linear algebra-based algorithm, such as an F_4-family including the (matrix-)F_5 algorithm and the XL family (cf. [10]), that follows Lazard's strategy [26] to reduce S-polynomials by the Gaussian elimination on Macaulay matrices, Caminata-Gorla [7] defined *another solving degree* in a different manner. Specifically, it is defined as the lowest degree d at which the reduced row echelon form (RREF) of the Macaulay matrix $M_{\leq d}(F)$ produces a Gröbner basis, see [7] for details. In this case, the complexity is estimated to be $O(mN^\omega)$ with $N = \binom{n+d}{d}$, where ω is the *matrix multiplication exponent* with $2 \leq \omega < 3$. For a given polynomial sequence $F \in R^m$ and a graded monomial ordering $\prec$, we denote by $\text{sd}_\prec^{\text{mac}}(F)$ this solving degree. In a series of works (cf. [5, 7, 8]) by Gorla et al., they provided a mathematical formulation for the relation between the solving degree $\text{sd}_\prec^{\text{mac}}(F)$ (or $\text{sd}_\prec^{\text{mut}}(F)$ described below) and algebraic invariants coming from F, such as the maximal Gröbner basis degree, the degree of regularity, the Castelnuovo–Mumford regularity, the first and last fall degrees, and so on. Here, the *maximal Gröbner basis degree* of the ideal $\langle F \rangle_R$ is the maximal degree of elements in the reduced Gröbner basis of $\langle F \rangle_R$ with respect to a fixed monomial ordering $\prec$, and is denoted by $\text{max.GB.deg}_\prec(F)$.

In the following, we recall some of the results of Caminata et al. We set $\prec$ as the DRL ordering on R with $x_n \prec \cdots \prec x_1$, and fix it throughout the rest of this subsection. Let y be an extra variable for homogenization as in the previous subsection, and $\prec^h$ the homogenization of $\prec$, so that $y \prec x_i$ for any i with $1 \leq i \leq n$. Then, we have

$$\text{max.GB.deg}_\prec(F) \leq \text{sd}_\prec^{\text{mac}}(F) = \text{sd}_{\prec^h}^{\text{mac}}(F^h) = \text{max.GB.deg}_{\prec^h}(F^h),$$

see [7] for a proof. Here, we also recall Lazard's bound for the maximal Gröbner basis degree of $\langle F^h \rangle_{R'}$ with $R' = R[y]$:

Theorem 4 (Lazard; [26, Theorem 2]) *With notation as above, we assume that the number of projective zeros of F^h is finite (and therefore $m \geq n$), and that $f_1^h = \cdots = f_m^h = 0$ has no non-trivial solution over the algebraic closure $\overline{K}$ with $y = 0$, i.e., F^{top} has no solution in $\overline{K}^n$ other than $(0, \ldots, 0)$. Then, supposing also* that $d_1 \geq \cdots \geq d_m$ and putting $\ell := \min\{m, n+1\}$, we have

$$\text{max.GB.deg}_{\prec^h}(\boldsymbol{F}^h) \leq d_1 + \cdots + d_\ell - \ell + 1. \tag{9}$$

Lazard's bound given in (9) is also referred to as the *Macaulay bound*, and it provides an upper bound for the solving degree of $\boldsymbol{F}$ with respect to a DRL ordering. As for the maximal Gröbner basis degree of $\langle \boldsymbol{F} \rangle$, if $R/\langle \boldsymbol{F}^{\text{top}} \rangle$ is Artinian, we have

$$\text{max.GB.deg}_{\prec'}(\boldsymbol{F}) \leq d_{\text{reg}}(\langle \boldsymbol{F}^{\text{top}} \rangle)$$

for any graded ordering $\prec'$ on R, see [7, Remark 15] or Lemma 4 for a proof. Both $d_{\text{reg}}(\langle \boldsymbol{F}^{\text{top}} \rangle)$ and $\text{sd}_{\prec}^{\text{mac}}(\boldsymbol{F})$ are greater than or equal to $\text{max.GB.deg}_{\prec}(\boldsymbol{F})$, whereas the degree of regularity (or the first fall degree) used in the cryptographic literature as a proxy (or a heuristic upper bound) for the solving degree. However, it is pointed out in [5, 7], and [8] by explicit examples that *any* of the degree of regularity and the first fall degree does *not* produce an estimate for the solving degree in general, even when $\boldsymbol{F}$ is an affine (cryptographic) semi-regular sequence. In [8], Caminata-Gorla provided yet another solving degree, denoted by $\text{sd}_{\prec'}^{\text{mut}}(\boldsymbol{F})$, with respect to algorithms based on the *mutant strategy* (see [22] for details), and they proved that it is nothing but the *last fall degree* if it is greater than the maximal Gröbner basis degree.

Theorem 5 ([8, Theorem 3.1]) *With notation as above, for any graded monomial ordering $\prec'$ on R, we have the following inequality:*

$$\text{sd}_{\prec'}^{\text{mut}}(\boldsymbol{F}) = \max\{d_{\boldsymbol{F}}, \text{max.GB.deg}_{\prec'}(\boldsymbol{F})\},$$

where $d_{\boldsymbol{F}}$ denotes the last fall degree of $\boldsymbol{F}$ defined in [8, Definition 1.5].

By this theorem, if $d_{\text{reg}}(\langle \boldsymbol{F}^{\text{top}} \rangle) < d_{\boldsymbol{F}}$, the degree of regularity is no longer an upper bound on the solving degree $\text{sd}_{\prec'}^{\text{mut}}(\boldsymbol{F})$. Salizzoni proved that, if $d_{\text{reg}}(\langle \boldsymbol{F}^{\text{top}} \rangle) \geq \max\{\deg(f) : f \in \boldsymbol{F}\}$, then $\text{sd}_{\prec'}^{\text{mut}}(\boldsymbol{F}) \leq d_{\text{reg}}(\langle \boldsymbol{F}^{\text{top}} \rangle) + 1$, see [35].

On the other hand, Semaev and Tenti state (see Tenti's thesis [37] for a proof) that the solving degree $\text{sd}_{\prec}^{\text{hsd}}(\boldsymbol{F})$ (in terms of the highest step degree) is linear in the degree of regularity, if K is a (large) finite field of order q, and if the input system contains polynomials related to the *field equations*, say $x_i^q - x_i$ for $1 \leq i \leq n$:

Theorem 6 ([36, Theorem 2.1], [37, Corollary 3.67]) *With notation as above, assume that $K = \mathbb{F}_q$, and that $\boldsymbol{F}$ contains $x_i^q - x_i$ for $1 \leq i \leq n$. Put $D = d_{\text{reg}}(\langle \boldsymbol{F}^{\text{top}} \rangle)$. If $D \geq \max\{\deg(f) : f \in \boldsymbol{F}\}$, then we have*

$$\text{sd}_{\prec}^{\text{hsd}}(\boldsymbol{F}) \leq 2D - 2. \tag{10}$$

In Sect. 5.2, we will prove the same inequality as (10) for the case where $\boldsymbol{F}$ not necessarily contains a field equation but is cryptographic semi-regular.

We also refer to our subsequent paper [25], where the several definitions on solving degree given as above are described and classified more conceptionally.

4 Hilbert–Poincaré Series of Affine Semi-regular Sequence

As in the previous sections, let K be a field, and $R = K[X] = K[x_1, \ldots, x_n]$ denote the polynomial ring of n variables over K. We denote by R_d the homogeneous part of degree d, that is, the set of homogeneous polynomials of degree d and 0. Recall Definition 7 for the definition of cryptographic semi-regular sequences.

The Hilbert–Poincaré series associated to a (homogeneous) cryptographic semi-regular sequence is given by (7). On the other hand, the Hilbert–Poincaré series associated to the homogenization $\boldsymbol{F}^h$ of $\boldsymbol{F} = (f_1, \ldots, f_m) \in R^m$ not necessarily homogeneous polynomials cannot be computed without knowing its Gröbner basis in general, but we shall prove that it can be computed up to the degree $D := d_{\mathrm{reg}}(\langle \boldsymbol{F}^{\mathrm{top}} \rangle)$ if $\boldsymbol{F}$ is affine cryptographic semi-regular, namely, $\boldsymbol{F}^{\mathrm{top}}$ is cryptographic semi-regular. Note that Theorem 7 and Corollary 1 can be proved also by directly applying Diem's results (Theorem 3 and Proposition 2) to $(f_1^h, \ldots, f_m^h, y)$: Since it is D-regular by $R'/\langle f_1^h, \ldots, f_m^h, y \rangle \cong R/\langle \boldsymbol{F}^{\mathrm{top}} \rangle$ (this isomorphy is described in the proof of Theorem 7), its subseqence $(f_1^h, \ldots, f_m^h)$ is D-regular.

Theorem 7 *Let $R = K[x_1, \ldots, x_n]$ and $R' = R[y]$, and let $\boldsymbol{F} = (f_1, \ldots, f_m)$ be a sequence of not necessarily homogeneous polynomials in R. Assume that $\boldsymbol{F}$ is affine cryptographic semi-regular. Then, for each d with $d < D := d_{\mathrm{reg}}(\langle \boldsymbol{F}^{\mathrm{top}} \rangle)$, we have*

$$\mathrm{HF}_{R'/\langle \boldsymbol{F}^h \rangle}(d) = \mathrm{HF}_{R/\langle \boldsymbol{F}^{\mathrm{top}} \rangle}(d) + \mathrm{HF}_{R'/\langle \boldsymbol{F}^h \rangle}(d-1), \tag{11}$$

and hence

$$\mathrm{HF}_{R'/\langle \boldsymbol{F}^h \rangle}(d) = \mathrm{HF}_{R/\langle \boldsymbol{F}^{\mathrm{top}} \rangle}(d) + \cdots + \mathrm{HF}_{R/\langle \boldsymbol{F}^{\mathrm{top}} \rangle}(0), \tag{12}$$

whence we can compute the value $\mathrm{HF}_{R'/\langle \boldsymbol{F}^h \rangle}(d)$ from the formula (7).

Proof Let $K_\bullet = K_\bullet(f_1^h, \ldots, f_m^h)$ be the Koszul complex on $(f_1^h, \ldots, f_m^h)$, which is given by (1). We endow R with the structure of a graded R'-module by the evaluation homomorphism $R' = R[y] \to R$ sending y to 0. By tensoring $K_\bullet$ with $R'/\langle y \rangle_{R'} \cong K[x_1, \ldots, x_n] = R$ over R', we obtain the following exact sequence of chain complexes:

$$0 \longrightarrow K_\bullet \xrightarrow{\times y} K_\bullet \xrightarrow{\pi_\bullet} K_\bullet \otimes_{R'} R \longrightarrow 0,$$

where $\times y$ is a graded homomorphism of degree 1 multiplying each entry of a vector with y, and where π_i is a canonical homomorphism sending $v \in K_i$ to $v_i \otimes 1 \in K_i \otimes_{R'} R$. Note that there is an isomorphism

$$K_i \otimes_{R'} R \cong \bigoplus_{1 \leq j_1 < \cdots < j_i \leq m} R(-d_{j_1 \cdots j_i}) \mathbf{e}_{j_1 \cdots j_i},$$

via which we can interpret $\pi_i : K_i \to K_i \otimes_{R'} R$ as a homomorphism that projects each entry of a vector in K_i modulo y. In particular, we have

$$K_0 \otimes_{R'} R = R'/\langle f_1^h, \ldots, f_m^h \rangle_{R'} \otimes_{R'} R'/\langle y \rangle_{R'}$$
$$\cong R'/\langle f_1^h, \ldots, f_m^h, y \rangle_{R'}$$
$$\cong R/\langle f_1^{\text{top}}, \ldots, f_m^{\text{top}} \rangle_R$$

for $i = 0$. This means that the chain complex $K_\bullet \otimes_{R'} R$ gives rise to the Koszul complex on $(f_1^{\text{top}}, \ldots, f_m^{\text{top}})$. We induce a long exact sequence of homology groups. In particular, for each degree d, we have the following long exact sequence:

$$H_{i+1}(K_\bullet)_{d-1} \xrightarrow{\times y} H_{i+1}(K_\bullet)_d \xrightarrow{\pi_{i+1}} H_{i+1}(K_\bullet \otimes_{R'} R)_d$$
$$\xrightarrow{\delta_{i+1}} H_i(K_\bullet)_{d-1} \xrightarrow{\times y} H_i(K_\bullet)_d \xrightarrow{\pi_i} H_i(K_\bullet \otimes_{R'} R)_d,$$

where δ_{i+1} is the connecting homomorphism produced by the Snake lemma. For $i = 0$, we have the following exact sequence:

$$H_1(K_\bullet \otimes_{R'} R)_d \longrightarrow H_0(K_\bullet)_{d-1} \xrightarrow{\times y} H_0(K_\bullet)_d \longrightarrow H_0(K_\bullet \otimes_{R'} R)_d \longrightarrow 0.$$

From our assumption that $\boldsymbol{F}^{\text{top}}$ is cryptographic semi-regular, it follows from Proposition 3 that $H_1(K_\bullet \otimes_{R'} R)_{\leq D-1} = 0$ for $D := d_{\text{reg}}(\langle \boldsymbol{F}^{\text{top}} \rangle)$. Therefore, if $d \leq D-1$, we have an exact sequence

$$0 \longrightarrow H_0(K_\bullet)_{d-1} \xrightarrow{\times y} H_0(K_\bullet)_d \longrightarrow H_0(K_\bullet \otimes_{R'} R)_d \longrightarrow 0$$

of K-linear spaces, so that

$$\dim_K H_0(K_\bullet)_d = \dim_K H_0(K_\bullet \otimes_{R'} R)_d + \dim_K H_0(K_\bullet)_{d-1}$$

by the dimension theorem. Since $H_0(K_\bullet) = R'/\langle \boldsymbol{F}^h \rangle$ and $H_0(K_\bullet \otimes_{R'} R) \cong R/\langle \boldsymbol{F}^{\text{top}} \rangle$, we have the equality (11), as desired. □

Remark 6 With notation as in Theorem 7, assume that $D < \infty$ (and thus $m \geq n$). In the proof of Theorem 7, the multiplication map $H_0(K_\bullet)_{d-1} \to H_0(K_\bullet)_d$ by y is injective for all $d < D$, whence $\text{HF}_{R'/\langle F^h \rangle}(d)$ is monotonically increasing for $d < D-1$. On the other hand, since $H_0(K_\bullet \otimes_{R'} R)_d = (R/\langle \boldsymbol{F}^{\text{top}} \rangle)_d = 0$ for all $d \geq D$ by the definition of the degree of regularity, the multiplication map $H_0(K_\bullet)_{d-1} \to H_0(K_\bullet)_d$ by y is surjective for all $d \geq D$, whence $\text{HF}_{R'/\langle F^h \rangle}(d)$ is monotonically decreasing for $d \geq D - 1$. By this together with [9, Theorem 3.3.4], the homogeneous ideal $\langle \boldsymbol{F}^h \rangle$ is zero-dimensional or trivial, i.e., there are at most a finite number of projective zeros of $\boldsymbol{F}^h$ (and thus there are at most a finite number of affine zeros of $\boldsymbol{F}$).

By Theorem 4, it can be proved that the Hilbert–Poincaré series of $R'/\langle \boldsymbol{F}^h \rangle$ satisfies the following equality (13), which may correspond to [3, Proposition 6]:

Corollary 1 *Let $D = d_{\mathrm{reg}}(\langle \boldsymbol{F}^{\mathrm{top}}\rangle)$. Then we have*

$$\mathrm{HS}_{R'/\langle \boldsymbol{F}^h\rangle}(z) \equiv \frac{\prod_{i=1}^{m}(1-z^{d_i})}{(1-z)^{n+1}} \pmod{z^D}. \tag{13}$$

Therefore, by Theorem 3 ([13, Theorem 1]), $\boldsymbol{F}^h$ is D-regular. Here, we note that $D = \deg(\mathrm{HS}_{R/\langle \boldsymbol{F}^{\mathrm{top}}\rangle}) + 1 = \deg\left(\left[\frac{\prod_{i=1}^{m}(1-z^{d_i})}{(1-z)^n}\right]\right) + 1$.

Proof Let $\mathrm{HS}'(z) = \frac{\prod_{i=1}^{m}(1-z^{d_i})}{(1-z)^{n+1}} \bmod z^D$ and let $\mathrm{HF}'(d)$ denote the coefficient of $\mathrm{HS}'(z)$ of degree d for $d < D$. First we remark that, as $\boldsymbol{F}^{\mathrm{top}}$ is a cryptographic semi-regular sequence, the Hilbert–Poincaré series of $R/\langle \boldsymbol{F}^{\mathrm{top}}\rangle$ satisfies the following:

$$\mathrm{HS}_{R/\langle \boldsymbol{F}^{\mathrm{top}}\rangle}(d) = \left[\frac{\prod_{i=1}^{m}(1-z^{d_i})}{(1-z)^n}\right] = \frac{\prod_{i=1}^{m}(1-z^{d_i})}{(1-z)^n} \bmod z^D,$$

since $\mathrm{HF}_{R/\langle \boldsymbol{F}^{\mathrm{top}}\rangle}(d) = 0$ for $d \geq D$. Then we have

$$\begin{aligned}
\mathrm{HS}'(z) \bmod z^D &= \frac{\prod_{i=1}^{m}(1-z^{d_i})}{(1-z)^{n+1}} \bmod z^D \\
&= \frac{\prod_{i=1}^{m}(1-z^{d_i})}{(1-z)^n} \times (1+z+\cdots+z^{D-1}) \bmod z^D \\
&= \mathrm{HS}_{R/\langle \boldsymbol{F}^{\mathrm{top}}\rangle}(z) \cdot (1+z+\cdots+z^{D-1}) \bmod z^D.
\end{aligned}$$

Therefore, for $d < D$, the equation (12) gives

$$\mathrm{HF}'(d) = \mathrm{HF}_{R/\langle \boldsymbol{F}^{\mathrm{top}}\rangle}(d) + \cdots + \mathrm{HF}_{R/\langle \boldsymbol{F}^{\mathrm{top}}\rangle}(0) = \mathrm{HF}_{R'/\langle \boldsymbol{F}^h\rangle}(d),$$

which implies the desired equality (13). □

To prove the following corollary, we use a fact that, for a homogeneous ideal I in R, the equality $\sum_{i=0}^{d} \dim_K I_i = \dim_K (IR')_d$ holds for each $d \geq 0$. Also we take a graded ordering $\prec$ (preferably a DRL ordering) on monomials in X and its homogenization on monomials in $X \cup \{y\}$.

Corollary 2 *With notation as above, assume that $\boldsymbol{F} = (f_1, \ldots, f_m) \in R^m$ is affine cryptographic semi-regular. Put $\overline{I} := \langle \boldsymbol{F}^{\mathrm{top}}\rangle_R$ and $\tilde{I} := \langle \boldsymbol{F}^h\rangle_{R'}$. Then, we have $(\langle \mathrm{LM}(\tilde{I})\rangle_{R'})_d = (\langle \mathrm{LM}(\overline{I})\rangle_{R'})_d$ for each d with $d < D := d_{\mathrm{reg}}(\overline{I})$.*

Proof We prove $(\langle \mathrm{LM}(\tilde{I})\rangle_{R'})_d \subset (\langle \mathrm{LM}(\overline{I})\rangle_{R'})_d$ by the induction on d. The case where $d = 0$ is clear, and so we assume $d > 0$. Any element in $(\langle \mathrm{LM}(\tilde{I})\rangle_{R'})_d$ is represented as a finite sum of elements in R' of the form $g \cdot \mathrm{LM}(h)$ with $g \in R'$, $h \in \tilde{I}$, and $\deg(gh) = d$. Hence, we can also write each $g \cdot \mathrm{LM}(h)$ as a K-linear combination of elements of the form $\mathrm{LM}(th)$ for a monomial t in R' of degree deg(g), where th is an element in $\tilde{I}$ of degree d. Therefore, it suffices for showing "⊂" to prove that $\mathrm{LM}(f) \in (\langle \mathrm{LM}(\overline{I}) \rangle_{R'})_d$ for any $f \in \tilde{I}$ with deg(f) = d. We may assume that f is homogeneous. It is straightforward that $f|_{y=0} \in \overline{I}_{\leq d}$. If $\mathrm{LM}(f) \in R = K[x_1, \ldots, x_n]$, then we have $\mathrm{LM}(f) = \mathrm{LM}(f|_{y=0}) \in \mathrm{LM}(\overline{I})$. Thus, we may also assume that $y \mid \mathrm{LM}(f)$. In this case, it follows from the definition of the DRL ordering that any other term in f is also divisible by y, so that $f \in \langle y \rangle_{R'}$. Thus, we can write $f = yh$ for some $h \in R'$, where h is homogeneous of degree $d - 1$. As in the proof of Theorem 7, the multiplication map

$$(R'/\tilde{I})_{d'-1} \to (R'/\tilde{I})_{d'} \ ; \ h' \bmod \tilde{I} \mapsto yh' \bmod \tilde{I}$$

is injective for any $d' < D = d_{\mathrm{reg}}(\overline{I})$, since F is cryptographic semi-regular. Therefore, it follows from $f \in \tilde{I}_d$ that $h \in \tilde{I}_{d-1}$, whence $f = yh \in y\tilde{I}_{d-1}$. By the induction hypothesis, there exists $g \in \overline{I}$ such that $\mathrm{LM}(g) \mid \mathrm{LM}(h)$, whence $\mathrm{LM}(f) \in (\langle \mathrm{LM}(\overline{I}) \rangle_{R'})_d$.

Here, it follows from Theorem 7 that

$$\dim_K (R')_d - \dim_K \tilde{I}_d = \sum_{i=0}^{d} \left(\dim_K R_i - \dim_K \overline{I}_i \right) = \sum_{i=0}^{d} \dim_K R_i - \sum_{i=0}^{d} \dim_K \overline{I}_i$$
$$= \dim_K (R')_d - \dim_K (\overline{I}R')_d,$$

and thus $\dim_K \tilde{I}_d = \dim_K (\overline{I}R')_d$. Hence, it follows from $\langle \mathrm{LM}(\overline{I}) \rangle_{R'} = \langle \mathrm{LM}(\overline{I}R') \rangle_{R'}$ that

$$\dim_K (\langle \mathrm{LM}(\tilde{I}) \rangle_{R'})_d = \dim_K (\langle \mathrm{LM}(\overline{I}) \rangle_{R'})_d,$$

whence $(\langle \mathrm{LM}(\tilde{I}) \rangle_{R'})_d = (\langle \mathrm{LM}(\overline{I}) \rangle_{R'})_d$, as desired. □

Example 1 We give a simple example. Let $p = 73$, $K = \mathbb{F}_p$, and

$$f_1 = x_1^2 + (3x_2 - 2x_3 - 1)x_1 + x_2^2 + (-2x_3 - 2)x_2 + x_3^2 + x_3,$$
$$f_2 = 4x_1^2 + (3x_2 + 4x_3 - 2)x_1 - x_2 + x_3^2 + 2x_3,$$
$$f_3 = 3x_1^2 - x_1 + 9x_2^2 + (-6x_3 + 1)x_2 + x_3^2 - x_3,$$
$$f_4 = x_1^2 + (-6x_2 + 2x_3 - 2)x_1 + 9x_2^2 + (-6x_3 + 1)x_2 + 2x_3^2.$$

Then, $d_1 = d_2 = d_3 = d_4 = 2$. As their top parts (maximal total degree parts) are

$$f_1^{\mathrm{top}} = x_1^2 + (3x_2 - 2x_3)x_1 + x_2^2 - 2x_3x_2 + x_3^2,$$
$$f_2^{\mathrm{top}} = 4x_1^2 + (3x_2 + 4x_3)x_1 + x_3^2,$$
$$f_3^{\mathrm{top}} = 3x_1^2 + 9x_2^2 - 6x_3x_2 + x_3^2,$$
$$f_4^{\mathrm{top}} = x_1^2 + (-6x_2 + 2x_3)x_1 + 9x_2^2 - 6x_3x_2 + 2x_3^2,$$

one can verify that F^{top} is a cryptographic semi-regular sequence. Moreover, its degree of regularity is equal to 3. Indeed, the reduced Gröbner basis G_{top} of the ideal $\langle F^{\text{top}} \rangle$ with respect to the DRL ordering $x_1 \succ x_2 \succ x_3$ is

$$\{x_3^2 x_2, x_3^3, x_1^2 + 68x_3 x_2 + 55x_3^2, x_2 x_1 + 27x_3 x_2 + 29x_3^2, x_2^2 + x_3 x_2 + 71x_3^2, x_3 x_1 + 3x_3 x_2 + 33x_3^2\}.$$

Then its leading monomials are $x_3^3, x_3^2 x_2, x_1^2, x_1 x_2, x_2^2, x_3 x_1$ and its Hilbert–Poincaré series satisfies

$$\text{HS}_{R/\langle F^{\text{top}} \rangle}(z) = 2z^2 + 3z + 1 = \left(\frac{(1-z^2)^4}{(1-z)^3} \mod z^3 \right),$$

whence the degree of regularity D of $\langle F^{\text{top}} \rangle$ is 3.

On the other hand, the reduced Gröbner basis G_{hom} of the ideal $\langle F^h \rangle$ with respect to the DRL ordering $x_1 \succ x_2 \succ x_3 \succ y$ is

$$\{y^3 x_1, y^3 x_2, y^3 x_3, 60y^2 x_1 + (x_3^2 + 22y^2)x_2 + 39y^2 x_3,$$
$$72y^2 x_1 + 14y^2 x_2 + x_3^3 + 56y^2 x_3, 16y^2 x_1 + (yx_3 + 55y^2)x_2 + 38y^2 x_3,$$
$$72y^2 x_1 + 66y^2 x_2 + yx_3^2 + 70y^2 x_3, x_1^2 + 72yx_1 + (68x_3 + 40y)x_2 + 55x_3^2 + 14yx_3,$$
$$(x_2 + 20y)x_1 + (27x_3 + 37y)x_2 + 29x_3^2 + 12yx_3,$$
$$57yx_1 + x_2^2 + (x_3 + 3y)x_2 + 71x_3^2 + 52yx_3,$$
$$(x_3 + 22y)x_1 + (3x_3 + 5y)x_2 + 33x_3^2 + 14yx_3\}$$

and its leading monomials are $y^3 x_1, y^3 x_2, y^3 x_3, x_3^2 x_2, x_3^3, yx_2 x_3, yx_3^2, x_1^2, x_1 x_2, x_2^2$, and $x_1 x_3$. Then the Hilbert–Poincaré series of $R'/\langle F^h \rangle$ satisfies

$$\left(\text{HS}_{R'/\langle F^h \rangle}(z) \mod z^3 \right) = \left(6z^2 + 4z + 1 \mod z^3 \right) = \left(\frac{(1-z^2)^4}{(1-z)^4} \mod z^3 \right).$$

We note that $\text{HF}_{R'/\langle F^h \rangle}(3) = 4$ and $\text{HF}_{R'/\langle F^h \rangle}(4) = 1$. We can also examine $\text{LM}(G_{\text{hom}})_{<D} = \text{LM}(G_{\text{top}})_{<D}$ and, for $g \in G_{\text{hom}}$, if $\text{LM}(g)$ is divided by y, then $\deg(g) \geq D = 3$. Thus, at the degree 3, there occurs a *degree-fall*. See [8, Sect. 2.1] for details. Also, the reduced Gröbner basis of $\langle F \rangle$ with respect to $\prec$ is $\{x, y, z\}$ and we can examine that the dehomogenization of G_{hom} is also a Gröbner basis of $\langle F \rangle$.

5 Application to Gröbner Bases Computation

We use the same notation as in the previous section, and assume that F is cryptographic semi-regular such that $D := d_{\text{reg}}(\langle F^{\text{top}} \rangle) < \infty$. Here we apply results in the previous section to the computation of Gröbner bases of the ideals $\langle F \rangle$ and $\langle F^h \rangle$.

Let G, G_{hom}, and G_{top} be the reduced Gröbner bases of $\langle F \rangle$, $\langle F^h \rangle$, and $\langle F^{\text{top}} \rangle$, respectively, where their monomial orderings are DRL $\prec$ or its extension $\prec^h$.

As to the computation of G, in special settings on F such as F containing field equations or F appearing in a multivariate polynomial cryptosystem, methods using the value D or those of the Hilbert function for degrees less than D were proposed. (See [34, 36].) Our results in the section can be considered as a *certain extension* and to *give exact mathematical proofs* for the correctness of the methods.

Here, we extend the notion of *top part* to a homogeneous polynomial h in $R' = R[y]$. We call $h|_{y=0}$ the *top part* of h and denote it by h^{top}. Thus, if h^{top} is not zero, it coincides with the top part $(h|_{y=1})^{\text{top}}$ of the dehomogenization $h|_{y=1}$ of h. We remark that $g^{\text{top}} = (g^h)^{\text{top}}$ for a polynomial g in R.

5.1 Gröbner Basis Elements of Degree Less Than D

Here we show relations between $(G_{\text{hom}})_{<D}$ and $(G_{\text{top}})_{<D}$ with proofs, which are useful for the computations of G_{hom} and G.

Since F^{top} is cryptographic semi-regular and F^h is D-regular by Corollary 1, $H_1(K_\bullet(F^{\text{top}}))_{<D} = H_1(K_\bullet(F^h))_{<D} = 0$. As $H_1(K_\bullet(F^h)) = \text{syz}(F^h)/\text{tsyz}(F^h)$ and $H_1(K_\bullet(F^{\text{top}})) = \text{syz}(F^{\text{top}})/\text{tsyz}(F^{\text{top}})$ (see (2)), we have the following corollary, where $\text{tsyz}(F^h)$ denotes the module of trivial syzygies (see Definition 1).

Corollary 3 ([13, Theorem 1]) *It follows that* $\text{syz}(F^{\text{top}})_{<D} = \text{tsyz}(F^{\text{top}})_{<D}$ *and* $\text{syz}(F^h)_{<D} = \text{tsyz}(F^h)_{<D}$.

This implies that, in the Gröbner basis computation G_{hom} with respect to a graded ordering $\prec^h$, if an S-polynomial $S(g_1, g_2) = t_1 g_1 - t_2 g_2$ of degree less than D is reduced to 0, it comes from some trivial syzygy, that is, $\sum_{i=1}^{m}(t_1 a_i^{(1)} - t_2 a_i^{(2)} - b_i)\mathbf{e}_i$ belongs to $\text{tsyz}(F^h)_{<D}$, where $g_1 = \sum_{i=1}^{m} a_i^{(1)} f_i^h$, $g_2 = \sum_{i=1}^{m} a_i^{(2)} f_i^h$, and $S(g_1, g_2) = \sum_{i=1}^{m} b_i f_i^h$ is obtained by Σ-reduction in the F_5 algorithm (or its variant such as the matrix F_5 algorithm) with the Schreyer ordering. Thus, since the F_5 algorithm (or its variant such as the matrix-F_5 algorithm) with the *Schreyer ordering* automatically discards an S-polynomial whose signature is the LM of some trivial syzygy, we can avoid unnecessary S-polynomials. See Sect. 3.1.2 for a brief outline of the F_5 algorithm and the syzygy criterion (Proposition 4).

In addition to the above facts, as mentioned (somehow implicitly) in [1, Sect. 3.5] and [3], when we compute a Gröbner basis of $\langle F^h \rangle$ for the degree less than D by the F_5 algorithm with respect to $\prec^h$, for each computed non-zero polynomial g from an S-polynomial, say $S(g_1, g_2)$, of degree less than D, its signature does not come from any trivial syzygy and so the reductions of $S(g_1, g_2)$ are done only at its top part. This implies that the Gröbner basis computation process of $\langle F^h \rangle$ corresponds exactly to that of $\langle F \rangle$ for each degree less than D, see [25] for details. Especially, the following lemma holds. Here we give a *concrete and easy* proof using Corollary 2. We also note that the argument and the proof of Lemma 2 can be considered as corrected versions for those of [34, Theorem 4].

Lemma 2 *For each degree $d < D$, we have*

$$\mathrm{LM}(G_{\mathrm{hom}})_d = \mathrm{LM}(G_{\mathrm{top}})_d. \tag{14}$$

Proof We can prove the equality (14) by the induction on d. The case where $d = 0$ is trivial. Assume that the equality (14) holds for $d < D - 1$.

Consider any $t \in \mathrm{LM}(G_{\mathrm{hom}})_{d+1}$. Then, there is a polynomial $g \in G_{\mathrm{hom}}$ such that $\mathrm{LM}(g) = t$. By Corollary 2, for $d + 1 < D$, we have

$$(\langle \mathrm{LM}(\langle \boldsymbol{F}^h \rangle_{R'}) \rangle_{R'})_{d+1} = (\langle \mathrm{LM}(\langle \boldsymbol{F}^{\mathrm{top}} \rangle_{R}) \rangle_{R'})_{d+1}$$

and $\mathrm{LM}(g)$ is divided by $\mathrm{LM}(g')$ for some $g' \in G_{\mathrm{top}}$. Since G_{hom} is reduced, $\mathrm{LM}(g)$ is not divisible by any monomial in $\mathrm{LM}(G_{\mathrm{hom}})_{\leq d} = \mathrm{LM}(G_{\mathrm{top}})_{\leq d}$, so that $\deg(g') = d + 1$. Then we have $\mathrm{LM}(g) = \mathrm{LM}(g')$, and so $\mathrm{LM}(G_{\mathrm{hom}})_{d+1} \subset \mathrm{LM}(G_{\mathrm{top}})_{d+1}$.

By the same argument, $\mathrm{LM}(G_{\mathrm{hom}})_{d+1} \supset \mathrm{LM}(G_{\mathrm{top}})_{d+1}$ can be shown. We note that for each $t \in \mathrm{LM}(G_{\mathrm{top}})_{d+1}$, there is a polynomial $g \in (G_{\mathrm{top}})_{d+1} \subset \langle \boldsymbol{F}^{\mathrm{top}} \rangle_{d+1}$ such that $t = \mathrm{LM}(g)$. In this case, there are homogeneous polynomials $a_1, \ldots, a_m$ such that $g = \sum_{i=1}^m a_i f_i^{\mathrm{top}}$. Then $g' = \sum_{i=1}^m a_i f_i^h$ in $\langle \boldsymbol{F}^h \rangle_{d+1}$ has t as its LM. □

Next we consider $(G_{\mathrm{hom}})_D$.

Lemma 3 *For each monomial t in X of degree D, there is an element g in $(G_{\mathrm{hom}})_{\leq D}$ such that $\mathrm{LM}(g)$ divides t. Therefore,*

$$\langle \mathrm{LM}((G_{\mathrm{hom}})_{\leq D}) \rangle_{R'} \cap R_D = R_D. \tag{15}$$

Moreover, for each element g in $(G_{\mathrm{hom}})_D$ with $g^{\mathrm{top}} \neq 0$, the top-part g^{top} consists of one term, that is, $g^{\mathrm{top}} = \mathrm{LT}(g)$, where LT denotes the leading term of g. (We recall $\mathrm{LT}(g) = \mathrm{LC}(g)\mathrm{LM}(g)$.)

Proof Since $\langle \boldsymbol{F}^{\mathrm{top}} \rangle_D = R_D$, for each monomial t in X of degree D, there are homogeneous $a_1, \ldots, a_m \in R$ with $t = \sum_{i=1}^m a_i f_i^{\mathrm{top}}$. Now consider $h = \sum_{i=1}^m a_i f_i^h$, which belongs to $\langle \boldsymbol{F}^h \rangle$. Then, as $f_i^h = f_i^{\mathrm{top}} + y h_i$ for some homogeneous h_i in R', we have

$$h = \sum_{i=1}^m a_i(f_i^{\mathrm{top}} + y h_i) = \sum_{i=1}^m a_i f_i^{\mathrm{top}} + y \sum_{i=1}^m a_i h_i = t + y \sum_{i=1}^m a_i h_i$$

and $\mathrm{LM}(h) = t$. As G_{hom} is the reduced Gröbner basis of $\langle \boldsymbol{F}^h \rangle$, there is some g in $(G_{\mathrm{hom}})_{\leq D}$ whose LM divides $\mathrm{LM}(h)$, as desired.

Next we prove the second assertion. Let $g_1, \ldots, g_k$ be all elements of $(G_{\mathrm{hom}})_D$ that have non-zero top parts, and set $\mathrm{LM}(g_1) \prec \cdots \prec \mathrm{LM}(g_k)$. We show that $g_i^{\mathrm{top}} = \mathrm{LT}(g_i)$ for all i. Suppose, to the contrary, that our claim does not hold for some g_i. Then, g_i^{top} can be written as $g_i^{\mathrm{top}} = \mathrm{LT}(g_i) + T_2 + \cdots + T_s$ for some terms $T_2, \ldots, T_s$ in R_D. Since $\mathrm{LM}(T_j) \prec \mathrm{LM}(g_i)$ for $2 \leq j \leq s$, it follows from equality (15) that each $\mathrm{LM}(T_j)$ is equal to $\mathrm{LM}(g_\ell)$ for some $\ell < i$ or is divisible by $\mathrm{LM}(g')$ for some $g' \in (G_{\mathrm{hom}})_{<D}$. This contradicts to the fact that G_{hom} is reduced. □

Remark 7 If we apply a signature-based algorithm such as the F_5 algorithm or its variant to compute the reduced Gröbner basis of $\langle F^h \rangle$, its Σ-Gröbner basis is a Gröbner basis, but is not always *reduced* in the sense of ordinary Gröbner basis, in general. In this case, we have to compute so-called *inter-reduction* among elements of the Σ-Gröbner basis for obtaining the reduced Gröbner basis.

5.2 Gröbner Basis Elements of Degree Not Less Than D

In this subsection, we shall extend the upper bound on solving degree given in [36, Theorem 2.1] to our case.

Remark 8 In [36], polynomial ideals over $\mathbb{F}_q[X]$ are considered. Under the condition where the generating sequence F contains the field equations $x_i^q - x_i$ for $1 \leq i \leq n$, recall from Theorem 6 ([37, Theorem 6.5 and Corollary 3.67]) that the solving degree $\mathrm{sd}_{\prec}^{\mathrm{hsd}}(F)$ with respect to a Buchberger-like algorithm for $\langle F \rangle$ is upper bounded by $2D - 2$, where $D = d_{\deg}(\langle F^{\mathrm{top}} \rangle)$. In the proofs of [37, Theorem 6.5 and Corollary 3.67], the property $\langle F^{\mathrm{top}} \rangle_D = R_D$ was essentially used for obtaining the upper bound. As the property also holds in our case, we may apply their arguments. Also in [5, Sect. 3.2], the case where F^h is cryptographic semi-regular is considered. The results on the solving degree and the maximal degree of the Gröbner basis are heavily related to our result in this subsection.

Now we show an upper bound on the solving degree of F by using the set $H := \{g|_{y=1} : g \in (G_{\mathrm{hom}})_{\leq D}\}$, that is, at the pre-process of the computation of G, we first compute $H = (G_{\mathrm{hom}})_{\leq D}$, and at the latter process, we continue the computation from H. We remark that, when we use the normal selection strategy on the choice of S-polynomials, the Gröbner basis computation of $\langle F \rangle$ proceeds along with the graded structure of R in its early stages. By Lemma 2 it simulates faithfully that of $\langle F^{\mathrm{top}} \rangle$ until the degree of computed polynomials becomes $D - 1$, that is, it produces $\{g|_{y=1} : g \in (G_{\mathrm{hom}})_{< D}\}$. Also, by Lemma 3, it may also produce $\{g|_{y=1} : g \in (G_{\mathrm{hom}})_D, g^{\mathrm{top}} \neq 0\}$ by carefully choosing S-polynomials, see [25] for details. We also note that the F_5 algorithm actually uses the normal selection strategy.

Lemma 4 *If $D \geq \max\{\deg(f) : f \in F\}$, then the maximal Gröbner basis degree and the solving degree $\mathrm{sd}_{\prec}^{\mathrm{hsd}}(F)$ (see Sect. 3.2 for the definition of $\mathrm{sd}_{\prec}^{\mathrm{hsd}}(F)$) are bounded as follows:*

$$\max.\mathrm{GB}.\deg_{\prec}(F) \leq D \text{ and } \mathrm{sd}_{\prec}^{\mathrm{hsd}}(F) \leq 2D - 2.$$

Proof Recall from Lemma 3 that $\langle \mathrm{LM}(H) \rangle$ contains all monomials in X of degree D. We continue the Gröbner basis computation from H. In this *latter process*, all polynomials generated from S-polynomials are reduced by elements of H. Therefore, their LMs are reduced with respect to any monomial (in X) of degree D, and thus their degrees are not more than $D - 1$. Thus, the maximal Gröbner basis degree is upper bounded by D, and the degree of S-polynomials dealt in the whole computation is upper bounded by $2D$.

Next we show that we can avoid any S-polynomial of degree $2D$ or $2D-1$.

- If an S-polynomial $S(g_1, g_2)$ has its degree $2D$, then we have $\deg(g_1) = \deg(g_2) = D$ and $\gcd(\mathrm{LM}(g_1), \mathrm{LM}(g_2)) = 1$. Then, Buchberger's criterion predicts that $S(g_1, g_2)$ is always reduced to 0.
- If an S-polynomial $S(g_1, g_2)$ has its degree $2D-1$, then one has $\deg(g_1) = \deg(g_2) = D$, $\deg(g_1) = D, \deg(g_2) = D-1$ or $\deg(g_1) = D-1, \deg(g_2) = D$. For the case where $\deg(g_1) = D, \deg(g_2) = D-1$ or $\deg(g_1) = D-1, \deg(g_2) = D$, we have $\gcd(\mathrm{LM}(g_1), \mathrm{LM}(g_2)) = 1$, and hence $S(g_1, g_2)$ is always reduced to 0 by Buchberger's criterion.
Finally, we consider the remaining case where $\deg(g_1) = \deg(g_2) = D$. In this case, g_1 and g_2 should belong to H and recall from Lemma 3 that both of $(g_1)^{\mathrm{top}}$ and $(g_2)^{\mathrm{top}}$ are single terms. Then $S(g_1, g_2)$ cancels the top parts of $t_1 g_1$ and $t_2 g_2$, where $S(g_1, g_2) = t_1 g_1 - t_2 g_2$ for some terms t_1 and t_2. Thus, the degree of $S(g_1, g_2)$ is less than $2D-1$.

Remark 9 We refer to [7, Remark 15] for another proof of $\max.\mathrm{GB}.\deg_{\prec}(F) \leq D$. We also note that, if $D = d_{\mathrm{reg}}(F^{\mathrm{top}})$ is finite, Lemmas 3 and 4 hold without the assumption that F^{top} is cryptographic semi-regular.

As to the computation of G_{hom}, we have a result similar to Lemma 4. Since $\langle \mathrm{LM}(G_{\mathrm{hom}})_{\leq D} \rangle$ contains all monomials in X of degree D, for any polynomial g generated in the middle of the computation of G_{hom} the degree of the X-part of $\mathrm{LM}(g)$ is less than D. Because g is reduced by $(G_{\mathrm{hom}})_{\leq D}$. Thus, letting $\mathcal{U}$ be the set of all polynomials generated during the computation of G_{hom}, we have

$$\{\text{The } X\text{-part of } \mathrm{LM}(g) : g \in \mathcal{U}\} \subset \{x_1^{e_1} \cdots x_n^{e_n} : e_1 + \cdots + e_n \leq D\}.$$

As different $g, g' \in \mathcal{U}$ cannot have the same X part in their leading terms, the size $\#\mathcal{U}$ is upper bounded by the number of monomials in X of degree not greater than D, that is, $\binom{n+D}{D}$. By using the F_5 algorithm or its efficient variant, under an assumption that every unnecessary S-polynomial can be avoided, the number of computed S-polynomials during the computation of G_{hom} coincides with the number $\#\mathcal{U}$ and is upper bounded by $\binom{n+D}{D}$.

Example 2 When $m = n+1$ and $d_1 = \cdots = d_m = 2$, the Hilbert–Poincáre series of $R/\langle F^{\mathrm{top}} \rangle$ is $\left[\frac{(1-z^2)^{n+1}}{(1-z)^n}\right]$. Since $\frac{(1-z^2)^n}{(1-z)^n} = (1+z)^n = \sum_{i=0}^{n} \binom{n}{i} z^i$, we have

$$\frac{(1-z^2)^{n+1}}{(1-z)^n} = (1+z)^n(1-z^2) = 1 + nz + \sum_{i=2}^{n}\left(\binom{n}{i} - \binom{n}{i-2}\right) z^i - nz^{n+1} - z^{n+2},$$

so that $D = d_{\mathrm{reg}}(\langle F^{\mathrm{top}} \rangle) = \min\left\{i : \binom{n}{i} - \binom{n}{i-2} \leq 0\right\} = \lfloor (n+1)/2 \rfloor + 1$, see [5, Theorem 4.1]. In this case, it follows from

$$2D - 2 = 2(\lfloor (n+1)/2 \rfloor + 1) - 2 = \begin{cases} n+1 & (n: \text{odd}), \\ n & (n: \text{even}) \end{cases}$$

that $\text{sd}_{\prec}^{\text{hsd}}(F) \leq n+1$ in Lemma 4; see [5, Theorems 4.2, 4.7] for the bound in the case where F^h is a generic sequence.

We note that, in the homogeneous case, the solving degree $\text{sd}_{\prec_h}^{\text{hsd}}(F^h)$ is equal to the maximal Gröbner basis degree of F^h (for an appropriate setting in the algorithm one adopts), so that we can apply Lazard's bound, see Theorem 4. It also follows (see [25] for details) that the solving degree $\text{sd}_{\prec}^{\text{hsd}}(F)$ can be upper bounded by $\max.\text{GB}.\deg_{\prec_h}(F^h) = \text{sd}_{\prec_h}^{\text{hsd}}(F^h)$, and we can apply Theorem 4, as our case satisfies its conditions. Then, for the case where $m = n+1$ and $d_1 = \cdots = d_{n+1} = 2$, Lazard's bound gives the bound $n+2$ for $\max.\text{GB}.\deg_{\prec_h}(F^h) = \text{sd}_{\prec_h}^{\text{hsd}}(F^h) \geq \text{sd}_{\prec}^{\text{hsd}}(F)$.

Acknowledgements The authors thank the anonymous referee for helpful comments. The authors are also grateful to Yuta Kambe and Shuhei Nakamura for helpful comments. This work was supported by JSPS Grant-in-Aid for Young Scientists 20K14301 and 23K12949, JSPS Grant-in-Aid for Scientific Research (C) 21K03377, and JST CREST Grant Number JPMJCR2113.

References

1. M. Bardet, Étude des systémes algébriques surdéterminés. Applications aux codes correcteurs et á la cryptographie. PhD thesis, Université Paris IV (2004)
2. M. Bardet, J.-C. Faugère, B. Salvy, On the complexity of Gröbner basis computation of semi-regular overdetermined algebraic equations (extended abstract), in *Proceedings of the International Conference on Polynomial System Solving* (2004), pp. 71–74
3. M. Bardet, J.-C. Faugère, B. Salvy, B.-Y. Yang, Asymptotic behaviour of the degree of regularity of semi-regular polynomial systems, in *Proceedings of MEGA 2005: Eighth International Symposium on Effective Methods in Algebraic Geometry* (2005)
4. T. Becker, V. Weispfenning, *Gröbner Bases: A Computational Approach to Commutative Algebra. GTM*, vol. 141 (Springer, NY, 1993)
5. M. Bigdeli, E. De Negri, M.M. Dizdarevic, E. Gorla, R. Minko, S. Tsakou, Semi-regular sequences and other random systems of equations, in *Women in Numbers Europe III*, vol. 24 (Springer, 2021)
6. B. Buchberger, Ein Algorithmus zum Auffinden der Basiselemente des Restklassenringes nach einem nulldimensionalen Polynomideal. Innsbruck, Univ. Innsbruck, Mathematisches Institut (Diss.) (1965)
7. A. Caminata, E. Gorla, Solving multivariate polynomial systems and an invariant from commutative algebra, in *WAIFI 2020*, pp. 3–36
8. A. Caminata, E. Gorla, Solving degree, last fall degree, and related invariants. J. Symbol. Comput. **114**, 322–335 (2023)
9. J.G. Capaverde, Gröbner bases: Degree bounds and generic ideals. Ph.D thesis, Clemson University (2014)
10. N. Courtois, A. Klimov, J. Patarin, A. Shamir, Efficient algorithms for solving overdefined systems of multivariate polynomial equations, in *EUROCRYPT 2000*, LNCS, vol. 1807 (Springer, 2000)

11. D.A. Cox, J. Little, D. O'Shea, *Ideals, Varieties, and Algorithms*, Undergraduate Texts in Mathematics, 4th edn. (Springer, NY, 2010)
12. W. Decker, C. Lossen, *Computing in Algebraic Geometry: A Quick Start Using SINGULAR. Algorithms and Computation in Mathematics*, vol. 16 (Springer, 2006)
13. C. Diem, Bounded regularity. J. Algebra **423**, 1143–1160 (2015)
14. C. Diem, The XL-algorithm and a conjecture from commutative algebra. *Asiacrypt'04*, LNCS, vol. 3329, pp. 323–337
15. J. Ding, D. Schmidt, Solving degree and degree of regularity for polynomial systems over a finite fields, in *Number Theory and Cryptography*, Lecture Notes in Computer Science, ed. by M. Fischlin, S. Katzenbeisser, vol. 8260 (Springer, Berlin, Heidelberg)
16. D. Eisenbud, Commutative algebra: with a view toward algebraic geometry. *GTM*, vol. 150 (Springer, 1995)
17. C. Eder, J.-C., Faugère, A survey on signature-based algorithms for computing Gröbner bases. J. Symbol. Comput. **80**, 719–784 (2017)
18. J.-C. Faugère, A new efficient algorithm for computing Gröbner bases (F4). J. Pure Appl. Algebra **139**, 61–88 (1999)
19. J.-C. Faugère, A new efficient algorithm for computing Gröbner bases without reduction to zero (F5). *Proceedings of ISSAC 2002* (ACM Press, 2002), pp. 75–82
20. J.-C. Faugère, P. Gianni, D. Lazard, T. Mora, Efficient computation of zero-dimensional Gröbner bases by change of ordering. J. Symbol. Comput. **16**(4), 329–344 (1993)
21. R. Fröberg, An inequality for Hilbert series of graded algebras. Math. Scand. **56**, 117–144 (1985)
22. G. Gaggero, E. Gorla, The complexity of solving a random polynomial system. arxiv: 2309.03855 (2023)
23. M. Kreuzer, L. Robbiano, *Computational Commutative Algebra*, vol. 2 (Springer, 2005)
24. M. Kudo, K. Yokoyama, Generalized cryptographic semi-regular sequences: a variant of Fröberg conjecture and a simple complexity estimation for Gröbner basis computation. arXiv:2410.23211 (2024)
25. M. Kudo, K. Yokoyama, The solving degrees for computing Groebner bases of affine semi-regular polynomial sequences. Accepted for presentation at *MEGA2024: Effective Methods in Algebraic Geometry* (arXiv:2404.03530 or IACR Cryptology ePrint Archive 2024/528) (2024)
26. D. Lazard, Gröbner bases, Gaussian elimination and resolution of systems of algebraic equations. *Computer Algebra* (London, 1983), pp. 146–156, Lecture Notes in Computer Science, vol. 162 (Springer, Berlin, 1983)
27. D. Lazard, Résolution des systèmes d'équations algébriques. Theor. Comput. Sci. **15**(1), 77–110 (1981)
28. E.W. Mayr, S. Ritscher, Dimension-dependent bounds for Gröbner bases of polynomial ideals. J. Symbol. Comput. **49**, 78–94 (2013)
29. G. Moreno-Socías, Autour de la fonction de Hilbert-Samuel (escaliers d'idéaux polynomiaux), Thèse, École Polytechnique (1991)
30. S. Nakamura, Formal power series on algebraic cryptanalysis, arxiv: 2007.14729 (2023)
31. M. Noro, K. Yokoyama, Usage of modular techniques for efficient Computation of ideal operations. Math. Comput. Sci. **12**, 1–32 (2018)
32. K. Pardue, Generic sequences of polynomials. J. Algebra **324**(4), 579–590 (2010)
33. S. Ritscher, Degree Bounds and Complexity of Gröbner Bases of Important Classes of Polynomial Ideals. Ph.D thesis, Technische Universität München Institut für Mathematik (2012)
34. Y. Sakata, T. Takagi, An Efficient Algorithm for Solving the MQ Problem using Hilbert Series, Cryptology ePrint Archive, 2023/1650
35. F. Salizzoni, An upper bound for the solving degree in terms of the degree of regularity. arXiv:2304.13485 (2023)
36. I. Semaev, A. Tenti, Probabilistic analysis on Macaulay matrices over finite fields and complexity constructing Gröbner bases. J. Algebra **565**, 651–674 (2021)

37. A. Tenti, Sufficiently overdetermined random polynomial systems behave like semiregular ones, PhD Thesis, University of Bergen (2019). https://hdl.handle.net/1956/21158
38. C. Traverso, Hilbert functions and the Buchberger algorithm. J. Symbol. Comput. **22**(4), 355–376 (1996)

Open Access This chapter is licensed under the terms of the Creative Commons Attribution 4.0 International License (http://creativecommons.org/licenses/by/4.0/), which permits use, sharing, adaptation, distribution and reproduction in any medium or format, as long as you give appropriate credit to the original author(s) and the source, provide a link to the Creative Commons license and indicate if changes were made.

The images or other third party material in this chapter are included in the chapter's Creative Commons license, unless indicated otherwise in a credit line to the material. If material is not included in the chapter's Creative Commons license and your intended use is not permitted by statutory regulation or exceeds the permitted use, you will need to obtain permission directly from the copyright holder.

Parallel DeepBKZ 2.0: Development of Parallel DeepBKZ Reduction with Large Blocksizes

Satoshi Nakamura, Nariaki Tateiwa, Masaya Yasuda, and Katsuki Fujisawa

Abstract Lattice basis reduction is a powerful tool for solving lattice problems, such as the shortest vector problem (SVP) that asks to find a non-zero shortest vector given a lattice basis. DeepBKZ is a practical refinement of the BKZ reduction algorithm that calls LLL with deep insertions every time after finding non-zero shortest vectors in a projected block. In this paper, we develop "parallel DeepBKZ 2.0", parallelized DeepBKZ reduction using large blocksizes. It can share multiple short lattice vectors among solvers to drastically accelerate the reduction process of DeepBKZ in every solver. We report experimental results of parallel DeepBKZ 2.0 for solving several instances of the Darmstadt SVP challenge.

Keywords Lattice basis reduction · SVP · DeepBKZ · Parallelization

1 Introduction

For a positive integer d, the set of integral linear combinations of d linearly independent vectors in $\mathbb{R}^d$ is called a (full-rank) *lattice* of dimension d, which defines a discrete subgroup of $\mathbb{R}^d$. The set of linearly independent vectors spanning a lattice L is called a *basis* of L. There exist infinitely many bases spanning any lattice of dimension $d \geq 2$. Given a basis of a lattice L, *lattice basis reduction* finds a basis of L

S. Nakamura
NTT Social Informatics Laboratories, Tokyo, Japan
e-mail: satoshi.nakamura.xn@hco.ntt.co.jp

N. Tateiwa
NTT Software Innovation Center, Tokyo, Japan
e-mail: nariaki.tateiwa@ntt.com

M. Yasuda (✉)
Department of Mathematics, Rikkyo University, Tokyo, Japan
e-mail: myasuda@rikkyo.ac.jp

K. Fujisawa
Institute of Mathematics for Industry, Kyushu University, Fukuoka, Japan
e-mail: fujisawa@imi.kyushu-u.ac.jp

consisting of relatively short vectors in L that are also nearly orthogonal to each other. It is a mandatory tool for solving lattice problems, whose hardness assures the security of lattice-based cryptography. In particular, the security of many lattice-based cryptosystems relies on the hardness of solving approximate-SVP with polynomial approximation factors. In cryptanalysis, the BKZ reduction algorithm [16] and its variants such as BKZ 2.0 [7] and pump and jump-BKZ (pnj-BKZ) [1] are the de facto standard to estimate the security level of lattice-based cryptosystems. Given a blocksize β, BKZ repeatedly calls an exact-SVP algorithm for every projected block of dimension β, and also LLL [12] for sub-bases. As a practical refinement of BKZ, DeepBKZ was proposed in [21] that calls LLL with deep insertions (DeepLLL) [16] as a subroutine alternative to LLL. Experimental results in [21, 23] showed that DeepBKZ can find short lattice vectors by smaller blocksizes than BKZ in practice.

In this paper, we parallelize DeepBKZ with large blocksizes, which we call "parallel DeepBKZ 2.0" like BKZ 2.0 [7]. (See [19, 20] for parallelization of DeepBKZ with small blocksizes.) Specifically, we utilize the framework of [20] for massive parallelization of a reduction algorithm. The framework enables each solver to race to generate a more reduced basis while all solvers share multiple short lattice vectors asynchronously. It is useful for solving approximate-SVP in a large-scale computing environment. On the other hand, we redevelop a core DeepBKZ reduction algorithm that each solver runs in the parallelization framework. Specifically, we control the timings of communications for each solver to obtain short lattice vectors from the parallelization system and also introduce a new measurement method for DeepBKZ-reduced bases. There are two choices for an exact-SVP algorithm over projected blocks of dimension β; enumeration and sieving (see [22]). The time complexity of sieving is $2^{O(\beta)}$ while that of enumeration is $2^{O(\beta^2)}$. Therefore sieving is much faster than enumeration for large β. In contrast, the space complexity of sieving is exponential in β while that of enumeration is polynomial. In our implementation, we adopt pruned enumeration of [11] due to the memory limitation of our computing systems. (Cf., the general sieve kernel (G6K) [1] supports various sieving algorithms that need expensive memory requirements.) We report experimental results of parallel DeepBKZ 2.0 for solving several instances of the Darmstadt SVP challenge [9].

Notation Throughout this paper, we represent all vectors in column format. For a vector $a \in \mathbb{R}^d$, let $\|a\|$ denote its Euclidean norm. Let $\langle u, v \rangle$ denote the inner product between two vectors $u, v \in \mathbb{R}^d$. Let $\lfloor z \rceil$ denote the rounded value to the nearest integer of a real number $z \in \mathbb{R}$.

2 Lattices and Their Basis Reduction

In this section, we summarize the basics of lattices and also recall typical algorithms of lattice basis reduction (e.g., see [5, 13, 22] for details).

2.1 Basics on Lattices

Lattices and their bases We fix a positive integer d. A (full-rank) *lattice* in $\mathbb{R}^d$ is the set of all integral linear combinations of d linearly independent vectors $\boldsymbol{b}_1, \ldots, \boldsymbol{b}_d \in \mathbb{R}^d$ as

$$L = \mathcal{L}(\boldsymbol{b}_1, \ldots, \boldsymbol{b}_d) := \left\{ \sum_{i=1}^{d} x_i \boldsymbol{b}_i : x_i \in \mathbb{Z}, \ 1 \le i \le d \right\}.$$

The set $\mathbf{B} = (\boldsymbol{b}_1, \ldots, \boldsymbol{b}_d)$ (resp., the number) of linearly independent vectors spanning L is called a *basis* (resp., the *dimension*) of L. We can regard every basis $\mathbf{B}$ as the $d \times d$ matrix whose column vectors are $\boldsymbol{b}_i$'s. There are infinitely many bases of a lattice of dimension d if $d \ge 2$. Specifically, two $d \times d$ matrices $\mathbf{B}$ and $\mathbf{C}$ span the same lattice of dimension d if and only if there exists a $d \times d$ unimodular matrix $\mathbf{T}$ satisfying $\mathbf{C} = \mathbf{BT}$. (An integral square matrix is said *unimodular* if its determinant is equal to ± 1.) The *volume* of a lattice L is defined by $\mathrm{vol}(L) := |\det \mathbf{B}|$ for any basis $\mathbf{B}$ of L, independent of the choice of bases of L since $|\det(\mathbf{BT})| = |\det \mathbf{B}|$ for any unimodular matrix $\mathbf{T}$.

Gram-Schmidt orthogonalization and projected lattices For a basis $\mathbf{B} = (\boldsymbol{b}_1, \ldots, \boldsymbol{b}_d)$ of a lattice L, its *Gram-Schmidt orthogonalization vectors* are defined recursively as $\boldsymbol{b}_1^\star := \boldsymbol{b}_1$ and

$$\boldsymbol{b}_i^\star := \boldsymbol{b}_i - \sum_{j=1}^{i-1} \mu_{i,j} \boldsymbol{b}_j^\star \in \mathbb{R}^d, \quad \mu_{i,j} := \frac{\langle \boldsymbol{b}_i, \boldsymbol{b}_j^\star \rangle}{\|\boldsymbol{b}_j^\star\|^2} \tag{1}$$

for $2 \le i \le d$. Then we have $\langle \boldsymbol{b}_i^\star, \boldsymbol{b}_j^\star \rangle = 0$ for any $i \ne j$. Let $\mathbf{B}^\star$ be the $d \times d$ matrix whose column vectors are Gram-Schmidt vectors $\boldsymbol{b}_i^\star$'s, and $\mathbf{U}$ the $d \times d$ lower triangular matrix with non-diagonal entries given by the Gram-Schmidt coefficients $\mu_{i,j}$'s and diagonal entries $\mu_{i,i} = 1$. Then we have $\mathbf{B} = \mathbf{U}\mathbf{B}^\star$ from Eq. (1), and hence

$$\mathrm{vol}(L) = |\det \mathbf{B}^\star| = \sqrt{\det({}^t\mathbf{B}^\star \mathbf{B}^\star)} = \prod_{i=1}^{d} \|\boldsymbol{b}_i^\star\| \tag{2}$$

by the orthogonality of Gram-Schmidt vectors $\boldsymbol{b}_i^\star$'s (i.e., it holds $\langle \boldsymbol{b}_i^\star, \boldsymbol{b}_j^\star \rangle = 0$ for any $i \ne j$). For $1 \le k \le d$, we define an orthogonal projection map as

$$\pi_k : \mathbb{R}^d \longrightarrow \mathrm{Span}_\mathbb{R}\left(\boldsymbol{b}_k^\star, \ldots, \boldsymbol{b}_d^\star\right), \quad \pi_k(\boldsymbol{v}) := \sum_{i=k}^{d} \frac{\langle \boldsymbol{v}, \boldsymbol{b}_i^\star \rangle}{\|\boldsymbol{b}_i^\star\|^2} \boldsymbol{b}_i^\star \ (\boldsymbol{v} \in \mathbb{R}^d),$$

where $\mathrm{Span}_\mathbb{R}(\boldsymbol{b}_k^\star, \ldots, \boldsymbol{b}_d^\star)$ is the sub-vector space of $\mathbb{R}^d$ spanned by $d - k + 1$ Gram-Schmidt vectors $\boldsymbol{b}_k^\star, \ldots, \boldsymbol{b}_d^\star$. (In particular, π_1 is the identity map.) The lattice in $\mathbb{R}^d$ spanned by $d - k + 1$ projected vectors $\pi_k(\boldsymbol{b}_k), \ldots, \pi_k(\boldsymbol{b}_d)$ is denoted by $\pi_k(L)$, called a *projected lattice*. The projected lattice $\pi_k(L)$ is of dimension $d - k + 1$, and its volume is equal to $\prod_{i=k}^{d} \|\boldsymbol{b}_i^*\|$ like Eq. (2).

The first successive minimum and Gaussian Heuristic Let L be a lattice in $\mathbb{R}^d$ of dimension d. The length of a shortest non-zero vector in L is called the first *successive minimum*, denoted by $\lambda_1(L)$. (More generally, the k-th successive minimum $\lambda_k(L)$ of L is defined as the minimum of $\max_{1 \le j \le k} \|\boldsymbol{v}_j\|$ over all k linearly independent vectors $\boldsymbol{v}_1, \ldots, \boldsymbol{v}_k$ in L for each $1 \le k \le d$.) Given a measurable set S in $\mathbb{R}^d$ of volume $\mathrm{vol}(S)$, *Gaussian Heuristic* predicts that the number of vectors in $L \cap S$ is approximately equal to $\mathrm{vol}(S)/\mathrm{vol}(L)$. We apply as S the ball of radius $\lambda_1(L)$ centered at the origin in $\mathbb{R}^d$ to obtain a prediction of $\lambda_1(L)$ from Gaussian Heuristic as

$$\lambda_1(L) \approx \nu_d^{-\frac{1}{d}} \cdot \mathrm{vol}(L)^{\frac{1}{d}} \sim \sqrt{\frac{d}{2\pi e}} \cdot \mathrm{vol}(L)^{\frac{1}{d}} =: \mathrm{GH}(L). \tag{3}$$

Here we let ν_d denote the volume of the unit ball in $\mathbb{R}^d$ given by

$$\nu_d = \frac{\pi^{d/2}}{\Gamma(1 + d/2)} \sim \frac{1}{\sqrt{\pi d}} \left(\frac{2\pi e}{d}\right)^{d/2},$$

where $\Gamma(s) = \int_0^\infty t^{s-1} e^{-t} dt$ denotes the Gamma function and the approximation of ν_d is by Stirling's formula. This heuristic roughly holds in practice for "random" lattices of high dimensions such as $d \ge 45$.

2.2 Lattice Basis Reduction and Its Practical Algorithms

Given an input basis of a lattice L, lattice basis reduction finds a new basis $\mathbf{B} = (\boldsymbol{b}_1, \ldots, \boldsymbol{b}_d)$ of the same lattice L with short and nearly-orthogonal basis vectors $\boldsymbol{b}_i$'s. Such a basis is said *good* or *reduced*, and most lattice problems are easier to be solved with a more reduced basis. Below we introduce several reduction notions and algorithms to achieve them.

Lenstra-Lenstra-Lovász (LLL) Let $\mathbf{B} = (\boldsymbol{b}_1, \ldots, \boldsymbol{b}_d)$ be a basis of a lattice L with Gram-Schmidt coefficients $\mu_{i,j}$'s and vectors $\boldsymbol{b}_k^*$'s. We say that $\mathbf{B}$ is *LLL-reduced* for a constant $\frac{1}{4} < \delta < 1$ if it satisfies two conditions;

(i) (Size-reduced) $|\mu_{i,j}| \le \frac{1}{2}$ for any $j < i$.
(ii) (Lovász' condition) $\delta \|\boldsymbol{b}_{k-1}^*\|^2 \le \|\pi_{k-1}(\boldsymbol{b}_k)\|^2$ for all $2 \le k \le d$.

The first basis vector of any LLL-reduced basis satisfies $\|\boldsymbol{b}_1\| \le \alpha^{\frac{d-1}{2}} \lambda_1(L)$ and $\|\boldsymbol{b}_1\| \le \alpha^{\frac{d-1}{4}} \mathrm{vol}(L)^{\frac{1}{d}}$ with $\alpha = \frac{4}{4\delta-1}$. Given any basis $\mathbf{B}$ of L, the LLL algorithm [12] repeatedly performs size-reduction like $\boldsymbol{b}_i \leftarrow \boldsymbol{b}_i - \lfloor \mu_{i,j} \rceil \boldsymbol{b}_j$ and swaps adjacent basis vectors $\boldsymbol{b}_{k-1}, \boldsymbol{b}_k$ that do not satisfy Lovász' condition to find an LLL-reduced basis. The LLL algorithm terminates when conditions (i) (ii) are both satisfied, and its time complexity is polynomial in d. The LLL algorithm is applicable for linearly dependent vectors to remove their linear dependency (see [8, 14, 18]).

LLL with deep insertions (DeepLLL) It is a simple generalization of LLL, which can change *non-adjacent* basis vectors. Specifically, given a lattice basis $\mathbf{B} = (\boldsymbol{b}_1, \ldots, \boldsymbol{b}_d)$ and a reduction parameter $\frac{1}{4} < \delta < 1$ as inputs, the m-th basis vector $\boldsymbol{b}_m$ is inserted before $\boldsymbol{b}_k$, defined as

$$\sigma_{k,m}(\mathbf{B}) := (\ldots, \mathbf{b}_{k-1}, \mathbf{b}_m, \mathbf{b}_k, \ldots, \mathbf{b}_{m-1}, \mathbf{b}_{m+1}, \ldots) \qquad (4)$$

for any indexes $k < m$ satisfying the condition $\|\pi_k(\boldsymbol{b}_m)\|^2 < \delta \|\boldsymbol{b}_k^\star\|^2$. Equation (4) is just a permutation among basis vectors of $\mathbf{B}$, called a *deep insertion*. After a deep insertion, the new k-th Gram-Schmidt vector is given by $\pi_k(\boldsymbol{b}_m)$, strictly shorter than the old k-th Gram-Schmidt vector $\boldsymbol{b}_k^\star$. The DeepLLL algorithm [16] often finds shorter basis vectors than LLL. However, the time complexity of DeepLLL is no longer polynomial in d.

Block Korkine-Zolotarev (BKZ) For a basis $\mathbf{B} = (\boldsymbol{b}_1, \ldots, \boldsymbol{b}_d)$ of a lattice L, we set $\mathbf{B}_{[j:k]} = (\pi_j(\boldsymbol{b}_j), \pi_j(\boldsymbol{b}_{j+1}), \ldots, \pi_j(\boldsymbol{b}_k))$ and $L_{[j:k]}$ the sub-lattice of $\pi_j(L)$ spanned by $k - j + 1$ column vectors of $\mathbf{B}_{[j:k]}$ for $j < k$. We say $\mathbf{B}$ to be β-*BKZ-reduced* for a blocksize $\beta \geq 2$ if it is size-reduced and it also satisfies $\|\boldsymbol{b}_j^\star\| = \lambda_1(L_{[j:k]})$ for every $1 \leq j \leq d - 1$ with $k = \min(j + \beta - 1, d)$. The BKZ algorithm [16] repeatedly calls an exact-SVP algorithm (such as enumeration and sieving) over $L_{[j:k]}$ for every j to find a β-BKZ-reduced basis. A sequence of performing exact-SVP over $L_{[j:k]}$ from $j = 1$ to $d - 1$ is called a *tour*. In the basic BKZ, LLL is also called as a subroutine to reduce $\mathbf{B}_{[j:k]}$ before calling an exact-SVP algorithm over $L_{[j:k]}$. For a large blocksize β, it is practical to terminate the BKZ algorithm after several tours. The time complexity of BKZ depends on that of an exact-SVP algorithm and the number of tours. A larger β makes the computational cost of BKZ more expensive, but it finds a more reduced basis. In practice, for a β-BKZ-reduced basis $(\boldsymbol{b}_1, \ldots, \boldsymbol{b}_d)$ of a random lattice of volume 1, its Gram-Schmidt lengths are roughly given by

$$\|\boldsymbol{b}_i^\star\| \approx c_\beta^{\frac{d-1}{2} - i} \quad \text{with} \quad c_\beta := \left(\frac{\beta}{2\pi e}\right)^{\frac{1}{\beta-1}} \qquad (5)$$

when $\beta > 50$ and $\beta \ll d$ (see [6, 7, 24] for experimental evidences).

DeepBKZ It is a practical refinement of BKZ proposed in [21] that performs DeepLLL as an alternative of LLL to reduce a basis $\mathbf{B} = (\boldsymbol{b}_1, \ldots, \boldsymbol{b}_d)$ before calling an exact-SVP algorithm over $L_{[j:k]}$ for every $1 \leq j \leq d - 1$. As well as BKZ, the time complexity of DeepBKZ depends on that of an exact-SVP algorithm and the number of tours. DeepBKZ often finds short lattice vectors by even small blocksizes

Algorithm 1 Basic DeepBKZ [21]

1: **procedure** DeepBKZ(**B**, β, δ)
                                       ▷ **B** $= (\boldsymbol{b}_1, \ldots, \boldsymbol{b}_d)$: a basis of a lattice L, β: a blocksize
                                       ▷ δ: a reduction parameter such as $\delta = 0.99$
2:     Compute Gram-Schmidt information $\mu_{i,j}$ and $B_j = \|\boldsymbol{b}_j^\star\|^2$ of **B**
3:     **B** ← DeepLLL(**B**, δ)   ▷ DeepLLL reduction together with update of $\mu_{i,j}$, B_j
4:     $z = 0, j = 0$
5:     **while** $z < d - 1$ **do**
6:        $j \leftarrow (j \bmod d - 1) + 1, k = \min(j + \beta - 1, d), h = \min(k + 1, d)$
7:        Find $\boldsymbol{v} \in L$ with $\|\pi_j(\boldsymbol{v})\| = \lambda_1(L_{[j:k]})$ by an exact-SVP algorithm over $L_{[j:k]}$
8:        **if** $\|\pi_j(\boldsymbol{v})\|^2 \geq B_j$ **then**
9:           $z \leftarrow z + 1$
10:       **else**
11:          $z = 0$
12:          $[\boldsymbol{b}_1, \ldots, \boldsymbol{b}_h] \leftarrow$ MLLL$([\boldsymbol{b}_1, \ldots, \boldsymbol{b}_{j-1}, \boldsymbol{v}, \boldsymbol{b}_j, \ldots, \boldsymbol{b}_h], \delta)$
                                     ▷ Insert $\boldsymbol{v}$ and remove its linear dependency by modified LLL
13:         **B** ← DeepLLL(**B**, δ)   ▷ DeepLLL reduction for the current basis **B**
14:       **end if**
15:     **end while**
16: **end procedure**

β in practice (see [21, 23] for experimental results). We present a basic procedure of DeepBKZ in Algorithm 1.

3 Development of Parallel DeepBKZ 2.0

In this section, we develop parallel DeepBKZ 2.0, an improvement of parallel Deep-BKZ reduction with large blocksizes. We utilize the framework of [20] for parallelization of lattice basis reduction to develop parallel DeepBKZ 2.0. Below we begin to describe the parallelization framework.

3.1 Parallelization Framework for Lattice Basis Reduction

The parallelization framework of [20] is a supervisor-worker style consisting of a supervisor process and multiple worker processes. Each worker handles its tasks, and the supervisor manages all the worker processes. We call a worker process a *solver* as in [20]. Given an input basis of a lattice L, the supervisor randomizes it to generate many different bases of L and then distributes them to solvers. Each solver executes a reduction algorithm for a received basis and communicates periodically with the supervisor to update its basis during reduction. In contrast, the supervisor keeps the input basis as a *global basis*, and updates it by communicating with solvers. We show the overall procedure of the parallelization framework in Fig. 1, and describe each procedure below (see [20, Sect. 3] for details of the framework):

(i) Given an input basis **B** of a lattice L, the supervisor sets it as an initial global basis **S**.
(ii) The supervisor randomizes the input basis **B** to generate n different bases $\mathbf{B}_1, \ldots, \mathbf{B}_n$ of L, where n is the number of solvers. Then the supervisor distributes them to solvers.
(iii) Each solver starts to execute a reduction algorithm for a received basis $\mathbf{B}_i$, and then periodically sends its reduced basis $\mathbf{B}'_i$ to the supervisor.
(iv) `Communicate()`: When a solver sends a reduced basis $\mathbf{B}'_i$, the supervisor compares $\mathbf{B}'_i = (\boldsymbol{b}'_1, \ldots, \boldsymbol{b}'_d)$ with the global basis $\mathbf{S} = (\boldsymbol{s}_1, \ldots, \boldsymbol{s}_d)$. We let m denote the number of sharing basis vectors with $1 \leq m \leq d$.

  (a) *Update global basis*: If the first m basis vectors $\boldsymbol{b}'_1, \ldots, \boldsymbol{b}'_m$ of $\mathbf{B}'_i$ are shorter than those of **S** in the lexicographic order of Gram-Schmidt lengths (see [20, Sect. 3] for the order), then the supervisor replaces the first m basis vectors of **S** with $\boldsymbol{b}'_1, \ldots, \boldsymbol{b}'_m$ to generate a new global basis using a modified LLL algorithm.
  (b) *Update local basis*: Otherwise, the supervisor sends the first m basis vectors $\boldsymbol{s}_1, \ldots, \boldsymbol{s}_m$ of **S** to the solver. The solver replaces the first m basis vectors of $\mathbf{B}'_i$ with $\boldsymbol{s}_1, \ldots, \boldsymbol{s}_m$ to obtain a new reduced basis using a modified LLL algorithm. After that, the solver restarts to execute a reduction algorithm for the new reduced basis.

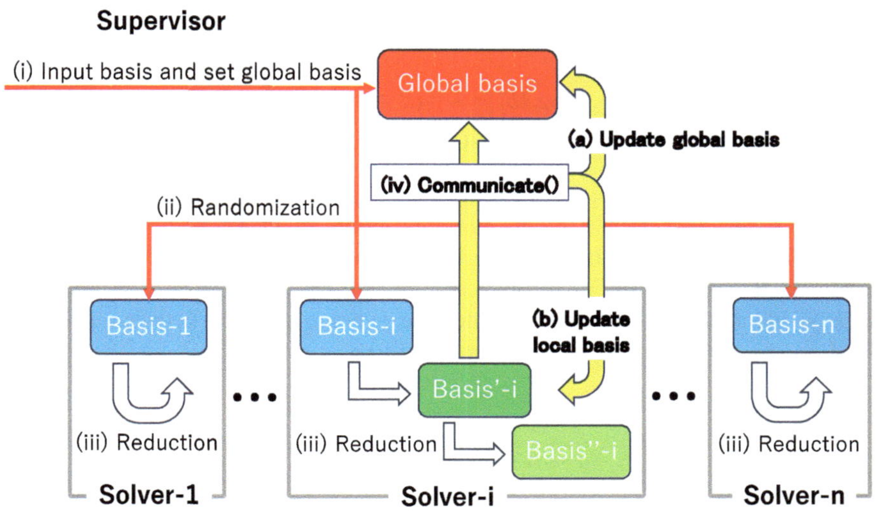

Fig. 1 Overall procedure of the parallelization framework of [20]

During the above processing of parallel reduction, the first m basis vectors of a global basis $\mathbf{S}$ are not larger than those of any solver's reduced basis $\mathbf{B}'_i$ in the lexicographic order of Gram-Schmidt lengths. In contrast, all solvers share the first m basis vectors of $\mathbf{S}$ in their own reduced basis. It is achieved through communications only between each solver and the supervisor. (In other words, it does not require any communication among solvers.)

3.2 Updates for Parallel DeepBKZ 2.0

Here we develop a core algorithm of parallel DeepBKZ 2.0, which every solver runs in the parallelization framework. Algorithm 2 is our algorithm of parallel DeepBKZ 2.0 with a large blocksize $\beta \geq 45$. Below we describe our updates from the basic DeepBKZ algorithm presented in Algorithm 1.

Timings of communications with supervisor We call Communicate() at Step 7 of Algorithm 2 for each solver to communicate with the supervisor at every loop before calling SVP enumeration over $L_{[j:k]} = \mathcal{L}(\mathbf{B}_{[j:k]})$ for $1 \leq j \leq d-1$ with $k = \min(j + \beta - 1, d)$, where $\mathbf{B}$ is an input basis of a lattice L of dimension d. Since it reduces $\mathbf{B}$ using the supervisor's global basis vectors as a pre-processing of enumeration over $L_{[j:k]}$, it reduces the enumeration cost of enumeration. In particular, it would be effective for projected lattices $L_{[j:k]}$ of high dimensions.

New measurement of reduced bases The geometric series assumption (GSA) in [15] states that Gram-Schmidt log-lengths of a reduced basis are roughly on a straight line. Given an input basis $\mathbf{B} = (\boldsymbol{b}_1, \ldots, \boldsymbol{b}_d)$, we let ρ denote a least squares fit of logarithms of squared Gram-Schmidt lengths $\log \|\boldsymbol{b}_i^\star\|^2$ for $1 \leq i \leq d$, defined by

$$\rho := \frac{\sum_{i=1}^d i \log \|\boldsymbol{b}_i^\star\|^2 - \frac{1}{2}(d+1)\sum_{i=1}^d \log \|\boldsymbol{b}_i^\star\|^2}{d(d^2-1)/12}. \tag{6}$$

The gradient of GSA slope (6) is popular in general to measure the quality of reduced bases. However, it seems not appropriate to measure the quality of β-DeepBKZ-reduced bases $\mathbf{B}$ with large blocksizes β since GSA does not hold accurately for $\mathbf{B}$ as shown in Fig. 3. For reducing $\mathbf{B} = (\boldsymbol{b}_1, \ldots, \boldsymbol{b}_d)$ by DeepBKZ with a large blocksize $\beta \geq 45$, we consider the average of gaps between $\|\boldsymbol{b}_j^\star\|$ and $\lambda_1(L_{[j:j+\beta-1]}) \approx \mathrm{GH}(L_{[j:j+\beta-1]})$ from $j = 1$ to $d - \beta + 1$ as

$$\gamma := \frac{1}{d-\beta+1} \sum_{j=1}^{d-\beta+1} \frac{\|\boldsymbol{b}_j^\star\|}{\mathrm{GH}\left(L_{[j:j+\beta-1]}\right)}. \tag{7}$$

Note that after every tour in DeepBKZ for a basis $\mathbf{B}$, the last projected block $\mathbf{B}_{[d-\beta+1:d]}$ of dimension β is HKZ-reduced and thus it holds $\|\boldsymbol{b}_j^\star\| = \lambda_1(L_{[j:k]}) \approx \mathrm{GH}(L_{[j:k]})$

Algorithm 2 A core algorithm of parallel DeepBKZ 2.0 (cf., Algorithm 1)

1: **procedure** Core_DeepBKZ($\mathbf{B}, \beta, \beta', \delta$)
                                        ▷ $\mathbf{B} = (\boldsymbol{b}_1, \ldots, \boldsymbol{b}_d)$: a basis of a lattice L of dimension $d \geq 100$
                                        ▷ β: a large blocksize $\beta \geq 45$, β': a small blocksize with $\beta' \approx \frac{\beta}{2}$
                                        ▷ δ: a reduction parameter for LLL and DeepLLL, such as $\delta = 0.99$
2:     Compute Gram-Schmidt information $\mu_{i,j}$ and $B_j = \|\boldsymbol{b}_j^\star\|^2$
3:     $\mathbf{B} \leftarrow$ DeepLLL$(\mathbf{B}, \delta)$
4:     Compute $\gamma = \dfrac{1}{d-\beta+1} \sum_{j=1}^{d-\beta+1} \dfrac{\|\boldsymbol{b}_j^\star\|}{\text{GH}(L_{[j:j+\beta-1]})}$ ▷ γ is the gap average (7)
5:     $z = 0, j = 0$
6:     **while** $z < d - 1$ **do**
7:        Communicate()   ▷ Communication with the supervisor
8:        **if** $j = d - 1$ **then**
9:           Compute $\gamma = \dfrac{1}{d-\beta+1} \sum_{j=1}^{d-\beta+1} \dfrac{\|\boldsymbol{b}_j^\star\|}{\text{GH}(L_{[j:j+\beta-1]})}$, and set $j = 0$
10:       **end if**
11:       $j \leftarrow j + 1, k = \min(j + \beta - 1, d), h = \min(k + 1, d)$
12:       **if** $\|\boldsymbol{b}_j^\star\| \geq \gamma \text{GH}(L_{[j:k]})$ **then**   ▷ Screening for enumeration
13:           $(v_j, \ldots, v_k) = $ Enum$(\mu_{[j:k]}, B_j, \ldots, B_k)$ ▷ SVP enumeration over $L_{[j:k]}$
14:           $\boldsymbol{v} = \sum_{i=j}^{k} v_i \boldsymbol{b}_i \in L$ ▷ $\|\pi_j(\boldsymbol{v})\| = \lambda(L_{[j:k]})$
15:           **if** $\|\pi_j(\boldsymbol{v})\|^2 \geq B_j$ **then**
16:              $z \leftarrow z + 1$
17:           **else**
18:              $z = 0$
19:              $[\boldsymbol{b}_1, \ldots, \boldsymbol{b}_h] \leftarrow$ MLLL$([\boldsymbol{b}_1, \ldots, \boldsymbol{b}_{j-1}, \boldsymbol{v}, \boldsymbol{b}_j, \ldots, \boldsymbol{b}_h], \delta)$
20:              $\mathbf{B} \leftarrow$ DeepBKZ$(\mathbf{B}, \beta', \delta)$ ▷ DeepBKZ with small blocksizes
21:           **end if**
22:       **end if**
23:     **end while**
24: **end procedure**

with $k = \min(j + \beta - 1, d)$ for all j with $d - \beta + 1 \leq j \leq d$. As the gap average γ of a basis is closer to 1, the basis is more reduced by DeepBKZ with β. In Step 9 of Algorithm 2, we compute the gap average of a basis after every tour to measure how reduced the basis is by DeepBKZ with β. In addition, we use the gap average of a basis $\mathbf{B}$ for screening of SVP enumeration over every projected lattice $L_{[j:k]} = \mathcal{L}(\mathbf{B}_{[j:k]})$. Specifically, we perform SVP enumeration over $L_{[j:k]}$ only if $\|\boldsymbol{b}_j^\star\| \geq \gamma \text{GH}(L_{[j:k]})$. It enables us to efficiently decrease the gap average γ in every tour since there might exist many vectors in $L_{[j:k]}$ shorter than $\boldsymbol{b}_j^\star$ for positions j with $\|\boldsymbol{b}_j^\star\| \geq \gamma \text{GH}(L_{[j:k]})$.

Calling DeepBKZ with small blocksizes as subroutines In Step 20 of Algorithm 2, we call basic DeepBKZ (Algorithm 1) with a small blocksize β' as a main subroutine alternative to DeepLLL (cf., Step 13 of Algorithm 1). During the procedure of DeepBKZ with β', short projected vectors can be pumped up to generate short lattice vectors. In our implementation, for a large blocksize $\beta \geq 45$, we set

$$\beta' = \left\lfloor \frac{\beta - 10}{2} \right\rceil \in \mathbb{Z}$$

as its corresponding small blocksize. (For examples, we have $\beta' = 20, 25, 30$, and 35 for $\beta = 50, 60, 70$, and 80, respectively.) In DeepBKZ with $\beta' \geq 20$, we call Communicate() at the beginning of every tour to get short basis vectors from the supervisor. We use such short vectors to reduce a basis in DeepBKZ with β', as for large blocksizes $\beta \geq 45$. It enables us to reduce an input basis drastically when we use many solvers in the parallelization system. For a small blocksize $\beta' \geq 30$, we use the gradient of GSA slope (6) of a reduced basis to force termination of the procedure of DeepBKZ with a small blocksize β'.

3.3 Implementation for Algorithm 2

We implemented Algorithm 2 in C++ programs with the NTL library [17]. (Cf., see [20] for implementation details of the parallelization framework described in Sect. 3.1.) As a class in our implementation, we set a triple of a basis $\mathbf{B} = (\mathbf{b}_1, \ldots, \mathbf{b}_d)$, its Gram-Schmidt coefficients $\mu = (\mu_{i,j})_{1 \leq j \leq i \leq d}$, and squared lengths $(B_j)_{1 \leq j \leq d}$ of its Gram-Schmidt vectors with $B_j = \|\mathbf{b}_j^*\|^2$. We also used the int data type for $\mathbf{B}$ and double for both μ and $(B_j)_{1 \leq j \leq d}$. For Step 13 of Algorithm 2, we implement [11, Algorithm 2] to enumerate short vectors in a projected lattice $L_{[j:k]}$ of dimension $n = k - j + 1$. For large dimensions $n \geq 45$, we adopted the extreme pruning technique of [11] with a limitation of the enumeration radius (cf., updates for BKZ 2.0 [7]). (We did not use any recent technique for the state-of-the-art enumeration of [2] in our implementation.) Specifically, we used a part of the open-source library of progressive-BKZ [4] to determine bounding functions $R_1^2 \leq R_2^2 \leq \cdots \leq R_n^2$ as inputs of [11, Algorithm 2]. (See the webpage in [4] for the progressive-BKZ library.) On the other hand, we did neither use any pruning technique nor limit the enumeration radius for small dimensions $n < 45$. For Step 19 of Algorithm 2, we adopted the LLL function implemented in the NTL library for the modified LLL algorithm.

3.4 Experiments for the Darmstadt SVP Challenge

We conducted experiments of parallel DeepBKZ 2.0 for solving several instances of the Darmstadt SVP challenge [9] that is a public contest for solving approximate-SVP. Below we report the experimental results.

The number of sharing basis vectors In the parallelization framework described in Sect. 3.1, all solvers can share the first m basis vectors $\mathbf{s}_1, \ldots, \mathbf{s}_m$ of a global basis $\mathbf{S}$ through communications with the supervisor. For our experiments, we set the number

m of sharing basis vectors as follows: For dimension $d = 100$, we set $m = 16$ that is suitable in both the diversity of lattice bases and the performance of parallel reduction from experimental results in [20, Sects. 4 and 5]. For higher dimensions d, we follow the technique of "dimensions for free" in [10] to increase m gradually. For solving the SVP in dimension d, the technique realizes free dimensions around $O\left(\frac{d}{\log d}\right)$ by pre-processing of lattice basis reduction and Babai's lifting (or size reduction). Therefore we set

$$m = \left\lfloor \frac{cd}{\log d} \right\rfloor \in \mathbb{Z}$$

with $c = 0.75$ so that we have $m = 16$ for $d = 100$. For examples, we have $m = 17$, 18, and 20 for $d = 110$, 120, and 130, respectively.

Experimental results The SVP challenge [9] is a public contest for finding non-zero shorter vectors in given a high-dimensional lattice L of length less than $1.05\text{GH}(L)$. We conducted experiments of parallel DeepBKZ 2.0 with blocksizes from $\beta = 50$ to 70 (using pruned enumeration of [11]) for solving instances of the SVP challenge in dimensions $100 \leq d \leq 130$ with seeds from 0 to 9. We used the following four computing systems with different numbers of cores (we assigned a core for the supervisor process and the other cores for solver processes):

1. A server with 8 cores of Intel(R) Xeon(R) Gold 5222 CPU @ 3.80 GHz with 32 GByte memory
2. A server with 36 cores of Intel(R) Xeon(R) Gold 6240 CPU @ 2.60 GHz with 128 GByte memory.
3. A PC cluster with a total of 200 cores (30 cores of Intel(R) Xeon(R) Silver 4214R CPU @ 2.40 GHz with 768 GByte memory, 80 cores of Intel(R) Xeon(R) Gold 5220R CPU @ 2.20 GHz with 768 GByte memory and 90 cores of AMD EPYC 7543 Processor with 1TByte memory)
4. The supercomputer ITO (Kyushu University) with a total of 4,608 cores (128 servers with 36 cores of Intel(R) Xeon(R) Gold 6154 CPU @ 3.0 GHz with 192 GByte memory)

In Fig. 2, we show the average running time of parallel DeepBKZ 2.0 for solving instances of the SVP challenge in each dimension $100 \leq d \leq 130$ with seeds from 0 to 9 (that is, the average running time for finding non-zero short vectors in d-dimensional lattices L of length less than $1.05\text{GH}(L)$). We see from Fig. 2 that the average running time increases about 5 times as the dimension d increases by 5, regardless of the number of processes of computing systems. We also see that the average running time can be reduced approximately by the increase in the number of processes of computing systems. For dimension $d = 135$, we conducted the same experiments using the supercomputer ITO system for several lattices L. We succeeded in finding a non-zero lattice vector of length less than $1.05\text{GH}(L)$ for a few instances within 24 h.

In Fig. 3, we show Gram-Schmidt log-lengths of a global basis **S** reduced by parallel DeepBKZ 2.0 for the SVP challenge instance in dimension $d = 130$ with seed

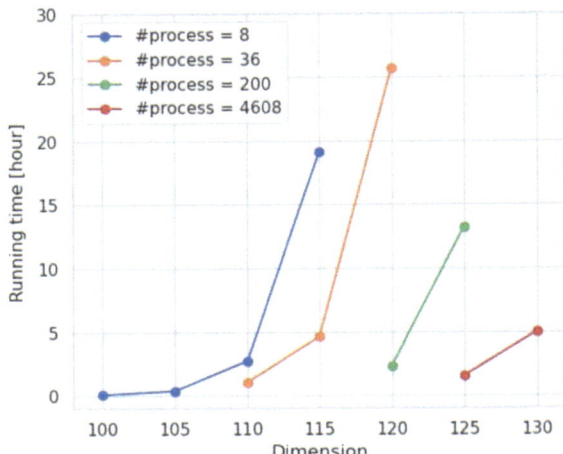

Fig. 2 Average running times (hours) of parallel DeepBKZ 2.0 with pruned enumeration of [11] for solving instances of the SVP challenge in dimensions $100 \leq d \leq 130$ (i.e., average running times for finding non-zero short vectors in lattices L of length less than $1.05\mathrm{GH}(L)$)

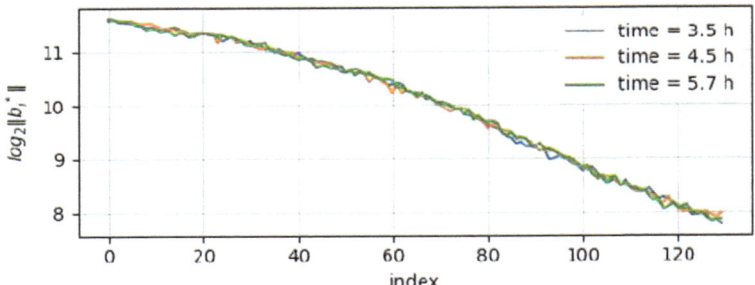

Fig. 3 Gram-Schmidt log-lengths of a global basis reduced by parallel DeepBKZ 2.0 for the SVP challenge instance in dimension $d = 130$ with seed 0

0. We see from Fig. 3 that Gram-Schmidt log-lengths of **S** form convex slightly. In particular, the gradient of Gram-Schmidt log-lengths of **S** is gentle at the first $m = 20$ indices, thanks to sharing m short basis vectors among solvers in the parallelization system. (Note that we set $m = 20$ as the number of sharing basis vectors for $d = 130$.)

Remark 1 Given a constant factor $\kappa \geq 1$, Albrecht et al. [3] estimated the cost of an approximate enumeration oracle for finding a non-zero short vector in a lattice L of dimension d of length less than $\kappa\mathrm{GH}(L)$. Specifically, they considered the relaxed enumeration of [2] with extreme cylinder pruning and extended preprocessing. Based on their simulations and implementation results, they counted the expected cost of the enumeration as the number T_d of nodes visited enumeration nodes, given by

$$\log_2(T_d) = \frac{d \log_2(d)}{2e} - 1.040d + 17.69$$

for $\kappa = 1.05$ from [3, Fig. 3] (with a parameter $c = 0.00$ for pre-processing), where it is assumed to take 64 CPU cycles per node. If we could parallelize the enumeration *perfectly* in a computing system with $p = 4,608$ cores of 3.0 GHz, then it is predicted to take

$$\frac{T_d \times 64}{p \times 3.0 \times 10^9 \times 3600} \approx 1.9\,\text{h} \tag{8}$$

for dimension $d = 130$. On the other hand, it took about 5 h on average by parallel DeepBKZ 2.0 from Fig. 2 for the same dimension d using the same number of processes p. However, the prediction (8) does not take into account the running time of pre-processing for reducing an input basis. Since pre-processing takes almost the same running time of enumeration in practice, it may be almost the same as the running time of parallel DeepBKZ 2.0 even though we did not use the relaxed enumeration of [2] in our implementation. (In other words, it may imply that our parallelization is nearly perfect.)

4 Conclusion

In this paper, we developed parallel DeepBKZ 2.0 with large blocksizes $\beta \geq 50$ using the parallelization framework of [20]. In particular, we gave several updates for a core reduction algorithm (Algorithm 2) that every solver runs, such as timings of communications and a new measurement method of reduced bases. In Fig. 2, we reported the average running time of parallel DeepBKZ 2.0 with the pruned enumeration of [11], implemented in the progressive-BKZ library [4], for solving several instances of the SVP challenge [9] by blocksizes from $\beta = 50$ to 70 in computing systems with different numbers of processes. In particular, we succeeded in finding non-zero short vectors in 130-dimensional lattices L of length less than $1.05 \text{GH}(L)$ within about 5 h on average in a computing system with 4,608 processes.

Acknowledgements The authors would like to thank Yoshinori Aono, a senior researcher at the National Institute of Information and Communications Technology in Japan, for teaching us how to set the success probability of pruned enumeration implemented in the progressive-BKZ library. This work was supported by JST CREST Grant Number JPMJCR2113 and JSPS KAKENHI Grant Number JP20H04142, Japan.

References

1. M. Albrecht, L. Ducas, G. Herold, E. Kirshanova, E.W. Postlethwaite, M. Stevens, The general sieve kernel and new records in lattice reduction, in *Advances in Cryptology–EUROCRYPT 2019*. Lecture Notes in Computer Science, vol. 11477 (Springer, 2019), pp. 717–746
2. M.R. Albrecht, S. Bai, P.A. Fouque, P. Kirchner, D. Stehlé, W. Wen, Faster enumeration-based lattice reduction: root Hermite factor $k^{1/(2k)}$ in time $k^{k/8+o(k)}$, in *Advances in Cryptology–CRYPTO 2020*, Part II. Lecture Notes in Computer Science, vol. 12171 (Springer, 2020), pp. 186–212
3. M.R. Albrecht, S. Bai, J. Li, J. Rowell, Lattice reduction with approximate enumeration oracles: practical algorithms and concrete performance, in *Advances in Cryptology–CRYPTO 2021*, Part II. Lecture Notes in Computer Science, vol. 12826 (Springer, 2021), pp. 732–759
4. Y. Aono, Y. Wang, T. Hayashi, T. Takagi, Improved progressive BKZ algorithms and their precise cost estimation by sharp simulator, in *Advances in Cryptology–EUROCRYPT 2016*. Lecture Notes in Computer Science, vol. 9665 (Springer, 2016), pp. 789–819. Progressive BKZ library is available from https://www2.nict.go.jp/security/pbkzcode/
5. M.R. Bremner, *Lattice Basis Reduction: An Introduction to the LLL Algorithm and its Applications* (CRC Press, 2011)
6. Y. Chen, Réduction de réseau et sécurité concrete du chiffrement completement homomorphe. Ph.D. thesis, Paris 7 (2013)
7. Y. Chen, P.Q. Nguyen, BKZ 2.0: better lattice security estimates, in *Advances in Cryptology–ASIACRYPT 2011*. Lecture Notes in Computer Science, vol. 7073 (Springer, 2011), pp. 1–20
8. H. Cohen, *A Course in Computational Algebraic Number Theory*, Graduate Texts in Mathematics, vol. 138 (Springer, 2013)
9. T. Darmstadt, SVP challenge, https://www.latticechallenge.org/svp-challenge/
10. L. Ducas, Shortest vector from lattice sieving: a few dimensions for free, in *Adavances in Cryptology–EUROCRYPT 2018*. Lecture Notes in Computer Science, vol. 10820 (Springer, 2018), pp. 125–145
11. N. Gama, P.Q. Nguyen, O. Regev, Lattice enumeration using extreme pruning, in *Advances in Cryptology–EUROCRYPT 2010*. Lecture Notes in Computer Science, vol. 6110 (Springer, 2010), pp. 257–278
12. A.K. Lenstra, H.W. Lenstra, L. Lovász, Factoring polynomials with rational coefficients. Mathematische Annalen **261**(4), 515–534 (1982)
13. P.Q. Nguyen, Hermite's constant and lattice algorithms, in *The LLL Algorithm* (Springer, 2009), pp. 19–69
14. M. Pohst, A modification of the LLL reduction algorithm. J. Symbol. Comput. **4**(1), 123–127 (1987)
15. C.P. Schnorr, Lattice reduction by random sampling and birthday methods, in *Symposium on Theoretical Aspects of Computer Science (STACS 2003)*. Lecture Notes in Computer Science, vol. 2607 (Springer, 2003), pp. 145–156
16. C.P. Schnorr, M. Euchner, Lattice basis reduction: improved practical algorithms and solving subset sum problems. Math. Program. **66**, 181–199 (1994)
17. V. Shoup, NTL: a library for doing number theory, http://www.shoup.net/ntl/
18. C.C. Sims, *Computation with Finitely Presented Groups*, vol. 48 (Cambridge University Press, 1994)
19. N. Tateiwa, Y. Shinano, S. Nakamura, A. Yoshida, S. Kaji, M. Yasuda, K. Fujisawa, Massive parallelization for finding shortest lattice vectors based on ubiquity generator framework, in *SC20: International Conference for High Performance Computing, Networking, Storage and Analysis* (IEEE, 2020), pp. 1–15
20. N. Tateiwa, Y. Shinano, M. Yasuda, S. Kaji, K. Yamamura, K. Fujisawa, Development and analysis of massive parallelization of a lattice basis reduction algorithm. Jpn. J. Ind. Appl. Math. 1–44 (2023)

21. J. Yamaguchi, M. Yasuda, Explicit formula for Gram-Schmidt vectors in LLL with deep insertions and its applications, in *Number-Theoretic Methods in Cryptology (NuTMiC 2017)*. Lecture Notes in Computer Science, vol. 10737 (Springer, 2017), pp. 142–160
22. M. Yasuda, A survey of solving SVP algorithms and recent strategies for solving the SVP challenge, in *International Symposium on Mathematics, Quantum Theory, and Cryptography* (Springer, 2021), pp. 189–207
23. M. Yasuda, S. Nakamura, J. Yamaguchi, Analysis of DeepBKZ reduction for finding short lattice vectors. Designs Codes Crypt. **88**, 2077–2100 (2020)
24. Y. Yu, L. Ducas, Second order statistical behavior of LLL and BKZ, in *Selected Areas in Cryptography (SAC 2017)*. Lecture Notes in Computer Science, vol. 10719 (Springer, 2017), pp. 3–22

Open Access This chapter is licensed under the terms of the Creative Commons Attribution 4.0 International License (http://creativecommons.org/licenses/by/4.0/), which permits use, sharing, adaptation, distribution and reproduction in any medium or format, as long as you give appropriate credit to the original author(s) and the source, provide a link to the Creative Commons license and indicate if changes were made.

The images or other third party material in this chapter are included in the chapter's Creative Commons license, unless indicated otherwise in a credit line to the material. If material is not included in the chapter's Creative Commons license and your intended use is not permitted by statutory regulation or exceeds the permitted use, you will need to obtain permission directly from the copyright holder.

Post-quantum Cryptography

A Survey on Middle-Product Learning with Errors Cryptography

Masayuki Tezuka and Keisuke Tanaka

Abstract Middle-product learning with errors (MP-LWE) introduced by Rosca et al. (CRYPTO 2017) is a variant of polynomial learning with errors (Poly-LWE). MP-LWE-based cryptosystems offer higher efficiency than LWE-based cryptosystems. Moreover, a reduction from the Poly-LWE problems to the MP-LWE problem exists, the MP-LWE problem is at least as hard as the Poly-LWE problems. After the seminal work by Rosca et al., MP-LWE-based cryptography has progressed. In particular, the MP-LWE-based public key encryption scheme called Titanium was submitted in the NIST post-quantum cryptography competition process. Moreover, the trapdoor technique for MP-LWE developed, identity-based encryption schemes were proposed. In this paper, we give a survey on MP-LWE cryptography.

Keywords Middle-product learning with errors · Public key encryption · Identity-based encryption · Identification · Digital signature

1 Introduction to MP-LWE Cryptography

Lattice-Based Cryptography. Lattice-based cryptography is the front-running candidate for post-quantum cryptography (PQC). The lattice-based public key encryption and key-establishment algorithm CRYSTALS-KYBER [10] and digital signature algorithm CRYSTALS-DILITHIUM [14] were selected for algorithms 2022 in the NIST PQC competition [31]. These algorithms are based on learning with errors (LWE) [35] in lattices with algebraic structures. Structured lattices offer efficient scheme compared to unstructured lattices.

M. Tezuka (✉) · K. Tanaka
Tokyo Institute of Technology, Tokyo, Japan
e-mail: tezuka.m.ac@m.titech.ac.jp

K. Tanaka
e-mail: keisuke@is.titech.ac.jp

Structured Lattices. Several LWE on various structured lattices have been proposed. Polynomial learning with errors (Poly-LWE) [39] (initially called Ideal-LWE) is defined by a ring $\mathbb{Z}_q[x]/(f)$ where q is a prime and $f \in \mathbb{Z}_q[x]$ is a monic and irreducible polynomial. Ring learning (Ring-LWE) [29] is defined by the full ring of integers O_K of a number field K. Module learning with errors (Module-LWE) [11] is an extension of Ring-LWE. Module-LWE is obtained by setting the polynomial ring dimension of Ring-LWE to a dimension greater than one. However, compared to schemes based on LWE in unstructured lattices, Poly-LWE/Ring-LWE/Module-LWE-based schemes do not provide a stronger security guarantee.

MP-LWE Cryptography. Rosca, Sakzad, Stehlé, and Steinfeld [36] introduced the middle-product learning with errors (MP-LWE) which is a variant of Poly-LWE that uses the middle product of polynomials modulo prime q. For a large class of polynomials $f \in \mathcal{F}$, they gave a reduction from the Poly-LWE$^{(f)}$ problems which are parametrized by a polynomial f to the MP-LWE problem. This result implies that the MP-LWE problems are at least as hard as the Poly-LWE$^{(f)}$ problem for a wide class of polynomials $f \in \mathcal{F}$. The use of the MP-LWE problem achieves lower security risk than the Poly-LWE$^{(f)}$ problem in a single fixed polynomial f. Let us consider a Poly-LWE$^{(f)}$-based scheme implemented in a single fixed polynomial f. If the Poly-LWE$^{(f)}$ problem is broken, we must change the scheme implementation from f to $f^* \in \mathcal{F}$ that the hardness of the Poly-LWE$^{(f^*)}$ problem still holds. By contrast, in a MP-LWE-based scheme, we do not need to change the implementation even if the Poly-LWE$^{(f)}$ problem is broken. Since as long as the hardness of the Poly-LWE$^{(f^*)}$ problem for $f^* \in \mathcal{F}$ holds, the hardness of the MP-LWE problem still holds. Thus, the use of the MP-LWE problem achieves lower security risk than reliance on the Poly-LWE$^{(f)}$ problem over a single fixed polynomial f.

PKE Schemes. Rosca et al. [36] proposed the public key encryption scheme based on the MP-LWE problem. After the seminal work by Rosca et al., MP-LWE-based cryptography has been progressed. The optimized parameters' variant of their scheme called Titanium [40] was submitted to the NIST PQC competition [31].

IBE/HIBE/IPE Schemes. Lombardi, Vaikuntanathan, and Vuong [25] developed the lattice trapdoor technique for MP-LWE and gave an adaptive-ID secure identity-based encryption (IBE) scheme in the random oracle model (ROM) [7] and a selective-ID secure IBE without the ROM. Fan, Lu, and Au [15] proposed an adaptive-ID secure IBE scheme without the random oracle model. Le, Duong, Susilo, and Pieprzyk [22] developed the trapdoor delegation technique for MP-LWE and gave a hierarchical identity-based encryption (HIBE) scheme. Yang, Yang, Zhao, and Wu [41] proposed an inner product encryption (IPE) scheme. Yang, Yang, Zhao, Wu, and Wang [42] improved their IPE scheme.

Digital Signature and Ring Signature Schemes. Hiromasa [20] proposed the first MP-LWE-based digital signature scheme. However, Bai, Das, Hiromasa, Rosca, Sakzad, Stehlé, Steinfeld, and Zhang [6] pointed out that the security proof of the signature scheme by Hiromasa is incorrect and mentioned that it is unclear how to fix it. Instead, they proposed alternative constructions for an identification scheme and a digital signature scheme. As for special type of signatures schemes, Das, Au, and Zhang [13] proposed a ring signature scheme. Lin, Sun, Wang, Liu, and Wang [23] proposed a linkable ring signature scheme.

MP-LWE Variants. Bai, Boudgoust, Das, Roux-Langlois, Wen, and Zhang [5] introduced a deterministic variant of MP-LWE called middle-product learning with rounding (MP-LWR). They constructed a MP-LWR-based PKE scheme. Ling and Mendelsohn [24] proposed a new variant of MP-LWE called skew polynomial learning with errors (Skew-Poly-LWE) and skew middle-product learning with errors (Skew-MP-LWE). They gave a reduction from the cyclic learning with errors (Cyclic-LWE) problem [18] to the Skew-Poly-LWE problem. They also gave a reduction from the Skew-Poly-LWE problem to the Skew-MP-LWE. Moreover, they constructed a Skew-MP-LWE-based PKE scheme. Yang and Huang [43] proposed a generalized variant of MP-LWE called universal-product learning with errors (UP-LWE). The UP-LWE is obtained by replacing the product $a \cdot s$ mod f in the Poly-LWE$^{(f)}$ problem by universal product. They gave a reduction from Poly-LWE problems to the UP-LWE problem.

General Framework for Algebraic Structured LWE. Peikert and Pepin [33] proposed general learning with errors (General-LWE) which captures LWE over commutative rings. General-LWE captures MP-LWE, Module-LWE, Ring-LWE, Poly-LWE, and order learning with errors (Order-LWE) [8]. The Order-LWE problem is parametrized by an order O. For a large class of orders O, Peikert and Pepin [33] gave a reduction from the Order-LWE$^{(O)}$ problems to the MP-LWE problem.

Related Works. The short integer solution (SIS) [3] is a counterpart of the LWE. Its polynomial ring variant called polynomial short integer solution (Poly-SIS) [34] was also proposed. Similar to the Poly-LWE$^{(f)}$ problem, the Poly-SIS$^{(f)}$ problem is parametrized by a polynomial f. Lyubashevsky [27] proposed a variant of the Poly-SIS which is not parametrized by a specific polynomial f, but parameterized by a degree n. We write this variant as Poly-SIS$^{(\emptyset)}$. The Poly-LWE$^{(f)}$ problems can be reduced to the Poly-SIS$^{(\emptyset)}$ problem for many polynomial f.

Paper Organization. In Sect. 2, we introduce notations and Poly-LWE. In Sect. 3, we review MP-LWE and reduction results from the Poly-LWE problems to the MP-LWE problem. In Sect. 4, we review MP-LWE-based PKE schemes. In Sect. 5, we review MP-LWE-based IBE schemes. In Sect. 6, we review a MP-LWE-based identification scheme and a MP-LWE-based digital signature scheme. In Sect. 7, we review MP-LWR cryptography. In Sect. 8, we review a general framework for LWE with an algebraic structure.

2 Preliminaries

Notations. Let λ be the security parameter. A function $f(\lambda)$ is negligible in λ if $f(\lambda)$ tends to 0 faster than $\frac{1}{\lambda^c}$ for every constant $c > 0$. For a positive integers a, b such that $a \leq b$, we write $[a] := \{1, \ldots, a\}$ and $[a, b] := \{a, a+1, \ldots, b\}$. For a finite set S, $s \xleftarrow{\$} S$ represents that an element s is sampled from S uniformly at random, and $|S|$ represents the number of elements in S. For a distribution χ on a finite set S, $s \xleftarrow{r} \chi$ represents that an element s is sampled from according to χ.

Let q be an integer. $\mathbb{Z}_q$ denote the ring of integers modulo q and $\mathbb{R}_q$ the set of real number modulo q. For a positive integer a, $\mathbb{Z}_{\leq a}$ represents the set of integers $\{-a, \ldots, 0, \ldots, a\}$. For a ring R, $R^{<n}[x]$ is the set of all polynomials in R with degree less than n. We extend this notation to any unstructured set S.

For a polynomial $s = \sum_{i=0}^{n-1} a_i x^i \in \mathbb{Z}^{<n}[x]$, we define ℓ_1-norm as $\|s\|_1 := \sum_{i=0}^{n-1} |a_i|$, ℓ_2-norm as $\|s\|_2 := \sqrt{\sum_{i=0}^{n-1} a_i^2}$, and ℓ_∞-norm as $\|s\|_\infty := \max_{i \in \{0, \ldots, n-1\}} |a_i|$. We extend these notations to any vector $a = (a_1, \ldots, a_n) \in \mathbb{R}^n$.

For an algorithm A, $y \leftarrow \mathsf{A}(x)$ denotes that the algorithm A outputs y on input x. We abbreviate probabilistic polynomial time as PPT.

Gaussian Distributions. We recall the definition of the Gaussian distribution.

Definition 1 (*Continuous Gaussian Distribution*) For a positive semidefinite matrix $\Sigma \in \mathbb{R}^{n \times n}$ and a positive integer n, the continuous Gaussian D_Σ is the probability distribution over $\mathbb{R}^n$ whose density is proportional to $\rho_\Sigma(x) := \exp(-\pi x^T \Sigma^{-1} x)$.

Definition 2 (*Discrete Gaussian Distribution*) For a positive integer n, a countable set $S \subset \mathbb{R}^n$, and $\sigma > 0$, the discrete Gaussian distribution $D_{S,\sigma,c}$ is the probability distribution over S whose density is proportional to $\rho_{\sigma,c}(x) = \rho_\sigma(x - c) = \exp\left(-\pi \frac{\|x-c\|_2^2}{\sigma^2}\right)$. We can write $D_{S,\sigma,c} = \rho_{\sigma,c}(x)/\rho_{\sigma,c}(S)$ where $\rho_{\sigma,c}(S) = \sum_{s \in S} \rho_{\sigma,c}(s)$. If $c = 0$, we simply write $D_{S,\sigma,c}$ as $D_{S,\sigma}$.

Poly-LWE Problem. We recall the definition of the Poly-LWE problem.

Definition 3 (*Poly-LWE Distribution*) Let $m > 0$ and $q \geq 2$ be positive integers, f a polynomial of degree m, and χ a distribution over $\mathbb{R}_q[x]/(f)$. We define the polynomial distribution $\mathsf{Poly\text{-}D}_{q,\chi}^{(f)}(s)$ as the distribution obtained by sampling $a \xleftarrow{\$} \mathbb{Z}_q[x]/(f)$, $e \xleftarrow{r} \chi$, and returning $(a, b \leftarrow a \cdot s + e)$.

Definition 4 (*Poly-LWE* [39]) Let $m > 0$, $q \geq 2$ be positive integers, f a polynomial of degree m, and χ_1, χ_2 are distributions over $\mathbb{R}_q[x]/(f)$ and $\mathbb{Z}_q[x]/(f)$, respectively. The $\mathsf{Poly\text{-}LWE}_{q,\chi_1,\chi_2}^{(f)}$ problem consists in distinguishing between polynomially many samples from $\mathsf{Poly\text{-}D}_{q,\chi_1}^{(f)}(s)$ and the same number of uniform samples in $\mathbb{Z}_q[x]/(f) \times \mathbb{R}_q[x]/(f)$ over the choice of $s \xleftarrow{r} \chi_2$. We say that the

Poly-LWE$_{q,\chi_1,\chi_2}^{(f)}$ assumption holds if the advantage of any PPT algorithm trying to solve the Poly-LWE$_{q,\chi_1,\chi_2}^{(f)}$ problem is negligible in λ. If χ_2 is a uniform distribution over $\mathbb{Z}_q[x]/(f)$, we denote Poly-LWE$_{q,\chi_1}^{(f)}$. Otherwise, we call the Poly-LWE assumption the Poly-LWE with small secrets (Poly-LWEwSS) assumption.

Definition 5 (*Expansion Factor* [28]) Let $f \in \mathbb{Z}[x]$ be a polynomial of degree m. The expansion factor of f is defined as

$$\mathsf{EF}(f) := \max_{g \in \mathbb{Z}^{<2m-1}[x] \setminus \{0\}} \left(\frac{\|g \bmod f\|_\infty}{\|g\|_\infty} \right).$$

3 Middle Product of Polynomials and MP-LWE

Middle Product of Polynomials. Middle Product of polynomials was studied to accelerate computations on polynomials in polynomial rings [19, 38]. We review a definition of the middle product of polynomials.

Definition 6 (*Middle Product* [36]) Let d_a, d_b, k, b be integers which satisfy that $d_a + d_b - 1 = 2k + d$. The middle-product $\odot$ of two polynomials $a \in \mathbb{Z}^{<d_a}[x]$ and $b \in \mathbb{Z}^{<d_b}[x]$ is defined as

$$a \odot_d b := \left\lfloor \frac{(a \cdot b) \bmod x^{k+d}}{x^k} \right\rfloor \in \mathbb{Z}^{<d}[x].$$

Lemma 1 (Associativity Property of Middle Product [36]) *Let d, k, n be positive integers. For all $r \in \mathbb{Z}^{<k+1}[x]$, $a \in \mathbb{Z}^{<n}[x]$, and $s \in \mathbb{Z}^{<n+d+k-1}[x]$, $r \odot_d (a \odot_{d+k} s) = (r \cdot a) \odot_d s$ holds.*

MP-LWE Problem. We review a definition of the MP-LWE problem, and its degree-parametrized variant problem.

Definition 7 (*MP-LWE Distribution* [36]) Let $n, d > 0$ and $q \geq 2$ be positive integers and χ a distribution over $\mathbb{R}_q^{<d}[x]$. We define the middle-product distribution MP-D$_{q,n,d,\chi}(s)$ as the distribution obtained by sampling $a \xleftarrow{\$} \mathbb{Z}_q^{<n}[x]$, $e \xleftarrow{r} \chi$, and returning $(a, b \leftarrow a \odot_d s + e)$.

Definition 8 (*MP-LWE* [6, 36]) Let $n, d > 0$ and $q \geq 2$ be positive integers and χ_1 and χ_2 distributions over $\mathbb{R}_q^{<d}[x]$ and $\mathbb{Z}_q^{<n+d-1}[x]$, respectively. The MP-LWE$_{q,n,d,\chi_1,\chi_2}$ problem consists in distinguishing between polynomially many samples from MP-D$_{q,n,d,\chi_1}(s)$ and the same number of uniform samples in $\mathbb{Z}_q^{<d}[x] \times \mathbb{R}_q^{<n+d-1}[x]$ over the choice of $s \xleftarrow{r} \chi_2$. We say that the MP-LWE$_{q,n,d,\chi_1,\chi_2}$

assumption holds if the advantage of any PPT algorithm trying to solve the MP-LWE$_{q,n,d,\chi_1,\chi_2}$ problem is negligible in λ. If χ_2 is a uniform distribution over $\mathbb{Z}_q^{<n+d-1}[x]$, we denote MP-LWE$_{q,n,d,\chi_1}$. Otherwise, we call the MP-LWE assumption the MP-LWE with small secrets (MP-LWEwSS) assumption.

The degree-parametrized variant of the MP-LWE problem is defined in a similar manner to the MP-LWE problem.

Definition 9 (*DPMP-LWE Distribution* [25]) Let $n > 0$ and $q \geq 2$ be positive integers, $\mathbf{d} = (d_1, \ldots, d_t) \in [\frac{n}{2}]^t$ and χ a distribution over $\mathbb{R}_q$. We define the degree-parametrized middle-product distribution DPMP-D$_{q,n,\mathbf{d},\chi}(s)$ as the distribution obtained as follows:

- For $i \in [t]$, sample $a_i \xleftarrow{\$} \mathbb{Z}_q^{<n-d_i}[x]$, $e \xleftarrow{r} \chi^{<d_i}[x]$, compute $b_i \leftarrow a_i \odot_{d_i} s + e_i$.
- Return $(\mathbf{a}, \mathbf{b}) \leftarrow ((a_i)_{i \in [t]}, (b_i)_{i \in [t]})$.

Definition 10 (*DPMP-LWE* [25]) Let $n > 0$ and $q \geq 2$ be positive integers, $\mathbf{d} = (d_1, \ldots, d_t) \in [\frac{n}{2}]^t$ and χ a distribution over $\mathbb{R}_q$. The DPMP-LWE$_{q,n,\mathbf{d},\chi}$ problem consists in distinguishing between polynomially many samples from DPMP-D$_{q,n,\mathbf{d},\chi}(s)$ and the same number of uniform samples in $\prod_{i \in [t]}(\mathbb{Z}_q^{<n-d_i}[x]) \times \prod_{i \in [t]}(\mathbb{R}_q^{<d_i}[x])$ over the choice of $s \xleftarrow{\$} \mathbb{Z}_q^{<n-1}[x]$.

We say that the DPMP-LWE$_{q,n,\mathbf{d},\chi}$ assumption holds if the advantage of any PPT algorithm trying to solve the DPMP-LWE$_{q,n,\mathbf{d},\chi}$ problem is negligible in λ.

Reduction from Poly-LWE to MP-LWE From the following theorems, the MP-LWE$_{q,n,d,\chi}$ problem is at least as hard as the Poly-LWE$^{(f)}$ problems.

Theorem 1 (*Reduction from Poly-LWE to MP-LWE* [36]) *Let $n, d > 0$, $q \geq 2$ be positive integers and $\alpha \in (0, 1)$. For $S > 0$, let $\mathcal{F}(S, d, n)$ be the set of polynomials $f \in \mathbb{Z}[x]$ which are a monic, have constant coefficient coprime with q, have degree $m \in [d, n]$, and $\mathsf{EF}(f) < S$. Then, there is a PPT reduction from the Poly-LWE$^{(f)}_{q,D_{\alpha q}}$ problem for any $f \in \mathcal{F}(S, d, n)$ to the MP-LWE$_{q,n,d,D_{\alpha q dS}}$.*

Theorem 2 (*Reduction from Poly-LWEwSS to MP-LWEwSS* [6]) *Let $n, d > 0$, $q \geq 2$ be positive integers $S > 0$ and $\frac{2\sqrt{n}}{q^S} \leq \alpha \leq 1$. Let $\mathcal{F}(S, d, n)$ be a set described in Theorem 1. Then, there is a PPT reduction from the Poly-LWE$^{(f)}_{q,D_{\mathbb{Z}^m,\alpha q},D_{\mathbb{Z}^m,\alpha q}}$ problem for any $f \in \mathcal{F}(S, d, n)$ to the MP-LWE$_{q,n,d,D_{\mathbb{Z}^d,\alpha' q},D_{\mathbb{Z}^d,\alpha'' q}}$ problem where $\alpha' = \alpha\sqrt{2d} \cdot \mathsf{EF}(f)$ and $\alpha'' = \alpha\sqrt{2d} \cdot \mathsf{EF}(f)^2$.*

Theorem 3 (*Reduction from Poly-LWE to DPMP-LWE* [25]) *Let $n \geq 2$ be positive integers, $\mathbf{d} = (d_1, \ldots, d_t) \in [\frac{n}{2}]^t$ and $\alpha \in (0, 1)$. For $S > 0$, let $\mathcal{F}(S, \mathbf{d}, n)$ be the set of polynomials $f \in \mathbb{Z}[x]$ which are a monic, have constant coefficient coprime with q, have degree $m \in \bigcap_{i \in [t]}[d_i, n - d_i]$, and $\mathsf{EF}(f) < S$. Then, there is a PPT reduction from the Poly-LWE$^{(f)}_{q,D_{\alpha q}}$ problem for any $f \in \mathcal{F}(S, \mathbf{d}, n)$ to the MP-LWE$_{q,n,\mathbf{d},D_{\alpha q\sqrt{n/2}S}}$ problem.*

4 Public Key Encryption from MP-LWE

Definition of PKE Scheme. We review a definition of a public key encryption scheme. A public key encryption scheme PKE with a plaintext space $\mathcal{M}$ is a tuple of PPT algorithms (KeyGen, Encrypt, Decrypt).

- KeyGen(1^λ): A key generation algorithm (probabilistic) takes as an input a security parameter 1^λ. It returns a public key pk and a secret key sk.
- Encrypt(pk, μ): An encryption algorithm (probabilistic) takes as an input a public key pk and a plaintext μ. It returns a ciphertext c.
- Decrypt(sk, c): A decryption algorithm (deterministic) takes as an input a secret key sk and a ciphertext c. It returns a plaintext μ.

Correctness. PKE has correctness error $\delta(\lambda)$, if for all $\lambda \in \mathbb{N}$, for all $\mu \in \mathcal{M}$,

$$\Pr[\text{Decrypt}(\text{sk}, c) \neq \mu] \leq \delta(\lambda)$$

holds where (pk, sk) $\leftarrow$ KeyGen(1^λ), $c \leftarrow$ Encrypt(pk, μ). The probability is taken over the randomness of KeyGen, Encrypt, and Decrypt. We require that correctness error $\delta(\lambda)$ is negligible in λ.

Definition 11 (*IND-CPA Security*) Let PKE = (KeyGen, Encrypt, Decrypt) be a public key encryption scheme and A a PPT algorithm. Indistinguishability under chosen plaintext attacks (IND-CPA) security for PKE is defined by the following IND-CPA game $\text{Game}_{\text{PKE,A}}^{\text{IND-CPA}}$ between the challenger C and A:

- C chooses $b \xleftarrow{\$} \{0, 1\}$, runs (pk, sk) $\leftarrow$ KeyGen(1^λ), and sends pk to A.
- A sends a challenge (μ_0^*, μ_1^*).
- C runs $c^* \leftarrow$ Encrypt(pk, μ_b^*) and sends c^* to A.
- A finally outputs a guess $b^* \in \{0, 1\}$.
- If $b^* = b$, $\text{Game}_{\text{PKE,A}}^{\text{IND-CPA}}$ returns 1. Otherwise, $\text{Game}_{\text{PKE,A}}^{\text{IND-CPA}}$ returns 0.

The advantage of an adversary A for the IND-CPA security game is defined by $\text{Adv}_{\text{PKE,A}}^{\text{IND-CPA}}(\lambda) := \Pr[\text{Game}_{\text{PKE,A}}^{\text{IND-CPA}} = 1]$. PKE satisfies IND-CPA security if for all PPT adversaries A, $\text{Adv}_{\text{PKE,A}}^{\text{IND-CPA}}(\lambda)$ is negligible in λ.

Regev Style PKE Scheme from MP-LWE. Rosca et al. [36] proposed the first MP-LWE-based PKE scheme. This construction is a variant of the Regev PKE scheme [35].

Let $n \geq \lambda$ be an integer, $q, d, t \geq 2$ integers such that q is odd and α is a noise rate. Let $\chi := \lfloor D_{\alpha \cdot q} \rceil$ be the distribution over $\mathbb{Z}$ where e is sampled from $D_{\alpha \cdot q}$ and then rounded to nearest integer and $\chi^{<d}[x]$ the distributions over $\mathbb{Z}^{<d}[x]$ where each element is sampled from χ. The PKE scheme PKE_{RSSS} by Rosca et al. is given in Fig. 1.

The IND-CPA security of PKE_{RSSS} is proven under the MP-LWE assumption [36].

```
KeyGen(1^λ) :
  s ←$ Z_q^{<n+d+k-1}[x].
  For i ∈ [t], a_i ←$ Z_q^{<n}[x], e_i ←r χ^{<d+k}[x], b_i ← a_i ⊙_{d+k} s + 2e_i.
  Return (pk, sk) ← ((a_i, b_i)_{i∈[t]}, s).
Encrypt(pk = (a_i, b_i)_{i∈[t]}, μ ∈ {0, 1}^{<d}[x]) :
  For i ∈ [t], r_i ←$ {0, 1}^{<k+1}[x].
  c_1 ← Σ_{i∈[t]} a_i · r_i, c_μ ← μ + Σ_{i∈[t]} r_i ⊙_d b_i.
  Return c = (c_μ, c_1).
Decrypt(sk = s, c = (c_μ, c_1)) :
  Return μ ← (c_μ − c_1 ⊙_d s  mod q)  mod 2.
```

Fig. 1 The Regev style PKE scheme PKE_RSSS [36]

Dual Regev Style PKE Scheme from MP-LWE. Lombardi et al. [25] proposed a variant of the Dual Regev PKE scheme [17] from MP-LWE. IBE and HIBE schemes are obtained by modifying their scheme.

Let $\lambda = n$ be an integer, $q = q(n)$ a prime, $\tau = \lceil \log_2 q \rceil$, d, k positive integers, σ a positive real number such that $\gamma = \frac{n+2d-2}{d} \in \mathbb{N}$, $2d + k \le n$, $\sigma = \omega(1)$, $dt/n = \Omega(\log n)$, $q = \omega(\log^{\frac{1}{2}} n)\sigma$ and $q = \Omega(\alpha^{-1} n^{1+c})$. Let $t > 0$, $t' = t + \gamma\tau$. Let $\chi := \lfloor D_{\alpha \cdot q} \rceil$ be the distribution over $\mathbb{Z}$ where e is sampled from $D_{\alpha \cdot q}$ and then rounded to nearest integer. The PKE scheme PKE_LVV with a message space $\mathcal{M} = \{0, 1\}^{<k+1}[x]$ is given in Fig. 2.

The IND-CPA security of PKE_LVV is proven under the DPMP-LWE assumption [25].

```
KeyGen(1^λ) :
  For i ∈ [t], a_i ←$ Z_q^{<n}[x], r_i ←r D_{Z^{2d-1},σ}[x].
  For i ∈ {t+1, ..., t'}, a_i ←$ Z_q^{<n+d-1}[x], r_i ←r D_{Z^d,σ}[x].
  a ← (a_i)_{i∈[t']}, u ← Σ_{i∈[t']} a_i · r_i.
  Return (pk, sk) ← ((a, u), (r_i)_{i∈[t']}).
Encrypt(pk = ((a_i)_{i∈[t']}, u), μ ∈ {0, 1}^{<k+1}[x]) :
  s ←$ Z_q^{<n+2d+k-1},
  For i ∈ [t], e_i ←$ χ^{<2d+k}[x], c_i ← a_i ⊙_{2d+k} s + 2e_i.
  For i ∈ {t+1, ..., t'}, e_i ←$ χ^{<d+k+1}[x], c_i ← a_i ⊙_{d+k+1} s + 2e_i.
  e' ←$ χ^{<k+1}[x], c_μ ← μ + u ⊙_{k+1} s + 2e'.
  Return c ← (c_μ, (c_i)_{i∈[t']}).
Decrypt(sk = (r_i)_{i∈[t']}, c = (c_μ, (c_i)_{i∈[t']})) :
  Return μ ← (c_μ − Σ_{i∈[t']} c_i ⊙_{k+1} r_i  mod q)  mod 2.
```

Fig. 2 The dual Regev style PKE scheme PKE_LVV [25]

5 Identity-Based Encryption from MP-LWE

Definition of IBE Scheme. We review a definition of an identity-based encryption scheme. An identity-based encryption scheme IBE with an identity space $\mathcal{I}$ and a plaintext space $\mathcal{M}$ is a tuple of algorithms (Setup, Extract, Encrypt, Decrypt).

- Setup(1^λ): A setup algorithm (probabilistic) takes as an input a security parameter 1^λ. It returns a master public key mpk and a master secret key msk.
- Extract(mpk, msk, id): An extraction algorithm (probabilistic) takes as an input a master public key mpk, a master secret key msk, and an identity id. It returns a secret key $\mathsf{sk}_{\mathsf{id}}$.
- Encrypt(mpk, id, μ): An encryption algorithm (probabilistic) takes as an input a master public key mpk, an identity id, and a plaintext μ. It returns a ciphertext c.
- Decrypt($\mathsf{sk}_{\mathsf{id}}$, c): A decryption algorithm (deterministic) takes as an input a secret key $\mathsf{sk}_{\mathsf{id}}$ and a ciphertext c. It returns a plaintext μ.

Correctness. IBE has correctness error $\delta(\lambda)$, if for all $\lambda \in \mathbb{N}$, for all id $\in \mathcal{I}$, $\mu \in \mathcal{M}$, we have
$$\Pr[\mathsf{Decrypt}(\mathsf{sk}_{\mathsf{id}}, c) \neq \mu] \leq \delta(\lambda),$$

where (mpk, msk) $\leftarrow$ Setup(1^λ), $\mathsf{sk}_{\mathsf{id}} \leftarrow$ Extract(mpk, msk, id), and $c \leftarrow$ Encrypt(pk, id, μ). The probability is taken over the randomness of Setup, Extract, Encrypt, and Decrypt. We require that correctness error $\delta(\lambda)$ is negligible in λ.

Definition 12 (*IND-ID-CPA Security*) Let IBE = (Setup, Extract, Encrypt, Decrypt) be an identity-based encryption scheme and A a PPT algorithm. Indistinguishability under chosen plaintext attacks (IND-ID-CPA) security for IBE is defined by the following IND-ID-CPA game $\mathsf{Game}_{\mathsf{IBE},\mathsf{A}}^{\mathsf{IND\text{-}ID\text{-}CPA}}$ between the challenger C and A.

- C choose $b \xleftarrow{\$} \{0, 1\}$, sets $Q^{\mathsf{Extract}} \leftarrow \{\}$, runs (mpk, msk) $\leftarrow$ KeyGen(1^λ), and sends mpk to A.
- A makes a number of key-extraction queries to the extraction oracle O^{Extract}. For a key-extraction query on id, O^{Extract} sets $Q^{\mathsf{Extract}} \leftarrow Q^{\mathsf{Extract}} \cup \{\mathsf{id}\}$, computes $\mathsf{sk}_{\mathsf{id}} \leftarrow$ Extract(mpk, msk, id), and returns $\mathsf{sk}_{\mathsf{id}}$.
- A sends a challenge (id*, μ_0^*, μ_1^*) where id* $\notin Q^{\mathsf{Extract}}$.
- C runs $c^* \leftarrow$ Encrypt(mpk, id, μ_b^*) and sends c^* to A.
- A makes a number of key-extraction queries to the extraction oracle O^{Extract} with a restriction that id* cannot be queried to O^{Extract}.
- A finally outputs a guess $b^* \in \{0, 1\}$.
- If $b^* = b$, $\mathsf{Game}_{\mathsf{IBE},\mathsf{A}}^{\mathsf{IND\text{-}ID\text{-}CPA}}$ returns 1. Otherwise, $\mathsf{Game}_{\mathsf{IBE},\mathsf{A}}^{\mathsf{IND\text{-}ID\text{-}CPA}}$ returns 0.

The advantage of an adversary A for the IND-ID-CPA security game is defined by $\mathsf{Adv}_{\mathsf{IBE},\mathsf{A}}^{\mathsf{IND\text{-}ID\text{-}CPA}}(\lambda) := \Pr[\mathsf{Game}_{\mathsf{IBE},\mathsf{A}}^{\mathsf{IND\text{-}ID\text{-}CPA}} = 1]$. IBE satisfies IND-ID-CPA security if for all PPT adversaries A, $\mathsf{Adv}_{\mathsf{IBE},\mathsf{A}}^{\mathsf{IND\text{-}ID\text{-}CPA}}(\lambda)$ is negligible in λ.

Algorithms Related to Lattice Trapdoor. The lattice trapdoor technique is proposed by Gentry, Peikert, and Vaikuntanathan [17]. Micciancio and Peikert [30] introduced the efficient gadget-based trapdoor framework. Lombardi et al. [25] applied these techniques to develop the lattice trapdoor technique for MP-LWE. We review algorithms for the lattice trapdoor technique.

Lemma 2 (TrapGen and SamplePre [25]) *Let $n = \lambda$ be an integer, q a polynomial in n, $d \leq n$, $dt/n = \Omega(\log n)$, $\sigma = \omega(\log^2 n \sqrt{ndt})$, $\gamma = \frac{n+2d-2}{d}$ an integer, and $\tau = \lceil \log_2 q \rceil$. Then, there exists a tuple of PPT algorithms* (TrapGen, SamplePre) *which satisfies the following properties:*

- TrapGen(1^n) *outputs polynomials* $(a_i)_{i \in [t+\gamma\tau]}$ *with the trapdoor* td. *The distribution* $(a_i)_{i \in [t+\gamma\tau]}$ *output by* TrapGen *is statistically closed to the uniform distribution on* $(\mathbb{Z}_q^{<n}[x])^t \times (\mathbb{Z}_q^{<n+d-1}[x])^{\gamma\tau}$.
- SamplePre($(a_i)_{i \in [t+\gamma\tau]}$, td, $u \in \mathbb{Z}_q^{<n+2d-2}[x]$, σ) *outputs polynomials* $(r_i)_{i \in [t+\gamma\tau]}$ *which satisfies* $\sum_{i \in [t+\gamma\tau]} a_i \cdot r_i = u$.
- *The distribution* $(r_i)_{i \in [t+\gamma\tau]}$ *output by* SamplePre($(a_i)_{i \in [t+\gamma\tau]}$, td, u, σ) *is exactly the conditional distribution* $(D_{\mathbb{Z}^{2d-1},\sigma}[x])^t \times (D_{\mathbb{Z}^d,\sigma}[x])^{\gamma\tau} \mid \sum_{i \in [t+\gamma\tau]} a_i \cdot r_i = u$.

Lemma 3 (SampleLeft [15]) *Let $n = \lambda$ be an integer, q a polynomial in n, $3d \leq n$, $dt/n = \Omega(\log n)$, $\sigma = \omega(\log^2 n \sqrt{ndt})$, $\gamma = \frac{n+2d-2}{d}$ an integer, and $\tau = \lceil \log_2 q \rceil$. Then, there exists a PPT algorithm* SampleLeft *which satisfies the following property:*

- SampleLeft($(a_i)_{i \in [t+\gamma\tau]}$, $(h_i \in \mathbb{Z}_q^{<n+d-1}[x])_{i \in [\gamma\tau]}$, td, u, σ) *outputs polynomials* $(r_i)_{i \in [t+2\gamma\tau]}$ *which satisfies* $\sum_{i \in [t+\gamma\tau]} a_i \cdot r_i + \sum_{i \in [\gamma\tau]} h_i \cdot r_{t+\gamma\tau+i} = u$. *The distribution* $(r_i)_{i \in [t+2\gamma\tau]}$ *output by* SampleLeft($(a_i)_{i \in [t+\gamma\tau]}$, $(h_i)_{i \in [\gamma\tau]}$, td, u, σ) *is exactly the conditional distribution* $(D_{\mathbb{Z}^{2d-1},\sigma}[x])^t \times (D_{\mathbb{Z}^d,\sigma}[x])^{2\gamma\tau} \mid \sum_{i \in [t+\gamma\tau]} a_i \cdot r_i + \sum_{i \in [\gamma\tau]} h_i \cdot r_{t+\gamma\tau+i} = u$.

Adaptive-ID Secure IBE Scheme from MP-LWE in the ROM. Lombardi et al. [25] constructed the adaptive-ID secure IBE scheme in the ROM. This scheme is based on the PKE scheme PKE$_{\text{LVV}}$.

Let $n = \lambda$ be an integer, $q = q(n)$ a prime, $\tau = \lceil \log_2 q \rceil$, d, k positive integers, σ a positive real number such that $\gamma = \frac{n+2d-2}{d} \in \mathbb{N}$, $2d + k \leq n$, $\sigma = \omega(1)$, $dt/n = \Omega(\log n)$, $q = \omega(\log^{\frac{1}{2}} n)\sigma$ and $q = \Omega(\alpha^{-1} n^{1+c})$. Let $t > 0$ be a positive integer, $t' = t + \gamma\tau$, and $\chi := \lfloor D_{\alpha \cdot q} \rceil$ be the distribution over $\mathbb{Z}$ where e is sampled from $D_{\alpha \cdot q}$ and then rounded to nearest integer. Let $H : \mathbb{Z}_q^{n+2d-2} \to \mathbb{Z}_q^{<n+2d-2}[x]$ be a hash function. The IBE scheme IBE$_{\text{LVV}}$ with an identity space $\mathcal{I} = \mathbb{Z}_q^{n+2d-2}$ and a message space $\mathcal{M} = \{0, 1\}^{<k+1}[x]$ is given in Fig. 3.

The IND-ID-CPA security of IBE$_{\text{LVV}}$ is proven under the DPMP-LWE assumption in the ROM.

Lombardi et al. [25] explained how to obtain the selective-ID (IND-sID-CPA) secure IBE without the ROM from the IBE scheme [2] in the middle-product setting.

```
Setup($1^\lambda$) :
  ($\mathbf{a} = (a_i)_{i \in [t']}$, td) ← TrapGen($1^n$).
  Return (mpk, msk) ← ($\mathbf{a}$, td).
Extract(mpk, msk = td, id) :
  $(r_i)_{i \in [t']}$ ← SamplePre(td, $H$(id)).
  Return sk$_{id}$ ← $(r_i)_{i \in [t']}$.
Encrypt(mpk = $(a_i)_{i \in [t']}$, id ∈ $\mathbb{Z}_p^{<n+2d-2}$, $\mu \in \{0,1\}^{<k+1}[x]$) :
  $u \leftarrow H$(id), $s \xleftarrow{\$} \mathbb{Z}_q^{<n+2d+k-1}$.
  For $i \in [t]$, $e_i \xleftarrow{\$} \chi^{<2d+k}[x]$, $c_i \leftarrow a_i \odot_{2d+k} s + 2e_i$.
  For $i \in \{t+1, \ldots, t'\}$, $e_i \xleftarrow{\$} \chi^{<d+k+1}[x]$, $c_i \leftarrow a_i \odot_{d+k+1} s + 2e_i$.
  $e' \xleftarrow{\$} \chi^{<k+1}[x]$, $c_\mu \leftarrow \mu + u \odot_{k+1} s + 2e'$.
  Return $c \leftarrow (c_\mu, (c_i)_{i \in [t']})$.
Decrypt(sk$_{id} = (r_i)_{i \in [t']}$, $c = (c_\mu, (c_i)_{i \in [t']})$) :
  Return $\mu \leftarrow (c_\mu - \sum_{i \in [t']} c_i \odot_{k+1} r_i \mod q) \mod 2$.
```

Fig. 3 The adaptive-ID secure IBE scheme IBE$_{\text{LVV}}$ [25] in the ROM

Adaptive-ID Secure IBE Scheme from MP-LWE without the ROM. Fan et al. [15] gave the adaptive-ID secure IBE scheme based on the MP-LWE problem without the ROM. This scheme is based on the IBE scheme by Agrawal, Boneh, and Boyen [1].

Let $q = q(n)$ be a prime, $\tau = \lceil \log_2 q \rceil$, n, d, k positive integers such that $\gamma = \frac{n+2d-2}{d} \in \mathbb{N}$, $2d + k \leq n$. Let $t > 0$ be a positive integer, $t' = t + \gamma\tau$, and $t'' = t' + \gamma\tau$. Let $\chi := \lfloor D_{\alpha \cdot q} \rceil$ and $\chi_1 := \lfloor D_{\alpha_1 \cdot q} \rceil$ be the distributions over $\mathbb{Z}$ where $e \xleftarrow{r} D_{\alpha \cdot q}$ or $e_1 \xleftarrow{r} D_{\alpha_1 \cdot q}$ is sampled and then rounded to nearest integer, respectively. Let s be an integer such that $0 < s \leq \frac{d}{2}$ and $\sigma = c\ell \cdot \omega(n \log^{\frac{3}{2}} n)$ for some constant c. The IBE scheme IBE$_{\text{FLA}}$ with an identity space $\mathcal{I} = \{-1, 1\}^\ell$ and a message space $\mathcal{M} = \{0, 1\}^{<k+1}[x]$ is given in Fig. 4.

The IND-ID-CPA security of IBE$_{\text{FLA}}$ is proven under the DPMP-LWE assumption [15].

6 Identification and Digital Signature from MP-LWE

Definition of Identification Scheme. We review a definition of an identification scheme. A (three-move) identification scheme ID with a challenge space C is a tuple of algorithms (KeyGen, $\mathsf{P} = (\mathsf{P}_1, \mathsf{P}_2)$, V_1, Verify).

- KeyGen(1^λ): A key generation algorithm (probabilistic) takes as an input a security parameter 1^λ. It returns a public key pk and a secret key sk.
- P_1(sk): A prover algorithm P_1 (probabilistic) which is run at the beginning of interaction with the verifier. It takes as an input a secret key sk and returns a commitment com and a state st.

```
Setup(1^λ) :
 (a = (a_i)_{i∈[t']}, td) ← TrapGen(1^n), u ←$ Z_q^{<n+2d-1}[x].
 For i ∈ {0,...,ℓ}, h^(i) ←$ (Z_q^{<n+d-1}[x])^{γτ}.
 Return (mpk, msk) ← ((a, (h^(i))_{i∈{0,...,ℓ}}, u), td).
Extract(mpk, msk = td, id ∈ {−1, 1}^ℓ) :
 h_id ← h^(0) + Σ_{i∈[ℓ]} id_i h^(i) where id_i is the i-th bit of id.
 (r_i)_{i∈[t'']} ← SampleLeft((a_i)_{i∈[t']}, h_id, td, u, σ).
 Return sk_id ← (r_i)_{i∈[t'']}.
Encrypt(mpk = (a_i)_{i∈[t']}, id ∈ {0, 1}^ℓ, μ ∈ {0, 1}^{<k+1}[x]) :
 h_id ← h^(0) + Σ_{i∈[ℓ]} id_i h^(i), s ←$ Z_q^{<n+2d+k-1}.
 For i ∈ [t], e_i ←$ χ^{<2d+k}[x], c_i ← a_i ⊙_{2d+k} s + 2e_i.
 For i ∈ {t+1,...,t'}, e_i ←$ χ^{<d+k+1}[x], c_i ← a_i ⊙_{d+k+1} s + 2e_i.
 For i ∈ {t'+1,...,t''}, e_i ←$ χ_1^{<d+k+1}[x], c_i ← a_i ⊙_{d+k+1} s + 2e_i.
 e' ←$ χ^{<k+1}[x], c_μ ← μ + u ⊙_{k+1} s + 2e'.
 Return c ← (c_μ, (c_i)_{i∈[t'']}).
Decrypt(sk_id = (r_i)_{i∈[t'']}, c = (c_μ, (c_i)_{i∈[t'']})) :
 Return μ ← (c_μ − Σ_{i∈[t'']} c_i ⊙_{k+1} r_i  mod q)  mod 2.
```

Fig. 4 The adaptive-ID secure IBE scheme IBE_FLA [15] without the ROM

- $V_1(\mathsf{pk}, \mathsf{com})$: A verifier algorithm V_1 (probabilistic) which takes as an input a public key and a commitment com. It returns a challenge ch $\overset{\$}{\leftarrow} C$.
- $P_2(\mathsf{sk}, \mathsf{com}, \mathsf{ch}, \mathsf{st})$: A prover algorithm P_2 (deterministic) which takes as an input a secret key, a commitment c, a challenge, and a state st. It returns a response resp or $\perp$.
- Verify(pk, com, ch, resp): A verification algorithm (deterministic) takes as an input a public key pk, a commitment c, a challenge ch, and a response resp. It returns a bit $b \in \{0, 1\}$.

We also define a transcript algorithm Trans which takes a secret key and returns a real interaction (com, ch, resp) where (com, st) ← $P_1(\mathsf{sk})$, ch ← $V_1(\mathsf{pk}, \mathsf{com})$, and resp ← $P_2(\mathsf{sk}, \mathsf{com}, \mathsf{ch}, \mathsf{st})$.

Correctness. ID has correctness error $\delta(\lambda)$, if for all $\lambda \in \mathbb{N}$, (pk, sk) ← KeyGen(1^λ) the followings holds:

- For all possible transcript (com, ch, resp) ← Trans(sk) with resp $\neq \perp$, Verify(pk, com, ch, resp) = 1 holds.
- Pr[resp = $\perp$ | (com, ch, resp) ← Trans(sk)] $\leq \delta(\lambda)$ holds where the probability is taken with respect to the randomness of KeyGen, P_1, V_1, and P_2.

We require that correctness error $\delta(\lambda)$ is negligible in λ.

Min-Entropy. We say ID has α bits of min-entropy if the following

$$\Pr_{(\mathsf{pk},\mathsf{sk})\leftarrow\mathsf{KeyGen}(1^\lambda)}[H_\infty(\mathsf{com}|(\mathsf{com},\mathsf{st})\leftarrow\mathsf{P}_1(\mathsf{sk}))] \geq 1 - 2^\alpha$$

holds where H_∞ represents the min-entropy.

Definition 13 (*Lossiness*) Let $\mathsf{ID} = (\mathsf{KeyGen}, \mathsf{P} = (\mathsf{P}_1, \mathsf{P}_2), \mathsf{V}_1, \mathsf{Verify})$ be an identification scheme. ID satisfies the lossiness if there exist a lossy key-generation algorithm LossyKeyGen which takes as an input a security parameter 1^λ and returns a public key pk such that for all quantum PPT algorithm A, the following advantage

$$\mathsf{Adv}_{\mathsf{ID},\mathsf{A}}^{\mathsf{Lossiness}}(\lambda) := |\Pr[\mathsf{A}(\mathsf{pk}) = 1|(\mathsf{pk},\mathsf{sk}) \leftarrow \mathsf{KeyGen}(1^\lambda)]$$
$$- \Pr[\mathsf{A}(\mathsf{pk}) = 1|\mathsf{pk} \leftarrow \mathsf{LossyKeyGen}(1^\lambda)]|$$

is negligible in λ.

Definition 14 (*Lossy Soundness*) Let $\mathsf{ID} = (\mathsf{KeyGen}, \mathsf{P} = (\mathsf{P}_1, \mathsf{P}_2), \mathsf{V}_1, \mathsf{Verify})$ be an identification scheme and LossyKeyGen the algorithm in Definition 13. ID satisfies ϵ-lossy soundness if for all quantum adversary A, the following holds:

$$\Pr\left[\begin{array}{l}\mathsf{Verify}(\mathsf{pk},\mathsf{com}^*,\\ \mathsf{ch}^*,\mathsf{resp}^*)=1\end{array}\middle|\begin{array}{l}\mathsf{pk} \leftarrow \mathsf{LossyKeyGen}(1^\lambda), (\mathsf{com}^*,\mathsf{st}) \leftarrow A(\mathsf{pk}),\\ \mathsf{ch}^* \xleftarrow{\$} C, \mathsf{resp}^* \leftarrow A(\mathsf{st}, c^*)\end{array}\right] \leq \epsilon.$$

Definition 15 (*No-Abort Honest-Verifier Zero-Knowledge*) Let $\mathsf{ID} = (\mathsf{KeyGen}, \mathsf{P} = (\mathsf{P}_1, \mathsf{P}_2), \mathsf{V}_1, \mathsf{Verify})$ be an identification scheme. ID satisfies ϵ-perfect no-abort honest-verifier zero-knowledge if there exists a PPT algorithm Sim which takes as an input a public key pk and outputs a transcript $(\mathsf{com}, \mathsf{ch}, \mathsf{resp})$ such that the followings hold: For all $(\mathsf{pk},\mathsf{sk}) \leftarrow \mathsf{KeyGen}(1^\lambda)$,

- The statistical distance between $(\mathsf{com}, \mathsf{ch}, \mathsf{resp}) \leftarrow \mathsf{Sim}(\mathsf{pk})$ and $(\mathsf{com}, \mathsf{ch}, \mathsf{resp}) \leftarrow \mathsf{Trans}(\mathsf{sk})$ is at most ϵ.
- The distribution of com from $(\mathsf{com}, \mathsf{ch}, \mathsf{resp}) \leftarrow \mathsf{Sim}(\mathsf{pk})$ conditioned on com $\neq \bot$ corresponds to uniform distribution on C.

Identification Scheme from MP-LWE. We describe the identification scheme by Bai et al. [6]. We introduce an extendable output function *Sam* that is a function on bit strings in which the output can be extended to any required length. We write $y \xleftarrow{\$} S := \mathsf{Sam}(x)$ when we want the deterministic output y of *Sam* on input to be uniformly distributed on the set S. The identification scheme $\mathsf{ID}_{\mathsf{BDH}^+}$ is given in Fig. 5. Due to the limitation of the paper we omit the concrete parameter settings. See [6] for concrete parameter settings for $n, d, k, q, \kappa, a', a'', A', A', \alpha', \alpha''$.

Under the MP-LWEwSS assumption, $\mathsf{ID}_{\mathsf{BDH}^+}$ satisfies lossiness, ϵ-lossy-soundness, 0-perfect no-abort HVZK and has $d \cdot \log(2a'' + 1)$ bits of mini-entropy [6].

```
KeyGen(1^λ) :
  ρ ←$ {0,1}^256, a ←$ Z_q^{<n}[x] := Sam(ρ).
  s ←^r D_{Z^{n+d+k-1}, α'q}, e ←^r D_{Z^{d+k}, α''q}, b ← a ⊙_{d+k} s + e.
  Return (pk, sk) ← ((b, ρ), (s, e, ρ)).
P_1(sk = (s, e, ρ)) :
  a ←$ Z_q^{<n}[x] := Sam(ρ), y_1 ←$ Z_{α'}^{<n+d-1}[x], y_2 ←$ Z_{α''}^{<d}[x], w ← a ⊙_d y_1 + y_2.
  Return (com, st) ← (w, (w, y_1, y_2)).
V_1(pk, com) :
  Return ch = c ←$ C where C := {p ∈ {-1, 0, 1}^{<k+1}[x] with ||p||_1 = κ}.
P_2(sk = (s, e, ρ), com = w, ch = c, st = (w, y_1, y_2)) :
  z_1 ← c ⊙_{n+d-1} s + y_1, z_2 ← c ⊙_d e + y_2.
  If ||z_1||_∞ > A' ∨ ||z_2||_∞ > A'', (z_1, z_2) ← ⊥.
  Return resp ← (z_1, z_2).
Verify(pk = (b, ρ), com = w, ch = c, resp = (z_1, z_2)) :
  a ←$ Z_q^{<n}[x] := Sam(ρ).
  If w = a ⊙_d z_1 + z_2 − c ⊙_d b ∧ ||z_1||_∞ ≤ A' ∧ ||z_2||_∞ ≤ A'', return 1.
  Otherwise, return 0.
```

Fig. 5 The identification scheme $\mathsf{ID}_{\mathsf{BDH}^+}$ [6]

Definition of Digital Signature Scheme. We review a definition of a digital signature scheme. A digital signature scheme DS with a message space $\mathcal{M}$ is a tuple of algorithms (KeyGen, Sign, Verify).

- KeyGen($1^λ$): A key generation algorithm (probabilistic) takes as an input a security parameter $1^λ$. It returns a public key pk and a secret key sk.
- Sign(sk, $μ$): A signing algorithm (probabilistic) takes as an input a secret key sk and a message $μ$. It returns a signature $σ$.
- Verify(pk, $μ$, $σ$): A verification algorithm (deterministic) takes as an input a public key pk, a message $μ$, and a signature $σ$. It returns a bit $b ∈ \{0, 1\}$.

Correctness. DS has correctness error $δ(λ)$, if for all $λ ∈ \mathbb{N}$, for all $μ ∈ \mathcal{M}$,

$$\Pr[\mathsf{Verify}(\mathsf{pk}, μ, σ) \neq 1] \leq δ(λ)$$

holds where (pk, sk) ← KeyGen($1^λ$), $σ$ ← Sign(sk, m). The probability is taken over the randomness of KeyGen, Sign, and Verify. We require that correctness error $δ(λ)$ is negligible in $λ$.

Definition 16 (*EUF-CMA Security*) Let DS = (KeyGen, Sign, Verify) be a signature scheme and A a PPT algorithm. Existential unforgetability under chosen message attack (EUF-CMA) security for DS is defined by the following EUF-CMA security game $\mathsf{Game}_{\mathsf{DS},\mathsf{A}}^{\mathsf{EUF-CMA}}$ between the challenger C and A.

```
KeyGen($1^\lambda$) :
  $\rho \xleftarrow{\$} \{0, 1\}^{256}, a \xleftarrow{\$} \mathbb{Z}_q^{<n}[x] := \mathsf{Sam}(\rho).$
  $s \xleftarrow{r} D_{\mathbb{Z}^{n+d+k-1}, \alpha'q}, e \xleftarrow{r} D_{\mathbb{Z}^{d+k}, \alpha''q}, b \leftarrow a \odot_{d+k} s + e.$
  Return (pk, sk) $\leftarrow ((b, \rho), (s, e, K, \rho)).$
Sign(sk = $(s, e, K, \rho), \mu$) :
  $i \leftarrow 0, a \xleftarrow{\$} \mathbb{Z}_q^{<n}[x] := \mathsf{Sam}(\rho).$
  While $(z_1, z_2) \wedge i \leq k_m$ do:
    $i \leftarrow i + 1,$
    $y_1 \xleftarrow{\$} \mathbb{Z}_{a'}^{<n+d-1}[x] := \mathsf{Sam}(K, m, i, 0), y_2 \xleftarrow{\$} \mathbb{Z}_{a''}^{<d}[x] := \mathsf{Sam}(K, m, i, 1).$
    $w \leftarrow a \odot_d y_1 + y_2, c \leftarrow H(w, m), z_1 \leftarrow c \odot_{n+d-1} s + y_1, z_2 \leftarrow c \odot_d e + y_2.$
    If $\|z_1\|_\infty > A' \vee \|z_2\|_\infty > A'', (z_1, z_2) \leftarrow \bot.$
  Return $\sigma \leftarrow (z_1, z_2, c).$
Verify(pk = $(b, \rho), \mu, \sigma = (z_1, z_2, c)$) :
  $a \xleftarrow{\$} \mathbb{Z}_q^{<n}[x] := \mathsf{Sam}(\rho), w \leftarrow a \odot_d z_1 + z_2 - c \odot_d b.$
  If $c = H(w, m) \wedge \|z_1\|_\infty \leq A' \wedge \|z_2\|_\infty \leq A''$, return 1.
  Otherwise, return 0.
```

Fig. 6 The digital signature scheme $\mathsf{DS}_{\mathsf{BDH}^+}$ in the QROM [6]

- C sets $Q^{\mathsf{Sign}} \leftarrow \emptyset$, runs (pk, sk) $\leftarrow$ KeyGen(1^λ), and sends pk to A.
- A makes a number of the signing queries to the signing oracle O^{Sign}. a signing query on μ, O^{Sign} sets $Q^{\mathsf{Sign}} \leftarrow Q^{\mathsf{Sign}} \cup \{\mu\}$, computes $\sigma \leftarrow$ Sign(sk, m), and returns σ.
- A finally outputs a forgery (m^*, σ^*).
- If Verify(pk, μ^*, σ^*) = $1 \wedge \mu^* \notin Q^{\mathsf{Sign}}$, $\mathsf{Game}_{\mathsf{DS,A}}^{\mathsf{EUF\text{-}CMA}}$ returns 1. Otherwise, $\mathsf{Game}_{\mathsf{DS,A}}^{\mathsf{EUF\text{-}CMA}}$ returns 0.

The advantage of an adversary A for the EUF-CMA security game is defined by $\mathsf{Adv}_{\mathsf{DS,A}}^{\mathsf{EUF\text{-}CMA}}(\lambda) := \Pr[\mathsf{Game}_{\mathsf{DS,A}}^{\mathsf{EUF\text{-}CMA}} = 1]$. DS satisfies EUF-CMA security if for all PPT adversaries A, $\mathsf{Adv}_{\mathsf{DS,A}}^{\mathsf{EUF\text{-}CMA}}(\lambda)$ is negligible in λ.

Digital Signature Scheme from MP-LWE in the ROM. Generally, a digital signature scheme can be obtained from an identification scheme via Fiat–Shamir transformation [16]. However, the original Fiat–Shamir transformation cannot be applied to $\mathsf{ID}_{\mathsf{BDH}^+}$, since the $\mathsf{ID}_{\mathsf{BDH}^+}$ scheme allows to abort. For this type of identification scheme, the Fiat–Shamir with Abort (FSwA) transformation [26] was proposed.

A MP-LWE-based digital signature scheme $\mathsf{DS}_{\mathsf{BDH}^+}$ is obtained by applying the FSwA transformation [26] to $\mathsf{ID}_{\mathsf{BDH}^+}$. Let $H : \{0, 1\}^* \to D_H$ be a hash function where $D_H := \{p \in \{-1, 0, 1\}^{<k+1}[x] \mid \|p\|_1 = \kappa\}$. The signature scheme $\mathsf{DS}_{\mathsf{BDH}^+}$ with a message space $\mathcal{M} = \{0, 1\}^*$ is given in Fig. 6. Due to the limitation of the paper we omit the concrete parameter settings. See [6] for concrete parameter settings for $n, d, k, q, \kappa, a', a'', A', A', \alpha', \alpha''$.

By the Fiat–Shamir transformation analysis result by Kiltz et al. [21] in the quantum random oracle model (QROM) [9], the EUF-CMA security of $\mathsf{DS}_{\mathsf{BDH}^+}$ is proven under the MP-LWEwSS assumption in the QROM [6].

7 Public Key Encryption from MP-LWR

MP-LWR Problem. We review the computational MP-LWR problem. To define the computational MP-LWR problem, we introduce some notations.

For a polynomial $s = \sum_{i \in \{0,\ldots,n+d-2\}} a_i x^i \in \mathbb{Z}^{<n+d-1}[x]$, we define the Hankel matrix as

$$\mathsf{Hank}(s) := \begin{pmatrix} a_0 & a_1 & \cdots & a_{d-1} & \cdots & a_{n-1} \\ a_1 & a_2 & \cdots & a_d & \cdots & a_n \\ & & \cdot^{\cdot^{\cdot}} & & & \\ a_{d-1} & a_d & \cdots & a_{2d-2} & \cdots & a_{n+d-2} \end{pmatrix} \in \mathbb{Z}_q^{d \times n}$$

and $\mathbb{Z}_q^{<n+d+k-1}[x]^\times := \{s \in \mathbb{Z}_q^{<n+d+k-1}[x] | \mathsf{Hank}(s) \text{ has full rank}\}$.

Let $\lfloor \cdot \rceil : \mathbb{R} \to \mathbb{Z}$ be the rounding function that rounds to the closest integer. In the case of equality, we take the floor. Let $p, q, n, d > 0$ be positive integers such that $n \geq d > 0$ and $q \geq p \geq 2$. We define the modular rounding function $\lfloor \cdot \rceil_p : \mathbb{Z}_q \to \mathbb{Z}_p$ as $\lfloor x \rceil_p := \left\lfloor \left(\frac{p}{q}\right) \cdot x \right\rceil \mod p$. We extend the above notations to the coefficients of a polynomial.

Now, we are ready to describe the computational MP-LWR problem. The computational MP-LWR problem is an adaption of the computational Ring-LWR problem [12] to the middle-product setting.

Definition 17 (*Computational MP-LWR* [5]) Let $p, q, n, d, t > 0$ be positive integers such that $n \geq d > 0$ and $q \geq p \geq 2$. For $s \xleftarrow{\$} \mathbb{Z}_q^{<n+d-1}[x]^\times$, denote by $\mathcal{X}_s$ the distribution of $(a, \lfloor a \cdot_d \odot s \rceil_p)$ where $a \xleftarrow{\$} \mathbb{Z}_q^{<n}[x]$ and denote by $\mathcal{U}$ the distribution of $(a, \lfloor b \rceil_p)$ where $a \xleftarrow{\$} \mathbb{Z}_q^{<n}[x]$ and $b \xleftarrow{\$} \mathbb{Z}_q^{<d}[x]$. We define distributions $\mathsf{var}_0 := (\mathcal{X}_s)^t$ and $\mathsf{var}_1 := (\mathcal{U})^t$. Let con be an arbitrary distribution over $\{0, 1\}^*$ which is independent from var_0 and var_1. For $i \in \{0, 1\}$, we define sources as $\mathsf{Source}_i := (\mathsf{var}_i, \mathsf{con})$. The computational MP-LWE$_{p,q,n,d,t}$ assumption is defined by the following experiments Exp_0 and Exp_1 between a challenger C and an adversary A in Fig. 7.

For $i \in \{0, 1\}$ and for fixed C, let $Q_{i,\mathsf{A},\mathsf{C}}$ be the probability that Exp_i returns 1. We say that the computational MP-LWE$_{p,q,n,d,t}$ assumption is that for any challenger C, if $Q_{1,\mathsf{A},\mathsf{C}}$ is negligible in λ for any PPT adversary A, then $Q_{0,\mathsf{A},\mathsf{C}}$ is also negligible in λ.

$\mathsf{Exp}_0(\mathsf{C}, \mathsf{A}, \mathsf{Source}_0)$:	$\mathsf{Exp}_1(\mathsf{C}, \mathsf{A}, \mathsf{Source}_1)$:
$(X_0, \mathsf{aux}) \xleftarrow{r} \mathsf{Source}_0$.	$(X_1, \mathsf{aux}) \xleftarrow{r} \mathsf{Source}_1$.
$(\mathsf{Input}_0, \mathsf{Target}_0) \leftarrow \mathsf{C}(X_0, \mathsf{aux})$,	$(\mathsf{Input}_1, \mathsf{Target}_1) \leftarrow \mathsf{C}(X_1, \mathsf{aux})$,
$\mathsf{Output}_0 \leftarrow \mathsf{A}(\mathsf{Input}_0)$.	$\mathsf{Output}_1 \leftarrow \mathsf{A}(\mathsf{Input}_1)$.
If $\mathsf{Target}_0 = \mathsf{Output}_0$, return 1.	If $\mathsf{Target}_1 = \mathsf{Output}_1$, return 1.
Otherwise, return 0.	Otherwise, return 0.

Fig. 7 The experiments Exp_0 and Exp_1 for the computational MP-LWE assumption

Bai et al. [5] gave a reduction from the (decisional) MP-LWE problem to the computational MP-LWR problem.

PKE Scheme from MP-LWR. Bai et al. [5] proposed the MP-LWR-based PKE scheme. This construction is a variant of the Ring-LWR-based PKE scheme by Chen, Zhang, and Zhang [12]. To describe the PKE scheme by Bai et al. [5], we introduce some notations.

Let p and q be integers such that $2 \leq p \leq q$. We define the reconciliation cross-rounding function $\langle \cdot \rangle_2 : \mathbb{Z}_q \to \mathbb{Z}_2$ as $\langle x \rangle_2 := \left\lfloor \left(\frac{4}{q} \right) \cdot x \right\rceil \mod 2$. We introduce functions Lift, DBL, and REC. The probabilistic lifting function Lift takes as an input $x \in \mathbb{Z}_p$ and outputs $u \in \mathbb{Z}_q$ where u is sampled uniformly at random on the set $\{u \in \mathbb{Z}_q | \lfloor u \rceil_p = x\}$. The probabilistic doubling function DBL takes as an input $x \in \mathbb{Z}_q$ and outputs $y = 2x - e \in \mathbb{Z}_{2q}$ where e is sampled from the distribution that $e = 1$ is sampled with probability $\frac{1}{4}$, $e = -1$ is sampled with probability $\frac{1}{4}$, and $e = 0$ is sampled with probability $\frac{1}{2}$. For a positive even integer $q \in \mathbb{N}$, the reconciliation function REC takes as an input two values $y \in \mathbb{Z}_q$ and $b \in \{0, 1\}$. It outputs $\lfloor x \rceil$ where x is the closest element to y such that $\langle x \rangle_2 = b$. We extend the above notations to the coefficients of a polynomial.

Let k, d, n, p, q, t be positive integers such that $d + k \leq n$, $q \geq p \geq 2$, and $t \geq (2\lambda + (k + d + n) \log q)/(k + 1)$. Let $H : \{0, 1\}^d \to \{0, 1\}^k$ be a hash function. The PKE scheme $\mathsf{PKE_{BBD^+}}$ by Bai et al. is given in Fig. 8.

The correctness of $\mathsf{PKE_{BBD^+}}$ can be proven by the following lemma.

Lemma 4 ([32]) *For a positive odd integer q and $x, y \in \mathbb{Z}_q$ such that $|x - y| < \frac{q}{8}$, $\mathsf{REC}(y, \langle \mathsf{DBL}(x) \rangle_2) = \lfloor x \rceil_2$ holds.*

The IND-CPA security of $\mathsf{PKE_{BBD^+}}$ is proven under the computational MP-LWR assumption in the ROM [5].

$\mathsf{KeyGen}(1^\lambda)$:
$\quad s \xleftarrow{\$} \mathbb{Z}_q^{<n+d+k-1}[x]^\times$.
$\quad$ For $i \in [t]$, $a_i \xleftarrow{\$} \mathbb{Z}_q^{<n}[x]$, $b_i \leftarrow \lfloor a_i \odot_{d+k} s \rceil_p$.
$\quad$ Return $(\mathsf{pk}, \mathsf{sk}) \leftarrow ((a_i, b_i)_{i \in [t]}, s)$.
$\mathsf{Encrypt}(\mathsf{pk} = (a_i, b_i)_{i \in [t]}, \mu \in \{0, 1\}^{<k}[x])$:
$\quad$ For $i \in [t]$, $r_i \xleftarrow{\$} \{0, 1\}^{<k+1}[x]$.
$\quad c_1 \leftarrow \sum_{i \in [t]} a_i \cdot r_i \mod p$, $v \leftarrow \sum_{i \in [t]} r_i \odot_d \mathsf{Lift}(b_i) \mod q$,
$\quad c_2 \leftarrow \langle \mathsf{DBL}(v) \rangle_2$, $c_\mu \leftarrow H(\lfloor \mathsf{DBL}(v) \rceil_2) \oplus \mu$.
$\quad$ Return $c = (c_\mu, c_1, c_2)$.
$\mathsf{Decrypt}(\mathsf{sk} = s, c = (c_\mu, c_1, c_2))$:
$\quad w \leftarrow c_1 \odot_d s$,
$\quad$ Return $\mu \leftarrow c_\mu \oplus H(\mathsf{REC}(w, c_2))$.

Fig. 8 The PKE scheme $\mathsf{PKE_{BBD^+}}$ [5] based on the computational MP-LWR

8 General Framework for Algebraically Structured LWE

Order, Ring of Integers, Fractional Ideal, and Tensor. Let $K = \mathbb{Q}(\theta)$ be an algebraic number field with degree n which is a $\mathbb{Q}$-vector space obtained by attaching a root θ of a monic, irreducible polynomial $f(x) \in \mathbb{Z}[x]$ of degree n. An order O in K is a subring and lattice in K. A ring of integers O_K is a set of all elements in a number field K that are the root of a monic polynomial in $\mathbb{Z}[x]$. That is $O_K := \{\alpha | \exists (monic) f(x) \in \mathbb{Z}[x] \text{ s.t. } f(\alpha) = 0\}$.

To define the dual of O_K, we review a notion of embedding. For a number field $K = \mathbb{Q}(\theta)$ of degree n, there are n field embeddings (injective homomorphisms) $\sigma_i : K \to \mathbb{C}$. Concretely, each σ_i maps θ to the complex root of the minimal polynomial f. The canonical embedding $\sigma : K \to \mathbb{C}^n$ is defined as $\sigma(x) := (\sigma_1(x), \ldots \sigma_n(x))$. A dual of the ring of O_K is defined as $O_K^\vee := \{x \in K | Tr_{K/\mathbb{Q}}(xO_K) \subset \mathbb{Z}\}$ where $Tr_{K/\mathbb{Q}}$ is the trace defined as $Tr_{K/\mathbb{Q}}(a) := \sum_{i=1}^{n} \sigma_i(a)$ for $a \in K$.

A fractional O_K-ideal $\mathcal{I}$ is an additive subgroup of K for which there exists $r \in O_K$ such that $r\mathcal{I} \subset O_K$. An order-three tensor T of local dimensions k_1, k_2, k_3 over a set S is defined as $T := (T_{i_1,i_2,i_3} \in S)_{i_1 \in [k_1], i_2 \in [k_2], i_3 \in [k_3]}$.

Ring-LWE and Module-LWE Problems. We review definitions of the Module-LWE, dual Ring-LWE, and primal Ring-LWE problems.

Definition 18 (*Module-LWE and Ring-LWE Distributions*) Let $q \geq 2$ and k be positive integers, K a degree n number field, O_K an its ring of integers, and χ a distribution over $K_\mathbb{R}$ where $K_\mathbb{R} = K \otimes_\mathbb{Q} \mathbb{R}$ is the field tensor product.

For $\mathbf{s} \in (O_K^\vee / q O_K^\vee)^k$, we define the dual module distribution dModule-$\mathsf{D}_{k,q,\chi}(\mathbf{s})$ as the distribution obtained by sampling $\mathbf{a} \xleftarrow{\$} (O_K/qO_K)^k, e \xleftarrow{r} \chi$, and returning $(\mathbf{a}, b \leftarrow \frac{1}{q} \sum_{i \in [k]} \mathbf{a}_i \cdot \mathbf{s}_i + e)$.

Similarly, for $s \in O_K/qO_K$, we define the primal ring distribution pRing-$\mathsf{D}_{q,\chi}(s)$ as the distribution obtained by sampling $a \xleftarrow{\$} O_K/qO_K, e \xleftarrow{r} \chi$, and returning $(a, b \leftarrow a \cdot s + e)$.

Definition 19 (*Module-LWE and Ring-LWE* [11, 29]) Let $q \geq 2$ and k be positive integers, K a degree n number field, O_K its ring of integers, and χ a distribution over $K_\mathbb{R}$ where $K_\mathbb{R} = K \otimes_\mathbb{Q} \mathbb{R}$ is the field tensor product. The dual Module-LWE$_{k,q,\chi}$ problem consists in distinguishing between polynomially many samples from dModule-$\mathsf{D}_{k,q,\chi}(\mathbf{s})$ and the same number of uniform samples in $(O_K^\vee/qO_K^\vee)^k \times K_\mathbb{R}/qO_K^\vee$ over the choice of $\mathbf{s} \in (O_K^\vee/qO_K^\vee)^k$. We say that the dual Module-LWE$_{k,q,\chi}$ assumption holds if the advantage of any PPT algorithm trying to solve the dual Module-LWE$_{k,q,\chi}$ problem is negligible in λ. We say that the dual Ring-LWE$_{q,\chi}$ assumption holds if the Module-LWE$_{1,q,\chi}$ assumption holds.

Similarly, the primal Ring-LWE$_{q,\chi}$ problem consists in distinguishing between polynomially many samples from pRing-$\mathsf{D}_{q,\chi}(s)$ and the same number of uniform samples in $O_K/qO_K \times K_\mathbb{R}/qO_K$ over the choice of $s \in O/qO$. We say that the primal Ring-LWE$_{q,\chi}$ assumption holds if the advantage of any PPT algorithm trying to solve the primal Ring-LWE$_{q,\chi}$ problem is negligible in λ.

General-LWE and $\mathcal{L}$-LWE Problems. Peikert and Pepin [33] proposed general learning with errors (General-LWE) which captures LWE over commutative rings. General-LWE is parameterized by the following factors:

- O: An order in a number field K.
- $\mathcal{I}_s, \mathcal{I}_a, \mathcal{I}_b$: Fractional O-ideals such that $\mathcal{I}_b = \mathcal{I}_a \mathcal{I}_b$, $\mathcal{I}_s, \mathcal{I}_a, \mathcal{I}_b$ have dimensions k_s, k_a, k_b, respectively, and a modulus q.
- T: An order-three tensor $T \in O_q^{k_s \times k_a \times k_b}$. T induces an O_q-bilinear map $T : (\mathcal{I}_s/q\mathcal{I}_s)^{k_s} \times (\mathcal{I}_a/q\mathcal{I}_s)^{k_a} \to (\mathcal{I}_b/q\mathcal{I}_b)^{k_b}$ defined by

$$T(\mathbf{s}, \mathbf{a})_{i_b} := \sum_{i_s \in k_s, i_a \in k_a} T_{i_s, i_a, i_b} \mathbf{s}_{i_s} \mathbf{a}_{i_a} \ (\text{for } i_b \in [k_b]).$$

- ψ: An error distribution over $K_{\mathbb{R}}^{k_b}$.

Definition 20 (*General-LWE Distribution* [33]) Let $T, \mathcal{I}_s, \mathcal{I}_a, \mathcal{I}_b, k_s, k_a, k_b, q, \psi$ be parameters described above. For $\mathbf{s} \in (\mathcal{I}_s/q\mathcal{I}_s)^{k_s}$, we define the generalized distribution General-$\mathsf{D}_{T,\mathcal{I}_s,\mathcal{I}_a,\psi,q}(\mathbf{s})$ as the distribution obtained by sampling $\mathbf{a} \xleftarrow{\$} (\mathcal{I}_a/q\mathcal{I}_a)^{k_a}$, $\mathbf{e} \xleftarrow{r} \psi$, and returning $(\mathbf{a}, \mathbf{b} \leftarrow T(\mathbf{s}, \mathbf{a}) + \mathbf{e})$.

Definition 21 (*General-LWE* [33]) The General-LWE$_{T,\mathcal{I}_s,\mathcal{I}_a,\psi,q}$ problem consists in distinguishing between polynomially many samples from General-$\mathsf{D}_{T,\mathcal{I}_s,\mathcal{I}_a,\psi,q}(\mathbf{s})$ and the same number of uniform samples in $(\mathcal{I}_a/q\mathcal{I}_a)^{k_a} \times (K_{\mathbb{R}}/q\mathcal{I}_b)^{k_b}$ over the choice of $\mathbf{s} \in (\mathcal{I}_s/q\mathcal{I}_s)^{k_s}$.

General-LWE captures MP-LWE, Module-LWE, Ring-LWE, Poly-LWE, MP-LWE, and Order-LWE. A special case of General-LWE called $\mathcal{L}$-LWE was proposed in [33]. $\mathcal{L}$-LWE$_{\psi,q}^k$ is General-LWE$_{T,\mathcal{L}^\vee,O^{\mathcal{L}},\psi,q}$ where $\mathcal{L}$ is a lattice in a number field K, $T \in O^{k \times k \times 1}$, $\mathcal{L}^\vee$ is the dual of $\mathcal{L}$, $O^{\mathcal{L}}$ is the coefficient ring of $\mathcal{L}$, and ψ is an error distribution over $K_{\mathbb{R}}$. $\mathcal{L}$-LWE^k captures Module-LWE, Ring-LWE, Poly-LWE, and Order-LWE. We provide the overview of existing reductions among the Module-LWE, Ring-LWE, Poly-LWE, and MP-LWE problems in Fig. 9.

$$\text{Module-LWE} \atop (O_K\text{-LWE}^k) \xrightleftharpoons[]{\text{dual } [4]} \text{Ring-LWE} \atop (O_K\text{-LWE}^1) \xrightleftharpoons[]{\text{dual } [37]} \text{Ring-LWE} \atop (O_K\text{-LWE}^1) \xrightleftharpoons[]{\text{primal } [37]} \text{Poly-LWE (dual)} \atop \mathbb{Z}[\alpha]\text{-LWE}^1) \xrightarrow{[36]} \text{MP-LWE}$$

$A \longrightarrow B$ represents that there is a reduction from the problem A to the problem B.

Fig. 9 Reductions among the Module-LWE, Ring-LWE, Poly-LWE, and MP-LWE problems. For $\mathcal{L}$-LWE^k, by taking $\mathcal{L} = O_K$ where O_K is the full ring of integers, we obtain dual Module-LWE when $k > 1$ and dual Ring-LWE when $k = 1$. By taking $\mathcal{L} = O^\vee$ where $O = \mathbb{Z}[\alpha]$ and $\mathbb{Z}[\alpha]$-LWE^k, we obtain primal Poly-LWE which does not involve any dual lattices when $k = 1$

References

1. S. Agrawal, D. Boneh, X. Boyen, Efficient lattice (H)IBE in the standard model, in *EUROCRYPT 2010*, LNSC, vol. 6110 (Springer, 2010), pp. 553–572
2. S. Agrawal, X. Boyen, Identity-based encryption from lattices in the standard model (2009)
3. M. Ajtai, Generating hard instances of lattice problems (extended abstract), in *ACM STOC 1996* (ACM, 1996), pp. 99–108
4. M.R. Albrecht, A. Deo, Large modulus ring-lwe $\geq$ module-lwe, in *ASIACRYPT 2017, Part I*, LNSC, vol. 10624 (Springer, 2017), pp. 267–296
5. S. Bai, K. Boudgoust, D. Das, A. Roux-Langlois, W. Wen, Z. Zhang, Middle-product learning with rounding problem and its applications, in *ASIACRYPT 2019, Part I*, LNSC, vol. 11921 (Springer, 2019), pp. 55–81
6. S. Bai, D. Das, R. Hiromasa, M. Rosca, A. Sakzad, D. Stehlé, R. Steinfeld, Z. Zhang, Mpsign: a signature from small-secret middle-product learning with errors, in *PKC 2020, Part II*, LNSC, vol. 12111 (Springer, 2020), pp. 66–93
7. M. Bellare, P. Rogaway, Random oracles are practical: a paradigm for designing efficient protocols, in *CCS '93* (ACM, 1993), pp. 62–73
8. M. Bolboceanu, Z. Brakerski, R. Perlman, D. Sharma, Order-lwe and the hardness of ring-lwe with entropic secrets, in *ASIACRYPT 2019, Part II*, LNCS, vol. 11922 (Springer, 2019), pp. 91–120
9. D. Boneh, Ö. Dagdelen, M. Fischlin, A. Lehmann, C. Schaffner, M. Zhandry, Random oracles in a quantum world, in *ASIACRYPT 2011*, LNSC, vol. 7073 (Springer, 2011), pp. 41–69
10. J.W. Bos, L. Ducas, E. Kiltz, T. Lepoint, V. Lyubashevsky, J.M. Schanck, P. Schwabe, G. Seiler, D. Stehlé, CRYSTALS - kyber: a cca-secure module-lattice-based KEM, in *IEEE EuroS&P 2018* (IEEE, 2018), pp. 353–367
11. Z. Brakerski, C. Gentry, V. Vaikuntanathan, (leveled) fully homomorphic encryption without bootstrapping, in *ITCS 2012* (ACM, 2012), pp. 309–325
12. L. Chen, Z. Zhang, Z. Zhang, On the hardness of the computational ring-lwr problem and its applications, in *ASIACRYPT 2018, Part I*, LNCS, vol. 11272 (Springer, 2018)
13. D. Das, M.H. Au, Z. Zhang, Ring signatures based on middle-product learning with errors problems, in *AFRICACRYPT 2019*, LNSC, vol. 11627 (Springer, 2019), pp. 139–156
14. L. Ducas, E. Kiltz, T. Lepoint, V. Lyubashevsky, P. Schwabe, G. Seiler, D. Stehlé, Crystals-dilithium: a lattice-based digital signature scheme. IACR Trans. Cryptogr. Hardw. Embed. Syst. **2018**(1), 238–268 (2018)
15. J. Fan, X. Lu, M.H. Au, Adaptively secure identity-based encryption from middle-product learning with errors, in *ACISP 2023*, LNSC, vol. 13915 (Springer, 2023), pp. 320–340
16. A. Fiat, A. Shamir, How to prove yourself: practical solutions to identification and signature problems, in *CRYPTO '86*, LNSC, vol. 263 (Springer, 1986), pp. 186–194
17. C. Gentry, C. Peikert, V. Vaikuntanathan, Trapdoors for hard lattices and new cryptographic constructions, in *ACM STOC 2008* (ACM, 2008), pp. 197–206
18. C. Grover, A. Mendelsohn, C. Ling, R. Vehkalahti, Non-commutative ring learning with errors from cyclic algebras. J. Cryptol. **35**(3), 22 (2022)
19. G. Hanrot, M. Quercia, P. Zimmermann, The middle product algorithm I. Appl. Algebra Eng. Commun. Comput. **14**(6), 415–438 (2004)
20. R. Hiromasa, Digital signatures from the middle-product LWE, in *ProvSec 2018*, LNSC, vol. 11192 (Springer, 2018), pp. 239–257
21. E. Kiltz, V. Lyubashevsky, C. Schaffner, A concrete treatment of fiat-shamir signatures in the quantum random-oracle model, in *EUROCRYPT 2018, Part III*, LNSC, vol. 10822 (Springer, 2018), pp. 552–586
22. H. Q. Le, D.H. Duong, W. Susilo, J. Pieprzyk, Trapdoor delegation and HIBE from middle-product LWE in standard model, in *ACNS 2020, Part I*, LNSC, vol. 12146 (Springer, 2020), pp. 130–149
23. H. Lin, S. Sun, M. Wang, J.K. Liu, W. Wang, Shorter linkable ring signature based on middle-product learning with errors problem. Comput. J. **66**(12), 2974–2989 (2023)

24. C. Ling, A. Mendelsohn, Middle-products of skew polynomials and learning with errors, in *IMA IMACC 2023*, LNCS, vol. 14421 (Springer, 2023), pp. 199–219
25. A. Lombardi, V. Vaikuntanathan, T. Vuong, Lattice trapdoors and IBE from middle-product LWE, in *TCC 2019, Part I*, LNSC, vol. 11891 (Springer, 2019), pp. 24–54
26. V. Lyubashevsky, Fiat-shamir with aborts: applications to lattice and factoring-based signatures, in *ASIACRYPT 2009*, LNCS, vol. 5912 (Springer, 2009), pp. 598–616
27. V. Lyubashevsky, Digital signatures based on the hardness of ideal lattice problems in all rings, in *ASIACRYPT 2016, Part II*, LNSC, vol. 10032 (2016), pp. 196–214
28. V. Lyubashevsky, D. Micciancio, Generalized compact knapsacks are collision resistant, in *ICALP 2006, Part II*, LNSC, vol. 4052 (Springer, 2006), pp. 144–155
29. V. Lyubashevsky, C. Peikert, O. Regev, On ideal lattices and learning with errors over rings, in *EUROCRYPT 2010*, LNSC, vol. 6110 (Springer, 2010), pp. 1–23
30. D. Micciancio, C. Peikert, Trapdoors for lattices: simpler, tighter, faster, smaller, in *EUROCRYPT 2012*, LNSC, vol. 7237 (Springer, 2012), pp. 700–718
31. NIST. NIST post-quantum competition. https://csrc.nist.gov/projects/post-quantum-cryptography
32. C. Peikert, Lattice cryptography for the internet, in *PQCrypto 2014*, LNCS, vol. 8772 (Springer, 2014), pp. 197–219
33. C. Peikert, Z. Pepin, Algebraically structured lwe, revisited, in *TCC 2019, Part I*, LNCS, vol. 11891 (Springer, 2019), pp. 1–23
34. C. Peikert, A. Rosen, Efficient collision-resistant hashing from worst-case assumptions on cyclic lattices, in *TCC 2006*, LNSC, vol. 3876 (Springer, 2006), pp. 145–166
35. O. Regev, On lattices, learning with errors, random linear codes, and cryptography, in *ACM STOC 2005* (ACM, 2005), pp. 84–93
36. M. Rosca, A. Sakzad, D. Stehlé, R. Steinfeld, Middle-product learning with errors, in *CRYPTO 2017, Part III*, LNSC, vol. 10403 (Springer, 2017), pp. 283–297
37. M. Rosca, D. Stehlé, A. Wallet, On the ring-lwe and polynomial-lwe problems, in *EUROCRYPT 2018, Part I*, LNSC, vol. 10820 (Springer, 2018), pp. 146–173
38. V. Shoup, Efficient computation of minimal polynomials in algebraic extensions of finite fields, in *ISSAC '99* (ACM, 1999), pp. 53–58
39. D. Stehlé, R. Steinfeld, K. Tanaka, K. Xagawa, Efficient public key encryption based on ideal lattices, in *ASIACRYPT 2009*, LNSC, vol. 5912 (Springer, 2009), pp. 617–635
40. R. Steinfeld, A. Sakzad, R.K. Zhao, Practical MP-LWE-based encryption balancing security-risk versus efficiency. Des. Codes Cryptogr. **87**(12), 2847–2884 (2019)
41. N. Yang, S. Yang, Y. Zhao, W. Wu, Inner product encryption from middle-product learning with errors, in *SocialSec 2022*, CCIS, vol. 1663 (Springer, 2022), pp. 94–113
42. N. Yang, S. Yang, Y. Zhao, W. Wu, X. Wang, Inner product encryption from middle-product learning with errors. Comput. Stand. Interfaces **87**, 103755 (2024)
43. S. Yang, X. Huang, Universal product learning with errors: a new variant of LWE for lattice-based cryptography. Theor. Comput. Sci. **915**, 90–100 (2022)

Open Access This chapter is licensed under the terms of the Creative Commons Attribution 4.0 International License (http://creativecommons.org/licenses/by/4.0/), which permits use, sharing, adaptation, distribution and reproduction in any medium or format, as long as you give appropriate credit to the original author(s) and the source, provide a link to the Creative Commons license and indicate if changes were made.

The images or other third party material in this chapter are included in the chapter's Creative Commons license, unless indicated otherwise in a credit line to the material. If material is not included in the chapter's Creative Commons license and your intended use is not permitted by statutory regulation or exceeds the permitted use, you will need to obtain permission directly from the copyright holder.

Expanded Lattices for Solving Ring-Based LWE and NTRU Problems

Satoshi Nakamura and Masaya Yasuda

Abstract In modern cryptography, structured lattices from ring-based LWE and NTRU are useful to construct efficient and compact cryptosystems. Search problems of ring-based LWE and NTRU can be reduced to particular cases of the shortest vector problem (SVP) on Kannan's embedding and the NTRU lattices, respectively. In this paper, we expand Kannan's embedding and the NTRU lattices for solving the ring-based LWE and the NTRU problems, respectively. Both expanded lattices include many short vectors that are amplified by using rotations of secret short vectors. Since many target short vectors are embedded in our expanded lattices, it could increase the success probability for solving the ring-based LWE and the NTRU problems by using the block Korkine–Zolotarev (BKZ) reduction algorithm. We demonstrate by experiments the efficacy of our expansions from viewpoints of the success probability and the running time.

Keywords Lattice problems · Embedding · Rotations · BKZ

1 Introduction

The security of most lattice-based cryptosystems relies on the hardness of solving the learning with errors (LWE) or the NTRU problem. An analog of LWE over rings is known as ring-based LWE [7], whose framework includes polynomial-LWE [24] and ring-LWE [17]. Both problems of ring-based LWE and NTRU use the structure

---

This is a fully revised version of the conference paper [20]. In this version, we extended our approach for solving the NTRU problem in Sect. 4.2. We also presented experimental results with discussion on trade-offs of our approach in Sect. 5.

S. Nakamura
NTT Social Informatics Laboratories, Tokyo, Japan
e-mail: satoshi.nakamura.xn@hco.ntt.co.jp

S. Nakamura · M. Yasuda (✉)
Department of Mathematics, Rikkyo University, Tokyo, Japan
e-mail: myasuda@rikkyo.ac.jp

of a ring $R = \mathbb{Z}[x]/(\phi)$ for some ϕ, and they are useful to build efficient various cryptosystems. In general, $\phi = x^n + 1$ and $\phi = x^n - 1$ are used with a 2-power integer n for ring-based LWE and NTRU, respectively. For an odd prime q, the search problem of ring-based LWE asks to find a secret $s(x) \in R_q = R/qR$ from ring-based LWE samples of the form $(a(x), t(x))$, where $a(x)$ is randomly chosen from R_q and $t(x) = a(x) \cdot s(x) + e(x)$ for some unknown error $e(x)$ with small coefficients. When ϕ has degree n, we express any element of R (resp., R_q) as $p(x) = p_0 + p_1 x + \cdots + p_{n-1} x^{n-1}$ with p_i's in $\mathbb{Z}$ (resp., $\mathbb{Z}_q$), and write its coefficient vector by $\boldsymbol{p} = (p_0, p_1, \ldots, p_{n-1})$. We denote the rotation of $\boldsymbol{p}$ as the coefficient vector of $xp(x)$ by $\text{rot}(\boldsymbol{p})$. Then every ring-based LWE sample is rewritten in the matrix form as $(\boldsymbol{A}, \boldsymbol{t})$ satisfying $\boldsymbol{t} \equiv \boldsymbol{sA} + \boldsymbol{e} \pmod{q}$, where $\boldsymbol{t}, \boldsymbol{s}, \boldsymbol{e}$ denote the coefficient vectors of $t(x), s(x), e(x)$, respectively, and $\boldsymbol{A}$ the matrix whose rows are rotations $\text{rot}^i(\boldsymbol{a})$ of the coefficient vector $\boldsymbol{a}$ of $a(x)$. We regard $(\boldsymbol{A}, \boldsymbol{t})$ as an instance of the closest vector problem over a so-called q-ary lattice with targeting $\boldsymbol{t}$ (see [19] for q-ary lattices). It is also transformed to a particular instance of SVP by Kannan's embedding [14]. Kannan's embedding lattice contains unusually short error vectors $\pm \boldsymbol{e}$, which might be found by using BKZ. Once $\boldsymbol{e}$ is found, the secret vector $\boldsymbol{s}$ can be recovered from the linear equation $\boldsymbol{t} - \boldsymbol{e} \equiv \boldsymbol{sA} \pmod{q}$ by using Gaussian elimination. On the other hand, the NTRU problem [12] asks to find two secret polynomials $f(x), g(x) \in R$ with small coefficients from $h(x) = g(x) \cdot f(x)^{-1} \in R_q$ under the assumption that $f(x)$ is invertible in R_q. The NTRU lattice is constructed from the coefficient vector of $h(x)$ to include the concatenated unusually short vector $(\boldsymbol{g} \mid \boldsymbol{f})$ of two coefficient vectors of $g(x)$ and $f(x)$ (e.g., see [2, 13]).

In this paper, we expand Kannan's embedding and the NTRU lattices for solving the ring-based LWE and the NTRU problems, respectively. For both problems, we amplify secret short vectors by using rotations, and embed them into our expanded lattices. For an instance $(\boldsymbol{A}, \boldsymbol{t})$ of ring-based LWE, we add k rotated vectors of $\boldsymbol{t}$ to a basis of Kannan's embedding lattice for an expansion parameter $k \geq 1$. Then our expanded Kannan's embedding lattice contains k rotated vectors of the error $\boldsymbol{e}$ since it holds $\text{rot}^i(\boldsymbol{t}) \equiv \text{rot}^i(\boldsymbol{s}) \cdot \boldsymbol{A} + \text{rot}^i(\boldsymbol{e}) \pmod{q}$ for any i. (Our approach can apply to Bai-Galbraith's embedding [5] that is a variant of Kannan's embedding for small $\boldsymbol{s}$.) It is analogous to the NTRU lattice that contains all rotated vectors of the secret vector $\left(\text{rot}^i(\boldsymbol{g}) \mid \text{rot}^i(\boldsymbol{f})\right)$. On the other hand, for an expansion parameter $\ell \geq 0$, we expand the NTRU lattice to include all the rotated vectors of the form $(\text{rot}^j(\boldsymbol{g}) \mid \boldsymbol{f})$ for $0 \leq j \leq \ell$. In other words, our expanded NTRU lattice has $\ell + 1$ times more secret short vectors than the original NTRU lattice. Since our expanded lattices include many secret short vectors of the same length, we expect that it could increase the success probability for solving the ring-based LWE and the NTRU problems by using BKZ over our expanded lattices. We demonstrate the efficacy of our expansions through experiments, and also discuss the trade-offs of our approach.

Notation Let $\mathbb{Z}_q = \mathbb{Z} \cap \left[-\frac{q}{2}, \frac{q}{2}\right)$ denote the set of representatives of integers modulo an odd prime q. Let $\lfloor z \rceil$ denote the rounded value to the nearest integer of a real number $z \in \mathbb{R}$. We write row vectors (resp., matrices) in bold lower case (resp., capital) letters as $\boldsymbol{u}$ (resp., $\boldsymbol{A}$). We denote by $\boldsymbol{u}^\top$ (resp., $\boldsymbol{A}^\top$) the transpose of $\boldsymbol{u}$

(resp., A). For $u = (u_1, \ldots, u_d)$, $v = (v_1, \ldots, v_d) \in \mathbb{R}^d$, we let $\langle u, v \rangle$ denote the inner product defined by $\sum_{i=1}^{d} u_i v_i$. We let $\|u\|$ denote the Euclidean length defined by $\|u\| = \sqrt{\langle u, u \rangle}$.

2 Lattices, Lattice Problems, and Lattice Basis Reduction

In this section, we recall the basics on lattices and lattice problems that are algorithmic problems for lattices. We also present lattice basis reduction and its practical algorithms that are a strong tool to solve lattice problems.

2.1 Lattices

For a positive integer d, a (full-rank) *lattice* in $\mathbb{R}^d$ is the set of all *integral* linear combinations of d linearly independent vectors $b_1, \ldots, b_d \in \mathbb{R}^d$ as

$$L = \mathcal{L}(b_1, \ldots, b_d) = \left\{ \sum_{i=1}^{d} x_i b_i : x_i \in \mathbb{Z}, \ 1 \leq i \leq d \right\}.$$

We denote by $B = (b_1, \ldots, b_d)$ the ordered set of b_i's, and also regard it as the square matrix of size d whose rows are given by b_i's. Such B is called a *basis* of L, and we simply write $L = \mathcal{L}(B)$, the lattice spanned by the rows of B. The number of linearly independent vectors spanning L is called the *dimension* of L. There exist infinitely many bases of a lattice of dimension $d \geq 2$. Precisely, two matrices B and C span the same lattice of dimension d if and only if there exists a unimodular matrix T of size d satisfying $C = TB$. (A square matrix with integral components is called *unimodular* if its determinant is equal to 1 in absolute.) The *volume* of a lattice L is defined by $\mathrm{vol}(L) = |\det B|$ for any basis B of L, which is independent of the choice of bases since $|\det(TB)| = |\det B|$ for any unimodular matrix T.

Given a basis $B = (b_1, \ldots, b_d)$ of a lattice L, its *Gram–Schmidt vectors* are defined recursively as $b_1^\star = b_1$ and

$$b_i^\star = b_i - \sum_{j=1}^{i-1} \mu_{ij} b_j^\star \in \mathbb{R}^d, \quad \mu_{ij} = \frac{\langle b_i, b_j^\star \rangle}{\|b_j^\star\|^2} \tag{1}$$

for $2 \leq i \leq d$. Then $\langle b_i^\star, b_j^\star \rangle = 0$ for any $i \neq j$. We denote by $B^\star$ the $d \times d$ matrix whose row vectors are Gram–Schmidt vectors $b_i^\star$'s, and U the $d \times d$ lower triangular matrix with non-diagonal entries given by Gram–Schmidt coefficients μ_{ij}'s and all diagonal entries equal to 1. Then $B = UB^\star$ from (1), and hence $\mathrm{vol}(L) = |\det B^\star| =$

$\sqrt{\det\left(\boldsymbol{B}^\star(\boldsymbol{B}^\star)^\top\right)} = \prod_{i=1}^{d} \|\boldsymbol{b}_i^\star\|$ from the orthogonality of Gram–Schmidt vectors. For each $1 \le k \le d$, define an orthogonal projection map as

$$\pi_k : \mathbb{R}^d \longrightarrow \mathrm{span}_{\mathbb{R}}\left(\boldsymbol{b}_k^\star, \ldots, \boldsymbol{b}_d^\star\right), \quad \pi_k(\boldsymbol{v}) = \sum_{i=k}^{d} \frac{\langle \boldsymbol{v}, \boldsymbol{b}_i^\star \rangle}{\|\boldsymbol{b}_i^\star\|^2} \boldsymbol{b}_i^\star \quad (\boldsymbol{v} \in \mathbb{R}^d),$$

where $\mathrm{span}_{\mathbb{R}}(\boldsymbol{b}_k^\star, \ldots, \boldsymbol{b}_d^\star)$ denotes the sub-vector space of $\mathbb{R}^d$ spanned by $d-k+1$ Gram–Schmidt vectors $\boldsymbol{b}_k^\star, \ldots, \boldsymbol{b}_d^\star$. (In particular, the map π_1 is the identity.) The lattice in $\mathbb{R}^d$ spanned by $d-k+1$ projected vectors $\pi_k(\boldsymbol{b}_k), \ldots, \pi_k(\boldsymbol{b}_d)$ is denoted by $\pi_k(L)$, called a *projected lattice*. The projected lattice $\pi_k(L)$ is of dimension $d-k+1$, and its volume is equal to $\prod_{i=k}^{d} \|\boldsymbol{b}_i^\star\|$.

2.2 Lattice Problems

Let L be a full-rank lattice in $\mathbb{R}^d$. For each $1 \le k \le d$, the k-th *successive minimum* of L, denoted by $\lambda_k(L)$, is defined as the minimum of $\max_{1 \le j \le k} \|\boldsymbol{v}_j\|$ over all k linearly independent vectors $\boldsymbol{v}_1, \ldots, \boldsymbol{v}_k$ in L. In particular, the first minimum $\lambda_1(L)$ is the length of a shortest non-zero vector in L. *Lattice problems* are algorithmic problems for lattices. Among various lattice problems, the *shortest vector problem (SVP)* is the most famous; "Given a basis $\boldsymbol{B}$ of a lattice L, find a shortest non-zero vector in L, that is, a lattice vector $\boldsymbol{s} \in L$ satisfying $\|\boldsymbol{s}\| = \lambda_1(L)$." SVP is proven NP-hard under randomized reductions [1]. Approximate factors relax SVP; "Given a basis $\boldsymbol{B}$ of a lattice L and an approximation factor $f \ge 1$, find a non-zero vector $\boldsymbol{v}$ in L satisfying $\|\boldsymbol{v}\| \le f\lambda_1(L)$." The *closest vector problem (CVP)* is another famous problem; "Given a basis $\boldsymbol{B}$ of a lattice L and a target vector $\boldsymbol{t}$, find a vector in L closest to $\boldsymbol{t}$, that is, a lattice vector $\boldsymbol{v} \in L$ such that its distance to $\boldsymbol{t}$ is minimized." It is known that SVP is not harder than CVP (see [18]). As in the case of SVP, approximate factors can relax CVP. Since Kannan's embedding [14] transforms approximate-CVP into approximate-SVP, both problems seem equally hard in practice.

Remark 1 (*Gaussian Heuristic*) We fix a full-rank lattice L in $\mathbb{R}^d$. Given a measurable set S in $\mathbb{R}^d$ of volume $\mathrm{vol}(S)$, Gaussian Heuristic predicts that the number of vectors in $L \cap S$ is approximately equal to $\mathrm{vol}(S)/\mathrm{vol}(L)$. We apply as S the ball of radius $\lambda_1(L)$ centered at the origin in $\mathbb{R}^d$ to obtain a prediction of $\lambda_1(L)$ from Gaussian Heuristic as

$$\lambda_1(L) \approx v_d^{-\frac{1}{d}} \cdot \mathrm{vol}(L)^{\frac{1}{d}} \sim \sqrt{\frac{d}{2\pi e}} \cdot \mathrm{vol}(L)^{\frac{1}{d}}. \tag{2}$$

Here we let v_d denote the volume of the unit ball in $\mathbb{R}^d$ given by

$$v_d = \frac{\pi^{d/2}}{\Gamma(1+d/2)} \sim \frac{1}{\sqrt{\pi d}} \left(\frac{2\pi e}{d}\right)^{d/2},$$

where $\Gamma(s) = \int_0^\infty t^{s-1} e^{-t} dt$ denotes the Gamma function and the approximation of v_d is due to Stirling's formula. This roughly holds in practice for random lattices of high dimensions such as $d \geq 45$.

2.3 Lattice Basis Reduction and Its Practical Algorithms

Given a basis of a lattice L as input, *lattice basis reduction* seeks a new basis $B = (b_1, \ldots, b_d)$ of L with short and nearly orthogonal vectors b_i's by a suitable combination of unimodular transformations. Such B is called *reduced*. Below we introduce two practical reduction algorithms.

Lenstra–Lenstra–Lovász (LLL) Let $B = (b_1, \ldots, b_d)$ be a basis of a lattice L. We say B *LLL-reduced* for a constant $\frac{1}{4} < \delta < 1$ if it satisfies the following two conditions: (i) (Size-reduced) $|\mu_{ij}| \leq \frac{1}{2}$ for any $j < i$. (ii) (Lovász' condition) $\delta \|b_{k-1}^*\|^2 \leq \|\pi_{k-1}(b_k)\|^2$ for all $2 \leq k \leq d$. Here μ_{ij}'s and b_k^*'s are Gram–Schmidt coefficients and vectors of B, respectively. The first basis vector of any LLL-reduced basis satisfies both $\|b_1\| \leq \alpha^{\frac{d-1}{2}} \lambda_1(L)$ and $\|b_1\| \leq \alpha^{\frac{d-1}{4}} \mathrm{vol}(L)^{\frac{1}{d}}$ for $\alpha = \frac{4}{4\delta - 1}$. Given any basis B of L as input, the LLL algorithm [16] repeatedly performs size reduction like $b_i \leftarrow b_i - \lfloor \mu_{ij} \rceil b_j$ and swaps adjacent basis vectors b_{k-1}, b_k that do not satisfy Lovász' condition to find an LLL-reduced basis.

Block Korkine–Zolotarev (BKZ) Let $B = (b_1, \ldots, b_d)$ be a basis of a lattice L. For $j < k$, set $B_{[j:k]} = (\pi_j(b_j), \pi_j(b_{j+1}), \ldots, \pi_j(b_k))$ and $L_{[j:k]}$ the sub-lattice of the projected lattice $\pi_j(L)$ spanned by $B_{[j:k]}$. We say B β-*BKZ-reduced* for a blocksize $\beta \geq 2$ if it is size-reduced and satisfies $\|b_j^*\| = \lambda_1(L_{[j:k]})$ for every $1 \leq j \leq d-1$ with $k = \min(j + \beta - 1, d)$. In particular, it is called *Hermite–Korkine–Zolotarev (HKZ)-reduced* when $\beta = d$. The BKZ algorithm [23] repeatedly calls an exact-SVP algorithm (such as enumeration and sieving) over $L_{[j:k]}$ for every j to find a β-BKZ-reduced basis. In the basic BKZ algorithm, LLL is called as a subroutine to reduce $B_{[j:k]}$ before calling an exact-SVP algorithm over $L_{[j:k]}$. A larger β makes the computational cost of BKZ much more expensive, but it finds a more reduced basis. Under the Gaussian Heuristic and the geometric series assumption (GSA) of [22], the Gram–Schmidt lengths of a β-BKZ-reduced basis B of a random lattice of volume 1 are estimated as

$$\|\boldsymbol{b}_i^\star\| \approx \delta_\beta^{d-1-2i}, \quad \delta_\beta = \left(\frac{\beta}{2\pi e}(\pi\beta)^{\frac{1}{\beta}}\right)^{\frac{1}{2(\beta-1)}} \tag{3}$$

for $\beta > 50$ with $\beta \ll d$. (The GSA states that the Gram–Schmidt log-lengths $\log \|\boldsymbol{b}_i^\star\|$'s of a reduced basis are roughly on a straight line.) There are experimental evidences in [8, 9, 26] to support Eq. (3), except for last β indexes since the last β-dimensional block $\boldsymbol{B}_{[d-\beta+1:d]}$ is HKZ-reduced.

3 Ring-Based LWE and NTRU Problems

In this section, we recall the ring-based and the NTRU problems and their reduction to lattice problems. We fix a base ring $R = \mathbb{Z}[x]/(\phi)$ with $\phi = x^n + 1$ or $\phi = x^n - 1$ for a 2-power integer n. Such a ring is commonly employed for structured lattice-based cryptography. Specifically, we shall consider the anti-cyclic case by $\phi = x^n + 1$ for the ring-based LWE problem, and the cyclic case by $\phi = x^n - 1$ for the NTRU problem.

Coefficient representation and rotations For a prime q, we let $R_q = R/qR$. Every element of R (resp., R_q) can be expressed as a polynomial $f(x) = f_0 + f_1 x + \cdots + f_{n-1} x^{n-1}$ of degree less than n with coefficients f_i's in $\mathbb{Z}$ (resp., $\mathbb{Z}_q$), and we write its coefficient vector as $\boldsymbol{f} = (f_0, f_1, \ldots, f_{n-1})$. We then define the rotation operation for $\boldsymbol{f}$ as

$$\mathrm{rot}(\boldsymbol{f}) = \begin{cases} (-f_{n-1}, f_0, f_1, \ldots, f_{n-2}) & \text{when } \phi = x^n + 1, \\ (f_{n-1}, f_0, f_1, \ldots, f_{n-2}) & \text{when } \phi = x^n - 1. \end{cases}$$

This rotated vector is equal to the coefficient vector of $xf(x)$. More generally, the i times rotated vector $\mathrm{rot}^i(\boldsymbol{f})$ is given as the coefficient vector of $x^i f(x)$ for every $1 \le i \le n$. We clearly have $\mathrm{rot}^n(\boldsymbol{f}) = -\boldsymbol{f}$ (resp., $\mathrm{rot}^n(\boldsymbol{f}) = \boldsymbol{f}$) when $\phi = x^n + 1$ (resp., $\phi = x^n - 1$) since $x^n f(x) = -f(x)$ (resp., $x^n f(x) = f(x)$).

3.1 Ring-Based LWE Problem and Its Reduction

We adhere to the ring-based LWE framework of [7] that incorporates the ring-LWE [17] and the polynomial-LWE [24]. We commonly consider the anti-cyclic case by $\phi = x^n + 1$ for the search ring-based LWE problem below.

Definition 1 (*Search ring-based LWE*) Fix a so-called secret $s(x) \in R_q$ for a prime q. We also fix an error distribution χ that produces small elements in R to construct a ring-based LWE distribution $A_{s,\chi}$, sampling a pair

$$(a(x), t(x)) \in R_q \times R_q \quad \text{satisfying} \quad t(x) = s(x) \cdot a(x) + e(x), \tag{4}$$

where $a(x)$ is uniformly chosen at random from R_q and $e(x)$ is sampled from the distribution χ. Given any independent samples from $A_{s,\chi}$, the search ring-LWE problem inquires us to identify the secret $s(x)$.

Remark 2 The number of samples (4) is not limited in the above definition. We assume that any number of samples can be acquired, or we may consider the number of samples as an additional parameter. We also assume that the distribution χ on R is centered at every coefficient in $\mathbb{Z}$, that is, every coefficient in $\mathbb{Z}$ of an element sampled from χ has expectation 0.

Reduction to standard LWE Let $(a(x), t(x))$ be a ring-based LWE sample from the distribution $A_{s,\chi}$. For the coefficient vector $\boldsymbol{a} = (a_0, a_1, \ldots, a_{n-1})$ of $a(x)$, we consider an $n \times n$ matrix

$$A = \begin{pmatrix} \boldsymbol{a} \\ \mathrm{rot}(\boldsymbol{a}) \\ \vdots \\ \mathrm{rot}^{n-1}(\boldsymbol{a}) \end{pmatrix} = \begin{pmatrix} a_0 & a_1 & \cdots & a_{n-1} \\ -a_{n-1} & a_0 & \cdots & a_{n-2} \\ \vdots & \vdots & \ddots & \vdots \\ -a_1 & -a_2 & \cdots & a_0 \end{pmatrix}. \tag{5}$$

Then we can reformulate the ring-based LWE relation $t(x) = s(x) \cdot a(x) + e(x)$ on R_q in the matrix form as

$$\boldsymbol{t} \equiv \boldsymbol{s} A + \boldsymbol{e} \pmod{q}, \tag{6}$$

where $\boldsymbol{t}$, $\boldsymbol{s}$, and $\boldsymbol{e}$ are three coefficient vectors of $t(x), s(x), and e(x)$, respectively. Indeed, for the variable vector $\boldsymbol{x} = (1, x, x^2, \ldots, x^{n-1}) \in R^n$, we rewrite the ring-based LWE relation on R_q as

$$\boldsymbol{t}\boldsymbol{x}^\top = t(x) = s(x) \cdot a(x) + e(x) = \boldsymbol{s}\boldsymbol{x}^\top \cdot a(x) + \boldsymbol{e}\boldsymbol{x}^\top$$

$$= \boldsymbol{s} \begin{pmatrix} a(x) \\ xa(x) \\ \vdots \\ x^{n-1}a(x) \end{pmatrix} + \boldsymbol{e}\boldsymbol{x}^\top = \boldsymbol{s} \begin{pmatrix} \boldsymbol{a} \\ \mathrm{rot}(\boldsymbol{a}) \\ \vdots \\ \mathrm{rot}^{n-1}(\boldsymbol{a}) \end{pmatrix} \boldsymbol{x}^\top + \boldsymbol{e}\boldsymbol{x}^\top \tag{7}$$

$$= (\boldsymbol{s} A + \boldsymbol{e})\boldsymbol{x}^\top \in R_q,$$

since we can express every element $f(x)$ in R_q as $\boldsymbol{f}\boldsymbol{x}^\top$ by its coefficient vector $\boldsymbol{f}$. Since $\boldsymbol{x}$ gives a basis of R_q as a $\mathbb{Z}_q$-module, we obtain Eq. (6).

For a positive integer m, let $(a_1(x), t_1(x)), \ldots, (a_m(x), t_m(x))$ be m independent ring-based LWE samples from $A_{s,\chi}$, satisfying $t_i(x) = s(x) \cdot a_i(x) + e_i(x)$ on R_q with some $e_i(x) \leftarrow \chi$ for every $1 \leq i \leq m$. For each $1 \leq i \leq m$, we let A_i be the matrix corresponding to $a_i(x)$ like Eq. (5). Then, from the previous paragraph, we obtain m relations $t_i \equiv sA_i + e_i \pmod{q}$ for every $1 \leq i \leq m$. Moreover, we concatenate the relations as

$$(\widetilde{A}, \widetilde{t}), \quad \widetilde{t} \equiv s\widetilde{A} + \widetilde{e} \pmod{q}, \tag{8}$$

where $\widetilde{A} = (A_1 \mid \cdots \mid A_m)$ is the $n \times d$ matrix concatenated with A_i's for $d = mn$, and $\widetilde{t} = (t_1 \mid \cdots \mid t_m), \widetilde{e} = (e_1 \mid \cdots \mid e_m)$ are concatenated target and error vectors of lengths d, respectively. This system can be viewed as an instance of standard LWE of size $n \times d$.

Reduction to lattice problems over q-ary lattices We say a full-rank lattice L in $\mathbb{R}^d$ to be q-ary for a prime q if $q\mathbb{Z}^d \subseteq L \subseteq \mathbb{Z}^d$. We can construct a q-ary lattice from the concatenated matrix $\widetilde{A}$ of size $n \times d$ in (8) as

$$\Lambda_q(\widetilde{A}) = \left\{ \widetilde{z} \in \mathbb{Z}^d \mid \widetilde{z} \equiv u\widetilde{A} \pmod{q}, \exists u \in \mathbb{Z}^n \right\}. \tag{9}$$

(See [19] for q-ary lattices constructed from LWE.) We have $\mathrm{vol}(\Lambda_q(\widetilde{A})) = q^{d-n}$ for almost all matrices $\widetilde{A}$. The q-ary lattice is spanned by the rows of the $(d+n) \times d$ matrix $\begin{pmatrix} \widetilde{A} \\ qI_d \end{pmatrix}$, where I_d is the identity matrix of size d. Moreover, we apply the Hermite normal form algorithm (or the LLL algorithm) for such a generating matrix to obtain a basis of $\Lambda_q(\widetilde{A})$.

The LWE instance (8) can be viewed as a CVP instance over $\Lambda_q(\widetilde{A})$ targeting $\widetilde{t}$. In the cryptographic setting, the concatenated error vector $\widetilde{e}$ is significantly shorter than the modulus prime q, and its length $\|\widetilde{e}\|$ is bounded by some public information. Under such setting, the unknown vector $s\widetilde{A}$ is the closest in $\Lambda_q(\widetilde{A})$ to the concatenated target vector $\widetilde{t}$ since the difference is equal to $\widetilde{e} = \widetilde{t} - s\widetilde{A}$. Technically, it is known as the *bounded distance decoding (BDD)* problem, which is a specific case of CVP with a promise about the minimal distance of a target from a lattice.

3.2 NTRU Problem and Its Reduction

Here we commonly consider the cyclic case by $\phi = x^n - 1$ for a 2-power integer n to define the base ring $R = \mathbb{Z}[x]/(\phi)$ for the NTRU problem below.

Definition 2 (*NTRU* [12]) For a prime q, let $f(x), g(x) \in R$ be two polynomials with small coefficients in $\mathbb{Z}$ such that $f(x)$ is invertible in R_q. We then set $h(x) = g(x) \cdot f(x)^{-1}$ in R_q. The NTRU problem is the problem of finding $f(x), g(x)$ from $h(x)$ (or any equivalent solution $(x^i f(x), x^i g(x))$ for some $i \in \mathbb{Z})$.

Remark 3 In NTRUEncrypt [12], a basic NTRU public-key cryptosystem, two secret polynomials in R are randomly selected as

$$f(x) \in \mathcal{T}(m+1, m), \quad g(x) \in \mathcal{T}(m, m) \tag{10}$$

for a public parameter m, where $\mathcal{T}(s, t)$ denotes the set of ternary polynomials in R with s coefficients equal to 1, t coefficients equal to -1, and all other coefficients equal to 0 for two positive integers s, t with $s + t \leq n$. We here consider two secret polynomials $f(x), g(x) \in R$ satisfying (10). In addition, in basic NTRU cryptosystems, a 2-power integer is commonly used as the modulus of ciphertext space q for efficiency. For simple discussion, we consider an odd prime for q in this paper.

The NTRU lattice Let $\boldsymbol{h} = (h_0, h_1, \ldots, h_{n-1})$ denote the coefficient vector of the public polynomial $h(x) \in R_q$ in the NTRU problem. We use rotated vectors of $\boldsymbol{h}$ to construct an $n \times n$ matrix

$$\boldsymbol{H} = \begin{pmatrix} \boldsymbol{h} \\ \mathrm{rot}(\boldsymbol{h}) \\ \mathrm{rot}^2(\boldsymbol{h}) \\ \vdots \\ \mathrm{rot}^{n-1}(\boldsymbol{h}) \end{pmatrix} = \begin{pmatrix} h_0 & h_1 & h_2 & \cdots & h_{n-1} \\ h_{n-1} & h_0 & h_1 & \cdots & h_{n-2} \\ h_{n-2} & h_{n-1} & h_0 & \cdots & h_{n-3} \\ \vdots & \vdots & \vdots & \ddots & \vdots \\ h_1 & h_2 & h_3 & \cdots & h_0 \end{pmatrix}. \tag{11}$$

Since the coefficient vector of $x^i h(x) \in R_q$ is given as $\mathrm{rot}^i(\boldsymbol{h})$ for $0 \leq i \leq n-1$, we have $\boldsymbol{H}\boldsymbol{x}^\top = h(x)\boldsymbol{x}^\top \in (R_q)^n$ for the variable vector $\boldsymbol{x} = (1, x, x^2, \ldots, x^{n-1})$. We now consider a $2n \times 2n$ matrix

$$\boldsymbol{B} = \begin{pmatrix} q\boldsymbol{I}_n & \boldsymbol{O}_n \\ \boldsymbol{H} & \boldsymbol{I}_n \end{pmatrix}, \tag{12}$$

where $\boldsymbol{O}_n$ denotes the square zero matrix of size n. We take the so-called *NTRU lattice* $L = \mathcal{L}(\boldsymbol{B})$ that is spanned by the rows of $\boldsymbol{B}$.

Short vectors in the NTRU lattice For two secret polynomials $f(x), g(x) \in R_q$, the NTRU relation is written as $f(x)h(x) = g(x) + qr(x)$ in R for some polynomial $r(x) \in R$. Then we have

$$\begin{aligned} \boldsymbol{g}\boldsymbol{x}^\top &= g(x) = f(x)h(x) - qr(x) \\ &= \boldsymbol{f}\boldsymbol{x}^\top h(x) - q\boldsymbol{r}\boldsymbol{x}^\top = (\boldsymbol{f}\boldsymbol{H} - q\boldsymbol{r})\boldsymbol{x}^\top \in R_q, \end{aligned} \tag{13}$$

where $\boldsymbol{f}, \boldsymbol{g}$, and $\boldsymbol{r}$ are coefficient vectors of $f(x), g(x)$, and $r(x)$, respectively. Since the variable vector $\boldsymbol{x}$ gives a basis of R_q as a $\mathbb{Z}_q$-module, we have $\boldsymbol{g} = \boldsymbol{f}\boldsymbol{H} - q\boldsymbol{r}$. This shows the concatenated vector $(\boldsymbol{g} \mid \boldsymbol{f})$ is included in L since we have $(\boldsymbol{g} \mid \boldsymbol{f}) = (\boldsymbol{f}\boldsymbol{H} - q\boldsymbol{r} \mid \boldsymbol{f}) = (-\boldsymbol{r} \mid \boldsymbol{f})\boldsymbol{B} \in L$ from the form (12). More generally, we have $(\mathrm{rot}^i(\boldsymbol{g}) \mid \mathrm{rot}^i(\boldsymbol{f})) \in L$ for every $0 \leq i \leq n-1$. These concatenated vectors have the same length $\|(\mathrm{rot}^i(\boldsymbol{g}) \mid \mathrm{rot}^i(\boldsymbol{f}))\| = \sqrt{4m+1}$ when two secret polynomials $f(x)$, $g(x)$ are chosen from Eq. (10). In the general NTRU setting, these concatenated vectors are extremely short in the NTRU lattice L, and some of them could be recovered by applying a reduction algorithm for a basis of L.

4 Expanded Lattices for Ring-Based LWE and NTRU

In this section, we propose expanded lattices for solving the ring-based LWE and the NTRU problems. For both problems, we amplify target short vectors by rotations, and then construct lattices including all of them.

4.1 Expansion of Embedding for Ring-Based LWE Problem

As described in Sect. 3.1, the search ring-based LWE problem can be reduced to the BDD problem over a q-ary lattice. An embedding technique, due to Kannan [14], converts the BDD problem to the unique-SVP problem that is a particular case of SVP with a promise about the gap between the first and the second successive minimum of a lattice. Here we expand Kannan's embedding for solving the search ring-based LWE problem using rotations.

Expansion of embedding lattices We consider the LWE instance $(\widetilde{\boldsymbol{A}}, \widetilde{\boldsymbol{t}})$ of size $n \times d$ from Eq. (8) that is constructed from m ring-based LWE samples. Let $\boldsymbol{B}$ be a basis of the q-ary lattice $\Lambda_q(\widetilde{\boldsymbol{A}})$. For an expansion parameter $k \geq 1$ and a small constant η, we construct a square matrix of size $d + k$ as

$$C_k = \begin{pmatrix} \boldsymbol{B} & 0 & 0 & \cdots & 0 \\ \widetilde{\boldsymbol{t}} & \eta & 0 & \cdots & 0 \\ \mathrm{rot}(\widetilde{\boldsymbol{t}}) & 0 & \eta & \cdots & 0 \\ \vdots & \vdots & \vdots & \ddots & \vdots \\ \mathrm{rot}^{k-1}(\widetilde{\boldsymbol{t}}) & 0 & 0 & \cdots & \eta \end{pmatrix}, \quad (14)$$

where $\mathrm{rot}^i(\widetilde{\boldsymbol{t}})$ is the i times rotated vector of the concatenated target vector $\widetilde{\boldsymbol{t}} = (\boldsymbol{t}_1 \mid \cdots \mid \boldsymbol{t}_m)$ as $\mathrm{rot}^i(\widetilde{\boldsymbol{t}}) = (\mathrm{rot}^i(\boldsymbol{t}_1) \mid \cdots \mid \mathrm{rot}^i(\boldsymbol{t}_m))$ for $1 \leq i \leq k - 1$. The lattice spanned by the rows of C_k is denoted by $\bar{\Lambda}_k$, whose dimension is $d + k$. In particular, $\bar{\Lambda}_k$ is the original embedding lattice when $k = 1$.

Expanded Lattices for Solving Ring-Based LWE and NTRU Problems 283

Short vectors in the expanded embedding lattice For $0 \leq i \leq k-1$, m ring-based LWE relations $x^i t_j(x) = x^i s(x) \cdot a_j(x) + x^i e_j(x)$ $(1 \leq j \leq m)$ multiplied by x^i are transformed as

$$\operatorname{rot}^i(\tilde{t}) \equiv \operatorname{rot}^i(s)\tilde{A} + \operatorname{rot}^i(\tilde{e}) \pmod{q} \tag{15}$$

from the same argument as for Eq. (7), where $\operatorname{rot}^i(\tilde{e})$ is the i times rotated vector of the concatenated error vector $\tilde{e} = (e_1 \mid \cdots \mid e_m)$ (cf., $\operatorname{rot}^i(s)$ is the i times rotated vector of the secret vector s). We see from Eq. (15) and the form of C_k that $\bar{\Lambda}_k = \mathcal{L}(C_k)$ contains k short vectors

$$\begin{cases} \bar{e} = (\tilde{e} \mid \eta, 0, \ldots, 0), \\ \operatorname{rot}(\bar{e}) = (\operatorname{rot}(\tilde{e}) \mid 0, \eta, \ldots, 0), \\ \quad \vdots \\ \operatorname{rot}^{k-1}(\bar{e}) = \left(\operatorname{rot}^{k-1}(\tilde{e}) \mid 0, \ldots, 0, \eta\right). \end{cases} \tag{16}$$

Since any rotation does not alter the length of a vector, these lattice vectors are all the same length. In other words, k unusual short vectors (16) are embedded in $\bar{\Lambda}_k$, on which it is no longer the unique-SVP for $k \geq 2$. We note that the number of rotation operations performed on the concatenated error vector $\tilde{e}$ is indicated by the position of η in the latter components of (8).

Remark 4 Several combinations of vectors in (16) with small integers might be non-zero shortest in the expanded embedding lattice $\bar{\Lambda}_k$. When such vectors are recovered, we can build a system of linear equations by looking at the positions of η with scalars to recover the non-rotated error $\tilde{e}$. However, when employing small k, it is unlikely for any combination of vectors in (16) to be shorter than those in $\bar{\Lambda}_k$. (Indeed, from preliminary experiments for $1 \leq k \leq 5$, the vectors in (16) seem to be the non-zero shortest in $\bar{\Lambda}_k$.)

Remark 5 *(Expansion of Bai-Galbraith's embedding)* Bai-Galbraith's embedding [5] is a refinement of Kannan's embedding in solving the LWE problem with a small secret vector. Here we will expand Bai-Galbraith's embedding in a similar manner to the above by utilizing rotated vectors. Given an LWE instance (8) of size $n \times d$ with small secret vector s, we construct a square matrix of size $d+n+k$ like Eq. (14) as

$$C'_k = \begin{pmatrix} qI_d & 0 & 0 & 0 & \cdots & 0 \\ -\tilde{A} & I_n & 0 & 0 & \cdots & 0 \\ \tilde{t} & 0 & \eta & 0 & \cdots & 0 \\ \operatorname{rot}(\tilde{t}) & 0 & 0 & \eta & \cdots & 0 \\ \vdots & \vdots & \vdots & \vdots & \ddots & \vdots \\ \operatorname{rot}^{k-1}(\tilde{t}) & 0 & 0 & 0 & \cdots & \eta \end{pmatrix}.$$

The lattice spanned by the rows of C'_k is denoted by $\bar{\Lambda}'_k$, whose dimension is $d + n + k$. In particular, $\bar{\Lambda}'_k$ is just the original embedding lattice due to Bai-Galbraith when $k = 1$. We see by Eq. (15) that the expanded embedding lattice $\bar{\Lambda}'_k = \mathcal{L}(C'_k)$ contains k short vectors

$$\begin{cases} \bar{e}' = (\widetilde{e} \mid s \mid \eta, 0, \ldots, 0), \\ \operatorname{rot}(\bar{e}') = (\operatorname{rot}(\widetilde{e})) \mid \operatorname{rot}(s) \mid 0, \eta, \ldots, 0), \\ \quad \vdots \\ \operatorname{rot}^{k-1}(\bar{e}') = (\operatorname{rot}^{k-1}(\widetilde{e})) \mid \operatorname{rot}^{k-1}(s) \mid 0, 0, \ldots, \eta). \end{cases}$$

Remark 6 (*Application to module-LWE*) A counterpart of LWE over modules was introduced in [6, 15] as Module-LWE that lies between LWE and ring-based LWE. For $r \geq 1$, it employs a free R_q-module of rank r. Specifically, a module-LWE sample is a pair $(a, t) \in R_q^r \times R_q$ with $a = (a_1(x), \ldots, a_r(x))$ that satisfies $t(x) = \sum_{i=1}^r a_i(x) \cdot s_i(x) + e(x)$ over R_q for some secret vector $s = (s_1(x), \ldots, s_r(x)) \in R_q^r$ and some error polynomial $e(x) \in R_q$. (Especially, the case $r = 1$ defines ring-based LWE.) As in Sect. 3.1, the module-LWE relation can be represented in the matrix form as

$$t \equiv (s_1 \mid \cdots \mid s_r) \begin{pmatrix} A_1 \\ \vdots \\ A_r \end{pmatrix} + e \pmod{q},$$

where t, e, and s_i's are coefficient vectors of $t(x)$, $e(x)$, and $s_i(x)$'s, respectively, and A_i's are matrices corresponding to $a_i(x)$'s like Eq. (5). With this matrix form, we can apply Kannan's or Bai-Galbraith's embedding for solving a module-LWE instance, and we can expand such an embedding as above.

Recovery of short lattice vectors by using BKZ We consider the procedure of the BKZ reduction algorithm for a basis of the expanded embedding lattice $\bar{\Lambda}_k$ to find any of k short vectors $v = \operatorname{rot}^h(\bar{e}) \in \bar{\Lambda}_k$ in (16) for $0 \leq h \leq k - 1$. Once such v is found, we can recover the concatenated error vector $\widetilde{e}$ by looking at the position of η in the components of v, and then identify the secret vector s from the LWE relation (8). For a blocksize $\beta \geq 50$, we consider an almost β-BKZ-reduced basis $B = (b_1, \ldots, b_{d+k})$ of $\bar{\Lambda}_k$ such that its basis vectors b_i's are not equal to any of short vectors in (16). For simple discussion, we assume from Eq. (3) that Gram–Schmidt vectors $(b_1^\star, \ldots, b_{d+k}^\star)$ of B have length

$$\|b_i^\star\| \approx \delta_\beta^{d+k-1-2i} \operatorname{vol}(\bar{\Lambda}_k)^{\frac{1}{d+k}}$$

for every $1 \leq i \leq d + k$, where δ_β is the constant in (3). Since the volume of the q-ary lattice $\Lambda_q(\widetilde{A})$ is equal to q^{d-n} for almost all $\widetilde{A}$, the volume of $\bar{\Lambda}_k$ is estimated as $\eta^k q^{d-n}$ from the form (14) of a basis of $\bar{\Lambda}_k$. As discussed in [2, 4], if the projected vector of v at index $d+k-\beta$ holds

$$\|\pi_{d+k-\beta}(v)\| < \|b^*_{d+k-\beta}\| \approx \delta_\beta^{2\beta-d-k-1} \mathrm{vol}(\bar{\Lambda}_k)^{\frac{1}{d+k}}, \tag{17}$$

then $\pi_{d+k-\beta}(v)$ is a non-zero shortest vector in the last projected block lattice $L_{[d+k-\beta:d+k]} = \mathcal{L}(B_{[d+k-\beta:d+k]})$. In the procedure of the BKZ algorithm, the projected vector $\pi_{d+k-\beta}(v)$ is found by (pruned) ENUM for the last block $B_{[d+k-\beta:d+k]}$, and then it becomes the Gram–Schmidt vector at index $d+k-\beta$ (that is, $\pi_{d+k-\beta}(v) = b^*_{d+k-\beta}$). Since every projected vector of v is the non-zero shortest in the other projected block lattices, the whole target vector v is recovered by performing ENUM over the remaining blocks. (See Fig. 1 for an illustration of such a procedure.) In particular, since there exists k candidates of v in the expanded embedding lattice $\bar{\Lambda}_k$, we expect that the success probability of finding $\pi_{d+k-\beta}(v)$ by pruned ENUM over the last block $B_{[d+k-\beta:d+k]}$ would grow as the expansion parameter k increases.

Remark 7 In the literature, the above explanation is referred to as "the 2016 estimate" by [4], and it was empirically verified in [3, 21] that a target short lattice vector can be retrieved with a high probability by the BKZ algorithm for a blocksize β satisfying (17).

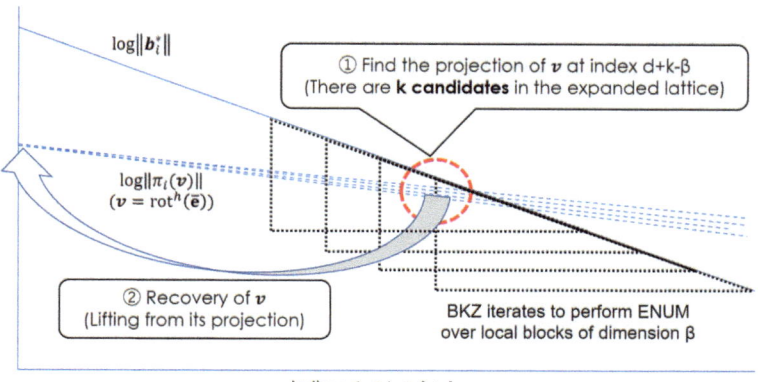

Fig. 1 An illustration of the Gram–Schmidt log-lengths $\log \|b^*_i\|$'s of a β-BKZ-reduced basis $B = (b_1, \ldots, b_{d+k})$ of the expanded lattice $\bar{\Lambda}_k$ to recover any of k short lattice vectors $v = \mathrm{rot}^h(\bar{e}) \in \bar{\Lambda}_k$ in (16) for $0 \leq h \leq k-1$

4.2 Expansion of the NTRU Lattice

Here we propose an expansion of the NTRU lattice for solving the NTRU problem. Similar to the previous subsection for the ring-based LWE problem, we amplify target short vectors by using rotations, and also construct an expanded NTRU lattice including those vectors.

Expanded NTRU lattice We use an expansion parameter $\ell \geq 0$ to expand the NTRU lattice L of dimension $2n$, defined in Sect. 3.2. Specifically, we use the rotated matrix H of the public vector h, defined by (11), to build a square matrix B_ℓ of size $2n + \ell$ as (cf., the form (12))

$$B_\ell = \begin{pmatrix} qI_n & O_n & O_{n\times\ell} \\ H & I_n & O_{n\times\ell} \\ h & 0_{1\times n} & 1\ 0\ 0\ \cdots\ 0 \\ \mathrm{rot}(h) & 0_{1\times n} & 0\ 1\ 0\ \cdots\ 0 \\ \mathrm{rot}^2(h) & 0_{1\times n} & 0\ 0\ 1\ \cdots\ 0 \\ \vdots & \vdots & \vdots\ \vdots\ \vdots\ \ddots\ \vdots \\ \mathrm{rot}^{\ell-1}(h) & 0_{1\times n} & 0\ 0\ 0\ \cdots\ 1 \end{pmatrix} = \begin{pmatrix} qI_n & O_{n\times(n+\ell)} \\ H_\ell & I_{n+\ell} \end{pmatrix}, \qquad (18)$$

where $0_{s\times t}$ denotes the zero matrix of size $s \times t$. We remark that H_ℓ is the $(n + \ell) \times n$ matrix in which the top ℓ rows of the original rotated matrix H are repeatedly added below the matrix. We denote by $L_\ell = \mathcal{L}(B_\ell)$ the lattice spanned by the $2n + \ell$ rows of B_ℓ to consider it as an expanded NTRU lattice, defining the original NTRU lattice when $\ell = 0$.

Short vectors in the expanded NTRU lattice Similar to Eq. (13), we see from the NTRU relation $f(x)h(x) = g(x) + qr(x)$ in the ring R with some $r(x) \in R$ that for any i, we have

$$\begin{aligned}
&\mathrm{rot}^i(g)x^\top \\
&= x^i g(x) = x^i f(x)h(x) - qx^i r(x) \\
&= f \begin{pmatrix} x^i \\ x^{i+1} \\ \vdots \\ x^{n+i-1} \end{pmatrix} h(x) - q\mathrm{rot}^i(r)x^\top = \left\{ f \begin{pmatrix} \mathrm{rot}^i(h) \\ \mathrm{rot}^{i+1}(h) \\ \vdots \\ \mathrm{rot}^{n+i-1}(h) \end{pmatrix} - q\mathrm{rot}^i(r) \right\} x^\top.
\end{aligned}$$

(Recall that for any element $p(x) \in R_q$ with its coefficient vector p, we have $x^j p(x) = \mathrm{rot}^j(p)x^\top$ for any j.) This shows

$$\mathrm{rot}^i(g) = f\mathrm{rot}^i(H) - q\mathrm{rot}^i(r). \qquad (19)$$

Here $\text{rot}^i(\boldsymbol{H})$ denotes the $n \times n$ matrix whose rows are i times rotated rows of $\boldsymbol{H}$. We remark that the $n \times n$ matrix $\text{rot}^i(\boldsymbol{H})$ is a sub-matrix of the $(n + \ell) \times n$ matrix $\boldsymbol{H}_\ell$ for each $0 \le i \le \ell$. Therefore we see from the form (18) of $\boldsymbol{B}_\ell$ that the expanded NTRU lattice $L_\ell = \mathcal{L}(\boldsymbol{B}_\ell)$ includes $\ell + 1$ short vectors

$$\begin{cases} (\boldsymbol{g} \mid \boldsymbol{f}, 0, 0, \ldots, 0), \\ (\text{rot}(\boldsymbol{g}) \mid 0, \boldsymbol{f}, 0, \ldots, 0), \\ (\text{rot}^2(\boldsymbol{g}) \mid 0, 0, \boldsymbol{f}, \ldots, 0), \\ \quad \vdots \\ (\text{rot}^\ell(\boldsymbol{g}) \mid 0, 0, 0, \ldots, \boldsymbol{f}). \end{cases} \qquad (20)$$

For example, by Eq. (19), we have

$$\begin{aligned} (\text{rot}(\boldsymbol{g}) \mid 0, \boldsymbol{f}, 0, \ldots, 0) &= (-q\text{rot}(\boldsymbol{r}) + \boldsymbol{f}\text{rot}(\boldsymbol{H}) \mid 0, \boldsymbol{f}, 0, \ldots, 0) \\ &= (-\text{rot}(\boldsymbol{r}) \mid 0, \boldsymbol{f}, 0, \ldots, 0)\boldsymbol{B}_\ell \in L_\ell. \end{aligned}$$

(We note from Eq. (18) that the one time rotated matrix $\text{rot}(\boldsymbol{H})$ lies in $\boldsymbol{H}_\ell$ from the second to the $n + 1$ rows.) As in the original NTRU lattice, all rotated vectors of (20) are also included in the expanded NTRU lattice L_ℓ. Specifically, the lattice L_ℓ includes $(\ell + 1)n$ short vectors

$$\begin{cases} (\text{rot}^i(\boldsymbol{g}) \mid \text{rot}^i(\boldsymbol{f}), 0, 0, \ldots, 0), & (0 \le i \le n - 1), \\ (\text{rot}^{i+1}(\boldsymbol{g}) \mid 0, \text{rot}^i(\boldsymbol{f}), 0, \ldots, 0), & (0 \le i \le n - 1), \\ (\text{rot}^{i+2}(\boldsymbol{g}) \mid 0, 0, \text{rot}^i(\boldsymbol{f}), \ldots, 0), & (0 \le i \le n - 1), \\ \quad \vdots \\ (\text{rot}^{i+\ell}(\boldsymbol{g}) \mid 0, 0, 0, \ldots, \text{rot}^i(\boldsymbol{f})), & (0 \le i \le n - 1). \end{cases} \qquad (21)$$

Since the length of a vector does not change by the rotation operation, lengths of $(\ell + 1)n$ vectors in (21) are all the same, and they are equal to $\|(\boldsymbol{g} \mid \boldsymbol{f})\|$, which is the length of concatenation of coefficients vectors of two secret polynomials $f(x), g(x)$ in the NTRU problem. In other words, the expanded NTRU lattice L_ℓ includes $(\ell + 1)n$ vectors with all lengths equal to $\|(\boldsymbol{g} \mid \boldsymbol{f})\|$.

Lattice basis reduction for the expanded NTRU lattice Here we consider applying the BKZ algorithm for a basis of the expanded NTRU lattice L_ℓ of dimension $2n + \ell$ to recover any of the short lattice vectors in (21). Let $(\boldsymbol{b}_1, \ldots, \boldsymbol{b}_{2n+\ell})$ be a β-BKZ-reduced basis of L_ℓ for a blocksize $\beta > 50$, and $(\boldsymbol{b}_1^\star, \ldots, \boldsymbol{b}_{2n+\ell}^\star)$ its Gram–Schmidt vectors. From a similar argument in Sect. 4.1 for the ring-based LWE, the blocksize β of BKZ requires to satisfy (cf., Eq. (17))

$$\|\pi_{2n+\ell-\beta}(\boldsymbol{v})\| < \|\boldsymbol{b}_{2n+\ell-\beta}^\star\| \approx \delta_\beta^{2\beta - 2n - \ell - 1} \text{vol}(L_\ell)^{\frac{1}{2n+\ell}} \qquad (22)$$

for recovering some vector $v \in L_\ell$ in (21), where δ_β denote the same constant as in Eq. (3). In particular, we see from the form of (18) that the volume of $L_\ell = \mathcal{L}(B_\ell)$ is given by q^n, independent of the extension parameter ℓ. When two secret polynomials $f(x)$ and $g(x)$ are chosen as (10), the norm of v is equal to $\sqrt{4m+1}$ and that of its projected vector at index $2n + \ell - \beta$ is estimated as $\|\pi_{2n+\ell-\beta}(v)\| \approx \frac{\sqrt{\beta}}{\sqrt{2n+\ell}}\|v\| = \frac{\sqrt{\beta(4m+1)}}{\sqrt{2n+\ell}}$ according to [3]. From a similar argument in Sect. 4.1, once the projected vector $\pi_{2n+\ell-\beta}(v)$ at $2n + \ell - \beta$ is recovered by (pruned) ENUM over the last β-dimensional projected lattice $\pi_{2n+\ell-\beta}(L_\ell)$ in the procedure of BKZ, the whole vector $v \in L_\ell$ can be restored by other ENUM subroutines. However, we see from the form of (18) that the expanded NTRU lattice $L_\ell = \mathcal{L}(B_\ell)$ includes 2ℓ trivial non-zero shortest vectors of length $\sqrt{2}$

$$\begin{cases} (\mathbf{0}_{1 \times n} \mid \pm 1, 0, 0, \ldots, 0, 0, \ldots, 0 \mid \mp 1, 0, 0, \ldots, 0), \\ (\mathbf{0}_{1 \times n} \mid 0, \pm 1, 0, \ldots, 0, 0, \ldots, 0 \mid 0, \mp 1, 0, \ldots, 0), \\ \vdots \\ (\mathbf{0}_{1 \times n} \mid 0, 0, 0, \ldots, \pm 1, 0, \ldots, 0 \mid 0, 0, 0, \ldots, \mp 1), \end{cases}$$

where the first, second, and third blocks of every vector have lengths n, n, and ℓ, respectively. Therefore in a β-BKZ-reduced basis of L_ℓ, the top ℓ vectors are given by the above ℓ linearly independent vectors excluding the sign difference, and several short vectors in (21) appear as basis vectors after them. Moreover, we can extract the first block of length n from some $v \in L_\ell$ in (21) to get $\mathrm{rot}^j(g)$, and also know the index j by looking at the last block of v to recover the coefficient vector g of the secret polynomial $g \in R_q$.

5 Experiments

In this section, we implemented our expanded embeddings for solving ring-based LWE and NTRU problems in **SageMath** [10], the Sage mathematics software. We conducted all experiments on one core of an Intel Xeon Gold 6240 CPU @ 2.60GHz with 128 GByte memory.

Ring-based LWE case Given parameters (n, q, m) of ring-based LWE with $d = mn$, we generated m ring-LWE samples $\{(a_i(x), t_i(x)\}_{i=1}^m$ with unknown common secret $s(x)$ and errors $e_i(x)$ by using the ring-LWE oracle generator in **SageMath**. (For each $1 \leq i \leq m$, it holds $t_i(x) = a_i(x) \cdot s(x) + e_i(x)$.) We set the generator so that every coefficient of $s(x)$ is chosen uniformly from $\mathbb{Z}_q$, and every coefficient of each $e_i(x)$ is sampled from the discrete Gaussian sampler with mean zero and standard deviation σ for a fixed constant $\sigma > 0$. We expressed the m ring-LWE samples in the form of (8) to construct the basis C_k of our expanded lattice $\bar{\Lambda}_k$ of dimension $d + k$ for an expansion parameter $k \geq 1$ (see Eq. (14) for the form of C_k). We set $\eta = \lfloor \sigma \rceil$

for C_k. We reduced C_k by using BKZ 2.0 implemented in fpylll [25] (available in SageMath) to find any of k short vectors (16) embedded in $\bar{\Lambda}_k = \mathcal{L}(C_k)$. We also adopted the loop_max option to limit the number of loop iterations of BKZ for efficiency. For our experiments, we chose the size of β smaller than the minimum blocksize satisfying Eq. (17). (We estimated the left-hand side of (17) as $\sigma\sqrt{\beta}$ according to [3].) We judged that our approach succeeded by confirming that the first basis vector of a BKZ-reduced basis has a length less than $1.2\sigma\sqrt{d}$ with all entries at most 4σ in absolute. (All short vectors in (16) have the same length that is roughly expected as $\sigma\sqrt{d}$.)

In Table 1, we summarize experimental results on the success probability and the average running time for solving ring-based LWE instances by our expanded embedding with BKZ. For each parameter setting in Table 1, we conducted 100 times experiments for randomly chosen instances. From Table 1, the success probability is the highest in case $k = 2$ or 3. On the other hand, as k increases, the running time also increases slightly since the dimension of $\bar{\Lambda}_k = d + k$ increases. In Fig. 2, we show transitions of success probabilities of our expanded embedding with expansion parameters $1 \leq k \leq 5$ by BKZ with different blocksizes $50 \leq \beta \leq 64$. Figure 2 demonstrates that for most of β, the success probabilities in cases $k = 2, 3, 4$ are higher than those in the case $k = 1$ that gives the original Kannan's embedding.

Table 1 The success probability and the average running time (seconds) for solving ring-based LWE instances by our expanded embedding using BKZ with blocksizes $\beta \geq 50$ (Here "loop" denotes the maximum number of loop iterations in BKZ)

Experimental parameters			Expansion parameter for $\bar{\Lambda}_k = \mathcal{L}(C_k)$				
Ring-based LWE		BKZ	$k = 1$	$k = 2$	$k = 3$	$k = 4$	$k = 5$
$n = 32$	$\sigma = 6.0$	$\beta = 50$	20%	**44%**	38%	32%	34%
$q = 257$		loop $= 2$	5.6 s	5.5 s	5.5 s	5.5 s	5.7 s
$m = 3$	$\sigma = 8.0$	$\beta = 65$	10%	**19%**	14%	12%	10%
($d = 96$)		loop $= 2$	79.3 s	82.4 s	83.5 s	88.1 s	89.2 s
$n = 32$	$\sigma = 10.0$	$\beta = 50$	38%	**53%**	46%	46%	37%
$q = 577$		loop $= 2$	4.4 s	4.6 s	4.7 s	4.9 s	4.9 s
$m = 3$	$\sigma = 11.0$	$\beta = 65$	69%	**75%**	**75%**	69%	65%
($d = 96$)		loop $= 2$	80.1 s	80.3 s	81.7 s	81.9 s	82.5 s
$n = 64$	$\sigma = 1.7$	$\beta = 50$	21%	21%	**27%**	18%	21%
$q = 257$		loop $= 4$	16.6 s	16.8 s	17.0 s	17.2 s	17.4 s
$m = 2$	$\sigma = 2.0$	$\beta = 60$	24%	36%	**38%**	20%	17%
($d = 128$)		loop $= 4$	74.7 s	74.5 s	74.0 s	74.7 s	74.8 s
$n = 64$	$\sigma = 4.0$	$\beta = 55$	18%	**22%**	21%	19%	17%
$q = 1153$		loop $= 4$	29.5 s	30.2 s	30.4 s	30.8 s	31.5 s
$m = 2$	$\sigma = 4.6$	$\beta = 65$	24%	**32%**	26%	24%	21%
($d = 128$)		loop $= 4$	323.4 s	327.9 s	332.2 s	340.7 s	341.1 s

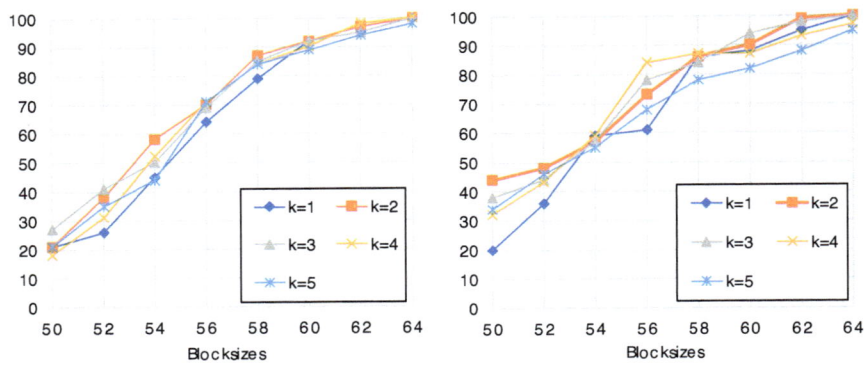

(a) Experimental parameters: $n = 32$, $q = 257$, $m = 3$, $d = 96$, $\sigma = 6.0$, loop $= 2$

(b) Experimental parameters: $n = 64$, $q = 257$, $m = 2$, $d = 128$, $\sigma = 1.7$, loop $= 4$

Fig. 2 Transitions of success probabilities of our expanded embedding for solving ring-based LWE instances by BKZ with different blocksizes $50 \leq \beta \leq 64$

Table 2 The success probability and the average running time (seconds) of our expanded attack using BKZ with $\beta = 60$ for solving NTRU instances

Parameters	Expansion parameter for $L_\ell = \mathcal{L}(\boldsymbol{B}_\ell)$			
(n, q, m)	$\ell = 0$	$\ell = 1$	$\ell = 2$	$\ell = 3$
(64, 31, 18)	31% (33.5 s)	**36%** (33.7 s)	32% (34.2 s)	31% (34.9 s)
(64, 41, 23)	46% (34.0 s)	**52%** (34.4 s)	38% (34.4 s)	42% (34.7 s)
(64, 53, 28)	65% (34.8 s)	71% (34.3 s)	**78%** (35.5 s)	67% (35.7 s)
(72, 31, 14)	71% (62.9 s)	**78%** (62.6 s)	68% (63.0 s)	74% (63.0 s)
(72, 41, 19)	52% (63.8 s)	**58%** (64.2 s)	48% (65.5 s)	51% (64.8 s)
(72, 53, 27)	18% (65.7 s)	15% (64.9 s)	13% (65.9 s)	**21%** (66.4 s)
(80, 67, 25)	41% (106.1 s)	**48%** (106.8 s)	42% (111.2 s)	45% (111.3 s)
(80, 89, 31)	69% (108.6 s)	**80%** (110.2 s)	75% (111.0 s)	70% (111.0 s)
(80, 101, 36)	66% (108.2 s)	**74%** (109.2 s)	62% (112.2 s)	69% (111.2 s)

NTRU case In Table 2, we summarize experimental results on the success probability and the average running time for solving NTRU instances by using BKZ for our expanded NTRU lattice L_ℓ for an expansion parameter $\ell \geq 0$. For NTRU parameters (n, q, m), we randomly chose two secret polynomials $f(x)$ and $g(x)$ satisfying Eq. 10 to generate an NTRU instance with the public polynomial $h(x) = g(x) \cdot f(x)^{-1} \in R_q$. Given each pair (n, q) in Table 2, we chose m so that a blocksize of BKZ is required at least $\beta = 70$ to satisfy Eq. (22) in the case $\ell = 0$. For our experiments, we used $\beta = 60$ to reduce the basis $\boldsymbol{B}_\ell$ of L_ℓ (see Eq. (18) for the form of $\boldsymbol{B}_\ell$). In addition, we set loop $= 2, 3, 4$ as the maximum number of loops in BKZ for $n = 64, 72, 80$, respectively. We conducted 100 times experiments for every parameter setting. As in the ring-based LWE case, Table 2 demonstrates that the success probability of solving NTRU instances is the highest by $\ell \neq 0$. (Recall that $\ell = 0$ defines the original NTRU lattice.) Moreover, as ℓ increases, the running time also increases slightly due to dim $L_\ell = 2n + \ell$.

Discussion A larger expansion parameter k (resp., ℓ) enables our expanded lattice $\bar{\Lambda}_k$ (resp., L_ℓ) to contain more target short vectors for solving the ring-based LWE (resp., NTRU) problem. As discussed in Sect. 4, we expect that a larger expansion parameter would increase the success probability of finding any of target short lattice vectors. When we use a large blocksize $\beta \geq 50$ of BKZ, a pruning strategy is adopted for ENUM on projected blocks of dimension β as the main subroutine of BKZ (see [11] for pruning strategies). Since the number of target short lattice vectors increases by k (resp., $\ell + 1$) times by our expansion for ring-based LWE (resp., NTRU), it might increase by at most k (resp., $\ell + 1$) times the success probability of finding a projected target vector on the last β-dimensional block by pruned ENUM. However, experimental results of Table 1 (resp., Table 2) show that the success probability is high at $k = 2, 3$ (resp., $\ell = 1, 2$) for solving the ring-based LWE (resp., NTRU) problem. In other words, a larger expansion parameter does not necessarily increase the success probability. This is due to that a larger expansion parameter increases the dimension of our expanded lattice, and thus it requires a larger blocksize β of BKZ for the success of finding a target short lattice vector (see Eqs. (17) and (22)). Thus $k = 2, 3$ (resp., $\ell = 1, 2$) are suitable when we use blocksizes from $\beta = 50$ to 65. In contrast, larger expansion parameters might be useful with a higher success probability in using larger β since the success probability of pruned ENUM is set to be lower for larger β.

6 Conclusion

For an expansion parameter $k \geq 1$ (resp., $\ell \geq 0$), we constructed an expanded Kannan's embedding lattice $\bar{\Lambda}_k$ (resp., expanded NTRU lattice L_ℓ) for solving the ring-based LWE (resp., NTRU) problem. In particular, the case $k = 1$ (resp., $\ell = 0$) defines the original embedding (resp., NTRU) lattice. Both expanded lattices contain many short vectors that are amplified by rotations of secret short vectors. Our experimental results demonstrated from Table 1 (resp., Table 2) that $k = 2, 3$ (resp., $\ell = 1, 2$) give a higher success probability around 10% than $k = 1$ (resp., $\ell = 0$) for solving ring-based LWE (resp., NTRU) instances by using BKZ with blocksizes from $\beta = 50$ to 65. (As discussed in the previous section, larger expansion parameters might give a higher success probability in using larger β.) In contrast, the running time of BKZ increases slightly for such small expansion parameters.

Acknowledgements This work was supported by JST CREST Grant Number JPMJCR2113 and JSPS KAKENHI Grant Number JP20H04142, Japan.

References

1. M. Ajtai, The shortest vector problem in L_2 is NP-hard for randomized reductions, in *Symposium on Theory of Computing (STOC 1998)* (1998), pp. 10–19
2. M.R. Albrecht, L. Ducas, Lattice attacks on NTRU and LWE: a history of refinements. IACR ePrint 2021/799 (2021)
3. M.R. Albrecht, F. Göpfert, F. Virdia, T. Wunderer, Revisiting the expected cost of solving uSVP and applications to LWE, in *Advances in Cryptology–EUROCRYPT 2017*. Lecture Notes in Computer Science, vol. 10624 (Springer, 2017), pp. 297–322
4. E. Alkim, L. Ducas, T. Pöppelmann, P. Schwabe, Post-quantum key exchange: a new hope, in *25th USENIX Security Symposium* (2016), pp. 327–343
5. S. Bai, S.D. Galbraith, Lattice decoding attacks on binary LWE, in *Australasian Conference on Information Security and Privacy (ACISP 2014)*. Lecture Notes in Computer Science, vol. 8544 (Springer, 2014), pp. 322–337
6. Z. Brakerski, C. Gentry, V. Vaikuntanathan, (Leveled) fully homomorphic encryption without bootstrapping. ACM Trans. Comput. Theory (TOCT) **6**(3), 1–36 (2014)
7. W. Castryck, I. Iliashenko, F. Vercauteren, On error distributions in ring-based LWE. LMS J. Comput. Math. **19**(A), 130–145 (2016)
8. Y. Chen, Réduction de réseau et sécurité concrete du chiffrement completement homomorphe. Ph.D. thesis, Paris 7 (2013)
9. Y. Chen, P.Q. Nguyen, BKZ 2.0: better lattice security estimates, in *Advances in Cryptology–ASIACRYPT 2011*. Lecture Notes in Computer Science, vol. 7073 (Springer, 2011), pp. 1–20
10. Developers, T Sage: Sagemath (2016), https://www.sagemath.org/
11. N. Gama, P.Q. Nguyen, O. Regev, Lattice enumeration using extreme pruning, in *Advances in Cryptology–EUROCRYPT 2010*. Lecture Notes in Computer Science, vol. 6110 (Springer, 2010), pp. 257–278
12. J. Hoffstein, J. Pipher, J.H. Silverman, NTRU: a ring-based public key cryptosystem, in *Algorithmic Number Theory Symposium (ANTS III)*. Lecture Notes in Computer Science, vol. 1423 (Springer, 1998), pp. 267–288
13. J. Hoffstein, J. Pipher, J.H. Silverman, An introduction to mathematical cryptography, vol. 1 (Springer, 2008)
14. R. Kannan, Minkowski's convex body theorem and integer programming. Math. Oper. Res. **12**(3), 415–440 (1987)
15. A. Langlois, D. Stehlé, Worst-case to average-case reductions for module lattices. Designs, Codes Crypt. **75**(3), 565–599 (2015)
16. A.K. Lenstra, H.W. Lenstra, L. Lovász, Factoring polynomials with rational coefficients. Mathematische Annalen **261**(4), 515–534 (1982)
17. V. Lyubashevsky, C. Peikert, O. Regev, On ideal lattices and learning with errors over rings, in *Advances in Cryptology–EUROCRYPT 2010*. Lecture Notes in Computer Science, vol. 6110 (Springer, 2010), pp. 1–23
18. D. Micciancio, S. Goldwasser, Complexity of lattice problems: a cryptographic perspective, vol. 671 (Springer, 2012)
19. D. Micciancio, O. Regev, Lattice-based cryptography. Post-Quantum Cryptography (2009), pp. 147–191
20. S. Nakamura, M. Yasuda, An extension of Kannan's embedding for solving ring-based LWE problems, in *IMA International Conference on Cryptography and Coding (IMACC 2021)*. Lecture Notes in Computer Science, vol. 13129 (Springer, 2021), pp. 201–219
21. E.W. Postlethwaite, F. Virdia, On the success probability of solving unique SVP via BKZ, in *Public-Key Cryptography–PKC 2021*. Lecture Notes in Computer Science, vol. 12710 (Springer, 2021), pp. 68–98
22. C.P. Schnorr, Lattice reduction by random sampling and birthday methods, in *Symposium on Theoretical Aspects of Computer Science (STACS 2003)*. Lecture Notes in Computer Science, vol. 2607 (Springer, 2003), pp. 145–156

23. C.P. Schnorr, M. Euchner, Lattice basis reduction: improved practical algorithms and solving subset sum problems. Math. Program. **66**, 181–199 (1994)
24. D. Stehlé, R. Steinfeld, K. Tanaka, K. Xagawa, Efficient public key encryption based on ideal lattices, in *Advances in Cryptology–ASIACRYPT 2009*. Lecture Notes in Computer Science, vol. 5912 (Springer, 2009), pp. 617–635
25. The FPLLL development team: fpylll, a Python wraper for the fplll lattice reduction library, Version: 0.5.6 (2021), https://github.com/fplll/fpylll, https://github.com/fplll/fpylll
26. Y. Yu, L. Ducas, Second order statistical behavior of LLL and BKZ, in *Selected Areas in Cryptography (SAC 2017)*. Lecture Notes in Computer Science, vol. 10719 (Springer, 2017), pp. 3–22

Open Access This chapter is licensed under the terms of the Creative Commons Attribution 4.0 International License (http://creativecommons.org/licenses/by/4.0/), which permits use, sharing, adaptation, distribution and reproduction in any medium or format, as long as you give appropriate credit to the original author(s) and the source, provide a link to the Creative Commons license and indicate if changes were made.

The images or other third party material in this chapter are included in the chapter's Creative Commons license, unless indicated otherwise in a credit line to the material. If material is not included in the chapter's Creative Commons license and your intended use is not permitted by statutory regulation or exceeds the permitted use, you will need to obtain permission directly from the copyright holder.

Analysis of (U,U+V)-Code Problem with Gramian Over Binary and Ternary Fields

Ichiro Iwata, Yusuke Yoshida, and Keisuke Tanaka

Abstract Debris-Alazard, Sendrier, and Tillich proposed SURF, which is a code-based signature scheme and enjoys efficient signature generation and verification (eprint in 2017). The security of this scheme is based on two problems: one is DOOM (Decoding One Out of Many), and the other is the plain (U,U+V)-code problem over $\mathbb{F}_2$. There are many studies on the former one but few studies on the latter one. Later the security of SURF was broken because the hardness of the plain (U,U+V)-code problem does not hold with considering a notion of the hull. Then Debris-Alazard et al. proposed Wave as a successor of SURF, which is known as one of the most promising quantum-resistant signature schemes (ASIACRYPT 2019). Wave is based on similar problems used in SURF. Wave uses DOOM and the normalized generalized (U,U+V)-code problem over $\mathbb{F}_3$. In this paper, we utilize a notion of the Gramian (the determinant of the Gram matrices) of public keys and analyze the plain (U,U+V)-code problem over $\mathbb{F}_2$. For this purpose, we compute the asymptotic probability distribution of Gramians of random matrices. Furthermore, we also show a way to analyze the normalized generalized (U,U+V)-code problem over $\mathbb{F}_2$. Finally, we apply our analysis to the normalized generalized (U,U+V)-code problem over $\mathbb{F}_3$ in a special case. By our analysis with Gramian, SURF is completely broken, however, Wave is not directly threatened.

Keywords Code-based cryptography · Digital signature scheme · (U,U+V)-code problem and Gramian

---

This paper is based on "Analysis of (U,U+V)-code Problem with Gramian over Binary and Ternary Fields" [10].

I. Iwata (✉)
Ministry of Defense, Tokyo, Japan
e-mail: sing.the.wrath@gmail.com

Y. Yoshida · K. Tanaka
Tokyo Institute of Technology, Tokyo, Japan
e-mail: yoshida.yusuke@c.titech.ac.jp

K. Tanaka
e-mail: keisuke@is.titech.ac.jp

1 Introduction

1.1 Code-Based Signature Schemes

Digital signature plays a significant role in modern cryptographic applications and recently it becomes necessary to be quantum-resistant. To build secure digital signature schemes, first, we have to find quantum-resistant cryptographic problems.

Decoding a linear code is counted as one of those problems and cryptosystems based on this problem are called code-based. Recently, NIST announced the first three quantum-resistant digital signature schemes for standardization. Two of them are based on lattices while the other is based on hash functions. A code-based digital signature scheme was proposed as a candidate [7], but unfortunately not selected. It is still an important problem to build a secure code-based digital signature scheme.

Courtois, Finiasz, and Sendrier gave the first code-based digital signature scheme [3]. Its security depends on two problems; one is distinguishing Goppa codes from random codes and the other is decoding a linear code. However, it was later discovered that the former problem is not as hard as originally thought. As a result, the unpractical size of public keys is required for practical security level. Later its variant was proposed [11] but failed [14].

Aragon, Blazy, Gaborit, Hauteville, and Zémor proposed a new code-based digital signature scheme recently, whose name is Durandal [1]. By adopting Lyubashevsky's approach [12], this scheme enjoys small sizes of signatures and public keys. Durandal is based on a novel assumption, namely PSSI+ (Product Spaces Subspaces Indistinguishability). However, this assumption is not studied adequately and this scheme could leak the secret key information [6].

Debris-Alazard, Sendrier, and Tillich proposed a code-based signature scheme SURF, which enjoys efficient signature generation and verification [4]. This scheme is based on the GPV construction, which is an improved hash-and-sign digital signature scheme with trapdoor functions [8]. SURF is based on two problems: one is DOOM (Decoding One Out of Many) and the other is the plain (U,U+V)-code problem over $\mathbb{F}_2$.

First, DOOM is a variation of Syndrome Decoding Problems (SDP). SDP is the most fundamental problem in code-based cryptography. DOOM is as follows:

Given $\mathbf{H} \in \mathbb{F}_2^{(n-k) \times n}$, $\mathbf{s}_1, \ldots, \mathbf{s}_q \in \mathbb{F}_2^{n-k}$ and a sufficiently small w ($\leq n$), find $(\mathbf{e}, i)$ such that $|\mathbf{e}| = w$ and $\mathbf{He}^t = \mathbf{s}_i^t$.

There are many studies on DOOM and its related problems [2, 15, 16].

Second, given linear codes U and V, we define a (U,U+V)-code as follows:

$$\{(\mathbf{u}, \mathbf{u} + \mathbf{v}) : \mathbf{u} \in U, \mathbf{v} \in V\}.$$

The plain (U,U+V)-code problem is a decisional problem such that deciding whether a linear code is a permuted (U,U+V)-code or a random code. This problem can be converted into the problem of distinguishing a parity check matrix of a permuted (U,U+V)-code from a random matrix. The parity check matrix is denoted by

$$\mathbf{SHP} = \mathbf{S} \begin{pmatrix} \mathbf{H}_U & \mathbf{O} \\ \mathbf{H}_V & \mathbf{H}_V \end{pmatrix} \mathbf{P}$$

where $\mathbf{S}$ is an invertible matrix, $\mathbf{P}$ is a permutation matrix, $\mathbf{H}_U$, $\mathbf{H}_V$ are parity check matrices of U and V, and $\mathbf{O}$ is the zero matrix.

In summary, the main idea of SURF is to use the following code-based function:

$$f_{\mathbf{H},w} : S_w \to \mathbb{F}_2^{n-k}$$
$$\mathbf{e} \mapsto \mathbf{eH}^t$$

for $\mathbf{H} \in \mathbb{F}_2^{(n-k) \times n}$, a sufficiently small w ($\leq n$) and $S_w = \{\mathbf{x} \in \mathbb{F}_2^n : |\mathbf{x}| = w\}$. Generally, this function is one way as far as $\mathbf{H}$ is random due to the DOOM problem, and in some cases invertible when $\mathbf{H}$ has a particular structure, for example, $\mathbf{H}$ is a parity check matrix of a (U,U+V)-code. Here $\mathbf{SHP}$ is random due to the plain (U,U+V)-code problem, therefore $f_{\mathbf{SHP},w}$ can be considered as a trapdoor function where $\mathbf{S}$, $\mathbf{H}$ and $\mathbf{P}$ are the trapdoors. We obtain a signature of a message m as $f_{\mathbf{SHP},w}^{-1}(h(m))$ with a hash function h.

Unfortunately, the security of SURF was broken because the hardness of the plain (U,U+V)-code problem does not hold with considering the hull of the code [5]. For a linear code C, the hull of C is defined by the intersection of C itself and its dual code. Generally, the dimension of the hull of a random code is not always 0, but that of a permuted (U,U+V)-code is 0 with an overwhelming probability. Hence if we compute the dimension of the hull of a code, then we can decide whether a permuted (U,U+V)-code or a random code. They showed an attack on SURF.

After that, Debris-Alazard, Sendrier, and Tillich proposed Wave [6] as a successor of SURF. Wave works over $\mathbb{F}_3$, unlike SURF works over $\mathbb{F}_2$. Therefore the security of Wave is based on the normalized generalized (U,U+V)-code problem over $\mathbb{F}_3$ instead of the plain (U,U+V)-code problem over $\mathbb{F}_2$. Given linear codes U and V, the normalized generalized (U,U+V)-code is as follows:

$$\{(\mathbf{a} \odot \mathbf{u} + \mathbf{b} \odot \mathbf{v}, \mathbf{c} \odot \mathbf{u} + \mathbf{d} \odot \mathbf{v}) : \mathbf{u} \in U, \mathbf{v} \in V\}$$

where $\odot$ denotes Hadamard product and $\mathbf{a}, \mathbf{b}, \mathbf{c}$, and $\mathbf{d}$ are random vectors which satisfy

$$\forall i \in \{1, \ldots, n/2\}, \quad \text{and} \quad a_i c_i \neq 0 \quad a_i d_i - b_i c_i = 1.$$

The normalized generalized (U,U+V)-code problem is a decisional problem such that deciding whether a linear code is a permuted normalized generalized (U,U+V)-code or a random code. As far as we know, there are few studies on this problem, and no efficient attack against this problem is found.

1.2 Our Contribution

First, we utilize a notion of the Gramian as an indicator for distinguishing the plain (U,U+V)-code problem. By considering the Gramian in case of $\mathbf{H}_{pk} = \mathbf{S}\mathbf{H}_{sk}\mathbf{P}$, the effect of the randomizing matrices $\mathbf{S}$ and $\mathbf{P}$ is canceled and we obtain the Gramian of a secret key matrix $\mathbf{H}_{sk}$ such that

$$\det(\mathbf{H}_{pk}\mathbf{H}^t_{pk}) = \det(\mathbf{H}_{sk}\mathbf{H}^t_{sk}).$$

We prove that if we instantiate $\mathbf{H}_{sk}$ with a parity check matrix of (U,U+V)-code, then we obtain $\det(\mathbf{H}_{sk}\mathbf{H}^t_{sk}) = 0$.

Second, we estimate the distribution of the Gramian of random matrices. In other words, for each $a \in \mathbb{F}_q$, we would like to know

$$\Pr[\det(\mathbf{H}\mathbf{H}^t) = a \mid \mathbf{H} \in \mathbb{F}_q^{m \times n}].$$

Though this distribution seems to be common, it has never been analyzed mathematically. Hence we show an asymptotical formula, and conclude that this probability approaches around 0.42 in case of $q = 2$ and $a = 1$ as n increases. With such an analysis, we can construct a polynomial-time algorithm that distinguishes the plain (U,U+V)-code problem over $\mathbb{F}_2$.

Third, we deal with the normalized generalized (U,U+V)-code problem over $\mathbb{F}_2$. In this problem, we have to consider the additional secret variables. However, we prove that this problem is distinguished as well.

Finally, we apply our analysis to the normalized generalized (U,U+V)-code problem over $\mathbb{F}_3$ which Wave is based on. We found some weak public key in Wave. This is a new argument added from an earlier version [10] of this paper. However such weak keys are rarely generated, so Wave is not directly threatened by our analysis.

2 Preliminaries

2.1 Notation

For a prime number q, we denote the finite field with q elements by $\mathbb{F}_q$, for example, the field $\mathbb{F}_3$ denotes $\{0, 1, -1\}$. The vector space $\mathbb{F}_q^n$ denotes an n-dimensional vector space on $\mathbb{F}_q$. Vectors are denoted by small bold letters (such as $\mathbf{a}$) and matrices by capital bold letters (such as $\mathbf{A}$).

Vectors are in row notation. Let $\mathbf{u}, \mathbf{v}$ be two vectors in $\mathbb{F}_q^n$. The product $(\mathbf{u}, \mathbf{v}) \in \mathbb{F}_q$ denotes their inner product. The product $\mathbf{u} \odot \mathbf{v} \in \mathbb{F}_q^n$ denotes their Hadamard product such that $(u_1v_1, \ldots, u_nv_n)$. The Hamming weight of $\mathbf{u}$ is denoted by $|\mathbf{u}|$. For a vector $\mathbf{a} \in \mathbb{F}_q^n$, the matrix $\mathbf{Diag}(\mathbf{a})$ denotes the diagonal matrix $\mathbf{A} \in \mathbb{F}_q^{n \times n}$ with its entries given by $\mathbf{a}$, i.e., for all $i, j \in \{1, \ldots, n\}$, $\mathbf{A}(i, i) = a_i$ and $\mathbf{A}(i, j) = 0$ for $i \neq j$.

The matrix **I** denotes the identity matrix and the matrix **O** denotes the zero matrices. For $\mathbf{X} \in \mathbb{F}_q^{m \times n}$, $\mathbf{X}^t \in \mathbb{F}_q^{n \times m}$ denotes the transpose of **X**. We also define $\dim(V)$ as the dimension of a linear space V and $\text{rank}(\mathbf{X})$ as the dimension of the vector space generated by the columns of **X**.

We define a notion of permutations and their signatures for the definition of the determinant. S_n denotes the set which is consisted of all permutations of the set $\{1,\ldots,n\}$. If σ is achieved by interchanging two entries an odd/even number of times, σ is called odd/even. The signature of σ is defined to be $+1$ if σ is even and -1 if σ is odd, and which is denoted by $\text{sgn}(\sigma)$. Given $\mathbf{A} \in \mathbb{F}_q^{n \times n}$, the determinant of **A** is defined as follows:

$$\det(\mathbf{A}) \triangleq \sum_{\sigma \in S_n} \left(\text{sgn}(\sigma) \prod_{i=1}^n a_{i,\sigma(i)} \right).$$

A linear code of length n and dimension k is denoted by $[n,k]$-code, which is defined as a linear subspace V with dimension k of the vector space $\mathbb{F}_q^n$.

In the following, let C be an $[n,k]$-code. $C^\perp$ denotes the dual of C which is defined as:

$$\{\mathbf{h} \in \mathbb{F}_q^n : \forall \mathbf{c} \in C, (\mathbf{c},\mathbf{h}) = 0\}.$$

We denote $\text{hull}(C)$ as a vector space such that $C \cap C^\perp$.

2.2 (U,U+V)-Codes and Problems

Given linear codes U and V of length $n/2$, we define a (U,U+V)-code as

$$\{(\mathbf{u}, \mathbf{u} + \mathbf{v}) : \mathbf{u} \in U, \mathbf{v} \in V\}.$$

The plain (U,U+V)-code problem is defined as deciding whether a certain linear code is a permuted (U,U+V)-code or a random code.

This problem over $\mathbb{F}_2$ is equivalent to the following problem in the light of a parity check matrix.

Problem 1 (*The plain (U,U+V)-code problem over $\mathbb{F}_2$*) For a random non-singular matrix $\mathbf{S} \in \mathbb{F}_2^{m \times m}$, a random permutation matrix $\mathbf{P} \in \mathbb{F}_2^{n \times n}$ and $\mathbf{H}_{sk} \in \mathbb{F}_2^{m \times n}$ given by

$$\begin{pmatrix} \mathbf{H}_U & \mathbf{O} \\ \mathbf{H}_V & \mathbf{H}_V \end{pmatrix}$$

where $\mathbf{H}_U \in \mathbb{F}_2^{l \times (n/2)}$, $\mathbf{H}_V \in \mathbb{F}_2^{m \times (n/2)}$ ($l < m$) are random, distinguish $\mathbf{H}_{pk} \triangleq \mathbf{S}\mathbf{H}_{sk}\mathbf{P}$ from a random matrix $\mathbf{H}_{rand} \in \mathbb{F}_2^{m \times n}$.

Remark 1 The notations above such that $\mathbf{H}_{pk}$ and $\mathbf{H}_{sk}$ imply that $\mathbf{H}_{pk}$ is used as a public key and $\mathbf{H}_{sk}$ is used as a secret key.

Also, we define a normalized generalized (U,U+V)-code as

$$\{(\mathbf{a} \odot \mathbf{u} + \mathbf{b} \odot \mathbf{v}, \mathbf{c} \odot \mathbf{u} + \mathbf{d} \odot \mathbf{v}) : \mathbf{u} \in U, \mathbf{v} \in V\}$$

where $\mathbf{a}, \mathbf{b}, \mathbf{c}$ and $\mathbf{d} \in \mathbb{F}_q^{n/2}$ are some random vectors which satisfy the following conditions:

$$\forall i \in \{1, \ldots, n/2\}, \quad \text{and} \quad a_i c_i \neq 0 \quad a_i d_i - b_i c_i = 1.$$

The normalized generalized (U,U+V)-code problem is defined as deciding whether a certain linear code is a permuted normalized generalized (U,U+V)-code or a random code.

This problem is equivalent to the following problem in the light of a parity check matrix.

Problem 2 (*The normalized generalized (U,U+V)-code problem*) Let n be an even integer and let $\mathbf{a}, \mathbf{b}, \mathbf{c}$ and $\mathbf{d} \in \mathbb{F}_q^{n/2}$ be some random vectors which satisfy the following conditions:

$$\forall i \in \{1, \ldots, n/2\}, \quad \text{and} \quad a_i c_i \neq 0 \quad a_i d_i - b_i c_i = 1.$$

In addition, we take $\mathbf{H}_U \in \mathbb{F}_q^{l \times (n/2)}$ and $\mathbf{H}_V \in \mathbb{F}_q^{m \times (n/2)}$ at random and define $\mathbf{H}_{sk} \in \mathbb{F}_q^{m \times n}$ as follows:

$$\mathbf{H}_{sk} \triangleq \begin{pmatrix} \mathbf{H}_U \mathbf{D} & -\mathbf{H}_U \mathbf{B} \\ -\mathbf{H}_V \mathbf{C} & \mathbf{H}_V \mathbf{A} \end{pmatrix}$$

where $\mathbf{A} \triangleq \mathrm{Diag}(\mathbf{a})$, $\mathbf{B} \triangleq \mathrm{Diag}(\mathbf{b})$, $\mathbf{C} \triangleq \mathrm{Diag}(\mathbf{c})$, and $\mathbf{D} \triangleq \mathrm{Diag}(\mathbf{d})$.

Then for a random non-singular matrix $\mathbf{S} \in \mathbb{F}_q^{m \times m}$ and a random permutation matrix $\mathbf{P} \in \mathbb{F}_q^{n \times n}$, distinguish $\mathbf{H}_{pk} \triangleq \mathbf{S} \mathbf{H}_{sk} \mathbf{P} \in \mathbb{F}_q^{m \times n}$ from a random matrix $\mathbf{H}_{rand} \in \mathbb{F}_q^{m \times n}$.

3 Gramian

We define a notion of the Gramian and prove some useful properties. In this section, K denotes $\mathbb{F}_2$ or $\mathbb{F}_3$.

3.1 Basic Formulae

Definition 1 Given $\mathbf{H} \in \mathbb{F}_q^{m \times n}$, the Gramian (the determinant of Gram matrix) of $\mathbf{H}$ is defined by $\det(\mathbf{H}\mathbf{H}^t)$.

Lemma 1 *For any non-singular matrix* $\mathbf{S} \in K^{n \times n}$, *the Gramian of* $\mathbf{S}$ *equals* 1.

Proof When $K = \mathbb{F}_2$, $\det(\mathbf{S}) = 1$. When $K = \mathbb{F}_3$, $\det(\mathbf{S}) = 1$ or -1. In any case, we obtain $\det(\mathbf{S}\mathbf{S}^t) = \det(\mathbf{S})^2 = 1$. □

Remark 2 In case of K is not $\mathbb{F}_2$ or $\mathbb{F}_3$, this lemma does not hold.

Theorem 1 *Suppose that* $\mathbf{S} \in K^{m \times m}$, $\mathbf{H}_{sk} \in K^{m \times n}$, $\mathbf{P} \in K^{n \times n}$, *and* $\mathbf{H}_{pk} \triangleq \mathbf{S}\mathbf{H}_{sk}\mathbf{P}$, *where* $\mathbf{S}$ *is a non-singular matrix and* $\mathbf{P}$ *is a permutation matrix. Then we have*

$$\det(\mathbf{H}_{pk}\mathbf{H}_{pk}^t) = \det(\mathbf{H}_{sk}\mathbf{H}_{sk}^t).$$

Proof Lemma 1 and a well-known fact that $\mathbf{P}\mathbf{P}^t = \mathbf{I}$ for any permutation matrix $\mathbf{P}$, we can obtain:

$$\begin{aligned}
\det(\mathbf{H}_{pk}\mathbf{H}_{pk}^t) &= \det((\mathbf{S}\mathbf{H}_{sk}\mathbf{P})(\mathbf{P}^t\mathbf{H}_{sk}^t\mathbf{S}^t)) \\
&= \det(\mathbf{S}\mathbf{H}_{sk}(\mathbf{P}\mathbf{P}^t)\mathbf{H}_{sk}^t\mathbf{S}^t) \\
&= \det(\mathbf{S})\det(\mathbf{H}_{sk}\mathbf{H}_{sk}^t)\det(\mathbf{S}^t) \\
&= \det(\mathbf{S}\mathbf{S}^t)\det(\mathbf{H}_{sk}\mathbf{H}_{sk}^t) \\
&= \det(\mathbf{H}_{sk}\mathbf{H}_{sk}^t).
\end{aligned}$$

□

We also have the following theorem:

Theorem 2 *Let the notations be the same as above, then we have*

$$\mathrm{rank}(\mathbf{H}_{pk}\mathbf{H}_{pk}^t) = \mathrm{rank}(\mathbf{H}_{sk}\mathbf{H}_{sk}^t).$$

Proof We can obtain like Theorem 1. □

3.2 Distribution of Gramian

As far as we know, the distribution of the Gramian of random matrices is not studied adequately. Here we show an asymptotic analysis over $\mathbb{F}_2$.

First, we prove the following lemma:

Lemma 2 For arbitarary $i, j \in \{1, \ldots, m\}$,

$$Pr\big[(\mathbf{v}_i, \mathbf{v}_j) = 1\big] = \begin{cases} \frac{1}{2} & i = j \\ \frac{1}{2} - \frac{1}{2^{n+1}} & i \neq j \end{cases}$$

Proof We can easily prove this lemma. □

Then we refer to a kind of counting symmetric matrices theorem. MacWilliams showed the following [13]:

Theorem 3 Let $N(t, r)$ denote the number of symmetric matrices of size $t \times t$, rank r, with entries in a finite field $GF(q)$, $q = p^n$.

$$N(t, 2s) = \prod_{i=1}^{s} \frac{q^{2i}}{q^{2i} - 1} \cdot \prod_{i=0}^{2s-1} (q^{t-i} - 1) \qquad (2s \leq t),$$

$$N(t, 2s+1) = \prod_{i=1}^{s} \frac{q^{2i}}{q^{2i} - 1} \cdot \prod_{i=0}^{2s} (q^{t-i} - 1) \qquad (2s + 1 \leq t).$$

Here we show the main theorem in this section as follows:

Theorem 4 For a random matrix $\mathbf{A} \in \mathbb{F}_2^{m \times n}$, the probability that the Gramian of A equals 1 approaches the following proportion asymptotically as n increases:

$$Pr\big[\det(\mathbf{A}\mathbf{A}^t) = 1\big] \xrightarrow[n \to \infty]{} \left(1 - \frac{1}{2}\right)\left(1 - \frac{1}{2^3}\right)\left(1 - \frac{1}{2^5}\right) \cdots \left(1 - \frac{1}{2^{m \text{ or } m-1}}\right).$$

Proof We prove this theorem by demonstrating that $\mathbf{A}\mathbf{A}^t$ approaches a random symmetric matrix asymptotically. In this proof, vectors are column notation. We can write

$$\mathbf{A} = \begin{pmatrix} \mathbf{v}_1 \\ \vdots \\ \mathbf{v}_m \end{pmatrix}$$

where $\mathbf{v}_1, \cdots, \mathbf{v}_m \in \mathbb{F}_2^n$ and

$$\det(\mathbf{A}\mathbf{A}^t) = \det \begin{pmatrix} (\mathbf{v}_1, \mathbf{v}_1) & (\mathbf{v}_1, \mathbf{v}_2) & \cdots & (\mathbf{v}_1, \mathbf{v}_m) \\ (\mathbf{v}_2, \mathbf{v}_1) & (\mathbf{v}_2, \mathbf{v}_2) & \cdots & (\mathbf{v}_2, \mathbf{v}_m) \\ \vdots & & & \\ (\mathbf{v}_m, \mathbf{v}_1) & (\mathbf{v}_m, \mathbf{v}_2) & \cdots & (\mathbf{v}_m, \mathbf{v}_m) \end{pmatrix}.$$

By Lemma 2, all of the elements of $\mathbf{A}\mathbf{A}^t$ approach to $\frac{1}{2}$ asymptotically as n increases.

Second, we consider the independence of the elements. Here we remark on their pairwise independence. Without loss of generality, we would like to prove the following three patterns where i, j, k are different indices:

$$\Pr[(v_i, v_i) = 1 \cap (v_j, v_j) = 1] \xrightarrow[n \to \infty]{} \Pr[(v_i, v_i) = 1] \cdot \Pr[(v_j, v_j) = 1] \quad (1)$$

$$\Pr[(v_i, v_i) = 1 \cap (v_i, v_j) = 1] \xrightarrow[n \to \infty]{} \Pr[(v_i, v_i) = 1] \cdot \Pr[(v_i, v_j) = 1] \quad (2)$$

$$\Pr[(v_i, v_j) = 1 \cap (v_i, v_k) = 1] \xrightarrow[n \to \infty]{} \Pr[(v_i, v_j) = 1] \cdot \Pr[(v_i, v_k) = 1]. \quad (3)$$

By the above lemma, the right sides of (1) to (3) approach $\frac{1}{4}$ asymptotically. We can check easily the left side of (1) equals $\frac{1}{4}$. As for (2), we have v_i has odd number 1s from the first condition of the left side. From the second condition, the probability of $(v_i, v_j) = 1$ is $\frac{1}{2}$, so the left side probability equals $\frac{1}{4}$. As for (3), considering the above lemma as well, we obtain the left side probability equals $\frac{1}{4}(1 - \frac{1}{2^n})$.

Finally, we have the following equation by Theorem 3,

$$\Pr\left[\det(\mathbf{X}) = 1 \mid \mathbf{X} \in \mathbb{F}_2^{2s \times 2s} \text{ is symmetric matrix}\right]$$

$$= \frac{N(2s, 2s)}{2 \cdot 2^2 \cdots 2^{2s-1} \cdot 2^{2s}}$$

$$= \frac{2^2 \cdot 2^4 \cdots 2^{2s-2} \cdot 2^{2s}}{(2^2 - 1)(2^4 - 1)(2^{2s-2} - 1)(2^{2s} - 1)}$$

$$\cdot \frac{(2^{2s} - 1)(2^{2s-1} - 1) \cdots (2^2 - 1)(2 - 1)}{2 \cdot 2^2 \cdots 2^{2s-1} \cdot 2^{2s}}$$

$$= \frac{(2^{2s-1} - 1)(2^{2s-3} - 1) \cdots (2^3 - 1)(2 - 1)}{2^{2s-1} \cdot 2^{2s-3} \cdots 2^3 \cdot 2}$$

$$= \left(1 - \frac{1}{2}\right)\left(1 - \frac{1}{2^3}\right) \cdots \left(1 - \frac{1}{2^{2s-3}}\right)\left(1 - \frac{1}{2^{2s-1}}\right),$$

and we obtain the following equation as well:

$$\Pr\left[\det(\mathbf{X}) = 1 \mid \mathbf{X} \in \mathbb{F}_2^{2s+1 \times 2s+1} \text{ is symmetric matrix}\right]$$

$$= \left(1 - \frac{1}{2}\right)\left(1 - \frac{1}{2^3}\right) \cdots \left(1 - \frac{1}{2^{2s-1}}\right)\left(1 - \frac{1}{2^{2s+1}}\right).$$

This concludes the proof. □

We would also like to prove the well-known lemma as follows:

Lemma 3

$$|\{\mathbf{X} \in \mathbb{F}_q^{m \times m} \mid \mathbf{X} \text{ is invertible}\}| = (q^m - q^{m-1}) \cdots (q^m - 1).$$

Table 1 $G(10, n)$

n	10	12	14	16	18	∞
Theoretic	0.28907	–	–	–	–	0.41969
Experimental	0.28907	0.38533	0.41099	0.41751	0.41915	–

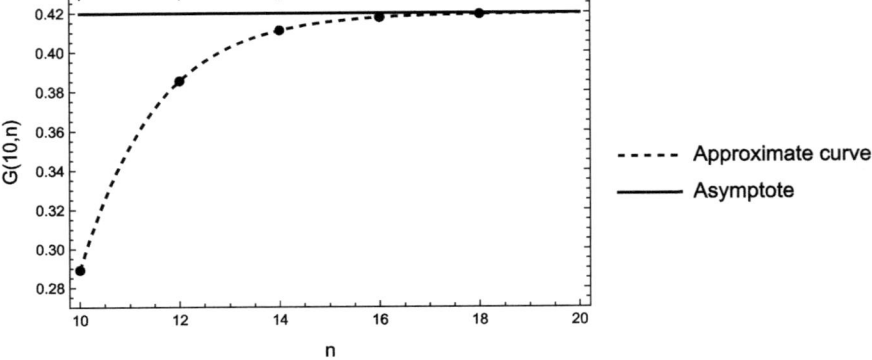

Fig. 1 $G(10, n)$

Proof This can be shown by counting the possible columns of the matrix: the first column can be anything but the zero vector; the second column can be anything but the multiples of the first column; and in general, the k-th column can be any vector not in the linear span of the first $k - 1$ columns. □

By this lemma, for $\mathbf{A} \in \mathbb{F}_2^{m \times m}$, the probability distribution of the Gramian of A equals the following proportion:

$$\Pr\left[\det(\mathbf{A}\mathbf{A}^t) = 1\right] = \Pr\left[\det(\mathbf{A}) = 1\right]$$
$$= \frac{(2^m - 2^{m-1}) \cdots (2^m - 1)}{(2^m)^m}$$
$$= \left(1 - \frac{1}{2}\right)\left(1 - \frac{1}{2^2}\right) \cdots \left(1 - \frac{1}{2^m}\right).$$

Let $G(m, n)$ be $\Pr\left[\det(\mathbf{A}\mathbf{A}^t) = 1 | \mathbf{A} \in \mathbb{F}_q^{m \times n}\right]$. We conduct some experiments in order to estimate $G(10, n)$ for $n = 10, 12, 14, 16, 18$ and $G(100, n)$ for $n = 100, 102, 104, 106, 108$. We compute the average of the determinants of one million random matrices for each (m, n).

The following tables and figures are Table 1 and Fig. 1 for $G(10, n)$ and Table 2 and Fig. 2 for $G(100, n)$. It shows that they converge to around 0.42 quickly. We use these results in the following section.

Table 2 $G(100, n)$

n	100	102	104	106	108	∞
Theoretic	0.28878	–	–	–	–	0.41942
Experimental	0.28878	0.38505	0.41072	0.41724	0.41887	–

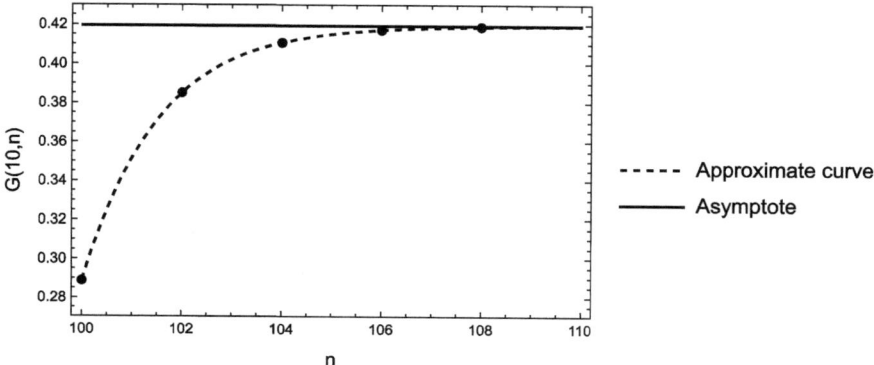

Fig. 2 $G(100, n)$

4 The Plain (U,U+V)-Code Problem Over $\mathbb{F}_2$

Debris-Alazard et al. analyze SURF with a notion of the hull [5]. Here, we produce another analysis with the Gramian.

Proposition 1 *For any block matrix* $\mathbf{X} \in \mathbb{F}_q^{(l+m) \times (l+m)}$ *given by*

$$\left(\begin{array}{c|c} \mathbf{A} & \mathbf{B} \\ \hline \mathbf{C} & \mathbf{O} \end{array} \right)$$

where $\mathbf{A} \in \mathbb{F}_q^{l \times l}$, $\mathbf{B} \in \mathbb{F}_q^{l \times m}$ *and* $\mathbf{C} \in \mathbb{F}_q^{m \times l}$, $\det(\mathbf{X})$ *equals 0 in case of* $l < m$.

Proof By the pigeonhole principle, for any $\sigma \in S_{l+m}$, $\{\sigma(l+1), \ldots, \sigma(l+m)\}$ contains at least one element of $\{l+1, \ldots, l+m\}$. For such an element $\sigma(i)$, we have $x_{i,\sigma(i)} = 0$. Hence, the definition of the determinant shows that all additive terms in the expansion of $\det(\mathbf{X})$ equal 0, and the sum of those is as well. □

Theorem 5 *There exists an algorithm that solves the plain (U,U+V)-code problem in polynomial time.*

Proof By Theorem 1 and Proposition 1, we have

$$\det(\mathbf{H}_{pk}\mathbf{H}^t_{pk}) = \det(\mathbf{H}_{sk}\mathbf{H}^t_{sk})$$
$$= \det\left(\begin{pmatrix}\mathbf{H}_U & \mathbf{O} \\ \mathbf{H}_V & \mathbf{H}_V\end{pmatrix}\begin{pmatrix}\mathbf{H}^t_U & \mathbf{H}^t_V \\ \mathbf{O} & \mathbf{H}^t_V\end{pmatrix}\right)$$
$$= \det\begin{pmatrix}\mathbf{H}_U\mathbf{H}^t_U & \mathbf{H}_U\mathbf{H}^t_V \\ \mathbf{H}_V\mathbf{H}^t_U & \mathbf{O}\end{pmatrix}$$
$$= 0.$$

On the other hand, $\det(\mathbf{H}_{rand}\mathbf{H}^t_{rand})$ is not always 0. □

We can show another proof using the two propositions: one is the following relationship between the hull and the Gramian [9].

Proposition 2 *Let* $\mathbf{H} \in \mathbb{F}_2^{m \times n}$ *be a parity check matrix of an arbitrary code* C*. Then we can obtain:*

$$\mathrm{rank}(\mathbf{H}\mathbf{H}^t) = m - \dim(\mathrm{hull}(C)).$$

The other is the proposition of the original analysis of SURF [5]:

Proposition 3 *For* $\mathbf{H}_U \in \mathbb{F}_2^{l \times (n/2)}$ *and* $\mathbf{H}_V \in \mathbb{F}_2^{m \times (n/2)}$*, let* $\mathbf{H}_{sk} \in \mathbb{F}_2^{m \times n}$ *be same as Problem 1 and* $l \le m$*. If* $\mathbf{H}_{sk}$ *is a parity check matrix of an arbitrary code* C*, then we have with probability* $1 - O(2^{l-m})$

$$\dim(\mathrm{hull}(C)) = m - l.$$

By Theorem 2 and the two propositions above, we have

$$\mathrm{rank}(\mathbf{H}_{pk}\mathbf{H}^t_{pk}) = \mathrm{rank}(\mathbf{H}_{sk}\mathbf{H}^t_{sk})$$
$$= m - \dim(\mathrm{hull}(C))$$
$$= l \quad < m.$$

Then we have $\det(\mathbf{H}_{pk}\mathbf{H}^t_{pk}) = 0$ which concludes the proof.

In the previous distinguishing method [5], the hull calculation itself is complex and difficult to understand intuitively, although it is completed in polynomial time. On the other hand, we note that the method presented here is intuitively easy to understand and can be easily extended to other problems as shown in the next section.

5 The Normalized Generalized (U,U+V)-Code Problem Over $\mathbb{F}_2$

In the binary case ($q = 2$), since $a_i c_i \neq 0$, we obtain $a_i = 1$ and $c_i = 1$ and $d_i - b_i = 1$. Thus, we have $\mathbf{A} = \mathbf{C} = \mathbf{I}$ and $\mathbf{B} + \mathbf{D} = \mathbf{I}$. In addition, since $\mathbf{B}$ and $\mathbf{D}$ are binary diagonal matrices, we also have $\mathbf{B}^2 = \mathbf{B}$ and $\mathbf{D}^2 = \mathbf{D}$. Hence, we obtain:

$$\det(\mathbf{H}_{pk}\mathbf{H}^t_{pk}) = \det(\mathbf{H}_{sk}\mathbf{H}^t_{sk})$$
$$= \det\left(\begin{pmatrix} \mathbf{H}_U \mathbf{D} & \mathbf{H}_U \mathbf{B} \\ \mathbf{H}_V & \mathbf{H}_V \end{pmatrix} \begin{pmatrix} \mathbf{D}\mathbf{H}^t_U & \mathbf{H}^t_V \\ \mathbf{B}\mathbf{H}^t_U & \mathbf{H}^t_V \end{pmatrix}\right)$$
$$= \det\begin{pmatrix} \mathbf{H}_U \mathbf{H}^t_U & \mathbf{H}_U \mathbf{H}^t_V \\ \mathbf{H}_V \mathbf{H}^t_U & \mathbf{O} \end{pmatrix}$$
$$= 0.$$

Therefore, we reach the following theorem by Theorem 5:

Theorem 6 *There exists an algorithm that solves the normalized generalized (U,U+V)-code problem over $\mathbb{F}_2$ in polynomial time.*

6 Special Case of Normalized Generalized (U,U+V)-Code Problem Over $\mathbb{F}_3$

The security proof of Wave depends on two problems [6]. One is DOOM (Decoding One Out of Many) and the other is the normalized generalized (U,U+V)-code problem over $\mathbb{F}_3$. Our analysis with the Gramian does not seem to be efficient as the previous section, since the determinant does not vanish, and even four secret variables $\mathbf{A}$, $\mathbf{B}$, $\mathbf{C}$, and $\mathbf{D}$ remain as follows:

$$\det(\mathbf{H}_{pk}\mathbf{H}^t_{pk}) = \det(\mathbf{H}_{sk}\mathbf{H}^t_{sk})$$
$$= \det\left(\begin{pmatrix} \mathbf{H}_U \mathbf{D} & -\mathbf{H}_U \mathbf{B} \\ -\mathbf{H}_V \mathbf{C} & \mathbf{H}_V \mathbf{A} \end{pmatrix} \begin{pmatrix} \mathbf{D}\mathbf{H}^t_U & -\mathbf{C}\mathbf{H}^t_V \\ -\mathbf{B}\mathbf{H}^t_U & \mathbf{A}\mathbf{H}^t_V \end{pmatrix}\right)$$
$$= \det\begin{pmatrix} \mathbf{H}_U(\mathbf{D}^2 + \mathbf{B}^2)\mathbf{H}^t_U & -\mathbf{H}_U(\mathbf{AB} + \mathbf{CD})\mathbf{H}^t_V \\ -\mathbf{H}_V(\mathbf{AB} + \mathbf{CD})\mathbf{H}^t_U & \mathbf{H}_V(\mathbf{A}^2 + \mathbf{C}^2)\mathbf{H}^t_V \end{pmatrix}.$$

Here, we can transform this into a simpler one by the following proposition:

Proposition 4 *Let a matrix Δ be $-(\mathbf{AB} + \mathbf{CD})$. Then we obtain:*

Table 3 Tuples of (a_i, b_i, c_i, d_i)

a_i	b_i	c_i	d_i	$a_i^2 + c_i^2$	$-(a_i b_i + c_i d_i)$	$b_i^2 + d_i^2$
1	0	1	1	-1	-1	1
1	1	1	-1	-1	0	-1
1	-1	1	0	-1	1	1
1	0	-1	1	-1	1	1
1	1	-1	0	-1	-1	1
1	-1	-1	-1	-1	0	-1
-1	0	1	-1	-1	1	1
-1	1	1	1	-1	0	-1
-1	-1	1	0	-1	-1	1
-1	0	-1	-1	-1	-1	1
-1	1	-1	0	-1	1	1
-1	-1	-1	1	-1	0	-1

$$\det(\mathbf{H}_{pk}\mathbf{H}_{pk}^t) = \det\begin{pmatrix} -\mathbf{H}_U(\Delta^2+\mathbf{I})\mathbf{H}_U^t & \mathbf{H}_U\Delta\mathbf{H}_V^t \\ \mathbf{H}_V\Delta\mathbf{H}_U^t & -\mathbf{H}_V\mathbf{H}_V^t \end{pmatrix}.$$

Proof In the ternary case ($q = 3$), among all $81 (= 3^4)$ tuples of (a_i, b_i, c_i, d_i), there are only 12 tuples which satisfies $a_i c_i \neq 0$ and $a_i d_i - b_i c_i = 1$. Table 3 shows these tuples and additional 3 values corresponding to $\mathbf{A}^2 + \mathbf{C}^2$, $\mathbf{AB} + \mathbf{CD}$ and $\mathbf{D}^2 + \mathbf{B}^2$.

From this table, we can easily check $a_i^2 + c_i^2 = -1$ and $-(-(a_i b_i + c_i d_i))^2 - 1 = b_i^2 + d_i^2$. Hence we can obtain $\mathbf{A}^2 + \mathbf{C}^2 = -\mathbf{I}$ and $-\Delta^2 - \mathbf{I} = \mathbf{D}^2 + \mathbf{B}^2$, then this concludes the proof. □

By this proposition, we can reduce the number of random variables. Here we consider a special case such that $\Delta = \mathbf{O}$. Then we obtain the following theorem:

Theorem 7 *There exists an algorithm that solves the normalized generalized $(U, U+V)$-code problem over $\mathbb{F}_3$ where $\Delta = \mathbf{O}$ in polynomial time.*

Proof By the proposition above, we obtain

$$\det(\mathbf{H}_{pk}\mathbf{H}_{pk}^t) = \det\begin{pmatrix} -\mathbf{H}_U\mathbf{H}_U^t & \mathbf{O} \\ \mathbf{O} & -\mathbf{H}_V\mathbf{H}_V^t \end{pmatrix}$$
$$= \det(-\mathbf{H}_U\mathbf{H}_U^t)\det(-\mathbf{H}_V\mathbf{H}_V^t).$$

By Theorem 4, there is approximately a 0.64 probability that $\det(\mathbf{HH}^t) \neq 0$ for a random $\mathbf{H}$ over $\mathbb{F}_3$. Therefore, for $n \to \infty$,

$$\Pr\left[\det(\mathbf{H}_{pk}\mathbf{H}_{pk}^t) \neq 0\right] = \Pr\left[\det(\mathbf{H}_U\mathbf{H}_U^t) \neq 0\right]\Pr\left[\det(\mathbf{H}_V\mathbf{H}_V^t) \neq 0\right]$$
$$\to 0.64^2 = 0.4096.$$

Table 4 Tuples of (a_i, b_i, c_i, d_i) where $\Delta = \mathbf{O}$

a_i	b_i	c_i	d_i	$a_i^2 + c_i^2$	$-(a_i b_i + c_i d_i)$	$b_i^2 + d_i^2$
1	1	1	−1	−1	0	−1
1	−1	−1	−1	−1	0	−1
−1	1	1	1	−1	0	−1
−1	−1	−1	1	−1	0	−1

On the other hand, as we see for $n \to \infty$ that

$$\Pr\left[\det(\mathbf{H}_{rand}\mathbf{H}_{rand}^t) \neq 0\right] \to 0.64$$

Then we can distinguish them with non-negligible probability. □

However, our results can be applied only in this case, and the condition $\Delta = \mathbf{O}$ is satisfied iff $\mathbf{A}, \mathbf{B}, \mathbf{C}$, and $\mathbf{D}$ are occupied with 4 tuples of all 12 tuples. Table 4 is an excerpt of such tuples from Table 3.

Then a probability that $\Delta = \mathbf{O}$ is $1/3^{\frac{n}{2}} (= 4^{\frac{n}{2}}/12^{\frac{n}{2}})$ and this is negligible. In other cases, the Gramian cannot be computed simply. Hence by our analysis with Gramian, SURF is completely broken, however, Wave is not directly threatened. However, our results indicate that using fixed matrix variable might be vulnerability of Wave.

7 Conclusion

In this paper, we have introduced another view on the (U,U+V)-code problem. First, we utilize a notion of Gramian as an indicator for distinguishing the plain (U,U+V)-code problem. By considering the Gramian in case of $\mathbf{H}_{pk} = \mathbf{S}\mathbf{H}_{sk}\mathbf{P}$, the effect of the randomizing matrices $\mathbf{S}$ and $\mathbf{P}$ is canceled and we obtain the Gramian of a secret key matrix $\mathbf{H}_{sk}$ such that $\det(\mathbf{H}_{pk}\mathbf{H}_{pk}^t) = \det(\mathbf{H}_{sk}\mathbf{H}_{sk}^t)$. We prove that if we instantiate $\mathbf{H}_{sk}$ with a parity check matrix of (U,U+V)-code, then we obtain $\det(\mathbf{H}_{sk}\mathbf{H}_{sk}^t) = 0$. Second, we estimate the distribution of the Gramian of random matrices asymptotically. In other words, we would like to know $\Pr[\det(\mathbf{H}\mathbf{H}^t) \neq 0 \mid \mathbf{H} \in \mathbb{F}_q^{m \times n}]$. Though this distribution seems to be common but has never been analyzed mathematically. Hence we show an asymptotical formula and conclude that this probability approaches around 0.42 in the case of $q = 2$ and around 0.64 in the case of $q = 3$ as n increases. With such an analysis, we can construct a polynomial-time algorithm that distinguishes the plain (U,U+V)-code problem over $\mathbb{F}_2$. Third, we deal with the normalized generalized (U,U+V)-code problem over $\mathbb{F}_2$. In this problem, we have to consider the additional secret variables. However, we prove that this problem is distinguished as well. Finally, we apply our analysis to the normalized generalized (U,U+V)-code problem over $\mathbb{F}_3$. In a special case of this problem, we can distinguish this problem with Gramian. By our analysis with Gramian, SURF is completely broken, but Wave is not directly threatened. However, our results indicate that using fixed matrix variable might be vulnerability of Wave.

There are three open questions related to this paper as follows: First, our analysis can distinguish only a special case of the (U,U+V)-code problem on which Wave is based. Hence the general case is still an open problem. Second, our approaches with Gramian can be used only over binary or ternary fields, because concerning integer fields larger than 3, the squared determinant of randomizing matrix S does not always vanish. Wave can be instantiated with such large integer fields like $\mathbb{F}_5$ and $\mathbb{F}_7$, so this remains an important problem. Third, We have shown the distribution of random Gramians only asymptotically. Hence deducing a formula of this value is an open problem. We conducted some experiments and have some conjecture but could not prove it. This problem is interesting not only cryptographically.

References

1. N. Aragon, O. Blazy, P. Gaborit, A. Hauteville, G. Zémor, Durandal: a rank metric based signature scheme, in *Advances in Cryptology - EUROCRYPT 2019 - 38th Annual International Conference on the Theory and Applications of Cryptographic Techniques, Darmstadt, Germany, May 19–23, 2019, Proceedings, Part III*, Lecture Notes in Computer Science, vol. 11478, ed. by Y. Ishai, V. Rijmen (Springer, 2019), pp. 728–758
2. R. Bricout, A. Chailloux, T. Debris-Alazard, M. Lequesne, Ternary syndrome decoding with large weight, in *Selected Areas in Cryptography - SAC 2019 - 26th International Conference, Waterloo, ON, Canada, Aug 12–16, 2019, Revised Selected Papers*, Lecture Notes in Computer Science, vol. 11959, ed. by K.G. Paterson, D. Stebila (Springer, 2019), pp. 437–466
3. N.T. Courtois, M. Finiasz, N. Sendrier, How to achieve a mceliece-based digital signature scheme, in *Advances in Cryptology - ASIACRYPT 2001, 7th International Conference on the Theory and Application of Cryptology and Information Security, Gold Coast, Australia, Dec 9–13, 2001, Proceedings*, Lecture Notes in Computer Science, vol. 2248, ed. by C. Boyd (Springer, 2001), pp. 157–174
4. T. Debris-Alazard, N. Sendrier, J. Tillich, A new signature scheme based on (u|u+v) codes. Cryptology ePrint Archive, Paper 2017/662 (2017)
5. T. Debris-Alazard, N. Sendrier, J. Tillich, The problem with the surf scheme. https://csrc.nist.gov/projects/post-quantum-cryptography/round-1-submissions (2017)
6. T. Debris-Alazard, N. Sendrier, J. Tillich, Wave: a new family of trapdoor one-way preimage sampleable functions based on codes, in *Advances in Cryptology - ASIACRYPT 2019 - 25th International Conference on the Theory and Application of Cryptology and Information Security, Kobe, Japan, Dec 8–12, 2019, Proceedings, Part I*, Lecture Notes in Computer Science, vol. 11921, ed. by S.D. Galbraith, S. Moriai (Springer, 2019), pp. 21–51
7. K. Fukushima, P.S. Roy, R. Xu, S. Kiyomoto, K. Morozov, T. Takagi, Racoss. First round submission to the NIST post-quantum cryptography call (2017)
8. C. Gentry, C. Peikert, V. Vaikuntanathan, Trapdoors for hard lattices and new cryptographic constructions, in *Proceedings of the 40th Annual ACM Symposium on Theory of Computing, Victoria, British Columbia, Canada, May 17-20, 2008*, ed. by C. Dwork (ACM, 2008), pp. 197–206
9. K. Guenda, S. Jitman, T.A. Gulliver, Constructions of good entanglement-assisted quantum error correcting codes. Des. Codes Cryptogr. **86**(1), 121–136 (2018)
10. I. Iwata, Y. Yoshida, K. Tanaka, Analysis of (u, u+v)-code problem with gramian over binary and ternary fields, in *Information Security and Cryptology - ICISC 2022 - 25th International Conference, ICISC 2022, Seoul, South Korea, Nov 30–Dec 2, 2022, Revised Selected Papers*,

Lecture Notes in Computer Science, vol. 13849, ed. by S. Seo, H. Seo (Springer, 2022), pp. 435–449
11. G. Kabatianskii, E.A. Krouk, B.J.M. Smeets, A digital signature scheme based on random error-correcting codes, in *Cryptography and Coding, 6th IMA International Conference, Cirencester, UK, Dec 17–19, 1997, Proceedings, Lecture Notes in Computer Science*, vol. 1355, ed. by M. Darnell (Springer, 1997), pp. 161–167
12. V. Lyubashevsky. Fiat-shamir with aborts: applications to lattice and factoring-based signatures, in *Advances in Cryptology - ASIACRYPT 2009, 15th International Conference on the Theory and Application of Cryptology and Information Security, Tokyo, Japan, Dec 6–10, 2009. Proceedings, Lecture Notes in Computer Science*, vol. 5912, ed. by M. Matsui (Springer, 2009), pp. 598–616
13. J. MacWilliams, Orthogonal matrices over finite fields. Am. Math. Mon. **76**(2), 152–164 (1969)
14. A. Otmani, J. Tillich, An efficient attack on all concrete KKS proposals, in *Post-Quantum Cryptography - 4th International Workshop, PQCrypto 2011, Taipei, Taiwan, Nov 29–Dec 2, 2011. Proceedings, Lecture Notes in Computer Science*, vol. 7071, ed. by B. Yang (Springer, 2011), pp. 98–116
15. N. Sendrier, Decoding one out of many, in *Post-Quantum Cryptography - 4th International Workshop, PQCrypto 2011, Taipei, Taiwan, Nov 29–Dec 2, 2011. Proceedings, Lecture Notes in Computer Science*, vol. 7071, ed. by B. Yang (Springer, 2011), pp. 51–67
16. D.A. Wagner, A generalized birthday problem, in *Advances in Cryptology - CRYPTO 2002, 22nd Annual International Cryptology Conference, Santa Barbara, California, USA, Aug 18–22, 2002, Proceedings, Lecture Notes in Computer Science*, vol. 2442, ed. by M. Yung (Springer, 2002), pp. 288–303

Open Access This chapter is licensed under the terms of the Creative Commons Attribution 4.0 International License (http://creativecommons.org/licenses/by/4.0/), which permits use, sharing, adaptation, distribution and reproduction in any medium or format, as long as you give appropriate credit to the original author(s) and the source, provide a link to the Creative Commons license and indicate if changes were made.

The images or other third party material in this chapter are included in the chapter's Creative Commons license, unless indicated otherwise in a credit line to the material. If material is not included in the chapter's Creative Commons license and your intended use is not permitted by statutory regulation or exceeds the permitted use, you will need to obtain permission directly from the copyright holder.

A Survey on Small Public Key Signature Schemes Derived from UOV Signature Scheme

Yasuhiko Ikematsu

Abstract Multivariate public key cryptosystems (MPKC) are constructed based on the computational difficulty of solving quadratic equations (MQ problem), and are being studied as promising candidates for post-quantum cryptography (PQC). UOV is a multivariate signature scheme and is a very fast scheme with a small signature size, however, its main drawback lies in its large public key. Recently, some variants of UOV have been proposed in order to reduce the public key size, and have been submitted to the additional NIST PQC standardization which started in 2022. It will be important for the future progress of the additional NIST PQC standardization to understand these variants. In this paper, we provide an overview of UOV and variants MAYO, QR-UOV submitted to the additional NIST PQC standardization.

Keywords PQC · MPKC

1 Introduction

In these decades, post-quantum cryptography (PQC), which aims to study cryptosystems that are resistant to quantum computer attacks, has been actively researched. One of the main candidates for PQC is multivariate public key cryptosystems (MPKC), which are based on the difficulty of the problem of finding a solution to the system of multivariate quadratic equations (called MQ problem).

In the NIST PQC standardization project [29] that began in 2016, Rainbow [12] and GeMMS [7] have advanced to the third round as multivariate signature schemes. The Rainbow scheme was proposed by Ding et al. [10] as a more efficient variant of the UOV scheme [25] using the multi-layer technique. The GeMMS scheme proposed by Casanova et al. [7] is an instance of the HFEv- scheme to be a more secure variant of the HFE scheme [32]. Unfortunately, fatal attacks [8, 9] were found before the end of the third round, and these schemes were not been selected as standardization

---

Y. Ikematsu (✉)
Institute of Mathematics for Industry, Kyushu University, Fukuoka, Japan
e-mail: ikematsu@imi.kyushu-u.ac.jp

schemes. In 2022, the lattice-based schemes Crystals-Dilithium, Falcon, and hash-based scheme SPHINCS+ were adopted as signature standardization schemes in the NIST PQC standardization.

The HFE scheme on which the GeMMS scheme is based is already known to be an inefficient scheme, while the UOV scheme [25] on which Rainbow is based has no fatal attacks. The UOV scheme has efficient algorithms and small signature size as its features, however, its drawback is a large public key. So far, several improvements have been proposed to reduce its public key size. For example, as schemes proposed after the NIST PQC standardization, BAC-UOV [37] in 2019, MAYO [3] in 2021, and QR-UOV [18] in 2021 are known. The BAC-UOV scheme accomplishes key size reduction using circulant matrices, however, it was broken by using its algebraic structure [19]. Afterward, the QR-UOV scheme was proposed by generalizing the algebraic structure of circulant matrices. There have been proposed no fatal attacks on MAYO and QR-UOV so far.

In order to ensure the variety of PQC signature algorithms, NIST announced to start the project of the PQC standardization of additional digital signature schemes in 2022 [30] (called the additional NIST PQC standardization in this paper). In particular, NIST specified in the call for proposal [30] that it desires the feature of small signature size, which is the exactly feature of the UOV scheme. The call for proposal of the additional NIST PQC standardization [30] was closed in June 2023 with a total of 50 submission. Out of them, 40 schemes were accepted, and are being analyzed for their security and efficiency. Among these, 10 are MPKC-based, and 7 of those 10 are UOV-based. The specification of each scheme was opened in July, and various discussions are currently taking place in the NIST PQC forum [31]. For example, some flaws have already been reported for three non-UOV-based schemes: 3WISE [35], DME-Sign [28], and HPPC [36]. In addition, it was pointed out in [31] that the proposed parameters of the UOV-based VOX signature scheme [33] are not secure due to an attack not considered in the specification of VOX. The additional NIST PQC standardization, which has just begun, had a number of schemes using new ideas, some of which were just born and therefore some flaws were found. Thus, it would be important for the future progress of the additional NIST PQC standardization to understand these schemes more.

In this paper, we survey UOV, MAYO, and QR-UOV submitted to the additional NIST PQC standardization. The UOV scheme is an important scheme that is receiving attention as stated above. We review the construction and security analysis of the UOV scheme. Moreover, given the recent result that the MinRank attacks of Beullens [2] and Tao et al. [8, 9], initially used in Rainbow and GeMMS respectively, can also be employed in UOV variants, we will provide a summary of their approach. MAYO and QR-UOV are variants of the UOV scheme, and have small public key size compared with the UOV scheme. We review the construction and the security analysis of MAYO and QR-UOV. For the MAYO scheme, we explicitly compute the representation matrices of its public key, and consider how the MAYO scheme is constructed in the view of the matrix representation. For the QR-UOV scheme, we examine how the choice of irreducible polynomials used in its construction affects its security.

This paper is organized as follows. In Sect. 2, we consider the construction and security analysis of the UOV scheme. In Sect. 3, we consider the MAYO scheme and its matrix representation. In Sect. 4, we consider the QR-UOV scheme and the effect of the choice of the irreducible polynomial.

2 Unbalanced Oil Vinegar Scheme (UOV)

The UOV scheme was proposed by Kipnis et al. [25] in 1999. It is considered a secure and basic multivariate signature scheme since there have not been found critical attacks. In this section, we explain the UOV scheme which will be a foundation for schemes treated in this paper. In Sect. 2.1, we describe the construction of the UOV scheme. In Sect. 2.2, we review mainly known attacks for the UOV scheme. In Sect. 2.3, we explain the MinRank attack for the UOV scheme which is recently known.

2.1 Construction of UOV

We describe the key generation, signature generation, and verification in this subsection.

Let v, o, m be three integers, $\mathbb{F}_q$ be the finite field with q elements and set $n := v + o$. We use two variable sets $\mathbf{x}_v = (x_1, \ldots, x_v)$, and $\mathbf{x}_o = (x_{v+1}, \ldots, x_n)$, and put $\mathbf{x} = (\mathbf{x}_v, \mathbf{x}_o)$. We call the first variables $\mathbf{x}_v$ the *vinegar variables* and the second variables $\mathbf{x}_o$ the *oil variables*. The integer m will represent the number of quadratic polynomials used in the UOV scheme.

We explain the construction of the key of the UOV scheme. Randomly choose m quadratic polynomials in $\mathbb{F}_q[x_1, \ldots, x_n]$ in the following form:

$$f_k(\mathbf{x}) = f_k(\mathbf{x}_v, \mathbf{x}_o) = \sum_{1 \le i \le j \le v} a_{ij}^{(k)} x_i x_j + \sum_{i=1}^{v} \sum_{j=v+1}^{n} a_{ij}^{(k)} x_i x_j, \quad (1 \le k \le m). \quad (1)$$

Here, each coefficient $a_{ij}^{(k)}$ is randomly chosen from the finite field $\mathbb{F}_q$. From the form of f_i, it is clear that f_i is a linear polynomial regarding variables $\mathbf{x}_o$ when $\mathbf{x}_v$ is fixed as scalars. For each $k = 1, \ldots, m$, we set the $n \times n$ upper triangular matrix $A_k := (a_{ij}^{(k)})_{ij}$, where $a_{ij}^{(k)} := 0$ if $i > j$ or $i > v+1$. Then it holds $f_k(\mathbf{x}) = \mathbf{x} \cdot A_k \cdot {}^t\mathbf{x}$. We define a quadratic map $\mathcal{F} = (\{_\infty\}, \ldots, \{_\updownarrow\}) : \mathbb{F}_{\text{II}}^{\setminus} \to \mathbb{F}_{\text{II}}^{\updownarrow}$, and randomly choose a linear invertible map $\mathcal{S} : \mathbb{F}_q^n \to \mathbb{F}_q^n$. The secret key consists of $(f_1, \ldots, f_m, \mathcal{S})$. The public key is given as follows. Since the composite $\mathcal{P} := \mathcal{F} \circ \mathcal{S} : \mathbb{F}_q^n \to \mathbb{F}_q^m$ is a quadratic map, we can put $\mathcal{P} = (p_1, \ldots, p_m)$, which a set of m quadratic polynomials. Then the public key is the set $\mathcal{P} = (p_1, \ldots, p_m)$.

Remark 1 In general, the UOV scheme satisfies the condition $m = o$, which is necessary in the signature generation. However, in order to state MAYO and QR-UOV in later sections, it is convenient not to assume the condition. We call the UOV scheme with $m = o$ the *regular* UOV scheme throughout this paper.

The public key and secret key are usually obtained via the matrix computation as follows. Let S be the $n \times n$ matrix such that $\mathcal{S}(\mathbf{x}) = \mathbf{x} \cdot S$. Also, for any $n \times n$ matrix X, we define the upper triangular matrix $\text{Upper}(X)$ such that

$$\text{Upper}(X)_{ij} := \begin{cases} X_{ij} + X_{ji} & (j > i) \\ X_{ii} & (i = j) \\ 0 & (i > j). \end{cases}$$

Then, for any $k = 1, \ldots, m$, the coefficient of $x_i x_j$ $(j \geq i)$ of $p_k(\mathbf{x})$ is equal to the (i, j)-component of the following triangular matrix:

$$B_k := \text{Upper}(S \cdot A_k \cdot {}^t S).$$

In particular, we have $p_k(\mathbf{x}) = \mathbf{x} \cdot B_k \cdot {}^t \mathbf{x}$. Then we can also consider that the public key is the $n \times n$ upper triangular matrices $\{B_1, \ldots, B_m\}$ and the secret key is $\{A_1, \ldots, A_m, S\}$. It is clear that the size of the public key is that of $\frac{1}{2}n(n+1)m$ elements of $\mathbb{F}_q$.

The signature generation and verification processes for the regular UOV scheme are done as follows. Let $\mathcal{H}$ be a hash function and $\mathbf{m}$ be a message to be signed. Then a corresponding signature of $\mathbf{m}$ is generated as a solution to $\mathcal{P}(\mathbf{x}) = \mathcal{H}(\mathbf{m})$, where $\mathcal{H}(\mathbf{m}) \in \mathbb{F}_q^o$. To do so, randomly choose an element $\mathbf{c} = (c_1, \ldots, c_v) \in \mathbb{F}_q^v$, and find a solution $\mathbf{d} \in \mathbb{F}_q^o$ to the following o linear equations in $\mathbf{x}_o$:

$$f_1(\mathbf{c}, \mathbf{x}_o) = m'_1, \cdots, f_o(\mathbf{c}, \mathbf{x}_o) = m'_o, \tag{2}$$

where $\mathcal{H}(\mathbf{m}) = (m'_1, \ldots, m'_o) \in \mathbb{F}_q^o$. Note that since $f_i(\mathbf{c}, \mathbf{x}_o)$ is a linear polynomial as stated above, (2) are certainly linear equations. If there is no solution, we choose another element $\mathbf{c}$. The obtained vector $(\mathbf{c}, \mathbf{d}) \in \mathbb{F}_q^n$ is a solution to $\mathcal{F}(\mathbf{x}) = \mathcal{H}(\mathbf{m})$. Moreover, by computing $\mathbf{s} := (\mathbf{c}, \mathbf{d}) \cdot S^{-1}$, we obtain a solution to $\mathcal{P}(\mathbf{x}) = \mathcal{H}(\mathbf{m})$, and $\mathbf{s}$ is a signature of $\mathbf{m}$. The verification process is done by checking whether $\mathcal{P}(\mathbf{s}) = \mathcal{H}(\mathbf{m})$ or not.

We explain some standard techniques, which are applied to all variants of the UOV scheme, to reduce the sizes of secret and public key. For the secret key, $S \in \text{GL}_n(\mathbb{F}_q)$ is taken as follows:

$$S = \begin{pmatrix} 1_v & 0_{v \times o} \\ S_0 & 1_o \end{pmatrix},$$

where S_0 is taken as a random $o \times v$ matrix. It is known that even if the form of S is restricted in this way, it does not affect the security of the UOV scheme. The

Table 1 The proposed parameters of the UOV scheme [6] in the additional NIST PQC standardization project

NIST security level	(q, v, o, m)	Public key (bytes)	Signature (bytes)
I	(256, 68, 44, 44)	43576	128
	(16, 96, 64, 64)	66576	96
III	(256, 112, 72, 72)	189232	200
V	(256, 148, 96, 96)	446992	260

reduction of the public key size, which was developed by Petzoldt et al. [34], is done as follows. From the above form of S, for any k we obtain

$$B_k = \mathrm{Upper}(S \cdot A_k \cdot {}^t S) = \begin{pmatrix} A_k^{(1)} & A_k^{(1)t} S_0 + {}^t A_k^{(1)t} S_0 + A_k^{(2)} \\ 0_{o \times v} & \mathrm{Upper}\left(S_0 A_k^{(1)t} S_0 + S_0 A_k^{(2)}\right) \end{pmatrix}, \quad (3)$$

where we set $A_k = \begin{pmatrix} A_k^{(1)} & A_k^{(2)} \\ 0_{o \times v} & 0_o \end{pmatrix}$. This matrix (3) is distributed as the public key. Here, $A_k^{(1)}$ is a random matrix generated by the distributor, and is sent as it to verifiers. Thus, $A_k^{(1)}$ can be send as a seed **seed**$_1^k$, and this does not affect the security of the UOV scheme. Moreover, since the $v \times o$ matrix $A_k^{(2)}$ is generated randomly, the matrix $A_k^{(2)} + A_k^{(1)t} S_0 + {}^t A_k^{(1)t} S_0$ is also generated randomly. Thus, if we generate a random $v \times o$ matrix $B_k^{(2)}$ from a seed **seed**$_2^k$ and put $A_k^{(2)} := B_k^{(2)} - A_k^{(1)t} S_0 - {}^t A_k^{(1)t} S_0$, then we can distribute $A_k^{(2)} + A_k^{(1)t} S_0 + {}^t A_k^{(1)t} S_0$ as the seed **seed**$_2^k$. Therefore, the public key is compressed as the two seeds **seed**$_1^k$, **seed**$_2^k$ and $\mathrm{Upper}\left(S_0 A_k^{(1)t} S_0 + S_0 A_k^{(2)}\right)$ ($k = 1, \ldots, m$). As a result, the size of the public key is that of $\frac{1}{2} o(o+1)m$ elements of $\mathbb{F}_q$ and that of two seeds.

Remark 2 The Rainbow signature scheme [12], which has advanced to the third round of the NIST PQC standardization [29], is obtained by using the idea of multi-layer, and was proposed by Ding et al. in 2004. However, a critical attack (called simple attack) was proposed by Beullens [4], and the Rainbow scheme was not selected as a standardization scheme. Beullens' simple attack is not valid for the UOV scheme, and there have not found critical attacks for the UOV scheme even at this point. As a result, after the end of the third round of the NIST PQC standardization, the UOV scheme is paid a lot of attention as a plausible multivariate signature scheme. Beullens et al. proposed new parameters (Table 1) of the UOV scheme [6] based on the latest MPKC security analysis.

In 2023, the UOV scheme with these parameters of Table 1 were submitted to the additional NIST PQC standardization [30]. PROV [21] and TUOV [13], which are variants of the regular UOV scheme having certain provable security, were submitted to the additional NIST PQC standardization, and these public keys have almost same size as that of the UOV scheme in Table 1. Note that, however, a drawback of the UOV scheme is a large public key size compared with other PQC such as lattice-based and isogeny-based cryptosystems. To reduce such a drawback, there have been submitted some schemes such as MAYO [5], QR-UOV [17], VOX [33] and SNOVA [38]. In later sections, we will survey MAYO and QR-UOV.

2.2 Security Analysis

In this subsection, we explain the security analysis of the UOV scheme. In particular, we mainly state the direct attack, KS attack. Regarding the MinRank attack, we will explain it in the next subsection.

2.2.1 Direct Attack

The direct attack means to directly and algebraically solve an instance of the MQ problem related to the public key $\mathcal{P} = (p_1, \ldots, p_m)$. Since the (general) direct attack is related to some attacks other than the direct attack itself, we start with explaining the direct attack against the MQ problem.

The MQ problem is the problem of solving the system of m multivariate quadratic equations $g_1(\mathbf{x}) = \cdots = g_m(\mathbf{x}) = 0$ in n variables $\mathbf{x}$. Then we consider two cases for n and m, namely, the overdetermined case ($n \leq m$) and the underdetermined case ($n > m$). In the underdetermined case, by considering that the solution space is of $n - m$ dimension, we can reduce the system $g_1(\mathbf{x}) = \cdots = g_m(\mathbf{x}) = 0$ to a system of m quadratic equations in m variables by fixing $n - m$ variables in $\mathbf{x}$. Thus, to summarize the two cases, it is enough to solve the system of m quadratic equations in n' variables, where $n' = m$ if $n > m$, otherwise $n' = n$. To solve such a system, Gröbner basis algorithms such as F_4 [14], F_5 [15], and XL [40] are often considered. In particular, the XL algorithm is often used to estimate the complexity of solving such a system. Then, the complexity of solving the system of m quadratic equations in n' variables using the XL Wiedemann algorithm with the hybrid approach [1] is given by

$$\min_k q^k \cdot 3 \left(\frac{n' - k + D_{reg}}{D_{reg}} \right)^2 \binom{n' - k}{2}, \qquad (4)$$

where $0 \leq k \leq m$ is the number of fixed variables in the hybrid approach, and D_{reg} is given by the smallest integer d for which the coefficient of t^d in the function $\frac{(1 - t^2)^m}{(1 - t)^{n'-k}}$ is non-positive.

For the regular UOV scheme, the direct attack finds a solution to the underdetermined system of o quadratic equations $\mathcal{P}(\mathbf{x}) = \mathcal{H}(\mathbf{m})$ in $n = v + o$ variables. Thus, the complexity of the direct attack is given by (4) with $n' = o$ and $m = o$.

Remark 3 For the underdetermined case, Thomae-Wolf [39] proposed the technique to reduce the size of the MQ problem (namely, the numbers of variables and equations). Moreover, their technique was improved by Furue et al. [20] and Hashimoto [22]. The precise complexity of the direct attack for the UOV scheme is given by using such techniques.

2.2.2 Kipnis-Shamir (KS) Attack

The KS attack was proposed by Kipnis and Shamir [26], and is a key recovery attack for the UOV scheme. This attack utilizes the special form of $f_1, \ldots, f_m$ in (1).

First, we recall the matrix representation of quadratic polynomials. Let $g \in \mathbb{F}_q[x_1, \ldots, x_n]$ be a homogeneous quadratic polynomial. Then there exists a unique symmetric matrix $G \in M_n(\mathbb{F}_q)$ such that

$$\mathbf{x} \cdot G \cdot {}^t\mathbf{y} = g(\mathbf{x} + \mathbf{y}) - g(\mathbf{x}) - g(\mathbf{y}) \quad \mathbf{x}, \mathbf{y} \in \mathbb{F}_q^n.$$

We call G the symmetric representation matrix of g.

Second, let $\mathcal{F} = (f_1, \ldots, f_m)$ and $\mathcal{S}$ be a secret key of the UOV scheme and $\mathcal{P} = (p_1, \ldots, p_m)$ the corresponding public key, namely we have $\mathcal{P} = \mathcal{F} \circ \mathcal{S}$. We set F_k to be the symmetric representation matrix of f_k, and P_k that of p_k. Moreover, let S be the $n \times n$ matrix such that $\mathcal{S}(\mathbf{x}) = \mathbf{x} \cdot S$. It is clear that $F_k = A_k + {}^t A_k$ ($k = 1, \ldots, m$) and

$$(P_1, \ldots, P_m) = \left(S \cdot F_1 \cdot {}^t S, \ldots, S \cdot F_m \cdot {}^t S \right) \tag{5}$$

By using this relation, some attacks for MPKC have been proposed so far, such as KS attack, MinRank attack, and so on.

Finally, let $\{e_1, \ldots, e_n\}$ be a standard basis of $\mathbb{F}_q^n$; i.e., $e_1 = (1, 0, \ldots, 0)$ and so on. We set the vinegar space $V := \mathrm{Span}\{e_1, \ldots, e_v\}$ and the oil space $O := \mathrm{Span}\{e_{v+1}, \ldots, e_n\}$ in $\mathbb{F}_q^n$. Then the KS attack tries to find vectors of the twisted oil space

$$O \cdot S^{-1} := \mathrm{Span}\{e_{v+1} S^{-1}, \ldots, e_n S^{-1}\} \tag{6}$$

by computing stable subspaces of XY^{-1} for various two invertible matrices $X, Y \in \mathrm{Span}\{P_1, \ldots, P_m\}$. If the KS attack succeeds, an attacker can find an invertible matrix S' such that $O \cdot S^{-1} S' = O$, which is an equivalent secret key. Namely, the attacker can forge a signature for any message using S'. The complexity of the KS attack is given by $q^{v-o} \cdot o^4$.

2.2.3 Other Attacks

Any element in the twisted oil space $O \cdot S^{-1}$ is a solution to the system of quadratic equations $p_1(\mathbf{x}) = \cdots = p_m(\mathbf{x}) = 0$. An attack that finds such a solution is called the reconciliation attack [11]. By using a feature of the public key $p_1, \ldots, p_m$, the system $p_1(\mathbf{x}) = \cdots = p_m(\mathbf{x}) = 0$ can be reduced to a system of m quadratic equations in v variables. However, for the regular UOV scheme, this attack is harder than the direct attack in general.

The intersection attack was proposed by Beullens [2], and is obtained by combining with the reconciliation attack and KS attack. In general, this attack might be inefficient compare with the direct attack.

As a cryptographic attack, the collision attack is often considered for the UOV scheme. Although omitted in Sect. 2.1, strictly speaking, the signature generation of the UOV scheme [6] is an algorithm such that for a given message $\mathbf{m}$, a signer randomly choose a salt r, finds a solution $\mathbf{x} = \mathbf{s}$ to $\mathcal{P}(\mathbf{x}) = \mathcal{H}(\mathbf{m}||r)$, and outputs $(\mathbf{s}, r)$ as a signature of $\mathbf{m}$. Then the collision attack is to try to find a pair (i, j) satisfying $\mathcal{P}(\mathbf{s}_i) = \mathcal{H}(\mathbf{m}||r_j)$ by collecting a lot of vectors $\{\mathbf{s}_i\}_i$ and salts $\{r_j\}_j$. This collision attack is usually a considerable attack for the UOV scheme and its variants. Regarding the complexity analysis, see the specification of each scheme.

In addition, there are attacks using quantum algorithms. All known quantum attacks for the UOV scheme and its variants are obtained by improving classical attacks using Grover's algorithm. In particular, the complexity of a quantum attack is at least square root of the complexity of the based classical attack. In general, it is easy to analyze the complexity of a quantum attack for the UOV scheme and its variants if the based classical attack is understood well.

2.3 MinRank Attack for UOV

The MinRank attack tries to recover a secret key by finding a low rank matrix generated by the symmetric representation matrices of the public key. Though the MinRank attack had often been considered against HFE variants and Rainbow and so on, it was not known until recently that the MinRank attack can be applied to the UOV scheme. However, the work of Beullens [2] and Tao et al. [8, 9] for Rainbow and GeMMS has revealed that the MinRank attack may be applicable to the UOV scheme. In this subsection, we explain the idea of the MinRank attack for the UOV schemes. Since the effectiveness of the attack is not as threatening as the direct attack or the KS attack, the complexity estimation will not be described in detail (see the specification [6] of the UOV scheme for it).

We prepare to state the idea of the MinRank attack for the UOV scheme. Let $(G_1, \ldots, G_m)$ be a set of $n \times n$ matrices over $\mathbb{F}_q$, and $\mathbf{g}_i^{(j)}$ denotes the j-th column vector of G_i, namely,

$$G_i = \begin{pmatrix} \mathbf{g}_i^{(1)} & \mathbf{g}_i^{(2)} & \cdots & \mathbf{g}_i^{(n)} \end{pmatrix} \in M_n(\mathbb{F}_q).$$

Then, we define the new set $(\tilde{G}_1, \ldots, \tilde{G}_n)$ of $n \times m$ matrices by deforming $(G_1, \ldots, G_m)$ as follows:

$$\tilde{G}_1 := \begin{pmatrix} \mathbf{g}_1^{(1)} & \mathbf{g}_2^{(1)} & \cdots & \mathbf{g}_m^{(1)} \end{pmatrix}, \ldots, \tilde{G}_n := \begin{pmatrix} \mathbf{g}_1^{(n)} & \mathbf{g}_2^{(n)} & \cdots & \mathbf{g}_m^{(n)} \end{pmatrix}.$$

When we apply this deformation to the symmetric representation matrices $(P_1, \ldots, P_m)$ and $(F_1, \ldots, F_m)$ of the public key $\mathcal{P} = (p_1, \ldots, p_m)$ and the secret key $\mathcal{F} = (f_1, \ldots, f_m)$, the following lemma is easily proven from (5):

Lemma 1 [24, Lemma 5]

$$\begin{pmatrix} \tilde{P}_1, \ldots, \tilde{P}_n \end{pmatrix} = \begin{pmatrix} S \cdot \tilde{F}_1, \ldots, S \cdot \tilde{F}_n \end{pmatrix} \cdot {}^t S.$$

From the special form of $f_1, \ldots, f_m$ in (1), it is shown that $\tilde{F}_i$ has the following form:

$$\tilde{F}_i = \begin{cases} \begin{pmatrix} *_{v \times m} \\ *_{o \times m} \end{pmatrix} & (1 \leq i \leq v), \\ \begin{pmatrix} *_{v \times m} \\ 0_{o \times m} \end{pmatrix} & (v+1 \leq i \leq n). \end{cases}$$

By Lemma 1 and the form of $\tilde{F}_i$, we can consider the MinRank attack. Its purpose is to find non-zero vectors of the twisted oil space $O \cdot S^{-1}$ using the MinRank problem associated with $\tilde{F}_1, \ldots, \tilde{F}_n$. More precisely, the MinRank attack for the UOV scheme is explained as follows. Since $\dim O \cdot S^{-1} = o$, there exist $N := \lceil \frac{v}{m} \rceil$ linear independent $n \times 1$ vectors in $O \cdot S^{-1}$ with the following form:

$$\mathbf{a}^{(1)} = (a_1^{(1)}, a_2^{(1)}, \ldots, a_{v+N}^{(1)}, 0, \ldots, 0),$$
$$\vdots$$
$$\mathbf{a}^{(N)} = (a_1^{(N)}, a_2^{(N)}, \ldots, a_{v+N}^{(N)}, 0, \ldots, 0).$$

From $\mathbf{a}^{(k)} \cdot S \in O$ for any $k = 1, \ldots, N$, it is shown that

$$\sum_{i=1}^{v+N} a_i^{(k)} \tilde{P}_i = (\tilde{P}_1, \ldots, \tilde{P}_n) \cdot {}^t \mathbf{a}^{(k)} = (S \cdot \tilde{F}_1, \ldots, S \cdot \tilde{F}_n) \cdot {}^t (\mathbf{a}^{(k)} \cdot S)$$

is a linear combination of $S \cdot \tilde{F}_{v+1}, \ldots, S \cdot \tilde{F}_n$. Thus, the following $n \times Nm$ matrix is of rank $\leq v$:

$$\begin{pmatrix} \sum_{i=1}^{v+N} a_i^{(1)} \tilde{P}_i & \sum_{i=1}^{v+N} a_i^{(2)} \tilde{P}_i & \cdots & \sum_{i=1}^{v+N} a_i^{(N)} \tilde{P}_i \end{pmatrix} \quad (7)$$

Namely, the vectors $\mathbf{a}^{(1)}, \ldots, \mathbf{a}^{(N)}$ are a solution to the MinRank problem (7) with the target rank v consisting of $(\tilde{P}_1, \ldots, \tilde{P}_{v+N})$. Moreover, since $\mathcal{F} = (f_1, \ldots, f_m)$ is zero on O, the public key $\mathcal{P} = (p_1, \ldots, p_m)$ is zero on $O \cdot S^{-1}$. Thus, for any $k = 1, \ldots, N$, we have

$$p_1(\mathbf{a}^{(k)}) = \cdots = p_m(\mathbf{a}^{(k)}) = 0. \tag{8}$$

As a result, the MinRank attack for the UOV scheme can be considered as the kind of attack that finds a common solution $\mathbf{a}^{(1)}, \ldots, \mathbf{a}^{(N)}$ of above problems (7) and (8).

Remark 4 Beullens [2] proposed the rectangular MinRank attack for the Rainbow scheme [12]. This attack is considered to correspond to the above attack for the case of $N = 1$. Using this attack, Beullens pointed out that the proposed parameter of the Rainbow scheme in the third round of NIST PQC standardization does not satisfy the security level. Moreover, Tao et al. broke the GeMMS scheme [7] in [8, 9]. Their attack is done by solving the above MinRank problem (7) associated with the representation matrices obtained by lifting the public key on a certain extension field. Note that the work of Beullens and Tao et al. was proposed independently at the same time. In [16], it is shown that the rectangular MinRank attack can be applied to variants of the UOV scheme such as MAYO [3] and QR-UOV [18].

3 MAYO

The MAYO scheme is a multivariate signature scheme by improving the UOV scheme proposed by Beullens in 2021 [3]. It uses the public key of a small UOV scheme and construct the public key of a large UOV scheme from it. As a result, the MAYO scheme succeeds in making the public key drastically smaller. In Sect. 3.1, we describe the construction of the MAYO scheme. In Sect. 3.2, we explain the proposed parameters and the security analysis. In Sect. 3.3, we consider how the MAYO scheme is constructed in the view of the matrix representation.

3.1 Construction of MAYO

We describe the key generation, signature generation, and verification in this subsection.

Let v, o, m, k be four positive integers, $\mathbb{F}_q$ be the finite field with q elements and set $n := v + o$. Let $\mathcal{F} = (f_1, \ldots, f_m)$ and $\mathcal{S} : \mathbb{F}_q^n \to \mathbb{F}_q^n$ be a secret key of the UOV scheme with the parameter (q, v, o, m), and $\mathcal{P} := \mathcal{F} \circ \mathcal{S} = (p_1, \ldots, p_m)$ be the corresponding public key. In addition to $\mathcal{P}$, the MAYO scheme has additional data as follows. We define the bilinear map $\mathcal{P}' : \mathbb{F}_q^n \times \mathbb{F}_q^n \to \mathbb{F}_q^m$ by

$$\mathcal{P}'(\mathbf{x}, \mathbf{y}) := \mathcal{P}(\mathbf{x}+\mathbf{y}) - \mathcal{P}(\mathbf{x}) - \mathcal{P}(\mathbf{y}) \quad (\mathbf{x}, \mathbf{y} \in \mathbb{F}_q^n).$$

Choose a linear map $\mathcal{E} : \mathbb{F}_q^m \to \mathbb{F}_q^m$ such that its characteristic polynomial is irreducible over $\mathbb{F}_q$. Let $\mathcal{E}_1, \ldots, \mathcal{E}_k, \mathcal{E}_{ij}$ ($1 \le i < j \le k$) be elements randomly generated by $1_m, \mathcal{E}, \ldots, \mathcal{E}^{m-1}$ such that they are linearly independent. Here, since they are generated randomly, they can be generated using a seed $\mathbf{seed}_E$. Then, the public key of the MAYO scheme is given by $\{\mathcal{P}, \mathbf{seed}_E\}$ and the secret key is $\{\mathcal{F}, \mathcal{S}\}$.

Before explaining the signature generation, we prepare some notations. Define the following quadratic maps $\mathcal{P}^*, \mathcal{F}^* : \mathbb{F}_q^{nk} \to \mathbb{F}_q^m$:

$$\mathcal{P}^*(\mathbf{x}_1, \ldots, \mathbf{x}_k) := \sum_{i=1}^{k} \mathcal{E}_i \circ \mathcal{P}(\mathbf{x}_i) + \sum_{1 \le i < j \le k} \mathcal{E}_{ij} \circ \mathcal{P}'(\mathbf{x}_i, \mathbf{x}_j),$$

$$\mathcal{F}^*(\mathbf{x}_1, \ldots, \mathbf{x}_k) := \sum_{i=1}^{k} \mathcal{E}_i \circ \mathcal{F}(\mathbf{x}_i) + \sum_{1 \le i < j \le k} \mathcal{E}_{ij} \circ \mathcal{F}'(\mathbf{x}_i, \mathbf{x}_j),$$

where we set $\mathbf{x}_i = (x_{i,1}, \ldots, x_{i,n})$ for any $i = 1, \ldots, k$. It is clear that

$$\mathcal{F}^*(\mathbf{x}_1, \ldots, \mathbf{x}_k) = \mathcal{P}^*(\mathcal{S}^{-1}(\mathbf{x}_1), \ldots, \mathcal{S}^{-1}(\mathbf{x}_k)).$$

Moreover, as stated in Sect. 2.2.2, $\mathcal{F} = \mathcal{P} \circ \mathcal{S}^{-1}$ vanishes on the oil space O. Therefore, if we set $\mathbf{x}_i = \mathbf{v}_i + \mathbf{o}_i$ ($i = 1, \ldots, k$) such that $\mathbf{v}_i \in V, \mathbf{o}_i \in O$, then we have

$$\mathcal{F}^*(\mathbf{v}_1 + \mathbf{o}_1, \ldots, \mathbf{v}_k + \mathbf{o}_k)$$
$$= \sum_{i=1}^{k} \mathcal{E}_i \left(\mathcal{F}'(\mathbf{v}_i, \mathbf{o}_i) + \mathcal{F}(\mathbf{v}_i) \right) + \sum_{1 \le i < j \le k} \mathcal{E}_{ij} \left(\mathcal{F}'(\mathbf{v}_i, \mathbf{v}_j) + \mathcal{F}'(\mathbf{v}_i, \mathbf{o}_j) + \mathcal{F}'(\mathbf{o}_i, \mathbf{v}_j) \right)$$

From this computation and the fact that $\mathcal{F}'$ is a bilinear map, the map $\mathcal{F}^*(\mathbf{v}_1 + \mathbf{o}_1, \ldots, \mathbf{v}_k + \mathbf{o}_k)$ is an affine map in variables $\mathbf{o}_1, \ldots, \mathbf{o}_k$ when $\mathbf{v}_1, \ldots, \mathbf{v}_k$ are fixed.

Now we explain the signature generation of the MAYO scheme. Let $\mathbf{m}$ be a message to be signed. Then a corresponding signature of $\mathbf{m}$ is generated as a solution to $\mathcal{P}^*(\mathbf{x}_1, \ldots, \mathbf{x}_k) = \mathcal{H}(\mathbf{m})$, where $\mathcal{H}(\mathbf{m}) \in \mathbb{F}_q^m$. To do so, randomly choose elements $\mathbf{v}_1, \ldots, \mathbf{v}_k \in \mathbb{F}_q^v$, and find a solution $\mathbf{o}_1, \ldots, \mathbf{o}_k \in \mathbb{F}_q^o$ satisfying the following m linear equations in ko variables:

$$\mathcal{F}^*(\mathbf{v}_1 + \mathbf{o}_1, \ldots, \mathbf{v}_k + \mathbf{o}_k) = \mathcal{H}(\mathbf{m}).$$

Moreover, by computing $\mathbf{s} := (\mathcal{S}^{-1}(\mathbf{v}_1 + \mathbf{o}_1), \ldots, \mathcal{S}^{-1}(\mathbf{v}_k + \mathbf{o}_k))$, we obtain a solution to $\mathcal{P}^*(\mathbf{x}_1, \ldots, \mathbf{x}_k) = \mathcal{H}(\mathbf{m})$, and $\mathbf{s}$ is a signature of $\mathbf{m}$.

The verification process is done by computing $\mathcal{P}^*$ from the public key $\{\mathcal{P}, \mathbf{seed}_E\}$ and checking whether $\mathcal{P}^*(\mathbf{s}) = \mathcal{H}(\mathbf{m})$, or not.

Table 2 The proposed parameters of the MAYO scheme [3] in the additional NIST PQC standardization

NIST security level	(q, v, o, m, k)	Public key (bytes)	Signature (bytes)
I	(16, 58, 8, 64, 9)	1168	321
	(16, 60, 18, 64, 4)	5488	180
III	(16, 89, 10, 96, 11)	2656	577
V	(16, 121, 12, 128, 12)	5008	838

The quadratic map $\mathcal{P}^*$ used in the verification process consists of m quadratic polynomials in kn variables. Since a verifier can construct this map from the system $\mathcal{P}$ of m quadratic equations in n variables and $\mathbf{seed}_E$. Thus, the MAYO scheme can change the public key from $\mathcal{P}^*$ to $\{\mathcal{P}, \mathbf{seed}_E\}$. As a result, the size of the public key of the MAYO scheme can be reduced drastically.

3.2 Security Analysis

Table 2 is the proposed parameters of the MAYO scheme in the additional NIST PQC standardization. These parameters are proposed based on basically the attacks stated in Sect. 2. The most effective attack is the collision attack (called claw finding attack in the specification [5]), and the next one is the reconciliation attack. Here, the reconciliation attack for the MAYO scheme is the key recovery attack that finding a non-zero element in the twisted oil space $O \cdot S^{-1}$ by solving $\mathcal{P}(\mathbf{x}) = 0$. From the condition $n < m$, this attack is effective compared with the case of the regular UOV scheme. Note that the MinRank attack stated in Sect. 2 was not considered in the specification of the MAYO scheme [5], and it was pointed out in [16] that the MinRank attack with $N = 1$ can be applied to the MAYO scheme while the attack does not affect the security of the proposed parameters.

Remark 5 In [23], Hashimoto proposed an improved technique of Thomae-Wolf [39], and applied it to the MAYO scheme. As a result, it is pointed out that some parameters are lower than the corresponding NIST security level. For example, it is shown that the security of the first level I parameter is 126 bits and does not satisfy the NIST security level I.

3.3 Matrix Representation

In this subsection, we consider the structure of the matrix representation of $\mathcal{P}^*$. By doing so, we understand the key compression idea of the MAYO scheme in the view of the matrix representation.

First, we compute the representation matrices of $\mathcal{P}'$. When we put $\mathcal{P}' = \{p'_1, \ldots, p'_m\}$, we obtain $p'_l(\mathbf{x}, \mathbf{y}) = p_l(\mathbf{x} + \mathbf{y}) - p_l(\mathbf{x}) - p_l(\mathbf{y})$ from the definition. Then, by the definitions of B_l and P_l in Sect. 2.1, we have the following computation:

$$p_l(\mathbf{x} + \mathbf{y}) - p_l(\mathbf{x}) - p_l(\mathbf{y}) = (\mathbf{x} + \mathbf{y}) \cdot B_l \cdot {}^t(\mathbf{x} + \mathbf{y}) - \mathbf{x} \cdot B_l \cdot {}^t\mathbf{x} - \mathbf{y} \cdot B_l \cdot {}^t\mathbf{y}$$
$$= \mathbf{x} \cdot B_l \cdot {}^t\mathbf{y} + \mathbf{y} \cdot B_l \cdot {}^t\mathbf{x}$$
$$= \mathbf{x} \cdot (B_l + {}^tB_l) \cdot {}^t\mathbf{y} = \mathbf{x} \cdot P_l \cdot {}^t\mathbf{y}.$$

Thus, we obtain $p'_l(\mathbf{x}, \mathbf{y}) = \mathbf{x} \cdot P_l \cdot {}^t\mathbf{y}$. Moreover, when we put $\mathcal{P}^* = (p^*_1, \ldots, p^*_m)$, each p^*_l is represented by using $p_1, p'_1, \ldots, p_m, p'_m$.

$$p^*_l(\mathbf{x}_1, \ldots, \mathbf{x}_k) = \sum_{i=1}^{k}\sum_{h=1}^{m} \alpha^{(l)}_{ih} p_h(\mathbf{x}_i) + \sum_{1 \leq i < j \leq k}\sum_{1 \leq h \leq m} \beta^{(l)}_{ijh} p'_h(\mathbf{x}_i, \mathbf{x}_j),$$

where these coefficients $\alpha^{(l)}_{ih}, \beta^{(l)}_{ijh}$ are determined from $\mathcal{E}_1, \ldots, \mathcal{E}_k, \mathcal{E}_{ij}$ ($1 \leq i < j \leq k$). Therefore, we can represent p^*_l as follows:

$$p^*_l(\mathbf{x}_1, \ldots, \mathbf{x}_k) =$$

$$\begin{pmatrix} \mathbf{x}_1 & \cdots & \mathbf{x}_k \end{pmatrix} \begin{pmatrix} \sum_{h=1}^{m} \alpha^{(l)}_{1h} B_h & \sum_{1 \leq h \leq m} \beta^{(l)}_{12h} P_h & \cdots & \sum_{1 \leq h \leq m} \beta^{(l)}_{1kh} P_h \\ 0 & \sum_{h=1}^{m} \alpha^{(l)}_{2h} B_h & \cdots & \sum_{1 \leq h \leq m} \beta^{(l)}_{2kh} P_h \\ 0 & 0 & \ddots & \vdots \\ 0 & 0 & 0 & \sum_{h=1}^{m} \alpha^{(l)}_{kh} B_h \end{pmatrix} \begin{pmatrix} \mathbf{x}_1 \\ \vdots \\ \mathbf{x}_k \end{pmatrix}. \quad (9)$$

Moreover, the symmetric matrix representation of p^*_l is given by

$$\begin{pmatrix} \sum_{h=1}^{m} \alpha^{(l)}_{1h} P_h & \sum_{1 \leq h \leq m} \beta^{(l)}_{12h} P_h & \cdots & \sum_{1 \leq h \leq m} \beta^{(l)}_{1kh} P_h \\ \sum_{1 \leq h \leq m} \beta^{(l)}_{12h} P_h & \sum_{h=1}^{m} \alpha^{(l)}_{2h} P_h & \cdots & \sum_{1 \leq h \leq m} \beta^{(l)}_{2kh} P_h \\ \vdots & \vdots & \ddots & \vdots \\ \sum_{1 \leq h \leq m} \beta^{(l)}_{1kh} P_h & \sum_{1 \leq h \leq m} \beta^{(l)}_{2kh} P_h & \cdots & \sum_{h=1}^{m} \alpha^{(l)}_{kh} P_h \end{pmatrix}.$$

The representation matrix of p^*_l is an $nk \times nk$ matrix, and each block matrix with size $n \times n$ is a linear combination of $B_1, \ldots, B_m$ or $P_1, \ldots, P_m$, where P_h is generated by B_h as $P_h = B_h + {}^tB_h$. As a result, p^*_l is generated by $B_1, \ldots, B_m$ and the coefficients $\alpha^{(l)}_{ih}, \beta^{(l)}_{ijh}$ that are computed from $\mathcal{E}_1, \ldots, \mathcal{E}_k, \mathcal{E}_{ij}$ ($1 \leq i < j \leq k$).

Remark 6 We consider the following another construction of the MAYO scheme via the matrix representation. Let $B_1, \ldots, B_{m'}$ be the public key of the UOV scheme with parameter (q, v, o, m'). Randomly choose m seeds $\mathbf{seed}^{(1)}_{\text{coeff}}, \ldots, \mathbf{seed}^{(m)}_{\text{coeff}}$. From the seed $\mathbf{seed}^{(l)}_{\text{coeff}}$ for any $l = 1, \ldots, m$, one generates coefficients $\{\alpha^{(l)}_{ih}\}_{1 \leq i \leq k, 1 \leq h \leq m'}, \{\beta^{(l)}_{ijh}\}_{1 \leq i < j \leq k, 1 \leq h \leq m'}$. Finally, define $p^*_l(\mathbf{x}_1, \ldots, \mathbf{x}_k)$ ($1 \leq l \leq m$) as in (9), where m has changed to m'. If we take $m' < m$, then the public key size might be smaller than that of the MAYO scheme stated in Sect. 3.1. In order to propose such a modified variant of the MAYO scheme, the security analysis is required, and it will be future work.

4 QR-UOV

The QR-UOV scheme is a variant of the UOV scheme proposed by Furue et al. [18]. By using the structure of a quotient ring (QR) of $\mathbb{F}_q[t]$, the QR-UOV scheme realizes small public keys compared with the UOV scheme. In Sect. 4.1, we recall the construction of the QR-UOV scheme. In Sect. 4.2, we explain the security analysis. In Sect. 4.3, we consider how the choice of quotient rings used in the key size reduction affects the security of the QR-UOV scheme.

4.1 Construction of QR-UOV

Let V, O, k be three positive integers and set

$$v := Vk, o := Ok, N := V + O, n := v + o = Nk, m = o.$$

The idea of the QR-UOV scheme is to construct the UOV scheme with the parameter (q, v, o, m) from the UOV scheme with the parameter (q^k, V, O, m) using the embedding of a quotient ring.

Before stating the construction, we prepare some notations and facts. For an irreducible polynomial $h(t) \in \mathbb{F}_q[t]$ with degree k, we define the embedding

$$\phi : \mathbb{F}_{q^k} = \mathbb{F}_q[t]/(h(t)) \to M_k(\mathbb{F}_q)$$

by $(1, t, \ldots, t^{k-1}) \cdot \phi(g) = (g, gt, \ldots, gt^{k-1})$ for $g \in \mathbb{F}_{q^k}$. We denote the image of the embedding ϕ by $\mathcal{A}_h$, which is a subring of the matrix ring. By Theorem 1 in [18], there exists an invertible symmetric matrix $W \in M_k(\mathbb{F}_q)$ such that $W \cdot \phi(g)$ is symmetric for any $g \in \mathbb{F}_{q^k}$, that is, $W\mathcal{A}_h$ consists of symmetric matrices. We extend this ϕ to the embedding between the matrix rings as follows:

$$\phi : M_N(\mathbb{F}_{q^k}) \ni (a_{ij}) \mapsto (\phi(a_{ij})) \in M_n(\mathbb{F}_q).$$

Then, from the property of W stated above, it is proven that

$$W^{(N)} \cdot \phi({}^t X) = {}^t\phi(X) \cdot W^{(N)} \quad \text{for any } X \in M_N(\mathbb{F}_{q^k}), \tag{10}$$

where $W^{(N)} := \begin{pmatrix} W & & \\ & \ddots & \\ & & W \end{pmatrix} \in M_n(\mathbb{F}_q)$, namely, $W^{(N)}$ is the matrix obtained by diagonally arranging W.

Now, the key generation is done as follows. Randomly choose m symmetric matrices $F_1, \ldots, F_m$ in $M_N(\mathbb{F}_{q^k})$ in the following form:

$$F_i = \begin{pmatrix} *_V & *_{V \times O} \\ *_{O \times V} & 0_O \end{pmatrix}. \tag{11}$$

Then, we define the quadratic map $\mathcal{F} = (f_1, \ldots, f_m) : \mathbb{F}_q^n \to \mathbb{F}_q^m$ by

$$f_i(\mathbf{x}) := \mathbf{x} \cdot W^{(N)} \cdot \phi(F_i) \cdot {}^t\mathbf{x} \quad (1 \le i \le m),$$

where $\mathbf{x} = (x_1, \ldots, x_n)$. Here, from the structure of $W^{(N)}$ and ϕ, the matrix $W^{(N)} \cdot \phi(F_i)$ has the form $\begin{pmatrix} *_v & *_{v \times v} \\ *_{o \times v} & 0_o \end{pmatrix}$. Thus, $f_i(\mathbf{x})$ is linear polynomial regarding variables $x_{v+1}, \ldots, x_n$. Moreover, from (10), the matrix $W^{(N)} \cdot \phi(F_i)$ is symmetric. Thus, the matrix representation of f_i is $2 \cdot W^{(N)} \cdot \phi(F_i)$. Next, randomly choose an invertible matrix $S \in M_N(\mathbb{F}_{q^k})$. The secret key consists of $\{F_1, \ldots, F_m, S\}$. Then the public key is given by $\{SF_1{}^tS, \ldots, SF_m{}^tS, W\}$. The quadratic map $\mathcal{P} = (p_1, \ldots, p_m)$ used in the verification process is given by

$$\begin{aligned} p_i(\mathbf{x}) &:= \mathbf{x} \cdot {}^t\phi({}^tS) \cdot W^{(N)} \cdot \phi(F_i) \cdot \phi({}^tS) \cdot {}^t\mathbf{x} \\ &= \mathbf{x} \cdot W^{(N)} \cdot \phi(S \cdot F_i \cdot {}^tS) \cdot {}^t\mathbf{x}. \end{aligned} \tag{12}$$

Though each p_i is a quadratic polynomial in n variables over $\mathbb{F}_q$, it can be computed from the $N \times N$ matrix $SF_i{}^tS$ over $\mathbb{F}_{q^k}$. Since p_i has $\frac{1}{2}n(n+1)$ coefficients of $\mathbb{F}_q$ and $SF_i{}^tS$ consists of $\frac{1}{2}N(N+1)$ elements of $\mathbb{F}_{q^k}$, $SF_i{}^tS$ is around k times smaller than p_i. This is the reason that the public key of the QR-UOV scheme is smaller than that of the UOV scheme.

The signature generation is the same as that of the UOV scheme since $f_1, \ldots, f_m$ have the property that they are linear polynomials regarding variables $x_{v+1}, \ldots, x_n$. The verification is done by computing $\mathcal{P}$ from the public key, and checking whether $\mathcal{P}(\mathbf{s}) = \mathcal{H}(\mathbf{m})$ for a message $\mathbf{m}$ and its signature $\mathbf{s}$.

Remark 7 Since f_i and p_i are computed from symmetric matrices, if the characteristic of $\mathbb{F}_q$ is 2 then almost the coefficients of f_i and p_i are zero, which might cause some problems in the security. To avoid such a situation, the characteristic of $\mathbb{F}_q$ is not 2 in the QR-UOV scheme.

4.2 Security Analysis

The following table is the proposed parameters of the QR-UOV scheme in the additional NIST PQC standardization [30] (Table 3).

There are two groups of attacks for the QR-UOV scheme. First one is the group of attacks against the quadratic map $\mathcal{P} = (p_1, \ldots, p_m)$, which is identified with the public key of the UOV scheme of the parameter (q, v, o, m). Thus, we can apply the attacks in Sect. 2 to $\mathcal{P}$. Second one is the group of key recovery attacks against the public key $\{SF_1{}^tS, \ldots, SF_m{}^tS\}$, which is considered as the public key of the UOV scheme with the parameter (q^k, V, O, m). We can apply the key recovery attacks in Sect. 2 to $\{SF_1{}^tS, \ldots, SF_m{}^tS\}$ in order to recover the secret key S or its equivalent key. Since the parameters v, o are quite large, the first group is not efficient except for the collision attack (see Table 8 in the specification [17]). Therefore, the choice of the parameter of the QR-UOV scheme is done using the collision attack and the second group (see Table 7 in the specification [17]).

Remark 8 VOX [33] was proposed as an improved variant of the UOV scheme by Patarin et al. in the additional NIST PQC standardization. This variant is obtained by removing some quadratic polynomials in the public key of the UOV scheme, and mixing some random quadratic polynomials with the remaining polynomials in the public key. By doing so, the VOX scheme enhances the high security compared with the UOV scheme. Moreover, by combining the structure of the QR-UOV scheme, the VOX scheme realized the compact public key compared with the QR-UOV scheme. Recently, Furue et al. pointed out in [31] that the proposed parameters in [33] do not satisfy the security level due to the MinRank attack with $N=1$ in Sect. 2.3. Therefore, new parameter set is currently in the process of being revised by the team of VOX.

Table 3 The proposed parameters of the QR-UOV scheme [17] in the additional NIST PQC standardization

NIST security level	(q, v, o, m, k)	Public key (bytes)	Signature (bytes)
I	(7, 740, 100, 100, 10)	20657	331
	(31, 165, 60, 60, 3)	23657	157
	(31, 600, 70, 70, 10)	12282	435
	(127, 156, 54, 54, 3)	24271	200
III	(7, 1100, 140, 140, 10)	55173	489
	(31, 246, 87, 87, 3)	71007	232
	(31, 890, 100, 100, 10)	34423	643
	(127, 228, 78, 78, 3)	71915	292
V	(7, 1490, 190, 190, 10)	135439	662
	(31, 324, 114, 114, 3)	158453	306
	(31, 1120, 120, 120, 10)	58564	807
	(127, 306, 105, 105, 3)	173708	392

4.3 Choice of Irreducible Polynomial

By choosing an irreducible polynomial h, the QR-UOV scheme reduces the public key size by representing the quadratic map $\mathcal{P}$ using the embedding ϕ. In this subsection, we consider how the choice of h affect the security of the QR-UOV scheme. In particular, we prove that the security of the QR-UOV scheme does not depend on the choice of h.

We choose an irreducible polynomial h, set the embedding ϕ and the subring $\mathcal{A}_h$, and choose a symmetric matrix W as in Sect. 4.1. We call the quadratic map $\mathcal{P}$ in (10) a verification map $\mathcal{P}$ of (h, W).

We choose another irreducible polynomial h'. The map ϕ' is the corresponding embedding to h'. Moreover, set the associated algebra $\mathcal{A}_{h'}$ and choose a symmetric matrix W'. Then, we prove that there exists a change of variables regarding $\mathbf{x}$ such that the set of verification maps of (h, W) bijectively corresponds that of (h', W').

First, from Skolem-Neother's theorem, we have the following.

Theorem 1 *There exists an element $g \in GL_k(\mathbb{F}_q)$ such that $\mathcal{A}_{h'} = g^{-1}\mathcal{A}_h g$.*

Next, we prove the following proposition regarding W and W'.

Proposition 1 *(i) Let W_0 be a non-singular symmetric matrix such that $W_0 \mathcal{A}_h$ consists of symmetric matrices. Then, there exists an element $a \in \mathcal{A}_h^\times$ such that $W = W_0 a$.*
(ii) There exists elements $g \in GL_k(\mathbb{F}_q)$ and $a \in \mathcal{A}_h^\times$ such that ${}^t g W = W' g^{-1} a$.

Proof (i) Let ϕ be the embedding stated in Sect. 4.1 and set $X := \phi(t)$. Then, X is a generator of the subring $\mathcal{A}_h$. Since WX and $W_0 X$ are symmetric, we have $WX = {}^t(WX) = {}^t X {}^t W = {}^t X W$ and $W_0 X = {}^t X W_0$. Therefore, $WXW^{-1} = {}^t X = W_0 X W_0^{-1}$ and $(W_0^{-1}W)X(W_0^{-1}W)^{-1} = X$. Since X commutes with $W_0^{-1} W$ and is a generator of the maximal subring $\mathcal{A}_h$ in $M_k(\mathbb{F}_q)$, we have $W_0^{-1} W \in \mathcal{A}_h$. This implies (i).
(ii) By Theorem 1, we have $W' \mathcal{A}_{h'} = W' g^{-1} \mathcal{A}_h g$. Since this space consists of symmetric matrices, so is ${}^t g^{-1} \left(W' g^{-1} \mathcal{A}_h g \right) g^{-1} = {}^t g^{-1} W' g^{-1} \mathcal{A}_h$. Thus, applying the result of (i) to $W_0 := {}^t g^{-1} W' g^{-1}$, it is shown that there exists $a \in \mathcal{A}_h^\times$ such that ${}^t g W = W' g^{-1} a$. □

Finally, we define the following matrices:

$$g^{(N)} := \begin{pmatrix} g & & \\ & \ddots & \\ & & g \end{pmatrix}, \quad a^{(N)} := \begin{pmatrix} a & & \\ & \ddots & \\ & & a \end{pmatrix} \in M_n(\mathbb{F}_q).$$

Since $a \in \mathcal{A}_h$, we can write as $a = \phi(\alpha)$ ($\alpha \in \mathbb{F}_{q^k}$). Then, we prove that by the change of variables defined as $g^{(N)}$, the set of verification map of (h, W) corresponds that of (h', W'). Let $\mathcal{P} = (p_1, \ldots, p_m)$ be a verification map of (h, W) in (12). Then, by Proposition 1 (ii), we have the following computation.

$$\begin{aligned} p_i(\mathbf{x} \cdot {}^t g^{(N)}) &= \mathbf{x} \cdot {}^t g^{(N)} \cdot W^{(N)} \cdot \phi(S \cdot F_i \cdot {}^t S) \cdot g^{(N)} \cdot {}^t \mathbf{x} \\ &= \mathbf{x} \cdot W'^{(N)} \cdot g^{(N),-1} a^{(N)} \cdot \phi(S \cdot F_i \cdot {}^t S) \cdot g^{(N)} \cdot {}^t \mathbf{x} \\ &= \mathbf{x} \cdot W'^{(N)} \cdot g^{(N),-1} \cdot \phi(S \cdot \alpha F_i \cdot {}^t S) \cdot g^{(N)} \cdot {}^t \mathbf{x}. \end{aligned}$$

By Theorem 1, $g^{(N),-1} \cdot \phi(S \cdot \alpha F_i \cdot {}^t S) \cdot g^{(N)}$ can be written as $\phi'(S' \cdot \alpha F_i' \cdot {}^t S')$ for some $S', F_i' \in M_N(\mathbb{F}_{q^k})$, where F_i' has the form in (11). Thus, $p_i(\mathbf{x} \cdot {}^t g^{(N)})$ is a quadratic polynomial in a verification map of (h', W'). Therefore, we proved that the set of verification maps of (h, W) bijectively corresponds to that of (h', W') by the change of variables ${}^t g^{(N)}$. Since g can be computed from h and h', an attacker can construct a verification map of (h, W) from that of (h', W'). This means that the choice of an irreducible polynomial h does not affect the security of the QR-UOV scheme. Note that, however, the choice of h might affect the efficiency, and its study might be important task for the QR-UOV scheme.

Acknowledgements This work was supported by JST CREST Grant Number JPMJCR2113 and JSPS KAKENHI Grant Number JP19K20266.

References

1. L. Bettale, J.C. Faugère, L. Perret, Hybrid approach for solving multivariate systems over finite fields. J. Math. Crypt. **3**, 177–197 (2009)
2. W. Beullens, Improved Cryptanalysis of UOV and Rainbow (Springer, EUROCRYPT. LNCS **12696**, 348–373 (2021)
3. W. Beullens, MAYO: Practical Post-quantum Signatures from Oil-and-Vinegar Maps (Springer, SAC. LNCS **13203**, 355–376 (2021)
4. W. Beullens, Breaking Rainbow Takes a Weekend on a Laptop (Springer, CRYPTO. LNCS **13508**, 464–479 (2022)
5. W. Beullens, F. Campos, S. Celi, B. Hess, M. Kannwischer, 'MAYO', *Specification document of NIST PQC Standardization of Additional Digital Signature Scheme* (2023)
6. W. Beullens, M.-S. Chen, J. Ding, B. Gong, M.J. Kannwischer, J. Patarin, B.Y. Peng, D. Schmidt, C.J. Shih, C. Tao, B.Y. Yang, 'UOV', *Specification document of NIST PQC Standardization of Additional Digital Signature Scheme* (2023)
7. A. Casanova, J.-C. Faugere, G. Macario-Rat, J. Patarin, L. Perret, J. Ryckeghem, 'GeMSS: A Great Multivariate Short Signature', Specification document of NIST PQC 3rd round submission package (2020)
8. C. Tao, A. Petzoldt, J. Ding, Improved key recovery of the HFEv- signature scheme. IACR Cryptol. ePrint Arch. **2020**, 1424 (2020)
9. C. Tao, A. Petzoldt, J. Ding, Efficient key recovery for all HFE signature variants. CRYPTO **1**, 70–93 (2021)
10. J. Ding, D.S. Schmidt, Rainbow, a New Multivariate Polynomial Signature Scheme (Springer, ACNS. LNCS **3531**, 164–175 (2005)
11. J. Ding, B.Y. Yang, C.H.O. Chen, M.S. Chen, C.M. Cheng, New Differential-algebraic Attacks and Reparametrization of Rainbow (Springer, ACNS. LNCS **5037**, 242–257 (2008)

12. J. Ding, M.S. Chen, A. Petzoldt, D.S. Schmidt, B.Y. Yang, 'Rainbow', technical report, national institute of standards and technology. Post-Quant. Crypt. https://csrc.nist.gov/Projects/Post-Quantum-Cryptography/Round-3-submissions
13. J. Ding, B. Gong, H. Guo, X. He, Y. Jin, Y. Pan, D. Schmidt, C. Tao, D. Xie, B.Y. Yang, Z. Zhao, *'TUOV', Specification document of NIST PQC Standardization of Additional Digital Signature Scheme* (2023)
14. J.C. Faugère, A new efficient algorithm for computing Gröbner bases (F4). J. Pure Appl. Algebra **139**, 61–88 (1999)
15. J.C. Faugère, A new efficient algorithm for computing Gröbner Bases without reduction to zero (F5) (ISSAC, 2002), pp. 75–83
16. H. Furue, Y. Ikematsu, *A New Security Analysis Against MAYO and QR-UOV Using Rectangular MinRank Attack* (Springer, IWSEC, 2023), pp.101–116
17. H. Furue, Y. Ikematsu, F. Hoshino, T. Takagi, K. Yasuda, T. Miyazawa, T. Saito, A. Nagai, *'QR-UOV', Specification document of NIST PQC Standardization of Additional Digital Signature Scheme* (2023)
18. H. Furue, Y. Ikematsu, Y. Kiyomura, T. Takagi, A new variant of unbalanced oil and vinegar using quotient ring: QR-UOV (Springer, ASIACRYPT. LNCS **13093**, 187–217 (2021)
19. H. Furue, K. Kinjo, Y. Ikematsu, Y. Wang, T. Takagi, *A Structural Attack on Block-anti-circulant UOV at SAC 2019* (Springer, PQC2020, 2020), pp. 323–339
20. H. Furue, S. Nakamura, T. Takagi, Improving Thomae-Wolf Algorithm for Solving Underdetermined Multivariate Quadratic Polynomial Problem (PQCrypto. LNCS **12841**, 65–78 (2021)
21. L. Goubin, B. Cogliati, J.C. Faugère, P.A. Fouque, R. Larrieu, M.R. Gilles, B. Minaud, J. Patarin, *PROV, Specification document of NIST PQC Standardization of Additional Digital Signature Scheme* (2023)
22. Y. Hashimoto, An improvement of algorithms to solve under-defined systems of multivariate quadratic equations. JSIAM Lett. **15**, 53–56 (2023)
23. Y. Hashimoto, *An Improvement of Algorithms to Solve Under-defined Systems of Multivariate Quadratic Equations* IACR Cryptology ePrint Archive, Report 2021/1045 (2021)
24. Y. Ikematsu, S. Nakamura, T. Takagi, *Recent Progress in the Security Evaluation of Multivariate Public-key Cryptography* (IET Information Security, 2022)
25. A. Kipnis, L. Patarin, L. Goubin, Unbalanced Oil and Vinegar Schemes (Springer, EUROCRYPT. LNCS **1592**, 206–222 (1999)
26. A. Kipnis, A. Shamir, *Cryptanalysis of the Hfe Public Key Cryptosystem by Relinearization* (Springer, CRYPTO 99, LNCS 1666), pp. 19-30
27. A. Kipnis, A. Shamir, *Cryptanalysis of the Oil and Vinegar Signature Scheme* (Springer, CRYPTO 98, LNCS 1462), pp. 257–266
28. I. Luengo, M. Avendaño, *DME-Sign, Specification document of NIST PQC Standardization of Additional Digital Signature Scheme* (2023)
29. National Institute of Standards and Technology, *Post-Quantum Cryptography Standardization* https://csrc.nist.gov/projects/post-quantum-cryptography
30. National Institute of Standards and Technology, *Call for Additional Digital Signature Schemes for the Post-Quantum Cryptography Standardization Process* https://csrc.nist.gov/csrc/media/Projects/pqc-dig-sig/documents/call-for-proposals-dig-sig-sept-2022.pdf
31. NIST PQC forum, https://groups.google.com/a/list.nist.gov/g/pqc-forum
32. J. Patarin, *Hidden Field Equations (HFE) and Isomorphisms of Polynomials (IP): Two New Families of Asymmetric Algorithms* (Springer, EUROCRYPT 1996, LNCS 1070), pp. 33–48
33. J. Patarin, B. Cogliati, J.C. Faugère, P.A. Fouque, L. Goubin, R. Larrieu, G. Macario-Rat, B. Minaud, *VOX, Specification document of NIST PQC Standardization of Additional Digital Signature Scheme* (2023)
34. A. Petzoldt, S. Bulygin, J.-A. Buchmann, CyclicRainbow-a Multivariate Signature Scheme with a Partially Cyclic Public Key (Springer. INDOCRYPT, LNCS **6498**, 33–48 (2010)
35. B.G. Rodríguez, *3WISE, Specification document of NIST PQC Standardization of Additional Digital Signature Scheme* (2023)

36. B.G. Rodríguez, *HPPC Specification document of NIST PQC Standardization of Additional Digital Signature Scheme* (2023)
37. A. Szepieniec, B. Preneel, Block-anti-circulant Unbalanced Oil and Vinegar (Springer. SAC **2019**, 574–588 (2020)
38. L.C. Wang, C.Y. Chou, J. Ding, Y.L. Kuan, M.S. Li, B.S. Tseng, P.E. Tseng, C.C. Wang, *SNOVA, Specification document of NIST PQC Standardization of Additional Digital Signature Scheme* (2023)
39. E. Thomae, C. Wolf, Solving Underdetermined Systems of Multivariate Quadratic Equations Revisited (PKC. LNCS **7293**, 156–171 (2012)
40. B.Y. Yang, J.M. Chen, All in the XL Family: Theory and Practice (Springer, ICISC. LNCS **3506**, 67–86 (2004)
41. D. Wiedemann, Solving sparse linear equations over finite fields. IEEE Trans. Inform. Theory **32**(1), 54–62 (1986)

Open Access This chapter is licensed under the terms of the Creative Commons Attribution 4.0 International License (http://creativecommons.org/licenses/by/4.0/), which permits use, sharing, adaptation, distribution and reproduction in any medium or format, as long as you give appropriate credit to the original author(s) and the source, provide a link to the Creative Commons license and indicate if changes were made.

The images or other third party material in this chapter are included in the chapter's Creative Commons license, unless indicated otherwise in a credit line to the material. If material is not included in the chapter's Creative Commons license and your intended use is not permitted by statutory regulation or exceeds the permitted use, you will need to obtain permission directly from the copyright holder.

A Survey of Attacks on Supersingular Isogeny with Torsion Problem and Its Variants

Hiroshi Onuki

Abstract The problem of computing an isogeny between two given supersingular elliptic curves over a finite field holds significant importance in cryptography. In particular, isogeny-based cryptography, which is a promising candidate for post-quantum cryptography, relies on the hardness of this problem. When constructing cryptosystems, it is often necessary to assume the hardness of a variant of the problem with additional information. One such problem is the supersingular isogeny with torsion (SSI-T) problem, where the degree of a secret isogeny and the image of a torsion subgroup under the isogeny are provided. SIDH is a key-exchange protocol founded on the hardness of the SSI-T problem. SIKE, a key encapsulation mechanism, is based on SIDH and was a round-4 candidate in the NIST post-quantum cryptography standardization process. However, in 2022, Castryck and Decru introduced a polynomial-time attack on the SSI-T problem in SIKE, thereby compromising the security of SIDH and SIKE. Subsequently, various works have emerged, which include extensions of this attack, countermeasures against this attack, and new cryptosystems using this attack. These efforts have explored different variants of the SSI-T problem. This paper provides a survey of the attacks on the SSI-T problem and its variants.

Keywords Isogeny-based cryptography · Supersingular isogeny problem · Supersingular elliptic curves

1 Introduction

Isogeny-based cryptography is one of the candidates for post-quantum cryptography (PQC). The security of isogeny-based cryptography relies on the problem to find an isogeny between two given isogenous abelian varieties over a finite field. This problem is called the *isogeny problem*. Especially, we use supersingular elliptic

---

H. Onuki (✉)
Department of Mathematical Informatics, The University of Tokyo, Tokyo, Japan
e-mail: hiroshi-onuki@g.ecc.u-tokyo.ac.jp

curves as abelian varieties for the efficiency of the computation. We call this problem the *supersingular isogeny (SSI) problem*.

The first cryptosystem using isogenies between supersingular elliptic curves is a hash function proposed by Charles, Goren, and Lauter [11]. The preimage resistance of this hash function is reduced to the hardness of the SSI problem. However, constructing more complex cryptosystems like public-key encryption and signature schemes needs stronger assumptions. In particular, we need to assume the hardness of the isogeny problem with additional information. One such problem is the *supersingular isogeny with torsion (SSI-T) problem*. In this problem, we are given the degree of a secret isogeny and the image of a torsion subgroup under the isogeny. Based on the hardness of the SSI-T problem, Jao, De Feo, and Plût [17, 28] proposed a key-exchange protocol SIDH (supersingular isogeny Diffie-Hellman protocol). SIKE (supersingular isogeny key encapsulation) [27] is a key encapsulation mechanism based on SIDH and was a round-4 candidate of NIST PQC standardization [1]. SIKE Cryptographic Challenge [2] was held in 2021 for the purpose of evaluating the security of SIKE. This challenge required to solve SSI-T problems in the SIKE setting of small parameters. The world record of this challenge at the beginning of 2022 [41] was a problem of a parameter less than half of the smallest SIKE parameter in the NIST proposal.

However, in 2022, Castryck and Decru [7] proposed a polynomial time attack on the SSI-T problem in SIKE. Their implementation of the attack by Magma [6] solved the SSI-T problem of the largest SIKE parameter in about 3 hours. Their attack needs the knowledge of the endomorphism ring of the domain of a secret isogeny. Independently, Maino, Martindale, Panny, Pope, and Wesolowski [33] proposed an attack on the SSI-T problem without the knowledge of the endomorphism ring, which has sub-exponential computational complexity. Later, Robert [38] proposed a polynomial time attack on the SSI-T problem without the knowledge of the endomorphism ring. These attacks are based on Kani's reducibility theorem [29], which connects isogenies between elliptic curves to those between abelian surfaces.

These attacks broke the security of some isogeny-based cryptosystems, e.g., SIDH, SIKE, B-SIDH [14], and Séta [16]. However, there are some isogeny-based cryptosystems that are not broken by these attacks, e.g., CSIDH [9] and SQIsign [18]. In addition, some countermeasures against these attacks [3, 24] and new cryptosystems using these attacks [4, 12, 15, 34, 35] were proposed.

Some of these new cryptosystems are based on the hardness of variants of the SSI-T problem, where the images of points are scaled by a matrix. Recently, Castryck and Vercauteren [10] advanced the analysis of the hardness of these variants of the SSI-T problem. In this paper, we explain the attacks on the SSI-T problem and its variants.

This paper is organized as follows. Section 2 gives the preliminaries. In Sect. 3, we define the problems we consider in this paper. In Sect. 4, we explain attacks on the SSI-T problem. In Sect. 5, we explain an attack on a variant of the SSI-T problem, where the images of one point are given instead of the image of a torsion subgroup. In Sect. 6, we explain attacks on another variant of the SSI-T problem, where the image of a torsion subgroup is scaled by a matrix.

2 Preliminaries

2.1 Notation

Throughout this paper, we let p be a prime of cryptographic size, i.e., we assume p is at least 100 bits in size. We say an algorithm has *polynomial time* if the algorithm terminates in polynomial in $\log p$ operations on $\mathbb{F}_p$. For real functions $f(x)$ and $g(x)$, we say $f(x)$ is in $O(g(x))$ if and only if there exists a positive constant C such that $|f(x)| \leq C|g(x)|$ for sufficiently large x. We say $f(x)$ is in $\tilde{O}(g(x))$ if and only if there exist positive constants C and k such that $|f(x)| \leq C|g(x)|(\log g(x))^k$ for sufficiently large x. We say $f(x)$ is in $\text{poly}(x)$ if and only if there exists a polynomial F such that $f(x)$ is in $O(F(x))$. For a positive integer n, we say n is *smooth* if and only if the largest prime factor of n is in $\text{poly}(\log n)$, and *power-smooth* if and only if the largest prime power dividing n is in $\text{poly}(\log n)$.

The varieties in this paper are principally polarized (p.p.) abelian varieties over a finite field of characteristic p. Let A be a p.p. abelian variety. We denote the neutral element of A by 0_A. We identify the dual of A with A by the polarization. For an isogeny φ between p.p. abelian varieties, we denote the dual of φ with respect to the polarizations by $\hat{\varphi}$. We say φ is a *d-isogeny* if $\hat{\varphi} \circ \varphi$ is the multiplication-by-d map on the domain of φ.

2.2 Computing Isogenies

Let A be a p.p. abelian variety, ℓ be a positive integer prime to p, and G be a maximal ℓ-isotropic subgroup of A. Then there exists an ℓ-isogeny φ with kernel G and φ is unique up to post-composition of an isomorphism.

Algorithms to compute an isogeny from its kernel are known. If the dimension of A is 1 then we can use Vélu's formulas [42]. Vélu's formulas give an algorithm to compute the codomain of φ in $O(\ell)$ operations on a field containing the points in G. For an additional input $P \in A$, we can compute $\varphi(P)$ in $O(\ell)$ operations on a field containing the points in G and P. These computational costs were improved to $\tilde{O}(\sqrt{\ell})$ by [5].

Algorithms to compute a 2-isogeny between p.p. abelian varieties of dimension 2 can be found in [40] and [26]. An algorithm for a general ℓ was given by [13]. The computational cost of this algorithm is $O(\ell^4)$ operations on a field containing the points in G. For an ℓ-isogeny of an arbitrary dimension g, there is an algorithm by [32] having complexity $O(\ell^g)$.

Let D be a positive integer prime to p having the prime factorization $D = \prod_i \ell_i$ and φ be a D-isogeny. Then we can compute φ as the composition of ℓ_i-isogenies. Therefore, if D is smooth and the points in $\ker \varphi$ are defined over a $\mathbb{F}_{p^k}$ of $k \in \text{poly}(\log p)$, then we can compute a D-isogeny in polynomial time.

2.3 Isogeny Diagram

Let d_1 and d_2 be positive integers prime to each other and p. Let $\varphi_1 : A \to A_1$ be a d_1-isogeny and $\varphi_2 : A \to A_2$ be a d_2-isogeny between p.p. abelian varieties. Then we say an isogeny with kernel $\varphi_1(\ker \varphi_2)$ a *push-forward* of φ_2 by φ_1 and denote it by $\varphi_{1*}\varphi_2$. Since $\ker(\varphi_{1*}\varphi_2 \circ \varphi_1) = \langle \ker \varphi_1, \ker \varphi_2 \rangle = \ker(\varphi_{2*}\varphi_1)$, the codomains of $\varphi_{1*}\varphi_2$ and $\varphi_{2*}\varphi_1$ are the same. In other words, we have the following isogeny diagram

$$\begin{array}{ccc} A & \xrightarrow{\varphi_1} & A_1 \\ \varphi_2 \downarrow & & \downarrow \varphi_{1*}\varphi_2 \\ A_2 & \xrightarrow{\varphi_{2*}\varphi_1} & B. \end{array}$$

The attacks in this paper are based on the following theorem from Kani's reducibility theorem [29, Theorem 2.3].

Theorem 1 *([38, Lemma 6])* We use the same notation as above and let $d = d_1 + d_2$. Suppose that we take the push-forwards so that the above diagram is commutative. We define an isogeny

$$F = \begin{pmatrix} \hat{\varphi}_1 & \hat{\varphi}_2 \\ -\varphi_{1*}\varphi_2 & \varphi_{2*}\varphi_1 \end{pmatrix} : A_1 \times A_2 \to A \times B,$$

i.e., $F(P_1, P_2) = (\hat{\varphi}_1(P_1) + \hat{\varphi}_2(P_2), -\varphi_{1*}\varphi_2(P_1) + \varphi_{2*}\varphi_1(P_2))$. *Then* F *is a d-isogeny with kernel* $\{(\varphi_1(P), \varphi_2(P)) \mid P \in A[d]\}$.

3 Isogeny Problems

This section gives the definitions of the problems we consider in this paper. In these problems, we assume that the degree D of isogenies and the order N of torsion points are smooth and prime to each other, and that $E[D]$ and $E[N]$ are contained in $E(\mathbb{F}_{p^k})$ for $k \in O(1)$. Therefore, the problems can be represented in $O(\log p)$ space.

First, we define two problems that are considered to be hard.

Problem 1 (Supersingular Isogeny (SSI) Problem) Given supersingular elliptic curves E and E' over $\mathbb{F}_{p^2}$, find an isogeny $\varphi : E \to E'$.

Problem 2 (Supersingular Isogeny with Degree (SSI-D) Problem) Let E and E' be supersingular elliptic curves over $\mathbb{F}_{p^2}$, D be a smooth integer. Assume that there exist an isogeny $\varphi : E \to E'$ of degree D. Given E, E', and D, find an isogeny $\varphi' : E \to E'$.

These two problems are considered to be hard if we choose a random E' or φ. More precisely, if we chose E' uniformly at random from the set of supersingular elliptic curves over $\mathbb{F}_{p^2}$, then the currently known best computational complexity to solve SSI problems is $O(p^{1/2} \log p)$ operations on $\mathbb{F}_{p^2}$ [21]. An attack on SSI-D problems can be found in [17, Sect. 5.1] and has computational complexity $O(D^{1/2} \log D)$ operations on $\mathbb{F}_{p^2}$.

Next, we define the SSI-T problem, on which the security of SIDH, SIKE, B-SIDH, and Séta is based.

Problem 3 (Supersingular Isogeny with Torsion (SSI-T) Problem) Let E and E' be supersingular elliptic curves over $\mathbb{F}_{p^2}$, D, N be smooth integers prime to each other, and (P, Q) be a basis of $E[N]$. Assume that there exists an isogeny $\varphi : E \to E'$ of degree D. Given E, E', D, N, P, Q, and $\varphi(P), \varphi(Q)$, find an isogeny $\varphi' : E \to E'$.

A natural generalization of the SSI-T problem is the following problem, in which the image of one point is given instead of the image of a basis.

Problem 4 (Supersingular Isogeny with One Torsion point (SSI-OT) Problem) Let E and E' be supersingular elliptic curves over $\mathbb{F}_{p^2}$, D, N be smooth integers prime to each other, and P be a point of E of order N. Assume that there exists an isogeny $\varphi : E \to E'$ of degree D. Given E, E', D, N, P, and $\varphi(P)$, find an isogeny $\varphi' : E \to E'$.

Another generalization of the SSI-T problem is the following problem, in which the images of points are scaled by a matrix.

Problem 5 (Supersingular Isogeny with Scaled Torsion (SSI-ST) Problem) Let E and E' be supersingular elliptic curves over $\mathbb{F}_{p^2}$, D, N be smooth integers prime to each other, and (P, Q) be a basis of $E[N]$. Let $\mathcal{M}$ be an abelian subgroup of $\mathrm{GL}_2(\mathbb{Z}/N\mathbb{Z})$, and $\mathbf{A} \in \mathcal{M}$. Assume that there exists an isogeny $\varphi : E \to E'$ of degree D. Given $E, E', D, N, P, Q, \mathcal{M}$, and $\mathbf{A}(\varphi(P), \varphi(Q))^\mathsf{T}$, find an isogeny $\varphi' : E \to E'$.

The security of FESTA [4], QFESTA [35], and IS-CUBE [34] is based on the hardness of the SSI-ST problem.

The following sections explain polynomial-time attacks on the SSI-T, SSI-OT, and SSI-ST problems. Table 1 shows the conditions on N and D and the additional requirements for the polynomial-time attacks on these problems. The exact requirements on P and Q are explained in Sect. 6.3.

Remark 1 We defined the problems to allow any isogeny between E and E' as a solution. In cryptographic setting, we often require that the solution is an isogeny satisfying the condition given in the problem. However, this difference does not matter in many cases. This is because the following holds in most cases:

- an attack is designed to find an isogeny satisfying the condition,
- or any isogeny between E and E' can be transformed to φ in polynomial time.

In the following sections, we assume that φ is a unique isogeny between E and E'.

Table 1 The conditions on N and D and the additional requirements for the polynomial-time attacks on the SSI-T, SSI-OT, and SSI-ST problems. See Sect. 6.3 for the definition of "special"

Problem	Condition on N and D	$\mathcal{M}$	Additional requirement
SSI-T	$N^2 > D$	None	None
SSI-OT	$N > D$	None	N is square
SSI-ST	$N^2 > D^2 \deg \omega$	scalar	$\exists \omega \in \text{End}(E) \setminus \mathbb{Z}$
	$N > D$	scalar	E is defined over $\mathbb{F}_p$
	$N^2 > D^2 \deg \omega$	diagonal	$\exists \omega \in \text{End}(E) \setminus \mathbb{Z}$ and P and Q are "special"
	$N > D^2 \deg \omega$	diagonal	$\exists \omega \in \text{End}(E) \setminus \mathbb{Z}$ and P is "special"

4 Attacks on SSI-T Problems

In this section, we consider attacks on SSI-T problems. The first polynomial-time attack on SSI-T problems was proposed by [36]. This attack works if $j(E) = 1728$, $N > D^4$, and $D > p$. This restriction was improved to $j(E) = 1728$, ($N > pD$ or $N > \sqrt{p}D^2$), and $p > D$ by [20]. These attacks use the fact that $\text{End}(E)$ contains a subring isomorphic to $\mathbb{Z}[\sqrt{-1}]$ and can be extended to the case that $\text{End}(E)$ has a non-integer endomorphism of small degree (the restriction needs to be slightly changed). However, these attacks cannot be applied to SIKE because SIKE uses the setting $N \approx D$.

A polynomial-time attack on SIKE was proposed by [7]. The attack works if $N > D$ and $\text{End}(E)$ is given. This restriction is enough to break SIKE. The attack by [7] uses a reduction from an SSI-T problem to the decisional version of it. Based on the same idea as in [7] (and independent of [7]), a direct attack on SSI-T problems was proposed by [33]. The attack by [33] does not need the restriction that $\text{End}(E)$ is known, but has sub-exponential computational time. Later, [38] proposed a polynomial time attack on the SSI-T problem without the knowledge of $\text{End}(E)$.

The attacks by [7, 33] are based on Kani's reducibility theorem (Theorem 1) with $g = 1$, thus use isogenies of dimension 2. On the other hand, the attack by [38] is based on Kani's reducibility theorem with $g = 2$ or 4, thus uses isogenies of dimension 4 or 8.

4.1 Dimension 2

We explain polynomial-time attacks on SSI-T problems where $N > D$ and $\text{End}(E)$ is given using an isogeny of dimension 2. These attacks are based on [7, 33]. Although attacks using dimension 4 or 8 can be applicable to stronger restriction, the attacks using dimension 2 are important in cryptography because the computation on dimension 2 is more efficient than that on higher dimension. Indeed, some cryptosystems using the attacks on dimension 2 were proposed (e.g., [4, 12, 34, 35]).

4.1.1 Case in SIKE

First, we consider an attack on the SSI-T problem in SIKE. In particular, we assume the following:

- $p = 2^a 3^b - 1$, where a, b are positive integers such that $2^a \approx 3^b$,
- $N = 2^a$ and $D = 3^b$,
- $j(E) = 66^3$.

In this case, $\text{End}(E)$ contains a subring isomorphic to $\mathbb{Z}[2\sqrt{-1}]$. We let τ be the endomorphism of E corresponding to $2\sqrt{-1}$. There exist exactly two degree-4 endomorphisms of E whose kernel is cyclic, which correspond to $\pm \tau$. Therefore, we can obtain τ or $-\tau$ by computing all degree-4 isogenies from E. Since the difference in sign corresponds to the way the isomorphism is taken, we can regard the obtained degree-4 endomorphism as τ.

In addition, we assume that there exist a nonnegative integer $\beta \in O(\log(\text{poly}(\log p)))$ and integers c, d such that $2^a - 3^{b-\beta} = c^2 + 4d^2$. Then $\gamma := c + d\tau \in \text{End}(E)$ has degree $2^a - 3^{b-\beta}$. Note that we can compute the image of a point in E by computing the 4-isogeny τ and scaler multiplications.

Let φ_1 be the first 3^β-isogeny of φ and φ_2 be the rest $3^{b-\beta}$-isogeny, i.e., $\varphi = \varphi_2 \circ \varphi_1$. Then we have the following diagram.

$$\begin{array}{ccccc}
E & \xrightarrow{\varphi_1} & E'' & \xrightarrow{\varphi_2} & E' \\
{\scriptstyle \gamma}\downarrow & & {\scriptstyle \varphi_{1*}\gamma}\downarrow & & \downarrow{\scriptstyle \varphi_{2*}(\varphi_{1*}\gamma)} \\
E & \xrightarrow{\gamma_*\varphi_1} & E_1 & \xrightarrow{(\varphi_{1*}\gamma)_*\varphi_2} & E_2.
\end{array}$$

In this diagram, $\deg \varphi_2 + \deg \varphi_{1*}\gamma = 2^a$. Therefore, from Theorem 1, an isogeny F with kernel $\{(\varphi_2(R), \varphi_{1*}\gamma(R)) \mid R \in E''[2^a]\}$ is a 2^a-isogeny $E' \times E_1 \to E'' \times E_2$ that represented by

$$\begin{pmatrix} \hat{\varphi}_2 & \widehat{\varphi_{1*}\gamma} \\ -\varphi_{2*}(\varphi_{1*}\gamma) & (\varphi_{1*}\gamma)_*\varphi_2 \end{pmatrix}$$

up to post-composition of an isomorphism.

In the SSI-T problem, the images $\varphi(P)$ and $\varphi(Q)$ of a basis (P, Q) of $E[2^a]$ are given. From this information and $\ker \varphi_1$, we can compute $\ker F$. Indeed,

$$\ker F = \{(\varphi_2(R), \varphi_{1*}\gamma(R)) \mid R \in E''[2^a]\}$$
$$= \{(\varphi_2(R), \varphi_{1*}\gamma(R)) \mid R \in \varphi_1(E[2^a])\}$$

Algorithm 1: Attack on SSI-T problem in SIKE

Input : The input of an SSI-T problem with $j(E) = 66^3$, $N = 2^a$, and $D = 3^b$, and $\beta \in \mathbb{Z}_{\geq 0}$ and $\gamma \in \mathrm{End}(E)$ such that $\deg \gamma = 2^a - 3^{b-\beta}$.
Output: A generator of $\ker \varphi$.

1 Let $\mathcal{G}$ be the set of all order-3^β cyclic subgroups of E.
2 **for** $G \in \mathcal{G}$ **do**
3   Compute an isogeny ψ with kernel G and let E'' be the codomain of ψ.
4   Compute a push-forward $\gamma_* \psi$ and let E_1 be the codomain of $\gamma_* \psi$.
5   Let $P_1 = \gamma_* \psi \circ \gamma(P)$ and $Q_1 = \gamma_* \psi \circ \gamma(Q)$.
6   Compute an isogeny F with kernel $\langle (P_1, \varphi(P)), (Q_1, \varphi(Q)) \rangle$.
7   **if** *the codomain of F isomorphic to $E'' \times E_2$ for some elliptic curve E_2* **then**
8   Let $\mathrm{pr}_{E''}$ be the projection from the codomain of F to E''.
9   Let (R, S) be a basis of $E'[3^b]$.
10   Let $R' = \hat{\psi} \circ \mathrm{pr}_{E''} \circ F(0_{E_1}, R)$ and $S' = \hat{\psi} \circ \mathrm{pr}_{E''} \circ F(0_{E_1}, S)$.
11   **if** *the order of R' is 3^b* **then**
12   **return** R'.
13   **else**
14   **return** S'.

$$= \langle (\varphi_2(R), \varphi_{1*}\gamma(R)) \rangle_{R \in \{\varphi_1(P), \varphi_1(Q)\}}$$
$$= \langle (\varphi_2(\varphi_1(R)), \varphi_{1*}\gamma(\varphi_1(R))) \rangle_{R \in \{P, Q\}}$$
$$= \langle (\varphi(R), \gamma_* \varphi_1(\gamma(R))) \rangle_{R \in \{P, Q\}}$$

and we know $\varphi(P)$, $\varphi(Q)$ and can compute γ and $\gamma_* \varphi_1$. Using F, we can compute the image of any point under $\hat{\varphi}_2$. We can obtain $\ker \varphi$ from this since $\ker \varphi = \hat{\varphi}(E'[3^b])$.

The attack is constructed by executing the above process for all candidates for $\ker \varphi_1$. Algorithm 1 shows a pseudo-code of this attack.

Theorem 2 *Algorithm 1 outputs the correct answer in polynomial time.*

Proof As explained in Sect. 2.2, the computation of isogenies in Algorithm 1 can be done in polynomial time. The other computation can be done by standard algorithms for the elliptic curve arithmetic in polynomial time. Therefore, each line in Algorithm 1 can be done in polynomial time. Since the cardinality of $\mathcal{G}$ is $4 \cdot 3^{\beta-1} \in \mathrm{poly}(\log p)$, Algorithm 1 terminates in polynomial time.

At least one of ψ is the correct guess of φ_1. From Theorem 1, $\hat{\psi} \circ \mathrm{pr}_{E''} \circ F = \hat{\varphi}$ for such ψ. Therefore, R' and S' generates $\ker \varphi$. Since the order of $\ker \varphi$ is a power of a prime, one of R' and S' is a generator of $\ker \varphi$. □

Remark 2 The fact that D is a power of prime is not essential. We can apply the same attack to the case that D is smooth.

The existence of $\beta \in O(\log \mathrm{poly}(\log p))$ is not guaranteed in general. The number of integers in $[1, x]$ that are of the form $c^2 + 4d^2$ is asymptotic to $C \frac{x}{\sqrt{\ln x}}$, where

$C = 0.57\ldots$, by a variation on a theorem of Landau [39]. Therefore, the probability that $2^a - 3^{b-\beta}$ is the form $c^2 + 4d^2$ when β varies is approximately linear with respect to $1/\sqrt{a}$. Since $a \in O(\log p)$, we expect that β is in $O(\sqrt{\log p})$.

For breaking the concrete SIKE parameters, this is not a problem since there are β's small enough to execute the attack for these parameters. Indeed, [7] reports that their implementation by Magma [6] breaks all SIKE parameters proposed to the NIST PQC standardization in several hours.

To claim (heuristic) polynomial time, there are two methods. The first is using other orders of imaginary quadratic fields. We explain this method in more general setting in the next subsection. The second method is adding small torsion images and a short isogeny. This was considered by [33] in the case that $\text{End}(E)$ is unknown. In this method, we use $e2^{a-\alpha} - f3^{b-\beta}$ instead of $2^a - 3^{b-\beta}$ for integers α, β, e, f. Since the number of $e2^{a-\alpha} - f3^{b-\beta}$ for $\alpha \in \mathbb{Z}_{\geq 0}, \beta \in O(\log \text{poly}(\log p)), e, f \in O(\log p)$ is approximately $(\log p)^2$, we can expect that $e2^{a-\alpha} - f3^{b-\beta}$ is of the form $c^2 + 4d^2$ for some α, β, e, f in these ranges.

4.1.2 General Case Where $\text{End}(E)$ is Known

We briefly explain the case that we only assume that $\text{End}(E)$ is known. In this case, we use a "special" supersingular elliptic curve E_0 and its endomorphism. We use the following two properties. The first is that E_0 has a computable endomorphism γ of degree $N - D$. The second is that we can compute an isogeny $\psi : E_0 \to E$ by using $\text{End}(E)$. Then we can execute the similar attack as in Algorithm 1 by using the push-forward $\psi_*\gamma$.

An overview of the attack is as follows.

1. Compute a supersingular elliptic curve E_0 whose endomorphism ring is "special".
2. Compute an isogeny $\psi : E_0 \to E$ of a power-smooth degree.
3. Compute an isogeny $\delta : E_0 \to E_3$ of a power-smooth degree such that there exists an isogeny $\gamma : E_0 \to E_3$ of degree $N - D$.
4. Compute a push-forward $\gamma_*\psi : E_3 \to E_1$ by using δ.
5. Compute the images of P, Q under $\psi_*\gamma$ by using $\hat{\psi}, \delta$, and $\gamma_*\psi$.
6. Apply Theorem 1 to the diagram of E, E_1, E' and $\varphi, \psi_*\gamma$, and recover φ as in Algorithm 1.

The following diagram shows the elliptic curves and the isogenies in the attack.

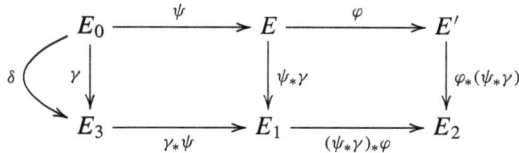

To explain the meaning of "special" and how to find E_0, we define the notation of quaternion algebras. Let $B_{p,\infty}$ be the quaternion algebra ramified only at p and ∞. $B_{p,\infty}$ is isomorphic to $\mathbb{Q} + \mathbb{Q}\mathbf{i} + \mathbb{Q}\mathbf{j} + \mathbb{Q}\mathbf{ij}$, where $\mathbf{i}^2 = -q$ and $\mathbf{j}^2 = -p$ for a prime $q \equiv 3 \pmod 4$ and $\left(\frac{p}{q}\right) = -1$ [37, Proposition 5.1]. Assuming the generalized Riemann hypothesis (GRH), we can find $q \in O((\log p)^2)$. In this notation, we have the following lemma.

Lemma 1 *(([43, Lemma 2.3]) Assuming GRH, there exists a maximal order O_0 in $B_{p,\infty}$ such that O_0 contains a suborder $R + R\mathbf{j}$ with index $O((\log p)^2)$, where R is the ring of integers of $\mathbb{Q}(\mathbf{i})$.*

Let O_0 be a maximal order in Lemma 1. By Deuring correspondence [22], there exists a supersingular elliptic curve E_0 whose endomorphism ring is isomorphic to O_0. We can compute $j(E_0)$ in polynomial time by [23, Algorithm 3] (see [23, Proposition 3] for the computational time). In the following, we fix an isomorphism between O_0 and $\text{End}(E_0)$ and identify these.

Next, we explain the step 2. Let O be a maximal order isomorphic to $\text{End}(E)$. The isogenies from E_0 to E correspond to the left-O_0 ideals that are right-O ideals (see [18]). One of these ideals can be computed by [30, Algorithm 3.5]. Let I be such an ideal. Then, by using [25, Algorithm 1 and 2], we can transform I to an isogeny $\psi : E_0 \to E$ whose degree is power-smooth and prime to N. This isogeny can be computed in polynomial time [25, Lemma 4].

For the step 3, we use the method in [19, Appendix E]. Let M be a power-smooth integer prime to $\deg \psi$ and N such that $M(N - D) > p$. Then we can compute $\alpha \in R + R\mathbf{j} \subset \text{End}(E_0)$ of degree $M(N - D)$ by using the method in [31, Sect. 3.2]. The step 3 is done by decomposing α as $\alpha = \hat{\delta} \circ \gamma$, where $\deg \delta = M$ and $\deg \gamma = N - D$.

For the step 4, we need to compute the image of $\ker \psi$ under γ. There exist generators $\{R_i\}_i$ of $\ker \psi$ such that each R_i is defined over an extension field of $\mathbb{F}_{p^2}$ of degree $O(\text{poly}(\log p))$ (see [25, the proof of Lemma 4] for the detail). Since $\deg \delta$ is power-smooth, we can compute δ and obtain its codomain E_3 in polynomial time as same as ψ. We can also compute the images under α since there exists an integral basis of R of degrees $O((\log p)^2)$ and $\mathbf{j}$ is the p-th power Frobenius endomorphism. Consequently, we can compute the image of a point of order prime to $\deg \delta$ under γ by $\gamma = \frac{1}{\deg \delta} \delta \circ \alpha$. This allows us to compute the push-forward $\gamma_* \psi$.

The step 5 can be executed by using the commutativity of the diagram. In particular, we have $\psi_* \gamma = \frac{1}{\deg \psi} \psi_* \gamma \circ \psi \circ \hat{\psi} = \frac{1}{\deg \psi} \gamma_* \psi \circ \gamma \circ \hat{\psi}$. Therefore, we can compute the images of P and Q under $\psi_* \gamma$.

The step 6 is the same as in the case of SIKE. This completes the attack for the case that $\text{End}(E)$ is known.

4.2 Dimension 4 or 8

The attacks using dimension 2 are extended by [38] to those using dimension 4 or 8. The advantage of using such dimensions is that the product E^g of the same elliptic curve has endomorphism represented by a $g \times g$ matrix over $\mathbb{Z}$. We can use this matrix to solve the SSI-T problem instead of the isogeny γ in the previous subsection.

In particular, if $g = 2$ then E^2 has an endomorphism α represented by $\begin{pmatrix} a_1 & a_2 \\ -a_2 & a_1 \end{pmatrix}$. Then α is an $(a_1^2 + a_2^2)$-isogeny. If $N - D = a_1^2 + a_2^2$ then, by applying Theorem 1 to the following diagram,

$$\begin{array}{ccc} E^2 & \xrightarrow{\varphi \mathrm{Id}} & E'^2 \\ \alpha \downarrow & & \downarrow \alpha \\ E^2 & \xrightarrow{\varphi \mathrm{Id}} & E'^2, \end{array}$$

where $\varphi \mathrm{Id} = \begin{pmatrix} \varphi & 0 \\ 0 & \varphi \end{pmatrix}$, we can obtain an N-isogeny $E^2 \times E'^2 \to E^2 \times E'^2$ of dimension 4 containing $\hat{\varphi}$ in its component.

Using $g = 4$, we do not need any restriction on $N - D$. In particular, an endomorphism of E^4 represented by

$$\begin{pmatrix} a_1 & -a_2 & -a_3 & -a_4 \\ a_2 & a_1 & a_4 & -a_3 \\ a_3 & -a_4 & a_1 & a_2 \\ a_4 & a_3 & -a_2 & a_1 \end{pmatrix}$$

is an $(a_1^2 + a_2^2 + a_3^2 + a_4^2)$-isogeny. Since any positive integer can be expressed as the sum of four squares, we can obtain an N-isogeny $E^4 \times E'^4 \to E^4 \times E'^4$ containing $\hat{\varphi}$ in its component by a similar diagram as above.

Note that these attack uses an endomorphism of E^g for $g = 2$ or 4, but does not use a non-integer endomorphism of E. Therefore, we do not need to know $\mathrm{End}(E)$ for these attacks. Consequently, the restriction is only $N > D$. In addition, this restriction was relaxed to $N^2 > D$ by [38].

Consider an SSI-T problem with $N^2 > D$. Let (P', Q') be a basis of $E[N^2]$ such that $P = NP'$ and $Q = NQ'$. If $N^2 - D$ is the sum of two squares then we let $g = 2$ and otherwise $g = 4$. Let α be an endomorphism of E^g that is an $(N^2 - D)$-isogeny defined by one of the above matrices. For $R_1, R_2 \in E$, we denote the set of points R in E^g such that one of the components of R is R_1 or R_2 and the others are 0_E as $\mathcal{S}(R_1, R_2)$. For example, if $g = 2$ then $\mathcal{S}(R_1, R_2) = \{(R_1, 0_E), (0_E, R_1), (R_2, 0_E), (0_E, R_2)\}$. Note that $E^g[N] = \langle \mathcal{S}(P, Q) \rangle$. As the above, Theorem 1 says that there exists an N^2-isogeny $F : E^g \times E'^g \to E^g \times E'^g$

Algorithm 2: Attack on SSI-T problem

Input : The input of an SSI-T problem with $N^2 > D$.
Output: Generators of ker φ.

1. Let (P', Q') be a basis of $E[N^2]$ such that $P = NP'$ and $Q = NQ'$.
2. Let $g = 2$ if $N^2 - D$ is the sum of two squares and $g = 4$ otherwise.
3. Let α be an $(N^2 - D)$-endomorphism of E^g.
4. Compute an N-isogeny $\hat{F}_2$ with kernel $\langle \alpha(R), \varphi(R) \rangle_{R \in S(P,Q)}$.
5. Let B be a basis of $(E^g \times E'^g)[N]$. // #$B = 4g$
6. Let B' be the image of B under $\hat{F}_2$.
7. Compute an N-isogeny F_2 with kernel $\langle B' \rangle$.
8. Compute an N-isogeny F_1 with kernel $\langle \alpha(R), \varphi(R) \rangle_{R \in S(P,Q)}$.
9. Let B'' be a basis of $E'^g[D]$.
10. Let $S = \{(0_{E^g}, R) \mid R \in B''\}$.
11. Let pr_{E^g} be the projection from $E^g \times E'^g$ to E^g.
12. Let $S' \subseteq E$ be the set of the components of the image of S under $\mathrm{pr}_{E^g} \circ F_2 \circ F_1$.
13. **return** S'.

represented by $\begin{pmatrix} \hat{\alpha} & \hat{\varphi}\mathrm{Id} \\ \varphi\mathrm{Id} & -\alpha \end{pmatrix}$ with kernel $\langle \alpha(R), \varphi(R) \rangle_{R \in S(P', Q')}$. In this case, we do not know the images of P' and Q' under φ. To solve this problem, we use the dual isogeny $\hat{F}$. From [38, Lemma 3], $\hat{F}$ is represented by $\begin{pmatrix} \alpha & \hat{\varphi}\mathrm{Id} \\ \varphi\mathrm{Id} & -\hat{\alpha} \end{pmatrix}$. Applying Theorem 1 to the diagram whose arrows are $\hat{\alpha}$ and $\varphi\mathrm{Id}$, we have ker $\hat{F} = \langle \hat{\alpha}(R), \varphi(R) \rangle_{R \in S(P', Q')}$. Although we do not know the image of $E[N^2]$ under φ, we know that of $E[N]$. Therefore, we can compute the halves of F and $\hat{F}$. More precisely, by decomposing F to two N-isogenies F_1 and F_2 as $F_2 \circ F_1$, we have ker $F_1 = N$ ker $F = \langle (\alpha(R), \varphi(R)) \rangle_{R \in S(P,Q)}$ and ker $\hat{F}_2 = N$ ker $\hat{F} = \langle (\hat{\alpha}(R), \varphi(R)) \rangle_{R \in S(P,Q)}$. Since we know the image of $E[N]$ under φ, we can compute F_1 and $\hat{F}_2$. From these, we can compute F and ker φ. Algorithm 2 shows a pseudo-code of this attack.

As in the proof of Theorem 2, the computation in Algorithm 2 can be done in polynomial time. Therefore, we have the following theorem.

Theorem 3 *SSI-T problems can be solved in polynomial time if $N^2 > D$.*

Remark 3 In the case $g = 1$, we can not use the attack using the dual isogeny even if we know End(E). In this case, we do not know E_2 in the diagrams in Sect. 4.1. Therefore, we do not know the domain of the dual isogeny.

5 Attack on SSI-OT Problems

De Feo et al. presented a reduction of an SSI-OT problem to an SSI-T problem at the KU Leuven isogeny days in 2022. This work has not been published, but is mentioned in [8]. The reduction is as follows.

Theorem 4 *SSI-OT problems where N is a square n^2 can be solved in polynomial-time by using one call of an oracle to solve SSI-T problems with $N = n$ and the same degree D.*

Proof Consider an SSI-OT problem with $N = n^2$. Let ψ be an isogeny with kernel $\langle nP \rangle$ and E_0 be its codomain. Let E_0' be the codomain of $\varphi_*\psi$. Then we have the following commutative diagram.

$$\begin{array}{ccc} E_0 & \xrightarrow{\psi_*\varphi} & E_0' \\ \psi \uparrow & & \uparrow \varphi_*\psi \\ E & \xrightarrow{\varphi} & E' \end{array}$$

In this diagram, we can compute E_0, E_0', ψ, and $\varphi_*\psi$ in polynomial time since $\ker \varphi_*\psi = \langle n\varphi(P) \rangle$. Let $P_0 = \psi(P)$, Q_0 be a generator of $\ker \psi$, $P_0' = \varphi_*\psi(\varphi(P))$, and Q_0' be a generator of $\ker \widehat{\varphi_*\psi}$. Then $\psi_*\varphi(P_0) = P_0'$ and $\psi_*\varphi(Q_0) = \lambda Q_0'$ for some $\lambda \in (\mathbb{Z}/n\mathbb{Z})^\times$ since $\psi_*\varphi(Q_0)$ generates $\ker \widehat{\varphi_*\psi}$. We can compute λ by using the Weil pairing. Let e_n be the n-th Weil pairing. Then we have

$$e_n(P_0', Q_0') = e_n(\psi_*\varphi(P_0), \lambda^{-1}\psi_*\varphi(Q_0)) = e_n(P_0, Q_0)^{D\lambda^{-1}}.$$

Since n is smooth, we can compute λ in polynomial time. Then, by calling the oracle with input $E_0, E_0', P_0, Q_0, \psi_*\varphi(P_0), \psi_*\varphi(Q_0)$, we obtain $\ker \psi_*\varphi$. This solves the SSI-OT problem we consider since $\ker \varphi = \hat{\psi}(\ker \psi_*\varphi)$. □

From this theorem and Theorem 3, we have the following corollary.

Corollary 1 *SSI-OT problems can be solved in polynomial time if N is square and $N > D$.*

6 Attacks on SSI-ST Problems

An attack on SSI-ST problems where $\mathcal{M}$ is the set of the scalar matrices was proposed by [24]. This attack requires that $\text{End}(E)$ has a non-integer endomorphism of small degree. This was extended by [10]. This section gives an overview of this extended attack.

6.1 Attack Diagram

Consider an SSI-ST problem. Suppose that there exist a supersingular elliptic curve E_1 and two isogenies $\sigma, \omega : E \to E_1$ such that a push-forward $\varphi_*\sigma$ can be computed without knowing φ and deg ω is small. Let E'_1 be the codomain of $\varphi_*\sigma$. Then we have the following diagram.

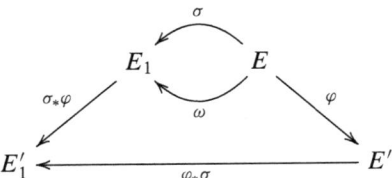

Let **M** be the matrix representation of the endomorphism $\hat{\sigma} \circ \omega$ on $E[N]$ with respect to the basis (P, Q), i.e., $\hat{\sigma} \circ \omega((P, Q)^\mathsf{T}) = \mathbf{M}(P, Q)^\mathsf{T}$. In the problem, the scaled images $\mathbf{A}(\varphi(P), \varphi(Q))^\mathsf{T}$ for unknown $\mathbf{A}$ are given. We denote these images as S, T. If **M** commutes with every matrix of $\mathcal{M}$ then we have ([10, Lemma 3])

$$(\deg \sigma)\sigma_*\varphi \circ \omega \circ \hat{\varphi}\begin{pmatrix} S \\ T \end{pmatrix} = D\mathbf{M}\varphi_*\sigma\begin{pmatrix} S \\ T \end{pmatrix}.$$

This means that we can compute the images of S, T under $\sigma_*\omega \circ \omega \circ \hat{\varphi}$. Therefore, SSI-ST problems satisfying the above assumptions can be reduce to the SSI-T problem of $\sigma_*\omega \circ \omega \circ \hat{\varphi}$ and (S, T).

The requirement that we can compute $\varphi_*\sigma$ without knowing φ restricts the possibility of σ. In particular, in [10], the following two candidates for σ are proposed.

1. The identity map on E, thus $E_1 = E$.
2. The p-th power Frobenius map π_p from E to the Frobenius conjugate $E^{(p)}$.

In the second case, we need an isogeny $\omega : E \to E^{(p)}$ of a small degree. This restricts the possibility of E that the attack can be applied to. In general, such ω does not exist. In this paper, we only consider the case that E is defined over $\mathbb{F}_p$.

6.2 Scalar Matrices

Suppose that $\mathcal{M}$ is the set of the scalar matrices in $\mathrm{GL}_2(\mathbb{Z}/N\mathbb{Z})$. In this case, any matrix in $\mathrm{GL}_2(\mathbb{Z}/N\mathbb{Z})$ commutes with every matrix of $\mathcal{M}$.

First, we consider the case that σ is the identity map on E. This is the case considered in [24]. In this case, ω is an endomorphism of E and $\varphi_*\sigma$ is the identity map on E', thus $E'_1 = E'$. We obtain a diagram of so-called a "lollipop isogeny"

$$\omega \,\begin{array}{c}\curvearrowleft\end{array} E \mathrel{\substack{\varphi \\ \rightleftarrows \\ \hat{\varphi}}} E'.$$

If the composition $\varphi \circ \omega \circ \hat{\varphi}$ can not be represented by the composition of a scalar multiplication and an isogeny, then $\ker(\varphi \circ \omega \circ \hat{\varphi}) \cap E'[D] = \ker \hat{\varphi}$. Therefore, we can obtain φ by $\varphi \circ \omega \circ \hat{\varphi}$ in this case. The exception occurs if $\ker \varphi \cap E[d]$ is an eigenspace of ω on $E[d]$ for a divisor d of D. We exclude this case, but the attack is applicable by guessing the missing part of $\ker \varphi$ in some cases.

As explained above, we know the image of $\varphi(P)$ and $\varphi(Q)$ under $\varphi \circ \sigma \circ \hat{\varphi}$. Therefore, using Theorem 3, we have the following.

Theorem 5 *SSI-ST problems can be solved in polynomial time if $\mathcal{M}$ is the set of the scalar matrices and E has an endomorphism ω such that ω can be evaluated in polynomial time, $\ker \varphi \cap E[d]$ is not any eigenspace of ω on $E[d]$ for any divisor d of D, and $N^2 > D^2 \deg \omega$.*

Next, we consider the case that σ is the p-the power Frobenius endomorphism π_p of E. In this case, we can take ω as the identity map on E and know the images of $\varphi(P)$ and $\varphi(Q)$ under $\varphi^{(p)} \circ \hat{\varphi} : E' \to E'^{(p)}$, where $\varphi^{(p)}$ is the isogeny defined by raising each coefficient of φ to the p power.

As above, we exclude the case that $\ker \varphi \cap E[d]$ is an eigenspace of the p-th power Frobenius endomorphism on $E[d]$ for a divisor d of D. We have the following.

Theorem 6 *SSI-ST problems can be solved in polynomial time if $\mathcal{M}$ is the set of the scalar matrices, E is defined over $\mathbb{F}_p$, $\ker \varphi \cap E[d]$ is not any eigenspace of π_p on $E[d]$ for any divisor d of D, and $N > D$.*

6.3 Diagonal Matrices

Suppose that $\mathcal{M}$ is the set of the diagonal matrices in $\mathrm{GL}_2(\mathbb{Z}/N\mathbb{Z})$. In this case, the matrix $\mathbf{M}$ representing $\hat{\sigma} \circ \omega$ should be a diagonal matrix for applying the attack on SSI-T problems. By applying the attack on SSI-OT problems, we can relax the requirement on $\mathcal{M}$. In particular, if P is an eigenvector of $\hat{\sigma} \circ \omega$ then we have

$$(\deg \sigma)\sigma_* \varphi \circ \omega \circ \hat{\varphi}(S) = D\mathbf{M}\varphi_* \sigma(S).$$

This allows us to compute the image of S under $\sigma_* \varphi \circ \omega \circ \hat{\varphi}$. Therefore, we have the following from Theorems 3 and 4.

Theorem 7 *SSI-ST problems can be solved in polynomial time if $\mathcal{M}$ is the set of the diagonal matrices, and E has an endomorphism ω such that $\ker \varphi \cap E[d]$ is not any eigenspace of ω on $E[d]$ for any divisor d of D and one of the following holds:*

1. *P and Q are eigenvectors of ω and $N^2 > D^2 \deg \omega$.*
2. *P is an eigenvector of ω, N is a square, and $N > D^2 \deg \omega$.*
3. *E is defined over $\mathbb{F}_p$, P and Q are eigenvectors of $\pi_p \circ \omega$, and $N^2 > D^2 \deg \omega$.*
4. *E is defined over $\mathbb{F}_p$, P is an eigenvector of $\pi_p \circ \omega$, N is a square, and $N > D^2 \deg \omega$.*

It rarely happens that P and/or Q are eigenvector(s) of ω or $\pi_p \circ \omega$. The probabilities that these happen for a randomly chosen P and Q in the case $j(E) = 66^3$ are estimated in [10]. The estimation shows that the probabilities are at most the order of $\frac{1}{N}$. In addition, for a fixed E whose endomorphism ring is known and a basis (P, Q), we can check whether there exists an endomorphism of E whose degree is sufficiently small to apply the attack in this subsection.

References

1. National Institute of Standards and Technology (NIST), *NIST Post-Quantum Cryptography Standardization* https://csrc.nist.gov/Projects/Post-Quantum-Cryptography
2. SIKE Cryptographic Challenge, https://www.microsoft.com/en-us/msrc/sike-cryptographic-challenge
3. A. Basso, T.B. Fouotsa, New SIDH countermeasures for a more efficient key exchange. Cryptology ePrint Archive, Paper 2023/791, 2023. https://eprint.iacr.org/2023/791
4. A. Basso, L. Maino, G. Pope, FESTA: Fast encryption from supersingular torsion attacks. Cryptology ePrint Archive, Paper 2023/660, 2023. https://eprint.iacr.org/2023/660. To appear in ASIACRYPT 2023
5. D.J. Bernstein, L.D. Feo, A. Leroux, B. Smith, Faster computation of isogenies of large prime degree, in *ANTS-XIV-14th Algorithmic Number Theory Symposium*. ed. by S. Galbraith, vol. 4 of Proceedings of the Fourteenth Algorithmic Number Theory Symposium (ANTS-XIV), (Mathematical Sciences Publishers, Auckland, New Zealand, 2020), pp.39–55
6. W. Bosma, J. Cannon, C. Playoust, The Magma algebra system. I. The user language. J. Symbolic Comput. **24**(3-4), 235–265 (1997). Computational algebra and number theory (London, 1993)
7. W. Castryck, T. Decru, An efficient key recovery attack on SIDH, in *Advances in Cryptology – EUROCRYPT 2023*, eds. by C. Hazay, M. Stam (Springer Nature, Cham, Switzerland, 2023), pp. 423–447
8. W. Castryck, M. Houben, S.-P. Merz, M. Mula, S. van Buuren, F. Vercauteren, Weak instances of class group action based cryptography via self-pairings, in *Advances in Cryptology – CRYPTO 2023*, eds. by H. Handschuh, A. Lysyanskaya (Springer Nature, Cham, Switzerland, 2023), pp. 762–792
9. W. Castryck, T. Lange, C. Martindale, L. Panny, J. Renes, CSIDH: An efficient post-quantum commutative group action, in *Advances in Cryptology – ASIACRYPT 2018*, eds. by T. Peyrin, S. Galbraith (Springer International Publishing, Cham, 2018), pp. 395–427
10. W. Castryck, F. Vercauteren, A polynomial-time attack on instances of M-SIDH and FESTA. Cryptology ePrint Archive, Paper 2023/1433 (2023). https://eprint.iacr.org/2023/1433. To apppear in ASIACRYPT 2023
11. D. Charles, E. Goren, K. Lauter, Cryptographic hash functions from expander graphs. Cryptology ePrint Archive, Report 2006/021 (2006). https://eprint.iacr.org/2006/021
12. M. Chen, A. Leroux, SCALLOP-HD: Group action from 2-dimensional isogenies. Cryptology ePrint Archive, Paper 2023/1488 (2023). https://eprint.iacr.org/2023/1488
13. C. Romain, R. Damien, Computing (l, l)-isogenies in polynomial time on Jacobians of genus 2 curves. Math. Comput. **84**(294), 1953–1975 (2015)
14. C. Costello, B-SIDH: Supersingular isogeny Diffie-Hellman using twisted torsion, in *Advances in Cryptology – ASIACRYPT 2020*, eds. by S. Moriai, H. Wang (Springer International Publishing, Cham, 2020), pp. 440–463
15. P. Dartois, A. Leroux, D. Robert, B. Wesolowski, SQISignHD: New dimensions in cryptography. Cryptology ePrint Archive, Paper 2023/436 (2023). https://eprint.iacr.org/2023/436

16. L. De Feo, C. Delpech de Saint Guilhem, T. Boris Fouotsa, P. Kutas, A. Leroux, C. Petit, J. Silva, B. Wesolowski, Séta: Supersingular encryption from torsion attacks, in *Advances in Cryptology – ASIACRYPT 2021*, eds. by M. Tibouchi, H. Wang (Springer International Publishing, Cham, 2021), pp. 249–278
17. D. Feo Luca, J. David, P. Jérôme, Towards quantum-resistant cryptosystems from supersingular elliptic curve isogenies. J. Math. Crypt. **8**(3), 209–247 (2014)
18. L. De Feo, D. Kohel, A. Leroux, C. Petit, B. Wesolowski, SQISign: Compact post-quantum signatures from quaternions and isogenies, in *Advances in Cryptology – ASIACRYPT 2020*, eds. by S. Moriai, H. Wang (Springer International Publishing, Cham, 2020), pp. 64–93
19. L. De Feo, D. Kohel, A. Leroux, C. Petit, B. Wesolowski, SQISign: Compact post-quantum signatures from quaternions and isogenies (extended version). Cryptology ePrint Archive, Report 2020/1240 (2020). https://ia.cr/2020/1240
20. V. de Quehen, P. Kutas, C. Leonardi, C. Martindale, L. Panny, C. Petit, K.E. Stange, Improved torsion-point attacks on SIDH variants, in *Advances in Cryptology – CRYPTO 2021*, eds. by T. Malkin, C. Peikert (Springer International Publishing, Cham, 2021), pp. 432–470
21. D. Christina, G.D. Steven, Computing isogenies between supersingular elliptic curves over $\mathbb{F}_p$. Designs Codes Cryptogr. **78**(2), 425–440 (2016)
22. D. Max, Die typen der multiplikatorenringe elliptischer funktionenkörper. Abhandlungen aus dem Mathematischen Seminar der Universität Hamburg **14**, 197–272 (1941)
23. K. Eisenträger, S. Hallgren, K. Lauter, T. Morrison, C. Petit, Supersingular isogeny graphs and endomorphism rings: Reductions and solutions, in *Advances in Cryptology – EUROCRYPT 2018*, eds. by J.B. Nielsen, V. Rijmen (Springer International Publishing, Cham, 2018), pp. 329–368
24. T.B. Fouotsa, T. Moriya, C. Petit, M-SIDH and MD-SIDH: Countering SIDH attacks by masking information, in *Advances in Cryptology – EUROCRYPT 2023*, eds. by C. Hazay, M. Stam (Springer Nature Switzerland, Cham, 2023), pp. 282–309
25. S.D. Galbraith, C. Petit, J. Silva, Identification protocols and signature schemes based on supersingular isogeny problems, in *Advances in Cryptology – ASIACRYPT 2017*, eds. by T. Takagi, T. Peyrin (Springer International Publishing, Cham, 2017), pp. 3–33
26. H.W. Everett, L. Franck, P. Bjorn, Large torsion subgroups of split Jacobians of curves of genus two or three. Forum Math. **12**(3), 315–364 (2000)
27. D. Jao, R. Azarderakhsh, M. Campagna, C. Costello, L.D. Feo, B. Hess, A. Jalali, B. Koziel, B. LaMacchia, P. Longa, M. Naehrig, G. Pereira, J. Renes, V. Soukharev, D. Urbanik, SIKE–Supersingular isogeny key encapsulation, Submission to the NIST Post-Quantum Cryptography Standardization project; https://sike.org
28. D. Jao, L. De Feo, Towards quantum-resistant cryptosystems from supersingular elliptic curve isogenies, in *Post-Quantum Cryptography*. ed. by B.-Y. Yang (Springer, Berlin, Heidelberg, 2011), pp.19–34
29. K. Ernst, The number of curves of genus two with elliptic differentials. J. für die reine und angewandte Mathematik **485**, 93–122 (1997)
30. K. Markus, V. John, Algorithmic enumeration of ideal classes for quaternion orders. SIAM J. Comput. **39**(5), 1714–1747 (2010)
31. D. Kohel, K. Lauter, C. Petit, J.-P. Tignol, On the quaternion ℓ-isogeny path problem. LMS J. Comput. Math. **17**(A), 418–432 (2014)
32. L. David, R. Damien, Computing isogenies between abelian varieties. Compositio Mathematica **148**(5), 1483–1515 (2012)
33. L. Maino, C. Martindale, L. Panny, G. Pope, B. Wesolowski, A direct key recovery attack on SIDH, in *Advances in Cryptology – EUROCRYPT 2023*, eds. by C. Hazay, M. Stam (Springer Nature Switzerland, Cham, 2023), pp. 448–471
34. T. Moriya, IS-CUBE: An isogeny-based compact KEM using a boxed SIDH diagram. Cryptology ePrint Archive, Paper 2023/1506 (2023). https://eprint.iacr.org/2023/1506
35. K. Nakagawa, H. Onuki, QFESTA: Efficient algorithms and parameters for FESTA using quaternion algebras. Cryptology ePrint Archive, Paper 2023/1468 (2023). https://eprint.iacr.org/2023/1468

36. C. Petit, Faster algorithms for isogeny problems using torsion point images, in *Advances in Cryptology – ASIACRYPT 2017*, eds. by T. Takagi, T. Peyrin (Springer International Publishing, Cham, 2017), pp. 330–353
37. P. Arnold, An algorithm for computing modular forms on $\Gamma_0(N)$. J. Algebra **64**(2), 340–390 (1980)
38. D. Robert, Breaking SIDH in polynomial time, in *Advances in Cryptology – EUROCRYPT 2023*, eds. by C. Hazay, M. Stam (Springer Nature, Cham, Switzerland, 2023), pp. 472–503
39. S. Daniel, S.P. Larry, Variations on a theorem of Landau. Part I. Math. Comput. **20**(96), 551–569 (1966)
40. B.A. Smith, *Explicit endomorphisms and correspondences*. PhD thesis, University of Sydney (2005)
41. A. Udovenko, G. Vitto, Revisiting Meet-in-the-Middle cryptanalysis of SIDH/SIKE with application to the $IKEp182 Challenge. Cryptology ePrint Archive, Paper 2021/1421 (2021). https://eprint.iacr.org/2021/1421, presented at SAC 2022
42. V. Jacques, Isogénies entre courbes elliptiques. Comptes-Rendues de l'Académie des Sciences **273**, 238–241 (1971)
43. B. Wesolowski, *The supersingular isogeny path and endomorphism ring problems are equivalent*, in *2021 IEEE 62nd Annual Symposium on Foundations of Computer Science (FOCS)* (IEEE Computer Society, Los Alamitos, CA, USA, 2022), pp.1100–1111

Open Access This chapter is licensed under the terms of the Creative Commons Attribution 4.0 International License (http://creativecommons.org/licenses/by/4.0/), which permits use, sharing, adaptation, distribution and reproduction in any medium or format, as long as you give appropriate credit to the original author(s) and the source, provide a link to the Creative Commons license and indicate if changes were made.

The images or other third party material in this chapter are included in the chapter's Creative Commons license, unless indicated otherwise in a credit line to the material. If material is not included in the chapter's Creative Commons license and your intended use is not permitted by statutory regulation or exceeds the permitted use, you will need to obtain permission directly from the copyright holder.

An Optimization for Efficient Computation of Multiradical (3, 3)-isogenies on Jacobians

Masahiro Ishii and Daiki Hayashida

Abstract In this paper, we focus on isogenies over abelian varieties for isogeny-based cryptography and efficient computation of the hash function using multiradical (3, 3)-isogenies. In particular, we performed a detailed analysis of the explicit formulae to compute multiradical (3, 3)-isogenies and optimized many parts for the efficient hash function. We optimized the formulae manually and achieved a 16.8% reduction in complexity arithmetic operations compared to the implementation of the previous work. In particular, we achieved an efficiency improvement of about 86.5% for the most complicated part of the hash function excluding the Gröbner basis computation. In addition, we provided a comparison of hash functions using isogenies on elliptic curves and Jacobians of genus 2 curves. We discuss further improvements and optimization of the multiradical (3, 3)-isogenies that make the hash function based on them faster than in the case of elliptic curves.

Keywords Isogeny-based cryptography · Abelian surfaces · Multiradical (3, 3)-isogeny · Hash function · Optimization

1 Introduction

Many researches on post-quantum cryptography computationally secure against general-purpose quantum computers are currently active and have attracted attention. In addition, various activities toward the standardization of such cryptosystems are underway in the Post-Quantum Cryptography Standardization Project by the National Institute of Standards and Technology (NIST) [19].

M. Ishii (✉)
Global Scientific Information and Computing Center, Tokyo Institute of Technology, Tokyo, Japan
e-mail: mishii@c.titech.ac.jp

D. Hayashida
Information Technology R&D Center, Mitsubishi Electric Corporation, Kanagawa, Japan
e-mail: Hayashida.Daiki@df.MitsubishiElectric.co.jp

The isogeny-based cryptography, whose security relies on the difficulty of computing isogenies between elliptic curves, is also attractive in addition to the cryptosystems based on lattices and code theory. There are several isogeny-based schemes, including the SIDH [11], one of the key exchange protocols, the SIDH-based KEM SIKE,[1] the signature scheme SQISign [12] using isogeny graph, and the isogeny graph-based hash function [7]. Although SIKE reached the 4th round of the NIST competition, several works [3, 17] have proposed attacks on it.

As in the case of the hyperelliptic curve cryptosystem [15], research on cryptographic techniques for abelian varieties of higher dimension and isogenies between them has been conducted as a natural extension to elliptic curve cryptosystems. The use of higher genus varieties allows for reduced key lengths with generally the same level of security as elliptic curve-based cryptography and is thus expected to be used in memory-saving cryptographic applications. In particular, the security of isogeny graph-based cryptographic schemes for abelian surfaces has been discussed, and SIDH-based key exchange schemes and hash functions using superspecial surfaces have been proposed [4, 8, 13].

In this paper, we focus on an explicit formula for computing isogenies between superspecial abelian surfaces.

1.1 Studies on Explicit Formulae Computing Isogenies

Here we present some related works that give explicit formulae to compute isogenies for elliptic curves and superspecial abelian surfaces.

Any separable isogeny of degree ℓ (denoted by ℓ-isogeny) between elliptic curves and its image can be calculated from the kernel subgroup of the isogeny (the cyclic group generated by the ℓ-torsion points) by using Vélu's formula [23]. An efficient method to compute isogenies of low degree during random walks of isogeny graphs in hash function has been proposed [22, 24]. In addition, when it is possible to utilize Montgomery curves different from the Weierstrass model to improve the efficiency of cryptographic protocols, it is useful to clarify the formula to compute an isogeny and its cost [18, 20]. To compute ℓ-isogenies, one can either use Vélu's formula after obtaining the ℓ-torsion points as described above or use the division polynomial or modular polynomial [1]. Recently, Castryck et al. [5] presented a new explicit formula for computing isogenies between elliptic curves. The generation of ℓ-torsion points is not needed in the computation of a sequence of isogenies (ℓ^2-isogeny map), and the formula is applied as an algebraic expression by power roots of Tate pairing values in the coefficients of the ℓ-torsion points generating the kernel subgroup.

The computation of isogenies between abelian surfaces is first studied in the low degree case. In particular formulae and applications to cryptographic primitives in the case of (2, 2)-isogeny (Richelot isogeny) and (3, 3)-isogeny have been discussed [2, 13, 22]. Recently, the multiradical isogeny [6], which extends [5] to the case of

---

[1] https://sike.org.

(polarized) abelian surfaces, was proposed. The authors showed that an explicit formula to compute $(N^2, \ldots, N^2)$-isogeny for a certain family of parametrized varieties could be given by input variety, coefficients of the points, and power roots of certain pairing values. Recently, several works dedicated to finding efficient algorithms for efficient computation of isogenies of abelian surfaces are reported [9, 10, 16]

1.2 Our Contribution

In this paper, we clarify the cost of formulae to compute isogenies between abelian surfaces and show optimization methods for implementing the formulae to reduce the computational cost. Our optimization is performed manually for the (3,3)-isogeny formula and the hash function based on it. We achieved an efficiency improvement of 16.8% for overall hash function and 86.5% for the most complicated part of the hash function excluding the Gröbner basis computation compared to the implementation shown in [6]. The Magma code can be found in our GitHub repository [14]. In addition, we provided a comparison of hash functions using isogenies on elliptic curves and Jacobians of genus 2 curves.

1.3 Organization

The rest of this paper is organized as follows. We review basic mathematical facts about isogenies over elliptic curves and abelian surfaces in Sect. 2. Section 3 is dedicated to describing Multiradical isogenies and their formulae. We propose our optimized (3, 3)-isogeny formulae in Sect. 4. Section 5 describes our experiments and comparisons to previous implementations. We remark future tasks and the conclusions in Sect. 6.

2 Preliminaries

In this section, we introduce the basic mathematical tools needed for hash functions that use isogenies such as elliptic curves, abelian varieties, and isogenies over them.

2.1 Abelian Varieties and Isogenies

An abelian variety is a smooth projective algebraic group variety. The abelian variety A is known to have a group structure, and its unit element is often written as

0_A. An abelian variety of dimension 1 is an elliptic curve, and the Jacobian variety constructed from a curve of genus 2 is an example of an abelian variety of dimension 2.

For two abelian varieties A_1 and A_2, the isogeny of the abelian varieties $\phi : A_1 \to A_2$ is a finite surjective morphism such that $\phi(0_{A_1}) = 0_{A_2}$. Note that this definition coincides with the definition of isogenies on elliptic curves when the dimension is 1.

It is often briefly explained that an abelian variety is a high-dimensional version of an elliptic curve, but strictly speaking, a principally polarized abelian variety can be said to be a high-dimensional generalization of an elliptic curve. The polarization of an abelian variety A is the isogeny $\lambda : A \to \widehat{A}$ which $\widehat{A} \cong \mathrm{Pic}^0(A)$ is a dual abelian variety of A. The polarization λ is principal if λ is an isomorphism as an abelian variety. When there exists a principally polarization λ for the abelian variety A, the set (A, λ) is called a principally polarized abelian variety (PPAV). When it is clear from the context, the principally polarized abelian variety is simply expressed as A.

Let (A_1, λ_1), (A_2, λ_2) be a PPAV. An isogeny between abelian varieties $\phi : A_1 \to A_2$ is an isogeny between PPAVs if there is an integer d and $\widehat{\phi} \circ \lambda_2 \circ \phi = [d]\lambda_1$ holds, where $\widehat{\phi}$ is a dual isogeny $\widehat{\phi} : \widehat{A_2} \to \widehat{A_1}$. Also, if $\phi^\dagger := \lambda_1^{-1} \circ \widehat{\phi} \circ \lambda_2$, then we have $\phi^\dagger \circ \phi = [d]$ and $\phi^\dagger$ is called Rosati dual. The isogenies between PPAVs and the Rosati dual are the generalizations of the isogenies and dual isogenies on elliptic curves, respectively.

Let A be a PPAV over $\overline{\mathbb{F}}_p$ of dimension g and ℓ an integer that is relatively prime to p. For ℓ, an $(\ell, \ldots, \ell)$-subgroup of $A[\ell]$ is a maximal isotropic subgroup with regard to the Weil pairing of $A[\ell]$. Let G be an $(\ell, \ldots, \ell)$-subgroup of $A[\ell]$. Then it is known that $G \cong (\mathbb{Z}/\ell\mathbb{Z})^g$ holds and we call the isogeny between PPAVs with kernel G the $(\ell, \ldots, \ell)$-isogeny.

Next we explain the properties of the sequence of isogenies. Given PPAVs A_1, A_2 and an $(\ell, \ldots, \ell)$-isogeny $\phi : A_1 \to A_2$, an $(\ell, \ldots, \ell)$-isogeny $\psi : A_2 \to A_3$ is a good extension of ϕ if the composite map

$$\psi \circ \phi : A_1 \to A_2 \to A_3$$

is an $(\ell^2, \ldots, \ell^2)$-isogeny.

An abelian variety $A/\overline{\mathbb{F}}_p$ of dimension g is supersingular if A is isogenous to $E_1 \times \cdots \times E_g$ and A is a superspecial abelian variety if A is isomorphic to $E_1 \times \cdots \times E_g$, where E_i is a supersingular elliptic curve.

The product $E_1 \times \cdots \times E_g$ is known to be isomorphic to the product of any supersingular elliptic curves. Therefore all superspecial abelian varieties are isomorphic to each other, however we remark that it is not necessarily to hold as a PPAV.

2.2 The Isogeny Graph of Superspecial PPAVs

In this section, we describe the isogeny graph of superspecial principally polarized abelian varieties.

Let $S_g(p)$ be the set of superspecial PPAVs defined over $\overline{\mathbb{F}}_p$ of dimension g. Then the size of $S_g(p)$ is known to be $\#S_g(p) = O(p^{g(g+1)/2})$. From this fact, it can be seen that the number of superspecial PPAVs increases quadratically as the dimension increases.

We define $\Gamma_2(\ell; p)$ as the graph with the vertex set superspecial principally polarized abelian surfaces (PPAS) up to isomorphism and the edge set the (ℓ, ℓ)-isogenies between superspecial PPASs. It is known that all PPASs are isomorphic to the Jacobian variety composed from the curve of genus 2 or the product of two elliptic curves as PPAVs. Hence, the vertices of the graph $\Gamma_2(\ell; p)$ are decomposed into the following set:

$$S_2(p)^J := \{A \in S_2(p) : A \cong J_C\},$$
$$S_2(p)^E := \{A \in S_2(p) : A \cong E_1 \times E_2 \text{ with } E_i \in S_1(p)\}.$$

In the graph $\Gamma_2(\ell; p)$, there is a large bias between the Jacobian points and the elliptic curve points, and for $p > 5$, the following holds ([4, Proposition 2]):

$$\#S_2(p)^J = \frac{1}{2880}p^3 + \frac{1}{120}p^2 + O(p), \quad \#S_2(p)^E = \frac{1}{288}p^2 + O(p).$$

As with the previous work of hash functions using isogenies [4, 8, 13], we mainly focus on the Jacobian points $S_2(p)^J$.

3 Multiradical Formula for (3, 3)-isogeny

In this section, we review the multiradical isogenies [6] and their computation formulae, in particular, the (3, 3)-isogeny and its implementation schemes presented in the previous studies.

For any field K with char $K \nmid 6$, we define a set S in space $\mathbb{A}^3$ with coordinates (r, s, t) as the joint complement of the zero loci of δ_i ($1 \leq i \leq 7$), Δ, $r - 1$, $r^2 - t$, $rs - st - 1$. δ_i ($1 \leq i \leq 7$), Δ is represented in r, s, t, and the detailed expressions can be found in [6, Sect. 5.1].

For such parameters $(r, s, t) \in S$, the genus-2 curve $C_{rst} : y^2 = F_{rst}(x)$ is defined where

$$F_{rst}(x) = G_1(x)^2 + \lambda_1 H_1(x)^3 = G_2(x)^2 + \lambda_2 H_2(x)^3$$

, and λ_i, $H_i(x)$, $G_i(x)$ are polynomials of degree 0, 2, 3 given by r, s, t, respectively. Let us denote the Jacobian of C_{rst} by J_{rst} and let $\alpha_{i1}, \alpha_{i2} \in \overline{K}$ be roots of $H_i(x)$ for $i = 1, 2$. We consider the divisor class $T_i \in J_{rst}$ of

$$(\alpha_{i1}, G_i(\alpha_{i1})) + (\alpha_{i2}, G_i(\alpha_{i2})) - \infty_1 - \infty_2,$$

with $\infty_1, \infty_2 \in C(\overline{K})$ the two points at infinity. Then $\langle T_1, T_2 \rangle \subset J_{rst}$ is a maximal isotropic subgroup and, $J_{rst}/\langle T_1, T_2 \rangle$ is isomorphic to $J_{r's't'}^{(-3)}$. Furthermore, the genus-2 curve is (3, 3)-isogenous to C_{rst}, and the explicit algebraic expressions of $r's't'$ by r, s, t is shown in [6, Proposition 11]. Therefore we want an explicit formula to represent all good extensions of (3, 3)-isogeny

$$J_{rst} \to J_{r's't'}^{(-3)} = J_{rst}/\langle T_1, T_2 \rangle \tag{1}$$

in this situation.

For a curve C/K of genus g, let t_N be the Tate pairing

$$t_N : \text{Pic}_K^0(C)[N] \times \text{Pic}_K^0(C)/N\text{Pic}_K^0(C) \to K^*/(K^*)^N.$$

About the radicands in the (conjectural) property [6, Sect. 3.2],

$$\mathfrak{r}_1 = t_3(T_1, T_1) t_3(T_1, T_2), \quad \mathfrak{r}_2 = t_3(T_1, T_2)^{-1}, \quad \mathfrak{r}_3 = t_3(T_1, T_2) t_3(T_2, T_2)$$

are taken to simplify the (3, 3)-isogeny formula instead of $\mathfrak{r}_{i,j} = t_3(T_i, T_j)$ ($1 \leq i \leq j \leq 2$) in [6]. Since the power roots of radicands are algebraically representable each other, formulae can be rewritten by both sets of radicals.

The good extensions of (3, 3)-isogeny (1) satisfy that the kernel subgroup intersects the kernel of the dual isogeny $\langle T_1', T_2' \rangle$ trivially. To find the kernel subgroup of the good extension, tuple $(b_1, \ldots, b_7)$ is calculated where

$$F_{r's't'} = (b_4 x^3 + b_3 x^2 + b_2 x + b_1)^2 + b_7 (x^2 + b_5 x + b_6)^3$$

as in [6, Sect. 5.2]. Indeed, 3-torsion points for divisor classes in $J_{r's't'}^{(-3)}[3]$ are given by $(b_1, \ldots, b_7)$, and vice versa. In [6], the authors compute $(b_1, \ldots, b_7)$ by using Gröbner basis with the monomial order where b_4 is the last term. They also compute two solutions for b_4 by observing the factors of minimal polynomial of b_4, and give the explicit expression with the radicals $\mathfrak{r}_i$ of pairings. All 27 good extensions are obtained any b_4 by varying the radicals $\sqrt[3]{\mathfrak{r}_i}$ with ζ_3 (primitive cube root of unity) and b_i.

From the above, we obtain a procedure for finding a good extension for a (3, 3)-isogeny. Furthermore, we consider to compute sequences of (3, 3)-isogenies to construct a hash function taking a random walk in the (3, 3)-isogeny graph. To do this, we need to compute parameters (R, S, T) corresponding to the kernel subgroup of the good extension shown above. Bruin, Flynn, and Testa use the Gröbner basis to compute these parameters [6, Sect. 5.2].

The followings show each step of random walks in the (3, 3)-isogeny graph.

get_pairings, action_on_cubic_roots, get_b4s, get_bi, get_new_rst, and get_isogenous_RST denote function names in the implementation[2] by Castryck et al. [6].

---

[2] https://github.com/KULeuven-COSIC/Multiradical-Isogenies.

1. For $J_{rst} \to J_{r's't'}^{(-3)} = J_{rst}/\langle T_1, T_2\rangle$, compute pairings $\mathfrak{t}_1 = t_3(T_1, T_1)t_3(T_1, T_2)$, $\mathfrak{t}_2 = t_3(T_1, T_2)^{-1}$, $\mathfrak{t}_3 = t_3(T_1, T_2)t_3(T_2, T_2)$ and radicals $\sqrt[3]{\mathfrak{t}_i}$ (get_pairings).
2. Take $\zeta_3^{j_i}$ acting on $\sqrt[3]{\mathfrak{t}_i}$ (action_on_cubic_roots).
3. Compute b_4 by $r, s, t, \zeta_3^{j_i}\sqrt[3]{\mathfrak{t}_i}$ (get_b4s).
4. Compute b_i ($i \neq 4$) sequentially using b_4 (get_bi).
5. Compute (R, S, T) corresponding to the kernel subgroup of good extension (get_bi, get_new_rst).
6. Compute parameters of next $(3, 3)$-isogenous curve from (R, S, T) (get_isogenous_RST).

4 Optimizations of Formulae to Compute (3, 3)-isogenies

In this section, we consider to optimize the hash function algorithm using the $(3, 3)$-isogeny computation formula explained in the Sect. 3. Specifically, we aim to reduce the number of operations by devising a computation circuit for each function.

We denote **M, S, A**, *and* **I** by the cost of multiplication, squaring, addition, and division in the base field $\mathbb{F}_{p^2}$ respectively.

In [6], they proposed the explicit algorithms but only naive implementations, leaving plenty of room for optimization in the algorithms. In this paper, we propose an optimization that can eliminate such useless computation process, although it is not an improvement of the algorithm itself. The hash function based on $(3, 3)$-isogenies in [6] deals with six functions get_pairings, action_on_cubic_roots, get_b4s, get_bi, get_new_rst, and get_isogenous_RST. We provide the optimized Magma code in our GitHub repository [14].

First, we optimize the computation of Δ that is common to the functions get_pairings, get_b4s, and get_isogenous_RST. In [6], the value Δ is computed by

$$\delta_1 = t, \quad \delta_2 = s, \quad \delta_3 = st + 1,$$
$$\delta_4 = r^3 - 3rt + t^2 + t, \quad \delta_5 = r^3s - 3rst + st^2 + st + t, \quad \delta_8 = r^2 - t,$$
$$\delta_9 = r - 1, \quad \delta_{10} = rs - st - 1,$$

$$\Delta = r^6s^2 - 6r^4s^2t - 3r^4s + 2r^3s^2t^2 + 2r^3s^2t + 3r^3st + r^3s + r^3 + 9r^2s^2t^2 \\ + 6r^2st - 6rs^2t^3 - 6rs^2t^2 - 9rst^2 - 3rst - 3rt + s^2t^4 + 2s^2t^3 + s^2t^2 \\ + 2st^3 + 3st^2 + t^2 + t. \tag{2}$$

Since the inputs r, s, t are the elements of $\mathbb{F}_{p^2}$, the sequential computation cost is $48\mathbf{M} + 40\mathbf{S} + 32\mathbf{A}$, where the cube power cost is $1\mathbf{M} + 1\mathbf{S}$, the fourth power cost is 2S, and the sixth power cost is $1M + 2S$, and the constant multiplication cost is ignored for simplicity.

For example, from the above equation (2), it is clear that we obtain $\delta_5 = \delta_4 s + t$, so recomputing δ_5 after computing δ_4 involves a lot of wasteful processing. Therefore, we propose the following computation procedure for δ_i.

$$\delta_1 = t, \quad \delta_2 = s, \quad A = \delta_1\delta_2, \quad \delta_3 = A + 1,$$
$$\delta_8 = r^2 - t, \quad \delta_4 = r(\delta_8 - 2\delta_1) + \delta_1^2 + \delta_1, \quad \delta_5 = \delta_2\delta_4 + \delta_1, \quad \delta_9 = r - 1,$$
$$\delta_{10} = r\delta_2 - \delta_3.$$

The computation cost is $4M + 2S + 8A$ when δ_i is computed in this way.

Similarly, it is inefficient to compute the value Δ sequentially from the front. In this paper, we give two expressions for the value Δ. We first consider the value Δ as a multivariate polynomial $\Delta(r, s, t)$ with variables r, s, t. $\Delta(r, s, t)$ is quadratic for the variable s. Hence we obtain

$$\Delta(r, s, t) = f_1(r, t)s^2 + f_2(r, t)s + f_3(r, t).$$

Then each coefficient $f_i(r, t)$ can be computed as

$$f_1(r, t) = \delta_4^2, \quad f_2(r, t) = (-3(r - t) + 1)\delta_4 + (\delta_9^3 - \delta_4)t, \quad f_3(r, t) = \delta_4.$$

Therefore we can represent Δ as

$$\Delta = (\delta_2\delta_4)^2 + (-3(r - \delta_1) + 1)(\delta_2\delta_4) + (\delta_9^3 - \delta_4)A + \delta_4. \tag{3}$$

The values $\delta_2\delta_4$, $\delta_9^3 - \delta_4$ in (3) can be said to be the elementary equation in this hash function, and it appears everywhere in the function get_b4s, so it is more efficient to precompute them before computing Δ. The computation cost of Δ is $2M + 5A$. Therefore, if $\delta_2\delta_4$, $\delta_9^3 - \delta_4$ is precomputed, the total computation cost before Δ is $6M + 2S + 13A$.

On the other hand, in get_pairings and get_isogenous_RST, it is not necessary to precompute $\delta_2\delta_4$, $\delta_9^3 - \delta_4$. If they are not precomputed, the computation cost is $4M + 2S + 6A$ in the expression (3). Then we give another expression for the value Δ.

We compute $\delta_6 = (1 - 3r + \delta_5)\delta_2 + \delta_3$, and obtain

$$\Delta = (\delta_6 + A)\delta_4 + (\delta_9^3 - \delta_4)A. \tag{4}$$

Then the computation cost of Δ is $4M + 1S + 6A$. The number of squaring operations can be reduced only once compared to using the expression (3).

4.1 Optimization of `get_pairings`

The naive computation cost of the function `get_pairings` in [6] is $86M + 67S + 50A + 3I$. We first need to compute r, s, t which determine the previous isogeny. In this part, three inversions $1/(\delta_{10}^2 \delta_4)$, $1/(\delta_1 \delta_2 \delta_9^3 \Delta)$, $1/(\delta_{10}^3 \delta_4^2)$ are performed and we multiply them by the values $-\delta_2 \delta_9 \delta_8 (\delta_5 - r)$, $\delta_{10}^3 \delta_4^2$, $\delta_2^2 \delta_9^3$. We performed the above to reduce the number of inversions as much as possible as follows:

$$A = \delta_{10}\delta_4, \quad B = A^2 \delta_{10}, \quad E = \delta_2 \delta_9, \quad F = E\delta_8, \quad G = E\delta_1 \delta_9^2,$$
$$P = G\Delta, \quad Q = 1/(BP), \quad C = PQ, \quad D = AC, \quad H = F^2 \delta_9 \delta_8,$$
$$r = -F(\delta_5 - r)D, \quad s = B^2 Q, \quad t = HC.$$

We then compute

$$d_1 = t, \quad d_2 = s, \quad d_3 = d_1 d_2 + 1,$$
$$d_4 = r(r^2 - 3d_1) + d_1^2 + d_1, \quad T = \delta_2 \delta_4, \quad d_5 = T + d_1,$$
$$d_6 = d_2(d_5 + 1 - 3r) + d_3, \quad d_7 = Td_3 + d_1(d_3 - d_2),$$

to obtain the Tate pairings. In the end, we need to compute the cube roots of pairings, which are not optimized in this paper.

Therefore, the computation cost of the function `get_pairings` excluding the cube root is $30M + 9S + 25A + 1I$.

4.2 Optimization of `get_isogenous_RST`

The naive computation cost of the function `get_isogenous_RST` in [6] is $64M + 49S + 33A + 3I$. After the computation of Δ, the same computation circuit as the optimization in `get_pairings` in the previous section is applied. Then the total cost is $24M + 7S + 15A + 1I$.

4.3 Optimization of `get_b4s`

The naive computation cost of the function `get_b4s` in [6] is $895M + 863S + 351A + 18I$.

This function computes the value Δ for the six input data R, S, T, a, b, c and outputs the value `b4ab` and the value `b4bc`. The computation up to the value Δ is the same as in the previous section, so it is omitted.

First, we explain the computation procedure of the b4ab. The b4ab is represented as

$$\text{b4ab} = \sqrt{-3}((\text{cofab9} + \text{cofab8} \cdot a + \text{cofab7} \cdot a^2)$$
$$+ (\text{cofab6} + \text{cofab5} \cdot a + \text{cofab4} \cdot a^2)b$$
$$+ (\text{cofab3} + \text{cofab2} \cdot a + \text{cofab1} \cdot a^2)b^2).$$

These coefficients cofab1, ..., cofab9 are represented by

$$\begin{aligned}
\text{cofab1} &= (-6ST - 2)(\delta_1^2 \delta_4^4 \delta_{10}^8 / (\delta_2^3 \delta_8^6 \delta_9^2 \Delta^2)), \\
\text{cofab2} &= -2(\delta_1^2 \delta_4^4 \delta_8 \delta_{10}^7 \delta_{11} / (\delta_2^2 \delta_8^6 \delta_9^2 \Delta^2)), \\
\text{cofab3} &= (6ST + 4)(\delta_1 \delta_4^4 \delta_8^2 \delta_{10}^6 / (\delta_2 \delta_8^6 \delta_9^2 \Delta^2)), \\
\text{cofab4} &= 2(\delta_1^2 \delta_4^2 \delta_{10}^5 \delta_{11} / (\delta_2^2 \delta_8^4 \delta_9 \Delta)), \\
\text{cofab5} &= (-6ST - 2)(\delta_1 \delta_4^2 \delta_8 \delta_{10}^4 / (\delta_2 \delta_8^4 \delta_9 \Delta)), \quad\quad (5) \\
\text{cofab6} &= -6(\delta_1 \delta_3 \delta_4^2 \delta_8^2 \delta_{10}^3 / (\delta_8^4 \delta_9 \Delta)), \\
\text{cofab7} &= 6(\delta_1^2 \delta_{10}^2 / \delta_8^2), \\
\text{cofab8} &= (6ST + 4)(\delta_1 \delta_{10} / \delta_8), \\
\text{cofab9} &= (2ST + 1)\delta_{11}.
\end{aligned}$$

These computations require 8 inversions, and we can reduce the number of them as follows:

$$\begin{aligned}
A_0 &= \delta_8^2, & A_1 &= A_0 \delta_9 \Delta, & A_2 &= A_1 \delta_2, & A_3 &= A_2^2, & A_4 &= A_0 A_3 \delta_2, \\
I_1 &= 1/A_4, & I_2 &= I_1 \delta_2 & I_3 &= I_2 \delta_2, & I_4 &= I_1 A_2, & I_5 &= I_4 \delta_2, \quad (6)\\
I_6 &= I_5 \delta_2, & I_7 &= I_6 A_1, & I_8 &= I_7 \delta_8.
\end{aligned}$$

We then compute

$$\begin{aligned}
B_0 &= \delta_1 \delta_4, & B_1 &= B_0 \delta_4, & B_2 &= B_0^2, & B_3 &= \delta_{10}^2, \\
B_4 &= B_3 \delta_{10}, & B_5 &= B_4^2, & B_6 &= B_1 B_4, & B_7 &= B_6 \delta_{10}, \\
B_8 &= B_6^2, & C_1 &= B_7^2, & C_2 &= B_8 \delta_8 \delta_{10} \delta_{11}, & C_3 &= B_1 \delta_4^2 A_1 B_5, \\
C_4 &= B_2 B_3 B_4 \delta_{11}, & C_5 &= B_7 \delta_8, & C_6 &= B_6 \delta_3 A_1, & C_8 &= \delta_1 \delta_{10}, \\
C_7 &= C_8^2,
\end{aligned}$$

and finally obtain

$$\text{cofab1} = C_1 I_1, \quad \text{cofab2} = C_2 I_2, \quad \text{cofab3} = C_3 I_3, \quad \text{cofab4} = C_4 I_4,$$

cofab5 $= C_5 I_5$, cofab6 $= C_6 I_6$, cofab7 $= C_7 I_7$, cofab8 $= C_8 I_8$, cofab9 $= \delta_{11}$.

Next we explain the b4bc. The b4bc is represented as

$$\begin{aligned} \text{b4bc} = \sqrt{-3}(&(\text{cofbc1} + \text{cofbc2} \cdot c + \text{cofbc3} \cdot c^2) \\ &+ (\text{cofbc4} + \text{cofbc5} \cdot c + \text{cofbc6} \cdot c^2)b \\ &+ (\text{cofbc7} + \text{cofbc8} \cdot c + \text{cofbc9} \cdot c^2)b^2). \end{aligned}$$

These coefficients cofbc1, ..., cofbc9 are represented by using δ_i, Δ as

$$\begin{aligned} \text{cofbc1} &= F_1(R, S, T) \cdot 1/(\delta_2 \delta_4^3), \\ \text{cofbc2} &= F_2(R, S, T) \cdot \delta_1^2 \delta_9 \delta_{10}/(\delta_2 \delta_4^3 \delta_8), \\ \text{cofbc3} &= F_3(R, S, T) \cdot \delta_1^3 \delta_9^2 \delta_{10}^2/(\delta_2 \delta_4^3 \delta_5 \delta_8^2), \\ \text{cofbc4} &= F_4(R, S, T) \cdot \delta_1 \delta_{10}^3/(\delta_2^2 \delta_4 \delta_8^3 \delta_9 \Delta), \\ \text{cofbc5} &= F_5(R, S, T) \cdot \delta_1^2 \delta_{10}^4/(\delta_2^2 \delta_4 \delta_8^3 \Delta), \quad (7) \\ \text{cofbc6} &= F_6(R, S, T) \cdot \delta_1^3 \delta_9 \delta_{10}^5/(\delta_2^2 \delta_4 \delta_5 \delta_8^4 \Delta), \\ \text{cofbc7} &= F_7(R, S, T) \cdot \delta_1 \delta_4 \delta_{10}^6/(\delta_2^3 \delta_8^4 \delta_9^2 \Delta^2), \\ \text{cofbc8} &= F_8(R, S, T) \cdot \delta_1^2 \delta_4 \delta_{10}^7/(\delta_2^3 \delta_8^5 \delta_9 \Delta^2), \\ \text{cofbc9} &= F_9(R, S, T) \cdot \delta_1^4 \delta_4 \delta_{10}^8/(\delta_2^3 \delta_5 \delta_8^6 \Delta^2). \end{aligned}$$

The explicit formulae to compute each coefficient are obtained by multiplying by the very complicated polynomials $F_i(R, S, T)$ consisting of inputs R, S, T. In order to compute the equations (7), 9 inversions are required in the naive way. We compute them by avoiding some inversions as follows:

$$\begin{aligned} &T_0 = A_0 \delta_2, \quad T_1 = T_0 \Delta, \quad T_2 = T_1 \delta_9, \quad E_0 = \delta_4^2, \quad E_1 = E_0 \delta_4, \\ &E_2 = E_1 \delta_5, \quad E_3 = 1/E_2, \quad E_4 = I_1 \delta_9^2, \quad E_5 = E_1 E_3, \quad E_6 = E_4 T_1, \\ &E_7 = E_0 E_3, \quad E_8 = E_6 \delta_8, \quad E_9 = E_7 \delta_5, \quad E_{10} = I_4 T_2, \quad J_9 = E_4 E_5, \quad (8) \\ &J_8 = I_1 \delta_8 \delta_9, \quad J_7 = I_1 A_0, \quad J_6 = E_6 E_7, \quad J_5 = E_8 E_9, \quad J_4 = J_7 T_2 E_9, \\ &J_3 = E_{10} E_3, \quad J_2 = J_3 \delta_5 \delta_8, \quad J_1 = J_2 \delta_8, \end{aligned}$$

where each J_i is the denominator of each in the equations (7).

For the equations (6) and (8), we can reduce an inversion by computing E_2 with simultaneously A_4 and performing $P = 1/A_4 E_2$. We then compute $I_1 = E_2 P$, $E_3 = A_4 P$.

The numerators in the equations (7) excluding $F_i(R, S, T)$ are represented as

$$F_0 = \delta_1\delta_9, \quad F_1 = F_0\delta_{10}, \quad F_2 = \delta_{10}^2, \quad F_3 = \delta_4\delta_{10}, \quad F_4 = \delta_1 F_2,$$
$$F_5 = F_2 F_3, \quad C_1 = 1, \quad C_2 = F1\delta_1, \quad C_3 = C_2 F_1, \quad C_4 = F_4\delta_{10}, \quad (9)$$
$$C_5 = F_4^2, \quad C_6 = C_5 F_1, \quad C_7 = C_4 F_5, \quad C_8 = C_4^2 F_3, \quad C_9 = C_5^2 \delta_4.$$

Then the equations (7) can be computed as follows:

$\text{cofbc}1 = C_1 J_1, \quad \text{cofbc}2 = C_2 J_2, \quad \text{cofbc}3 = C_3 J_3, \quad \text{cofbc}4 = C_4 J_4,$
$\text{cofbc}5 = C_5 J_5, \quad \text{cofbc}6 = C_6 J_6, \quad \text{cofbc}7 = C_7 J_7, \quad \text{cofbc}8 = C_8 J_8,$
$\text{cofbc}9 = C_9 J_9.$

Next, we consider the very complicated polynomials $F_i(R, S, T)$. Before this, we compute the relations

$$A_1 = \delta_9^3 - \delta_4, \quad A_2 = R - T, \quad A_3 = 2\delta_9 A_2, \quad A_4 = A_2 A_3,$$
$$A_5 = T A_1, \quad A_6 = 2A_5, \quad A_7 = 3A_6, \quad A_8 = 2A_7,$$
$$A_9 = A_7 + A_8, \quad A_{10} = T A_9, \quad A_{11} = \delta_2 \delta_4, \quad A_{12} = A_{11}^2,$$
$$A_{13} = R^2, \quad A_{14} = 4T, \quad A_{15} = A_{13} + A_{14}, \quad A_{16} = A_{14} + A_{15},$$
$$A_{19} = 2A_{15}, \quad A_{20} = A_{14} + A_{19}, \quad A_{21} = 2A_{20}, \quad A_{22} = 2R,$$
$$A_{23} = A_{22} T, \quad A_{24} = 3R, \quad A_{25} = A_{24} T, \quad A_{26} = A_{15} - A_{22},$$
$$A_{27} = A_{15} - A_{23}, \quad A_{28} = A_{16} - A_{24}, \quad A_{29} = A_{16} - A_{25},$$
$$A_{30} = A_{21} - 10R, \quad A_{31} = A_{21} - 10RT,$$
$$A_{32} = (T+2)(-A_{13} - T) + (5T+1)R, \quad A_{33} = T A_{32},$$
$$A_{34} = 6A_{33}, \quad A_{35} = 2A_{34}, \quad A_{36} = 3A_{34}, \quad A_{37} = T A_{33},$$
$$A_{38} = 6T A_{26}\delta_4 + 18A_{37}, \quad A_{39} = 2(9T(1 - A_2) - \delta_4)\delta_4,$$
$$A_{40} = T((2T+1)(A_{13} + T) - T(T+5)R), \quad A_{41} = 6A_{40}, \quad A_{42} = 2A_{41},$$
$$A_{43} = 3A_{41}, \quad A_{44} = T A_{43}, \quad A_{45} = 6T A_{27}\delta_4 - A_{44}, \quad A_{46} = A_{10} + A_{39}.$$
$$(10)$$

We use these values A_i to simplify $F_i(R, S, T)$.

4.3.1 The Detail of $F_1(R, S, T)$

The explicit formulae of $F_1(R, S, T)$ is given by (11). The polynomial $F_1(R, S, T)$ is quadratic with respect to S. Thus, it can be expressed as $F_1(R, S, T) = f_1(R, T)S^2 + f_2(R, T)S + f_3(R, T)$.

Especially since the relations $A_{11} = \delta_2\delta_4$, $A_{12} = (\delta_2\delta_4)^2$, $A_1 = \delta_9^3 - \delta_4$ appear in the formulae frequently, we can use these relations for simplification. The coefficient $f_1(R, T)$ can be expressed as $f_1(R, T) = ((T + 1)\delta_4 + A_6)\delta_4^2$, therefore

$$f_1(R, T)S^2 = ((T + 1)\delta_4 + A_6)A_{12}$$

holds.

Similarly, we obtain that

$$f_2(R, T)S = ((2T(8T - 7R + 8) - \delta_4)\delta_4 + A_{10})A_{11}, \quad f_3(R, T) = TA_{46}.$$

Thus we obtain the following proposition for `cofbc1`.

Proposition 1 *The polynomial $F_1(R, S, T)$ can be simplified as*

$$((T + 1)\delta_4 + A_6)A_{12} + ((2T(8T - 7R + 8) - \delta_4)\delta_4 + A_{10})A_{11} + TA_{46},$$

where each notation is as described above.

Although the polynomial $F_1(R, S, T)$ is a very complicated multivariate polynomial, it can be greatly reduced by focusing on the variable S and examining frequently appearing algebraic expressions.

$$\begin{aligned}
F_1(R, S, T) = {} & R^9 S^2 T + R^9 S^2 - R^9 S - 6R^8 S^2 T - 3R^7 S^2 T^2 \\
& - 3R^7 S^2 T - 5R^7 ST + R^6 S^2 T^3 + 40R^6 S^2 T^2 + R^6 S^2 T \\
& + 13R^6 ST^2 + 13R^6 ST - 2R^6 T - 21R^5 S^2 T^3 - 21R^5 S^2 T^2 \\
& + 3R^5 ST^2 + 6R^4 S^2 T^4 - 54R^4 S^2 T^3 + 6R^4 S^2 T^2 - 52R^4 ST^3 \\
& - 52R^4 ST^2 - 6R^4 T^2 - R^3 S^2 T^5 + 64R^3 S^2 T^4 + 64R^3 S^2 T^3 \\
& - R^3 S^2 T^2 + 11R^3 ST^4 + 103R^3 ST^3 + 11R^3 ST^2 + 14R^3 T^3 \\
& + 14R^3 T^2 - 33R^2 S^2 T^5 - 48R^2 S^2 T^4 - 33R^2 S^2 T^3 - 15R^2 ST^4 \\
& - 15R^2 ST^3 - 18R^2 T^3 + 9RS^2 T^6 + 15RS^2 T^5 + 15RS^2 T^4 \\
& + 9RS^2 T^3 + 7RST^5 - 40RST^4 + 7RST^3 - 6RT^4 \\
& - 6RT^3 - S^2 T^7 - 2S^2 T^6 - 2S^2 T^5 - 2S^2 T^4 \\
& - S^2 T^3 - 3ST^6 + 9ST^5 + 9ST^4 - 3ST^3 \\
& - 2T^5 + 14T^4 - 2T^3.
\end{aligned}$$

(11)

4.3.2 The Detail of $F_2(R, S, T)$

The explicit formulae of $F_2(R, S, T)$ is given by

$$\begin{aligned}
F_2(R, S, T) = &-2R^7S + 8R^6S - 6R^5S + 6R^5 + 2R^4ST^2 - 22R^4ST \\
&- 12R^4T + 22R^3ST^2 + 28R^3ST + 6R^3T - 18R^2ST^3 \\
&- 24R^2ST^2 - 6R^2ST + 6R^2T^2 - 12R^2T + 4RST^4 \quad (12) \\
&+ 20RST^3 - 2RST^2 + 6RT^3 + 6RT^2 - 4ST^4 \\
&- 2ST^3 + 2ST^2 - 12T^3 + 6T^2.
\end{aligned}$$

We simplify $F_2(R, S, T)$ with respect to S and write as $F_2(R, S, T) = f_1(R, T)S + f_2(R, T)$. Expressing $f_i(R, T)$ using A_j gives the following proposition.

Proposition 2 *The polynomial $F_2(R, S, T)$ can be simplified as*

$$2\delta_9(3A_2^2 - \delta_4)A_{11} + 6A_{27}\delta_4 + 18T(A_1 - \delta_9(T-1)(2R - T - 1)),$$

where each notation is as described above.

4.3.3 Other $F_i(R, S, T)$

In this paper, the details of $F_3(R, S, T), \ldots, F_9(R, S, T)$ corresponding to other `cofbc3, ..., cofb9` are omitted. See [6] for details. Here, only the result of simplifying $F_3(R, S, T), \ldots, F_9(R, S, T)$ is described as the following proposition.

Proposition 3 $F_3(R, S, T), \ldots, F_9(R, S, T)$ *can be simplified as*

$$\begin{aligned}
F_3(R, S, T) &= (2A_{28}\delta_4 + A_{35})A_{11} + A_{38}, \\
F_4(R, S, T) &= \delta_9 A_3 A_{12} + (A_{30}\delta_4 + A_{36})A_{11} + A_{38}, \\
F_5(R, S, T) &= (2(-R + 2T + 2)\delta_4 + A_7)A_{11} + A_{46}, \\
F_6(R, S, T) &= (2A_{29}\delta_4 - A_{42})A_{11} + A_{45}, \\
F_7(R, S, T) &= A_4 A_{12} + (A_{31}\delta_4 - A_{43})A_{11} + A_{45}, \\
F_8(R, S, T) &= (2(-R + 4T)\delta_4 + A_{34})A_{11} + 6T A_{26}\delta_4 + 18A_{37}, \\
F_9(R, S, T) &= (2(-4R + 5T + 5)\delta_4 + A_8)A_{11} + A_{46},
\end{aligned}$$

where each notation is as described above.

Overall, the total cost is $158\mathbf{M} + 20\mathbf{S} + 82\mathbf{A} + 1\mathbf{I}$.

5 Implementation

In this section, we sort out the computation costs described in the previous section, implement the simplified algorithm with Magma, and compare the processing time. As for the implementing environment, Magma (V2.26-10) was used on a machine equipped with Intel (R) Xeon (R) CPU E5-2630 v4 @ 2.20GHz CPU and 128GB memory.

We use the parameter $p = 2^{86} + 163$, which is proposed in [6] and is equivalent to 128-bit security for classical computers and 86-bit security for quantum computers.

Table 1 compares the number of operations in our get_pairings, get_b4s, get_isogenous_RST with those of the conventional methods.

Table 2 shows the processing time required for each of the six functions that consist of the hash function and the entire hash function per 1-bit input data.

We optimized the functions get_pairings, get_b4s, and get_isogenous_RST that achieved efficiency improvement of about 17.1%, 86.5%, and 69.0%, respectively, compared to the [6]. The total cost is reduced by about 16.8% for the entire hash function.

We also compare the performance of hash functions using isogenies on elliptic curves and Jacobians of genus 2 curves. In the case of using elliptic curves, the hash function based on 3-isogenies between elliptic curves is more efficient than the 2-isogeny hash function [21, 22]. Table 3 lists the performance of hash functions at 128 bits of classical security level.

Table 1 The computation cost of each function

Function name	[6] (reran)	This work
get_pairings	86M + 67S + 50A + 3I	30M + 9S + 25A + 1I
get_b4s	895M + 863S + 351A + 18I	158M + 20S + 82A + 1I
get_isogenous_RST	64M + 49S + 33A + 3I	24M + 7S + 15A + 1I

Table 2 The processing time of each function

Function name	[6] (reran) (/bit) (ms)	This work (/bit) (ms)
get_pairings	0.315	0.261
action_on_cubic_roots	0.002	0.002
get_b4s	0.838	0.113
get_bi	0.788	0.757
get_new_rst	2.978	2.956
get_isogenous_RST	0.058	0.018
hash function	4.930	4.101

Table 3 The processing time (/bit) of the hash functions based on isogenies

Hash function	Processing time (/bit) (ms)	Characteristic p
Based on (3, 3)-isogenies [6] (reran)	4.930	$2^{86} + 163$
Based on 3-isogenies[21] (reran)	2.413	$2^{255} + 141$
This work (an optimization of [6])	4.101	$2^{86} + 163$

Our results are still inefficient compared to the hash function in the genus-1 case. We are currently working on improving the functions get_bi and get_new_rst to achieve a faster hash function based on (3, 3)-isogenies.

6 Conclusions

In this work, we focused on the hash function, which is being actively studied as isogeny-based cryptography on abelian surfaces, and optimized its computation algorithm. In the explicitly shown multiradical (3, 3)-isogeny formulae, the complex polynomials are simplified to eliminate extra computation in [6]. We improved the efficiency by about 16.8% for the implementation method shown in [6]. In particular, we achieved an efficiency improvement of about 86.5% for the most complicated part get_b4s of the hash function excluding the Gröbner basis computation.

For further speed up the hash function based on (3, 3)-isogeny, we need to explore

- more efficient formulae to compute (3, 3)-isogeny that are simplified and optimized well,
- explicit formulae for isomorphic transformations to avoid Gröbner basis computations.

To solve the above issues, we are currently working with the following tasks. We should derive the explicit formula of get_bi more clearly to optimize it. Specifically, we need to simplify algebraic relations by the curve parameter r, s, t and the radicals of pairing values. Furthermore, we should explore other representations of the parameter b_4 and the other b_i and then simplify the formulae using computational algebra and solving optimization problems like [10].

In addition, we should find an explicit formula of get_new_rst for reparametrization of r, s, t to avoid the Gröbner basis computation in each step of the hash function. Note that the current implementation of get_new_rst costs most of the overall hash function. These optimizations would result in a more efficient construction of the hash function based on (3, 3)-isogenies compared to the hash function based on elliptic curve isogenies.

Acknowledgements This work was supported by JST CREST Grant Number JPMJCR21l3 and JSPS KAKENHI Grant Number JP21K17740, Japan.

References

1. D.J. Bernstein, T. Lange, C. Martindale, L. Panny, Quantum circuits for the CSIDH: optimizing quantum evaluation of isogenies, in *Advances in Cryptology–EUROCRYPT 2019 - 38th Annual International Conference on the Theory and Applications of Cryptographic Techniques, Darmstadt, Germany, May 19–23, 2019, Proceedings, Part II, Lecture Notes in Computer Science*, eds. by Y. Ishai, V. Rijmen, vol. 11477 (Springer, 2019), pp. 409–441. https://doi.org/10.1007/978-3-030-17656-3_15
2. N. Bruin, E.V. Flynn, D. Testa, Descent via (3,3)-isogeny on jacobians of genus 2 curves. Acta Arithmet. **165**(3), 201–223 (2014). https://doi.org/10.4064/aa165-3-1
3. W. Castryck, T. Decru, An efficient key recovery attack on sidh, in *Advances in Cryptology–EUROCRYPT 2023*, eds. by C. Hazay, M. Stam (Springer Nature Switzerland, Cham), pp. 423–447 (2023)
4. W. Castryck, T. Decru, B. Smith, Hash functions from superspecial genus-2 curves using richelot isogenies. J. Math. Cryptol. **14**(1), 268–292 (2020). https://doi.org/10.1515/jmc-2019-0021
5. W. Castryck, T. Decru, F. Vercauteren, Radical isogenies, in *Advances in Cryptology–ASIACRYPT 2020–26th International Conference on the Theory and Application of Cryptology and Information Security, Daejeon, South Korea, December 7-11, 2020, Proceedings, Part II, Lecture Notes in Computer Science*, eds. by S. Moriai, H. Wang, vol. 12492 (Springer, 2020), pp. 493–519. https://doi.org/10.1007/978-3-030-64834-3_17
6. C. Wouter, D. Thomas, Multiradical isogenies, in *Arithmetic, Gometry, Cryptography, and Coding Theory 2021: 18th International Conference Arithmetic, Geometry, Cryptography, and Coding Theory May 31-June 4, 2021 Centre International de Rencontres Mathématiques, Marseille, France, Contemporary mathematics*, eds. by A. Samuele, K. Valentijn, L. García Elisa, vol. 779 (American Mathematical Society, 2022), pp. 57–89. http://dx.doi.org/10.1090/conm/779
7. D.X. Charles, K.E. Lauter, E.Z. Goren, Cryptographic hash functions from expander graphs. J. Cryptol. **22**(1), 93–113 (2009). https://doi.org/10.1007/s00145-007-9002-x
8. C. Costello, B. Smith, The supersingular isogeny problem in genus 2 and beyond, in *Post-Quantum Cryptography–11th International Conference, PQCrypto 2020, Paris, France, April 15-17, 2020, Proceedings, Lecture Notes in Computer Science*, eds. by J. Ding, J. Tillich, vol. 12100 (Springer, 2020), pp. 151–168. https://doi.org/10.1007/978-3-030-44223-1_9
9. P. Dartois, L. Maino, G. Pope, D. Robert, An algorithmic approach to (2, 2)-isogenies in the theta model and applications to isogeny-based cryptography. Cryptology ePrint Archive, Paper 2023/1747 (2023). https://eprint.iacr.org/2023/1747
10. T. Decru, S. Kunzweiler, Efficient computation of $(3^n, 3^n)$-isogenies, in *Progress in Cryptology–AFRICACRYPT 2023–14th International Conference on Cryptology in Africa, Sousse, Tunisia, July 19-21, 2023, Proceedings, Lecture Notes in Computer Science*, eds. by N.E. Mrabet, L.D. Feo, S. Duquesne, vol. 14064 (Springer, 2023), pp. 53–78.https://doi.org/10.1007/978-3-031-37679-5_3
11. L.D. Feo, D. Jao, J. Plût, Towards quantum-resistant cryptosystems from supersingular elliptic curve isogenies. J. Math. Cryptol. **8**(3), 209–247 (2014). https://doi.org/10.1515/jmc-2012-0015
12. L.D. Feo, D. Kohel, A. Leroux, C. Petit, B. Wesolowski, Sqisign: Compact post-quantum signatures from quaternions and isogenies, in *Advances in Cryptology–ASIACRYPT 2020–26th International Conference on the Theory and Application of Cryptology and Information Security, Daejeon, South Korea, December 7-11, 2020, Proceedings, Part I, Lecture Notes in Computer Science*, eds. by S. Moriai, H. Wang, vol. 12491 (Springer, 2020), pp. 64–93. https://doi.org/10.1007/978-3-030-64837-4_3
13. E.V. Flynn, Y.B. Ti, Genus two isogeny cryptography, in *Post-Quantum Cryptography–10th International Conference, PQCrypto 2019, Chongqing, China, May 8-10, 2019 Revised Selected Papers, Lecture Notes in Computer Science*, eds. by J. Ding, R. Steinwandt, vol. 11505 (Springer, 2019), pp. 286–306.https://doi.org/10.1007/978-3-030-25510-7_16

14. M. Ishii, D. Hayashida, An optimization of the hash function based on (3, 3)-isogenies (2023). https://github.com/masahiro13/3_3_isogenies
15. N. Koblitz, Hyperelliptic cryptosystems. J. Cryptol. **1**(3), 139–150 (1989). https://doi.org/10.1007/BF02252872
16. S. Kunzweiler, Efficient computation of (2^n, 2^n)-isogenies. IACR Cryptol. ePrint Arch. p. 990 (2022). https://eprint.iacr.org/2022/990
17. L. Maino, C. Martindale, L. Panny, G. Pope, B. Wesolowski (2023) A direct key recovery attack on sidh, in *Advances in Cryptology–EUROCRYPT*, eds. by C. Hazay, M. Stam (Springer Nature, Switzerland, Cham, 2023), pp. 448–471
18. T. Moriya, H. Onuki, Y. Aikawa, T. Takagi, The generalized montgomery coordinate: A new computational tool for isogeny-based cryptography. Cryptology ePrint Archive, Paper 2022/150 (2022). https://eprint.iacr.org/2022/150
19. National Institute of Standards and Technology (NIST), Post-Quantum Cryptography Standardization. https://csrc.nist.gov/projects/post-quantum-cryptography/post-quantum-cryptography-standardization. Accessed 16 February 2022
20. H. Onuki, T. Moriya, Radical isogenies on montgomery curves, in *Public-Key Cryptography–PKC 2022–25th IACR International Conference on Practice and Theory of Public-Key Cryptography, Virtual Event, March 8-11, 2022, Proceedings, Part I, Lecture Notes in Computer Science*, eds. by G. Hanaoka, J. Shikata, Y. Watanabe, vol. 13177 (Springer, 2022), pp. 473–497. https://doi.org/10.1007/978-3-030-97121-2_17
21. H. Tachibana, K. Takashima, T. Takagi, Constructing an efficient hash function from 3-isogenies. JSIAM Letters **9**, 29–32 (2017).https://doi.org/10.14495/jsiaml.9.29
22. K. Takashima, Efficient algorithms for isogeny sequences and their cryptographic applications, in *Mathematical Modelling for Next-Generation Cryptography: CREST Crypto-Math Project, Mathematics for Industry*, eds. by T. Takagi, M. Wakayama, K. Tanaka, N. Kunihiro, K. Kimoto, D.H. Duong (Springer Singapore, 2017), pp. 97–114. https://doi.org/10.1007/978-981-10-5065-7_6
23. J. Vélu, Isogénies entre courbes elliptiques. Comptes Rendus de l'Acaémie des Sciences de Paris **273**, 238–241 (1971)
24. R. Yoshida, K. Takashima, Simple algorithms for computing a sequence of 2-isogenies, in *Information Security and Cryptology–ICISC 2008, 11th International Conference, Seoul, Korea, December 3-5, 2008, Revised Selected Papers, Lecture Notes in Computer Science*, eds. by P.J. Lee, J.H. Cheon, vol. 5461 (Springer, 2008), pp. 52–65.https://doi.org/10.1007/978-3-642-00730-9_4

Open Access This chapter is licensed under the terms of the Creative Commons Attribution 4.0 International License (http://creativecommons.org/licenses/by/4.0/), which permits use, sharing, adaptation, distribution and reproduction in any medium or format, as long as you give appropriate credit to the original author(s) and the source, provide a link to the Creative Commons license and indicate if changes were made.

The images or other third party material in this chapter are included in the chapter's Creative Commons license, unless indicated otherwise in a credit line to the material. If material is not included in the chapter's Creative Commons license and your intended use is not permitted by statutory regulation or exceeds the permitted use, you will need to obtain permission directly from the copyright holder.

Hash Function, Graph Theory, and Applications

Hashing by Walking Over Expanders: A Recipe for Constructing Provably Secure Hash Functions

Yusuke Aikawa

Abstract As a method to construct provably secure hash functions, the use of families of expander graphs is known. Two approaches are employed in their instantiation: one using the Cayley graph of a finite group and the other utilizing the isogeny graphs between supersingular elliptic curves (more generally, superspecial abelian varieties). This paper provides explanations for these constructions and summarizes recent developments.

Keywords Cryptographic hash functions · Expander graphs · Cayley graphs · Isogeny graphs

1 Introduction

Families of expander graphs (or expander graphs, simply) are families of sparse, highly connected graphs with a crucial property that random walks on them rapidly converge to a uniform distribution. Expander graphs have been explored for various applications in computer science, including applications in cryptography. Specifically, by corresponding inputs to walks on a graph and using the terminal of the walk as the output, it is expected to construct functions with high randomness, and their one-wayness is reduced to path finding problem of the graph.

In the explicit construction of expander graphs, Cayley graphs of groups play a significant role. As is well-known, if a group possesses property (T), its Cayley graph becomes an expander graph. In fact, Zémor realized the above idea using special linear groups $SL_2(\mathbb{F}_p)$ in his seminal work [50], marking the beginning of the study of Cayley hash functions. Since then, various approaches, including those using Ramanujan graphs, have been proposed ([6, 9, 12, 43, 45] et al.), and security analysis has progressed ([19, 22, 36, 37] et al.). In the explicit construction of expander graphs, Cayley graphs of groups play a significant role. As is well-known,

Y. Aikawa (✉)
Department of Mathematical Informatics, The University of Tokyo, Tokyo, Japan
e-mail: aikawa@mist.i.u-tokyo.ac.jp

© The Author(s) 2026
T. Takagi et al. (eds.), *Mathematical Foundations for Post-Quantum Cryptography*, Mathematics for Industry 40, https://doi.org/10.1007/978-981-96-1218-5_19

if a group possesses property (T), its Cayley graph becomes an expander graph. In fact, Zémor realized the above idea using special linear grouaikawa@mist.i.u-tokyo.ac.jpps $SL_2(\mathbb{F}_p)$ in his seminal work [50], marking the beginning of the study of Cayley hash functions. Since then, various approaches, including those using Ramanujan graphs, have been proposed ([6, 9, 43, 45], [12] et al.), and security analysis has progressed ([19, 22, 36, 37] et al.).

On the other hand, Charles, Lauter, and Goren [9] introduced a number-theoretic approach by using isogeny graphs of supersingular elliptic curves. It was proven by Pizer [34] that these graphs are Ramanujan graphs. This study opened a new direction in cryptography, relying on the hardness of isogeny problems as the security assumptions, widely known as isogeny-based cryptography today. Furthermore, in recent years, considering higher dimensional anaogue [8], specifically using superspecial abelian varieties, has led to developments in the study of isogeny graphs of superspecial abelian varieties ([2, 17] at al.).

This paper provides a concise overview of the current state of these studies from a graph-theoretical perspective. Due to space limitations, it was not possible to comprehensively address all aspects of this research topic, especially regarding security. Readers who desire a more detail on this subject are encouraged to refer to the references cited in this paper and their related citations.

2 Preliminaries

Notation. For a set S, we denote the number of elements in S by $|S|$. If S is a finite set, we denote $s \leftarrow S$ to sample s uniformly at random from S. For a positive integer d, we set $[d] := \{0, 1, \ldots, d-1\}$. The set of d-ary strings of any length is denoted by $[d]^*$. Let $p \in \mathbb{Z}$ denote a prime, and q represent its power. We denote the finite field of order q by $\mathbb{F}_q$. A prime distinct from p is denoted by ℓ. Let $X = (V, E)$ be a graph, where V is the set of vertices, and E is the set of edges. In this paper, an edge connecting x and y is denoted as $(x, y) \in E \subset V^2$. Furthermore, unless otherwise specified, graphs are assumed to admit for multiple edges and loops. For a group G, the unit element is denoted by e.

2.1 Hash Functions

Hash functions constitute a fundamental primitive in cryptography. A hash function H is a function that takes an arbitrary length bit string as input and produces a fixed length bit string as output, i.e., $H : \{0, 1\}^* \to \{0, 1\}^k$ for some a positive integer k. To maintain consistency with later discussion, more generally, for a positive integer d and a finite set S, the domain and codomain of a hash function may be written $H : [d]^* \to S$ via some correspondences $\{0, 1\}^* \to [d]^*$ and $S \to \{0, 1\}^k$. These functions are often used for purposes such as enhancing the security of cryptography. Thus, they are required to meet the security requirements as below:

Preimage Resistance Informally, a hash function H is preimage resistant if, for given $y \leftarrow \{0, 1\}^k$, it is computationally hard to find $x \in \{0, 1\}^*$ such that $H(x) = y$.

Target-collision Resistance Informally, a hash function H is target-collision resistant if, for given $x_1 \leftarrow \{0, 1\}^*$, it is computationally hard to find $x_2 \in \{0, 1\}^*$ such that $H(x_1) = H(x_2)$.

Collision Resistance Informally, a hash function H is collision resistant if it is computationally hard to find $x_1, x_2 \in \{0, 1\}^*$ such that $H(x_1) = H(x_2)$.

For a formal definition of these properties, see [26] or [31]. The aforementioned requirements are well-known, but the additional conditions regarding the security of hash functions exist. In this paper, we address universality as follows, in addition to the above, and focus on constructing universal provably secure collision resistant hash functions. While the notion of universality of hash functions is first introduced in [7], the following definition is drawn from [35].

Universality Informally, a hash function $H : \{0, 1\}^* \to \{0, 1\}^k$ is universal if the distribution of outputs of H is nearly equal to the uniform distribution on $\{0, 1\}^k$.

While functions satisfying these conditions are sometimes referred to as *cryptographic hash functions*, their construction is a hard task for cryptographers. Currently, SHA-2 and SHA-3 are practically employed, and these functions are heuristically believed to satisfy these conditions, although their security is not provably secure. Hence, research into the construction and security analysis of schemes whose security is based on computational hardness of a mathematical problem has been progressing.

One such approach involves the use of expander graphs, and the idea behind its construction is as follows: let X be a d-regular graph along with a fixed vertex v_0. Suppose there are efficiently computable invariants for each vertex of the graph. Additionally, we assume that any bit string corresponds to a non-backtracking path on the graph starting from v_0. Then, a function can be constructed by computing the invariant of the endpoint of the path corresponding to an input bit string and outputting it.

For example, let us consider the case of $d = 3$. Initially, from the three edges outgoing from v_0, we select two and label them as "0" and "1", respectively. From the vertex v_1 connected to v_0 via the edge labeled by "0", there are three edges, but one of them corresponds to the backtracking edge connected to v_0, so we label the remaining two edges with 0 and 1. Similarly, for the vertex v_2 connected to v_0 through the edge labeled "1," we follow the same procedure. By repeating this process, it is possible to label each edge of the graph with "0" and "1". Then, based on these labels, one can associate a path starting from v_0 with a bit string of any length.

When considering the construction of hash functions in this manner, various conditions are required for a based graph X. These conditions include existence of efficiently computable invariants for vertices, efficient computation of walks on the graph, a distribution of terminals of paths that approximates a uniform distribution, absence of short cycles to prevent collisions, and more. In the subsequent discussions in this paper, we will examine the explicit design of these, initiated by Zémor [50]. While explicit construction of such graphs is a challenging problem, one of the candidates for this purpose is families of expander graphs.

2.2 Expanders

In this section, we explain basics of families of expander graphs. Expander families are relevant to a variety of research fields and topics, but we will limit our discussion here to those necessary for cryptographic applications. We refer to [20, 27] for some general facts on expander graphs and [29] for general theory of Markov chains.

Families of Expander Graphs. Let $X = (V, E)$ be an undirected, connected, finite graph, where V (resp. E) denotes the set of vertices (resp. edges). In this section, we assume that graphs do not have multiedges and loops; however, later on, we will also consider graphs that have them. We consider quantifying how easily information over X can be expanded. Let F be a subset of V. Paths of propagation of F to its exterior are given by the edge boundary $\partial F = \{e = (x, y) \in E \mid x \in F, y \notin F\}$. Then, the quantity reflecting energy of propagation outward from F can be defined as $\frac{|\partial F|}{|F|}$, and we define the *expansion constant* or *Cheeger constant* of X as follows;

$$h(X) := \min \left\{ \frac{|\partial F|}{|F|} \mid F \subset V, 0 < |F| \leq \frac{|V|}{2} \right\}.$$

The larger $h(X)$, the more expandable the graph is. For instance, the complete graph K_n of order n has the expansion constant $h(K_n) = n - \lfloor \frac{n}{2} \rfloor$. However, complete graphs become inefficient as the number of edges increases quadratically as $n \to \infty$. On the other hand, a graph with large expansion constant and fewer edges, in other words, a sparse graph, can be considered an efficient graph for information transmission. Therefore, we define such efficient graphs as follows.

Definition 1 Let d be a fixed positive integer. Consider a family of d-regular connected, undirected graphs denoted as X_k for $k \in \mathbb{N}$, where X_k is also simply referred as X. The family $\{X_k\}_k$ is called a *family of expander graphs*, or simply *expander graphs*, if it satisfies the following conditions:

1. The number of vertices in X_k, tends to infinity as k approaches infinity.
2. There exists a positive real number $\epsilon > 0$ such that $h(X_k) > \epsilon$ for all k.

Since there are $2^{n-1} - 1$ possible choices of subsets F for a graph X with n-vertices, computing the expansion constant is generally computationally hard problem. On the other hand, it is known that various properties of X can be obtained from the eigenvalues. Let A_X be the adjacency matrix of X, which is an $n \times n$ matrix. The (i, j) entry represents the number of edges in X between vertices v_i and v_j when we denote $V = \{v_1, v_2, \ldots, v_n\}$. Since the matrix A_X is real and symmetric, we denote its real eigenvalues as $d = \mu_1 \geq \mu_2 \geq \cdots \geq \mu_n$. The expansion property of X is captured by:

Proposition 1 *Notation as above. We have*

$$\frac{d - \mu_2}{2} \leq h(X) \leq \sqrt{2d(d - \mu_2)}.$$

Based on the above inequality, the condition 2 for expander graphs can be expressed as there exists a positive real number $\epsilon > 0$ such that $d - \mu_{2,k} > \epsilon$ for all k, where $\mu_{2,k}$ denotes the second largest eigenvalue of X_k. The difference between the largest eigenvalue and the second largest eigenvalue $d - \mu_2$ is of significant importance, referred to as the *spectral gap* of X. Furthermore, this inequality informs us that a smaller second eigenvalue μ_2 implies higher expansion property of the graph. Hence, we are interested in whether there exists a lower bound for μ_2, it is known that there is a limit to bound this, given by Alon-Boppana's theorem. Here, we set $\mu := \{|\mu_2|, |\mu_n|\}$.

Proposition 2 *Let d be a fixed positive integer. Let $\{X_k\}_{k \in \mathbb{N}}$ be a family of d-regular connected, undirected graphs, with $|V_k| \to \infty$ as $k \to \infty$, where V_k denotes the set of vertices of X_k. Then, we have $\mu(X_k) \geq 2\sqrt{d-1} - o(1)$.*

By virtue of this theorem, we can define an optimal expander graphs, known as Ramanujan graphs, as follows.

Definition 2 A finite connected d-regular undirected graph X is a *Ramanujan graph* if, for any nontrivial eigenvalue $\tilde{\mu}$ of X, we have $|\tilde{\mu}| \leq 2\sqrt{d-1}$.

It is shown that random construction of graphs yields families of expander graphs, see [3]. However, providing explicit construction is of paramount importance in practical applications of families of expander graphs to computer science, including cryptography, error correcting code, derandamization, and so on. In the application to cryptography, two constructions are employed, as we will see later. One approach is to use isogeny graphs of supersingular elliptic curves, or more generally superspecial abelian varieties, while the other is employing Cayley graphs of linear groups over a finite field.

Random Walks on a Graph. Let $X = (V, E)$ be an undirected, connected, finite d-regular graph with n vertices. A *walk* on the graph X of length k is a sequence of vertices $v_0, v_1, \ldots, v_{k-1} \in V$ with $(v_i, v_{i+1}) \in E$ for $i = 0, 1, \ldots, k-2$. A *random walk* on X is a (discrete-time) stochastic process $(x_t)_{t \geq 0}$ taking values in V such that the initial vertex $x_0 = v_0$ is sampled from some initial distribution on V, and we have a transition probability $\Pr[x_{t+1} = v_j \mid x_t = v_i] = \frac{|\{\text{edges between } v_i \text{ and } v_j\}|}{|\{\text{edges outgoing from } v_i\}|}$. For simple d-regular graph, this transition probability is equal to $1/d$ if v_i and v_j are connected. We define the *transition matrix* of X as $M_X = \frac{1}{d} A_X$, where A_X is the adjacency matrix of X. Let us arrange the eigenvalues of M_X as $1 = \lambda_1 \geq \lambda_2 \cdots \geq \lambda_n$, bearing in mind that $\lambda_i = \frac{1}{d}\mu_i$. The difference $1 - \lambda_2$ between the largest eigenvalue and the second largest eigenvalue is also called the spectral gap of X. Moreover, we put $\lambda = \max\{|\lambda_2|, |\lambda_n|\}$ We remark that the random walk on X can be referred to as a Markov chain on the set of states V and the transition matrix M_X.

Theorem 1 *Notation as above. We assume that X is non-bipartite and simple. Consider a random walk on X starting from a fixed vertex $v_0 \in V$. We suppose that the walk reaches $v_k \in V$ after k steps. For any vertex $v \in V$, we have*

$$\left| Pr[v_k = v] - \frac{1}{n} \right| \leq \sqrt{n}\lambda^k.$$

For the proof of this theorem, we refer to [42]. This theorem implies that random walks on graphs converge rapidly to a uniform distribution when the absolute values of eigenvalues are small, and equivalently, when the spectral gap is large. Therefore, expander graphs can be characterized as families of graphs that achieve such rapid mixing, with Ramanujan graphs representing the optimal case. The utilization of this uniform randomness has been considered in applications of expander graphs to cryptography. On the other hand, when considering the construction of hash functions as explained in the previous subsection, it is essential to explicitly construct families of expander graphs. Generating such graphs in large quantities is a challenging problem, but it can be accomplished, for instance, by using isogeny graphs or Cayley graphs, as we will discuss below. In this way, schemes constructed explicitly from families of expander graphs are collectively called *expander hash functions*.

3 Isogeny-Based Construction

One instantiation of constructing a hash function from a family of expander graphs was proposed by Charles et al. in [9], who utilized isogeny graphs of supersingular elliptic curves. This construction not only opened doors to isogeny-based cryptography but also advanced studies of superspecial abelian varieties by considering their higher dimensional analogue. In this section, we provide an overview of these studies.

3.1 Abelian Varieties and Isogeny Graphs

Here, we provide a brief overview of abelian varieties and isogenies connecting them, and then introduce graphs constructed using these concepts, say *isogeny graphs*. For further details, refer to [40, 48] for elliptic curves and [33] for general theory of abelian varieties.

Definition 3 A complete connected group variety is called an *abelian variety*. In particular, a one dimensional abelian variety is called a *elliptic curve*.

We denote dimension of abelian varieties by g.

For an abelian variety A, a couple $(A, \mathcal{L})$ together with a principal polarization $\mathcal{L}$ is called a principally polarized abelian variety (PPAV, for short). While this paper exclusively deals with PPAVs, details such as definitions of polarization and more are omitted. For detailed information, we encourage to refer to [33].

In this paper, we always consider varieties over a finite field $\mathbb{F}_q$. For an abelian variety A over $\mathbb{F}_q$ and an extension K of $\mathbb{F}_q$, we denote the set of rational points of A valued in K by $A(K)$. By definition, $A(K)$ carries group structure, and the unit element is denoted by O_A. The kernel of N-multiplication map for $N \in \mathbb{Z}$ on an abelian variety A of dimension g is denoted by $A[N]$; $A[N] = \{P \in A(\overline{\mathbb{F}}_q) \mid NP = O_A\}$, which is isomorphic to $(\mathbb{Z}/N\mathbb{Z})^{2g}$ for $p \nmid N$. Subgroups of $A[\ell]$ for a prime $\ell \neq p$ which are maximally isotropic with respect to the ℓ-Weil pairing are called *Lagrangians*. If A is a PPAV of dimension g, Lagrangians in $A[\ell]$ are isomorphic to $(\mathbb{Z}/\ell\mathbb{Z})^g$. Since abelian varieties are algebraic varieties and groups, we define morphisms between them as follows.

Definition 4 Let A_1 and A_2 be abelian varieties over $\mathbb{F}_q$. A surjective homomorphism $A_1 \to A_2$ having a finite kernel is called an *isogeny*.

Abelian varieties A_1 and A_2 are isomorphic if there exist isogenies $\phi_1 : A_1 \to A_2$ and $\phi_2 : A_2 \to A_1$ such that $\phi_1 \circ \phi_2 = id$ on E_2 and $\phi_2 \circ \phi_1 = id$ on E_1. Isogenies and isomorphisms of PPAVs are defined in an appropriate manner compatible with principal polarizations of abelian varieties.

Lemma 1 *Let p and ℓ be distinct primes. Let A be a PPAV over $\overline{\mathbb{F}}_p$ and C a Lagrangian in $A[\ell]$. Then, there exists a PPAV $A_C := A/C$ which is unique up to isomorphisms of PPAVs and an isogeny $A \to A_C$ whose kernel is C.*

In particular, for g-dimensional PPAVs A_1 and A_2, an $(\ell)^g$-isogeny is defined to be an isogeny $A_1 \to A_2$ whose kernel is a Lagrangian of $A_1[\ell]$. We note that $(\ell)^g$-isogenies are compatible with the principally polarizations.

Let $\mathrm{Aut}(A)$ be the automorphism group of a PPAV A and $\mathscr{C}(A, \ell)$ the set of Lagrangians of A in $A[\ell]$. For $\sigma \in \mathrm{Aut}(A)$ and a Lagrangian $C \subset A[\ell]$, we have $\sigma(C) \in \mathscr{C}(A, \ell)$; thus, the group $\mathrm{Aut}(A)$ acts on the set $\mathscr{C}(A, \ell)$. Every PPAV has a nontrivial involution $[-1]$. Here, we consider $\mathrm{RA}(A) := \mathrm{Aut}(A)/\{\pm 1\}$ instead of $\mathrm{Aut}(A)$ since we have $[-1]C = C$ for every $C \in \mathscr{C}(A, \ell)$, which means that $[-1]$ commutes with any isogeny. We call the group $\mathrm{RA}(A)$ the *reduced automorphism group* of A.

Here, we remark that the number of Lagrangians in $A[\ell]$ for a g-dimensional PPAV A is

$$N_g(\ell) := \prod_{i=1}^{g}(\ell^i + 1), \tag{1}$$

which is equal to the number of $(\ell)^g$-isogenies outgoing from A.

Isogeny Graphs. An elliptic curve E in characteristic p is said to be supersingular if we have $E[p] = \{O_E\}$. An abelian variety A is superspecial if A is isomorphic to $E_1 \times \cdots \times E_g$ as an abelian variety, where E_i are supersingular elliptic curves ($i = 1, \ldots, g$). We refer to [28] for some general facts on supersingular/superspecial abelian varieties and [5] especially for the case $g = 2$ and 3.

We denote the set of isomorphism classes of g-dimensional principally polarized superspecial abelian varieties over $\overline{\mathbb{F}}_q$ by $SS_g(p)$. It is known that $SS_g(p)$ is a finite set and, we have $|SS_g(p)| = O(p^{\frac{g(g+1)}{2}})$, see [15].

Definition 5 Let us fix an integer g. The (superspecial) $(\ell)^g$-isogeny graph $\mathcal{G}_g(\ell, p)$ is a directed graph having weighted multiple edges defined by

Vertices: The set of vertices is $SS_g(p)$. For a PPAV A, we simply write A for the isomorphism class to which it belongs.

Edges: The set of directed edges consists of pairs $(A_1, A_2) \in SS_g(p) \times SS_g(p)$ such that there exists an $(\ell)^g$-isogeny between them. We note that if (A_1, A_2) is an edge, then (A_2, A_1) is an edge due to the existence of the dual isogeny. For each edge e whose kernel is $C \in \mathscr{C}(A, \ell)$ for some representative A of the class corresponding to the initial vertex of e, we define a weight of e as the number of the orbits of C under the action of RA(A).

We note that the graph $\mathcal{G}_g(\ell, p)$ is $N_g(\ell)$-regular. For a fixed integer g and a prime ℓ, we consider these graphs as a family of regular graphs $\{\mathcal{G}_g(\ell, p)\}_p$ indexed by primes $p \neq \ell$. It is known that these graphs are connected. For details, we refer to [34] for $g = 1$ and [2, 23, 24, 49] for $g \geq 2$.

Theorem 2 *Let p and ℓ be distinct primes and g a positive integer. Then, the superspecial isogeny graphs $\mathcal{G}_g(\ell, p)$ are connected.*

3.2 The Case of Elliptic Curves

In this subsection, we consider the case $g = 1$. Let E be a supersingular elliptic curve in characteristic p, given by Weierstrass model $Y^2 = X^3 + aX + b$. The j-invariant of E is defined by $j_E = -1728\frac{4a^3}{4a^3+27b^2}$. It is well known that every isomorphism class over $\mathbb{F}_q$ of elliptic curves is uniquely identified by j-invariants. One of the remarkable properties in the case of supersingular elliptic curves is that their j-invariants always belong to $\mathbb{F}_{p^2}$; thus the vertices of $\mathcal{G}_1(\ell, p)$ are labeled by elements in $\mathbb{F}_{p^2}$, which can be efficiently computed.

The properties of isogeny graphs $\mathcal{G}_1(\ell, p)$ are well investigated. The order of the set of vertices of $\mathcal{G}_1(\ell, p)$ is equal to $\lfloor \frac{p}{12} \rfloor + \epsilon$, where $\epsilon = 0$ (if $p \equiv 1$ mod 12), 1 (if $p \equiv 5, 7$ mod 12), 2 (if $p \equiv 11$ mod 12). Since most of the vertices of $\mathcal{G}_1(\ell, p)$ have trivial reduced automorphism group, this graph can be considered almost undirected. Elliptic curves with nontrivial reduced automorphism group correspond to the curves with j-invariants of 0 or 1728, and these are supersingular under the following conditions: if $p \equiv 2$ mod 3, $j = 0$ is supersingular, and if $p \equiv 3$ mod 4, $j = 1728$ is supersingular. Hence, if $p \equiv 1$ mod 12, the graph can be considered as an undirected graph. However, loops and multiple edges are possible, as indicated by the potential for modular polynomials to have multiple roots, for example.

From the spectral view point, a significant property of $\mathcal{G}_1(\ell, p)$ is the following, proved by Pizer.

Theorem 3 ([34]) *If $p \equiv 1 \mod 12$, the isogeny graph $\mathcal{G}_1(\ell, p)$ is a Ramanujan graph for any $\ell \neq p$.*

As mentioned above, the assumption on p in this theorem guarantees undirectedness of the graph $\mathcal{G}_1(\ell, p)$. This fact is one of the compelling reasons for using walks on supersingular isogeny graphs in construction of cryptography. The expansion property of expander graphs ensures uniformity of outputs of the function for inputs of length $O(\log p)$.

CGL Hash Functions Let p and ℓ be distinct primes. Moreover, we impose $p \equiv 1 \mod 12$. Charles, Goren. and Lauter [9] proposed construction of hash functions based on $\mathcal{G}_1(\ell, p)$. This hash function is often referred to as CGL hash function named after the proposer.

For the sake of simplicity, we consider the case of $\ell = 2$ below. In this case, the graph $\mathcal{G}_1(2, p)$ is 3-regular.[1] Let $E_0 \in SS_1(p)$ be a fixed curve given by $Y^2 = f(X)$ for a monic cubic $f(X)$. A hash function $H_{E_0} : \{0, 1\}^* \to \mathbb{F}_{p^2}$ is computed for an input bit string by computing a chain of 2-isogenies starting from E_0 as follows, and a pseudocode is given in Algorithm 1. The nontrivial 2-torsion points on E_0 are given by $P_i^0 = (x_i, 0)$ where x_i are three distinct roots of the cubic $f(X)$ for $i = 0, 1, 2$. Here, the points are numbered by some order of $\mathbb{F}_{p^2}$. Three 2-isogenies outgoing from E_0 correspond to the subgroups $\langle P_i^0 \rangle$. These are the setup.

Let $m = (m_0, m_1, \ldots, m_{n-1}) \in \{0, 1\}^n$ be a n-bit input. Firstly, we compute an isogeny $\phi_0 : E_0 \to E_1 := E_0/\langle P_{m_0}^0 \rangle$, for example, by using Vélu's formula [18, 47]. Secondly, we take non trivial 2-torsion points on E_1, and the point corresponding to the dual of ϕ_0 is labeled P_2^1. For the remaining two points, we label them P_0^1 and P_1^1 in accordance with the order of $\mathbb{F}_{p^2}$. Then, we compute 2-isogeny $\phi_1 : E_1 \to E_2 := E_1/\langle P_{m_1}^1 \rangle$. By repeating this computation, the endpoint $E_n \in SS_1(p)$ is obtained by the sequence of elliptic curves $(E_0, E_1, \ldots, E_n)$, which gives non-backtracking walk on $\mathcal{G}_1(2, p)$ by construction. Finaly, we define the output for the input m as j_{E_n}, i.e., $H_{E_0}(m) = j_{E_n} \in \mathbb{F}_{p^2}$.

Security of CGL Hash Functions. In this way, constructing a hash function allows us to reduce its security to the hardness of computational problems related to isogenies between elliptic curves. We define such computational problems below.

Problem 1 (ℓ-**Isogeny Path Finding Problem**) *Let p and ℓ be distinct primes, and n a positive integer. Let us assume that supersingular elliptic curves E_0 and E_1 in characteristic p are given. Then, find a chain of ℓ-isogeny of length n in $\mathcal{G}_1(\ell, p)$.*

---

[1] Even when ℓ is an arbitrary prime such that $\ell \neq p$, a similar argument can be applied by modifying the domain of hash functions to $[\ell]^* = \cup_{n=1}^{\infty} [\ell]^n$.

Algorithm 1 CGL hash function H_{E_0} (the case of $\ell = 2$)

Parameters: E_0: a fixed supersingular elliptic curve
Input : $m = (m_0, \ldots, m_{n-1}) \in \{0, 1\}^*$: a bit string of any length
Output : $j_n \in \mathbb{F}_{p^2}$: the j-invariant of the terminal of a walk
1: Find 2-torsion points of E_0: $\{P_0^0, P_1^0, P_2^0\}$
2: Compute 2-isogeny $\phi_0 : E_0 \to E_1 := E_0/\langle P_{m_1} \rangle$
3: **for** $1 \le i < n$:
4:     Find 2-torsion points of E_i: $\{P_0^i, P_1^i, P_2^i\}$
5:     Compute 2-isogeny $\phi_i : E_i \to E_{i+1} := E_i/\langle P_{m_{i+1}} \rangle$
6: **end for**
7: **return** j-invariant $j_{E_n} \in \mathbb{F}_{p^2}$ of E_n

---

Problem 2 (Cycle Finding Problem) *Let p and ℓ be distinct primes, and n a positive integer. Let us assume that a supersingular elliptic curve E_0 in characteristic p is given. Then, find a cycle in $\mathcal{G}_1(\ell, p)$ of length $2n$.*

The relationship between the hardness of these problems and the security of CGL hash function can be stated as follows.

Theorem 4 *([9], Theorem 2) Let n be a positive integer. Let E_1 be a supersingular elliptic curve corresponding to j-invariant $H_{E_0}(m)$ for an input of length n. An inverse image of the value $H_{E_0}(m)$ gives a solution of Problem 1 for an instance (E_0, E_1).*

Theorem 5 *([9], Theorem 1) Let n be a positive integer. A collision pair for inputs of length n of CGL hash function H_{E_0} gives a solution for Problem 2.*

The path finding on supersingular isogeny graphs is investigated from various points of view. For example, see [14, 25] and so on.

3.3 The Case of Higher Dimensional Abelian Varieties

Similarly to classical elliptic curves cryptography, it is natural to pursue research that generalizes the construction in one dimension to arbitrary dimensions. This not only whips up theoretical interest but also offers a motivation, for example, to construct cryptography utilizing arithmetic in smaller finite fields. In fact, for dimension g, as characteristic p become large, the size of graphs grows in $O(p^{\frac{g(g+1)}{2}})$ asymptotically; thus, in higher dimension, one can explicitly realize graphs of sufficient size with smaller values of p.

Construction of a higher dimensional analog of CGL hash functions was initiated by Takashima in [41] by using superspecial abelian surfaces and $(2)^2$-isogenies between them, which is computed by Richelot's formula. However, Flynn and Ti [16] showed that short cycles exist inherently in the isogeny graphs of superspecial abelian surfaces. This implies that Takashima's hash function is not collision-resistant. In contrast, Castryck, Decru, and Smith [8] addressed this vulnerability in the hash function by selecting a specific class of paths that avoid short cycles in the isogeny graphs of superspecial abelian surfaces. Furthermore, their construction motivated investigation on properties of the superspecial isogeny graphs $\mathcal{G}_g(\ell, p)$ for $g \geq 2$.

When considering a generalization of CGL hash functions to higher dimension, it is a crucial task to investigate the mixing properties of $\mathcal{G}_g(\ell, p)$ for $g \geq 2$. It is known that the eigenvalues of the transition matrix are real numbers; $1 = \lambda_1 > \lambda_2 \geq \cdots \geq \lambda_{|SS_g(p)|}$, and we set $\lambda := \max\{|\lambda_2|, |\lambda_{|SS_g(p)|}|\}$. While it is known that isogeny graphs of supersingular elliptic curves satisfy the Ramanujan bound, this does not hold for the case for $g \geq 2$ [23, 24, 49]. In [17], Florit and Smith investigate the stationary distribution and convergence rate of superspecial isogeny graphs.

Theorem 6 *([17]) Let p and ℓ be distinct primes and g a positive integer. The stationary distribution of random walks on $\mathcal{G}_g(\ell, p)$ is given by $\pi = \frac{\tilde{\pi}}{\|\tilde{\pi}\|_1}$ with $\tilde{\pi}(A) = \frac{\deg(A)}{|RA(A)|}$ for any vertex A, where $\|\cdot\|_1$ denotes L^1-norm and $\deg(A)$ denotes the number of isogenies outgoing from A.*

Moreover, we assume that a random walk starting from a fixed vertex A_0 reaches a vertex A_n after n steps. For any vertex A, we have

$$|\Pr[A_n \simeq A] - \pi(A)| \leq \lambda^n \sqrt{\frac{\deg(A)}{\deg(A_0)} \frac{|RA(A_0)|}{|RA(A)|}}.$$

It was also proven that there exists a positive bound for the spectral gap of the transition matrix of $\mathcal{G}_g(\ell, p)$ by Aikawa, Tanaka, and Yamauchi. Moreover, this bound was explicitly computed as follows:

Theorem 7 *([2]) Let p and ℓ be distinct primes and g a positive integer greater than 2. Then, we have*

$$1 - \lambda_2 \geq \frac{1}{4(g+2)} \left(\frac{\ell - 1}{2(\ell - 1) + 3\sqrt{2\ell(\ell + 1)}} \right)^2.$$

The proof differs entirely from the case $g = 1$, which employs the Jacquet-Langrands correspondence to prove the Ramanujan bound. For $g \geq 2$, the proof is accomplished by connecting superspecial isogeny graphs $\mathcal{G}_g(\ell, p)$ to special 1-simplexes of Bruhat-Tits buildings via ℓ-marking isogeny graphs, which is defined in [2], and using property (T) of the symplectic groups. While this theorem may seem to state that the families of graphs $\{\mathcal{G}_g(\ell, p)\}_p$ are families of expander graphs, we remark that the term expander is often used for undirected graphs. We also should note that this bound does not seem to be tight. Numerical examples of eigenvalues for small primes can be found in [17].

Similarly to CGL hash functions, the security of functions constructed by Castryck et al. is reduced to the path-finding problem of superspecial abelian varieties, which is a higher dimensional analogue of Problem 1 and Problem 2. Security analysis for these problem is also progressed as in [11, 39].

4 Group-Based Construction

Many families of expander graphs are constructed using Cayley graphs. Hence, it is natural to investigate properties of expander hash functions instantiated by Cayley graphs, often referred to as *Cayley hash functions*. In fact, many studies have been conducted in this direction, and here, we provide a brief overview of these researches.

4.1 Cayley Hash Functions

Let G be a finite group and S a subset of G. We say that S is *inverse closed* if, for any $s \in S$, we have $s^{-1} \in S$.

Definition 6 Let G be a finite group and S be a subset of $G \setminus \{e\}$. The *left Cayley graph*, denoted as $Cay_L(G, S)$, is a graph with the set of vertices as G and the set of edges as $\{(g, sg) \mid g \in G, s \in S\}$. In contrast, the *right Cayley graph*, denoted as $Cay_R(G, S)$, is a graph with the set of vertices as G and the set of edges as $\{(g, gs) \mid g \in G, s \in S\}$. When simply referred to as the Cayley graph, denoted as $Cay(G, S)$, it specifically refers to the left Cayley graph.

There are a few remarks. If G is abelian, the left and right Cayley graphs coincide. When S is inverse closed, the Cayley graph $Cay(G, S)$ becomes an undirected graph.

Hash functions based on Cayley graphs were initially introduced by Zémor [50]. The following outlines the construction of these functions. Let G be a finite group and S a subset in $G \setminus \{e\}$. Let $g_0 \in G$ be a fixed element. A *labeling map* is a map $\sigma : [d] \times S \to S$ satisfying $\sigma([d] \times \{s\}) = S \setminus \{s^{-1}\}$ for any $s \in S$. A function $H_{G,S} : [d]^* \to G$ is given by: for any input $m = (m_1, m_2, \ldots, m_\ell) \in [d]^*$, we define $s_i := \sigma(m_i, s_{i-1})$ ($i = 1, 2, \ldots, \ell$), where $s_0 := g_0$ and set

$$H_{G,S}(m) = s_\ell \cdots s_2 s_1 g_0.$$

Note that this function is simply computing walks on $Cay(G, S)$. A pseudocode for computing a Cayley hash function is described in Algorithm 2.

Algorithm 2 Cayley hash function $H_{G,S}$

Parameters: (G, S) as above, $g_0 \in G$: a fixed element, σ: a labeling map
Input : $m = (m_1, \ldots, m_\ell) \in [d]^\ell$: a string
Output : $g \in G$: an element
1: Set $g_0 = s_0$, $s_i = \sigma(m_i, s_{i-1})$ ($i = 1, 2, \ldots, \ell$)
2: $g \leftarrow g_0$
3: **for** $1 \leq i \leq \ell$:
4:      $g \leftarrow s_{m_i} g$
5: **end for**
6: **return** $g \in G$

Table 1 Parameters of Cayley hash functions

References	Cayley graphs of (G, S)	Broken?
Zémor [50]	$\left(\mathrm{SL}_2(\mathbb{F}_p), \{\begin{pmatrix}1 & 1\\ 1 & 0\end{pmatrix}, \begin{pmatrix}1 & 0\\ 1 & 1\end{pmatrix}\}\right)$	Yes [44]
Tillich and Zémor [45]	$\left(\mathrm{SL}_2(\mathbb{F}_{2^m}), \{\begin{pmatrix}X & 1\\ 1 & 0\end{pmatrix}, \begin{pmatrix}X & X+1\\ 1 & 1\end{pmatrix}\}\right)$ where $\mathbb{F}_{2^n} = \mathbb{F}_2[X]/(P(X))$ for some irreducible polynomial $P(X)$ of degree n	Yes [19, 37]
Charles et al. [9]	LPS graph [30]	Yes [36, 46]
Petit et al. [35]	Morgenstein graph [32]	Yes [36]
Jo et al. [22]	Cubic Ramanujan graph [10]	Yes [22]
Bromberg et al.[6]	$\left(\mathrm{SL}_2(\mathbb{F}_p), \{\begin{pmatrix}k_1 & 1\\ 1 & 0\end{pmatrix}, \begin{pmatrix}1 & 0\\ k_2 & 1\end{pmatrix}\}\right)$, where $k_1 k_2 \geq 4$	No
Tomkins et al.[43]	$\mathrm{GL}_2(\mathbb{F}_{p^n})$, see [43] for S	No
Coz et al. [12]	$\mathrm{SL}_n(\mathbb{F}_p)$, see [12] for S	No

From the construction of Cayley hash functions, the collision resistance property is reduced to a type of word problem known as the balancing problem, as follows.

Problem 3 (Balance Problem) *Let G be a finite group and $S = \{s_0, \ldots, s_{d-1}\}$ a subset of $G \setminus \{e\}$. And let $L \in \mathbb{Z}$ be a small integer. Then, find two messages $m = m_1 \cdots m_\ell \in [d]^*$ and $m' = m'_1 \cdots m'_{\ell'} \in [d]^*$ with ℓ, $\ell' \leq L$, such that $\prod_i s_{m_i} = \prod_j s_{m'_j}$.*

We note that the computational condition on the length is needed because we have $g^{|G|} = e$ for any $g \in G$.

A Cayley hash function $H_{G,S}$ possesses a certain homomorphic property. That is, when considering messages m and m', and denoting their concatenation as $m||m'$, the equation $H_{G,S}(m||m') = H_{G,S}(m) g_0^{-1} H_{G,S}(m')$ holds. This property leads to malleability [4], which stands as a drawback of Cayley hash functions.

Since the instances by Zémor [50] using $G = \mathrm{SL}_2(\mathbb{F}_p)$ were provided, various constructions and security analyses of these instances have been progressed. Table 1 summarizes these studies. Particularly, instantiations using Ramanujan graphs are comprehensively summarized in [21].

At present, the most effective cryptanalysis against Cayley hash functions is so-called the lifting attack. Several schemes have been cryptanalyzed through the lifting attack, but this method is only effective for specific generating sets S. Indeed, a general constructive solution for the word problem of groups, including balance problem, has not yet been provided. We note that relevance of cryptanalysis of Cayley hash functions to the Babai's conjecture is discussed in [38]. Therefore, making the exploration of highly secure parameters and the development of generally effective attack methods is still an open problem.

4.2 Left-Right Cayley Hash Functions

Left-right Cayley (cubical) complexes were introduced by Dinur et al. [13] as a higher dimensional analogue of Cayley graphs. Aikawa, Jo, and Satake [1] developed novel hash functions based on this concept, and demonstrated their universality when underlying Cayley graphs are expanders.

Definition 7 Let G be a finite group and let A and B be inverse closed subsets of $G \setminus \{e\}$ satisfying $|A| = |B| = d$. The *left-right Cayley complex*, denoted as $Cay^2(G, A, B)$, is a cubical complex consisting of:

Vertices: The set of vertices is $X^{(0)} := G$.
Edges: The set of edges $X^{(1)}$ is defined as the union of the set of edges of $Cay_L(G, A)$, denoted as $X_A^{(1)}$ and called A-edges, and that of $Cay_R(G, B)$, denoted as $X_B^{(1)}$ and called B-edges.
Squares: The set of squares $X^{(2)}$ is defined as $\{[a, g, b] := (g, gb, agb, gb) \mid (a, g, b) \in A \times G \times B\}$, with identification between $[a, g, b]$, $[a^{-1}, ag, b]$, $[a^{-1}, agb, b^{-1}]$ and $[a, gb, b^{-1}]$.

The *total no-conjugacy condition* (TNC, for short) on $Cay^2(G, A, B)$ requires that, for any $a \in A, b \in B, g \in G$, we have $ag \neq gb$.

Hash functions based on left-right Cayley complexes, referred to as *left-right Cayley hash functions*, are constructed with the following parameters:

- Let G be a finite (non abelian) group and $g_0 \leftarrow G \setminus \{e\}$ a fixed element.
- For an integer $d \geq 2$, choose inverse closed subsets A and B in $G \setminus \{e\}$ such that they generate G with $|A| = |B| = d + 1$, and also choose a fixes element $b_0 \leftarrow B$. Assuming that TNC holds for (G, A, B).
- Define labeling maps $\sigma_A : [d] \times A \to A$ and $\sigma_B : [d] \times B \to B$.
- For any positive integer k, define a division $k = k_L + k_R$; for instance, by setting $k_L = \lfloor \frac{k}{2} \rfloor$ and $k_R = \lceil \frac{k}{2} \rceil$.

Then, hashing $H_{G,A,B} : [d]^* \to G$ is performed as follows: Given a string $m = (m_1, m_2, \ldots, m_k) \in [d]^k$, divide m into $(m_1, m_2, \ldots, m_{k_L}, m'_1, m'_2, \cdots, m'_{k_R})$ according to the division parameter. Set $a_i := \sigma_A(m_i, a_{i-1}) \in A$ and $b_j = \sigma_B(m'_j, b_{j-1})$, where $a_0 := g_0$. Finally, compute $g := a_{k_L} \cdots a_2 a_1 g_0 b_1 b_2 \cdots b_{k_R}$, and output $H_{G,A,B}(m) = g$. We note that TNC condition prevents trivial collisions caused by specific backtracking walks. A pseudocode is presented in Algorithm 3.

The target of a left-right Cayley function is G; however, unlike traditional Cayley hash functions, G can be viewed simply as a set. In other words, the equations $H_{G,S}(m||m') = H_{G,S}(m) g_0^{-1} H_{G,S}(m')$ that hold for conventional Cayley hash functions $H_{G,S}$ do not apply here. This construction disrupts the algebraic structure of the target G, and left-right Cayley hash functions are expected to exhibit a good randomness. In fact, the following holds regarding the distribution of outputs.

Algorithm 3 Left-right Cayley hash function $H_{G,A,B}$

Parameters: (G, A, B) as above, $g_0 \in G$ and $b_0 \in B$: fixed elements, σ_A, σ_B: labeling maps, (k_L, k_R): a division
Input: $m = (m_1, \ldots, m_k) \in [d]^*$: a string, k: the length of m
Output: $g_0 \in G$: an element
1: Divide m into $m_1 m_2 \cdots m_{k_L} m'_1 m'_2 \cdots m'_{k_R}$
2: Set $a_0 := g_0$, $a_i := \sigma_A(m_i, a_{i-1}) \in A$ and $b_j = \sigma_B(m'_j, b_{j-1})$
3: $g \leftarrow g_0$
4: **for** $1 \le i \le k_L$:
5:    $g \leftarrow a_i g$
6: **end for**
7: **for** $1 \le j \le k_R$:
8:    $g \leftarrow g b_j$
9: **end for**
10: **return** $g \in G$

Theorem 8 *Notation as above. A left-right Cayley hash function $H_{G,A,B}$ has universality if underlying Cayley graphs $Cay_L(G, A)$, $Cay_R(G, B)$ are expander graphs.*

We refer to [1] for the rigorous definition of uniformity of hash functions and the precise statements of Theorem 8. Additionally, it is easy to see that if one can find a solution of the problem below, this leads to a collision of a left-right Cayley hash function.

Problem 4 (Simultaneous Balance Problems) *Let G, A and B as above. And let $L \in \mathbb{Z}$ be a small integer. Then, find two messages $m^1 = m_1^1 \cdots m_{k_L}^1 m_1'^1 \cdots m_{k_R}'^1 \in [d]^k$ and $m^2 = m_1^2 \cdots m_{n_L}^2 m_1'^2 \cdots m_{n_R}'^2 \in [d]^n$ with $k = k_L + k_R$, $n = n_L + n_R \le L$, such that $\prod_{i=1}^{k_L} a_{m_i^1} = \prod_{j=1}^{n_L} a_{m_j^2}$ and $\prod_{i=1}^{k_R} b_{m_i'^1} = \prod_{j=1}^{n_R} b_{m_j'^2}$.*

Instantiations using the special linear group over a finite field for left-right Cayley hash function are presented in [1]. On the other hand, due to the difference of mathematical structure between these functions and traditional Cayley hash functions, it appears that the lifting attack may not be applicable. Study on cryptanalysis for left-right Cayley hash functions remains an important future work.

5 Summary

In this paper, we discussed how to construct provably secure universal hash functions utilizing families of expander graphs. Although we couldn't delve into detailed discussions on security, we illustrate these constructions using instances derived from isogeny graphs and Cayley graphs.

While this paper has discussed the instantiation of expander hash functions using isogenies and groups, explicit constructions of families of expander graphs through alternative methods may potentially lead to development of new cryptographic schemes. This is a challenging but important problem.

References

1. Y. Aikawa, H. Jo, S. Satake, Left-right cayley hashing: A new framework for provably secure hash functions. Math. Cryptol. **3**(2), 53–65 (2023). Retrieved from https://journals.flvc.org/mathcryptology/article/view/134667
2. Y. Aikawa, R. Tanaka, T. Yamauchi, *Isogeny Graphs on Superspecial Abelian Varieties: Eigenvalues and Connection to Bruhat-Tits buildings*. to appear in Canadian Journal of Mathematics
3. N. Alon, Eigenvalues of expanders. Combinatorica **6**(2), 83–96 (1986)
4. A. Boldyreva, D. Cash, M. Fischlin, B. Warinschi, *Foundations of Non-malleable Hash and One-Way Functions. Advances in Cryptology – ASIACRYPT 2009. ASIACRYPT 2009. Lecture Notes in Computer Science*, vol. 5912. (Springer, Berlin, Heidelberg, 2009), pp. 524–541. https://doi.org/10.1007/978-3-642-10366-7_31
5. B.W. Brock, *Superspecial Curves of Genera Two and Three*, Ph.D. thesis (Princeton University, 1993)
6. L. Bromberg, V. Shpilrain, A. Vdovina, Navigating in the Cayley graph of and applications to hashing. Semigroup Forum **94**, 314–324 (2017). https://doi.org/10.1007/s00233-015-9766-5
7. J.L. Carter, M.N. Wegman, Universal classes of hash functions. J. Comput. Syst. Sci. **18**(2), 143–154 (1979)
8. W. Castryck, T. Decru, B. Smith, Hash functions from superspecial genus-2 curves using Richelot isogenies. J. Math. Cryptol. **14**(1), 268–292 (2020). https://doi.org/10.1515/jmc-2019-0021
9. D.X. Charles, K.E. Lauter, E.Z. Goren, Cryptographic hash functions from expander graphs. J. Cryptol. **22**, 93–113 (2009). https://doi.org/10.1007/s00145-007-9002-x
10. P. Chiu, Cubic Ramanujan graphs. Combinatorica **12**, 275–285 (1992). https://doi.org/10.1007/BF01285816
11. C. Costello, B. Smith, *The Supersingular Isogeny Problem in Genus 2 and Beyond. PQCrypto 2020. Lecture Notes in Computer Science*, vol. 12100 (Springer, Cham, 2020). https://doi.org/10.1007/978-3-030-44223-1_9
12. C.L. Coz, C. Battarbee, R. Flores, T. Koberda, D. Kahrobaei, Post-quantum hash functions using $SL_n(\mathbb{F}_p)$. IACR Cryptol. ePrint Arch. **2022**, 896 (2022)
13. I. Dinur, S. Evra, R. Livne, A. Lubotzky, S. Mozes, Locally testable codes with constant rate, distance, and locality. *Proceedings of the 54th Annual ACM SIGACT Symposium on Theory of Computing* (2022), pp. 357–374. https://doi.org/10.1145/3519935.3520024
14. C. Delfs, S.D. Galbraith, Computing isogenies between supersingular elliptic curves over $\mathbb{F}_p$. Designs Codes Cryptogr. **78**, 425–440 (2016)
15. T. Ekedahl, On supersingular curves and Abelian varieties. Math. Scand. **60**, 151–178 (1987). https://doi.org/10.7146/math.scand.a-12178
16. E.V. Flynn, Y.B. Ti, Genus two isogeny cryptography. *Post-Quantum Cryptography. PQCrypto 2019. Lecture Notes in Computer Science*, vol. 11505 (Springer, Cham, 2019). https://doi.org/10.1007/978-3-030-25510-7_16
17. E. Florit, B. Smith, Automorphisms and isogeny graphs of abelian varieties, with applications to the superspecial Richelot isogeny graph. Arithmet. Geomet. Cryptogr. Coding Theory 779, 103–32, *Math* (Amer. Math. Soc, Contemp, 2021), p. 2022
18. S.D. Galbraith, *Mathematics of Public Key Cryptography* (Cambridge University Press, 2012)

19. M. Grassl, I. Ilić, S. Magliveras, R. Steinwandt, Cryptanalysis of the Tillich-Zémor hash function. J. Cryptol. **24**(1), 148–156 (2011)
20. S. Hoory, V. Linial, A. Wigderson, Expander graphs and their applications. Bull. Amer. Math. Soc. (N.S.) **43**(4), 439–561 (2006)
21. H. Jo, *Hash Functions Based on Ramanujan Graphs. Mathematical Modelling for Next-Generation Cryptography. Mathematics for Industry*, vol. 29 (Springer, Singapore, 2018). https://doi.org/10.1007/978-981-10-5065-7_4
22. H. Jo, C. Petit, T. Takagi, Full cryptanalysis of hash functions based on cubic Ramanujan graphs. IEICE Trans. **E100-A**(9), 1891–1899 (2017)
23. B.W. Jordan, Isogeny graphs of superspecial abelian varieties. RIMS Kôkyûroku Bessatsu **B90**, 131–144 (2022)
24. B.W. Jordan, Y. Zaytman, Isogeny graphs of superspecial abelian varieties and Brandt matrices. arXiv:2005.09031
25. Y. Kambe, A. Katayama, Y. Aikawa, Y. Ishihara, M. Yasuda, K. Yokoyama, Computing supersingular endomorphism rings and its application to finding isogeny paths via the Deuring correspondence. submitted (2023)
26. J. Katz, Y. Lindell, *Introduction to Modern Cryptography*, 3rd edn. CRC Press (2020)
27. E. Kowalski, An Introduction to expander graphs (COURS SPÉECIALISÉS COLECTION SMF). Societe Mathematique De France (2019)
28. K.-Z. Li, F. Oort, *Moduli of Supersingular Abelian Varieties, Lecture Notes in Math.*, vol. 1680 (Springer-Verlag, 1998)
29. D.A. Levin, Y. Peres, *Markov Chains and Mixing Times*, 2nd edn. (American Mathematical Society, 2017)
30. A. Lubotzky, R. Phillips, P. Sarnak, Ramanujan graphs. Combinatorica **8**, 261–277 (1988). https://doi.org/10.1007/BF02126799
31. A. Mittelbach, M. Fischlin, *The Theory of Hash Functions and Random Oracles: An Approach to Modern Cryptography* (Springer, 2021)
32. M. Morgenstern, Existence and explicit constructions of q+1-regular Ramanujan graphs for every prime power q. J. Combinatorial Theory Ser. B **62**(1), 44–62 (1994)
33. D. Mumford, *Abelian Varieties (Tata Institute of Fundamental Research)* (American Mathematical Society, 2012)
34. A. Pizer, Ramanujan graphs. AMS/IP Stud. Adv. Math. **7**, 159–178 (1998)
35. C. Petit, K. Lauter, J.-J. Quisquater, *Cayley Hashes: A Class of Efficient Graph-based Hash Functions*. Preprint (2007)
36. C. Petit, K. Lauter, J.-J. Quisquater, *Full cryptanalysis of LPS and Morgenstern hash functions*. International Conference on Security and Cryptography for Networks (Springer, 2008), pp. 263–277
37. C. Petit, J.-J. Quisquater, *Preimages for the Tillich-Zémor Hash Function. Selected Areas in Cryptography. SAC 2010. Lecture Notes in Computer Science*, vol. 6544 (Springer, Berlin, Heidelberg, 2011), pp. 282–301. https://doi.org/10.1007/978-3-642-19574-7_20
38. C. Petit, J.-J. Quisquater, Rubik's for cryptographers. American Math. Soc. Not. **60**(6), 733–739 (2013)
39. M.C.-R. Santos, C. Costello, S. Frengley, An algorithm for efficient detection of (N, N)-splittings and its application to the isogeny problem in dimension 2. IACR Cryptol. ePrint Arch. **2022**, 1736 (2022)
40. J.H. Silverman, *The Arithmetic of Elliptic Curves*, 2nd edn. (Springer, New York, 2009)
41. K. Takashima, *Efficient Algorithms for Isogeny Sequences and Their Cryptographic Applications. Mathematical Modelling for Next-Generation Cryptography. Mathematics for Industry*, vol. 29. (Springer, Singapore, 2018). https://doi.org/10.1007/978-981-10-5065-7_6
42. A. Terras, *Fourier Analysis on Finite Groups and Applications.* (Cambridge University Press, 2010)
43. H. Tomkins, M. Nevins, H. Salmasian, New Zémor–Tillich type hash functions over $GL_2(\mathbb{F}_{p^n})$. J. Math. Cryptol. **14**(1), 236–253 (2020). https://doi.org/10.1515/jmc-2019-0033

44. L.-P. Tillich, G. Zémor, *Group-theoretic Hash Functions. Algebraic Coding. Algebraic Coding 1993. Lecture Notes in Computer Science*, vol. 781. (Springer, Berlin, Heidelberg, 1994), pp. 90–110. https://doi.org/10.1007/3-540-57843-9_12
45. L.-P. Tillich, G. Zémor, *Hashing with* SL_2 . *Advances in Cryptology – CRYPTO '94. Lecture Notes in Computer Science*, vol. 839. (Springer, Berlin, Heidelberg, 1994), pp. 40–49. https://doi.org/10.1007/3-540-48658-5_5
46. L.-P. Tillich, G. Zémor, *Collisions for the LPS Expander Graph Hash Function. Advances in Cryptology – EUROCRYPT 2008. Lecture Notes in Computer Science*, vol. 4965. (Springer, Berlin, Heidelberg, 2008), pp. 254–269. https://doi.org/10.1007/978-3-540-78967-3_15
47. J. Vélu, Isogénies entre courbes elliptiques. C.R. Acad. Sc. Paris, S'erie A. **273**, 238–241 (1971)
48. L.C. Washington, *Elliptic Curves: Number Theory and Cryptography*, 2nd edn. (CRC Press, 2008)
49. Y. Zaytman, Proving connectedness of isogeny graphs with strong approximation. RIMS Kôkyûroku Bessatsu **B90**, 145–151 (2022)
50. G. Zémor, *Hash Functions And Graphs With Large Girths, Advances in Cryptology – EUROCRYPT '91, Lecture Notes in Computer Science*, vol. 547 (1992), pp. 508–511. https://doi.org/10.1007/3-540-46416-6_44

Open Access This chapter is licensed under the terms of the Creative Commons Attribution 4.0 International License (http://creativecommons.org/licenses/by/4.0/), which permits use, sharing, adaptation, distribution and reproduction in any medium or format, as long as you give appropriate credit to the original author(s) and the source, provide a link to the Creative Commons license and indicate if changes were made.

The images or other third party material in this chapter are included in the chapter's Creative Commons license, unless indicated otherwise in a credit line to the material. If material is not included in the chapter's Creative Commons license and your intended use is not permitted by statutory regulation or exceeds the permitted use, you will need to obtain permission directly from the copyright holder.

Toward Hash Functions Based on Group-Subgroup Pair Graphs

Cid Reyes-Bustos

Abstract Cryptographic hash functions are fundamental elements in cryptography, used as building blocks for cryptographic systems. In recent years, there have been attempts to define hash functions using different mathematical structures, including finite groups and graphs. In particular, Cayley graphs have received considerable attention due to the existence of known families of graphs with good properties. In this paper, we give an overview of graph hash functions, focusing on Cayley hash functions, and discuss possible extensions to non-regular graphs using as a basis a non-regular generalization of Cayley graphs called group-subgroup pair graphs.

Keywords Hash functions · Cayley graphs · Graph hash functions · Group and subgroup pairs

1 Introduction

Informally speaking, a cryptographic hash function $\mathcal{H}$ is a function that takes an arbitrary input and produces a fixed-length output. The function $\mathcal{H}$ must be easy to compute, but finding preimages of $\mathcal{H}$ should be computationally infeasible [12]. Hash functions are used for ensuring data integrity, and with the addition of security assumptions they become essential elements of cryptographic systems. Indeed, hash functions are used in digital signatures, public-key cryptography, message authentication, key agreement protocols, password storage, and data structures such as Merkle trees, among other applications [9]. In Sect. 2.2 we recall the definition and the desirable properties of cryptographic hash functions.

As a result of technological and theoretical advances in quantum computing, there has been considerable work on developing cryptographic constructions that are resistant to attacks under the assumption of quantum computers. For cryptographic

---

C. Reyes-Bustos (✉)
NTT Institute for Fundamental Mathematics, NTT Communication Science Laboratories, Nippon Telegraph and Telephone Corporation, Tokyo, Japan
e-mail: cid.reyes@ntt.com

hash functions, new constructions have been proposed based on various mathematical structures [23, 28].

In particular, we note that in recent years, there have been attempts to construct cryptographic hash functions based on finite graphs [8, 18, 24]. The model of graph hash functions we consider in this paper is one where the input of the hash function corresponds to a path in the graph, and the properties of the hash correspond to the structure of the graph, for example, the existence of cycles of short length. In Sect. 2.2 we give a short introduction to graph hash functions, including the corresponding assumptions of hash functions in the graph context.

Due to the fast convergence of random walks to the uniform distribution, the so-called mixing property, expander graphs have been proposed as ideal graphs for use in the construction of graph hash functions [8, 18, 19]. Among these, cryptographic hash functions based on Cayley expander graphs have received considerable attention due to the nice properties afforded by the underlying group structure, related to deep mathematical theory in geometry and number theory. Unfortunately, up to this point, due to the discovery of attacks, a successful model of graph hash functions for practical use has not been found [19]. This makes the search for new graph hash functions beyond Cayley hash functions an interesting research prospect (see, e.g., [1, 21] for some recent proposals).

In this paper, we propose cryptographic hash functions based on non-regular graphs, using group-subgroup pair graphs (a generalization of Cayley graphs) as a basis [19]. For this purpose, we give a description of the structure of the group-subgroup pair graphs and the challenges needed for the construction of non-regular hash functions. Since further studies are needed on the properties of group-subgroup pair graphs, we only focus on the theoretical details of the definition and leave the possible implementation and cryptanalysis for future investigations.

2 Preliminaries

In this section we introduce the basic notions and notations used in the paper.

For $n \in \mathbb{Z}_{\geq 1}$ denote by $[n]$ the set of integers $\{1, \ldots, n\}$. We always denote by G a finite group with identity element $e \in G$ and order $|G|$. For a subset $S \subset G$, we denote by S^{-1} the set of inverses of elements of S. If $S^{-1} = S$, then we say S is a symmetric subset of G.

We use the notation H for a subgroup of G of index $[G : H] = k + 1$ for $k \in \mathbb{Z}_{\geq 0}$, we fix a set of representatives $\{x_0, x_1, x_2, \ldots, x_k\}$ of right cosets of H in G, and we always assume that $x_0 = e$. Necessarily, we have $0 \leq k \leq |G| - 1$. Note that when $k = 0$, H is equal to G and when $k = |G| - 1$, H is the trivial subgroup $\{e\}$.

2.1 Graph Theory

A graph $\mathcal{G} = (V, E)$ is determined by the set of vertices V (an arbitrary finite set) and a set of edges $E \subset V \times V$. For two vertices $v, w \in V$ we say that they are adjacent, denoted by $v \sim w$, if $(v, w) \in E$. The vertices $w \in V$ adjacent to v, that is the neighborhood of v, is denoted by $N(v)$.

In this paper we consider only simple undirected graphs, that is, we do not distinguish the elements (v, w) and (w, v) in E, and E is assumed to be a set, not a multiset (that is, any given edge appears only once). In general, we do not consider graphs with loops, that is, edges of the type (v, v) for $v \in V$. We denote by $\deg(v)$ the degree of $v \in V$, that is, the number of vertices $x \in V$ such that $x \sim v$. A graph is a d-regular graph if all vertices have degree d for an integer $d \geq 0$.

A path $\mathcal{P}$ is an (ordered) sequence of edges

$$\{(v_0, v_1), (v_1, v_2), \ldots, (v_{n-1}, v_n)\}$$

and we say that it is a path of length n starting at v_0 and ending at v_n. A closed path, or cycle, is a path $\mathcal{P}$ where $v_0 = v_n$. We say that a path $\mathcal{P}$ has backtracking if a given edge appears consecutively in the path (that is, if the sequence $(v_i, v_{i+1}), (v_{i+1}, v_i)$ appears in $\mathcal{P}$ for $i < n$).

A graph is connected if for any two vertices $v, w \in V$, there is path starting at v and ending at w. In general, the graphs here are assumed to be connected. The girth of $\mathcal{G}$ is the length of the shortest cycle contained in $\mathcal{G}$. We refer the reader to any standard reference on graph theory (e.g., [4]) for more details.

A measure of connectivity of a graph is given by the *isoperimetric constant*. The isoperimetric constant, or Cheeger constant, $h(\mathcal{G})$ of a graph $\mathcal{G} = (V, E)$ is defined as

$$h(\mathcal{G}) = \min\left\{ \frac{|\partial F|}{\min\{|F|, |V - F|\}} \,\bigg|\, F \subset V, 0 < |F| < |V| \right\}.$$

Here, ∂F is the *boundary* of F, that is, the set of edges with one endpoint in F and one in $V - F$.

A graph $\mathcal{G}$ is said to be a c-expander if $h(\mathcal{G}) > c$ for some constant $c > 0$. In general, finding an infinite family of c-expanders is an important problem in graph theory. The expansion of a graph is closely related to the spectral properties of the graph and the speed of convergence of random walks. We refer the reader to [7, 15, 16, 22] for an overview of the theory of expander graphs.

Next, we recall the definition of Cayley graph, graphs defined from finite groups. For a systematic treatment, we refer the reader to [16].

Definition 1 (**Cayley graph**) For a group G with a symmetric subset $S \subset G$, the *Cayley graph* $\mathcal{G}(G, S) = (V, E)$ is the graph with vertex set $V = G$ and where there is an edge $(h, g) \in E$ between two vertices $h, g \in G$ if

$$g = hs,$$

for $s \in S$.

It is immediate from the definition that a Cayley graph $\mathcal{G}(G, S)$ is a $|S|$-regular graph on $|G|$ vertices. Since we are only considering graphs without loops, we assume $e \notin S$.

Example 1 Let G be the symmetric group $\mathfrak{S}_3$, that is

$$G = \{e, (1, 2), (2, 3), (1, 3), (1, 2, 3), (1, 3, 2)\}$$

and let $S = \{(1, 2), (1, 2, 3), (1, 3, 2)\}$. The corresponding 3-regular Cayley graph $\mathcal{G}(G, S)$ is shown in Fig. 1.

A Cayley graph is connected if $\langle S \rangle = G$, that is S is a generating set for the group. We also note that Cayley graphs are vertex transitive, and a nontrivial graph automorphism is given by the left multiplication by any element $g \in G$ with $g \neq e$.

Next, we introduce a generalization of Cayley graphs called group-subgroup pair graphs. In this case, in addition to a group G, we have a subgroup H. We refer the reader to [20] for the basic properties of group-subgroup pair graphs.

Definition 2 (Group-subgroup pair graph) Let G be a group with a subgroup H of index $[G : H] = k + 1$. For $S \subset G$ such that $S \cap H$ is a symmetric set, the *group-subgroup pair graph*, or *pair-graph*, $\mathcal{G}(G, H, S) = (V, E)$ is the graph with vertex set $V = G$ and where there is an edge $(h, g) \in E$ between two vertices $h \in H$ and $g \in G$ if

$$g = hs,$$

for $s \in S$.

Clearly, when $H = G$, we have $\mathcal{G}(G, H, S) = \mathcal{G}(G, S)$, that is, the pair-graphs are generalizations of Cayley graphs. We note that the graph $\mathcal{G}(G, H, S)$ is, in general, not regular. In fact, $\mathcal{G}(G, H, S)$ can only be regular when H is a subgroup of index 1 or 2.

The degree of the vertices of $\mathcal{G}(G, H, S)$ depends on the coset they belong to. Define $S_i = S \cap Hx_i$ for $i = 0, 1, 2, \ldots, k$. Then, vertices on H have degree $|S|$ and the degree of the vertices in the coset Hx_i has degree $|S_i|$ for $1 \leq i \leq k$.

We say that $\mathcal{G}(G, H, S)$ is a $(|S|, |S_1|, \ldots, |S_k|)$-multiregular graph.

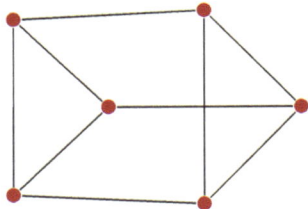

Fig. 1 The Cayley graph $\mathcal{G}(G, S)$

Example 2 Let n, m be positive integers. Take $G_1 = \mathbb{Z}/n(m+1)\mathbb{Z}$, $H_1 \simeq \mathbb{Z}/n\mathbb{Z}$ and

$$S_B = \{1, 2, \ldots, m, m+1, (n-1)(m+1), (n-1)(m+1)+1, \ldots, (n-1)(m+1)+m\},$$

then the resulting graph $\mathcal{G}(G_1, H_1, S_B)$ is a $(4, 2, 2, \ldots, 2)$-multiregular with girth 3 as shown in Fig. 2a.

On the other hand, for $n \geq 1$, take $G_2 = \mathbb{Z}/4n\mathbb{Z}$, $H_2 \simeq \mathbb{Z}/2n\mathbb{Z}$ (note that H has an element of order 2), and

$$S_C = \{1, 2n, 4n-1\},$$

then the resulting graph $\mathcal{G}(G_2, H_2, S_C)$ is a $(3, 2)$-multiregular graph with girth 5 for $n = 2$ and girth 6 for $n \geq 3$, shown in Fig. 2b.

Similar to Cayley graphs, the connectedness of pair-graphs depends on a generating property of the set S with respect to the subgroup H. Concretely, a pair-graph $\mathcal{G}(G, H, S)$ is connected if

$$\langle S_0 \cup H \cap (S - S_0)(S - S_0)^{-1} \rangle = H, \tag{1}$$

and S contains a representative of the cosets. The particular structure of the generating set in (1) is explained by Theorem 1 in Sect. 3.1.

We also note that since pair-graphs are in general not regular, they cannot be vertex transitive. However, as with Cayley graphs, left multiplication by elements of the subgroup H gives a graph automorphism of $\mathcal{G}(G, H, S)$.

As noted before, the expansion properties of graphs are determined by the graph spectrum. In Cayley graphs and group-subgroup pair graphs, the spectrum is controlled by the irreducible representations of the group G or the subgroup H, thus allowing exact computation or estimation (i.e., bounds) of the expansion based on character sums or functions of character sums (see [16] and [13] for the case of pair-graphs).

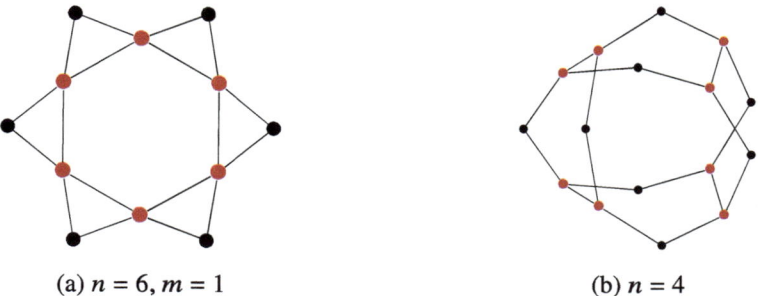

(a) $n = 6, m = 1$ (b) $n = 4$

Fig. 2 The pair-graphs $\mathcal{G}(G_1, H_1, S_B)$ and $\mathcal{G}(G_2, H_2, S_C)$. Vertices of H are shown in red

2.2 Hash Functions

In this section we introduce cryptographic hash functions. Here, we consider a simple setting that is convenient to our purposes. For a formal and more detailed definition of cryptographic hash functions, we refer the reader to [12]. We note that *cryptographic* is used to differentiate from hash functions used in contexts that do not require security assumptions (databases, data structures, etc.).

A *cryptographic hash function* $\mathcal{H}$ is a function

$$\mathcal{H} : \{0, 1\}^* \to \{0, 1\}^N$$

for fixed $N \in \mathbb{Z}_{>0}$ with the property that for a given input $m \in \{0, 1\}^*$ the cost of computing $\mathcal{H}(m)$ is relatively small. In general, N is assumed to be large but the specific details depend on the implementation. Hereafter, we refer to cryptographic hash functions simply as hash functions.

Clearly, such a function $\mathcal{H}$ cannot be injective. When two elements $m_1, m_2 \in \{0, 1\}^*$ have the same image, that is $\mathcal{H}(m_1) = \mathcal{H}(m_2)$, we say there is a collision.

The desirable properties for a hash function $\mathcal{H}$ are

- **(PR) Preimage resistance:** For $h \in \text{Im}(\mathcal{H})$ it should be computably infeasible to find $m \in \{0, 1\}^*$ such that $\mathcal{H}(m) = h$.
- **(SPR) Second-preimage resistance:** For a given input $m \in \{0, 1\}^*$, it should be infeasible to find $m' \neq m$ such that $\mathcal{H}(m) = \mathcal{H}(m')$.
- **(CR) Collision resistance:** It should be infeasible to find pairs of input m, m' such that $\mathcal{H}(m) = \mathcal{H}(m')$.

Depending on the actual applications, the hash function may not need to satisfy all the above properties. Examples of hash functions currently in use are the SHA family (SHA-2 and SHA-3) and the MD family (MD5, MD6), among others. We refer the reader to [9] for an introduction to these and other commonly used hash functions.

In addition to the properties above, another desirable property of hash functions is that the distribution of outputs is similar to a uniform random distribution (see, e.g., [19]). For the types of graph hash functions to be introduced in the next section, this property has a direct interpretation in terms of graph expansion.

2.2.1 Graph Hash Functions

In this section we give an introduction to graph hash functions (see, e.g., [8]). The basic idea is that to every input m we assign a path in the graph. Regular graphs are usually used for hash functions, but here we consider a general case. We fix a graph $\mathcal{G} = (V, E)$ with a designated vertex $v \in V$, and we assume that $\mathcal{G}$ has no degree 1 vertices.

Let $d = \max\{\deg(v) \mid v \in V\}$ and assume that each input $M \in \{0, 1\}^*$ is codified as

$$M = x_1 x_2 \ldots x_k,$$

for $k \in \mathbb{Z}_{\geq 1}$ and $x_i \in [d]$ for $i \in \{1, 2, \ldots, k\}$, then consider a function

$$\alpha : V \times [d] \to V,$$

such that $\alpha(v, x) \in N(v)$ for any $x \in [d]$. To simplify the notation, we denote an auxiliary function P as

$$P(v, x) = \alpha(v, x)$$

for $x \in [d]$ and

$$P(v, Mx_{k+1}) = P(P(v, M), x_{k+1}),$$

for $M = x_1 \cdots x_k$. Here, Mx denotes the concatenation of x to M.

Definition 3 (**Graph hash function**) Let $\mathcal{G}$ be a graph as described in the foregoing discussion. The graph hash function $\mathcal{H}(\cdot \, ; \mathcal{G}) : \{0, 1\}^* \to \{0, 1\}^N$ is defined for input $M = x_1 x_2 \cdots x_k$ as

$$\mathcal{H}(M; \mathcal{G}) = P(v, M).$$

The idea behind the hash function is that to each input M there is a path on $\mathcal{G}$ starting on v and ending in vertex $P(v, M)$

$$(v, p(v, x_1)), (p(v, x_1), p(v, x_1 x_2)), \ldots, (p(v, x_1 x_2 \cdots x_{k-1}), p(v, M))$$

as shown in Fig. 3. In general, we assume that the resulting path has no backtracking.

As mentioned before, a desirable property of hash functions is that the outputs resemble a uniform distribution. In general, random walks converge to the uniform distribution on the graph vertices (when the graph is not bipartite). For this reason, graphs with good expansion properties (expanders and Ramanujan graphs) have been considered as candidates for graph hash functions [8].

Finding a collision is equivalent to finding inputs M, M' such that

$$P(v, M) = P(v, M').$$

Note that concatenation of the two paths gives a closed cycle C in $\mathcal{G}$ containing the vertex v. Therefore, the desirable property is that $\mathcal{G}$ has no cycles of small length, or in other words that $\mathcal{G}$ has a large girth with respect to the number of vertices.

Fig. 3 The path associated with $M = x_1 x_2 \cdots x_n$

Finding graphs with large girth and expansion is not a trivial problem, but we note that the Ramanujan graphs by Lubotzky-Phillips-Sarnak have this property [17]. These graphs are actually Cayley graphs, and the resulting hash functions are called Cayley hash functions that we discuss in the next section.

We also remark in the discussion above we have tacitly assumed that there is a function $C : V \to \{0, 1\}^N$ to obtain the final output of the hash. In this paper we omit the description of this detail to simplify the notation.

2.3 Hash Function from Cayley Graphs

Cayley hash functions are graph hash functions where the associated graph is a Cayley graph $\mathcal{G}(G, S)$, for some group G and S, and where the function α is simply multiplication of a vertex by one of the elements of the generating set S.

Definition 4 (**Cayley hash function**) For a group G and set $S = \{s_1, s_2, \ldots, s_d\}$ consider a fixed element $v \in G$. For a plain text $M \in \{0, 1\}^*$ written as $x_1 x_2 \cdots x_n$ with $x_i \in [d]$, the output $\mathcal{H}(M; G, S)$ of the Cayley hash functions $\mathcal{H}(\cdot\,; G, S) : \{0, 1\}^* \to \{0, 1\}^N$ is given by the vertex

$$v s_{x_1} s_{x_2} \cdots s_{x_n}.$$

One of the advantages of using Cayley graphs is scalability. In other words, if we define families $\{G_i, S_i\}_{i \geq 1}$ such that $|G_i| \to \infty$ as $i \to \infty$, then it is possible to use the corresponding Cayley graph $\mathcal{G}(G_i, S_i)$ of the required size according to the application of the hash function.

The first practical proposal for Cayley hash functions was given by Zémor [27], and after cryptanalysis a new version was proposed by Tillich and Zémor [24] using the group $SL_2(\mathbb{F}_q)$ as a basis. A similar construction was proposed by [8] using the Lubotzky-Phillips-Sarnak Ramanujan graphs [17]. Several extensions of these constructions have appeared since.

It is important to note that the choice of vertex is not important for Cayley hash functions, since we can apply the graph isomorphism given by left multiplication by v^{-1} and we have

$$v s_{x_1} s_{x_2} \cdots s_{x_k} \mapsto s_{x_1} s_{x_2} \cdots s_{x_k}.$$

Therefore, we can always assume that $v = e$.

In this setting, preimage resistance reduces to a problem of group presentations in terms of generating sets. Concretely, for $w \in G$ the problem is to find a presentation

$$w = s_{x_1} s_{x_2} \cdots s_{x_n},$$

with $s_{x_i} \in S$ with $i \in [d]$. Similarly, for collision, we observe that if

$$s_{x_1}s_{x_1}\cdots s_{x_n} = s_{x'_1}s_{x'_2}\cdots s_{x'_m}$$

then we must have

$$e = s_{x_1}s_{x_1}\cdots s_{x_n}(s_{x'_1}s_{x'_2}\cdots s_{x'_m})^{-1}.$$

In particular, since

$$s_{x_1}s_{x_1}\cdots s_{x_n}(s_{x'_1}s_{x'_2}\cdots s_{x'_m})^{-1}$$

is a word in G with elements of S, the desired property for G and S is that there are no short words in G with elements in S that are equivalent to the identity.

Using the naming proposed in [19], these properties are given by the following theoretical problems:

- **(C1) Balance problem:** Find $x_i, x'_i \in [d]$ such that $\prod_{i=1}^{\ell} s_{x_i} = \prod_{i=1}^{\ell} s_{x'_i}$ for $\ell \geq 0$.
- **(C2) Representation problem:** Find $x_i \in [d]$ such that $e = \prod_{i=1}^{\ell} s_{x_i}$ for $\ell \geq 0$.
- **(C3) Factorization problem:** Given $h \in G$, find $x_i \in [d]$ such that $h = \prod_{i=1}^{\ell} s_{x_i}$ with ℓ small.

Unfortunately, to date there has not been a successful Cayley hash proposal resistant to attacks. We refer the reader to [19] for a summary of the situation and the current challenges. In the next section, we study the possibility of defining hash functions with group-subgroup pair graphs and describe the main challenges for such constructions.

3 Hash Functions from Pair-Graphs

In this section, we consider graph hash functions using group-subgroup pair graphs and we highlight some of the differences with respect to the case of Cayley hash functions.

For simplicity, we consider the case $|S_i| = d_1$ for all $i, j \geq 1$, but we note that the construction can easily be extended to the general case. Due to the difference of the degrees, we have to make an additional consideration compared to the Cayley hash case. We say that x is a prefix of $M \in \{0, 1\}^*$ if we can write $M = xM'$ for some (possibly empty) $M' \in \{0, 1\}^*$.

Definition 5 (Pair-graph hash functions) Suppose $\mathcal{G}(G, H, S)$ is a group-subgroup pair graph for group G, subgroup H, and subset S. We denote the elements of S by $S_0 = \{s_1, s_2, \ldots, s_{d_0}\}$ and $S_i = \{s_1^{(i)}, \ldots, s_{d_1}^{(i)}\}$ and take a fixed element $v \in H$.

The pair-graph hash function $\mathcal{H}(M; G, H, S)$ is the function

$$\mathcal{H}(\cdot; G, H, S) : \{0, 1\}^* \to \{0, 1\}^N$$

computed from the input M by the following algorithm:

1. Set
$$v_2 = vs_{y_1}$$
where $y_1 \in [d_0]$ is a prefix of M of length $log_2(d_0)$. Set M_2 to be M with the prefix y_1 removed.

2. Next, if $v_i \in H$, take $y_i \in [d_0]$ as the prefix of M_i and set
$$v_{i+1} = v_i s_{y_i}$$
and M_{i+1} to be M_i with y_i removed.
If $v_i \in Hx_i$, take $y_i \in [d_1]$ as the prefix of M_i and set
$$v_{i+1} = v_i (s_{y_i}^{(i)})^{-1},$$
and M_{i+1} to be M_i with the prefix y_i removed.

3. Repeat 2 until M_n becomes the empty word.

The output $\mathcal{H}(M; G, H, S)$ of the hash function is the vertex v_{n+1}.

We note that the construction above can be easily modified to allow $v \in Hx_i \neq H$. In the case described above, the choice of vertex $v \in H$ has no effect on the hash, due to the invariance of the graph $\mathcal{G}(G, H, S)$ by left-actions of H. By the same argument, in the general case the choice of vertex within a coset is not relevant; however the choice of starting coset is important as there may not be any graph isomorphism between vertices x_i, x_j belonging to different cosets.

Next, let us note that if the output vertex w is in H, we necessarily must have a sequence
$$s_{m_1}, s_{m_2} \cdots, s_{m_n} \in S$$
that we can arrange into elements
$$s_{m'_1}, s_{m'_2} \cdots, s_{m'_{n'}} \in H$$
where for $i \leq n'$, we have $s_{m'_i} = s_{m_j}$ or $s_{m'_i} = s_{m_j} s_{m_{j+1}}^{-1}$ for some $1 \leq j \leq n$. On the other hand, if $w \in Hx_i$ for some $i \in [k]$, then the sequence can be arranged as
$$s_{m'_1}, s_{m'_2} \cdots, s_{m'_n}, s_{m'_{n+1}}$$
with $s_{m'_i}$ for $i = 1, \ldots, n$ given as above and $s_{m'_{n+1}} = s_{m_n} \in Hx_i$. We illustrate the way paths are constructed for an input as in Fig. 4.

To unify the analysis of the two cases, for an input M we define the corresponding normalized sequence
$$s_{m'_1}, s_{m'_2} \cdots, s_{m'_n} \in H, \quad s_{m'_{n+1}} \in G$$

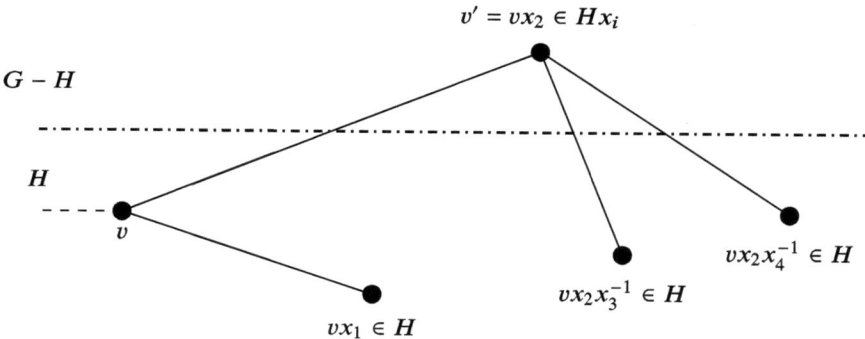

Fig. 4 Construction of paths in $\mathcal{G}(G, H, S)$ starting from $v \in H$. Here, all $x_1 \in S_0$ and $x_2, x_3, x_4 \in S_i$. Note that the possible vertices in the path depend on whether the current vertex $v \in H$ or $v' \in G - H$

such that for $i \leq n$, $s_{m'_i} \in S_0$ or $s_{m'_i} \in S_j S_j^{-1}$ for $1 \leq j \leq k$, and $s_{m'_{n+1}} = e$ or $s_{m'_{n+1}} \in S_j$ for $1 \leq j \leq k$.

Let us now consider a collision: for M and M' the corresponding normalized sequences must be

$$s_{m_1}, s_{m_2} \cdots, s_{m_n}, s_{m_{n+1}} = s_{m'_1}, s_{m'_2} \cdots, s_{m'_{n'}}, s_{m'_{n'+1}},$$

so that, in general, the computational problem is given by the following variation of the group presentation problems for Cayley graphs.

- **(P1) Balance problem:** Find normalized sequences $\{s_{m_i}\}, \{s_{m'_i}\} \subset G$ such that $\prod_{i=1}^{\ell+1} s_{m_i} = \prod_{i=1}^{\ell+1} s_{m'_i}$.
- **(P2) Representation problem:** Find a normalized sequence $\{s_{m_i}\} \subset G$ such that $e = \prod_{i=1}^{\ell} s_{m_i}$.
- **(P3) Factorization problem:** Given $g \in G$, find a normalized sequence $\{s_{m_i}\} \subset G$, such that $g = \prod_{i=1}^{\ell} s_{m_i}$ with small ℓ.

Let us make some remarks on the presentation problems for the case of pair-graphs. First, note that if both $s_{m_{\ell+1}} = s_{m'_{\ell+1}} = e$ then (P1) reduces to the balance problem for Cayley graphs (C1) for $\mathcal{G}(H, S_0 \bigcup_{i=1}^{k} (S_i S_i^{-1}))$. Since $e \in H$, (P2) corresponds to (C2) for $\mathcal{G}(H, S_0 \bigcup_{i=1}^{k} (S_i S_i^{-1}))$.

Clearly, the presentation problems above are intimately related to the structure of pair-graphs. As a first step to get a better understanding of the difficulty of these problems and therefore of the possibility for practical analysis of pair-graph hash functions, in the next section we study the general structure of group-subgroup pair graphs.

To conclude this section, we make some general considerations on other challenges needed to obtain good hash functions from pair-graphs.

Question 1 Open problem

- Find a family $\{\mathcal{G}(G_i, H_i, S_i)\}_{i=1}^{\infty}$ of multiregular c-expanders for some constant $c > 0$.
- Find multiregular $\mathcal{G}(G, H, S)$ graphs of arbitrary size with large girth.

Both of these questions remain open at this point. As mentioned before, the eigenvalues of pair-graphs are controlled by the irreducible representations of the subgroup H, so it is possible to attempt to generalize the constructions used for families of graph expanders directly.

A simpler approach is to consider a known expander or graph with large girth (e.g., [3, 5, 17]) given by a Cayley graph $\mathcal{G}(H, S_0)$ and find a supergroup G, with a subset S satisfying the hypothesis of pair-graphs such that $S \cap H = S_0$ and extend the construction. For the girth, it is important to avoid simple short cycles of length 3 and 4 as in Fig. 2a, in general, an analysis of the structure of the graph is needed for the construction of pair-graphs with large girth.

In Sect. 3.2 we give an idea of a different approach for this problem using a generalization of pair-graphs and leave the precise details of possible constructions of pair-graphs with large expansion and girth for other occasions.

Remark 1 There may be other ways to define the hash function from pair-graphs. Of particular importance is the choice of the sequence of generators from the plain text M, which ultimately determines the security of the hash function. A deeper study of the choice functions, technical details of the implementation, and the cryptanalysis of multiregular graph hash functions are the topics of a future paper of the author with Hyungrok Jo.

3.1 Structure of Pair-Graphs

In this section, we describe the structure of group-subgroup pair graphs to give further insight both to the general properties of the resulting graphs and the hash functions defined on them. We refer the reader to [11] for some of the details of graph theory used in this section.

Recall that the *generalized edge sum* of two graphs $\mathcal{G}_1 = (V_1, E_1)$ and $\mathcal{G}_2 = (V_2, E_2)$ is given by the graph

$$\mathcal{G}_1 \oplus \mathcal{G}_2 = (V_1 \cup V_2, E_1 \cup E_2).$$

Since the structure of group-subgroup pair graphs is complicated in general, we first consider a case where it is described as a graph covering via the union of Cayley graphs. This particular case corresponds to the case where there are no degree 1 vertices on the graph (that is $|S_i| > 1$ for all $i > 0$). We note that the case of degree

1 vertices is not useful for hash functions since arbitrary paths with no backtracking are not possible for graphs with such vertices.

For each $i \leq k$, define a corresponding symmetric multiset $\hat{S}_i$ by

$$\hat{S}_i = S_i S_i^{-1} = \{sr^{-1} | s, r \in S_i, s \neq r\},$$

and the union

$$\hat{S} = \bigcup_{i=1}^{k} \hat{S}_i.$$

We note that $\hat{S} \subset H$ and that $\hat{S}_i$ has cardinality $|S_i|(|S_i| - 1)$.

Next, we consider the Cayley (multi-)graph $\mathcal{G}(H, \hat{S})$ and its barycentric division $\mathcal{G}^{(2)}(H, \hat{S})$, that is, the graph obtained by dividing each edge into two edges by adding one intermediate vertex.

Then, we construct a strong graph morphism from $\mathcal{G}^{(2)}(H, \hat{S})$ to the pair-graph $\mathcal{G}(G, H, S)$. Let $\phi : \mathcal{G}^{(2)}(H, \hat{S}) \to \mathcal{G}(G, H, S)$ be defined on the vertices by

$$\phi(h) = h,$$
$$\phi((h, hs_i s_j^{-1})) = hs_i,$$

where $h \in H$ and $s_i s_j^{-1} \in \hat{S}$. Since the pair-graph is connected, by the definition of $\hat{S}$, it follows that this is a surjective map. Now, the induced map on the edges satisfies

$$\phi((h, (h, hs_i s_j^{-1}))) = (\phi(h), \phi((h, hs_i s_j^{-1}))) = (h, hs_i),$$

and the right-hand side is a vertex of the pair-graph $\mathcal{G}(G, H, S)$; therefore ϕ is a graph homomorphism. Conversely, let (h, hs_i) be an edge of $\mathcal{G}(G, H, S)$ with $s_i \in S_\ell$ for $\ell \leq k$. Since by assumption there are no vertices of degree one, there is a $s_j \in S_\ell$ such that $s_i s_j^{-1} \in \hat{S}_\ell \subset \hat{S}$. Now, since $(h, s_i s_j^{-1})$ is an edge of $\mathcal{G}(H, \hat{S})$ and

$$\phi((h, (h, hs_i s_j^{-1}))) = (h, hs_i),$$

therefore the map ϕ is a surjective *strong graph morphism*. By the homomorphism theorem for graphs, we have

$$\mathcal{G}(G, H, S) \simeq \mathcal{G}^{(2)}(H, \hat{S}) / \sim_\phi,$$

where $\sim_\phi$ is the congruence induced by ϕ.

Theorem 1 *Let G be a finite group, H a subgroup, and S a subset of G such that the corresponding pair-graph $\mathcal{G}(G, H, S)$ is connected and has no degree one vertices. Then, with the foregoing notation,*

$$\mathcal{G}(G, H, S) \simeq \mathcal{G}(H, S_0) \oplus \mathcal{G}^{(2)}(H, \hat{S})/ \sim_\phi .$$

As an example of the use of the structure theorem, let us note that the normalized form of paths given in the previous section is natural in this setting. First, we note that
$$\mathcal{G}^{(2)}(H, \hat{S}) = \oplus_{i=1}^{k} \mathcal{G}^{(2)}(H, \hat{S}_i),$$
then we see that the normalized sequence for a path corresponding to an input M is
$$s_{m_1}, s_{m_2} \cdots, s_{m_n}, s_{m_{n+1}},$$
then $s_{m_i} \in S_0$ or $s_{m_i} \in \hat{S}_j$ for $1 \leq j \leq k$. Therefore, for the analysis of hash functions it is convenient to use the graph structure description in Theorem 1.

For completeness, we consider the case including degree 1 vertices. Define the subset $\bar{S}$ of $S - H$ by
$$\bar{S} = \bigcup_{|S_i|=1} S_i,$$
and set $\mathcal{G}_1 = \mathcal{G}(G, H, \bar{S})$. The pair-graph $\mathcal{G}_1$ contains all the degree one vertices of the pair-graph $\mathcal{G}(G, H, S)$ and is necessarily a disconnected graph.

Corollary 1 *Let G be a finite group, H a subgroup, and S a subset of G such that the corresponding pair-graph (G, H, S) is connected. Then,*
$$\mathcal{G}(G, H, S) \simeq \mathcal{G}(H, S_0) \oplus \mathcal{G}_1 \oplus \mathcal{G}^{(2)}(H, \hat{S})/ \sim_\phi .$$
$\square$

3.2 Some Possible Extensions

As described in this paper, finding group-subgroup pair graphs suitable for hash functions is not a simple task. In this section, by considering an extension by Kazufumi Kimoto we extend the possible graphs that may be obtained as (generalized) pair-graphs for use in cryptography.

The *Petersen graph* $\mathcal{P}$ (see Fig. 5) is a 3-regular graph in 10 vertices and 15 edges that appears in many examples of exceptional behavior of graphs. For instance, it is known to be a Moore graph, that is, it is a graph whose girth is more than twice the diameter of the graph. In particular, it is the graph with the longest girth for its number of vertices and degree. We refer to [10] for a book devoted to the properties of this graph.

In particular, we recall the following result. The proof is done by direct verification.

Theorem 2 *The Petersen graph $\mathcal{P}$ is not a Cayley graph for any group.*

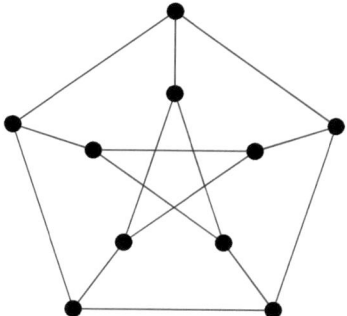

Fig. 5 The Petersen graph $\mathcal{P}$

As mentioned in previous sections, one of the main challenges for hash functions based on group-subgroup pair graphs is to find families of graphs with large girth, including the aforementioned Moore graphs or cage graphs (see [10] for the definitions).

Interestingly, the Petersen graph may be obtained in the context of group-subgroup pair graphs by considering a generalization of Kazufumi Kimoto that we recall here [14]. Note added in Proof: An equivalent notion was introduced by Majid Arezoomand and Bijan Taeri in [2].

Definition 6 Let $H \leq G$ be as in Sect. 2.1. Consider a family $\mathcal{S} = \{S_{i,j}\}_{i,j=0}^k$ of subsets of G such that

- $S_{i,j} \subset x_i^{-1} H x_j$,
- $S_{i,j}^{-1} = S_{j,i}$,
- $e \notin S_{i,j}$

Then, the *generalized group-subgroup pair graph* $\mathcal{G}(G, H, \mathcal{S}) = (V, E)$ is the graph with vertex set $V = G$ where $h \in Hx_i$ and $g \in Hx_j$

$$g = hs,$$

for $s \in S_{i,j}$.

Example 3 Let $D_n = \{s, t\}$ be the Dihedral group of degree $2n$. Let $H = \langle s \rangle$ and set

$$S_{0,0} = \{s, s^{-1}\}, \quad S_{1,0} = S_{0,1} = \{t\}, \quad S_{1,1} = \{s^2, s^{-2}\}.$$

Then, $\mathcal{G}(G, H, \mathcal{S})$ is the Petersen graph $\mathcal{P}$.

Obviously, following a similar construction to the one given in Sect. 3 it is possible to define a hash function from generalized pair-graphs.

Generally speaking, the generalized pair-graph allows the construction of a wide range of multiregular graphs. Indeed, when H is the trivial group, any connected pair-graph $\mathcal{G}(G, H, S)$ is a star graph, but in the generalized case any connected graph of $|G|$ vertices may be expressed as $\mathcal{G}(G, H, S)$ for appropriately chosen S. However, we note that the analysis of such graphs with $H = \{e\}$ is not different from that of graphs with no structure.

Therefore, a natural question is whether it is possible to construct Moore or cage graphs (or in general, graphs with large girth) as $\mathcal{G}(G, H, S)$ for certain group G, subgroup H with $1 < [G : H] < |G|$, and family of sets S. By the foregoing discussion, the answer of the above question for $[G : H] = |G|$ is yes. Therefore, when working with the generalized group-subgroup pair graphs, it is important to find a balance between choosing a subgroup with large index (that is, more flexibility in the construction but more complicated construction and less structure) or a subgroup with small index (simpler construction but less flexibility).

For instance, if the subgroup is abelian, all the eigenvalues can be computed explicitly for ordinary group-subgroup pair graphs, but in the generalized case this in only the case for certain choices of groups, subgroups, and generating sets. The generalized version of the graphs is a promising candidate for applications (including hash functions) but further study on their structure is needed.

Acknowledgements This work was supported by JST CREST Grant Number JPMJCR2113 including AIP challenge program, Japan. This work was supported by Institute of Mathematics for Industry, Joint Usage/Research Center in Kyushu University (FY2023 Short-term Joint Research (2023a017) "Toward a new method for constructing expander graphs and their applications 2").

References

1. Y. Aikawa, H. Jo, S. Satake, Left-right Cayley hashing: A new framework for provably secure hash functions. Math. Cryptol. **3**(2) (2023)
2. M. Arezoomand, B. Taeri, On the characteristic polynomial of n-Cayley digraphs. Electron. J. Combin. **20**(3), Paper 57 (2013)
3. Arzhantseva G, Biswas A (2022) Logarithmic girth expander graphs of $SL_n(\mathbb{F}_p)$. J. Algebr. Comb. 56:691–723
4. B. Bollobás, *Graph Theory. An Introductory Course. Graduate Texts in Mathematics*, vol. 63. (Springer, 1979)
5. J. Bourgain, A. Gamburd, New results on expanders. C. R. Math. Acad. Sci. Paris no.10 **342** 717–721 (2006)
6. L. Bromber, V. Shpilrain, A. Vdovina, Navigating in the Cayley graph of $SL_2(\mathbb{F}_p)$ and applications to hashing. Semigroup Forum **94**(2) (2017)
7. T. Ceccherini-Silberstein, F. Scarabotti, F. Tolli, *Discrete Harmonic Analysis. Representations, Number Theory, Expanders, and the Fourier Transform*. (Cambridge University Press, 2018)
8. Charles DX, Lauter KE, Goren EZ (2009) Cryptographic hash functions from expander graphs. J. Cryptol. 22(1):93–113
9. C. Easttom, *Modern Cryptography. Applied Mathematics for Encryption and Information Security*, 2nd edn. (Springer, 2022)
10. D. Holton, J. Sheehan, *The Petersen Graph. Australian Mathematical Society Lecture Series*. (Cambridge University Press, Cambridge, 1993)
11. U. Kanuer, *Algebraic Graph Theory. Morphisms, Monoids and Matrices. De Gruyter Studies in Mathematics*, vol. 41. (De Gruyter, 2011)

12. J. Katz, *Digital Signatures*. (Springer, 2010)
13. K. Kimoto, Spectra of group-subgroup pair graphs, in *Mathematical Modelling for Next-Generation Cryptography. Mathematics for Industry*, ed. by T. Takagi et al, vol. 29 (Springer, 2017), pp 139–157
14. K. Kimoto, Generalized group-subgroup pair graphs, in *International Symposium on Mathematics Quantum Theory and Crytography. Mathematics for Industry*, ed. by T. Takagi et al, vol. 33 (Springer, 2021), pp. 169–185
15. E. Kowalski, *An Introduction to Expander Graphs* (Société Mathématique de France, 2019)
16. M. Krebs, A. Shaheen, *Expander Families and Cayley Graphs : A Beginner's Guide*. (Oxford University Press, 2011)
17. Lubotzky A, Phillips R, Sarnak P (1998) Ramanujan graphs. Combinatorica 8(3):261–277
18. C. Petit, K.E. Lauter, J.-J. Quisquater, Cayley hashes: a class of efficient graph-based hash functions. Preprint (2007)
19. C. Petit, J.J. Quisquater, Cryptographic hash functions and expander graphs: The end of the story?, in *The New Codebreakers. Lecture Notes in Computer Science*, eds. by P. Ryan, D. Naccache, J.J. Quisquater, vol 9100. (Springer, Berlin, Heidelberg, 2016)
20. Reyes-Bustos C (2016) Cayley-type graphs for group-subgroup pairs. Linear Algebra Appl. 488(1):320–349
21. Satake S, Jo H (2022) On cryptographic hash functions from arc-transitive graphs. Math. Cryptol. 2(1):2–20
22. A. Terras, *Zeta functions of graphs. A stroll through the garden. Cambridge Studies in Advanced Mahtematics*, vol. 128 (Cambridge University Press, 2011)
23. Tillich JP, Zémor G (2011) Group theoretical hash functions. J. Algebraic Combinat. 33(1):95–109
24. J.P. Tillich, G. Zémor, Hashing with SL_2. Annual International Cryptology Conference. (Springer, 1994), pp. 40–49
25. Tomkins H, Nevins M, Salmasian H (2020) New Zémor-Tillich type hash functions over $GL_2(\mathbb{F}_p)$. J. Math. Cryptol. 14(1):236–253
26. P.C. van Oorschot, *Computer Security and the Internet*. (Springer, 2020)
27. G. Zémor, Hash functions and graphs with large girths, in *Workshop on the Theory and Applications of Cryptographic Techniques*. (Springer, 1991), pp. 508–511
28. Zémor G (1994) Hash functions and Cayley graphs. Designs Codes Cryptogr. 4(3):381–394

Open Access This chapter is licensed under the terms of the Creative Commons Attribution 4.0 International License (http://creativecommons.org/licenses/by/4.0/), which permits use, sharing, adaptation, distribution and reproduction in any medium or format, as long as you give appropriate credit to the original author(s) and the source, provide a link to the Creative Commons license and indicate if changes were made.

The images or other third party material in this chapter are included in the chapter's Creative Commons license, unless indicated otherwise in a credit line to the material. If material is not included in the chapter's Creative Commons license and your intended use is not permitted by statutory regulation or exceeds the permitted use, you will need to obtain permission directly from the copyright holder.

Improving Hash-Based Signature Schemes: From Theory to Practice

Quan Yuan

Abstract Hash-based signature schemes (HBS) are considered most secure and robust candidates in post-quantum public-key cryptographic schemes. For instance, SPHINCS+ has been chosen in the finalist of NIST post-quantum standardization. Although HBS has a long history in cryptographic research, the early versions of HBS are not practical enough. In this paper, we summarize the main challenges of applying HBS in practice and introduce solutions to improving the early versions to the modern ones.

Keywords Signature schemes · Hash functions · Post-quantum cryptography · Hash-based signatures

1 Introduction

A (computationally) secure cryptographic scheme is always based on assumptions about hard problems, such as the hardness of computing factorization and discrete logarithm. To design a secure scheme, we always try to use *minimal* assumptions: a stronger assumption implies a higher possibility to be broken in the future. For example, the above two problems are not hard anymore in the post-quantum world: one can solve them by Shor's algorithm [30] by a quantum machine with enough quantum circuits. Thus, in post-quantum cryptography, we need to avoid the usage of the above assumptions and try to construct schemes based on weaker assumptions.

For a secure signature scheme, the minimal assumption is the existence of one-way functions [29]: Let $\Gamma = (\mathsf{KeyGen}, \mathsf{Sig}, \mathsf{Ver})$ be a secure signature scheme, and $(pk, sk) \leftarrow \mathsf{KeyGen}(1^\lambda; s)$, where s denotes the randomness. Let f be the function mapping s to pk. Then, f is a one-way function, since any adversary breaking the one-wayness of f can recover the sk from pk.

Indeed, a one-way function is also *sufficient* to construct a *one-time* signature scheme (OTS), which is called Lamport's scheme [23]. Let f be a one-way function

Q. Yuan (✉)
Graduate School of Information Science and Technology, the University of Tokyo, Tokyo, Japan
e-mail: yuanquan@gmail.com

as a part of parameters. To sign a one-bit message b, the secret key includes random x_0 and x_1, and the public key includes $(y_0, y_1) = (f(x_0), f(x_1))$. The signature σ_b of b is x_b and the verification algorithm outputs 1 if and only if $f(\sigma_b) = y_b$. It can be extended to sign an n-bit message by instead picking $2n$ random x's in KeyGen and including n of them in a signature. In practice, the one-way function f can be instantiated by preimage-resistant hash functions, such as SHA. Thus, the construction is called *hash-based signature schemes* (HBS).[1]

The HBS is considered one of candidates in post-quantum cryptographic schemes. The main advantage is robustness and minimal assumptions. In post-quantum cryptosystems, hash functions are frequently used and always supposed to be post-quantum secure. Even if the security of hash functions may be broken in the future, we can then replace it with a secure one, and the main structure of the HBS is not modified. The replacing price is thus much lower than others (thus as replacing the lattice in lattice-based schemes).

However, the HBS is not widely used in practice due to the following challenges:

- **Signature size** In Lamport's scheme, the signature includes n preimages of the one-way function, where n is the message length. Thus, to achieve λ-bit classical security, a signature has at least $n\lambda$-bits, which is not practical for OTS.

In the finalist of NIST post-quantum standardization, there are three candidates for (many-time) signature schemes: CRYSTAL-Dilithium, Falcon, and SPHINCS+. SPHINCS+ is the only one based on hash functions, and the signature (with level-5 security) is approximately 5 times larger than CRYSTAL-Dilithium and 20 times larger than Falcon with the same security level.

- **Statefulness** We note that Lamport's scheme provides CMA-security in the one-time case. If the secret key is used to sign two messages, say 0^n and 1^n, then the secret key is totally revealed and thus the adversary can forge signature for any message. A natural approach to deal with many signatures is using Merkle tree structure [25] with collision-resistant hash functions. However, it means that the signer needs to maintain a dynamic state (or evolves a dynamic secret key) in signing operations. It does not fit the standard definition of a signature scheme.
- **Security Loss** In a HBS, we need to run a number of hash computations, which will lead to high security loss. For example, Lamport's scheme leads to $(\log n + 1)$-bit security loss when signing an n-bit message. The security loss gets much higher in many-time instances.

In this paper, we review the above challenges of HBS and introduce advanced solutions. Indeed, by reviewing these solutions, we can find the history from the basic HBS (such as Lamport's scheme) to the modern ones (such as SPHINCS+). The structure of practical HBS is somewhat complicated: it is always built from various lower level blocks based on hash functions. We introduce how the blocks

---

[1] Of course, f can also be instantiated by others based on mathematical hard problems, but hash functions are usually the most efficient choices.

are constructed and optimized, and then how they are used in the main body of the structure. Finally, we demonstrate other challenges of HBS, which we believe potentially interesting in the future research.

2 Eliminating the Signature Size

In this section, we introduce the approaches to eliminate the size of HBS. We only focus on the OTS in this section.

2.1 Improving Lamport's Scheme from Sperner Family

Let us get back to Lamport's scheme with n-bit messages. The main reason for the one-time security is that any two distinct messages have at least one-bit difference. In detail, the key generation picks $2n$ secret strings, and a signature will only leak n of them. Then, a forgery from the adversary (with a signing query) must contain at least one secret string such that it is not covered by the known signature. In other words, each message m in the message space $\mathcal{M}$ is *encoded* to a subset S_m of the secret strings $X = \{x_{b,i}\}_{b \in \{0,1\}, i \in [n]}$. Then, it must hold that $S_m \not\subseteq S_{m'}$ for any $m \neq m'$. Otherwise, a signature of m can be forged after querying m' to the signing oracle.

We observe that the above encoding in Lamport's scheme is not optimal. Fix X with $2n$ elements. Then, X has $\binom{2n}{n} = \Omega(2^{2n-\frac{1}{2}\log n})$ number of n-subsets, and apparently each of them is not covered by any other one. It implies that if we encode each m to an n-bit subset of X, then the message space size reaches $O(2^{2n-\frac{1}{2}\log n})$ rather than 2^n with the same key and signature size.

Indeed, the above question implies a Sperner family. Roughly, let $\mathcal{W}$ be a set of subsets of X. Then, we say $(X, \mathcal{W})$ is a Sperner family if for any distinct $A, B \in \mathcal{W}$, it holds that $A \cap B \neq \emptyset$.

Then, the question is how to pick an optimal Sperner family $(X, \mathcal{W})$ such that $|\mathcal{W}| \geq 2^n$ and thus the message space can be encoded to $\mathcal{W}$. Sperner's theorem [24] states that if $|X| = n$, then it holds that $|\mathcal{W}| \leq \binom{n}{\lfloor n/2 \rfloor}$ in a Sperner family. Thus, Lamport's scheme with n-bit messages can be theoretically improved as follows.

Construction 1 (*Improved Lamport's Scheme*) *Let* $\mathcal{F} : \{0, 1\}^\lambda \to \{0, 1\}^\lambda$ *be a one-way function family. Let t be the smallest integer such that* $\binom{t}{\lfloor t/2 \rfloor} \geq 2^n$, *and $\mathcal{W}$ be the optimal Sperner family of* $[t] := \{1, \ldots, t\}$.

- KeyGen(1^λ): *Pick* $f \leftarrow \mathcal{F}$ *and* $x_i \leftarrow \{0, 1\}^\lambda$ *for* $i \in [t]$. *Compute* $y_i = f(x_i)$ *for each i. The secret key is* $(f, x_1, \ldots, x_t)$ *and the public key is* $(f, y_1, \ldots, y_t)$.
- Sign(sk, m): *Parse* $sk = (f, x_1, \ldots, x_t)$. *Encode m to* $S_m \in \mathcal{W}$. *Return* $\sigma := \{x_i : i \in S_m\}$.
- Ver(pk, m, σ): *Parse* $pk = (f, y_1, \ldots, y_t)$ *and* $\sigma = \{x_i : i \in S_m\}$. *Return 1 if and only if* $f(x_i) = y_i$ *holds for each* $i \in S_m$.

We note that $|X|$ denotes the secret key size and $|S_m|$ denotes the signature size. The above improvement optimized the former one. It is possible to adjust the choice of Sperner family, and it results in different sizes. For example, we can pick larger $|X|$ and smaller $|S_m|$ such that $|\mathcal{W}| = \binom{|X|}{|S_m|} \geq 2^n$ still holds. See more details in [2] about an example of adjusting the parameters.

2.2 Winternitz Scheme and Its Variant

The above improvement only shortens the size by no more than one half. In this subsection, we introduce another variant called Winternitz one-time signature scheme (WOTS), which greatly compresses the signature size. The essence of WOTS is the same as the previous one: it also implies a Sperner family, while the keys and signatures are compressed by *hash chains*.

Construction 2 *(Winternitz Scheme) Denote w as the Winternitz parameter (usually as a power of 2). Let*

$$l_1 = \left\lceil \frac{n}{\log w} \right\rceil, \quad l_2 = \left\lfloor \frac{\log(l_1(w-1))}{\log w} \right\rfloor + 1,$$

and $l = l_1 + l_2$. Let h be a secure hash function. Denote $h^0(x) := x$ and $h^k(x) := h\bigl(h^{k-1}(x)\bigr)$ for any integer $k \geq 1$.

- **KeyGen**(1^λ): *Pick $x_i \leftarrow \{0, 1\}^\lambda$ for $i \in [l]$. Compute $y_i = h^{w-1}(x_i)$ for each i. The secret key is $sk := (x_1, \ldots, x_l)$ and the public key is $pk := h(y_1, \ldots, y_l)$.*
- **Sign**(sk, m): *Parse $sk = (x_1, \ldots, x_l)$. Let $(c_1, \ldots, c_{l_1})$ be the base-w representation of m, where $c_i \in \{0, \ldots, w-1\}$ for each $i \in [l_1]$. Compute the checksum $s = \sum_{i=1}^{l_1}(w - 1 - c_i)$. Let $(c_{l_1+1}, \ldots, c_l)$ be the base-w representation of s. Then, compute $\sigma_i := h^{c_i}(x_i)$ for each $i \in [l]$ and return $\sigma := (\sigma_1, \ldots, \sigma_l)$.*
- **Ver**(pk, m, σ): *Parse $\sigma = (\sigma_1, \ldots, \sigma_l)$. Compute $c_1, \ldots, c_l$ as in Sign and $y_i = h^{w-1-c_i}(\sigma_i)$. Return 1 if and only if $pk = (y_1, \ldots, y_l)$ holds.*

Observe that the base-w presentation of m implies a subset of elements in hash chains. For a simplified (but inefficient) example, let $w = 2^n$ and thus $l_1 = 1$. There is only one hash chain of x, and c_1 represents the algebraic expression of m. Then, S_m includes all integer i for $c_1 \leq i \leq w - 1$ and σ reveals all $h^i(x)$ for $i \in S_m$. Apparently, these S_m are not a Sperner family: for any $c_1 < c_1'$, $S_{m'} \subset S_m$ holds. This is why a checksum $c_2 := 2^n - 1 - c_1$ is introduced. If $c_1 < c_1'$, then $c_2 > c_2'$ must hold. Let the subset for checksum $\overline{S}_m = \{i : 0 \leq i \leq c_2 - 1\}$. Then, we have $\overline{S}_m \subset \overline{S}_{m'}$. Thus, $\{\{S_m, \overline{S}_m\}_m\}$ is a Sperner family of $\{0, \ldots, w-1\}$. Due to the hash chains, it is unnecessary to include all the elements of $\{S_m, \overline{S}_m\}$ in a signature. Instead, we can use two elements to compute all the others. As a result, the signature size is greatly compressed.

Generally, a message $m \in \mathcal{M}$ is encoded to some $\mathbf{c} = (c_1, \ldots, c_l) \in [w]^l$. For $\mathbf{c}, \mathbf{c}' \in [w]^l$, we define $\mathbf{c} \leq \mathbf{c}'$ if and only if $c_i' \leq c_i$ holds for any $i \in [l]$. We can see that $\leq$ is an anti-symmetric, transitive, and reflexive relation of set $[w]^l$. If $\mathbf{c} \leq \mathbf{c}'$ or $\mathbf{c}' \leq \mathbf{c}$ holds, then we say $\mathbf{c}$ and $\mathbf{c}'$ are comparable or otherwise they are incomparable. Let C be a subset of $\mathbf{c} \in [w]^l$. We say C is incomparable if any pair of distinct elements in C is incomparable (which implies a Sperner family). Then, the essence of WOTS is to encode the message space to a comparable C.

Similar to Lamport's scheme, a natural question is whether the encoding in WOTS is optimal. In other words, we need to find the largest incomparable subset of $\mathbf{c} \in [w]^l$, implying the largest message space. (Here, w and l are fixed. l implies the signature size, and the lw implies the key generation running time.)

A better solution is to use *constant-sum* encoding. Fix some $s \in [lw]$. Denote $C_s := \{\mathbf{c} \in [w]^l : \sum_{i=1}^{l} c_i = s\}$. Then, $\mathcal{M} \to C_s$ is a constant-sum encoding. When $s = \lfloor \frac{l(w-1)}{2} \rfloor$, $|C_s|$ reaches the largest one. It is recently proven that constant-sum encoding achieves the optimal encoding rate in Winternitz-style OTS [37]. The constant-sum strategy has been used to compress the signature size of SPHINCS+ [19] in a probabilistic manner.

Finally, we turn to the required assumptions in (variants of) WOTS: one may notice that we use a secure hash function h in WOTS instead of a one-way function f. Indeed, the assumption is implicitly enhanced. Recall that one-way property of f means that given $y = f(x)$ for a *uniform* x, it is hard to return x' such that $f(x') = y$. Intuitively, a secure WOTS requires that given $y = h^{w-1}(x)$ for uniform x, it is hard to return any x' such that $y = h^i(x')$ for *any* $i \in [w-1]$. It is a strictly stronger assumption, since the preimages in a hash chain may not always be uniformly distributed.

Thus, the hash function in WOTS is additionally required to be undetectable [13], which means that it is (computationally) hard to distinguish x from $h(x)$, where x is uniformly distributed in the range of h.[2] Undetectable functions are implied from pseudorandom functions (PRFs), which can be constructed from one-way functions [17]. Thus, the assumption of WOTS is still minimal.

3 Getting Rid of Signing Limitations

In the previous section, we only discuss one-time signature schemes. In this section, we show how to extend it to more general cases.

---

[2] In WOTS, the range of h should cover the domain.

3.1 From One-Time to Few-Time

We show in Sect. 2.1 that a Sperner family implies a OTS. A Sperner family is essentially a 1-cover-free family (1-CFF). Roughly, $(X, \mathcal{W})$ is a r-CFF if for any $A_1, \ldots, A_r \in \mathcal{W}$ and any other $B \in \mathcal{W}$, it holds that $B \not\subseteq \bigcup_{i=1}^{r} A_i$. Thus, we observe that the improved Lamport's scheme can be simply extended to an r-time signature scheme by replacing the Sperner family with an r-CFF [27]. Unfortunately, the construction is rather expensive (especially in terms of the running time of KeyGen) due to the lower bounds of constructing an r-CFF [32].

Next, we try to find other solutions for higher efficiency by weakening the requirement of r-CFF. For $(X, \mathcal{W})$ and an encoding $\mathsf{E} : m \to \mathcal{W}$, we instead require for any *polynomial-time* adversary $\mathcal{A}$, it is *computationally* (rather than information-theoretically) hard to find $m, m_1, \ldots, m_r \in \mathcal{M}$ such that $\mathsf{E}(m) \not\subseteq \bigcup_{i=1}^{r} \mathsf{E}(m_i)$. The encoding can be simply implemented by a secure hash function. That is, let H be a hash function mapping to $\{0, 1\}^l$, where $l = k\tau$. X contains $t := 2^\tau$ random elements x_i as the secret key. Let $\mathcal{W}$ be the family of all i-subsets of X for all $i \leq k$. For any message m, split $\mathsf{H}(m) := h_1(m) || h_2(m) || \ldots || h_k(m)$. Then, m is encoded to $\mathsf{E}(m) := \{x_{h_1(m)}, \ldots, x_{h_k(m)}\}$.[3]

Construction 3 *(HORS [28]) Let $\mathcal{F} : \{0, 1\}^\lambda \to \{0, 1\}^\lambda$ be a one-way function family, and $H : \{0, 1\}^* \to \{0, 1\}^{k \cdot \tau}$ be a hash function. Let $t = 2^\tau$.*

- KeyGen(1^λ): *Pick $f \leftarrow \mathcal{F}$ and $x_i \leftarrow \{0, 1\}^\lambda$ for $i \in [t]$. Compute $y_i = f(x_i)$ for each i. The secret key is $(f, x_1, \ldots, x_t)$ and the public key is $(f, y_1, \ldots, y_t)$.*
- Sign(sk, m): *Parse $sk = (f, x_1, \ldots, x_t)$ and split $H(m) = h_1(m) || \ldots || h_k(m)$, where $h_j(m)$ is the algebraic expression of the τ-bit string. Let $S_m = \{h_j(m) + 1\}_{j \in [k]}$. Return $\sigma := \{x_j\}_{j \in S_m}$.*
- Ver(pk, m, σ): *Parse $pk = (f, y_1, \ldots, y_t)$ and $\sigma = \{x_j\}_{j \in S_m}$. Return 1 if and only if $f(x_{h_i(m)+1}) = y_{h_j(m)+1}$ holds for each $j \in [k]$.*

The above scheme is called Hash to Obtain Random Subsets (HORS) [28]. To guarantee the r-time security, we need to additionally require H to be *subset-resilient*. Formally, for any polynomial-time adversary $\mathcal{A}$, it is computationally hard to find $(m, m_1, \ldots m_r)$ such that $\{h_j(m)\}_{j \in [k]} \subseteq \{h_j(m_i)\}_{j \in [k], i \in [r]}$ and $m \notin \{m_i\}_{i \in [r]}$. Unlike undetectability, subset resilience cannot be implied from one-way functions [35]. Thus, the assumption of HORS is not minimal.

In practice, the complexity of r is small. If r becomes larger, then the security of r-subset resilience will be rapidly decreased. Thus, HORS is called a few-time signature scheme (FTS).

The encoding in HORS has a potential drawback [1]. Note that $\mathcal{W}$ contains all subsets with elements *at most* k. It implies that there may be some messages whose signature contains less than k elements and thus is easier to be forged (with less signing queries). It is natural to restrict the subset to exactly contain k (labeled) elements. A modified version of HORS is depicted as follows.

---

[3] Since there may exist $i \neq j$ such that $h_i(m) = h_j(m)$, $\mathsf{E}(m)$ may contain less than k elements.

Construction 4 *(Simplified FORS [35])* Let $\mathcal{F} : \{0, 1\}^\lambda \to \{0, 1\}^\lambda$ be a one-way function family and $H : \{0, 1\}^* \to \{0, 1\}^{k \cdot \log t}$ be a hash function.

- **KeyGen**(1^λ): Pick $f \leftarrow \mathcal{F}$ and $x_{i,j} \leftarrow \{0, 1\}^\lambda$ for $i \in [t]$ and $j \in [k]$. Compute $y_{i,j} = f(x_{i,j})$ for each $i \in [t]$ and $j \in [k]$. The secret key is $(f, (x_{i,j})_{i \in [t], j \in [k]})$ and the public key is $(f, (y_{i,j})_{i \in [t], j \in [k]})$.
- **Sign**(sk, m): Parse $sk = (f, (x_{i,j})_{i \in [t], j \in [k]})$. Return $\sigma := (x_{h_j(m)+1, j})_{j \in [k]}$.
- **Ver**(pk, m, σ): Parse $pk = (f, (y_{i,j})_{i \in [t], j \in [k]})$ and $\sigma = (x'_j)_{j \in [k]}$. Return 1 if and only if $f(x'_j) = y_{h_j(m)+1}$ holds for each $j \in [k]$.

Correspondingly, the requirement of H should be enhanced to *restricted subset-resilient*: for any polynomial-time adversary $\mathcal{A}$, it is computationally hard to find $(m, m_1, \ldots m_r)$ such that $h_j(m) \in \{h_j(m_i)\}_{i \in [r]}$ for each $j \in [k]$.

Restricted subset resilience (rSR) is a stronger notion than the original one and thus impossible to be implied from one-way functions either. The resulting scheme, which is indeed the simplified version of Forest Of Random Subsets (FORS [6]), has larger signature sizes. An advantage is that the signature size of FORS is a constant ($k\lambda$ bits) while HORS is not.

3.2 From One-Time to Many-Time with States

The above approach to construct a many-time scheme (MTS) is rather inefficient. We instead have a more direct and efficient way: to construct an N-time signature scheme, we simply generate N OTS key pairs, and each of the secret keys is only used once. Even if N is few, the resulting scheme is more efficient than the FTS in the previous subsection.

A problem is that the resulting public key needs to contain all the N OTS public keys.[4] It is inefficient if N is large. A well-known solution is to use a (binary) Merkle tree structure: Without loss of generality, suppose $N = 2^\tau$ for some integer τ and H be a collision-resistant hash function $\{0, 1\}^* \to \{0, 1\}^\lambda$. Generate a binary tree with depth τ, where each leaf contains the hash value of a public key. A non-leaf node is labeled by the hash function value of the two children. Finally, we have a tree whose root's label is determined by all the N public keys. Denote the MTS public key by the root of the Merkle tree. A MTS signature needs to additionally contain the corresponding OTS public key, and the related τ nodes that can be used to calculate the root, which is called the *authentication path* (see details in Fig. 1). In verification, it first runs the original verification algorithm given the OTS signature and the public key in the MTS signature. If it outputs 1, calculate the root from the OTS public key and the authentication path. Return 1 if it is equal to the MTS public key.

For example, Fig. 1 implies a 2^3-time signature scheme. When we use the third OTS key pair (labeled 010) to sign a message m, the MTS signature contains (1) the

---

[4] The secret key can be compressed by a random seed and a PRF.

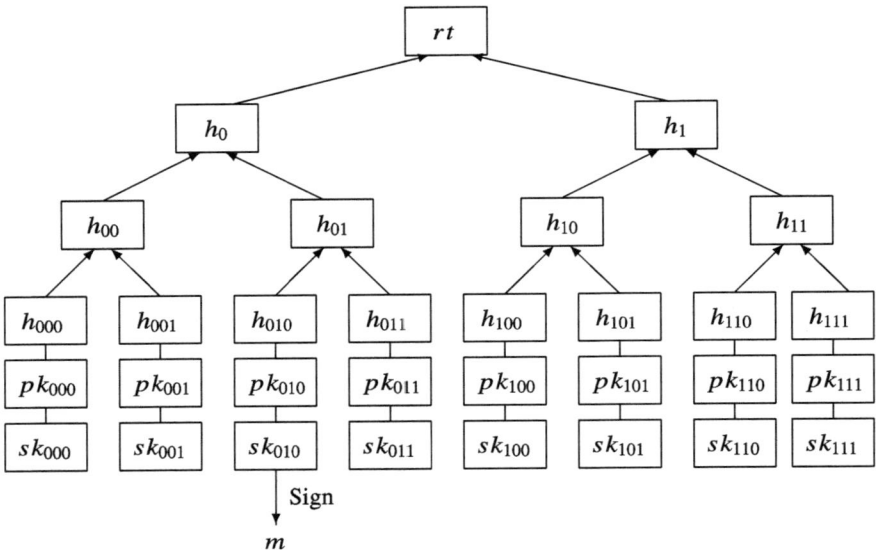

Fig. 1 A Merkle tree of a many-time signature scheme with $N = 2^3$. The nodes colored in gray denote the authentication path of leaf 010

OTS signature of $\sigma_{\mathsf{OTS}} = \mathsf{OTS.Sig}(sk_{010}, m)$, (2) pk_{010}, and (3) the authentication path $\mathsf{Auth} = (h_{011}, h_{00}, h_1)$. If we use WOTS as the OTS, where the public key can be recovered from the message and the corresponding signature, then the second part pk_{010} can be omitted. Instead, the verification algorithm recovers it from $(m, \sigma_{\mathsf{OTS}})$ and uses it to calculate the Merkle tree root with Auth. The MTS signature has additional $\lambda \log N$ bits to the OTS signature.

In a long-time protocol, a public key may be used to issue a great number of signatures, and it is impractical to prepare all of the sub-keys in the key generation. To solve this, we introduce a *hyper-tree structure* to expand the signing times.

Suppose we have had an N-time signature scheme with a Merkle tree, say Γ, with state $t \in [N]$. Here, a state t is an additional part of the input to the signing algorithm, meaning that it uses the t-th node to issue the signature. Then, we construct an N^2-time hyper-tree signature scheme (HTS) from Γ with state $(t_1, t_2) \in [N]^2$. Roughly, the key pair of HTS is the same as Γ. In each signing operation, it deterministically generates a sub-tree for state t_1, and uses the sub-tree to sign the message with state t_2. Then, use the main secret key to sign the root of the sub-tree. The formal description is as follows.

Construction 5 *(A hyper-tree signature scheme with 2 layers) Let* PRF *be a pseudorandom function with key space* $\{0, 1\}^\kappa$.

- $\mathsf{KeyGen}(1^\lambda)$: $Run\ (pk, sk) \leftarrow \Gamma.\mathsf{KeyGen}(1^\lambda).\ k \leftarrow \{0, 1\}^\kappa.\ Return\ (pk, (sk, k))$.
- $\mathsf{Sign}(sk, m, t)$: $Parse\ t = (t_1, t_2).\ Run\ (pk', sk') \leftarrow \Gamma.\mathsf{KeyGen}(1^\lambda; \mathsf{PRF}(k, t_1))$. $\sigma_2 \leftarrow \Gamma.\mathsf{Sign}(sk', m, t_2).\ \sigma_1 \leftarrow \Gamma.\mathsf{Sign}(sk, pk', t_1).\ Return\ (\sigma_1, \sigma_2, pk')$.

- Ver(pk, m, σ): Parse $\sigma = (\sigma_1, \sigma_2, pk')$. Return 1 if and only if

$$\Gamma.\text{Ver}(pk', m, \sigma_2) = \Gamma.\text{Ver}(pk, pk', \sigma_1) = 1.$$

Similar to WOTS, the public key of Γ can be recovered from the message and the signature. Thus, pk' can be omitted in σ.

The above construction can be further extended to d layers, and we eventually have an N^d-time signature scheme. The signature size is approximately d times of the original N-time signature. The signing algorithm contains $d - 1$ times of Γ.KeyGen and d times of Γ.Sign.

3.3 From Statefulness to Statelessness: Goldreich's Scheme and SPHINCS

The HT signature has an obvious drawback: every time a signature is issued, the signer needs to mark the OTS secret key so that it will not be used twice. Thus, the resulting scheme is called a *stateful* signature scheme [34] (or a *key-evolving* signature scheme [3]), since the signer needs to maintain a dynamic state in signing operations, and a failure of the evolving will break the security. It does not even fit the original definition of a signature scheme, where the signing algorithm should only take as input sk and m. Thus, we prefer a *stateless* signature scheme rather than a stateful one.

Goldreich [15] proposes a stateless signature scheme based on stateful one. First, generate a 2^λ-time HT signature pair (pk, sk) as above, where the state is in $\{0, 1\}^\lambda$. Then, pick a collision-resistant hash function H. To sign a message m, it simply runs Sign($sk, m, \text{H}(m)$), where H(m) is the state. Due to the collision resistance of H, it is hard to find (m_1, m_2) such that H(m_1) = H(m_2). It implies that the adversary cannot query two messages with the same state in two queries, and thus the security can be guaranteed.

However, we note that the size and the running time of the HT signature scheme grow linearly to the logarithm of the state space. In practice, a λ-bit hash function can only supply at most $\lambda/3$-bit quantum security [9]. That is, the size of state space should be at least $\Omega(2^{3\lambda})$ to achieve λ-bit quantum security, which is not practical enough.[5]

Another approach is to replace H(m) with a fresh random state. The scheme is secure if the signer never chooses the same state in two signatures. However, when the number of issued signatures grows up, the security will be degraded rapidly due to the "birthday attack": if the size of the state space is N, then an adversary is expected to break the security after knowing $O(\sqrt{N})$ signatures. Again, N is supposed to be exponentially large, and it thus leads to an inefficient construction.

---

[5] By the approach in the next section, the size can be reduced to $\Omega(2^{3\lambda})$, but it is still impractical.

As an improvement, SPHINCS framework [4] combines Goldreich's scheme and few-time signature schemes. As in Goldreich's scheme, it first generates a 2^h-time HT signature key pair (pk_{HT}, sk_{HT}), where h is a parameter. Then, each leaf of the hypter-tree corresponds to a few-time signature key pair. In each signing operation, it (pseudo-)randomly picks a leaf from the hyper-tree, say idx, generates the corresponding FTS key pair (pk_{idx}, sk_{idx}), and uses sk_{idx} to sign the message m. Then, use sk_{HT} to sign pk_{idx}. The final signature contains (1) idx, (2) the FTS signature σ_{FTS} of the message m, and (3) the HT signature σ_{HT} of pk_{idx}. Similar to the HT construction, pk_{idx} is finally omitted if it can be recovered from sk_{idx} and σ_{FTS}.

Construction 6 *(SPHINCS framework [4]) Let* HT *be a stateful hyper-tree signature scheme with state* $\{0, 1\}^h$ *and* FTS *be a few-time signature scheme.*

- KeyGen(1^λ): Run $(pk, sk) \leftarrow$ HT.KeyGen(1^λ). $k \leftarrow \{0, 1\}^\kappa$. Return $(pk, (sk, k))$.
- Sign(sk, m): Pick $idx \leftarrow \{0, 1\}^h$. $(pk_{idx}, sk_{idx}) :=$ FTS.KeyGen(1^λ; PRF(k, idx)). $\sigma_{FTS} \leftarrow$ FTS.Sign(sk_{idx}, m). $\sigma_{HT} \leftarrow \Gamma$.Sign($sk, pk_{idx}, idx$). Then, return ($idx, \sigma_{FTS}, \sigma_{HT}$).
- Ver(pk, m, σ): Parse $\sigma = (idx, \sigma_{FTS}, \sigma_{HT})$. Recover pk_{idx} from m and σ_{FTS}. Return 1 if and only if HT.Ver(pk_{HT}, m, σ_{HT}) = 1.

The construction ensures that a leaf of the hyper-tree never signs two different messages: it only signs a deterministic pk_{idx}. The potential risk comes from the FTS. If an index is chosen too many times, which exceed the limitation of FTS, then the scheme may be insecure. Fortunately, in the FTS introduced in Sect. 3.1, the limitation is not restricted by a constant number. For example, if we use HORS as the FTS,[6] then the security after r-time signing queries can be reduced to r-subset resilience. If r grows larger, then the security bits of r-subset resilience will go lower. Simultaneously, the probability becomes lower that a leaf is chosen more than r times. Let q_s be the total number of signing operations. The above probability can be bounded by

$$\frac{1}{2^{h(r-1)}} \binom{q_s}{r} \leq 2^{r(\log q_s - h) + h}.$$

Let $\mathsf{InSec}_H^{r\text{-SR}}(\xi)$ denote the maximum advantage of breaking r-subset resilience in ξ-time. Then, the risk from subset resilience is bounded by

$$\sum_{r=1}^{q_s} \min\left\{2^{r(\log q_s - h) + h}, 1\right\} \cdot \mathsf{InSec}_H^{r\text{-SR}}(\xi).$$

By carefully choosing the parameter h and the output length of H, we can bound the above probability by a negligible function.

---

[6] In HORS, the public key cannot be recovered from the secret key and the message. Thus, in SPHINCS, it instead uses a HORS with a Merkle tree (say HORST), and the signature additionally contains an authentication path.

We observe an attack on SPHINCS framework. After sending q_s number of signing queries, the adversary chooses the index idx^* that appears the most. It implies that the FTS secret key with index idx^* exposes the most information. Then, the adversary can aim to find a message m^* whose encoding is covered by those subsets and forge a signature for it with index idx^*. Although the success probability has been considered above, we cannot ignore attacks that can arbitrarily determine the index used in forgeries.

To avoid this attack, an improvement called SPHINCS+ [6] modified the strategy of choosing the index. In a SPHINCS+ signature, the index is not directly contained. Instead, it picks a (pseudo-)random nonce R, and the index is calculated by $idx = H'(R, m)$. Then, R is contained in the signature. Now, to forge a signature for some message with a target index idx^*, the adversary needs additional hash computations to find the corresponding R. This modification implicitly improves the bit security of the scheme.

4 Tightening the Security Bound

4.1 Concrete Quantum Bit Security

We first introduce how to evaluate the quantum bit security of a HBS. In the NIST standardization, the security level is evaluated by "the required computations to break the (EUF-CMA) security". For example, the level-5 security requires that breaking the scheme is at least as hard as key search on 256-bit block ciphers such as AES-256. By Grover's algorithm [16], it can be done by approximately 2^{128} quantum queries to AES. In bit security analysis, we say that the scheme has n-bit quantum security if breaking it requires at least 2^n computations of a symmetric primitive (such as AES or SHA).

As we can see, a level-5 signature scheme requires 128-bit quantum security. We note that if a hash function (or a block cipher) behaves as a random function (or a random permutation), then Grover's algorithm is optimal to break the one-wayness. In security analysis of a HBS, the hash functions are usually modeled as a random oracle. The optimality of Grover's search ensures that the quantum bit security can be reduced to the output length of the hash functions.

Let the one-way function in Lamport's scheme be an ideal λ-bit hash function. Does it mean that the scheme has $\frac{\lambda}{2}$-bit quantum security? Unfortunately, it is false. Let S be the set of the public keys. The adversary only needs to search x such that $f(x) \in S$. Since $|S| = 2n$, Grover's search algorithm is expected to succeed with $O(\sqrt{2^\lambda/2n})$ quantum queries to f. Indeed, the security proof of Lamport's scheme does cause $2n$ security loss since the adversary needs to randomly inject the one-way challenge into one of the public keys. It implies that Lamport's scheme only achieves $\frac{1}{2}(\lambda - \log n - 1)$-bit quantum security.

It seems not a big issue for Lamport's scheme, but it differs in others. For example, in a HTS with 2^h leaves, there are more than 2^h (potential) OTS key pairs. If the OTS is implemented by WOTS with parameters w and l, then the security loss reaches (at least) $2^h l w^2$, which could be a huge number. Instantiated by the parameters in SPHINCS-256, it causes over 80-bit quantum security loss.

Another problem appears in the requirement of collision resistance. BHT algorithm [9] can find a collision with $O(2^{n/3})$ number of hash queries. Thus, it is necessary to implement a 384-bit hash function to achieve 128-bit quantum security for Merkle trees. A similar problem appears in subset resilience [35].

Fortunately, the security of PRF in HBS does not cause explicit security loss, although a lot of pseudorandomness is picked in a HBS. It is proven that if PRF is constructed by a random oracle, then the advantage of breaking PRF is independent of the number of function queries [33] (but quadratic to the number of offline computations).

In the following subsections, we try to improve the scheme to avoid high security loss and have tighter security bounds.

4.2 Improving Collision Resistance: Bitmarks and Tweaks

We first try to remove the requirement of collision resistance, Indeed, collision resistance seems not necessary. A collision only implies that the adversary may have two different hash trees with the same root, but the hash tree is probably not the *real* tree corresponding to the secret key. Then, the collision is not helpful for a forgery.

Thus, what we really need for a hash function is the following property: given x_0, x_1, it is impossible to get $(x_0', x_1') \neq (x_0, x_1)$ such that $H(x_0'||x_1') = H(x_0||x_1)$. It is indeed quite close to a stronger property called *target collision resistance* (TCR). For an ideal and compressing hash function, breaking TCR requires an exhaustive (quantum) search, which leads to a much tighter bound than collision resistance.

However, we have two fatal problems in the reduction to TCR.

First, in a hash tree, the node is the hash of the compensation of two hash values, while the target of TCR is supposed to be a hash function of a *uniform* string. To make the reduction work, we can additionally add bitmarks to the hash trees [10]. Formally, to calculate the label of node i whose children are $x_{0,i}$ and $x_{1,i}$, pick a pseudorandom string c_i with the same length as $x_{0,i}||x_{1,i}$, say a *bitmark*. The label of node i is $H\big((x_{0,i}||x_{1,i}) \oplus c_i\big)$, and the bitmark is included in the signature. Then, the reduction can arbitrarily inject the challenge x to any nodes i. It simulates the node with label $H(x)$ and $c_i := (x_0||x_1) \oplus x$. The simulation cannot be detected if bitmarks are pseudorandom.

Second, it causes security loss. The reduction now needs to guess at which node the adversary's tree diverges from the real tree. The security loss is the total number of the nodes in the tree, which can be huge in practice. We can solve it by adding a *hash key* to the hash computations [20]. Formally, for each node i, pseudorandomly pick a hash key k_i. Then, the label of node i is modified to $H_{k_i}\big((x_{0,i}||x_{1,i}) \oplus c_i\big)$.

Correspondingly, the notion for TCR should be modified to a keyed version, called *multi-function, multi-target, target-collision resistance* (MM-TCR) with a parameter p. Here, the adversary is instead given p pairs of uniform random (k_i, x_i), and is required to output (k^*, x^*) such that $k^* = k_i$ for some $i \in [p]$ and $x^* \neq x_i$. It is interesting that if H is modeled as a random oracle mapping to $\{0, 1\}^\lambda$, then the probability of breaking p-MM-TCR is $O((q + 1)/2^\lambda)$ and $O((q + 1)^2/2^\lambda)$ for any classical and quantum adversary with q queries to H, respectively, which is independent of p. The reason for the independence will be explained in Sect. 4.4.

Then, we can tightly reduce the security to p-MM-TCR, where p is the number of nodes: all the nodes can be replaced with the challenge, and an arbitrary diverging node can lead to a successful reduction. We thus get rid of the high security loss p.

The idea can be further improved by replacing the keyed hash function with a *tweakable hash function* [6]: modify the label of node i with $H(i, x_{0,i}||x_{1,i})$, where i is called the tweak. An obvious improvement is that the signature does not contain any hash keys or bitmarks, and thus the size is smaller. The bitmarks, indeed, make no sense of avoiding attacks: it is only used for provable security. Thus, we remove them and modify the experiment of multi-target TCR to an *interactive* version: At the beginning, the adversary *adaptively* sends p queries to an oracle **Th**. Each query contains a tweak i and a target x_i, and each tweak can only be queried once. After that, the adversary is required to return (i^*, x'_{i^*}) such that i^* has been queried and $x'_{i^*} \neq x_{i^*}$. Then, the reduction can generate the hash tree by querying **Th**. A provable tweakable hash function can be constructed by several approaches (see [6]). Practically, if the tweakable hash function is defined as **Th**(i,x)=$H(i||x)$, where H is an original hash function modeled as a quantum random oracle, the probability of breaking p-MM-TCR is still independent of p.

4.3 Improving Subset Resilience: Nonce and Interleaves

Similar to collision resistance, subset resilience can also be weakened by the same approaches.

We first modify FORS signature as follows: the signing algorithm picks a (pseudo-)random nonce R and hashes $(R||m)$ instead of m. Now, we observe that an r-restricted subset cover $(x, x_1, \ldots, x_r)$ does not imply a forgery, since R cannot be controlled by the adversary. Then, the scheme can be reduced to a weakened rSR, say *extended target subset resilience* (eTSR, [35]). In the experiment, the adversary can (adaptively) query an oracle O with m_i and obtain a random R_i. Then, the adversary is required to output $(m, m_1, \ldots, m_r)$ such that $h_i(m) \in \{h_i(R_j, m_j)\}_{j \in [r]}$ holds for each i.

As a target version of rSR, eTSR has higher bit security. For a random function H : $\{0, 1\}^* \to \{0, 1\}^{k \cdot \log t}$. In the quantum setting, breaking r-rSR and r-eTSR approximately requires $O(t^{k/2r})$ and $O((t/r)^{k/2})$ computations of H.[7]

---

[7] The two quantities are approximated. See the concrete upper/lower bounds in [8, 35].

As mentioned in Sect. 3.3, in SPHINCS+, the index is replaced with a hash value of a nonce and the message. Thus, we can compute them together with a single hash function. That is, let $\mathsf{H} : \{0,1\}^* \to \{0,1\}^{h+k\log t}$ and separate $\mathsf{H}(\cdot) = h_0(\cdot)||h_1(\cdot)||\ldots||h_k(\cdot)$, where h_0 maps to $\{0,1\}^h$ and h_i maps to $\{0,1\}^{\log t}$. The resulting H is required to provide *interleaved target subset resilience* (ITSR). Formally, the adversary can also adaptively send q_s queries to oracle $\mathsf{O}(m_j)$ and obtain the corresponding (pseudo-)randomness R_j. Finally, the adversary is required to output (R^*, m^*) such that for each $i \in [k]$,

$$\big(h_0(R^*, m^*), h_i(R^*, m^*)\big) \in \big\{\big(h_0(R_j, m_j), h_i(R_j, m_j)\big)\big\}_{j \in [q_s]}.$$

We note that here the number of queries to O is q_s, the number of signing queries. Here, the probability of breaking ITSR is related to both q_s and q_h (the number of hash queries) [6, 18, 36]. If H is modeled as a random oracle, the probability of breaking ITSR is bounded by

$$O\left(q_s\sqrt{\frac{q_s + q_h + 1}{2^n}} + (q_s + q_h + 2)^2 X\right),$$

where $X = \sum_{r=1}^{q_s}\left(1 - \left(1 - \frac{1}{t}\right)^r\right)^k \binom{q_s}{r}\left(1 - \frac{1}{2^h}\right)^{q_s-r}\frac{1}{2^{hr}}$. By carefully adapting h, k, and r, the above probability can be bounded by a negligible function.

4.4 Improving One-Wayness: Multi-Targets and DSPR

Finally, we remove the high security loss from the reduction to one-wayness.

It can also be solved by replacing the one-way function with a tweakable hash function. Observe Lamport's scheme as an example. We replace $y_{b,i} = f(x_{b,i})$ with $y_{b_i} = H((b,i), x_{b,i})$, where (b,i) is considered a tweak.

Let us explain why we can get rid of the security loss by introducing the tweak. Similar to the attack in Sect. 4.1, let $S = \{y_{b,i}\}_{b,i}$. Now, an adversary needs to find x such that $H((b,i), x) \in S$. Here is the problem. Although the adversary also needs $\sqrt{2^\lambda/|S|} = \sqrt{2^\lambda/2n}$ queries by Grover's search, in each query, the adversary needs to check whether $H((b,i), x) \in S$ for *some* $(b,i) \in \{0,1\} \times [n]$. Thus, each query requires $2n$ computations of H. Thus, the multi-target quantum search does not have any advantage.

Then, what is the corresponding security notion for the tweakable hash function? Intuitively, we can give a definition of a "multi-target one-wayness with tweaks". The adversary can first adaptively query a tweak i to the oracle O, and is replied with $y_i = H(i, x_i)$ for a uniformly random x_i. Finally, it is required to return (i^*, x'_{i^*}) such that $H(i^*, x'_{i^*}) = y_i$ for some queried i^*.

Unfortunately, this definition is not sufficient for security proof. An EUF-CMA reduction needs to simulate the signing oracle. However, since x_i is not given from

O, it is unable to reply a signing query. A natural solution to modify the definition is as follows. After the queries to O, the adversary sends additional queries i to Open, and is replied with the corresponding x_i. Finally, the adversary is instead required to output (i^*, x'_{i^*}) for some i^* that have been queries to O but never to Open. This definition is called multi-target open preimage resistance (Open-PRE, [5]).

Now, the reduction works. However, Open-PRE experiment is too complicated to have a provable bound due to the existence of Open oracle. We prefer to eliminate the stateful interactive queries in experiments. A solution is to replace PRE with TCR: in the response of $O(i)$, x_i is directly given to the adversary. Finally, the adversary is required to return (i^*, x'_{i^*}) such that $H(i^*, x'_{i^*}) = H(i^*, x_{i^*})$ for some queried i and $x'_{i^*} \neq x_{i^*}$. The reduction is able to simulate the signing oracle using the responses from O.

However, for an adversary of Lamport's scheme, it is possible to return the *original* preimage in his forgery instead of the second preimage. In this case, the reduction fails. If we additionally suppose that each image has at least two preimages with the same tweak, then the reduction succeeds with probability at least $\frac{1}{2}$. However, not all tweakable hash functions meet the requirement, especially for length-preserving hash functions [5]. We can solve this by choosing a compressing hash, but as a price, the signature size will become larger.

A solution to this problem is to introduce a new notion for the hash function called *decisional second-preimage resistance*. Roughly, given a random x, it is hard for any adversary to decide whether there exists a second-preimage of x on the hash function. See details in [5, 7]. Then, the Open-PRE can be tightly reduced to multi-target TCR and multi-target DSPR.

Unlike TCR, it is an open question whether the security bound of DSPR for a random function is independent of the number of targets. It is conjectured [6] that the best strategy of solving DSPR is to find a second preimage, and thus as hard as solving TCR. If it is true, then it leads to a tight bound.

5 Other Challenges

Beyond what we have covered in the previous section, there are some interesting challenges for HBS that we haven't discussed yet. We believe they are worth looking into further for future research.

- **Attacks Other than CMAs** In previous sections, we only consider the security under *chosen message attacks* (CMAs). In practice, there are several other attacks in complex networks, such as side-channel attacks. Fortunately, many classical side-channel attacks (such as time or energy attacks) are not available for HBS. However, recent work shows potential risks against other attacks, such as superposition attacks [36], fault attacks [11, 14], state attacks [34], and so on.

- **Actual Notions for Hash Functions** In Sect. 4, we analyze the security on several notions in the case that the hash functions are modeled as (quantum) random oracles. However, the real-world hash functions do have their own special structures. For example, if the hash function uses SHA-256 with Merkle-Damård structure, then the resulting SPHINCS+ cannot reach level 5 security [26]. Their attack aims to the multi-target TCR. It is also an open problem whether a hash function with concrete structures can provide the required bit security on multi-target notions.
- **Limitations of Hash-Based Cryptography in PKC** In public-key cryptography, we only construct hash-based signature schemes but not public-key encryption schemes (PKE). Indeed, it is provably impossible to construct a secret key agreement protocol (implying PKE) or a blind signature scheme from one-way permutations [21, 22]. Of course, it does not completely rule out the possibility of constructing PKE, since hash functions can also provide other properties such as collision resistance. (Note that collision-resistant functions are not implied from one-way permutations [31].) However, one-way permutations can be used to construct most of the hash-based primitives, such as target-collision-resistant hash functions and pseudorandom functions. It implies that hash-based cryptography is not always friendly to all public-key cryptosystems.

Another branch to solve the problem is MPC-in-the-head cryptography. Unlike the HBS we introduced above, a MPC-in-the-head scheme does not treat the symmetric-key primitives (such as hash function) as black boxes. Instead, it builds zero-knowledge arguments related to the hash function by splitting it into circuits, and then uses them in public-key constructions. For example, Picnic [12] is a MPC-in-the-head based signatures with Fiat-Shamir heuristic.

References

1. J.-P. Aumasson, G. Endignoux, Improving stateless hash-based signatures, in *Topics in Cryptology–CT-RSA 2018*, vol. 10808 of *LNCS*. (Springer, 2018), pp. 219–242
2. B. Barak, M. Mahmoody-Ghidary, Lower bounds on signatures from symmetric primitives, in *48th Annual IEEE Symposium on Foundations of Computer Science (FOCS'07)*. (IEEE, 2007), pp. 680–688
3. M. Bellare, S.K. Miner, A forward-secure digital signature scheme, in *Annual International Cryptology Conference*. (Springer, 1999), pp. 431–448
4. D.J. Bernstein, D. Hopwood, A. Hülsing, T. Lange, R. Niederhagen, L. Papachristodoulou, M.S. Schwabe, Z. Wilcox-O'Hearn, SPHINCS: Practical stateless hash-based signatures, in *Advances in Cryptology–EUROCRYPT 2015*, vol. 9056 of *LNCS*. (Springer, 2015), pp. 368–397
5. D.J. Bernstein, A. Hülsing, Decisional second-preimage resistance: when does SPR imply PRE? In *International Conference on the Theory and Application of Cryptology and Information Security*. (Springer, 2019), pp. 33–62
6. D.J. Bernstein, A. Hülsing, S. Kölbl, R. Niederhagen, J. Rijneveld, P. Schwabe, The SPHINCS+ signature framework, in *Proceedings of the 2019 ACM SIGSAC Conference on Computer and Communications Security*. (Association for Computing Machinery, 2019), pp. 2129–2146

7. X. Bonnetain, A. Hosoyamada, M. Naya-Plasencia, Y. Sasaki, A. Schrottenloher, Quantum attacks without superposition queries: the offline Simon's algorithm, in *International Conference on the Theory and Application of Cryptology and Information Security*. (Springer, 2019), pp. 552–583
8. S. Bouaziz-Ermann, A.B. Grilo, D. Vergnaud, Quantum security of subset cover problems, in *4th Conference on Information-Theoretic Cryptography (ITC 2023)*. Schloss Dagstuhl-Leibniz-Zentrum für Informatik (2023)
9. G. Brassard, P. Høyer, A. Tapp, Quantum cryptanalysis of hash and claw-free functions, in *Latin American Symposium on Theoretical Informatics*. (Springer, 1998), pp. 163–169
10. J. Buchmann, E. Dahmen, A. Hülsing, XMSS–a practical forward secure signature scheme based on minimal security assumptions, in *PQCrypto 2011: Post-Quantum Cryptography*, vol. 7071 of *LNCS*. (Springer, 2011), pp. 117–129
11. L. Castelnovi, A. Martinelli, T. Prest, Grafting trees: a fault attack against the SPHINCS framework, in *International Conference on Post-Quantum Cryptography*. (Springer, 2018), pp. 165–184
12. M. Chase, D. Derler, S. Goldfeder, C. Orlandi, S. Ramacher, C. Rechberger, D. Slamanig, G. Zaverucha, Post-quantum zero-knowledge and signatures from symmetric-key primitives, in *Proceedings of the 2017 ACM SIGSAC conference on computer and communications security* (2017), pp. 1825–1842
13. C. Dods, N. Smart, M. Stam, Hash based digital signature schemes, in *Cryptography and Coding*, vol. 3796 of *LNCS*. (Springer Verlag, 2005), pp. 98–115
14. A. Genêt, On protecting SPHINCS+ against fault attacks. Cryptol. ePrint Archive (2023)
15. O. Goldreich, *Foundations of Cryptography: Basic Applications*, vol. 2 (Campbridge University Press, Cambridge, UK, 2004)
16. L.K. Grover, A fast quantum mechanical algorithm for database search, in *Proceedings of the twenty-eighth annual ACM symposium on Theory of computing* (1996), pp. 212–219
17. J. Håstad, R. Impagliazzo, L.A. Levin, M. Luby, A pseudorandom generator from any one-way function. SIAM J. Comput. **28**(4), 1364–1396 (1999)
18. A. Hülsing, M. Kudinov, Recovering the tight security proof of sphincs+, in *Advances in Cryptology–ASIACRYPT 2022: 28th International Conference on the Theory and Application of Cryptology and Information Security, Taipei, Taiwan, December 5–9, 2022, Proceedings, Part IV*. (Springer, 2023), pp. 3–33
19. A. Hülsing, M. Kudinov, E. Ronen, E. Yogev, SPHINCS+ C: Compressing SPHINCS+ with (almost) no cost, in *2023 IEEE Symposium on Security and Privacy (SP)*. (IEEE, 2023), pp. 1435–1453
20. A. Hülsing, J. Rijneveld, F. Song, Mitigating multi-target attacks in hash-based signatures, in *Public-Key Cyptography–PKC 2016*, volume 9614 of *LNCS*. (Springer, 2016), pp. 387–416
21. R. Impagliazzo, S. Rudich, Limits on the provable consequences of one-way permutations, in *Proceedings of the 21st annual ACM symposium on Theory of computing* (1989), pp. 44–61
22. J. Katz, D. Schröder, A. Yerukhimovich, Impossibility of blind signatures from one-way permutations, in *Theory of Cryptography: 8th Theory of Cryptography Conference, TCC 2011, Providence, RI, USA, March 28-30, 2011. Proceedings 8*. (Springer, 2011), pp. 615–629
23. L. Lamport, Constructing digital signatures from a one way function. Technical report, Technical Report SRI-CSL-98, SRI International Computer Science Laboratory (1979)
24. D. Lubell, A short proof of sperner's lemma. *Classic Papers in Combinatorics* (1987), pp. 402–402
25. R.C. Merkle, A certified digital signature, in *Advances in Cryptology–CRYPTO '89*, ed. by G. Brassard, vol. 435 of *LNCS*. (Springer, 1989), pp. 218–238
26. R. Perlner, J. Kelsey, D. Cooper, Breaking category five SPHINCS+ with sha-256, in *International Conference on Post-Quantum Cryptography*. (Springer, 2022), pp. 501–522
27. J. Pieprzyk, H. Wang, C. Xing, Multiple-time signature schemes against adaptive chosen message attacks, in *SAC 2003: Selected Areas in Cryptography*, vol. 3006 of *LNCS*. (Springer, 2003), pp. 88–100

28. L. Reyzin, N. Reyzin, Better than biba: Short one-time signatures with fast signing and verifying, in *ACISP 2002: Information Security and Privacy*, volume 2384 of *LNCS*. (Springer, 2002), pp. 144–153
29. J. Rompel, One-way functions are necessary and sufficient for secure signatures, in *Proceedings of the twenty-second annual ACM symposium on Theory of computing* (1990), pp. 387–394
30. P.W. Shor, Algorithms for quantum computation: discrete logarithms and factoring, in *Proceedings 35th annual symposium on foundations of computer science*. (IEEE, 1994), pp. 124–134
31. D.R. Simon, Finding collisions on a one-way street: Can secure hash functions be based on general assumptions? in *Advances in Cryptology–EUROCRYPT'98*, vol. 1403 of *LNCS*. (Springer, 1998), pp. 334–345
32. D.R. Stinson, R. Wei, L. Zhu, Some new bounds for cover-free families. J. Comb. Theory Ser. A **90**(1), 224–234 (2000)
33. K. Xagawa, T. Yamakawa, (Tightly) QCCA-secure key-encapsulation mechanism in the quantum random oracle model, in *Post-Quantum Cryptography: 10th International Conference, PQCrypto 2019, Chongqing, China, May 8–10, 2019 Revised Selected Papers 10*. (Springer, 2019), pp. 249–268
34. Q. Yuan, M. Tibouchi, M. Abe, Security notions for stateful signature schemes, in *IET Information Security* (IET, 2021)
35. Q. Yuan, M. Tibouchi, M. Abe, On subset-resilient hash function families. Designs Codes Cryptogr. 1–40 (2022)
36. Q. Yuan, M. Tibouchi, M. Abe, Quantum-access security of hash-based signature schemes, in *Australasian Conference on Information Security and Privacy*. (Springer, 2023), pp. 343–380
37. K. Zhang, H. Cui, Y. Yu, Revisiting the constant-sum winternitz one-time signature with applications to SPHINCS+ and XMSS, in *Annual International Cryptology Conference*. (Springer, 2023), pp. 455-483

Open Access This chapter is licensed under the terms of the Creative Commons Attribution 4.0 International License (http://creativecommons.org/licenses/by/4.0/), which permits use, sharing, adaptation, distribution and reproduction in any medium or format, as long as you give appropriate credit to the original author(s) and the source, provide a link to the Creative Commons license and indicate if changes were made.

The images or other third party material in this chapter are included in the chapter's Creative Commons license, unless indicated otherwise in a credit line to the material. If material is not included in the chapter's Creative Commons license and your intended use is not permitted by statutory regulation or exceeds the permitted use, you will need to obtain permission directly from the copyright holder.

A Formal Approach for Secured Risk Assessment

Kengo Zenitani

Abstract Information security management is a managerial activity that keeps the information security risk of an enterprise at some acceptable level. The process requires risk assessment iterations and non-trivial resources. At the same time, business executives compete for finite resources to invest in general business activities. The executives are motivated to intentionally reduce the resource assignment for risk assessment to broaden the estimated risk variance and then choose the lower limit to underestimate the actual risks. This study proposes a model-based risk assessment approach; it guarantees that an additional, thus expected to be costly, fine-grained risk assessment always gives a refined subset of the possible incidents compared to that of the relatively coarse-grained, therefore cheaper, risk assessment. In other words, estimated incidents roughly decrease with the growth of assessment cost. We expect this property to secure the risk management process from conflicts between the business executives and the security practitioners. Our approach introduces the concept named aggregation and segregation of facts appearing in attack graph analysis, a prominent approach in model-based network security risk analysis. This study contributes to expanding the attack graph analysis to enable secured risk assessment.

Keywords Information security management · Information security risk assessment · Attack graph analysis · Secured risk assessment

1 Introduction

Information security management is a managerial activity that controls the information security risk level of an enterprise. It is a kind of risk management and requires the repetition of risk assessment to identify weak parts of the enterprise to apply security controls and evaluate the degree of relevant risks. Generally, security practitioners perform the assessment, and business executives consider the security

K. Zenitani (✉)
Independent Researcher, Kawasaki, Japan
e-mail: k_zenitani@phd.jpn.org

investments in need. However, while security practitioners try to lower the security risks, business executives want to maximize profit—both share finite resources. Business executives have the potential motivation to underestimate the security risks as long as doing so enables their business investments by excluding several security investments.

The tendency above is a real phenomenon seen in actual cybersecurity incidents. The Office of Personnel Management (OPM), a US federal government agency, was attacked in 2014, and over twenty million personnel records of government officials, including detailed background investigation reports of counter-terrorism officers, were breached [5]. It is pointed out by the investigation committee that the OPM inappropriately prioritized its daily operation over information security. For example, they had failed to introduce secure authentication systems due to strong resistance from business staff reluctant to change their daily operations. Another similar case is the data breach in the Japan Pension Service in which 1.25 million citizen records were stolen [8, 11, 23]. The postmortem report criticizes the executives for loosening the security rules deliberately to keep their daily business not disturbed by them. Though it is not a cybersecurity case, we can see an analogous catastrophic case in the Fukushima nuclear reactors' failure. The investigating commissioners [7] "believe that the tsunami risk was underestimated .... Meanwhile, with regard to severe accident countermeasures, the probabilistic safety assessment of tsunamis was deemed uncertain, so the consideration itself was pushed back and measures were not taken. As such, Tokyo Electric Power Co. used probabilistic evaluations in ways that suited them in order to avoid clarifying the tsunami risk."

The Fukushima case indicates a critical point in risk management. Every risk assessment holds uncertainty up to some degree due to the nature of risk. Roughly speaking, the result of a less costly, coarse-grained risk assessment accompanies larger statistical variation; thus, the lower limit of the estimated risk is also lower than that of the costly, fine-grained risk assessment. The authority determining the level of security investment can choose the level of assessed risk at his or her discretion, for instance, by intentionally varying the level of risk assessment investments. In both the cases of the OPM and the Japan Pension Service, the business executives were under strong political pressure to perform their business as swiftly as possible. It is natural for such people to tweak the risk assessment process to secure their daily operations even when it degrades information security.

The problem here is the vulnerability of risk assessment to the tendency mentioned above. The preferable risk assessment method should be robust against intentionally constrained resource assignments. More precisely, such a method should estimate greater risks for coarse-grained assessment. At the same time, the method has to clarify the impact of security incidents on business to convince the authority to invest in security. In short, we need a robust, business-oriented security risk assessment method. Unfortunately, the existing risk assessment approaches do not satisfy these requirements.

This theoretical study proposes an information security risk assessment approach based on mathematically proven robustness. At the core, our model-based risk analysis method guarantees that an additional, thus expected to be costly, fine-grained risk analysis always gives a refined subset of the possible incidents compared to the relatively coarse-grained analysis. That is, estimated incidents roughly decrease with the growth of assessment cost. Moreover, our approach has powerful expressiveness in describing the dependency of business processes on IT infrastructure components. These characteristics contribute to the robust, business-oriented security risk assessment we want to realize.

Section 2 refers to prior studies and points out the lack of the robustness in the existing approaches. Section 3 defines our approach after introducing several prior studies as the foundation of our approach. The central tool in our approach is the set-theoretical segregation and aggregation of facts used for the risk analysis. Section 4 explains our approach with two imaginary examples. In the examples, we model the situation coarsely at first. Then, we segregate facts and filter several of the segregated facts to refine the analysis. The latter fine-grained analysis is guaranteed to give a refined subset of the possible incidents. The first example elaborates on the proposed concept in Sect. 3, and the second clarifies our approach's characteristics and issues. We conclude after discussing several relevant topics, including the obstacles in implementing our approach.

This article is the extended version of the prior report by the author [25]. We supplement mathematical proofs and a more realistic imaginary case of secured risk analysis. Especially the latter detailed case gives us further research topics that should be handled before the practical use of our approach.

2 Related Work

Many information security management practices in the field follow global standards such as ISO/IEC 27000 [9] and NIST SP800 [12] series. The standards define the management cycle and the governance structure in information security management. Though each standard has some unique characteristics, they are all risk management at their heart. Nevertheless, these standards do not provide specific approaches for risk assessment. Risk assessment practices in the field are performed with the personal expertise of analysts and have no formal, objective foundation.

On the other hand, various risk analysis approaches are proposed in the research field. We can divide them into two groups: quantitative and qualitative.

The representative quantitative approach is the use of numerical security metrics. One can find a variety of them in thorough surveys such as [10, 16, 17]. Many of these metrics are defined with a specific algorithm to calculate them. In theory, non-expert analysts can identify the level of the security risk by working out the metrics. Also, quantitative security metrics are fully ordered and thus easy to compare. Unfortunately, quantitative evaluation of risks requires reliable input data, which is usually unavailable. Verendel [22] has examined a number of said quantitative metrics and has pointed out their immaturity.

On the contrary, qualitative approaches are relatively superior regarding the reliability and objectiveness of assessment results. Qualitative approaches algorithmically analyze some formal models depicting the IT infrastructure and guess logically possible attacks and their characteristics qualitatively. Attack graph analysis [26] is a prominent example. It takes as input the network topology, system configuration, and the distribution of known vulnerabilities over the network. All these inputs are described in a formal language such as Datalog [3], and an algorithm generates a mathematical graph called an attack graph; a kind of AND/OR graph, or a merged set of fault trees. We can extract the logically possible incidents mechanically from the graph. It is well suited for use by non-experts and for enumerating the possible incidents. However, since the resultant attack graph is qualitative information, it is difficult to estimate the degree of risks or the probability of its implied incidents.

Though the approach still requires further development, this study focuses on extending attack graph analysis for the following two reasons.

First, several researchers have proposed quantitative metrics deriving from attack graphs. Wang et al. [24] propose a set of intuitive calculation rules of probabilities considering the AND/OR combination of nodes within an essentially acyclic attack graph. Sawilla and Ou [18] define an algorithm applicable to cyclic attack graphs to calculate a pseudo-probabilistic metric. Also, Zenitani [27] gives an algorithm for computing probabilistic metrics over cyclic attack graphs. The work by Noel and Jajodia [13] includes a suite of non-probabilistic attack-graph-based metrics. These additional metrics help us leverage attack graphs as the basis of quantitative risk assessment, putting aside the problem of statistical data availability.

Second, there is also assistance for business risk assessment based on attack graphs. Bacic et al. [2] expand the default rules given by Ou [14] to model the multiple aspects of risks, such as confidentiality, integrity, and availability. Froh and Henderson [6] designed a Datalog library to accommodate the dependency between services and their infrastructure. Sun et al. [19] propose a Datalog-based framework depicting the effect of infrastructure infringement on enterprise missions. These prior efforts indicate the powerful extensibility of attack graph analysis to support business risk assessment.

Currently, attack graph analysis lacks the robustness we need in this study. The intrinsic limitation of attack graph analysis is that it requires the distribution of vulnerabilities to be known beforehand. Zero-day exploits easily break this assumption. In practice, the lack of accurate information on system configuration also inhibits the analysis. Due to the nature of attack graph analysis, lack of information results in the smaller size of the generated attack graph, i.e., the underestimation of incidents. Intentional constraints on analysis resources and information quickly deteriorate the analysis result by losing sight of the possible incidents.

This study proposes a robust approach based on the monotonicity of Datalog-based inference. Our approach guarantees that the limited input always produces a greater estimation of incidents. We expect our contribution to be a building block that makes attack graph analysis practical and helps realize the secured risk assessment concept.

3 Approach

We introduce attack graph analysis and Datalog as our starting point. Then, we extend it with two new concepts; segregation and aggregation of facts. Based on these settings, a robust risk assessment process model is proposed.

3.1 Preliminary Settings

At first, let us define the *attack graphs*. We follow the definition in [1].

Definition 1 (Attack graph) Given a set of *exploits* E, a set of *security conditions* C where $E \cap C = \emptyset$, a *require* relation $R_r \subseteq C \times E$, and an *imply* relation $R_i \subseteq E \times C$, an *attack graph* G is the directed graph $G = (E \cup C, R_r \cup R_i)$, where $E \cup C$ is the vertex set and $R_r \cup R_i$ is the edge set. For each attack graph $G = (E \cup C, R_r \cup R_i)$, *initial conditions* refer to the subset $C_i \subseteq C$ composed of every condition $c \in C$ satisfying $\nexists e \in E$ such that $(e, c) \in R_i$, i.e., the set of all the leaves in the graph, whereas *intermediate conditions* refer to the subset $C \setminus C_i$.

An attack graph is a bipartite, directed graph and is also an AND/OR graph. The initial conditions are the leaves indicating the facts assumed unconditionally. Usually, we use them to represent the hosts' configuration and vulnerabilities. Each exploit is the node depicting the AND combination of facts; all the facts corresponding to its parent condition nodes are required for the exploit to happen. Each intermediate condition is the node corresponding to the OR nodes; at least one of its parent exploits is required to realize the corresponding intermediate fact. We can regard an attack graph as a merged set of attack paths. Each attack path is a subgraph where every OR node in it has only one parent exploit node.

Several algorithms generate attack graphs from a given input, such as the network topology, system configuration, vulnerability distribution, and exploit inference rules called *attack templates*. Many researchers use Datalog to describe those inputs. Datalog is a simplified subset of Prolog. It can handle the inference based on propositional logic without negation. An attack graph is the full trace—or the complete set of deducible facts—of a given Datalog program [15]. The following code is a minimal example of a Datalog program that can generate an attack graph shown in Fig. 1.

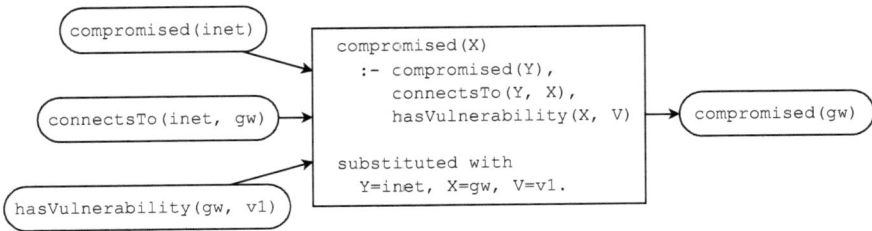

Fig. 1 A minimal attack graph

```
compromised(inet).
connectsTo(inet, gw).
connectsTo(inet, www).
hasVulnerability(gw, v1).

compromised(X)
  :- compromised(Y), connectsTo(Y,X),
     hasVulnerability(X,V).
```

A Datalog program is a set of *facts* and *rules*. Both facts and rules are composed of *terms*. A term is a *predicate* symbol followed by a list of *arguments*. An argument is either a *constant* or a *variable*. We denote constants by lower-cased names and variables by capitalized names. For example, "connectsTo(gw, H)" is a term with the predicate "connectsTo" with a constant "gw" and a variable "H". This term represents a situation in which a host named "gw" connects to some host referenced by "H". A *fact* is a term without variables. A *rule* is a pair of *head* and *body*; the last line in the code above is an example. Each rule must satisfy a constraint that every variable in the head must also appear in the body. The exemplified rule implies that host X can be compromised if there exists a compromised host Y connecting to X and X has some vulnerability. An appropriate assignment of values to the variables in a rule generates an *instance* that is a realized rule where the head and body are replaced with corresponding facts as follows:

```
compromised(gw)
  :- compromised(inet), connectsTo(inet,gw),
     hasVulnerability(gw,v1).
```

The realized head of an instance gives a new fact. Thus it deduces other value assignments and the realization of another fact. In this way, the execution of a Datalog program generates instances and facts until no valid assignment generates new facts.

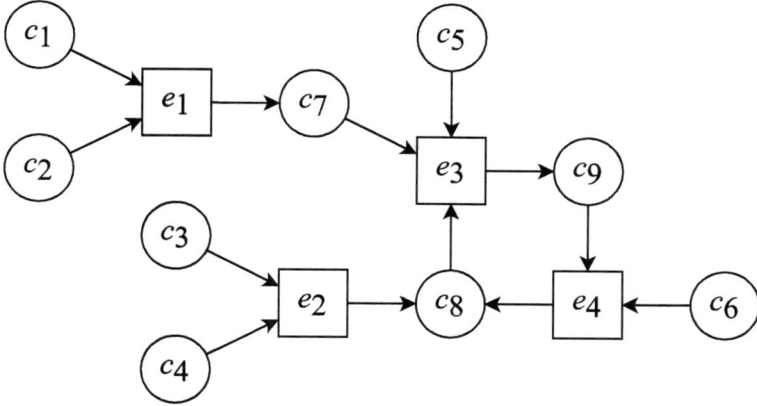

Fig. 2 Another small example of attack graph

Then, mapping instances to exploit nodes and facts to condition nodes identifies one unique attack graph. Figure 1 is the result of this process and shows a minimal attack graph. Rounded nodes are the conditions, and the square node is the exploit. By combining these minimal attack graphs, we can obtain attack graphs like shown in Fig. 2.

As Sect. 2 discusses, we can use attack graph analysis as a qualitative risk assessment tool in conjunction with additional quantitative metrics. Moreover, some prior studies leverage the powerful expressiveness of Datalog to handle the risk-relevant situation beyond the network incidents, including businesses' dependency on the infrastructure. Attack graph analysis provides a valuable foundation for business-oriented risk assessment.

3.2 Monotonicity, Aggregation and Segregation

The execution of a Datalog program takes a set of facts as input and then generates the deduced set of facts. Let P be a Datalog program without facts, S_0 be the input facts for P, and $P(S_0)$ be the set of all the deduced facts by executing P on S_0. Then, the relationship $S_0 \subseteq S'_0 \Rightarrow P(S_0) \subseteq P(S'_0)$ always holds in Datalog. Considering that the input facts for attack graph generation indicate the set of vulnerabilities, this *monotonicity* implies a natural tendency that the increased vulnerabilities trigger increased exploits.

The monotonicity has further implications. First, it indicates the lack of robustness in attack graph analysis, i.e., the less-sized input facts S_0 gives a smaller output $P(S_0)$ than $P(S'_0)$. It means the restriction on fact investigation by the analyst deteriorates the estimation of incidents. Second, it also implies that we can reduce the incident estimation by removing several input facts inconsistent with the actual situation.

For example, if we confirm the non-existence of concerned vulnerabilities, we can remove the input facts corresponding to them and legitimately obtain a reduced estimation of incidents.

The key idea behind our approach is to introduce coarse-grained opaque facts instead of removing input facts when we have insufficient information. Suppose the analyst is prohibited from accessing several client machines and identifying the list of installed applications. In that case, the analyst does not know the accurate reality and might remove several input facts corresponding to possible vulnerabilities. It brings a lower estimation of incidents. Instead, in such cases, we propose introducing a hypothetical fact expressing the set of every potential application and its vulnerabilities. Although this coarse-grained fact lacks detailed information, we can design it to allow the estimation of every might-be-possible incident. Introducing such opaque facts countervails the attack graph shrinkage coming from information shortage. If the assessment refinement is required, the analyst can ask for additional resources or authorization to clarify the details of the coarse-grained opaque facts and then implement them as fine-grained input facts.

Based on the intuition above, let us define the following one operator and the two operations applicable to Datalog programs:

Definition 2 (Constant replacement operator). Let C be the set of constants used in Datalog programs, and $\gamma : C \to C$ be a function satisfying $\gamma(c) = \gamma \circ \gamma(c)$ for any $c \in C$. Then, we call γ a *constant replacement operator*.

Definition 3 (Aggregation of facts). Let C be the set of constants used in Datalog programs, and $\gamma : C \to C$ be a constant replacement operator. The *aggregation of facts* is the operation of replacing every constant $c \in C$ appearing in a given Datalog program with $\gamma(c)$.

Replacing constants sometimes maps two different facts to the same single fact. This brings the aggregation of facts. Note that the constant replacement also applies to the constants in rules in the given program. Additionally, we define the following reversed operation of the aggregation of facts:

Definition 4 (Segregation of facts). Let $\gamma : C \to C$ be a constant replacement operator, $\gamma^{-1} : \text{Im } \gamma \to 2^C$ be the inverse mapping of γ such that $\gamma^{-1}(c) = \{c' \mid c' \in C, \gamma(c') = c\}$, P be a Datalog program as the set of facts and rules where every constant $c \in P$ satisfies $c \in \text{Im } \gamma$, and $\{c_1, c_2, ..., c_n\}$ be the multiset of all constants in P satisfying $\{c_k\} \subsetneq \gamma^{-1}(c_k)$ where each k corresponds to the order of appearance in the program P. Also, $P[c \to c']$ denotes the program obtained by replacing only one constant c appearing in P with c'. With these settings, we define the following recurring formula:

$$P_0 = P,$$
$$P_{k+1} = \bigcup_{c' \in \gamma^{-1}(c_{k+1}) \setminus \{c_{k+1}\}} P_k [c_{k+1} \to c'].$$

A Formal Approach for Secured Risk Assessment

We call the operation to obtain P_n from the pair of P and γ as the *segregation of facts*.

Here, let us see a tiny example to understand the behavior of aggregation and segregation. Suppose we have the following three facts in Datalog:

```
connectsTo(inet, gw).
connectsTo(gw, httpd).
connectsTo(gw, smtpd).
```

Additionally, let γ be a constant replacement operator such that

$$\gamma(\text{gw}) = \gamma(\text{httpd}) = \gamma(\text{smtpd}) = \gamma(\text{dmz}) = \text{dmz}.$$

This replacement operator implies that the three machines, gw, httpd, and smtpd, belong to the same subnet named dmz. The result of the aggregation of the facts above by γ is as follows:

```
connectsTo(inet, dmz).           connectsTo(dmz, dmz).
```

The second and third facts become the same fact by the aggregation and are merged into one. On the contrary, the segregation of the two aggregated facts shown above is as follows:

```
connectsTo(inet, gw).            connectsTo(gw, gw).
connectsTo(inet, httpd).         connectsTo(gw, httpd).
connectsTo(inet, smtpd).         connectsTo(gw, smtpd).
connectsTo(httpd, gw).           connectsTo(smtpd, gw).
connectsTo(httpd, httpd).        connectsTo(smtpd, httpd).
connectsTo(httpd, smtpd).        connectsTo(smtpd, smtpd).
```

In essence, dmz is expanded to three constants, gw, httpd, and smtpd. Accordingly, the fact "connectsTo(dmz, dmz)" is expanded to nine facts based on the product {gw, httpd, smtpd} × {gw, httpd, smtpd}. Note that this segregated form contains the three facts before the aggregation.

If we regard the first three facts as the true nature of the network to be analyzed, the aggregated form corresponds to the coarse grained, opaque representation since it lacks the detailed structure within the subnet. In contrast, the segregated form is the most conservative assumption about the subnet when we decompose it into more fine-grained components. This way, aggregation and segregation of facts can be tied to the concept of analysis granularity.

The following claim holds about aggregation and segregation. The proof is given in Sect. 3.3.

Theorem 5 *Let P be a Datalog program, S_0 be a set of facts, and γ be a constant replacement operator satisfying that $\gamma [S_0] = S_0$ and $\gamma [P] = P$. Here, $\gamma [P]$ denotes the aggregated form of P, and $\gamma^{-1} [P]$ denotes the segregated form of P. That is, we assume S_0 and P equal the aggregated forms of themselves. Then, the following equations hold:*

$$\gamma \left[\gamma^{-1} [P]\right] (S_0) = \gamma \left[\gamma^{-1} [P] \left(\gamma^{-1} [S_0]\right)\right] = P(S_0).$$

This theorem means that the aggregation after segregation preserves the set of deducible facts semantically. This property and the monotonicity of Datalog imply that if we segregate a given initial program and remove facts unmatched with reality, the deducible facts are guaranteed to be the subset of that of the initial program. In other words, combining segregation and facts removal could decrease the estimated risk events. It corresponds to the refinement of the given coarse-grained assessment expressed in P and S_0. Carefully designed constant replacement operator γ gives an interpretation that regards the facts contained in P and S_0 as the set of every possible application and its vulnerabilities, as mentioned previously. We use this guaranteed decrease of deducible incidents to secure the risk assessment.

3.3 Proof of the Main Theorem

This subsection shows the proof for Theorem 5. It is twofold. We start our proof with the elaboration of aggregation and segregation properties. Then, we discuss the Datalog program segregation and its invariance.

3.3.1 Aggregation and Segregation Functions

Definition 6 (Aggregation function and segregation function). Let X be an arbitrary finite set. When a function $\alpha : X \to X$ satisfies that $\forall x \in X, \alpha \circ \alpha (x) = \alpha (x)$, we call α an *aggregation function*. For any aggregation function α, X_α denotes the image $\{\alpha (x) \mid x \in X\}$ of α, and we call the inverse mapping $\alpha^{-1} : X_\alpha \to 2^X$ defined as follows a *segregation function*:

$$\alpha^{-1} (x) = \{w \mid w \in X, \alpha (w) = x\}.$$

We also define the application of them to an arbitrary set $W \subseteq X$ as follows:

$$\alpha (W) = \{\alpha (w) \mid w \in W\},$$
$$\alpha^{-1} (W) = \{x \mid x \in X, \alpha (x) \in W\}.$$

Definition 7 (Narrowed segregation function). Let $\alpha : X \to X$ be an aggregation function, and $\alpha^{-1} : X_\alpha \to 2^X$ be the coupled segregation function. When a function $\beta^{-1} : X_\alpha \to 2^X$ satisfies $\emptyset \neq \beta^{-1}(x) \subseteq \alpha^{-1}(x)$ for any $x \in X_\alpha$, we call β^{-1} a *narrowed segregation function* coupled with α.

Proposition 8 *Let $\alpha : X \to X$ be an aggregation function, and $\alpha^{-1} : X_\alpha \to 2^X$ be the coupled segregation function. Then, the following property holds:*

$$\forall W \subseteq X_\alpha, \ \alpha\left(\alpha^{-1}(W)\right) = W.$$

$\square$

Proof By definition,

$$\alpha^{-1}(\{w\}) = \{x \mid x \in X, \alpha(x) \in \{w\}\} = \{x \mid x \in X, \alpha(x) = w\}$$

for any $w \in X_\alpha$. This set, $\alpha^{-1}(\{w\})$, can never be empty since $w \in X_\alpha$. Thus,

$$\alpha\left(\alpha^{-1}(\{w\})\right) = \{\alpha(x) \mid x \in X, \alpha(x) = w\} = \{w\}.$$

Accordingly, for any $W \subseteq X_\alpha$,

$$\begin{aligned}\alpha\left(\alpha^{-1}(W)\right) &= \{\alpha(x) \mid x \in X, \alpha(x) \in W\} \\ &= \bigcup_{w \in W} \{\alpha(x) \mid x \in X, \alpha(x) = w\} \\ &= \bigcup_{w \in W} \{w\} = W.\end{aligned}$$

$\square$

Corollary 9 *Let $\alpha : X \to X$ be an aggregation function, and $\beta^{-1} : X_\alpha \to 2^X$ be a coupled narrowed segregation function. Then, the following holds:*

$$\forall W \subseteq X_\alpha, \ \alpha\left(\beta^{-1}(W)\right) = W.$$

Lemma 10 *Let α be an aggregation function, and β^{-1} be a narrowed segregation function coupled with α. Then, for any two sets $S, S' \subseteq X_\alpha$, the following claim holds:*

$$S \subseteq S' \Leftrightarrow \beta^{-1}(S) \subseteq \beta^{-1}(S').$$

Proof $\Rightarrow$) Due to the definition of β^{-1}, $\beta^{-1}(S_1 \cup S_2) = \beta^{-1}(S_1) \cup \beta^{-1}(S_2)$ holds for any two sets $S_1, S_2 \subseteq X_\alpha$. Also, $S' = (S' \setminus S) \cup S$ since $S \subseteq S'$. Therefore,

$$\begin{aligned}\beta^{-1}(S') &= \beta^{-1}\left((S' \setminus S) \cup S\right) \\ &= \beta^{-1}(S' \setminus S) \cup \beta^{-1}(S) \supseteq \beta^{-1}(S).\end{aligned}$$

⇐) Suppose $S \nsubseteq S'$, i.e., $S \setminus S'$ is not empty. On the other hand, $S = (S \setminus S') \cup (S \cap S')$, and $S' = (S' \setminus S) \cup (S \cap S')$. Thus,

$$\beta^{-1}(S) = \beta^{-1}((S \setminus S') \cup (S \cap S'))$$
$$= \beta^{-1}(S \setminus S') \cup \beta^{-1}(S \cap S'),$$
$$\beta^{-1}(S') = \beta^{-1}((S' \setminus S) \cup (S \cap S'))$$
$$= \beta^{-1}(S' \setminus S) \cup \beta^{-1}(S \cap S').$$

Be cautious that $\beta^{-1}(S \setminus S') \cap \beta^{-1}(S \cap S') = \beta^{-1}(S' \setminus S) \cap \beta^{-1}(S \cap S') = \emptyset$ since β^{-1} is defined as a narrowed inverse mapping of α. Therefore, under the assumptions that $\beta^{-1}(S) \subseteq \beta^{-1}(S')$ and $S \setminus S' \neq \emptyset$, it implies that $\emptyset \neq \beta^{-1}(S \setminus S') \subseteq \beta^{-1}(S' \setminus S)$. Then, there must be at least two elements $s \in S \setminus S'$ and $s' \in S' \setminus S$ such that $\beta^{-1}(s) \cap \beta^{-1}(s') \neq \emptyset$. Note that $s \neq s'$ since they are taken from $S \setminus S'$ and $S' \setminus S$, respectively. However, due to the nature of β^{-1} derived from α, it cannot be possible, i.e., s and s' must be equivalent whenever $\beta^{-1}(s) \cap \beta^{-1}(s') \neq \emptyset$. Therefore, $S \setminus S'$ is required to be empty, and it means $S \subseteq S'$. □

3.3.2 Datalog Program Segregations

Let P be a Datalog program, S be a set of facts, and γ be a constant replacement operator. In the following discussion, $\gamma[P]$ denotes the aggregated form of P by γ, and $\gamma^{-1}[P]$ denotes the segregated form of P by γ.

Lemma 11 *Let "body" be the body part of a rule in a Datalog program P, $S \subseteq X$ be a set of facts given to the program, and $\gamma : X \to X$ be a constant replacement operator satisfying that $\gamma[S] = S$ and $\gamma[P] = P$. The set of legitimate substitutions of constants for variables in "body" on S is denoted by $\Sigma(S, body)$. For any $\sigma \in \Sigma(S, body)$, we write the set of facts determined by the substitution as $\sigma(body)$, i.e., $\sigma(body) \subseteq S$. With these settings, the following two claims hold:*

1. $\forall \sigma \in \Sigma(S, body), \forall body' \in \gamma^{-1}[body],$
   $\exists \sigma' \in \Sigma(\gamma^{-1}[S], body')$ s.t. $\sigma'(body') \subseteq \gamma^{-1}[\sigma(body)]$.
2. $\forall body' \in \gamma^{-1}[body], \forall \sigma' \in \Sigma(\gamma^{-1}[S], body'),$
   $\exists \sigma \in \Sigma(S, body)$ s.t. $\sigma'(body') \subseteq \gamma^{-1}[\sigma(body)]$.

Proof In the following discussion, we use the property that γ and γ^{-1} applied to a set of facts work as an aggregation and a narrowed segregation function, respectively. For example, $\gamma[\gamma^{-1}[S']] = S'$ holds for any fact sets $S' \subseteq S$ because Corollary 9 applies.

1) Since σ defines a substitution of constants for variables in *body*, we can equate σ as a set of pairs of variables and constants. For example, if we have $\sigma = \{(X, c), (Y, d)\}$, let σ denotes the operation that substitutes c for X and

d for Y where c and d are constants, and X and Y are the variables in *body*. We write $\sigma = \{(X_1, c_1), \ldots, (X_n, c_n)\}$ where $X_1, \ldots, X_n$ are all the variables appearing in *body*, and $c_1, \ldots, c_n$ are the constants that will be assigned to the paired variables.

By definition, segregation of facts does not affect the variables in rules. It replaces constants in a rule while duplicating the rule to enumerate all the possible combinations of the replaced constants. Therefore, the set of variables in $body' \in \gamma^{-1}[body]$ is the same as that in *body*, i.e., σ and σ' have the same set of variables. Based on this observation, we define σ' as $\{(X_1, d_1), \ldots, (X_n, d_n)\}$ where d_k for $1 \leq k \leq n$ is a constant arbitrarily taken from $\gamma^{-1}(c_k)$. Note that $\gamma^{-1}(c_k)$ is always non-empty due to that $\gamma[S] = S$ and c_k is taken from the constants in S.

Then, $\sigma'(body')$ is a set of facts where each fact is determined by replacing constants in *body* by γ^{-1} or assigning constants to variables in *body* according to σ' where the constants taken from $\gamma^{-1}(c_k)$ for $1 \leq k \leq n$. In other words, $\sigma'(body')$ is a segregated form of $\sigma(body)$, i.e., $\sigma'(body') \subseteq \gamma^{-1}[\sigma(body)]$. At the same time, $\sigma(body) \subseteq S$ holds due to the nature of σ. Therefore, Lemma 10 guarantees $\gamma^{-1}[\sigma(body)] \subseteq \gamma^{-1}[S]$, i.e., $\sigma' \in \Sigma(\gamma^{-1}[S], body')$.

2) We write $\sigma' = \{(X_1, d_1), \ldots, (X_n, d_n)\}$ where $X_1, \ldots, X_n$ are all the variables appearing in *body*, and $d_1, \ldots, d_n$ are the constants that will be assigned to the paired variables. Since any d_k is taken from the constants in $\gamma^{-1}[S]$, there exists c_k contained in S such that $d_k \in \gamma^{-1}(c_k)$. For these $c_1, \ldots, c_n$, we define $\sigma = \{(X_1, c_1), \ldots, (X_n, c_n)\}$. Considering that $body' \in \gamma^{-1}[body]$, and every constant d in $body'$ satisfies $d \in \gamma^{-1}(c)$ where c is the corresponding constant in *body*, the definition of σ implies that $\sigma'(body') \subseteq \gamma^{-1}[\sigma(body)]$. On the other hand, for every $body'$, $\sigma'(body') \subseteq \gamma^{-1}[S]$ is guaranteed by the definition of σ'. By applying γ to both sides, we obtain $\gamma[\sigma'(body')] \subseteq \gamma[\gamma^{-1}[S]] = S$. Additionally, $\gamma[\sigma'(body')] = \sigma(body)$ holds since $\gamma[\sigma'] = \sigma$ and $\gamma[body'] = body$. Therefore, the substitution is legitimate, i.e., $\sigma \in \Sigma(S, body)$.

□

Lemma 12 *Let P be a Datalog program, S be a set of facts, and γ be a constant replacement operator satisfying that $\gamma[S] = S$ and $\gamma[P] = P$. Also, let $P * S$ denote the set of facts that is the expanded S by adding the facts determined by the non-transitive unification of P with S. Note that this notation implies that $S \subseteq P * S$. Then, the following equations hold:*

1. $\gamma^{-1}[P * S] = \gamma^{-1}[P] * \gamma^{-1}[S]$.   2. $P * S = \gamma[\gamma^{-1}[P] * \gamma^{-1}[S]]$.

Proof The first equation almost immediately follows from Lemma 11. Because the lemma implies that the body parts determined by the left-hand side and that of the right-hand side are equal. Therefore, their corresponding heads also match with each other. Also, we can obtain the second equation by applying γ to both sides of the first equation.

□

Based on the preparations above, Theorem 5 concludes as follows:

Proof (Proof for Theorem 5) At first, γ works as an aggregation function when applied to the set of facts since it maps a fact to another fact while satisfying $\gamma = \gamma \circ \gamma$. Similarly, $\gamma^{-1}[S]$ works as a (narrowed) segregation function when S is a set of facts since every modification by γ^{-1} is reversible by γ.

It is easy to see that $\gamma[\gamma^{-1}[P]] = P$. We focus on the equation, $\gamma[\gamma^{-1}[P](\gamma^{-1}[S_0])] = P(S_0)$. Let us think about the process to obtain the evaluation result of $P(S_0)$. Then, it implies that there exists some $K \in \mathbb{N}$ such that $S_K = P(S_0)$ where $S_{k+1} = P * S_k$ for $k = 0, \ldots, K-1$. Note that the binary operation "$*$" is right-associative, i.e., $P * P * S = P * (P * S)$. With this binary operation and Lemma 12, we can use the following properties:

(P1) $\gamma[\gamma^{-1}[P] * \gamma^{-1}[S]] = P * S$.
(P2) $\gamma^{-1}[\gamma[\gamma^{-1}[P] * \gamma^{-1}[S]]] = \gamma^{-1}[P * S]$.
(P3) $\gamma^{-1}[P] * \gamma^{-1}[S] = \gamma^{-1}[P * S]$.

Accordingly,

$$\begin{aligned}
P * P * S &\\
&= P * \gamma[\gamma^{-1}[P] * \gamma^{-1}[S]] & (\because \text{P1}) \\
&= \gamma[\gamma^{-1}[P] * \gamma^{-1}[\gamma[\gamma^{-1}[P] * \gamma^{-1}[S]]]] & (\because \text{P1}) \\
&= \gamma[\gamma^{-1}[P] * \gamma^{-1}[P * S]] & (\because \text{P2}) \\
&= \gamma[\gamma^{-1}[P] * \gamma^{-1}[P] * \gamma^{-1}[S]] & (\because \text{P3}).
\end{aligned}$$

This implies that

$$\begin{aligned}
P(S_0) &= P * \cdots * P * S_0 \\
&= \gamma[\gamma^{-1}[P] * \cdots * \gamma^{-1}[P] * \gamma^{-1}[S_0]] \\
&= \gamma[\gamma^{-1}[P](\gamma^{-1}[S_0])].
\end{aligned}$$

□

3.4 Secured Risk Assessment

The discussion above enables us to define a risk assessment process guaranteed to stay the same or decrease the estimated risk events when we spend more effort on it. It is an iterative process shown below:

1. Describe the situation to be analyzed as a coarse-grained (an aggregated) Datalog program. It is safe to start this first step with limited information.
2. Execute the program, obtain the attack graph, and then analyze it to see the risk level.

A Formal Approach for Secured Risk Assessment 439

3. Finish the assessment if the analysis result is satisfactory, e.g., the risk level is acceptable, or it is sufficient to consider security control options.
4. If not, choose the coarse-grained facts that should be elaborated, segregate them, and remove inappropriate facts that do not match the situation.
5. Return to the second step.

The iteration of this process incurs cost while reducing the qualitatively estimated risk events. If we use a quantitative risk metric proportional to the partial order of estimated risk events, the metric also decreases. The active incorporation of coarse-grained modeling also complements the lack of input information. These are the unique characteristics of this approach not seen in the past studies.

4 Example

This section shows two example cases of the segregation of facts in the context of secured risk assessment. We omit detail and focus only on the essential parts to clarify the points. The first example aims at the elaboration of the procedure of secured risk analysis. It helps us comprehend the step-by-step detail of our approach. The second example handles a more realistic scenario based on an enterprise architecture description to clarify the characteristics and issues in our approach to suggest further research topics.

4.1 Tiny Example

Figure 3 is a diagram representation of Datalog facts. It depicts a model composed of a network infrastructure, applications working on it, and business processes enabled by them. Be cautious that it is not an attack graph. Attack graphs are implicitly referred to determine the facts exploited for potential attacks.

For instance, the part "business-[enabledBy]-app" corresponds to the term "enabledBy(business, app)" in a Datalog program. The program corresponding to this diagram describes a situation where a "business" is enabled by some "app," and the "app" depends on a network infrastructure with vulnerabilities. We use the following rules to analyze the situation:

```
compromised(X)
  :- compromised(Y), connectsTo(Y,X),
     hasVulnerability(X,V).

compromised(S)
  :- hostedOn(S,H), compromised(H).

compromised(B)
  :- enabledBy(B,S), compromised(S).
```

As the red-colored part in the diagram indicates, the attack graph analysis applied to this situation infers the possible attack path from the "attacker" to the "business."

However, this analysis is rough since we do not distinguish the business's specific processes and the various applications enabling them. Figure 3 implies that every business will be affected by the potential attack. It might be too conservative an estimation. To refine the analysis, we introduce a constant replacement operator γ satisfying the following properties:

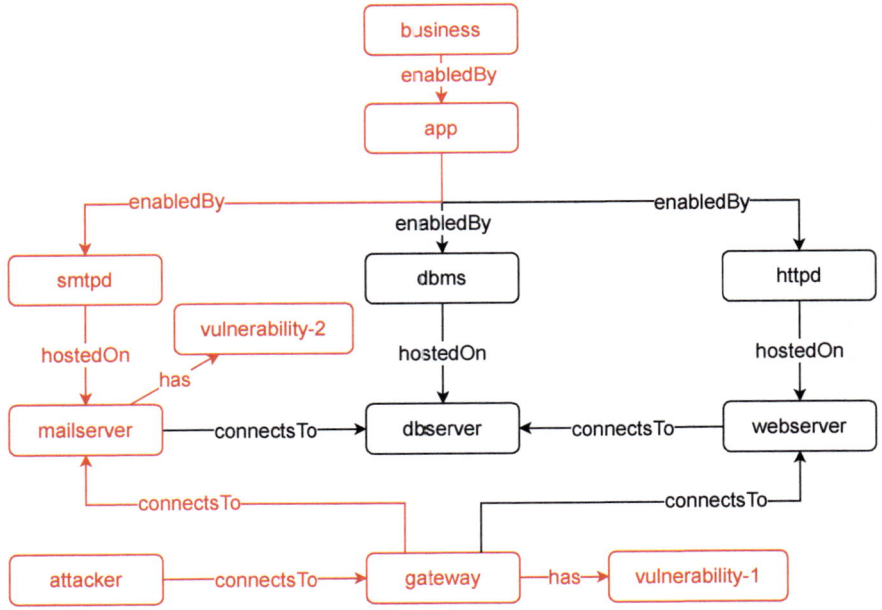

Fig. 3 Initial coarse-grained program diagram

A Formal Approach for Secured Risk Assessment

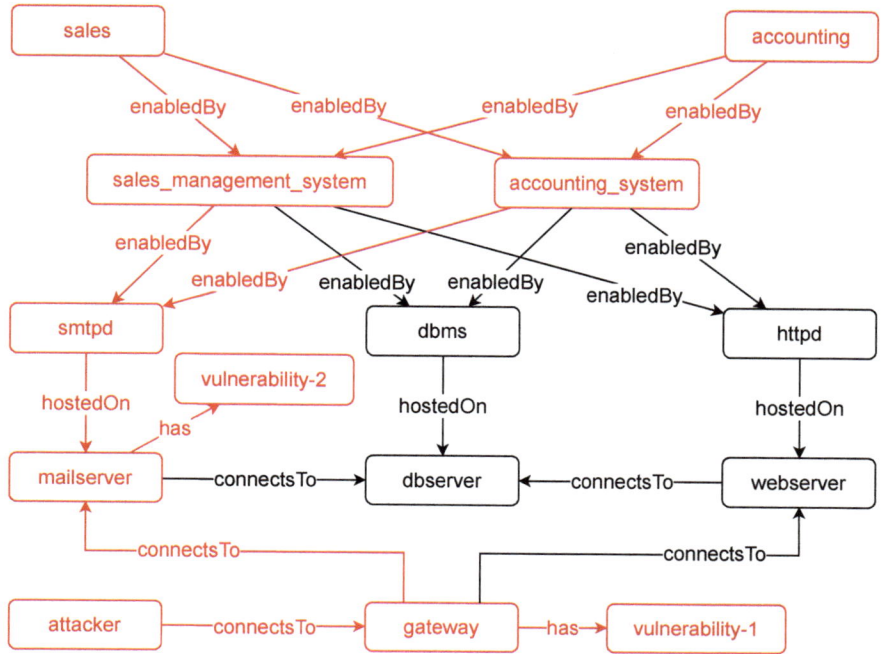

Fig. 4 Segregated form

$$\begin{aligned}
\text{business} &= \gamma(\text{accounting}) & \text{app} &= \gamma(\text{accounting\_system}) \\
&= \gamma(\text{sales}) & &= \gamma(\text{sales\_management\_system}) \\
&= \gamma(\text{business}), & &= \gamma(\text{app}).
\end{aligned}$$

With this operator, we can segregate Fig. 3 into the fine-grained form as shown in Fig. 4. Note that this transform corresponds to the "$\gamma^{-1}[S_0]$" in Theorem 5, and $\gamma^{-1}[P] = P$ since the rules used here have no constants.

This segregated form contains every possible relation between the segregated facts. If we examine the actual state of the business and its use of applications, we could remove several of them. For example, if we know that only the "sales_management_system" uses "smtpd," we can remove the "enabledBy" relation between the "accounting_system" and the "smtpd." It corresponds to the removal of a fact in Datalog programs. Figure 5 is the exemplified diagram of the facts to which this sort of reduction has been applied.

The attack graph analysis applied to this refined form identifies that "sales" could be affected by "attacker" while "accounting" is not. Considering that the "business" aggregates both "sales" and "accounting" in the coarse-grained Fig. 3, we can say the fine-grained analysis dismisses the risk event to "accounting." That is, the refinement of the analysis decreases the estimated incidents.

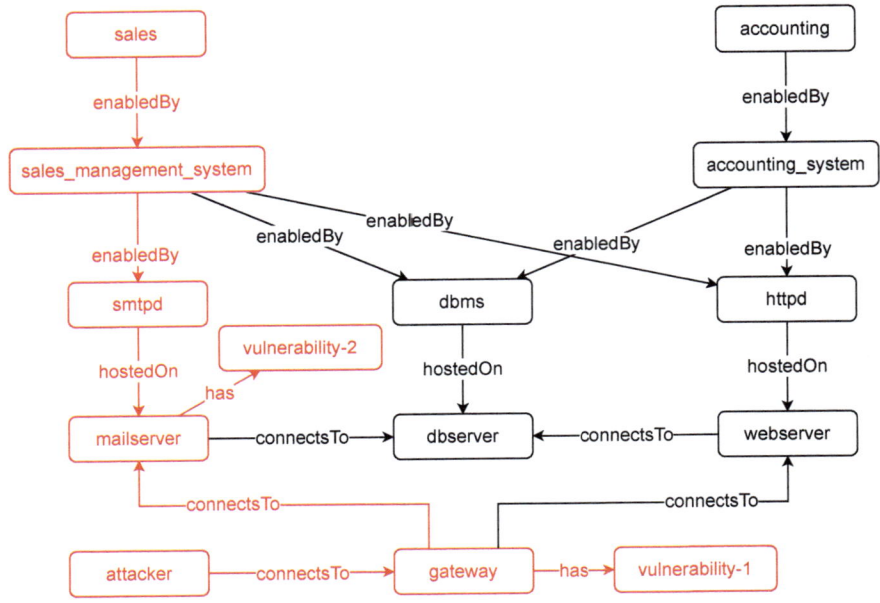

Fig. 5 Fine-grained program diagram

4.2 More Realistic Example

Let us examine a more realistic example to identify research issues that should be handled for the practical use of our approach. Here, we mainly focus on identifying issues in Datalog vocabulary design and clarifying the practical implications of the segregation of facts. The following explanations omit details due to the volume of the model implementation and the resultant attack graph. The reader can verify the explanation and the discussion by replicating the attack graph analysis with the analysis tool and the implemented model attached to this article.[1]

4.2.1 Motivations

The vocabulary design in Datalog is a fundamental prerequisite for business impact analysis by attack graph. As the tiny example above shows, our approach can handle the dependencies of business processes on IT infrastructures thanks to Datalog's powerful expressiveness. However, it requires a carefully designed set of predicates and constants, and such design is based on significant effort. At this point, enterprise architecture (EA) practices can be a valuable source for the design. EA is the whole

---

[1] https://github.com/kengo-zenitani/2023-secured-risk-assessment.

picture of an enterprise, including physical infrastructure, application services, business processes, and strategic objectives. Various ontologies, such as TOGAF [20] and ArchiMate [21], are dedicated to EA description. We can use these already tested frameworks as our basis. Notably, ArchiMate has a formal definition suitable for expression in Datalog; the diagrams in ArchiMate closely resemble the diagrams in this study. This section introduces a small EA description to examine the feasibility of EA in the vocabulary design for our approach.

In addition, we need some guidelines for selecting the facts to segregate. The motivation for secured risk assessment is to protect the analysis process from intentional risk underestimation. Therefore, the presumed primary target of the segregation is a department, a section, or a team in a corporation that are not cooperative with cybersecurity management. As the OPM case tells us, it is not uncommon to see such sections in the field. An uncooperative attitude is expected to make the details of the section opaque to security analysts. Secured risk analysis urges those people to disclose information in exchange for potentially decreased incident estimation. We verify this intuition by examining the effect of facts segregation in the realistic example. We also expect to obtain some practical implications from the verification.

4.2.2 EA Description

For the two aims above, the following discussion analyses a model based on an EA described in ArchiMate. Figure 6[2] shows the whole diagram of the EA.

The EA is composed of three layers: the business layer, the application layer, and the technology layer. Roughly speaking, the business layer depicts people's layer performing business processes, the application layer depicts application software, and the technology layer depicts hardware infrastructure. ArchiMate defines the vocabulary for depicting nodes and edges in EAs. For example, the cyan-colored box labeled with "Accounting" in the middle of the figure is an application component, and the edges running from the component to yellow boxes represent "serves" relations to business processes. Such a relation implies a fact like "`serves(Accounting, AccountingOffice)`" in Datalog. In this way, an EA written in ArchiMate gives fundamental information to model the business in Datalog.

4.2.3 Datalog Implementation

Based on the EA in Fig. 6, we implement a Datalog program for attack graph analysis. The EA descriptions are converted into facts in Datalog. Since ArchiMate models are a kind of Entity-Relationship model, the converted facts indicate the entities and their relationships composing an enterprise.

---

[2] PDF is available at: https://github.com/kengo-zenitani/2023-secured-risk-assessment/blob/main/Fig6.pdf.

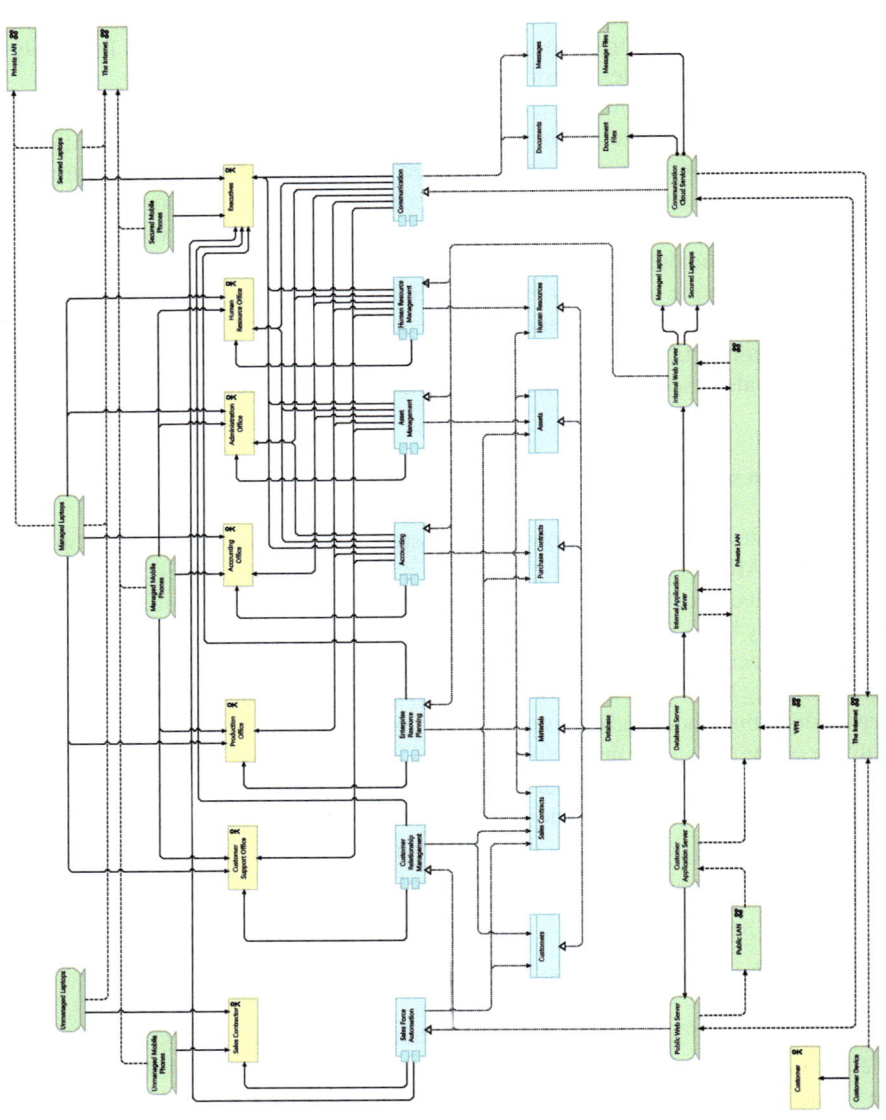

Fig. 6 An enterprise architecture described in ArchiMate

The fundamental limitation in using ArchiMate description for attack graph analysis is the lack of security-relevant information. For example, the ArchiMate specification does not define the vulnerability description. Attack templates are also out of the scope of ArchiMate. We need to supplement this missing information to enable attack graph analysis. In this example, we model the worst-case situation, where various common attacks are assumed, such as remote-code injection, phishing-triggered malware infection, distributed denial of service, and insider threats. We do not dig into the details of the supplemented information here. Section 5.1 discusses them.

4.2.4 Facts Segregation and Incident Estimations

We assume that an uncooperative section does not disclose security-relevant information while the security analyst can access the network configurations and communication logs serving the section. It means that the analyst can guess the technology layer of the architecture, but it is difficult for the analyst to grasp the relations between the applications used on the network and the actors and business processes performing in the section.

The assumption above makes some parts of the EA opaque to the analyst. Suppose the EA in Fig. 6 is the fine-grained form. Then, the list of "DataObjects," such as "Customers," "Sales Contracts," and "Materials," is opaque to the analyst and is to be aggregated into one entity like "Application Data." Similarly, it is unknown to the analyst who uses which applications. It leads to an aggregation of "BusinessActors".

We follow the implication above and define the following two constant mapping operators γ_1 and γ_2 to represent the aggregation and the segregation:

```
internal_office                     application_data
  = γ₁(accounting_office)             = γ₂(assets)
  = γ₁(administration_office)         = γ₂(customers)
  = γ₁(customer)                      = γ₂(human_resources)
  = γ₁(customer_support_office)       = γ₂(materials)
  = γ₁(human_resource_office)         = γ₂(purchase_contracts)
  = γ₁(production_office),            = γ₂(sales_contracts).
```

With these two operators, we define the following three models and generate attack graphs from them:

1. **Model-0**: the fine-grained original model as shown in Fig. 6.
2. **Model-1**: the coarse-grained model generated by applying γ_1 to Model-0.
3. **Model-2**: the coarse-grained model generated by applying both γ_1 and γ_2 to Model-0.

Table 1 Estimated incidents by the models

Estimated incidents	Model-0	Model-1	Model-2
accidentsOccur()	✓	✓	✓
moneySentToWrongAccount()	✓	✓	✓
productsSentToWrongPlace()	✓	✓	✓
interrupted(executives)	✓	✓	✓
interrupted(internal_office)		✓$_1$	✓$_1$
interrupted(accounting_office)	✓$_1$		
interrupted(administration_office)	✓$_1$		
interrupted(customer_support_office)	✓$_1$		
interrupted(human_resource_office)	✓$_1$		
interrupted(production_office)	✓$_1$		
interrupted(sales_contractor)	✓	✓	✓
leaked(documents)	✓	✓	✓
leaked(messages)	✓	✓	✓
leaked(database)	✓	✓	✓
leaked(application_data)			✓$_2$
leaked(assets)	✓$_2$	✓$_2$	
leaked(customers)	✓$_2$	✓$_2$	
leaked(human_resources)	✓$_2$	✓$_2$	
leaked(materials)	✓$_2$	✓$_2$	
leaked(purchase_contracts)	✓$_2$	✓$_2$	
leaked(sales_contracts)	✓$_2$	✓$_2$	

Model-1 is an aggregated form of Model-0, and Model-2 is an aggregated form of Model-1. In our hypothetical scenario, the uncooperative section gives the coarse-grained models to the analyst, and the analyst uses those models as the segregation target and then persuades the uncooperative section to disclose more information to remove segregated facts. In those cases, the analyst's implicit expectation is that removing segregated facts can decrease the estimated incidents. Note that Model-0 corresponds to the fine-grained model after the segregation and facts removal applied to Model-1 and Model-2 since Model-1 and Model-2 are obtained by aggregating facts in Model-0.

Table 1 shows the actual lists of the estimated incidents based on the models above. We examine these analysis results in the following section.

5 Discussions

We demonstrated the concept of our approach with small examples. At its heart, our approach is based on a risk analysis algorithm that guarantees the analysis refinement stays the same or decreases the estimated incidents. Limited information leads to a coarse-grained assessment with a greater estimation of incidents. This property encourages negatively motivated business executives to invest in risk assessment proactively. This secured monotonicity is the unique characteristic of our approach in this field. On the other hand, it is conceptual and is still in its early stage of development. We discuss the issues requiring further research in our approach here.

5.1 Limited Expressiveness in ArchiMate

As is mentioned in Sect. 4.2.2, the EA description in ArchiMate lacks the information required for attack graph analysis. The supplemented information that is implemented for the attack graph analysis in Datalog is as follows:

1. **Distribution of vulnerabilities and threats**: we added several types of facts denoting which entity has what kind of vulnerabilities or malicious property, such as the possibility that the targeted business actor is an insider threat.
2. **Access control list**: the Datalog implementation contains two types of facts, "readFrom" and "writeTo," each defining the access control, denoting which entity can access what data through which application service.
3. **Attack templates**: we also implemented the rules for inferring exploits based on available facts.

The information above is out of the scope of ArchiMate specification, and the analyst is required to write it down at present manually. It is a practical need to design the vocabulary and the standardized library of the supplemental information not limited to the above.

Another notable point is the form of access control list; the added facts "readFrom" and "writeTo" have three arguments: who, what, and by which. In contrast, every fact converted from the ArchiMate description has only one or two arguments since it is an Entity-Relationship model. This difference in the meta-structure of the description also affects the network topology representation in the EA. In ArchiMate-based ordinary EAs, only the connections between physically adjacent nodes are captured. We cannot identify how a network router controls the flow of packets. However, it is critical information to see the accessible scope of a node. In the example, we added some redundant "serves" relations to denote the logical connection among devices. The clarification of techniques like this should be another research topic.

5.2 Appropriate Target of Segregation

The aggregation that is applied in Sect. 4.2.4 identifies the segregation targets as the list of actors in the business layer and internal components in the application layer. This targeting naturally derives from the assumption that the analyst can access the network infrastructure information while the uncooperative section keeps its private information secret. However, the covered information is a typical target of access control and configuration management. The analyst can access them if the organization has implemented full-fledged access control and configuration management. It means that the appropriate choice of segregation target depends on the case, and the exemplified segregation in this article is just one possibility.

Additionally, compared to a typical attack graph analysis that refers to details of system configuration and vulnerability-specific properties, the examples in this article largely ignore the technical details of exploits. While it makes our discussion on segregation business-centric, there should be another segregation focusing on the technology layer. On the other hand, from the perspective of business impact analysis, it may be arguable whether conducting analysis at a granularity finer than Model-0 is necessary.

Since access controls and exploits are closely related, it is necessary to examine how we can choose segregation targets reasonably. We need validation with more detailed and larger scale models.

5.3 Masking Effect of Risks

If we segregate the aggregated constants in Model-1 and Model-2, the incidents shown in Table 1 become the same across models. For example, the fact "interrupted(internal_office)" is equivalent to the set of "interrupted(accounting_office)" and others. Remember that, compared to Model-1 and Model-2, Model-0 corresponds to the segregated form with facts removal. In this example, the segregation and removal of facts do not reduce the incident estimate. This contradicts our expectations, and there are two major reasons for this phenomenon.

First, the example in Sect. 4.2 addresses the worst-case scenario where a wide range of entities can be harmed in multiple ways. The removal of facts reduces incidents only when every attack path leading to the incidents is blocked. That is, simple qualitative evaluation ignores the value of attack path reduction, which should be preferred as a quantitative improvement.

Second, the example assumes that compromises in the technology layer cause most incidents. For example, if a database server is compromised, almost all applications are affected, and business processes are harmed transitively. Database server compromise is a technology layer event unrelated to segregation in the business and application layers.

The combination of the above two reasons makes the results of the risk analysis essentially the same, regardless of the granularity of the analysis. We can interpret this phenomenon as the *masking effect of risks*. In the example, the incidents in the technology layer mask the effect of facts segregation and removal in the business layer. Since introducing access control and other measures should deter some incidents, we have to suppress the masking by the technology layer incidents to make explicit the effect of such measures in the business layer. Such treatment is another topic for future research.

5.4 Need for Granularity Free Quantitative Metrics

The implication of Theorem 5 is that the segregation of facts itself does not change the level of risks inferred from the model to be analyzed. However, such invariance of segregation applies only qualitatively. If we have two models, the original one and its segregated form, the theorem guarantees the semantic equivalence of the inferred facts. It does not mean the invariance of metrics computed based on them. For example, as Table 1 indicates, the number of incidents in an attack graph is a metric increasing in the latter form. We need security metrics to have a property guaranteeing no increase in metric value in the segregated form. Let us call that property the *segregation stability*.

Another implication of the theorem and our approach is that we expect security metrics to improve when we remove the segregated facts. Let us call that property the *monotonicity of metrics*. The number of incidents in an attack graph satisfies this property, though it lacks stability. Several other attack-graph-based metrics satisfy the monotonicity, such as [4, 24, 27]. It is unclear whether those metrics satisfy the stability. The process in Sect. 3.4 implicitly assumes using such a metric to evaluate the risk level.

This study does not identify a metric satisfying both the segregation stability and the monotonicity. The assurance of stability is the focal point. The concept of segregation is pretty new to this field, and there is no consideration for it in the past metric design. We need further research.

6 Conclusion

This study proposed an information risk assessment approach guaranteeing that the refined assessment always gives a lower estimation of incidents. This property secures the risk assessment process from the conflicts between stakeholders; each has a different motivation and competes for finite resources. Our approach is based on the aggregation and the segregation of facts appearing in Datalog programs. Datalog programs are used to generate attack graphs, and attack graph analysis is one of the model-based risk assessment approaches with qualitative inference. We can use the powerful expressiveness of Datalog and quantitative metrics designed for attack graphs to leverage our approach as the basis of secured business risk assessment. However, its practical use requires carefully designed ontology in Datalog and appropriate treatments of the masking effects of risks. The unguaranteed improvement in quantitative metrics for refined assessment is an issue yet to be solved. Including the field testing, we need further research effort to make our approach a practical tool.

Acknowledgements The author wishes to acknowledge Dr. Keisuke Tanaka, Professor in the Department of Mathematical and Computing Science, School of Computing, Tokyo Institute of Technology, for constructive advice on this study.

References

1. M. Albanese, S. Jajodia, S. Noel, Time-efficient and cost-effective network hardening using attack graphs, in *IEEE/IFIP International Conference on Dependable Systems and Networks (DSN 2012)* (IEEE, 2012), pp. 1–12. ISBN 978-1-4673-1625-5. https://doi.org/10.1109/DSN.2012.6263942
2. E. Bacic, M. Froh, G. Henderson, MulVAL extensions for dynamic asset protection, in *Defence R&D Canada* (2006)
3. S. Ceri, G. Gottlob, L. Tanca, What you always wanted to know about Datalog (and never dared to ask). IEEE Trans. Knowl. Data Eng. 1(1), 146–166 (1989). ISSN 10414347. https://doi.org/10.1109/69.43410
4. P. Cheng, L. Wang, S. Jajodia, A. Singhal, Aggregating CVSS base scores for semantics-rich network security metrics, in *2012 IEEE 31st Symposium on Reliable Distributed Systems* (IEEE, 2012), pp. 31–40. ISBN 978-1-4673-2397-0. https://doi.org/10.1109/SRDS.2012.4
5. Committee on Oversight and Government Reform, *The OPM Data Breach: How the Government Jeopardized Our National Security for More than a Generation* (U.S. House of Representatives, Technical report, 2016)
6. M. Froh, G. Henderson, *MulVAL Extensions II Defence R&D Canada-Ottawa* (2009)
7. Commission Fukushima Nuclear Accident Independent. Investigation, *The official report of the Fukushima nuclear accident independent investigation commission* (Technical report, The National Diet of Japan, 2012)
8. Investigation Committee on Information Leakage due to Unauthorized Access, Report on the Results of the Investigation into the Incident of Information Leakage due to Unauthorized Access. Technical report, Japan Pension Service (2015)
9. ISO/IEC, Information technology–Security techniques–Information security management systems–Overview and vocabulary (2016)
10. B. Kordy, L. Piètre-Cambacédès, P. Schweitzer, DAG-based attack and defense modeling: Don't miss the forest for the attack trees. Comput. Sci. Rev. 13–14, 1–38 (2014). ISSN 15740137. https://doi.org/10.1016/j.cosrev.2014.07.001
11. National center of Incident readiness and Strategy for Cybersecurity, Results of the investigation into the cause of the personal information leak at the Japan Pension Service. Technical report, Cybersecurity Strategic Headquarters (2015)
12. National Institute of Standards and Technology, Guide for conducting risk assessments (2012). ISSN 01641212
13. S. Noel, S. Jajodia, *Metrics Suite for Network Attack Graph Analytics*, vol. 10 (ACM Press, 2014), pp. 5–8. ISBN 9781450328128. https://doi.org/10.1145/2602087.2602117. http://dl.acm.org/citation.cfm?doid=2602087.2602117

14. X. Ou, *A Logic-programming Approach to Network Security Analysis*. PhD thesis (Princeton University, USA, 2005)
15. X. Ou, W.F. Boyer, M.A. McQueen, A scalable approach to attack graph generation, in *Proceedings of the 13th ACM conference on Computer and communications security–CCS '06* (ACM Press, New York, USA, 2006), pp. 336–345. ISBN 1595935185. https://doi.org/10.1145/1180405.1180446
16. M. Pendleton, R. Garcia-Lebron, J.-H. Cho, S. Xu, A survey on systems security metrics. ACM Comput. Surv. **49**(4), 1–35 (2017). ISSN 0360-0300. https://doi.org/10.1145/3005714
17. A. Ramos, M. Lazar, R.H. Filho, J.J.P.C. Rodrigues, Model-based quantitative network security metrics: A survey. IEEE Commun. Surv. Tutor. **19**(4), 2704–2734 (2017). ISSN 1553-877X. https://doi.org/10.1109/COMST.2017.2745505
18. R.E. Sawilla, X. Ou, Identifying critical attack assets in dependency attack graphs, in *European Symposium on Research in Computer Security* (Springer, 2008), pp. 18–34. https://doi.org/10.1007/978-3-540-88313-5_2
19. X. Sun, A. Singhal, P. Liu, Towards actionable mission impact assessment in the context of cloud computing, in *Data and Applications Security and Privacy XXXI*, by G. Livraga, S. Zhu (Springer International Publishing, Cham, 2017), pp. 259–274. ISBN 978-3-319-61176-1
20. The Open Group, *The TOGAF® Standard, Version 9.2* (Van Haren Publishing, 2018). ISBN 1-947754-11-9
21. The Open Group, *ArchiMate® 3.1 Specification* (Van Haren Publishing, 2019)
22. V. Verendel, Quantified security is a weak hypothesis, in *Proceedings of the 2009 workshop on New security paradigms workshop–NSPW '09* (ACM Press, New York, USA, 2009), p. 37. ISBN 9781605588452.https://doi.org/10.1145/1719030.1719036
23. Verification Committee for the Incident of Information Leakage due to Unauthorized Access at Japan Pension Service, Verification Report. Technical report, Ministry of Health, Labour and Welfare (2015)
24. L. Wang, T. Islam, T. Long, A. Singhal, S. Jajodia, An attack graph-based probabilistic security metric, in *IFIP Annual Conference on Data and Applications Security and Privacy* (2008), pp. 283–296. ISBN 978-3-540-70567-3. https://doi.org/10.1007/978-3-540-70567-3_22
25. K. Zenitani, A formal approach for secured risk analysis in information security management, in *2022 9th International Conference on Behavioural and Social Computing (BESC)* (IEEE, Matsuyama, Japan, 2022), pp. 1–7. ISBN 979-8-3503-9814-4.https://doi.org/10.1109/BESC57393.2022.9995464
26. K. Zenitani, Attack graph analysis: An explanatory guide. Comput. Secur. 126, 103081 (2023a). ISSN 0167-4048. https://doi.org/10.1016/j.cose.2022.103081
27. K. Zenitani, A scalable algorithm for network reachability analysis with cyclic attack graphs. J. Comput. Secur. **31**(1), 29–55 (2023b). ISSN 18758924. https://doi.org/10.3233/JCS-210103

Open Access This chapter is licensed under the terms of the Creative Commons Attribution 4.0 International License (http://creativecommons.org/licenses/by/4.0/), which permits use, sharing, adaptation, distribution and reproduction in any medium or format, as long as you give appropriate credit to the original author(s) and the source, provide a link to the Creative Commons license and indicate if changes were made.

The images or other third party material in this chapter are included in the chapter's Creative Commons license, unless indicated otherwise in a credit line to the material. If material is not included in the chapter's Creative Commons license and your intended use is not permitted by statutory regulation or exceeds the permitted use, you will need to obtain permission directly from the copyright holder.

SoK: A Taxonomy for Layer-2 Scalability Related Protocols for Cryptocurrencies

Maxim Jourenko, Mario Larangeira, Kanta Kurazumi, and Keisuke Tanaka

Abstract Blockchain-based systems, in particular cryptocurrencies, face a serious limitation: scalability. This holds, especially, in terms of the number of transactions per second. Several alternatives are currently being pursued by both the research and practitioner communities. One venue for exploration is on protocols that do not constantly add transactions on the blockchain and therefore do not consume the blockchain's resources. This is done using off-chain transactions, *i.e.*, protocols that minimize the interaction with the blockchain, also commonly known as *Layer-2* approaches. This work relates several existing off-chain channel methods, also known as payment and state channels, channel network construction methods, and other components such as channel and network management protocols, *e.g.*, routing nodes. All these components are crucial to keep the usability of the channel and are often overlooked. In this work, we propose a taxonomy for all the components of Layer-2. We provide extensive coverage of the state-of-the-art protocols available outline their respective approaches, and discuss their advantages and disadvantages.

Keywords Blockchain · Systemization of knowledge · Payment channels · Layer 2

M. Jourenko (✉) · M. Larangeira · K. Kurazumi · K. Tanaka
Department of Mathematical and Computing Sciences, School of Computing, Tokyo Institute of Technology, Meguro-ku, Japan
e-mail: jourenko.m.ab@m.titech.ac.jp

M. Larangeira
e-mail: mario@c.titech.ac.jp; mario.larangeira@iohk.io

K. Tanaka
e-mail: keisuke@is.titech.ac.jp

M. Larangeira
Input Output Global (IOG), Singapore, Singapore

© The Author(s) 2026
T. Takagi et al. (eds.), *Mathematical Foundations for Post-Quantum Cryptography*, Mathematics for Industry 40, https://doi.org/10.1007/978-981-96-1218-5_23

1 Introduction

Blockchains are the main technology behind the successful rebirth of digital cash from its first attempts [28, 29], firstly by Bitcoin [82] and now with several decentralized cryptocurrencies [42, 46, 49, 101, 105]. Although Bitcoin's relative success is undeniable in offering worldwide payment alternatives to the more traditional mechanisms, like the VISA Network [84] and Paypal [83], it still has a long way ahead in terms of handling a larger number of transactions. Blockchains maintain a data structure, namely the ledger, that consists of an ordered and immutable list of transactions through a *consensus protocol*. Taking Bitcoin [82] as an example, transactions are added to the ledger in batches called blocks which, by design have a size limit of 1 MB and are issued every 10 minutes on average. This implies an upper bound for the potential transactions per second (TPS) that can be added to the ledger [31]. The technical challenge of increasing the number of transactions per second (TPS) of a blockchain system is urgent, and it is closely related to the inner workings of the system itself, *i.e.*, it's *consensus protocol*. Let alone that, in order to confirm a transaction, it is required to wait for a minimum number of blocks to be added in the main chain, which gives the *confirmation time*. Approaches to improve the TPS of a blockchain system can be roughly structured into three categories.

- **Layer-0** (the network infra-structure): Here, optimization and special servers are employed to decrease the latency of the network thus allowing for more frequent and larger blocks. An example of such an approach is BloXroute [65];
- **Layer-1** (the consensus protocol): Different consensus architectures [20, 42, 46, 49, 64, 86, 87, 101], where a complete taxonomy can be found in [47], different data-structures for blockchains [44, 98–100], valid chain criteria [67], sharding [66, 71, 106], federation [75], etc.;
- **Layer-2** (the off-chain channel/protocol): Protocols that perform minimal interaction, i.e., a constant number, with the blockchain. These are based on constructions called *channels*. While opening, closing, and disputing a channel require off-chain transactions, individual payments do not. Concrete examples are given by Decker and Wattenhofer [37], and the Lighting Network (LN) [90], among others. This layer encompasses many more protocols and algorithms than the channel construction alone, as we review in later sections. A somewhat similar survey on Layer-2 protocols exists [51].

Overview of Technical Aspects. Protocols in Layer-2 operate by coordinating their participants to create a series of *transactions*, which are objects that atomically change a ledger's state, for instance, payments. Transactions can stay in a complex relation to one another, effectively forming a transaction graph which is directed and acyclic. Here, a transaction can only be committed to the ledger if all its ancestors were committed to the ledger, and, furthermore, different branches of this graph can be mutually exclusive. Especially in the case of ledgers without smart contract capability, the graph formed by a protocol's transactions represents its functionality and *graph theoretical* consideration assists in its analysis. Moreover, Canetti's Universal Composability (UC) framework [24] can be considered the state-of-the-art rigorous security analysis of cryptographic protocols and finds frequent use in the related work. On the other side, payment channel networks can be modeled as *flow networks* where each payment requires the computation of a flow in the network and algorithms, such as Ford-Fulkerson [45], find application. Lastly, maintaining payment channel networks on a higher level involves rebalancing this flow network which can be approached as an optimization problem.

Contributions. We provide extensive coverage of Layer-2 solutions for scalability, namely off-chain channel constructions, payment network constructions, and network management protocols. We introduce a single framework (Table 1), which easily highlights and structures the different approaches. Moreover, we focus on the general description and the functionalities of the protocols. Therefore our contribution somewhat differs from the work in [51]. Our single framework is a taxonomy with further classification for the three cited levels: off-chain construction, channel network, and network management. The levels compose a stack of protocols for Layer-2, as illustrated by Table 1 in Sect. 2. As expected, different levels and their respective set of functions define specific technical challenges. A quick look at Table 1 reveals intense research in some areas and few works in others, or even no works for specific problems. We believe these observations are of strong interest to the research community, which otherwise would have to collect these works through numerous internet forums and repositories, in addition to scientific journals. Surely a time-consuming task even to distinguish among very distinct functionalities, let alone more subtle differences in implementation. Hence we expect the research and practitioners communities alike to appreciate this work.

2 A Taxonomy for Layer-2

We organize the different sets of protocols in three levels.[1] Each level is subdivided into the "functions" it performs. The intuition is that a protocol deployed on Layer-2 is expected to fulfill a certain function within a level. In order to illustrate and justify this organization, take for example the most popular Layer-2 Payment Channel Network (PCN) technique: Hashed Timelock Contract (HTLC) [90]. Briefly, HTLC allows PCN nodes to establish connections between themselves in order to transfer funds. A major challenge in this setting is to find a suitable route among the nodes, thus the function required is *Routing*. A routing protocol for the LN fits into the "Routing" function within the "Network Management" level. Later, Table 1 summarizes our taxonomy and the levels we identify in the literature. First, consider the following levels:

---

[1] We justify the choice of the term "level", instead of the more natural "layer", as an alternative to avoid ambiguity with the terms Layer-0, Layer-1, and Layer-2 which permeate this work.

Table 1 The channels stack and concrete protocols. The stack covers the "Off-chain Channels" Level, the very fundamental types of channel constructions, up to the "Network Management" Level, where the core function is more closely related with the guaranteeing safe and efficient networks. In the middle level are the protocols that construct the networks

Level	Function	Protocols
Network management	Routing	Silent whispers [72]
		Speedy murmur [95]
		Spider routing [97]
		Flare routing [91]
		Splitting payments [88]
		Hoenisch and weber [53]
		Atomic multi-path [2]
	Re-balancing channels	REVIVE [60]
		Hide & seek [16]
		Proactive rebalancing [89]
	Channel stability	PISA [76]
		Avarikioti et al. [14]
		Cerberus [17]
		Garrison [81]
		Fail-safe watchtowers [70]
		FPPW [80]
	Anonymity/privacy	Fulgor & Rayo [73]
		Tumblebit [52]
Network Sect. 4	Network payments	HTLC [90]
		Sprites [79]
		Payment trees [57]
		State assertion [22]
		Counterfactual [30]
	Virtual channels channels	Lightweight virtual payment channels [56]
		Elmo [63]
		Bitcoin-compatible virtual channels [11]
		Interhead hydra [55, 58]
	Post-quantum	Adaptor signatures (Esgin et al.) [43]
		Adaptor signatures (Tasiri et al.) [102]
Off-chain channels Sect. 3	Two-Party State	Z-Channel [107]
		Perun [40]
		NoCUST [61]
	Multi-party state	Hydra [26]
		Multi-party virtual state channels (MPVSC) [39]

(continued)

Table 1 (continued)

Level	Function	Protocols
	Duplex payment	Raiden [4]
		Lightning [90]
		Decker et al. [37]
		BOLT [50]
		Teechan [68]
		Burchert et al. [23]
		TumbleBit [52]
		Eltoo [36]
		Sleepy channels [12]
		Brick [15]
	Simplex payment	Simplex [7]
		Dimitrienko et al. [38]
		Takahashi et al. [103]
	Probabilistic payment	Pass et. al. [85]
		Hu and Zhang [54]

- **Off-chain Channels:** Here are the channel constructions. In other words, how the nodes, relying on the transaction design of a cryptocurrency, can construct single channels. These are the building blocks of the upper levels.
- **Network:** Here are the techniques employed to create the networks themselves. Typically by concatenating existing pairwise channels, or establishing a *network* of more than two nodes right from the start.
- **Network Management:** This is the level for the protocols that maintain the channels and network of channels, and allow their efficient use while keeping them "alive" and usable.

Off-Chain Channels. The terminology in the literature is not standard. However, it is possible to observe two major functionalities: (1) making payments and (2) keeping state. The first is the simplest, since it does not require to keep a state in order to perform a payment. Whereas the second is more general and it does require the aid of smart contracts. We further break down (1) into the three existing types: *Duplex*, *Simplex*, and *Probabilistic* and break down (2) into two-party -/ and Multiparty state channels.

Network. In order to establish PCNs, special techniques are required, in addition to carrying out the payments. There are off-chain channels that do not support network construction capabilities, therefore this level depends on the capabilities of the channels. Methods to form channel networks are split up into facilitation of *payments*, creation of *virtual channels* and we focus separately on *Post-Quantum* constructions. We further discuss this topic in Sect. 4.

Network Management. Over the network of channels, as further discussed in Sect. 5, several functionalities are required, and we identify the following main ones in the existing literature: (1) **Routing:** The payments can be carried over several nodes, however, it is necessary to select the set of nodes as the route, and several may be available. (2) **Re-Balancing:** A difference from regular computer networks is that each pairwise channel has a capacity which may exhaust during the course of the exchange of transactions, and therefore it needs to be rebalanced. (3) **Stability:** Another difference from the regular computer networks, is that some constructions require the nodes to be online in order to perform the dispute phase of the protocols. Therefore there are protocols that mitigate or solve this requirement. (4) **Anonymity/Privacy:** Information of nodes within a network, can be leaked as well as information about the performed transactions. Typically, the nodes in the middle of the channel can see the flow of funds.

Comparisons. Briefly, given the similarities with computer networks, researchers realized that techniques can be borrowed from currently known network algorithms, albeit the need to adapt them. This trend is promoted by Hoenisch and Weber [53], for *routing*, a crucial functionality within the *Network Management* Level. This trend contrasts with the technicalities of the *off-chain channel* and *network* levels. Our taxonomy decouples these different issues, and we further discuss them in Sects. 3 (channels), 4 (networks) and 5 (management). As expected the off-chain channels body of work, *i.e.*, described in Sect. 3, presents a greater number of works, in comparison, for example, with the *Network* level, *i.e.*, described in Sect. 4. The reason is that most of the protocols in Sect. 4 derive from a single technique to realize *conditional payments*, the HTLC, which was recently fully formalized by Kiayias et al. [62]. An example of a derived similar technique is Sprites [79] which employs a derived technique but with smart contracts. On the other hand, the amount of work on off-chain channels is comparable to that of the *Network Management* level, because of the significant amount of effort put into the functionality of *routing* among the nodes of the network by researchers and developers as later illustrated by Table 1 and also described in Sect. 5.

Scope. While there is a great body of work for scalability solutions for Blockchains, in this work we focus on Layer-2 solutions in the "pure" sense, *i.e.*, except for opening, closing, and dispute, interactions through Layer-2 stay off-chain or, more formally, all interactions on a Layer-2 construction result in at most a constant number of on-chain transactions. Respectively we do not cover scalability solutions such as side-chains or rollups. The reader should also note that our definition of the channel does not cover, for example, secure multi-party computation protocols, *i.e.*, several users interact off-chain and provide *correctness proof* of the computation. Here, purposely we focus only on protocols that realize a channel between nodes, and PCNs through the concatenation of channels. In other words, we leave out of our classification protocols with the multiuser distribution of funds, as used in [19, 33–35].

3 Off-Chain Channel Level

Preliminaries: Ledger Paradigms There are two paradigms that are most often used for blockchain design. For one, the *Unspent Transaction Output (UTxO) model* introduced in Bitcoin [82] revolves around UTxOs which are tuples of the form (c, s) where $c \in \mathbb{N}$ is an amount of coins and $s \in \{0, 1\}^*$ is a script. UTxOs represent the coins c that are in circulation and define their ownership through s by requiring that c can only be claimed by providing a witness $w \in \{0, 1\}^*$ s.t. $s(w) = \text{TRUE}$. Commonly, s is used to require w to be the signature of the UTxOs owner. Then, a transaction is of the form (I, O, τ) where I is a set of inputs, O is a list of outputs, and τ is a *timelock*, i.e., a point in time s.t. the transaction is only valid if time τ has passed. If a transaction is added to the ledger, all UTxO in its inputs are consumed and all UTxO in its outputs are added to the ledger's state. Δ is the maximum time required for a transaction to be committed by a party and subsequently be included in the ledger. The Extended UTxO (EUTxO) model [25] adds a field $\delta \in \{0, 1\}^*$ to UTxOs and extends the script evaluation function which effectively allows EUTxOs to execute state machines. Alternatively the *account based model* as used in Ethereum [105] revolves around the party's accounts and assigns coins directly to them. Notably, *smart contracts* are a type of account that operates according to program code, contains state and coins, and can be interacted with using messages. Note that most of the interaction in a pairwise channel happens only between the players in the channel. Therefore, the players need to keep the balance through all the transactions performed in the channel, in order to offer *balance security*. As we see next, very often, in order to have balanced security, the nodes need to be online to take action in case the partner in the channel misbehaves.

Overview. Layer-2 protocols are based on structures called *channels*. Originally, these are set up by two parties who move coins into it on *opening*, i.e., the channel's creation, with a transaction on the ledger. The channel then stores these coins and maintains the coin distribution between its owners which can be modified through an off-chain protocol effectively performing *payments* and without further interaction with the ledger. When closing a channel, another transaction is sent to the ledger by both parties, returning its coins to their owners according to the latest coin distribution. Lastly, in the case of *dispute*, the channel can be closed unilaterally and coins are returned based on how the dispute resolution is handled. *State Channels* allow storage and modification of arbitrary data in their state and allow the closure of the channel to be affected by it allowing the execution of state machines and smart contracts off-chain. Lastly, more recent constructions allow for arbitrary sets of parties to create *multi-party* state channels. Most channels are constructed according to a similar structure. A channel is opened through a *Funding transaction* that takes coins of both parties and makes them spendable with a multisignature by both parties. Then, a channel is closed using a *Refund transaction* that spends the Funding transaction's coins and returns it to the parties but is held off-chain. Parties can perform payments by creating alternative Refund transactions while *invalidating* all older ones. Note that the initial Refund transaction has to be held by all parties before the Funding transaction is created to prevent one party from aborting the protocol locking the channel's coins inside it indefinitely.

Channel constructions for ledgers with limiting scripting capabilities often form transaction graphs that are held off-chain, thus *graph theoretical* reasoning, together with protocol design, can be used to limit what transactions can be sent to the ledger and for analysis of the construction. On the other side, smart contracts implemented in the EUTxO model can be defined using Contraint Emitting Machines (CEMs) which are derived from Mealy Automata and it has been shown that there exists a weak bi-simulation between CEMs and EUTxO smart contracts. A run in a CEM equals a smart contract execution, thus analysis of all potential runs of a CEM is a valuable tool for analyzing a given construction.

Simplex Payment Channels. The earliest proposal of payment channels that we could find was posted on the Bitcoin Wiki [7]. Note that the first proposal mentioned here makes use of a sequence number to replace previous transactions, however, the respective, and already cited, Bitcoin field, nSequence has been disabled on 20th August 2010 [1, 7].[2] An updated version of this protocol has been created since: The channel is set up between two parties, a *Payer* and a *Payee*. The initial Refund transaction is set up with a timelock t. Then, newly created Refund transactions are signed by the Payer and sent to the Payee, however, the Payer does not receive the Payee's signature in return. Doing so, the Payee does not risk the channel to be closed in an old state as long as they send the latest transaction they receive before time $t - \Delta$. As the Payee is incentivized to only submit the transaction where they receive most coins, this effectively restricts the channel to one-way, i.e., simplex payments.

Probabilistic Simplex. Pass et al. [85] propose a probabilistic payment system to reduce the number of transactions on the blockchain and subsequently reduce the amount of transaction fees enabling micropayments. Moreover their solution provides near-instantaneous payments without requiring confirmation delays. Their work is based on the work of Wheeler [104], Rivest [93] as well as subsequent work [69, 78], and proposes three protocols. The first is naive, therefore vulnerable, construction for two parties. The second and third protocols are similar, with both relying on a verifiable third party with the difference that the latter is optimistic, and relies on the verifiable third party in case of a dispute. In these protocols the payer sets up an escrow address and puts an amount of coins into it. After the payment is performed the payer can potentially use the same escrow to pay another merchant. This work was improved by Hu and Zhang [54] using a time-locked deposit.

Duplex Payment. In a duplex payment channel both parties can allocate funds into their channel and redistribute them arbitrarily. The constructions for duplex channels [36, 37, 90] extend the idea of simplex payment channel[3] by adjusting Refund transaction invalidation, which differs in Duplex [37], Lightning [90] and Eltoo [36] Channels. While Duplex Channels uses an *invalidation* tree of transactions where by design the most recent Refund transaction is on a path within the tree where

---

[2] https://github.com/bitcoin/bitcoin/commit/05454818dc7ed92f577a1a1ef6798049f17a52e7#diff-118fcbaaba162ba17933c7893247df3aR522.

[3] A comparison can be found in [77].

transactions have the lowest timelocks compared to their siblings. Lightning on the other hand lets parties invalidate old transactions by having them exchange secret *invalidation* keys that allow them to claim the other party's coins within a dispute phase in case they submit a transaction that was previously invalidated this way. Eltoo on the other side allows parties to spend an invalidated Refund transaction with a more recent Refund transaction within a dispute phase. This allows forgoing punishing parties by taking their coins as in Lightning. A few channel constructions attempt to make them secure without the inherent requirement for all participants to stay online and enforce disputes. Brick [15] utilizes a set of $2f + 1$ wardens, which will enforce the dispute, where f wardens can be byzantine. The wardens are incentivized to behave through the submission of collateral. Sleepy Channels [12] adjust the Lightning Channel construction by changing the duration of its Dispute phase from being relative to submission of a Refund transaction to being an absolute point in time t, allowing a party to go offline until time $t - \Delta$ to perform a potential dispute in time. However, upon submission of a Refund transaction by a party to the ledger, both parties can obtain their coins early by having the counterparty claim their coins from the channel, thereby confirming a valid channel closure. This behavior is incentivized through collateral.

Hardware Based Approaches. Teechan is a protocol introduced by Lind et al. [68] which introduces a full duplex payment channel framework between two mutually distrusting parties, assuming that they are equipped with trusted execution environments (TEE), particularly on Intel's Software Guard Extensions (SGX), which provides a secure region of memory, *i.e.*, a *enclave*. By making only specific SGX instructions able to execute code or access data in an enclave, it guarantees confidentiality and integrity. Both private keys of the two parties interacting in Teechan are securely shared by their TEEs. In execution, only each TEE generates and encrypts every transaction, and then provides them to the party. Therefore each party is only responsible for the setup, payment, or close requests for its TEE, and forwards them to the counterpart. Lind et al. [68] measured 10 million transactions to be exchanged in optimal network conditions,*i.e.*, network bandwidth and latency, and observed that the protocol achieves 2,480 TPS on average. A similar work to [68] is given by Takahashi and Otsuka [103] which improves the work of Dmitrienko et al. [38] by selecting a different technique to validate funds of a hardware wallet. Both [38] and [103] rely on loading funds in advance into the wallet. However, whereas in [38] the payer wallet verifies the pre-loaded funds, in [103], the payee performs the verification. In [103], differently from [68], it does not need the signature of both sides of the channel, *i.e.*, payer and payee 2-out-of-2 signature.

Channel Factories. The general framework for channels imposes restrictions on the creation of channels. Namely, the confirmation time for the **Setup** Phase may still be unacceptable for micropayments, without an advance funding transaction. Furthermore, upon channel creation, the funds are locked for only that channel. The relaxation in these requirements is the motivation for the creation of *channel factories* proposed by Burchert et al. [23]. The core idea is to introduce a layer between the blockchain and the payments. This translates into a step where a group of collaborators jointly fund a factory. This first still step requires blockchain interaction, therefore still is subject to the limitations of the consensus algorithm. However, any new pairwise channel can be created, among the initial users, from this point, upon communication between the collaborators, hence creating channels. Although the factory creation still requires time and funds locking into the blockchain, the advantage of this design is that it allows the reallocation of funds between the channels.

Privacy Preserving Channels. Anonymity focused approach has been proposed by Green et. al. [50] with BOLT. It assumes anonymity of the underlying blockchain, *i.e.*, either by an anonymous blockchain as Zcash [8] or means to anonymize it, *e.g.*, by using a mixer. Then, assuming that one of the parties is well known, *e.g.*, it is a merchant, it provides additional anonymity guarantees for the respective other parties, *e.g.*, a customer. The merchant knows that they received a payment, however, does not know from which customer. However, their construction does not consider a definition of balance security when a malicious customer stops cooperating with the merchant after doing a couple of payments.[4] As the customer is required to start channel closure the merchant, *i.e.*, one of the participants of the channel, cannot close it arbitrarily, therefore it cannot claim the funds paid to him. Recently, Zhang et al. [107] introduced Z-Channel, which provides anonymity for micropayments adapting the ZeroCash. Another protocol that minimizes the payer information leakage is TumbleBit [52], which presents a unidirectional unlinkable payment channel, where the payer pays to payee via an untrusted payment hub (called Tumbler) that plays a role of a mixer. In TumbleBit, unlinkability is achieved for the Tumbler itself. Moreover, the protocol prevents the theft by a malicious user of any other user's funds without requiring trust from the users. A remarkable feature of the protocol is compatibility with the Bitcoin protocol before SegWit was adopted.

Two-Party State Channels. Until now, in this work, the channels are built based on special instructions for transactions, as described in Sect. 3. Smart contracts, as in Ethereum, enable more complex off-chain structures which have been investigated in [40, 79].[5] A particular use for smart contracts aims to enforce fairness and correctness on distributed systems. The approach of relying on penalties to guarantee correctness and fairness of computation, in particular, on *secure multi-party computation protocols*, has been independently explored for efficient protocols on the top of blockchain. For the general case, this idea has been introduced by Andrychowics et al. [9, 10], and later by Bentov and Kumaresan [18]. For specific purposes, *e.g.*, card games, it was further explored by Bentov et al. [19] and David et al. [33–35]. Compared to the previously mentioned constructions for payment channels, the state channel can be implemented more straightforwardly. They are opened by committing the respective smart contract onto the blockchain, whose state can be changed using a message signed by both parties as well as a sequence number. Parties change the state off-chain by computing a message to the smart contract that would make transition into that state including a sequence number and exchanged signatures for this message but without committing the message onto the blockchain. NoCUST [61] is another protocol which relies heavily on smart contracts. Here two parties wishing

---

[4] Although the authors acknowledge it by explicitly assuming the customer follows the protocol.

[5] We discuss [40] and [79] in more detail in Sect. 4.

to exchange small amounts create a channel by making their deposits into a smart contract, therefore two on-chain operations. All the payments from that point on, are executed off-chain via a third trusted node which intermediates the off-chain operations between the two participants, and each payment requires issuing request payments and receipts. This design has the advantage, in comparison to regular payment hubs over off-chain channels, of not requiring the hub node to allocate a large amount of funds, hence *no custodian*, while intermediating payments between large groups of pairs of nodes.

Multi-Party State Channels. Multi-Party Virtual State Channels allow an arbitrary set of parties which are connected through a State Channel Network, to create a Multi-Party State Channel. Transactions are confirmed through a 4 round synchronous protocol. Hydra [26] utilizes the EUTxO model to create an isomorphic multi-party state channel called *Hydra Heads* which stores a set of EUTxO in its state. Here parties can move EUTxO between the ledger and the Hydra Head using on-chain interactions and issue off-chain transactions that modify its stored EUTxO set through an asynchronous protocol.

4 Network Level

Network Payments. The main function of this level is the construction of a payment route. More concretely, how the pairwise "inner" channels can be concatenated and used to provide a medium for payments between two nodes that are more than one hop apart.

The payment networks and related concepts had been previously studied: trust networks [48, 59, 92], credit network [32], path-based transaction (PBT) [96] and privacy in PCN [73]. Here we are interested in construction techniques for pairwise channels, in order to concatenate several channels into a *channel network*. Such networks can enhance even more the scalability of a system, because two nodes do not need to contact each other to open a channel. Instead, they can create a "virtual channel", in the sense of Dziembowski et al. [41] via a mutually connected node (more on this, later in the section). The most known technique is the early cited HTLC [90], which is currently being used in the Bitcoin network and paved the way to new propositions for different cryptocurrencies, as in the Raiden Network [4] for the Ethereum [105].

HTLC Overview and Conditional Transfers. Nodes can execute a payment atomically on a set of channels using HTLC. A payment can be routed across a sequence of channels without payer and payee having to create a channel between themselves that would require committing transactions onto the blockchain. Briefly, this is done by setting up a conditional payment on each channel on a path from Payer to Payee. This payment is either executed if a secret x is presented, or aborted after the expiration of a timelock. After completion of the setup, the secret x will be revealed by a node to its predecessor on the path, allowing them to resolve the payment and making the predecessor learn the secret in turn. This ensures that a payment can be executed atomically across all channels. Note that for security, the timelocks have to be increased by Δ for each hop on the path. Because of this payments are potentially locked in conditional payments for a time linearly to the path's length. The Sprites Protocol [79] introduced by Miller et al. is designed for Ethereum's smart contracts and has the goal of reducing the time complexity of resolving a dispute by having the secret be broadcast through a smart contract in case of a dispute. Payment Tree's [57] are Bitcoin compatible and optimize a virtual channel construction [56] for a one-time payment to reduce the time complexity for an individual party to be logarithmic to a path's length whereas the average summed up time complexity across the whole path is constant. An approach introduced by Buckland and McCorry [22] named *State Assertion Channels*, and it is analogous to [79] but for computational costs. The goal of [22] is to guarantee that during the Dispute Phase, which can also be triggered by closing a channel, the honest party always can be paid back regardless of the cost of performing the verification of the full state of the application in the blockchain deployed smart contract.

The Wormhole Attack and Improved HTLC. The attack introduced by Malavolta et al. [74] affects all the PCN constructions based on 2-step interaction between the nodes, therefore HTLC based PCNs are in general, vulnerable to it. The colluding nodes aim to capture the transaction fee which the inner nodes on a payment route would expect to receive in order to carry the payments in the network. Despite the gravity of the attack, the authors of [74] succeeded in proposing a fix for the main protocol and even warned the LN team about the protocol weakness, which triggered new developments. In particular the implementation of the two-party ECDSA construction from [74], and its later incorporation into the system [6].

Virtual Channels. Instead of using a PCN to perform payments, Virtual channel constructions aim to create channels on top of two adjacent channels, allowing their participants to interact with one another directly and without further interaction with intermediaries, e.g., for individual payments. Dziembowski et al. [40] extend their previous work [41] to create *virtual state channels*. These operate based on a management smart contract that stores the channel's state, balance distribution, and a sequence number. State updates can be performed by providing a new by both parties signed state with a higher sequence number. The construction can be performed recursively to allow for virtual channels across multiple hops. Furthermore, Lightweight Virtual Payment Channels [56], Elmo [63], and Bitcoin-Compatible Virtual Channels [11] all propose virtual channel constructions that are compatible with Bitcoin. More recently Interhead Hydra [55, 58] proposes two constructions for virtual multi-party isomorphic state channels as well as a light-weight ad-hoc construction. These are implemented using the EUTxO model and allow EUTxO to be moved from the ledger or isomorphic state channels in and out of the virtual channel. Moreover, the latter construction allows for concatenation of an arbitrary amount of adjacent channels. In a sense, Coleman et al. [30] also aim to build a richer structure on the top of the off-chain channels. They introduced the notion of *counterfactual*, which is, in a nutshell, all the events of the channel, which can, or cannot, be committed to the blockchain. In this paradigm, a payment, in the off-chain channel, changes a so-called *counterfactual state* of the system. In addition, it is also possible to create *counterfactual contracts* via commitments and signatures, which generalizes the channels even further. The work in [30] presents a framework, focused on practical implementation. Similar to channel constructions, virtual channel constructions for ledgers without expressive smart contracts are utilizing transactions graphs and protocol design such that *graph theoretical* considerations are used for security and performance analysis.

(Post-Quantum) Adaptor Signatures / Scriptless Scripts. An idea to preserve space in the blockchains is to embed more instructions into the signature issuing procedure by relying on a more sophisticated, *i.e.*, one that offers more properties, signature scheme. That is the approach introduced by Poelstra [3, 5] denoted *scriptless script* for the Schnorr signatures and LN, which was recently extended by Malavolta et al. [74] to the more suitable, for cryptocurrencies, ECDSA signature scheme. Alternatively, [3] outlines an approach with the Schnorr signature scheme to embed the value of the pre-image x, *i.e.*, for the HTLC hash challenge, into the signature computation algorithm, jointly carried by the payer and the payee of the channel. Although the formalization effort displayed is in [74], the authors point out it is not the case for other proposals. Furthermore, the main cryptocurrencies, *e.g.*, Bitcoin and Ethereum, may not be compatible with the Schnorr scheme. Recently, Esgin et al. [43] and Tasiri et al. [102] proposed post-quantum adaptor signatures and proposed their use for PCNs. The construction of Esgin et al. [43] uses lattice assumptions Module-SIS and Module-LWE whereas Tasiri et al. [102] use assumptions on isogenies.

5 Network Management Level

When it comes to the Network Management Level, we observe that there is a significant number of works regarding "Routing" and "Stability" functions. For the remaining section, we group them respectively under "Node Routing" and "Network Maintenance, Stability and Privacy."

5.1 Node Routing

Routing. Similarly to a regular computer network, in order to find a payment path, it is necessary to probe the network for nodes available to route the payments. On the other hand, in the case of channel networks, other variables can influence the construction of such paths, in particular availability and *fees* to intermediate the payments can heavily influence the routing. A potential model for PCNs are *flow networks* where the *Ford-Fulkerson* algorithm can be used to identify potential payment paths. However, this approach has high complexity, and instead more practical approaches are used instead such as those in Mobile Ad Hoc Networks (MANET). Nevertheless, approaches using *Ford-Fulkerson* [45] can serve as a benchmark in an experimental setting.

Dynamic Networks. Routing in PCNs faces distinct challenges such as a highly dynamic network where nodes can spontaneously appear or go offline, channel capacities limit which routes payments can take and are ever-changing and multiple routing attempts can happen concurrently. Briefly, there are two ways to approach routing in these networks: through *landmarks*, and *embeddings*. In *landmark routing* all nodes know the route to a set of *landmarks*. In such a protocol payments are, first, routed from the payer to a landmark and, second, routed from the landmark to the payee. This approach promotes the centralization of payment channel networks with nodes close to a landmark having an advantageous position. Another approach is routing using *embeddings*, where nodes, among themselves, decide, on an address space of the network and then they can find routes using these addresses which requires more communication for maintaining routing tables.

Ad hoc Networks and Gossip Approach. The Lightning Network [90] relies on a *gossip protocol* for channel maintenance and recovery[6] and onion style routing for privacy.[7] However it is not clear how nodes can find routes, and an attempt to tackle this is the routing proposal algorithm for LN: Flare [91]. The protocol in [91] is probabilistic and proceeds in two ways: (1) each node stores the topology of its neighborhood, and (2) each node chooses a set of beacon nodes globally and at random according to a uniform distribution. Routing between two nodes is done by first checking both peers' neighborhoods for intersections, and if this does not work checking whether a beacon node is within the other peers neighborhood. The nodes continue by using the neighborhoods of a few beacon nodes when searching for mutually known nodes to route through. Hoenisch and Weber [53] pioneered the adaptation of techniques from Mobile Ad Hoc Network (MANET), by compiling a list of requirements and arguing with off-chain channel networks. An interesting feature of their adapted protocol, named On-Demand Distance Vector (AODV), is that it takes into account the balance and fees of intermediate nodes, a feature not present in MANET but highly relevant given that these values can change arbitrarily and without coordination in the Layer-2 scenario.

Routing Privacy. SilentWhispers Protocol [72] proposes using landmark routing using shortest or, where all landmarks are publicly known and semi-honest or, with a proposed extension malicious. Routing tables are updated in epochs. The SpeedyMurmurs routing protocol [95] is an embedding-based routing protocol, for path-based transaction networks based on the work in [94]. In [95] some nodes are designated landmarks and are used as roots of respective spanning trees are computed to create an address space within the network. A route can be computed using a distance function on nodes' addresses to find the next hop-on route to the target.

Routing by Splitting Payment. Orthogonal to the problem of routing, AMP [2] provides a protocol for splitting payments through multiple routes to improve routing success for large payments or in the face of low-capacity channels. The Split Payment

---

[6] A description can be found in https://github.com/lightningnetwork/lightning-rfc/blob/master/07-routing-gossip.md.

[7] A description can be found in https://github.com/lightningnetwork/lightning-rfc/blob/master/04-onion-routing.md.

Protocol [88] tackles a similar problem as in [2], however [88] allows for payments to complete only partially. Another approach is the Spider Network [97] which employs packet-based routing including congestion control for multi-path payments. Each payment is split up into smaller packets which can be routed through separate routes and claimed atomically by the sender.

5.2 Network Maintenance, Stability, and Privacy

The payment networks, similar to regular networks, require maintenance in order to ensure the channels are available for payments. This can be translated to make sure enough funds are available for transactions and also that a timely reaction is available when the misbehaving of parts happens.

Rebalancing Channel. Given the limited capacity of a channel, it may be necessary to rebalance a set of payment channels to avoid the channel turning into a simplex one because of exhausted capacity. In order to circumvent the skewness in funds, the REVIVE protocol [60] relies on an untrusted third party that creates a block of transactions that rebalance funds. Each peer will lose money on one or more payment channels but gain an equal amount on others. After each peer verifies this on the rebalancing transactions they can apply the changes off-chain. Nodes can set constraints on how their channels can be rebalanced which results in algorithms that attempt rebalancing to solve an *optimization problem* and *linear programming* has been used in this context.

Channel Stability. The current techniques for channel constructions very often rely on the assumption that the users of the channel, and therefore of a PCN, will be online during the lifespan of the channel, which is crucial in the case of a Dispute Phase when the parties need to act timely. An approach to address this is to assume that the receiver can assign a *watchtower* [13, 14], sometimes also referred to as *custodian node*, to monitor their payment channel while they are offline. Similarly, this is the idea of the PISA [76]. The custodian can dispute malicious commitments of the other peer while its customer is offline. The watchtower has to commit collateral which is slashed if they fail to dispute a channel and acts as an incentive for their honest behavior.

Anonymity and Privacy. The network management layer may leak crucial pieces of information from nodes belonging to a route to the origin node of a payment. Malavolta et al. [73] investigate privacy notions on PCN and how to enforce them, and propose routing protocols Fulgor and Rayo [73] that consider concurrent routing processes and attempt to avoid deadlocks in a sense that at least one payment completes. Unfortunately, their privacy notions rely on assumptions that the underlying routing protocol does not store the state (capacity) of payment channels within the network, which if done could be used to circumvent their attempts for privacy. Since each node in the network can charge an arbitrary fee in order to conduct the payment along the path, an interesting question to investigate is the game-theory behavior of rational nodes within the network and other economic questions. A framework to study network topologies and economic aspects is given by Brânzei et al. [21].

6 Conclusion

We provided a major review, as shown in Table 1, of the body of work on off-chain channels, networks, and related protocols for Layer-2 solutions, an emerging area targeting the scalability of cryptocurrencies. We believe it provides significant value for the community because it contains a wide overview of the landscape of the ideas and approaches present in the scientific literature and Internet repositories and forums. We focused this work on providing a thorough overview of the landscape with emphasis on problems and approaches of the available protocols.

We highlight the rich literature on routing protocols, for the Network Management Level, in comparison to attention given to, for example, channel Re-Balancing protocols. That difference can be explained by the similarities of the off-chain channel and computer networks, which can be seen as a source of already tested ideas to find routes. Similarly to state channel research, within the Off-Chain Level in Table 1. More recently, we can also note a trend in network construction protocols for the Network Level, in particular, the works on Virtual Channels [40] and Counterfactual [30]. In both cases, they distinguish themselves by providing a richer framework for the constructions on the top of the off-chain channels. Furthermore, we also highlight that the most established technique, *i.e.*, HTLC, was recently fully thoroughly studied [62], likewise the one in [40].

Taking these works as a basis we see a few aspects that we consider to be underrepresented or even to be *open problems*. Works on the economic aspects of the channels, *i.e.*, fees charged by nodes, and *reputation* of nodes seem to be underreported in the literature. Furthermore, while off-chain transactions do not appear on the ledger and thus enhance the participant's privacy, means for auditability and accountability are not available and are defined as Gap 6 in [27]. We consider both aspects critical to pave the way for more practical systems. Moreover, *adaptor signatures* are a comparatively new concept and while constructions and suggestions on their use exist, only a little work has been done to utilize them in off-chain constructions. Lastly, while some Layer-2 protocols have been treated within the UC framework, more complex protocols often opt for proving a few properties of their protocol and we argue that more formal security treatment can be provided in this area. Especially Layer-2 protocols based on the EUTxO model have not yet received any such full formal treatment.

Acknowledgements This work was supported by JST CREST Grant Number JPMJCR2113, Japan and Input Output Global.

References

1. Bip 68, https://github.com/bitcoin/bips/blob/master/bip-0068.mediawiki. Accessed 28 November 2018
2. [lightning-dev] amp: Atomic multi-path payments over lightning. https://lists.linuxfoundation.org/pipermail/lightning-dev/2018-February/000993.html. Accessed 18 October 2018
3. Lightning in scriptless script, Mailing list port: https://github.com/lightningnetwork/lnd. Accessed 26 March 2019
4. Raiden network, raiden.network. Accessed 3 September 2018
5. Scriptless script, Presentation slides: https://download.wpsoftware.net/bitcoin/wizardry/mw-slides/2017-05-milan-meetup/slides.pdf. Accessed 26 March 2019
6. tpec: 2p-ecdsa signatures, Github repository, https://github.com/cfromknecht/tpec. Accessed 27 March 2019
7. Rapidly-adjusted (micro)payments to a pre-determined party. https://en.bitcoin.it/wiki/Contract (2011). Accessed 3 September 2018
8. Zcash, https://z.cash (2016). Accessed 26 September 2018
9. M. Andrychowicz, S. Dziembowski, D. Malinowski, L. Mazurek, Fair two-party computations via bitcoin deposits, in *FC 2014 Workshops, LNCS*, eds. by R. Böhme, M. Brenner, T. Moore, M. Smith, vol. 8438 (Springer, Heidelberg, 2014), pp. 105–121. https://doi.org/10.1007/978-3-662-44774-1_8
10. M. Andrychowicz, S. Dziembowski, D. Malinowski, L. Mazurek, Secure multiparty computations on bitcoin, in *2014 IEEE Symposium on Security and Privacy*. (IEEE Computer Society Press, 2014), pp. 443–458.https://doi.org/10.1109/SP.2014.35
11. L. Aumayr, M. Maffei, O. Ersoy, A. Erwig, S. Faust, S. Riahi, K. Hostáková, P. Moreno-Sanchez, Bitcoin-compatible virtual channels, in *2021 IEEE Symposium on Security and Privacy*. (IEEE, 2021), pp. 901–918
12. L. Aumayr, S.A. Thyagarajan, G. Malavolta, P. Moreno-Sanchez, M. Maffei, Sleepy channels: Bi-directional payment channels without watchtowers, in *2022 ACM SIGSAC Conference on Computer and Communications Security* (2022), pp. 179–192
13. G. Avarikioti, F. Laufenberg, J. Sliwinski, Y. Wang, R. Wattenhofer, Incentivizing payment channel watchtowers, in *Scaling Bitcoin* (2018)
14. G. Avarikioti, F. Laufenberg, J. Sliwinski, Y. Wang, R. Wattenhofer, Towards secure and efficient payment channels. arXiv preprint (2018)
15. Z. Avarikioti, E. Kokoris-Kogias, R. Wattenhofer, D. Zindros, B rick: Asynchronous incentive-compatible payment channels, in *Financial Cryptography and Data Security 2021*. (Springer, 2021), pp. 209–230
16. Z. Avarikioti, K. Pietrzak, I. Salem, S. Schmid, S. Tiwari, M. Yeo, Hide & seek: Privacy-preserving rebalancing on payment channel networks, in *Financial Cryptography and Data Security*. (Springer, 2022), pp. 358–373
17. Z. Avarikioti, O.S. Thyfronitis Litos, R. Wattenhofer, Cerberus channels: Incentivizing watchtowers for bitcoin, in *Financial Cryptography and Data Security 2020*. (Springer, 2020), pp. 346–366
18. I. Bentov, R. Kumaresan, How to use bitcoin to design fair protocols, in *CRYPTO 2014, Part II, LNCS*, eds. by J.A. Garay, R. Gennaro, vol. 8617 (Springer, Heidelberg, 2014), pp. 421–439. https://doi.org/10.1007/978-3-662-44381-1_24
19. I. Bentov, R. Kumaresan, A. Miller, Instantaneous decentralized poker, in *ASIACRYPT 2017, Part II, LNCS*, eds. by T. Takagi, T. Peyrin, vol. 10625 (Springer, Heidelberg, 2017), pp. 410–440.https://doi.org/10.1007/978-3-319-70697-9_15
20. I. Bentov, R. Pass, E. Shi, Snow white: Provably secure proofs of stake. Cryptology ePrint Archive, Report 2016/919 (2016)
21. S. Brânzei, E. Segal-Halevi, A. Zohar, How to charge lightning. arXiv preprint arXiv:1712.10222 (2017)

22. C. Buckland, P. McCorry, Two-party state channels with assertions. Third Workshop on Trusted Smart Contracts (2019)
23. C. Burchert, C. Decker, R. Wattenhofer, Scalable funding of bitcoin micropayment channel networks-regular submission, in *SSS* (2017), pp. 361–377
24. R. Canetti, Universally composable security: A new paradigm for cryptographic protocols, in *IEEE Symposium on Foundations of Computer Science* (IEEE, 2001), pp. 136–145
25. M.M. Chakravarty, J. Chapman, K. MacKenzie, O. Melkonian, M. Peyton Jones, P. Wadler, The extended utxo model, in *Financial Cryptography and Data Security* (Springer, 2020), pp. 525–539
26. M.M. Chakravarty, S. Coretti, M. Fitzi, P. Gazi, P. Kant, A. Kiayias, A. Russell, Hydra: Fast isomorphic state channels, in *Financial Cryptography and Data Security*. Springer (2021)
27. P. Chatzigiannis, F. Baldimtsi, K. Chalkias, Sok: Auditability and accountability in distributed payment systems, in *Applied Cryptography and Network Security* (Springer, 2021), pp. 311–337
28. D. Chaum, Blind signature system, in *CRYPTO'83*, ed. by D. Chaum (Plenum Press, New York, USA, 1983), p. 153
29. D. Chaum, A. Fiat, M. Naor, Untraceable electronic cash, in *CRYPTO'88, LNCS*, ed. by S. Goldwasser, vol. 403 (Springer, Heidelberg, 1990), pp. 319–327.https://doi.org/10.1007/0-387-34799-2_25
30. J. Coleman, L. Horne, L. Xuanji, Counterfactual: Generalized state channels (2018)
31. K. Croman, C. Decker, I. Eyal, A.E. Gencer, A. Juels, A. Kosba, A. Miller, P. Saxena, E. Shi, E.G. Sirer, et al., On scaling decentralized blockchains, in *Financial Cryptography and Data Security* (Springer, 2016), pp. 106–125
32. P. Dandekar, A. Goel, R. Govindan, I. Post, Liquidity in credit networks: A little trust goes a long way, in *ACM Conference on Electronic Commerce, EC '11* (ACM, 2011), pp. 147–156
33. B. David, R. Dowsley, M. Larangeira, Kaleidoscope: An efficient poker protocol with payment distribution and penalty enforcement. Cryptology ePrint Archive, Report 2017/899 (2017). http://eprint.iacr.org/2017/899
34. B. David, R. Dowsley, M. Larangeira, 21 - bringing down the complexity: Fast composable protocols for card games without secret state, in *ACISP 18, LNCS*, eds. by W. Susilo, G. Yang, vol. 10946 (Springer, Heidelberg, 2018), pp. 45–63.https://doi.org/10.1007/978-3-319-93638-3_4
35. B. David, R. Dowsley, M. Larangeira, ROYALE: A framework for universally composable card games with financial rewards and penalties enforcement. Cryptology ePrint Archive (2018)
36. C. Decker, R. Russel, O. Osuntokun, eltoo: A simple layer2 protocol for bitcoin. Whitepaper, https://blockstream.com/eltoo.pdf Accessed: 2019-03-29
37. C. Decker, R. Wattenhofer, A fast and scalable payment network with bitcoin duplex micropayment channels, in *Symposium on Self-Stabilizing Systems* (Springer, 2015), pp. 3–18
38. A. Dmitrienko, D. Noack, M. Yung, Secure wallet-assisted offline bitcoin payments with double-spender revocation, in *ASIACCS 17*, eds. by R. Karri, O. Sinanoglu, A.R. Sadeghi, X. Yi (ACM Press, 2017), pp. 520–531
39. S. Dziembowski, L. Eckey, S. Faust, J. Hesse, K. Hostáková, Multi-party virtual state channels, in *Theory and Applications of Cryptographic Techniques* (Springer, 2019), pp. 625–656
40. S. Dziembowski, L. Eckey, S. Faust, D. Malinowski, PERUN: Virtual payment channels over cryptographic currencies. Cryptology ePrint Archive, Report 2017/635 (2017). http://eprint.iacr.org/2017/635
41. S. Dziembowski, S. Faust, K. Hostakova, Foundations of state channel networks. Cryptology ePrint Archive (2018)
42. EOS, EOS. https://eos.io/ (2018). [Online; accessed 6-May-2018]
43. M.F. Esgin, O. Ersoy, Z. Erkin, Post-quantum adaptor signatures and payment channel networks, in *European Symposium on Research in Computer Security* (Springer, 2020), pp. 378–397

44. I. Eyal, A.E. Gencer, E.G. Sirer, R. Van Renesse, Bitcoin-ng: A scalable blockchain protocol, in *Usenix Conference on Networked Systems Design and Implementation, NSDI'16* (USENIX Association, 2016), pp. 45–59
45. L.R. Ford, D.R. Fulkerson, Maximal flow through a network. Canadian J. Math. **8**, 399–404 (1956)
46. C. Foundation, Cardano Hub. https://www.cardano.org/ (2018). [Online; accessed 28-March-2018]
47. J. Garay, A. Kiayias, Sok: A consensus taxonomy in the blockchain era. Cryptology ePrint Archive (2018)
48. A. Ghosh, M. Mahdian, D.M. Reeves, D.M. Pennock, R. Fugger, Mechanism design on trust networks, in *Internet and Network Economics*, eds. by Deng X, Graham FC (Springer, Berlin Heidelberg, 2007), pp 257–268
49. Y. Gilad, R. Hemo, S. Micali, G. Vlachos, N. Zeldovich, Algorand: Scaling byzantine agreements for cryptocurrencies. Cryptology ePrint Archive, Report 2017/454 (2017). http://eprint.iacr.org/2017/454
50. M. Green, I. Miers, Bolt: Anonymous payment channels for decentralized currencies, in *ACM CCS 2017*, eds. by B.M. Thuraisingham, D. Evans, T. Malkin, D. Xu (ACM Press, 2017), pp. 473–489.https://doi.org/10.1145/3133956.3134093
51. L. Gudgeon, P. Moreno-Sanchez, S. Roos, P. McCorry, A. Gervais, Sok: Layer-two blockchain protocols, in *Financial Cryptography and Data Security* (Springer, 2020), pp. 201–226
52. E. Heilman, L. Alshenibr, F. Baldimtsi, A. Scafuro, S. Goldberg, Tumblebit: An untrusted bitcoin-compatible anonymous payment hub, in *Network and Distributed System Security Symposium* (2017)
53. P. Hoenisch, I. Weber, Aodv-based routing for payment channel networks, in *Blockchain - ICBC 2018*, eds. by Chen S, Wang H, Zhang LJ (Springer International Publishing, Cham, 2018), pp. 107–124
54. K. Hu, Z. Zhang, Fast lottery-based micropayments for decentralized currencies, in *ACISP 18, LNCS*, eds. by W. Susilo, G. Yang, vol. 10946 (Springer, Heidelberg, 2018), pp. 669–686. https://doi.org/10.1007/978-3-319-93638-3_38
55. M. Jourenko, M. Larangeira, State machines across isomorphic layer 2 ledgers, in *Financial Cryptography and Data Security*. Springer (2023)
56. M. Jourenko, M. Larangeira, K. Tanaka, Lightweight virtual payment channels, in *Cryptology and Network Security* (Springer, 2020), pp. 365–384
57. M. Jourenko, M. Larangeira, K. Tanaka, Payment trees: Low collateral payments for payment channel networks, in *Financial Cryptography and Data Security 2021* (Springer, 2021), pp. 189–208
58. M. Jourenko, M. Larangeira, K. Tanaka, Interhead hydra: Two heads are better than one, in *Mathematical Research for Blockchain Economy* (Springer, 2022), pp. 187–212
59. D. Karlan, M. Mobius, T. Rosenblat, A. Szeidl, Trust and social collateral. Quart. J. Econom. **124**(3), 1307–1361 (2009)
60. R. Khalil, A. Gervais, Revive: Rebalancing off-blockchain payment networks, in *2017 ACM SIGSAC Conference on Computer and Communications Security* (ACM, 2017), pp. 439–453
61. R. Khalil, A. Gervais, NOCUST–A non-custodial 2nd-layer financial intermediary. Cryptology ePrint Archive, Report 2018/642 (2018). https://eprint.iacr.org/2018/642
62. A. Kiayias, O.S.T. Litos, A composable security treatment of the lightning network. Cryptology ePrint Archive (2019)
63. A. Kiayias, O.S.T. Litos, Elmo: Recursive virtual payment channels for bitcoin. Cryptology ePrint Archive (2021)
64. A. Kiayias, A. Russell, B. David, R. Oliynykov, Ouroboros: A provably secure proof-of-stake blockchain protocol, in *CRYPTO 2017, Part I, LNCS*, eds. by J. Katz, H. Shacham, vol. 10401 (Springer, Heidelberg, 2017), pp. 357–388.https://doi.org/10.1007/978-3-319-63688-7_12
65. U. Klarman, S. Basu, A. Kuzmanovic, E.G. Sirer, bloxroute: A scalable trustless blockchain distribution network whitepaper

66. E. Kokoris-Kogias, P. Jovanovic, L. Gasser, N. Gailly, E. Syta, B. Ford, OmniLedger: A secure, scale-out, decentralized ledger via sharding, in *2018 IEEE Symposium on Security and Privacy* (IEEE Computer Society Press, 2018), pp. 583–598.https://doi.org/10.1109/SP.2018.000-5
67. Y. Lewenberg, Y. Sompolinsky, A. Zohar, Inclusive block chain protocols, in *FC 2015, LNCS*, eds. by R. Böhme, T. Okamoto, vol. 8975 (Springer, Heidelberg, 2015), pp. 528–547. https://doi.org/10.1007/978-3-662-47854-7_33
68. J. Lind, I. Eyal, P. Pietzuch, E.G. Sirer, Teechan: Payment channels using trusted execution environments. arXiv preprint arXiv:1612.07766 (2016)
69. R.J. Lipton, R. Ostrovsky, Micro-payments via efficient coin-flipping, in *Financial Cryptography* (Springer, 1998), pp. 1–15
70. B. Liu, P. Szalachowski, S. Sun, Fail-safe watchtowers and short-lived assertions for payment channels, in *ACM Asia Conference on Computer and Communications Security* (2020), pp. 506–518
71. L. Luu, V. Narayanan, C. Zheng, K. Baweja, S. Gilbert, P. Saxena, A secure sharding protocol for open blockchains, in *ACM CCS 2016*, eds. by E.R. Weippl, S. Katzenbeisser, C. Kruegel, A.C. Myers, S. Halevi (ACM Press, 2016), pp. 17–30.https://doi.org/10.1145/2976749.2978389
72. G. Malavolta, P. Moreno-Sanchez, A. Kate, M. Maffei, Silentwhispers: Enforcing security and privacy in credit networks. NDSS (2017)
73. G. Malavolta, P. Moreno-Sanchez, A. Kate, M. Maffei, S. Ravi, Concurrency and privacy with payment-channel networks, in *ACM CCS 2017*, eds. by B.M. Thuraisingham, D. Evans, T. Malkin, D. Xu (ACM Press, 2017), pp. 455–471. https://doi.org/10.1145/3133956.3134096
74. G. Malavolta, P. Moreno-Sanchez, C. Schneidewind, A. Kate, M. Maffei, Multi-hop locks for secure, privacy-preserving and interoperable payment-channel networks. Cryptology ePrint Archive, Report 2018/472 (2018). https://eprint.iacr.org/2018/472
75. D. Mazieres, The stellar consensus protocol: A federated model for internet-level consensus. Stellar Development Foundation (2015)
76. P. McCorry, S. Bakshi, I. Bentov, A. Miller, S. Meiklejohn, Pisa: Arbitration outsourcing for state channels. Cryptology ePrint Archive (2018)
77. P. McCorry, M. Möser, S.F. Shahandasti, F. Hao, Towards bitcoin payment networks, in *Australasian Conference on Information Security and Privacy* (Springer, 2016), pp. 57–76
78. S. Micali, R.L. Rivest, Micropayments revisited, in *Cryptographers' Track at the RSA Conference* (Springer, 2002), pp. 149–163
79. A. Miller, I. Bentov, S. Bakshi, R. Kumaresan, P. McCorry, Sprites and state channels: Payment networks that go faster than lightning, in *Financial Cryptography and Data Security* (Springer, 2019), pp. 508–526
80. A. Mirzaei, A. Sakzad, J. Yu, R. Steinfeld, Fppw: A fair and privacy preserving watchtower for bitcoin, in *Financial Cryptography and Data Security 2021* (Springer, 2021), pp. 151–169
81. A. Mirzaei, A. Sakzad, J. Yu, R. Steinfeld, Garrison: a novel watchtower scheme for bitcoin, in *Australasian Conference on Information Security and Privacy* (Springer, 2022), pp. 489–508
82. S. Nakamoto, Bitcoin: A peer-to-peer electronic cash system (2008)
83. P. Network, Paypal Network. https://www.paypal.com/jp/home (2018). [Online; accessed 17-October-2018]
84. V. Network, Visa Network. https://www.visa.ca/ (2018). [Online; accessed 17-October-2018]
85. R. Pass, A. shelat, Micropayments for decentralized currencies, in *ACM CCS 2015*, eds. by I. Ray, N. Li, C. Kruegel (ACM Press, 2015), pp. 207–218. https://doi.org/10.1145/2810103.2813713
86. R. Pass, E. Shi, The sleepy model of consensus, in *ASIACRYPT 2017, Part II, LNCS*, eds. by T. Takagi, T. Peyrin, vol. 10625 (Springer, Heidelberg, 2017), pp. 380–409. https://doi.org/10.1007/978-3-319-70697-9_14
87. R. Pass, E. Shi, Thunderella: Blockchains with optimistic instant confirmation, in *EUROCRYPT 2018, Part II, LNCS*, eds. by J.B. Nielsen, V. Rijmen, vol. 10821 (Springer, Heidelberg, 2018), pp. 3–33.https://doi.org/10.1007/978-3-319-78375-8_1

88. D. Piatkivskyi, M. Nowostawski, Split payments in payment networks, in *Data Privacy Management, Cryptocurrencies and Blockchain Technology* (Springer, 2018), pp. 67–75
89. R. Pickhardt, M. Nowostawski, Imbalance measure and proactive channel rebalancing algorithm for the lightning network, in *2020 IEEE International Conference on Blockchain and Cryptocurrency* (IEEE, 2020), pp. 1–5
90. J. Poon, T. Dryja, The bitcoin lightning network: Scalable off-chain instant payments. See https://lightning.network/lightning-network-paper.pdf (2016)
91. P. Prihodko, S. Zhigulin, M. Sahno, A. Ostrovskiy, O. Osuntokun, Flare: An approach to routing in lightning network. White Paper (bitfury. com/content/5-white-papers-research/whitepaper_flare_an_approach_to_routing_in_lightning_n etwork_7_7_2016. pdf) (2016)
92. P. Resnick, R. Sami, Sybilproof transitive trust protocols, in *ACM Conference on Electronic Commerce* (ACM, 2009), pp. 345–354
93. R.L. Rivest, Electronic lottery tickets as micropayments, in *Financial Cryptography* (Springer, 1997), pp. 307–314
94. S. Roos, M. Beck, T. Strufe, Anonymous addresses for efficient and resilient routing in f2f overlays, in *Computer Communications, IEEE INFOCOM 2016* (IEEE, 2016), pp. 1–9
95. S. Roos, P. Moreno-Sanchez, A. Kate, I. Goldberg, Settling payments fast and private: Efficient decentralized routing for path-based transactions. arXiv preprint arXiv:1709.05818 (2017)
96. S. Roos, P. Moreno-Sanchez, A. Kate, I. Goldberg, Settling payments fast and private: Efficient decentralized routing for path-based transactions. arXiv preprint arXiv:1709.05818 (2017)
97. V. Sivaraman, S.B. Venkatakrishnan, M. Alizadeh, G. Fanti, P. Viswanath, Routing cryptocurrency with the spider network. arXiv preprint arXiv:1809.05088 (2018)
98. Y. Sompolinsky, Y. Lewenberg, A. Zohar, SPECTRE: A fast and scalable cryptocurrency protocol. Cryptology ePrint Archive (2016)
99. Y. Sompolinsky, A. Zohar, Accelerating Bitcoin's transaction processing. Fast money grows on trees, not chains. Cryptology ePrint Archive (2013)
100. Y. Sompolinsky, A. Zohar, PHANTOM: A scalable BlockDAG protocol. Cryptology ePrint Archive (2018)
101. Steemit, Steemit. https://steemit.com/ (2018). [Online; accessed 6-May-2018]
102. E. Tairi, P. Moreno-Sanchez, M. Maffei, Post-quantum adaptor signature for privacy-preserving off-chain payments, in *Financial Cryptography and Data Security* (Springer, 2021), pp. 131–150
103. T. Takahashi, A. Otsuka, Short paper: Secure offline payments in bitcoin. Third Workshop on Trusted Smart Contracts (2019)
104. D. Wheeler, Transactions using bets, in *International Workshop on Security Protocols* (Springer, 1996), pp. 89–92
105. G. Wood, Ethereum: A secure decentralised generalised transaction ledger. Ethereum project yellow paper pp. 1–32 (2014)
106. M. Zamani, M. Movahedi, M. Raykova, RapidChain: Scaling blockchain via full sharding, in *ACM CCS 2018*, eds. by D. Lie, M. Mannan, M. Backes, X. Wang (ACM Press, 2018), pp. 931–948.https://doi.org/10.1145/3243734.3243853
107. Y. Zhang, Y. Long, Z. Liu, Z. Liu, D. Gu, Z-channel: Scalable and efficient scheme in zerocash, in *ACISP 18, LNCS*, eds. by W. Susilo, G. Yang, vol. 10946 (Springer, Heidelberg, 2018), pp. 687–705. https://doi.org/10.1007/978-3-319-93638-3_39

Open Access This chapter is licensed under the terms of the Creative Commons Attribution 4.0 International License (http://creativecommons.org/licenses/by/4.0/), which permits use, sharing, adaptation, distribution and reproduction in any medium or format, as long as you give appropriate credit to the original author(s) and the source, provide a link to the Creative Commons license and indicate if changes were made.

The images or other third party material in this chapter are included in the chapter's Creative Commons license, unless indicated otherwise in a credit line to the material. If material is not included in the chapter's Creative Commons license and your intended use is not permitted by statutory regulation or exceeds the permitted use, you will need to obtain permission directly from the copyright holder.

A Survey on Private Information Retrieval: Information-Theoretic Constructions and Error-Correction Techniques

Reo Eriguchi and Koji Nuida

Abstract Private Information Retrieval (PIR) is a fundamental cryptographic primitive to enable a client to retrieve a data item of his choice from a database without revealing any information on his query. Due to its wide variety of applications, PIR has been the subject of a large body of research and many efficient constructions and extensions have been proposed. In this survey, we review developments in PIR and related problems especially focusing on the information-theoretically secure schemes. We then provide the technical details of several state-of-the-art PIR schemes and their extensions to active security.

Keywords Private information retrieval · Information-theoretic security · Error correction

1 Introduction

As networking becomes a popular tool, there emerges a wide variety of services which allow a client to search a database held by a single or multiple servers for desired contents. Since a client's query may be confidential, we need to realize *secure database search*, enabling a client to obtain desired contents without revealing his or her query. As a motivating example, an investor may want to get the value of a certain stock from a stock-market database hiding the identity of the stock he is interested in. A trivial solution to this problem is that a client downloads the whole database and searches it locally. However, it results in high communication complexity that is

---

R. Eriguchi (✉)
National Institute of Advanced Industrial Science and Technology (AIST),Tokyo, Japan
e-mail: eriguchi-reo@aist.go.jp

K. Nuida
Institute of Mathematics for Industry (IMI), Kyushu University, Fukuoka, Japan
e-mail: nuida@imi.kyushu-u.ac.jp

National Institute of Advanced Industrial Science and Technology (AIST), Tokyo, Japan

© The Author(s) 2026
T. Takagi et al. (eds.), *Mathematical Foundations for Post-Quantum Cryptography*,
Mathematics for Industry 40, https://doi.org/10.1007/978-981-96-1218-5_24

linear in the size of the database. Can we construct secure database search protocols whose communication complexity is sufficiently smaller than the database size?

Private Information Retrieval (PIR) is a fundamental cryptographic primitive offering an efficient solution to the above problem. Specifically, a PIR scheme allows a client to retrieve a data item a_τ from a database $\boldsymbol{a} = (a_1, \ldots, a_n)$ without revealing the index τ to servers. In this survey, we review the brief history of developments in PIR and related problems, and introduce the currently best-known results, especially focusing on the information-theoretically secure schemes (i.e., the ones keeping security against adversaries with unbounded computational power).

The concept of PIR was first introduced in the seminal paper by Chor et al. [16]. They showed an impossibility result that when only a single server is available, there is no better solution than downloading the whole database; namely, a client should communicate at least n bits to guarantee privacy against computationally unbounded adversaries. To get around that difficulty, they assumed $\ell \geq 2$ non-colluding servers holding copies of a database and proposed a PIR scheme with sublinear communication complexity $O(n^{1/\ell})$.[1] Since then, a line of works have proposed various PIR schemes to improve communication complexity. The schemes in [1, 8] have communication complexity $O(n^{1/(2\ell-1)})$ and the scheme in [7] has communication complexity $n^{O(\log \log \ell/(\ell \log \ell))}$. The breakthrough result was established by Yekhanin [47], who gave the first 3-server PIR scheme whose communication complexity is a sub-polynomial function $n^{o(1)}$ (i.e., smaller than n^ϵ for any constant $\epsilon > 0$). Although the scheme in [47] assumed a number-theoretic conjecture, the assumption was then removed by Efremenko [23]. Dvir and Gopi [22] showed that a similar communication complexity can be achieved even in the 2-server setting. While these schemes assume non-colluding servers, we can consider a more general notion of *t-secure* PIR schemes in which any coalition of t servers learn no information on a client's query. Woodruff and Yekhanin [46] proposed an efficient construction of t-secure $O(td)$-server schemes with $O(n^{1/d})$ communication for any $d \geq 2$, and Barkol, Ishai and Weinreb [5] devised a general amplification technique to transform 1-secure schemes to t-secure ones at the cost of an exponential number of servers in t. In summary, the currently best-known t-secure PIR schemes are:

- The 3^t-server scheme with $n^{o(1)}$ communication obtained by combining [5] and [23];
- The $O(td)$-server scheme with $O(n^{1/d})$ communication in [46] for any $d \geq 2$.

We will provide more detailed descriptions of these schemes in Sect. 3.

- Is there any passively/actively t-secure PIR scheme which achieves sub-polynomial communication complexity in the database size n and polynomial number of servers in t simultaneously? The currently best-known PIR scheme with $n^{o(1)}$ communication in [23] requires an exponential number of servers 3^t.

---

[1] Here, we focus on the dependency on the database size n and omit a factor of the number of servers ℓ.

- Can we derive a tight lower bound on the communication complexity of PIR schemes? In the 2-server setting, the currently best-known lower bound is $5 \log n$ [45] while the trivial lower bound is $\log n$.
- Can we improve communication complexity if multiple rounds of interaction between a client and servers are allowed? Regarding the download rate (see Sect. 1 for the definition), Sun and Jafar [43] showed that the efficiency of multi-round PIR is no better than single-round PIR.

Traditionally, most of works have considered *passively secure* PIR schemes, i.e., the privacy and correctness are guaranteed only if servers follow the protocol specifications. On the other hand, in real-world scenarios, servers may try to let a client accept an incorrect result or compute responses from an out-of-date copy of a database. Motivated by this, Beimal and Stahl [10] introduced a notion of *actively secure* PIR schemes. Namely, such PIR schemes should keep the privacy of queries and guarantee the correctness of results even if some servers deviate from the protocols arbitrarily. Beimel and Stahl [10] proposed a generic compiler which transforms passively t-secure ℓ-server PIR schemes into actively t-secure $(\ell + 2t)$-server schemes, and Eriguchi, Kurosawa, and Nuida [24] showed that the number of servers in resulting schemes can be reduced to $\ell + t$ at the cost of allowing a small soundness error. Since the above generic compilers incur a large computational overhead that is exponential in t, Kurosawa [31] obtained a computationally efficient error correction technique tailored to the scheme in [46], and Zhang, Wang and Wang [48] proposed an error correction technique for the scheme in [23] assuming non-colluding servers. In Sect. 4, we will provide the technical details of the compilers in [10, 24].

Related Problems. Since PIR was introduced in [16], it has been the subject of a large body of research. Here, we give a short overview of other related problems on PIR which are omitted to limit the survey.

Computationally secure constructions. A way to get around the impossibility of achieving information-theoretic security in the single-server scenario is to construct computationally secure PIR schemes (i.e., the ones keeping security only against computationally bounded adversaries) based on cryptographic assumptions. Kushilevitz and Ostrovsky [32] proposed the first single-server PIR scheme with sub-polynomial communication complexity, assuming a number-theoretic public-key encryption. And then, more communication-efficient schemes with polylogarithmic communication complexity were proposed under different number-theoretic assumptions [13, 15, 35]. There are also computationally secure PIR schemes based on post-quantum assumptions [17, 18, 26, 49].

Doubly efficient PIR. Beimel, Ishai, and Malkin [9] showed that in any PIR scheme, the server-side computation must involve all data items per query, which results in $\Omega(n)$ server-side computation. This is roughly because if a server does not touch a data item, then it learns that a client's query is independent of that item, which partially reveals the content of the query. Beimel et al. [9] introduced the preprocessing model to get around this difficulty and achieve sublinear server-side computation by shifting the bulk of computation into off-peak hours. Although there were some candidate constructions based on non-standard assumptions or in a restricted model [11, 14], the first doubly efficient PIR scheme under standard assumptions was recently constructed by Lin, Mook, and Wichs [33].

Download cost of PIR. The problem of PIR was also considered in a special setting where the length of each entry of a database is large. An efficiency measure considered there is *download rate*, which is defined as the asymptotic ratio between the total length of servers' answers to a query and the length of each entry. We refer the interested reader to [2, 41, 42] and references therein.

Symmetric PIR. It is essential to keep the privacy of data contents besides the privacy of a client's query, in some practical scenarios such as a commercial database which sells information to clients. Symmetric PIR [27] is a variant of PIR which additionally guarantees that a client does not learn database records that he does not query.

List-decodable PIR. If at least half of the servers are malicious, it is necessarily impossible to achieve active security since there is an obvious attack that malicious servers substitute the database by a different one. To handle more than $\ell/2$ errors, Goldberg [28] proposed a notion of *list-decodable PIR*, which allows a client to output a list of L possibilities one of which is correct. It can be shown that list-decodable PIR is possible as long as the number of errors is at most $\ell L/(L+1)$ [20, 28].

Dealing with general queries. To allow more complex queries such as partial match search, Barkol and Ishai [3] generalized the setting of PIR as follows: A client has a private input x and servers share a function f, and the goal of the client is to obtain $f(x)$ by communicating with servers. A rich line of works proposed efficient schemes computing various classes of functions f, e.g., polynomials [4, 19, 30, 46] and circuits of bounded or unbounded depth [12, 21, 36, 39].

Lower bounds. There is very little known about lower bounds on the communication complexity of PIR. In the 2-server setting, the currently best-known lower bound is $5 \log n$ [45] while the trivial lower bound is $\log n$. An $\Omega(n^{1/3})$ lower bound is derived if PIR schemes have a certain algebraic property [38] and an $\Omega(n)$ lower bound is derived if the length of servers' answers is limited [6].

Other surveys. Finally, we list some other surveys on PIR. Gasarch [25] provides a survey on general topics on PIR. The survey [37] focuses on single-server PIR. The recent survey [44] introduces PIR schemes with low download rate and their applications to other problems.

2 Definitions

Notations. For $m \in \mathbb{N}$, define $[m] = \{1, \ldots, m\}$. For a set X, we denote the set of all subsets of X of size m by $\binom{X}{m}$. For a vector $\boldsymbol{x}$, let x_i denote the i-th entry of $\boldsymbol{x}$. Let $f \in \mathbb{F}_q[X_1, \ldots, X_m]$ be an m-variate polynomial over a finite field $\mathbb{F}_q$ of size q. We say that f is a degree-d polynomial if its total degree is at most d.

We write $u \leftarrow_\$ \mathcal{U}$ if u is randomly chosen from a set $\mathcal{U}$. For two vectors $\boldsymbol{x} = (x_i)_{i \in [m]}$, $\boldsymbol{y} = (y_i)_{i \in [m]}$ over a ring, we define $\langle \boldsymbol{x}, \boldsymbol{y} \rangle = \sum_{i \in [h]} x_i y_i$. Throughout the paper, we use the following notations:

- ℓ denotes the total number of servers.
- t denotes the number of corrupted servers.

We follow the client-servers model in [3]. In this model, there is a client C who holds a private input $\tau \in [n]$, and ℓ servers $\mathsf{S}_1, \ldots, \mathsf{S}_\ell$ who all hold the same database $\boldsymbol{a} = (a_1, \ldots, a_n) \in \{0, 1\}^n$. Roughly speaking, the goal of *private information retrieval (PIR)* is:

Correctness. The client learns the data item a_τ;
Privacy. The client keeps his search index τ hidden from any collusion of t servers.

We call a message from a client to servers a *query* and a message from servers to the client an *answer*. More formally, a PIR scheme is defined as follows.

Definition 1 (**Syntax**) An ℓ-server PIR scheme Π for n-sized databases is a tuple of three algorithms $\Pi = (\mathcal{Q}, \mathcal{A}, \mathcal{D})$, where $\mathcal{Q}$ is randomized while $\mathcal{A}$ and $\mathcal{D}$ are deterministic:

- $\mathcal{Q}(\tau; r) \rightarrow (\mathsf{que}_1, \ldots, \mathsf{que}_\ell; \mathsf{aux})$: A query algorithm $\mathcal{Q}$ takes a search index $\tau \in [n]$ as input, samples a random string $r \leftarrow_\$ \mathfrak{R}_\mathcal{Q}$, and outputs $\mathsf{que}_i \in \{0, 1\}^{c_{\mathsf{que}}}$ for $i \in [\ell]$ and $\mathsf{aux} \in \{0, 1\}^{c_{\mathsf{aux}}}$.
- $\mathcal{A}(i, \mathsf{que}_i, \boldsymbol{a}) \rightarrow \mathsf{ans}_i$: An answer algorithm $\mathcal{A}$ takes $i \in [\ell]$, $\mathsf{que}_i \in \{0, 1\}^{c_{\mathsf{que}}}$ and $\boldsymbol{a} = (a_1, \ldots, a_n) \in \{0, 1\}^n$ as input, and outputs $\mathsf{ans}_i \in \{0, 1\}^{c_{\mathsf{ans}}}$.
- $\mathcal{D}(\mathsf{ans}_1, \ldots, \mathsf{ans}_\ell; \mathsf{aux}) \rightarrow y$: A reconstruction algorithm $\mathcal{D}$ takes $\mathsf{ans}_i \in \{0, 1\}^{c_{\mathsf{ans}}}$ for $i \in [\ell]$ and $\mathsf{aux} \in \{0, 1\}^{c_{\mathsf{aux}}}$ as input, and outputs $y \in \{0, 1\}$;

satisfying the following property:

Correctness. For any $\boldsymbol{a} \in \{0, 1\}^n$ and any $\tau \in [n]$,

$$\Pr[r \leftarrow_\$ \mathfrak{R}_\mathcal{Q} : \mathcal{D}(\mathsf{ans}_1, \ldots, \mathsf{ans}_\ell; \mathsf{aux}) = a_\tau] = 1,$$

where $(\mathsf{que}_1, \ldots, \mathsf{que}_\ell; \mathsf{aux}) = \mathcal{Q}(\tau; r)$ and $\mathsf{ans}_i = \mathcal{A}(i, \mathsf{que}_i, \boldsymbol{a})$ for $i \in [\ell]$.

The (total) communication complexity of Π is given by $\mathrm{Comm}(\Pi) = \ell(c_{\mathsf{que}} + c_{\mathsf{ans}})$.

Definition 2 (**Passive security**) An ℓ-server PIR scheme $\Pi = (\mathcal{Q}, \mathcal{A}, \mathcal{D})$ is said to be passively t-secure if for any $X \in \binom{[\ell]}{t}$ and any $\tau, \tau' \in [n]$, the distributions of $(\mathsf{que}_i)_{i \in X}$ and $(\mathsf{que}'_i)_{i \in X}$ are perfectly identical, where $r, r' \leftarrow_\$ \mathfrak{R}_\mathcal{Q}$, $(\mathsf{que}_1, \ldots, \mathsf{que}_\ell; \mathsf{aux}) = \mathcal{Q}(\tau; r)$ and $(\mathsf{que}'_1, \ldots, \mathsf{que}'_\ell; \mathsf{aux}') = \mathcal{Q}(\tau'; r')$.

A PIR scheme directly implies a one-round protocol in the above client-servers setting. Indeed, when a client C has a search index $\tau \in [n]$ and ℓ servers $\mathsf{S}_1, \ldots, \mathsf{S}_\ell$ have a common database $\boldsymbol{a} \in \{0, 1\}^n$, we obtain the following protocol:

Query. On input $\tau \in [n]$, C chooses $r \leftarrow_\$ \mathfrak{R}_Q$ and runs Q on input τ, r to obtain $\mathsf{que}_1, \ldots, \mathsf{que}_\ell$ and aux. Then, C sends que_i to S_i for $i \in [\ell]$.
Answer. On input $a \in \{0,1\}^n$, each S_i returns $\mathsf{ans}_i = \mathcal{A}(i, \mathsf{que}_i, a)$ to C.
Reconstruction. C outputs $y = \mathcal{D}(\mathsf{ans}_1, \ldots, \mathsf{ans}_\ell; \mathsf{aux})$.

We can translate the above security notions of the PIR scheme Π into the relevant notions in the language of protocols. The correctness means that the client always obtains the correct result a_τ if servers correctly compute answers, and the passive security ensures that any (possibly computationally unbounded) semi-honest adversary corrupting t servers learns no information on the client's input τ. We will use the terminologies interchangeably for the sake of readability.

3 Information-Theoretic Constructions of PIR

In this section, we provide the currently best-known PIR schemes satisfying information-theoretic security.

3.1 Constructions Based on Matching Vector Codes

First, we show PIR schemes whose communication complexity is sub-polynomial in n (i.e., smaller than n^ϵ for any constant $\epsilon > 0$). These schemes are based on a combinatorial object called *matching vector codes*.

Definition 3 Let $m \in \mathbb{Z}$ and $S \subseteq \mathbb{Z}_m \setminus \{0\}$. We say that $\mathcal{U} = (u_1, \ldots, u_n)$ is an S-matching vector code over $\mathbb{Z}_m^h$ if $u_i \in \mathbb{Z}_m^h$ for all $i \in [n]$ and the following conditions are satisfied:

- $\langle u_i, u_i \rangle = 0$ for every $i \in [n]$;
- $\langle u_i, u_j \rangle \in S$ for every $i \neq j$.

There exists an explicit construction of a matching vector code such that $h = n^{o(1)}$.

Proposition 1 ([29]) *Let $r \geq 2$ and $p_1 < p_2 < \cdots < p_r$ be r primes, and set $m = p_1 p_2 \cdots p_r$. For any integer $n > 1$, there exists an S-matching vector code $\mathcal{U} = (u_1, \ldots, u_n)$ over $\mathbb{Z}_m^h$ such that*

- $h = \exp\left(\theta_m (\log n)^{1/r} (\log \log n)^{1-1/r}\right)$, *where θ_m a constant depending on m only;*
- $S = \{s \in \mathbb{Z}_m \setminus \{0\} : \forall j \in [r],\ s \bmod p_j \in \{0,1\}\}$. *In particular we have $|S| = 2^r - 1$.*

Definition 4 Let $S \subseteq \mathbb{Z}_m \setminus \{0\}$ and $\mathbb{F}_q$ be a field containing an element γ of multiplicative order m. We say that a polynomial $P \in \mathbb{F}_q[x]$ is an S-decoding polynomial if the following conditions are satisfied:

- $P(\gamma^0) = P(1) = 1$;
- $P(\gamma^s) = 0$ for all $s \in S$.

For any m, we can see that the above field $\mathbb{F}_q$ and an S-decoding polynomial P actually exist. A prime q satisfying $q \equiv 1 \bmod m$ can be chosen as $q = m^{O(1)}$ from Linnik's theorem [34]. Let γ_0 be the multiplicative generator of $\mathbb{F}_q$ and set $\gamma = \gamma_0^{(q-1)/m}$. Then, the multiplicative order of γ is m. Furthermore, if we define $P_0(x) = \prod_{s \in S}(x - \gamma^s)$, then $P(x) := P_0(x)/P_0(1)$ is an S-decoding polynomial. Since the degree of P is $|S|$, P has at most $|S| + 1$ monomials.

Theorem 1 ([23]) *Let $m \in \mathbb{Z}$, $S \subseteq \mathbb{Z}_m \setminus \{0\}$ and $\mathbb{F}_q$ be a field containing an element γ of multiplicative order m. Assume that $\mathcal{U} = (\mathbf{u}_1, \ldots, \mathbf{u}_n)$ is an S-matching vector code over $\mathbb{Z}_m^h$ and $P \in \mathbb{F}_q[x]$ is an S-decoding polynomial with ℓ monomials. Then, there exists a passively 1-secure ℓ-server PIR scheme Π for n-sized databases such that $\text{Comm}(\Pi) = O(\ell(h \log m + \log q))$.*

Proof The polynomial $P(x)$ can be written as $P(x) = c_1 x^{e_1} + c_2 x^{e_2} + \cdots + c_\ell x^{e_\ell}$ for some $c_i \in \mathbb{F}_q$ and $e_i \in \mathbb{Z}$. Consider the following PIR scheme: Each server encodes a database $\mathbf{a} = (a_1, \ldots, a_n) \in \{0, 1\}^n$ into $a_1 \gamma^{\mathbf{u}_1} + \cdots + a_n \gamma^{\mathbf{u}_n}$, where for a vector $\mathbf{v} = (v_1, \ldots, v_h)$, we denote $(\gamma^{v_1}, \ldots, \gamma^{v_h})$ by $\gamma^{\mathbf{v}}$. On input $\tau \in [n]$, a client chooses a vector $\mathbf{r}$ uniformly at random from $\mathbb{Z}_m^h$ and sends each server S_i

$$\mathsf{que}_i := e_i \cdot \mathbf{u}_\tau + \mathbf{r} \in \mathbb{Z}_m^h.$$

In response, each server S_i returns

$$\mathsf{ans}_i := \sum_{\sigma \in [n]} a_\sigma \gamma^{\langle \mathbf{u}_\sigma, \mathsf{que}_i \rangle} \in \mathbb{F}_q.$$

Note that $\gamma^{\langle \mathbf{u}, \mathbf{v} \rangle}$ can be computed from $\gamma^{\mathbf{u}} = (\gamma^{u_1}, \ldots, \gamma^{u_h})$ and $\mathbf{v} = (v_1, \ldots, v_h)$ by $(\gamma^{u_1})^{v_1} \cdots (\gamma^{u_h})^{v_h}$. Finally, a client computes $y = c_1 \cdot \mathsf{ans}_1 + \cdots + c_\ell \cdot \mathsf{ans}_\ell$, and outputs 0 if $y = 0$ and outputs 1 otherwise.

First, we see the correctness. Since $\langle \mathbf{u}_\sigma, \mathbf{u}_\tau \rangle = 0$ if and only if $\sigma = \tau$, it holds that

$$\mathsf{ans}_i = \sum_{\sigma \in [n]} a_\sigma \gamma^{e_i \langle \mathbf{u}_\sigma, \mathbf{u}_\tau \rangle + \langle \mathbf{u}_\sigma, \mathbf{r} \rangle} = a_\tau \gamma^{\langle \mathbf{u}_\tau, \mathbf{r} \rangle} + \sum_{s \in S} b_s (\gamma^s)^{e_i},$$

where $b_s = \sum_{\sigma:\langle u_\sigma, u_\tau \rangle = s} a_\sigma \gamma^{\langle u_\sigma, r \rangle}$. Therefore, we have that

$$\begin{aligned}
y &= \sum_{i \in [\ell]} c_i \cdot \text{ans}_i \\
&= a_\tau \gamma^{\langle u_\tau, r \rangle} \sum_{i \in [\ell]} c_i + \sum_{s \in S} b_s \sum_{i \in [\ell]} c_i (\gamma^s)^{e_i} \\
&= a_\tau \gamma^{\langle u_\tau, r \rangle} P(1) + \sum_{s \in S} b_s P(\gamma^s) \\
&= a_\tau \gamma^{\langle u_\tau, r \rangle},
\end{aligned}$$

from which the client obtains a_τ since $\gamma^x \neq 0$ for any x.

The 1-privacy can be easily seen since a client's input τ is perfectly masked by a uniformly random vector r.

If we instantiate $\mathcal{U}$ with the matching vector code in Proposition 1 and P with the aforementioned S-decoding polynomial with $|S| + 1$ monomials, then, for any $r \geq 2$, we obtain a passively 1-secure 2^r-server PIR scheme Π such that

$$\text{Comm}(\Pi) = \exp\left(O((\log n)^{1/r} (\log \log n)^{1-1/r})\right) = n^{o(1)},$$

omitting any constant depending on r only. This general template can only yield PIR schemes with at least 4 servers since the trivial construction of S-decoding polynomials results in polynomials with $\ell = 2^r \geq 4$ monomials.

Efremenko [23] constructed a 3-server PIR scheme with $n^{o(1)}$ communication by finding an S-decoding polynomial with fewer monomials by an exhaustive search. Concretely, let $m = 511 = 7 \cdot 73$ and $S = \{1, 147, 365\}$. Set

$$\mathbb{F}_q = \mathbb{F}_{2^9} = \mathbb{F}_2[\gamma]/(\gamma^9 + \gamma^4 + 1).$$

Then, γ is a multiplicative generator of $\mathbb{F}_q$, i.e., the multiplicative order of γ is $q - 1 = 511 = m$. We can see that

$$P(x) = \gamma^{423} \cdot x^{65} + \gamma^{257} \cdot x^{12} + \gamma^{342}$$

is an S-decoding polynomial with 3 monomials. In summary, we have the following corollary.

Corollary 1 *There exist passively 1-secure PIR schemes Π for n-sized databases such that*

- *the number of servers is 3 and the communication complexity is*

$$\text{Comm}(\Pi) = \exp\left(O\left(\sqrt{(\log n)(\log \log n)}\right)\right);$$

- the number of servers is 2^r and the communication complexity is

$$\text{Comm}(\Pi) = \exp\left(O\left((\log n)^{1/r}(\log\log n)^{1-1/r}\right)\right)$$

for any constant $r \geq 2$.

There is a technique to amplify the privacy threshold in [5]. Specifically, if we have a passively t_0-secure ℓ_0-server scheme whose query length is c_{que} and whose answer length is c_{ans}, then the scheme can be generically transformed into a passively $t_0 t$-secure ℓ_0^t-server scheme whose query length is $O(c_{\text{que}} t \ell_0^t)$ and whose answer length is $O(c_{\text{ans}}^t)$. By applying this technique, we obtain the following corollary.

Corollary 2 *There exist passively t-secure PIR schemes Π for n-sized databases such that*

- the number of servers is 3^t and the communication complexity is

$$\text{Comm}(\Pi) = \exp\left(O\left(\sqrt{(\log n)(\log\log n)}\right)\right) \cdot 2^{O(t)};$$

- the number of servers is 2^{rt} and the communication complexity is

$$\text{Comm}(\Pi) = \exp\left(O\left((\log n)^{1/r}(\log\log n)^{1-1/r}\right)\right) \cdot 2^{O(t)}$$

for any constant $r \geq 2$.

Remark 1 Dvir and Gopi [22] successfully devised a technique to optimize the 3-server scheme [23], and obtain a 2-server PIR scheme with $n^{o(1)}$ communication. However, since the answer length c_{ans} is not constant, the passively t-secure scheme obtained by applying the above amplification technique has larger communication complexity $\exp(O(t\sqrt{(\log n)(\log\log n)}))$.

3.2 Constructions Based on Polynomial Interpolation

The previous construction of PIR schemes with sub-polynomial communication requires exponentially many servers in the privacy threshold t. In this section, we show a PIR scheme whose number of servers is linear in t although its communication complexity is n^c for a constant $c < 1$. Woodruff and Yekhanin [46] proposed a PIR scheme based on *polynomial interpolation*. Specifically, they used Hermite interpolation [40], a technique to recover a polynomial using its values and derivatives on given points.

We start with some additional notations. Assume that $\mathbb{F}_q$ is a prime field such that $q \geq \ell + 1$. Let $f \in \mathbb{F}_q[X_1, \ldots, X_m]$ be an m-variate polynomial. Define the partial derivative of f with respect to X_j as

$$\frac{\partial f}{\partial X_j} = \sum_{e=(e_i)_{i\in[m]}\in I} c_e e_j X_j^{e_j-1} \prod_{i\in[m]\setminus\{j\}} X_i^{e_i}$$

if $f = \sum_{e\in I} c_e \prod_{i\in[m]} X_i^{e_i}$, where $c_e \in \mathbb{F}_q$ and I is a finite set of m-tuples of non-negative integers. For a univariate polynomial $f(X)$, we denote by df/dX the derivative of f with respect to its unique variable.

Let $y_{j,w} \in \mathbb{F}_q$ for each $j \in [\ell]$ and $w \in \{0, 1\}$. Then, there exists an explicit formula for finding a unique polynomial $g \in \mathbb{F}_q[X]$ of degree at most $2\ell - 1$ such that

$$g(j) = y_{j,0} \text{ and } \frac{dg}{dx}(j) = y_{j,1} \ (\forall j \in [\ell]).$$

Theorem 2 *([46]) Let $d \geq 1$ and $t \geq 1$. Suppose that*

$$\ell \geq \frac{td+1}{2}.$$

Then, there exists a passively t-secure ℓ-server PIR scheme for n-sized databases such that $\mathrm{Comm}(\Pi) = O(n^{1/d} d\ell \log \ell)$.

Proof To begin with, we prepare some notations. Let $\mathbb{F}_q$ be a prime field such that $q \geq \ell + 1$. Let h be the smallest integer such that $\binom{h}{d} \geq n$. Note that such h can be chosen as $h = O(dn^{1/d})$. We can then define an injective function $E : [n] \to \{0, 1\}^h$ such that for all $\tau \in [n]$, the Hamming weight of $E(\tau)$ is d. For a vector $z = (z_1, \ldots, z_h)$ and $E(\tau) = (E(\tau)_1, \ldots, E(\tau)_h)$, we define

$$z^{E(\tau)} = \prod_{j:E(\tau)_j=1} z_j.$$

Consider the following PIR scheme: Each server encodes a database $\boldsymbol{a} = (a_1, \ldots, a_n) \in \{0, 1\}^n$ into a polynomial

$$F(z_1, \ldots, z_h) = \sum_{\sigma \in [n]} a_\sigma z^{E(\tau)} \in \mathbb{F}_q[z_1, \ldots, z_h].$$

On input $\tau \in [n]$, a client chooses t vectors $\boldsymbol{r}_1, \ldots, \boldsymbol{r}_t$ uniformly at random from $\mathbb{F}_q^h$, and then sends each server S_i

$$\mathrm{que}_i := E(\tau) + \sum_{k=1}^{t} i^k \cdot \boldsymbol{r}_j = E(\tau) + i \cdot \boldsymbol{r}_1 + i^2 \cdot \boldsymbol{r}_2 + \cdots + i^t \cdot \boldsymbol{r}_t \in \mathbb{F}_q^h.$$

In response, each server S_i computes

$$y_i^{(0)} = F(\mathrm{que}_i), \ u_{ij}^{(1)} = \frac{\partial F}{\partial z_j}(\mathrm{que}_i) \ (\forall j \in [h])$$

and returns $\text{ans}_i = (y_i^{(0)}, \boldsymbol{u}_i^{(1)})$, where $\boldsymbol{u}_i^{(1)} = (u_{ij}^{(1)})_{j \in [h]}$. To obtain a_τ, the client computes

$$v_i := \sum_{k=1}^{t} k i^{k-1} \cdot \boldsymbol{r}_k = \boldsymbol{r}_1 + 2i \cdot \boldsymbol{r}_2 + \cdots + t i^{t-1} \cdot \boldsymbol{r}_t$$

from the $\boldsymbol{r}_k$'s he chose when computing queries, and then computes $y_i^{(1)} = \langle \boldsymbol{u}_i^{(1)}, v_i \rangle$. Finally, he computes a polynomial $\tilde{g}(x)$ of degree at most td such that

$$y_i^{(0)} = \tilde{g}(i) \text{ and } y_i^{(1)} = \frac{d\tilde{g}}{dx}(i) \ (\forall i \in [\ell])$$

using Hermite interpolation, and outputs $\tilde{g}(0)$.

To see the correctness, we define a polynomial $g(x)$ as

$$g(x) = F(E(\tau) + x \cdot \boldsymbol{r}_1 + x^2 \cdot \boldsymbol{r}_2 + \cdots + x^t \cdot \boldsymbol{r}_t).$$

The degree of g is at most $td \leq 2\ell - 1$ since the degree of F is d. It holds that $y_i^{(0)} = F(\text{que}_i) = g(i)$. Due to the chain rule, it also holds that

$$\frac{dg}{dx}(i) = \langle \nabla F(\text{que}_i), \boldsymbol{r}_1 + 2x \cdot \boldsymbol{r}_2 + \cdots + t x^{t-1} \cdot \boldsymbol{r}_t \rangle = \langle \boldsymbol{u}_i^{(1)}, v_i \rangle = y_i^{(1)},$$

where

$$\nabla F(\text{que}_i) = \left(\frac{\partial F}{\partial z_1}(\text{que}_i), \ldots, \frac{\partial F}{\partial z_h}(\text{que}_i) \right).$$

Therefore, we have that $\tilde{g}(x) = g(x)$ from the correctness of Hermite interpolation. The output of the client is thus $\tilde{g}(0) = g(0) = F(E(\tau)) = a_\tau$.

The t-privacy easily follows from the fact that t random vectors $\boldsymbol{r}_k$'s make any set of t queries independent of a client's input τ.

4 Dealing with Malicious Servers

In this section, we consider a malicious adversary who can make corrupted servers deviate from protocols arbitrarily. Specifically, a (possibly computationally unbounded) malicious adversary corrupts at most t servers and lets them return answers to a client which may be incorrect. Note that since we consider one-round PIR schemes, the adversary will not learn any new information as a result of sending wrong answers. The main technical problem is thus how to guarantee the correctness of a client's output.

Definition 5 An ℓ-server PIR scheme $\Pi = (Q, \mathcal{A}, \mathcal{D})$ is said to be actively t-secure if it is passively t-secure and for any adversary $\mathcal{B}$, any $B \subseteq [\ell]$ of size at most t, any $\boldsymbol{a} \in \{0, 1\}^n$, and any $\tau \in [n]$,

$$\Pr[r \leftarrow_\$ \mathfrak{R}_Q, y \leftarrow \mathcal{D}(\mathsf{ans}_1, \ldots, \mathsf{ans}_\ell; \mathsf{aux}) : y = a_\tau] = 1, \tag{1}$$

where $(\mathsf{que}_1, \ldots, \mathsf{que}_\ell; \mathsf{aux}) = Q(\tau; r)$, $\mathsf{ans}_i = \mathcal{A}(i, \mathsf{que}_i, \boldsymbol{a})$ for $i \in [\ell] \setminus B$, and $(\mathsf{ans}_i)_{i \in B} \leftarrow \mathcal{B}(\boldsymbol{a}, (\mathsf{que}_i)_{i \in B})$.

For $\epsilon > 0$, we consider a weaker notion of *actively* (t, ϵ)-*secure* PIR, which satisfies the requirements in Definition 5 except that the condition (1) is replaced with

$$\Pr[r \leftarrow_\$ \mathfrak{R}_Q, y \leftarrow \mathcal{D}(\mathsf{ans}_1, \ldots, \mathsf{ans}_\ell; \mathsf{aux}) : y = a_\tau] \geq 1 - \epsilon.$$

4.1 Perfect Error Correction

First, we consider the case of actively secure PIR schemes with $\epsilon = 0$ and show the generic construction proposed by Beimel and Stahl [10]. Along the way, they introduced a notion of k-*robust PIR*, which guarantees that a client can compute a_τ from answers of any k out of ℓ servers.

Definition 6 An ℓ-server PIR scheme $\Pi = (Q, \mathcal{A}, \mathcal{D})$ is said to be k-robust if

- $\mathcal{D}$ takes $X \in \binom{[\ell]}{k}$, $(\mathsf{ans}_i)_{i \in X} \in (\{0, 1\}^{c_{\mathsf{ans}}})^k$ and $\mathsf{aux} \in \{0, 1\}^{c_{\mathsf{aux}}}$ as input, and outputs $y \in \{0, 1\}$;
- For any $\boldsymbol{a} \in \{0, 1\}^n$, any $\tau \in [n]$ and any $X \in \binom{[\ell]}{k}$,

$$\Pr[r \leftarrow_\$ \mathfrak{R}_Q : \mathcal{D}(X, (\mathsf{ans}_i)_{i \in X}; \mathsf{aux}) = a_\tau] = 1,$$

where $(\mathsf{que}_1, \ldots, \mathsf{que}_\ell; \mathsf{aux}) = Q(\tau; r)$ and $\mathsf{ans}_i = \mathcal{A}(i, \mathsf{que}_i, \boldsymbol{a})$ for $i \in [\ell]$.

It was shown in [10] that if a passively t-secure ℓ-server PIR scheme is $(\ell - 2t)$-robust, then the scheme is actively t-secure.

Proposition 2 *Assume that there exists a passively t-secure k-robust ℓ-server PIR scheme $\Pi_1 = (Q_1, \mathcal{A}_1, \mathcal{D}_1)$ such that $\ell = k + 2t$. Then, there exists an actively t-secure ℓ-server PIR scheme Π such that $\mathsf{Comm}(\Pi) = \mathsf{Comm}(\Pi_1)$.*

Proof We construct an error correction algorithm $\mathcal{D}$ as follows. Suppose that a client receives (possibly incorrect) ℓ answers $\widetilde{\mathsf{ans}}_1, \ldots, \widetilde{\mathsf{ans}}_\ell$ and auxiliary information aux, which is computed along with queries.

1. Let $\mathcal{Y} = \emptyset$.
2. For each $S \in \binom{[\ell]}{\ell - t}$, do the following:

a. For each $T \in \binom{S}{k}$, let

$$x_{S,T} = \mathcal{D}_1(T, (\widetilde{\mathsf{ans}}_i)_{i \in T}; \mathsf{aux}).$$

b. If $\{x_{S,T} : T \in \binom{S}{k}\} = \{y\}$ for some $y \in \{0, 1\}$, then add y to $\mathcal{Y}$.

3. Output $\mathcal{Y}$.

To see that $\Pi = (\mathcal{Q}_1, \mathcal{A}_1, \mathcal{D})$ is actively secure, let $\boldsymbol{a} \in \{0, 1\}^n$ be a database and $\tau \in [n]$ be a client's index. Let $H \subseteq [\ell]$ be a set of servers who returned a correct answer $\widetilde{\mathsf{ans}}_i = \mathsf{ans}_i$. Clearly, it holds that $|H| \geq \ell - t$. We have that $a_\tau \in \mathcal{Y}$ with probability 1 since for any $S \in \binom{[\ell]}{\ell-t}$ such that $S \subseteq H$, it holds that

$$x_{S,T} = \mathcal{D}_1(T, (\mathsf{ans}_i)_{i \in T}; \mathsf{aux}) = a_\tau$$

for any subset $T \in \binom{S}{k}$. To show that $\mathcal{Y} = \{a_\tau\}$, assume otherwise that $y \in \mathcal{Y}$ for some $y \neq a_\tau$. Then, there must exist $S \in \binom{[\ell]}{\ell-t}$ such that $x_{S,T} = \mathcal{D}_1(T, (\widetilde{\mathsf{ans}}_i)_{i \in T}; \mathsf{aux}) = y$ for all $T \in \binom{S}{k}$. On the other hand, since the number of incorrect answers is at most t, the number of correct answers in $(\widetilde{\mathsf{ans}}_i)_{i \in S}$ is at least $|S| - t \geq \ell - 2t = k$. This implies that there exists $T \in \binom{S}{k}$ such that $(\widetilde{\mathsf{ans}}_i)_{i \in T}$ are all correct, and hence that $x_{S,T} = \mathcal{D}_1(T, (\widetilde{\mathsf{ans}}_i)_{i \in T}; \mathsf{aux}) = a_\tau$, which is a contradiction. Therefore, it holds with probability 1 that $\mathcal{Y} = \{a_\tau\}$.

Beimel and Stahl [10] showed generic constructions of k-robust ℓ-server PIR schemes from any passively secure k-server PIR scheme. Their constructions incur a $\min\{\binom{\ell}{k}, 2^{O(k)} \ell \log \ell\}$ multiplicative overhead to communication complexity. We thus obtain the following theorem.

Theorem 3 *Assume that there exists a passively t-secure k-server PIR scheme Π_0 with $k > t$. Let $\ell = k + 2t$. Then, there exists an actively t-secure ℓ-server PIR scheme Π such that*

$$\mathrm{Comm}(\Pi) = O\left(\mathrm{Comm}(\Pi_0) \cdot \min\left\{\binom{\ell}{k}, 2^{O(k)} \ell \log \ell\right\}\right).$$

By combining Corollary 2 and Theorem 3, we have concrete actively secure PIR schemes with $n^{o(1)}$ communication.

Corollary 3 *There exist actively t-secure PIR schemes Π such that*

- *the number of servers is $3^t + 2t$ and the communication complexity is*

$$\mathrm{Comm}(\Pi) = \exp\left(O\left(\sqrt{(\log n)(\log \log n)}\right)\right) \cdot 2^{O(t^2)};$$

- the number of servers is $2^{rt} + 2t$ and the communication complexity is

$$\mathrm{Comm}(\Pi) = \exp\left(O\left((\log n)^{1/r}(\log\log n)^{1-1/r}\right)\right) \cdot 2^{O(t^2)}$$

for any constant $r \geq 2$.

We can see that the PIR scheme in Theorem 2 is k-robust for

$$\frac{td+1}{2} \leq k \leq \ell.$$

Therefore, we can apply Proposition 2 and obtain the following corollary.

Corollary 4 *Let $d \geq 1$ and*

$$\ell \geq \left(\frac{d}{2}+2\right)t + \frac{1}{2}.$$

Then, there exists an actively t-secure ℓ-server PIR scheme Π such that

$$\mathrm{Comm}(\Pi) = O(n^{1/d} d\ell \log \ell).$$

4.2 Non-Perfect Error Correction

Next, we consider the case of actively secure PIR schemes with a non-zero probability of failure $\epsilon > 0$. Eriguchi, Kurosawa, and Nuida [24] showed that allowing a small probability of failure allows us to obtain actively secure schemes with fewer servers than the construction in [10]. To this end, they introduced an intermediate notion of *error-detecting PIR*, which guarantees that a client detects it if servers behave maliciously.

Definition 7 An ℓ-server PIR scheme $\Pi = (Q, \mathcal{A}, \mathcal{D})$ is said to be (t, ϵ)-error-detecting if it is passively t-secure and satisfies the following properties:

- $\mathcal{D}$ is allowed to output a special symbol $\perp \notin \{0, 1\}$;
- For any adversary $\mathcal{B}$, any $B \subseteq [\ell]$ of size at most t, any $a \in \{0, 1\}^n$, and any $\tau \in [n]$,

$$\Pr[r \leftarrow_\$ \mathfrak{R}_Q, y \leftarrow \mathcal{D}(\mathsf{ans}_1, \ldots, \mathsf{ans}_\ell; \mathsf{aux}) : y \in \{a_\tau, \perp\}] \geq 1 - \epsilon,$$

where $(\mathsf{que}_1, \ldots, \mathsf{que}_\ell; \mathsf{aux}) = Q(\tau; r)$, $\mathsf{ans}_i = \mathcal{A}(i, \mathsf{que}_i, a)$ for $i \in [\ell] \setminus B$, and $(\mathsf{ans}_i)_{i \in B} \leftarrow \mathcal{B}(a, (\mathsf{que}_i)_{i \in B})$;

- For any $a \in \{0, 1\}^n$ and any $\tau \in [n]$,

$$\Pr[r \leftarrow_\$ \mathfrak{R}_Q, y \leftarrow \mathcal{D}(\mathsf{ans}_1, \ldots, \mathsf{ans}_\ell; \mathsf{aux}) : y = a_\tau] \geq 1 - \epsilon,$$

where $(\mathsf{que}_1, \ldots, \mathsf{que}_\ell; \mathsf{aux}) = Q(\tau; r)$ and $\mathsf{ans}_i = \mathcal{A}(i, \mathsf{que}_i, a)$ for $i \in [\ell]$.

The third property is necessary to rule out the trivial scheme whose reconstruction algorithm $\mathcal{D}$ just outputs $\bot$ regardless of inputs.

There is a generic construction of $(\ell - 1, \epsilon)$-error-detecting ℓ-server PIR schemes from any ℓ-server PIR scheme [24].

Proposition 3 *If there exists a passively t-secure ℓ-server PIR scheme $\Pi_0 = (Q_0, \mathcal{A}_0, \mathcal{D}_0)$, then for any $\epsilon > 0$, there exists a (t, ϵ)-error-detecting t-private ℓ-server PIR scheme Π such that*

$$\mathrm{Comm}(\Pi) = O(\mathrm{Comm}(\Pi_0) \cdot (\ell^2 \log \ell) \log \epsilon^{-1}).$$

Proof (Proof(sketch)) Here, we only provide a basic construction of $(\ell - 1, \epsilon)$-error-detecting PIR schemes with non-negligible ϵ. The full construction for negligible ϵ can be obtained by running the non-negligible scheme sufficiently many times in parallel. The interested reader is referred to [24].

Given $\Pi_0 = (Q_0, \mathcal{A}_0, \mathcal{D}_0)$, we consider the following two query algorithms Π_0^{Compute} and $\Pi_0^{\mathsf{Verify},(i,j)}$. Π_0^{Compute} is used to actually compute a_τ and $\Pi_0^{\mathsf{Verify},(i,j)}$ is used to verify whether a server S_j correctly computes her answer assuming that S_i is honest.

Π_0^{Compute}: On input $\tau \in [n]$, C chooses $r \leftarrow_\$ \mathfrak{R}_{Q_0}$ and computes $Q_0(\tau; r) = (\mathsf{que}_1, \ldots, \mathsf{que}_\ell; \mathsf{aux})$. Then, he sends (m, que_m) to S_m for $m \in [\ell]$.

$\Pi_0^{\mathsf{Verify},(i,j)}$: On input $\tau \in [n]$, C chooses $r \leftarrow_\$ \mathfrak{R}_{Q_0}$ and computes $Q_0(\tau; r) = (\mathsf{que}_1, \ldots, \mathsf{que}_\ell; \mathsf{aux})$. Then, he sends (m, que_m) to S_m for $m \in [\ell] \setminus \{i\}$, and (j, que_j) to S_i.

Now we consider an ℓ-server PIR scheme $\Pi = (Q, \mathcal{A}, \mathcal{D})$, where the client C chooses $i \neq j \in [\ell]$ uniformly at random, randomly permutes two instances of Π_0^{Compute} and $\Pi_0^{\mathsf{Verify},(i,j)}$, and executes them in parallel. C verifies that S_j correctly computes her answer using $\Pi_0^{\mathsf{Verify},(i,j)}$ and then he runs $\mathcal{D}_0$ on the answers obtained during the execution of Π_0^{Compute}. Formally,

> **PIR scheme Π**
>
> **Query.** On input $\tau \in [n]$, C chooses $i \neq j \in [\ell]$ uniformly at random and executes Π_0^{Compute}, $\Pi_0^{\mathsf{Verify},(i,j)}$ in a random order. Specifically, C does the following:
>
> 1. C chooses $i \neq j \in [\ell]$ uniformly at random.
> 2. C randomly permutes two protocols Π_0^{Compute}, $\Pi_0^{\mathsf{Verify},(i,j)}$. Formally, C chooses $\alpha \leftarrow_{\$} \{0, 1\}$ and sets $\Pi^{(\alpha)} = \Pi_0^{\mathsf{Verify},(i,j)}$ and $\Pi^{(1-\alpha)} = \Pi_0^{\mathsf{Compute}}$.
> 3. C generates queries for $\Pi^{(0)}$, $\Pi^{(1)}$. Let $\mathsf{que}_{m,\alpha}$ and aux_α denote the query sent to S_m and auxiliary information for Π_α, respectively.
> 4. C sends $\mathsf{que}_m = (\mathsf{que}_{m,0}, \mathsf{que}_{m,1})$ to each S_m.
>
> **Answer.** On input $a \in \{0, 1\}^n$, each server S_m does the following:
>
> 1. For each $\mathsf{que}_{m,\alpha} = (x, \mathsf{que}_x)$, S_m computes $\mathsf{ans}_{m,\alpha} = \mathcal{A}_0(x, \mathsf{que}_x, a)$.
> 2. S_m returns $(\mathsf{ans}_{m,0}, \mathsf{ans}_{m,1})$ to C.
>
> **Error detection.** After receiving $(\widetilde{\mathsf{ans}}_{m,0}, \widetilde{\mathsf{ans}}_{m,1})$ from S_m as $(\mathsf{ans}_{m,0}, \mathsf{ans}_{m,1})$, C does the following:
>
> 1. For $\alpha \in \{0, 1\}$ with $\Pi^{(\alpha)} = \Pi_0^{\mathsf{Compute}}$, C sets $z \leftarrow \mathcal{D}_0(\widetilde{\mathsf{ans}}_{1,\alpha}, \ldots, \widetilde{\mathsf{ans}}_{\ell,\alpha}; \mathsf{aux}_\alpha)$.
> 2. For $\alpha \in \{0, 1\}$ with $\Pi^{(\alpha)} = \Pi_0^{\mathsf{Verify},(i,j)}$, C verifies whether $\widetilde{\mathsf{ans}}_{j,\alpha} = \widetilde{\mathsf{ans}}_{i,\alpha}$ holds. If it holds, then C outputs z. Otherwise, C outputs $\perp$.

It is easy to see that Π is passively t-secure. We will see that Π is $(\ell - 1, \epsilon)$-error-detecting for $\epsilon = 1 - 1/(2\ell(\ell - 1))$. Without loss of generality, we suppose that S_1 is honest and a malicious adversary $\mathcal{B}$ corrupts $S_2, \ldots, S_\ell$. Clearly, if $S_2, \ldots, S_\ell$ return correct answers, the client C obtains the correct value a_τ. We may assume that at least one malicious server, say S_2, modifies her answer.

Consider the case where C chooses $(i, j) = (1, 2)$ at Step 1 of **Query**, which occurs with probability $1/(\ell(\ell - 1))$. To make C output the incorrect value $1 - a_\tau$, S_2 needs to behave honestly in the instance $\Pi_0^{\mathsf{Verify},(1,2)}$ and to modify her answer in the other instance Π_0^{Compute}. Note that $\mathcal{B}$ cannot distinguish between two instances Π_0^{Compute}, $\Pi_0^{\mathsf{Verify},(1,2)}$ since the distributions of queries that $S_2, \ldots, S_\ell$ receive are the same in both cases. Hence, the distribution of an answer returned by S_2 is independent of the permutation chosen by C at Step 2. With probability at least $1/2$, S_2 fails to guess the instance $\Pi_0^{\mathsf{Verify},(1,2)}$, in which she has to behave honestly. Therefore, C can detect errors with probability $1/(2\ell(\ell - 1)) = 1 - \epsilon$. We conclude that Π is $(\ell - 1, \epsilon)$-error-detecting.

It is straightforward to transform a (t, ϵ')-error-detecting k-server PIR scheme into an actively (t, ϵ)-secure ℓ-server PIR scheme for $\ell = k + t$. Indeed, consider the scheme in which a client runs $\binom{\ell}{\ell-t}$ independent instances of an error-detecting scheme for all sets of $\ell - t$ servers. The client obtains a correct result in an instance in which all involved servers are honest. On the other hand, in all instances, he does not accept an incorrect result except with negligible probability thanks to the error-detecting property. We thus obtain the following theorem. Note that compared to Theorem 3, the number of servers of a resulting protocol is reduced from $\ell = k + 2t$ to $\ell = k + t$, when assuming a passively secure k-server PIR scheme.

Theorem 4 *Assume that there exists a passively t-secure k-server PIR scheme Π_0 such that $k > t$. Let $\ell = k + t$ and $\epsilon > 0$. Then, there exists an actively (t, ϵ)-secure ℓ-server PIR scheme Π such that*

$$\mathrm{Comm}(\Pi) = \mathrm{Comm}(\Pi_0) \cdot \ell^{O(t)} \log \epsilon^{-1}.$$

By combining Corollary 2 and Theorem 4, we have concrete actively secure PIR schemes with $n^{o(1)}$ communication.

Corollary 5 *Let $\epsilon > 0$. There exist actively (t, ϵ)-secure PIR schemes Π such that*

- *the number of servers is $3^t + t$ and the communication complexity is*

$$\mathrm{Comm}(\Pi) = \exp\left(O\left(\sqrt{(\log n)(\log \log n)}\right)\right) \cdot 2^{O(t^2)} \log \epsilon^{-1};$$

- *the number of servers is $2^{rt} + t$ and the communication complexity is*

$$\mathrm{Comm}(\Pi) = \exp\left(O\left((\log n)^{1/r}(\log \log n)^{1-1/r}\right)\right) \cdot 2^{O(t^2)} \log \epsilon^{-1}$$

for any constant $r \geq 2$.

By combining Theorems 2 and 4, we have a concrete actively secure PIR scheme with fewer servers.

Corollary 6 *Let $d \geq 1$ and $\epsilon > 0$. Suppose that*

$$\ell \geq \left(\frac{d}{2} + 1\right) t + \frac{1}{2}.$$

Then, there exists an actively (t, ϵ)-secure ℓ-server PIR scheme Π such that

$$\mathrm{Comm}(\Pi) = n^{1/d} d \ell^{O(t)} \log \epsilon^{-1}.$$

5 Open Problems

We conclude the paper with some interesting open problems.

- Is there any passively/actively t-secure PIR scheme which achieves subpolynomial communication complexity in the database size n and polynomial number of servers in t simultaneously? The currently best-known PIR scheme with $n^{o(1)}$ communication in [23] requires an exponential number of servers 3^t.
- Can we derive a tight lower bound on the communication complexity of PIR schemes? In the 2-server setting, the currently best-known lower bound is $5 \log n$ [45] while the trivial lower bound is $\log n$.
- Can we improve communication complexity if multiple rounds of interaction between a client and servers are allowed? Regarding the download rate (see Sect. 1 for the definition), Sun and Jafar [43] showed that the efficiency of multi-round PIR is no better than single-round PIR.

Acknowledgements This work is partially supported by JST AIP Acceleration Research JPMJCR22U5, Japan, JST CREST Grant Number JPMJCR22M1, Japan, and JSPS KAKENHI Grant Number JP19H01109, Japan.

References

1. A. Ambainis, Upper bound on the communication complexity of private information retrieval, in *Automata, Languages and Programming* (1997), pp. 401–407
2. K. Banawan, S. Ulukus, The capacity of private information retrieval from byzantine and colluding databases. IEEE Trans. Inf. Theory **65**(2), 1206–1219 (2019)
3. O. Barkol, Y. Ishai, Secure computation of constant-depth circuits with applications to database search problems, in *Advances in Cryptology — CRYPTO 2005* (2005), pp. 395–411
4. O. Barkol, Y. Ishai, E. Weinreb, On d-multiplicative secret sharing. J. Cryptol. **23**(4), 580–593 (2010)
5. O. Barkol, Y. Ishai, E. Weinreb, On locally decodable codes, self-correctable codes, and t-private PIR. Algorithmica **58**(4), 831–859 (2010)
6. R. Beigel, L. Fortnow, W. Gasarch, A tight lower bound for restricted PIR protocols. Comput. Complex. **15**, 82–91 (2006)
7. A. Beimel, Y. Ishai, E. Kushilevitz, J.F. Raymond, Breaking the o(n/sup 1/(2k-1)/) barrier for information-theoretic private information retrieval, in *The 43rd Annual IEEE Symposium on Foundations of Computer Science, 2002. Proceedings* (2002), pp. 261–270
8. A. Beimel, Y. Ishai, E. Kushilevitz, General constructions for information-theoretic private information retrieval. J. Comput. Syst. Sci. **71**(2), 213–247 (2005)
9. A. Beimel, Y. Ishai, T. Malkin, Reducing the servers' computation in private information retrieval: PIR with preprocessing. J. Cryptol. **17**(2), 125–151 (2004)
10. A. Beimel, Y. Stahl, Robust information-theoretic private information retrieval. J. Cryptol. **20**(3), 295–321 (2007)
11. E. Boyle, Y. Ishai, R. Pass, M. Wootters, Can we access a database both locally and privately? in *Theory of Cryptography* (2017), pp. 662–693
12. E. Boyle, L. Kohl, P. Scholl, Homomorphic secret sharing from lattices without FHE, in Advances in Cryptology - EUROCRYPT 2019. Part **II**, 3–33 (2019)
13. C. Cachin, S. Micali, M. Stadler, Computationally private information retrieval with polylogarithmic communication, in *Advances in Cryptology – EUROCRYPT '99* (1999), pp. 402–414

14. R. Canetti, J. Holmgren, S. Richelson, Towards doubly efficient private information retrieval, in *Theory of Cryptography* (2017), pp. 694–726
15. Y.C. Chang, Single database private information retrieval with logarithmic communication, in *Information Security and Privacy* (2004), pp. 50–61
16. B. Chor, O. Goldreich, E. Kushilevitz, M. Sudan, Private information retrieval. J. ACM **45**(6), 965–982 (1998)
17. H. Corrigan-Gibbs, A. Henzinger, D. Kogan, Single-server private information retrieval with sublinear amortized time. Advances in Cryptology - EUROCRYPT **2022**, 3–33 (2022)
18. H. Corrigan-Gibbs, D. Kogan, Private information retrieval with sublinear online time. Advances in Cryptology - EUROCRYPT **2020**, 44–75 (2020)
19. Q. Dao, Y. Ishai, A. Jain, H. Lin, Multi-party homomorphic secret sharing and sublinear MPC from sparse LPN. Advances in Cryptology - CRYPTO **2023**, 315–348 (2023)
20. C. Devet, I. Goldberg, N. Heninger, Optimally robust private information retrieval, in *21st USENIX Security Symposium (USENIX Security 12)* (2012), pp. 269–283
21. Y. Dodis, S. Halevi, R.D. Rothblum, D. Wichs, Spooky encryption and its applications, in Advances in Cryptology - CRYPTO 2016. Part **III**, 93–122 (2016)
22. Z. Dvir, S. Gopi, 2-server PIR with subpolynomial communication. J. ACM **63**(4), 1–15 (2016)
23. K. Efremenko, 3-query locally decodable codes of subexponential length. SIAM J. Comput. **41**(6), 1694–1703 (2012)
24. R. Eriguchi, K. Kurosawa, K. Nuida, On the optimal communication complexity of error-correcting multi-server PIR, in *Theory of Cryptography* (2022), pp. 60–88
25. W. Gasarch, A survey on private information retrieval https://citeseerx.ist.psu.edu/viewdoc/summary?doi=10.1.1.9.8246
26. C. Gentry, Z. Ramzan, Single-database private information retrieval with constant communication rate, in *Automata, Languages and Programming* (2005), pp. 803–815
27. Y. Gertner, Y. Ishai, E. Kushilevitz, T. Malkin, Protecting data privacy in private information retrieval schemes. J. Comput. Syst. Sci. **60**(3), 592–629 (2000)
28. I. Goldberg, Improving the robustness of private information retrieval, in *2007 IEEE Symposium on Security and Privacy (SP'07)* (2007), pp. 131–148
29. V. Grolmusz, Superpolynomial size set-systems with restricted intersections mod 6 and explicit ramsey graphs. Combinatorica **20**(1), 71–86 (2000)
30. Y. Ishai, R.W.F. Lai, G. Malavolta, A geometric approach to homomorphic secret sharing. Public-Key Cryptography - PKC **2021**, 92–119 (2021)
31. K. Kurosawa, How to correct errors in multi-server PIR. Advances in Cryptology - ASIACRYPT **2019**, 564–574 (2019)
32. E. Kushilevitz, R. Ostrovsky, Replication is not needed: single database, computationally-private information retrieval, in *Proceedings 38th Annual Symposium on Foundations of Computer Science* (1997), pp. 364–373
33. W.K. Lin, E. Mook, D. Wichs, Doubly efficient private information retrieval and fully homomorphic RAM computation from ring LWE, in *Proceedings of the 55th Annual ACM Symposium on Theory of Computing* (2023), pp. 595–608
34. U. Linnik, On the least prime in an arithmetic progression. II. The Deuring–Heilbronn phenomenon. Rec. Math. [Mat. Sbornik] N.S. **15**(3), 347–368 (1944)
35. H. Lipmaa, An oblivious transfer protocol with log-squared communication, in *Information Security* (2005), pp. 314–328
36. C. Orlandi, P. Scholl, S. Yakoubov, The rise of paillier: Homomorphic secret sharing and public-key silent ot. Advances in Cryptology - EUROCRYPT **2021**, 678–708 (2021)
37. R. Ostrovsky, W.E. Skeith, A survey of single-database private information retrieval: Techniques and applications. Public Key Cryptography - PKC **2007**, 393–411 (2007)
38. A.A. Razborov, S. Yekhanin, An $\omega(n^{1/3})$ lower bound for bilinear group based private information retrieval, in *2006 47th Annual IEEE Symposium on Foundations of Computer Science (FOCS'06)* (2006), pp. 739–748
39. L. Roy, J. Singh, Large message homomorphic secret sharing from DCR and applications. Advances in Cryptology - CRYPTO **2021**, 687–717 (2021)

40. A. Spitzbart, A generalization of Hermite's interpolation formula. Am. Math. Mon. **67**(1), 42–46 (1960)
41. H. Sun, S.A. Jafar, The capacity of private information retrieval. IEEE Trans. Inf. Theory **63**(7), 4075–4088 (2017)
42. H. Sun, S.A. Jafar, The capacity of robust private information retrieval with colluding databases. IEEE Trans. Inf. Theory **64**(4), 2361–2370 (2018)
43. H. Sun, S.A. Jafar, Multiround private information retrieval: Capacity and storage overhead. IEEE Trans. Inf. Theory **64**(8), 5743–5754 (2018)
44. S. Vithana, Z. Wang, S. Ulukus, Private information retrieval and its applications: An introduction, open problems, future directions (2023)
45. S. Wehner, R. de Wolf, Improved lower bounds for locally decodable codes and private information retrieval, in *Automata, Languages and Programming* (2005), pp. 1424–1436
46. D. Woodruff, S. Yekhanin, A geometric approach to information-theoretic private information retrieval. SIAM J. Comput. **37**(4), 1046–1056 (2007)
47. S. Yekhanin, Towards 3-query locally decodable codes of subexponential length. J. ACM (JACM) **55**(1), 1–16 (2008)
48. L.F. Zhang, H. Wang, L.P. Wang, Byzantine-robust private information retrieval with low communication and efficient decoding, in *Proceedings of the 2022 ACM on Asia Conference on Computer and Communications Security* (ASIA CCS '22, 2022), pp. 1079–1085
49. M. Zhou, W. Lin, Y. Tselekounis, E. Shi, Optimal single-server private information retrieval. Advances in Cryptology - EUROCRYPT **2023**, 395–425 (2023)

Open Access This chapter is licensed under the terms of the Creative Commons Attribution 4.0 International License (http://creativecommons.org/licenses/by/4.0/), which permits use, sharing, adaptation, distribution and reproduction in any medium or format, as long as you give appropriate credit to the original author(s) and the source, provide a link to the Creative Commons license and indicate if changes were made.

The images or other third party material in this chapter are included in the chapter's Creative Commons license, unless indicated otherwise in a credit line to the material. If material is not included in the chapter's Creative Commons license and your intended use is not permitted by statutory regulation or exceeds the permitted use, you will need to obtain permission directly from the copyright holder.

Index

A
Abc inequalities conjectures, 3
Arithmetic geometry, 3

B
Binary field, 127
Binary GCD algorithm, 163
Blockchain, 453

C
Cayley graph, 371, 389
CGL hash function, 379
Code-based cryptography, 296
Continued fraction, 97
Coppersmith algorithm, 164
Cryptocurrencies, 453
CRYSTALS-Dilithium, 251
CRYSTALS-Kyber, 251

D
Dagger categories, 15
DeepBKZ, 233

E
Elliptic Curve Discrete Logarithm Problem (ECDLP), 125
Entanglement, 35
Error correction, 6, 164, 477
Expander graphs, 371, 390

F
Factorization, 125, 335, 397
Falcon, 408
Farey graph, 89
Fault attack, 183
Field with one element, 89
Fuchsian, 51

G
Gaussian heuristic, 166, 236, 276
Gramian, 295
Gröbner basis, 205
Group-subgroup pair graphs, 392
Grover's algorithm, 145, 320, 417

H
Hash-based signature, 422
Hash function, 296, 316, 334, 351, 371, 372, 389, 408
Heun differential equation, 51
Hilbert–Poincaré series, 205

I
Isogeny-based cryptography, 333, 352, 372
IUT theory, 3

J
Jacobians of genus 2 curve, 351
Jaynes-Cummings model, 63, 76

K
Kani's reducibility theorem, 338
Kannan's embedding, 273

L
Lattice-based cryptography, 234, 278
Learning with errors, 251, 273
Lie algebra, 52
LLL algorithm, 166, 237

M
Magma, 334, 353
Middle-product learning with errors, 252
MQ problem, 184, 206, 318
Multiradical isogeny, 352
Multivariate public-key cryptography, 184, 206, 313

N
Non-commutative harmonic oscillator, 51
NP-hard, 36, 276
NTRU, 273

P
Prime number, 127, 164, 298
Private information retrieval, 475

Q
Quantum circuit, 127, 145

Quantum optics, 75
Quantum proof, 36
Quantum Rabi model, 63, 85
Quantum walks, 15
Qubit, 85, 128

R
Ramanujan graph, 375
Representation theory, 8, 51
Riemann hypothesis, 89, 342
Risk assessment, 425
Rivest–Shamir–Adleman (RSA), 125, 163

S
Semi-regular polynomial sequences, 207
Shor's algorithm, 126
Shortest vector problem, 166, 233, 273
Side channel attacks, 163, 421
SPHINCS+, 407
Supersingular elliptic curve, 334, 354, 376
Supersingular Isogeny Diffie-Hellman protocol (SIDH), 333, 352

U
Unbalanced oil and vinegar, 184
(U,U+V)-code problem, 295

If you have any concerns about our products,
you can contact us on
ProductSafety@springernature.com

In case Publisher is established outside the EU,
the EU authorized representative is:
**Springer Nature Customer Service Center GmbH
Europaplatz 3, 69115 Heidelberg, Germany**

Printed by Libri Plureos GmbH
in Hamburg, Germany